Membrane Biological Reactors

Membrane Biological Reactors

Theory, Modeling, Design, Management and Applications to Wastewater Reuse

Second Edition

Edited by
Faisal I. Hai, Kazuo Yamamoto and Chung-Hak Lee

IWA PUBLISHING

Published by **IWA Publishing**
Alliance House
12 Caxton Street
London SW1H 0QS, UK
Telephone: +44 (0)20 7654 5500
Fax: +44 (0)20 7654 5555
Email: publications@iwap.co.uk
Web: www.iwapublishing.com

First published 2019

British Library Cataloguing in Publication Data
A CIP catalogue record for this book is available from the British Library

ISBN: 9781780409160 (paperback)
ISBN: 9781780409177 (eBook)
ISBN: 9781780409184 (ePub)

Cover image: www.istockphoto.com

Contents

List of abbreviations .... **xv**

Nomenclature .... **xxiii**

About the editors .... **xxix**

Preface .... **xxxi**

Dedication .... **xxxiii**

Chapter 1
Introduction to membrane biological reactors ***1***
Faisal I. Hai, Kazuo Yamamoto and Chung-Hak Lee

1.1 Membrane Biological Reactors – Definition and Application 1
1.2 Historical Development of Biosolids Separation in MBRs 2
1.3 Process Comparison with Cas Process 5
  1.3.1 Treatment efficiency 6
  1.3.2 Sludge production and handling 6
  1.3.3 Space requirements 6
  1.3.4 Wastewater treatment cost 7
  1.3.5 Comparative energy usage 7
  1.3.6 Relative advantages of MBR 7
1.4 Factors Influencing Performance/Design Considerations 8
  1.4.1 Pre-treatment 8
  1.4.2 Membrane maintenance 8
  1.4.3 Membrane selection and applied flux 9
  1.4.4 SRT and MLSS concentration 9
1.5 Municipal vs Industrial MBRs 9

1.5.1 Membrane types and materials ... 9
1.5.2 Permeate flux ... 10
1.5.3 Treatment performance ... 11
1.5.4 Capital cost ... 11
1.6 Market Drivers/Restraints and Development Trend ... 12
1.6.1 Current status and typical drivers ... 12
1.6.2 Challenges ... 14
1.6.3 The way forward ... 16
1.7 MBR Market ... 16
1.7.1 Global market overview ... 16
1.7.2 Regional key drivers and constraints and market trend ... 17
1.8 Worldwide Research Trend ... 24
1.9 Summary and Future Outlook ... 26
References ... 27

Chapter 2
Process fundamentals: from conventional biological wastewater treatment to MBR ... ***35***
V. Jegatheesan and C. Visvanathan

2.1 Introduction ... 35
2.2 Need for Biological Treatment ... 36
2.3 Microbial Communities, their Environments and Degradation Pathways of Pollutants ... 36
2.4 Biological Treatment Fundamentals ... 38
2.4.1 Conventional activated sludge (CAS) process basics ... 40
2.4.2 Nitrogen removal ... 41
2.4.3 Phosphorus removal ... 42
2.4.4 Combined biological nutrient removal (BNR) ... 43
2.4.5 Operational requirements ... 44
2.5 Membrane Fundamentals ... 46
2.5.1 Membrane performance parameters ... 47
2.5.2 Membrane classifications ... 47
2.5.3 Membrane interactions with suspended impurities ... 49
2.5.4 Membrane materials, system configurations, and operating modes ... 50
2.5.5 Membrane resistance ... 51
2.6 Fundamentals of MBR ... 53
2.6.1 History of MBR technology ... 53
2.6.2 Differences between CAS and MBR processes ... 53
2.6.3 Design of MBR systems ... 54
2.6.4 Process overview ... 57
2.6.5 Biology in MBR ... 58
2.6.6 Operation of the membrane system in MBR ... 59
2.6.7 Energy utilization in MBR ... 63
2.7 Summary and Future Outlook ... 64
References ... 64

Chapter 3
Membrane bioreactors design, operation, maintenance ***67***
Christoph Brepols

3.1 Introduction 67
3.2 Technical Concepts 68
3.3 Reference Data on Design and Operation 69
3.3.1 Municipal/Urban applications 70
3.3.2 Industrial applications 74
3.3.3 Groundwater replenishment 76
3.4 MBR Design 78
3.4.1 Design workflow 78
3.4.2 General plant layout 78
3.4.3 Wastewater composition, volume and temperature 81
3.4.4 Process units: Inflow equalization 82
3.4.5 Process units: Mechanical pre-treatment 83
3.4.6 Process units: Biological treatment 84
3.4.7 Process units: Membrane unit design 89
3.4.8 Process units: Aeration 95
3.4.9 Process units: Automation 98
3.4.10 Energy consumption 99
3.4.11 Cost evaluations 100
3.4.12 Alternative MBR concepts 102
3.5 Operation and Plant Management 102
3.5.1 Membrane cleaning and maintenance 102
3.5.2 Process reliability 103
3.5.3 Residuals and sludge management 104
3.5.4 Energy improvements in existing MBRs 105
3.5.5 Personnel and qualification 105
3.6 Summary 106
References 106

Chapter 4
Monitoring, characterization and control of membrane biofouling in MBR ***115***
Katsuki Kimura, Fangang Meng, In-Soung Chang and Chung-Hak Lee

4.1 Introduction 116
4.2 Monitoring 116
4.2.1 Importance of monitoring 116
4.2.2 Methods used for assessment of filterability of mixed liquor 117
4.2.3 Identification of dominant parameters in filterability of mixed liquor 119
4.2.4 Problems to be addressed in monitoring of the filterability of mixed liquor 121

4.3 Characterization of Membrane Foulants in MBRs ... 122
4.3.1 Approaches to morphological visualization ... 122
4.3.2 Approaches to componential characterization ... 126
4.3.3 Approaches to microbiological identification ... 128
4.3.4 Summary of approaches to characterization ... 130
4.4 Biofouling Control ... 133
4.4.1 Membrane development ... 133
4.4.2 Chemical approaches ... 134
4.4.3 Physical (hydrodynamic, mechanical) approaches ... 136
4.4.4 Biological approaches ... 139
4.5 Conclusion and Future Outlook ... 145
References ... 145

Chapter 5
Nutrient removal and disinfection by MBR ... ***155***
So-Ryong Chae, Yongtae Ahn, Yuhoon Hwang, Duksoo Jang, Fangang Meng, Jeffrey Shi, Sang-Hyup Lee and Hang-Sik Shin

5.1 Introduction ... 156
5.2 Reuse and Recycling of Reclaimed Wastewater ... 157
5.2.1 Urban reuse ... 159
5.2.2 Agricultural reuse ... 159
5.2.3 Impoundments ... 160
5.2.4 Environmental reuse ... 160
5.2.5 Industrial reuse ... 160
5.2.6 Groundwater recharge – nonpotable reuse ... 161
5.2.7 Potable reuse ... 161
5.3 Advanced Designs of MBRs for Nutrient Removal ... 161
5.3.1 Design of MBRs for removal of organic matter and nitrogen ... 163
5.3.2 Design of MBRs for simultaneous removal of nitrogen and phosphorus ... 163
5.4 Effects of the Microbial Community on Nutrient Removal in MBRs ... 164
5.5 Case Studies: Reuse and Recycling of MBR Effluents ... 166
5.6 Nutrient Recovery from MBR Effluents ... 169
5.7 Challenges Associated with Pathogen Removal by MBRs ... 170
5.8 Post-treatments for Disinfection of the MBR Effluents ... 173
5.8.1 Chlorination ... 173
5.8.2 Ultraviolet irradiation ... 174
5.8.3 Ozonation ... 174
5.8.4 Other post-treatments for MBR effluents ... 174
5.8.5 Applications of AOPs for MBR effluents ... 175
5.9 Summary and Future Outlook ... 175
References ... 176

Chapter 6
Wastewater reuse applications and MBR ***183***
Muhammad B. Asif and Faisal I. Hai

6.1 Introduction 184
6.2 Overview of Current Wastewater Reuse Practices and Trends 186
6.3 Performance of MBR Technology for Pollutant Removal 188
6.3.1 Removal of bulk organics and nutrients 188
6.3.2 Removal of bacteria and viruses 190
6.3.3 Removal of emerging TrOCss 192
6.4 Wastewater Reuse Application Examples Involving MBR 196
6.5 Future Perspective and Conclusion 197
References 200

Chapter 7
Impacts of hazardous events on performance of membrane bioreactors ***207***
Trang Trinh, Amos Branch, Ben van den Akkerc, Pierre Le-Clech, Jörg E. Drewes and Stuart J. Khan

7.1 Introduction – Hazardous Events in Risk Assessment 208
7.2 Characterisation of Potential Hazardous Events and their Impact on MBR Operation 210
7.2.1 Deviation from normal operation 210
7.3 Expected Consequences of Key Hazardous Event Types 212
7.3.1 Impact on the removal of bulk organic matter and nutrients 212
7.3.2 Impact on the removal of pathogens and microbial indicators 216
7.4 Assessing Likelihoods of MBR Hazardous Events 221
7.5 Management of Hazardous Events Through Engineered Redundancy and Multiple Barrier Treatment Systems 222
7.6 Conclusions and Future Outlook 223
References 223

Chapter 8
Cost benefit and environmental life cycle assessment ***227***
Gonzalo Rodriguez-García, María Molinos-Senante, Sara Gabarron, Carolina Alfonsín, Almudena Hospido, Lluis Corominas, Francesc Hernandez-Sancho, Francisco Omil, Gumersindo Feijoo, Ramón Sala-Garrido, Ignasi Rodriguez-Roda and Maria Teresa Moreira

8.1 Introduction 228
8.2 Cost–Benefit Analysis 228
8.2.1 Modeling of operational costs in WwTP and membrane technologies 229
8.2.2 Calculation of the environmental benefits associated with WwTP: the shadow prices methodology 230

8.3 Life Cycle Assessment 231
8.3.1 Life cycle assessment methodology 231
8.3.2 Life cycle assessment in WwTP and membrane technologies 232
8.4 Economic and Environmental Profile of Full-scale MBR 237
8.4.1 Economic profile 238
8.4.2 Environmental profile 241
8.5 Environmental Profile of Pilot Plant MBR 251
8.5.1 Goal and scope 251
8.5.2 Life cycle inventory analysis 252
8.5.3 Life cycle impact assessment 257
8.5.4 Result interpretation 261
8.6 Conclusions and Future Outlook 262
References 262

Chapter 9
MBR modeling studies* *269
G. Andreottola, G. Guglielmi and M. Langone

9.1 Introduction 269
9.2 Biological Models 270
9.2.1 Introduction to ASM models 270
9.2.2 ASMs to MBR modeling 273
9.2.3 Application of unmodified/conventional ASMs to MBR 274
9.2.4 Application of modified/integrated ASMs models to MBR 285
9.3 Filtration Models 291
9.4. CFD and Hydrodynamics – Modeling of MBR Tanks and Fluid Dynamics 295
9.4.1 Module design 295
9.4.2 Process design and operation 295
9.5 Control and Operational Strategies 296
9.6 Conclusions and future outlook 297
References 297

Chapter 10
Gas-diffusion, extractive, biocatalytic, and electrochemical membrane biological reactors* *305
Oskar Modin, Faisal I. Hai, Long D. Nghiem, Angelo Basile and Kensuke Fukushi

10.1 Introduction 306
10.2 Membrane Biofilm Reactors 307
10.2.1 Overview 307
10.2.2 Membrane materials and configurations 308
10.2.3 Aeration MBfRs 308
10.2.4 Hydrogen MBfRs 314
10.2.5 Methane MBfRs 314
10.3 Extractive MBRs for Corrosive/Toxic Wastewater Treatment 315

10.4 Biocatalytic MBRs 317
10.4.1 Types and applications of biocatalytic MBRs 317
10.4.2 Membranes for biocatalytic MBRs 319
10.4.3 EMRs for xenobiotics removal 319
10.4.4 Membrane fouling in EMRs for xenobiotics removal 322
10.4.5 Inhibition of enzymatic activity in EMRs for xenobiotics removal 323
10.4.6 ICMRs for xenobiotics removal 324
10.5 Electrochemical MBRs 327
10.6 Summary and Future Outlook 330
References 331

Chapter 11
Anaerobic MBRs ***343***
Weijue Gao, Bao-Qiang Liao and Faisal I. Hai

11.1 Introduction 344
11.2 History 345
11.3 System Configurations 345
11.4 Applications of AnMBRs 347
11.4.1 Municipal wastewater treatment 347
11.4.2 Industrial wastewater treatment 349
11.5 Membrane Fouling 354
11.5.1 Membrane fouling mechanisms 355
11.5.2 Membrane fouling characterization 355
11.6 Factors Affecting the Treatment Performance and Membrane Fouling 358
11.6.1 Membrane properties 358
11.6.2 Effects of operating and environmental conditions 364
11.6.3 Hydrodynamic conditions 366
11.6.4 Sludge properties 367
11.6.5 Strategies for performance stability and membrane fouling control 370
11.7 Commercial Potential of AnMBRs 372
11.7.1 Water reuse and energy production 372
11.7.2 Reduced energy consumption 373
11.7.3 Economic analysis 374
11.8 Conclusion and Future Outlook 374
References 375

Chapter 12
Hybrid processes, new generation membranes and novel MBR designs ***387***
Olivier Lefebvre, Kok Kwang Ng, Kai Yin Tang, Tze Chiang Albert Ng and How Yong Ng

12.1 Introduction 388
12.2 Integrated MBR Systems for Water Reclamation 388

12.2.1 Biofilm MBR 388
12.2.2 Granular sludge MBR 391
12.2.3 MBR integrated with physico-chemical processes 391
12.3 Innovative Membrane Design for MBR 394
12.3.1 CNT-doped membranes 394
12.3.2 TiO_2-doped membranes 394
12.3.3 Grafted polymer membranes 395
12.3.4 Graphene oxide membranes 396
12.3.5 Electrospun nanofiber membranes 396
12.3.6 Metal-organic framework membranes 397
12.4 Innovative MBR Designs 397
12.4.1 NF-MBR 397
12.4.2 FO-MBR 398
12.4.3 MD-MBR 399
12.4.4 Anammox-MBR 400
12.4.5 Bioaugmented MBR 401
12.4.6 Bioaugmented MBR for fouling control 401
12.4.7 Aeration imposed shear for fouling control 402
12.4.8 Mechanically imposed shear for fouling control 403
12.5 Innovative Concepts 403
12.5.1 Mechanical recovery of energy in MBR 403
12.5.2 PRO-MBR 405
12.5.3 MFC-MBR 405
12.6 Conclusion and Future Outlook 407
References 407

Chapter 13
Removal of emerging trace organic contaminants (TrOC) by MBR* *413
Faisal I. Hai, Long D. Nghiem, Stuart J. Khan, Muhammad B. Asif, William E. Price and Kazuo Yamamoto

13.1 Introduction 414
13.2 TrOC in Water and their Potential Impact on Reuse 414
13.3 Relative Performance of MBR and Other Biological Processes 415
13.3.1 Conceptual expectations 415
13.3.2 Reported comparative performance of CAS and MBR 416
13.4 Effect of Troc Presence in Wastewater on Basic Performance of MBR 419
13.5 Factors Affecting Troc Removal by MBR 420
13.5.1 Characteristics of the TrOC 420
13.5.2 Operating parameters 427
13.6 Correlation of TrOC Removal with Nitrification and Denitrification 436
13.7 Effect of MBR-effluent Disinfection on TrOC Removal 437
13.8 Overall Fate and Metabolic Pathways 438
13.9 Post Treatments and MBR-based Hybrid Systems 439
13.9.1 Combination with physicochemical processes 439

13.9.2 Bioaugmented and enzymatic MBR for TrOC removal 444
13.9.3 High retention (HR)-MBR for TrOC removal 446
13.10 Conclusion and Future Outlook 453
References 453

Chapter 14
Commercial MBR technologies and selected case studies ***469***
Graeme Pearce

14.1 Introduction to Commercial Products 470
14.1.1 Background 470
14.1.2 Membrane materials 471
14.1.3 Module format 474
14.1.4 System configuration 477
14.1.5 Product nomenclature 479
14.2 Manufacturers' Review 479
14.2.1 Overview 479
14.2.2 Immersed hollow fibre 483
14.2.3 Immersed flat sheet 505
14.2.4 External 512
14.3 Case Studies 517
14.3.1 Immersed hollow fibre case studies 517
14.3.2 Flat sheet case studies 524
14.3.3 External case studies 525
14.4 Summary and Future Outlook 528
References 530
A1: Standard Product Literature and Website Downloads 533
A1.1 Introduction 533
A1.1.1 Overview 533
A1.1.2 Immersed (submerged) hollow fibre 533
A1.1.3 Immersed (submerged) flat sheet 534
A1.1.4 External 534

Chapter 15
MBR design calculations ***535***
Muhammad B. Asif and Faisal I. Hai

References 556

Index 559

List of abbreviations

3CBA	3-Chlorobenzoate
3D	EEM Three-dimensional excitation-emission matrix fluorescence spectroscopy
A/O MBR	Anaerobic–oxic membrane bioreactor
AD	Anaerobic digestion
ADBI	Celestolide
ADF	Average daily flow
ADUF	Anaerobic digestion ultrafiltration
AEM	Anion exchange membrane
AeMBR	Aerobic MBR
AFM	Atomic force microscopy
AFMBR	Anaerobic fluidized-bed MBR
AHL	acylated homoserine lactone
AHTN	Tonalide
AniMBR	Anaerobic immersed membrane bioreactor
AnMBR	Anaerobic membrane bioreactor
AnsMBR	Anaerobic sidestream membrane bioreactor
AOB	Ammonia-Oxidizing Bacteria
AOBF/MF	Anaerobic–oxic–anoxic biofilm filtration followed by membrane filtration
AOI	Adsorbable organo-iodine
AOP	Advanced oxidation process
AP	Acidification potential
AS	Activated sludge
ASM	Activated sludge model
ASM1	Activate Sludge Model N.1

ASM2	Activate Sludge Models N.2
ASM2d	Extension of ASM2
ASM2d2N	ASM2d with 2 step nitrification model
ASM3	Activate Sludge Models N.3
ASM3-SMP	ASM3 combined with SMP model
ASMs	Activate Sludge Models
BAC	Biological activated carbon
BAPs	Biomass-associated products
BEMBR	bio-entrapped MBR
BES	Bioelectrochemical system
BNR	Biological Nutrient Removal
BOD	Biochemical Oxygen Demand
BOD_5	Five-day biochemical oxygen demand
BOW	Beijing Origin Water
BSM	Benchmark Simulation Model
C/N	Carbon to Nitrogen ratio
CAS	Conventional Activated Sludge
CAS-MF/UF	Microfiltration or ultrafiltration following CAS process
CBA	Cost–benefit analysis
CBZ	Carbamazepine
CD	Centrifuge dewatering
CEL	Citic Environment Ltd
CEM	Cation exchange membrane
CF	Characterization factor
CFD	Computational Fluid Dynamics
CFU	Colony Forming Unit
CFV	Cross-flow velocity
CIP	Cleaning in place
CLSM	Confocal Laser Scanning Microscopy
CMF	Continuous microfiltration
CNT	carbon nanotube
COD	Chemical Oxygen Demand
CoF	Coagulation/Flocculation
COP	Cleaning-out-of-place
CPCB	Central pollution control board
CSTR	Completely stirred tank reactor
CT	Computer Tomography
CTA	cellulose triacetate
CTL	Citalopram

DAF	Dissolved air flotation
DBPs	Disinfection By-products
DCF	Diclofenac
DEET	*N, N*-Diethyl-meta-toluamide
DGGE	Denaturing Gradient Gel Electrophoresis
DNA	Deoxyribonucleic Acid
DO	Dissolved Oxygen
DOC	Dissolved organic carbon
DOM	Dissolved organic matter
DPR	Direct potable reuse
DZP	Diazepam
E1	Estrone
E2	Estradiol
EA	Extended aeration
EBPR	Enhanced Biological Phosphorus Removal
ECA	Enhanced capital allowance
EDC	Endocrine disrupting chemicals
EDCs	Endocrine Disrupting Compounds
EDCs	Endocrine disrupting chemicals
EDG	Electron-donating groups
EDX	Energy-dispersive X-ray spectroscopy
EE2	17α-Ethinylestradiol
EGSB	Expended granular sludge-bed reactor
EMBR	Extractive membrane bioreactor
EMR	Enzymatic membrane reactor
EOP	Electrochemical Oxidation Potential
EP	Eutrophication potential
EPC	Engineering, procurement, and construction contactor
EPS	Extracellular Polymeric Substances
ERY	Erythromycin
ETP	Ecotoxicity potential
EV	Enteric Virus
EWG	Electron-withdrawing groups
F	Filtration
F/M	Food to microorganism ratio
F/M ratio	Food to-Microorganism ratio
FBMBR	fixed-bed membrane bioreactor
FC	Fecal Coliforms
FISH	Fluorescence in Situ Hybridization

FLX	Fluoxetine
FO	Forward osmosis
FO-MBR	forward osmosis membrane bioreactor
FS	Flat sheet membrane
FT	Flotation thickener
FTIR	Fourier transform infrared spectroscopy
FU	Functional unit
GAC	Granular activated carbon
GFC	Gel filtration Chromatography
GHG	Greenhouse gases
GO	graphene oxide
GravT	Gravity thickener
GWP	Global warming potential
HAdV	Human Adenovirus
HDPE	high-density polyethylene
HF	Hollow fibre
HHCB	Galaxolide
HMI	Human–machine interface
HRAS	High rate activated sludge
HRT	hydraulic retention time
HTP	Human toxicity potential
IA	intermittent aeration
IBP	Ibuprofen
ICMR	Immobilized-cell membrane reactor
ICP	Inductively coupled plasma
IFAS	Integrated fixed film activated sludge
iMBR	Immersed (submerged) MBR
iMBR	Immersed membrane bioreactor
IMLR	Internal Mixed Liquor Recycle
IPR	Indirect potable reuse
ISO	International Standards Organization
IWVA	International Water Volleyball Association
LAS	Linear alkylbenzene sulfonates
LCA	Life cycle assessment
LCC	Life cycle costing
LCC	Life cycle costs
LCI	Life cycle inventory
LCIA	Life cycle impact assessment
LF	Landfilling

LMH	Litres per square metre per hour (L/m^2 h)
LRV	Log Removal Value
LS	Lime stabilization
MBBR	Moving-bed bioreactor
MBfR	Membrane biofilm reactor
MBMBR	moving-bed membrane bioreactor
MBR	Membrane bioreactors
MCDA	Multi-criteria decision analysis
MCR	Membrane chemical reactors
MD	membrane distillation
MD-MBR	membrane distillation bioreactor
MEC	Microbial electrolysis cell
MechT	Mechanical thickening
MF	Microfiltration
MFC	Microbial fuel cell
MFC-MBR	microbial fuel cell membrane bioreactor
MFE	Membrane flux enhancer
MFZ	Multi-family apartment building zones
MIET	Missing inventory estimation tool
MLD	Mega liter per Day
MLE	Modified Ludzacke-Ettinger
MLE	Modified Lutzak-Ettinger
MLIT	Ministry of Land, Infrastructure and Transport
MLSS	Mixed Liquor Suspended Solids
MLTS	Mixed liquor total solids
MLVSS	Mixed liquor volatile suspended solids
MOF	metal-organic framework
MRC	Mitsubishi Rayon Corporation
MSBR	membrane sequencing batch reactor
MSS	Mineral Suspended Solid
MTBF	Mean time between failures
MTTR	Mean time to repair
MW	Molecular weight
MWCO	Molecular weight cut-off
NDSA	Naphthalene disulfonates
NEB	Net environmental benefit
NF	Nanofiltration
NIPS	Non-solvent-induced phase separation
NMR	Solid-state 13C-nuclear magnetic resonance spectroscopy

NOB	Nitrite-Oxidizing Bacteria
NP	Net profit
NP	Nonylphenol
NPX	Naproxen
NSAID	Nonsteroidal anti-inflammatory drugs
NTU	Nephelometric Turbidity Unit
OLR	Organic loading rate
oNP	Ortho-nitrophenol
P	Phosphorus
PAC	Powdered activated carbon
PACl	polyaluminum chloride
PAOs	Polyphosphate Accumulating Organisms
PAOs	Phosphorus accumulating organisms
PCP	Pentachlorphenol
PCR	Polymerase Chain Reaction
PCR–DGGE	Polymerase chain reaction-denaturing gradient gel electrophoresis
PDF	Peak daily flow
PDF	Peak daily flow capacity
PDT	Pressure decay test
PE	Polyethylene
PES	Polyethersulphone
PET	Polyethylene terephthalate
PFBA	Perfluorobutanoic acid
PFHxA	Perfluorohexanoic acid
PFHxS	Perfluorohexane sulfonate
PFOA	Perfluorooctanoic acid
PFOS	Perfluorooctane sulfonate
PFPeA	Perfluoropentanoic acid
PHF	Peak hourly flow
PLC	Programmable logic controller
PN/PC	Proteins/polysaccharides ratio
PP	Polypropylene
PPCPs	Pharmaceutical and Personal Care Products
PRO	pressure retarded osmosis
PRO-MBR	pressure retarded osmosis membrane bioreactor
PTFE	Polytetrafluoroethylene
PU	Polyurethane
PUB	Public Utilities Board
PVC	Polyvinyl chloride

PVDF	Polyvinylidene fluoride
PVDF	Polyvinylidene difluoride
PVDF-g-POEM	polyvinylidene fluoride-graft-polyoxyethylene methacrylate
PVP	Polyvinyl pyrrolidone
QMC-D	Quartz Crystal Microbalance with Dissipation monitoring
QQ	quorum quenching
RAS	Return Activated Sludge
RED	Reverse electrodialysis
RIS	Resistance-in-series
RO	Reverse Osmosis
ROX	Roxithromycin
rRNA	ribosomal Ribonucleic Acid
RSF	Relative Sensitivity Function
SAC	Spectra Absorption Coefficient
SADm	Specific aeration demand per unit of membrane area
SADp	Specific aeration demand per unit of permeate or filtrate flow
SBR	sequencing batch reactors
SCADA	Supervisory control and data acquisition
SEC	Size-exclusion chromatography
SEM	Scanning Electron Microscopy
SFZ	Single-family house zone
SGD_m	Specific gas demand
SMA	Simple moving average
sMBR	Separate MBR
sMBR	Sidestream MBR
SMP	Soluble Microbial Products
SMX	Sulfamethoxazole
SND-MBR	Simultaneous nitrification–denitrification membrane bioreactor
SNDN	simultaneous nitrification and denitrification
SRT	Sludge Retention Time
SRT	Solids retention time
SS	Suspended Solids
SSSG-ASM3-SMP	ASM3 combined with SMP model and simultaneous soluble substrate and storage
ST	Settling tank
STP	Sewage treatment plants
T	Temperature
T-RFLP	Terminal Restriction Fragment Length Polymorphism
TAN	Total Ammonia Nitrogen
TCE	Trichloroethylene

TDS	Total Dissolved Solids
TGGE	Temperature Gradient Gel Electrophoresis
THMs	Trihalomethanes
TIPS	Thermal-induced phase separation
TKN	Total Kjeldahl Nitrogen
TMBR	Tertiary membrane bioreactor
TMP	Trans-membrane pressure
TN	Total Nitrogen
TOC	Total Organic Carbon
TP	Total Phosphorus
TRI	Trimethoprim
TrOC	Trace organic contaminants
TS	Total solids
TSS	Total Suspended Solids
UAP	Substrate-utilization-associated products
UAPs	Utilization associated products
UASB	Upflow anaerobic sludge blanket
UASB	Upflow anaerobic sludge blanket bioreactor
UCT	University of Cape Town
UF	Ultrafiltration
UFO-MBR	Hybrid ultrafiltration osmotic MBR
UG	Superficial gas velocity
UNICEF	United Nations Children's Fund
USEPA	United States Environmental Protection Agency
UV	Ultraviolet
UV	Ultraviolet radiation
UV-AOP	Ultraviolet advanced oxidation process
V_{AER}/V_{AX}	Aeration volume and anoxic volume ratio
VFA	Volatile fatty acids
VOC	Volatile organic compound
VSS	Volatile Suspended Solids
WHO	World Health Organization
WRF	Water reclamation facility
WRP	Water reclamation plant
WTP	Water treatment plant
WwSP	Wastewater stabilization pond
WWTP	Wastewater treatment plant
XPS	X-ray photoelectron spectroscopy
ZS	Zirconium Sulphate

Nomenclature

Greek Letters

β	Salinity-surface tension correction factor
Δp	Transmembrane pressure
η_{20}	Viscosity of water at 20°C
η_T	Temperature-related viscosity of water
$\mu_{Adesign}$	Design growth rate, nitrifying (autotroph) bacteria
ω	Alpha-factor exponent coefficient
φ	Temperature correction factor for oxygen transfer
ρ_{sludge}	Density of sludge
ρ_A	Density of air
θ_{xaer}	Aerobic SRT
θ_X	Sludge retention time
μ_{max}	Maximum specific growth rate, 1/d
θ	Hydraulic retention time
θ_c	Sludge retention time
A	specific cake resistance, m/kg
$\mu_{ANO,max}$	Maximum growth rate of X_{ANO}
μ_{EBAP}	Maximum growth rate of EBAP degradation
μ_{GBAP},	Maximum growth rate of GBAP degradation
$\mu_{OHO,max}$	Maximum growth rate of X_{OHO}
$\mu_{PAO,max}$	Maximum growth rate of X_{PAO}
μ_{SMP}	Maximum specific growth rate of SMP for heterotrophs
μ_{UAP}	Maximum growth rate of UAP degradation
μ_P	Permeate dynamic viscosity

A	Age of the wastewater treatment plant, year
A2/O	Anaerobic/Anoxic/Oxic
A_M	Membrane surface area
AOTE	Actual oxygen transfer efficiency
b_{ANO}	Decay rate for X_{ANO}
B_i	Benefits
b_{OHO}	Decay rate for X_{OHO}
b_{PAO}	Decay rate for X_{PAO}
BOD	Biochemical oxygen demand removal efficiency
BOD	Biochemical oxygen demand, mg/L
BOD_5	Biological oxygen demand in five days
C_b	bulk MLSS concentration, kg/m^3
C_i	Costs
CAGR	Compound annual growth rate, %
COD	Chemical oxygen demand, mg/L
CS	Total operating costs for secondary treatment, €/m^3
CT	Total operating costs for tertiary treatment, €/m^3
DO	Dissolved oxygen
F	Correction factor for fouling of air diffusors
F/M	Food to microorganism ratio, g BOD/ (g MLVSS.d)
f_{BAP}	Fraction of BAP generated in biomass decayed
f_{SMP}	Fraction of SMP generated in biomass decay
f_{UAP}	Fraction of UAP generated in biomass decay
$f_{XU\_Bio,lys}$	fraction of X_U generated in biomass decay
F_S	Specific footprint of membrane module
F_{SPBio}	Biological sludge production
FSP	Sludge production
g	Gravitational acceleration
H_2O_2	Hydrogen peroxide
HCl	Hydrochloric acid
HNO_3	Nitric acid
$[HM]_e$	Concentration of a heavy metal in the effluent, mg/m^3
$[HM]_i$	Concentration of a heavy metal in the influent, mg/m^3
$[HM]_S$	Concentration of a heavy metal in the sludge, mg/kg
HRT	Hydraulic retention time, h
J'	Membrane flux
J_P	Permeate flow through the single pore
$J_{net\ Design}$	Design net membrane flux
K'_{EPS}	Fraction of electron shunted to EPS formation

K2*	Coefficient of BAP degradation referred to the biomass Xa
k_{BAP}	BAP formation coefficient
k_{biol}	Kinetic biodegradation constant (L/(g $_{SS}$ d)).
k_d	Endogenous decay rate constant, 1/d
K_d	Solid water partitioning coefficient
k_H	Henry's law constant (atm m^3/mol)
K_{ow}	Octanol–water partitioning coefficient
K_p	Sludge–water partition coefficient
K_S	Half-saturation constant, mg/L
k_{SMP}	SMP formation coefficient
k_{UAP}	UAP formation coefficient
K_1	Carbonaceous oxygen demand
K_1	Coefficient of EPS formation in the growth-associated term
K_2	Coefficient of EPS formation in the non-growth term
K_2	Respiration oxygen demand
K_{BAP}	Half-saturation constant for BAP formation
K_{EBAP}	Biomass affinity constant for EBAP
K_{EPS}	EPS formation rate
k_f	Coefficient of fouling propensity
K_{GBAP}	biomass affinity constant for GBAP
$K_{hyd,BAP}$	Hydrolysis rate for BAP
$K_{hyd,UAP}$	Hydrolysis rate for UAP
k_{hyd1}	EPS hydrolysis rate for growth-associated BAP
k_{hyd2}	EPS hydrolysis rate for endogeny-associated BAP
K_{hyd}	Hydrolysis rate of EPS
$K_{NHx,ANO}$	Half saturation parameter for S_{NHx} for ANO
$K_{NOx,OHO}$	Half saturation parameter for S_{NOx} for OHO
$K_{O2,ANO}$	Half saturation parameter for S_{O2} for ANO
$K_{O2,OHO}$	Half saturation parameter for S_{O2} for OHO
K_S	External substrate affinity constant (half-saturation constant)
K_{SMP}	SMP half-saturation constant for heterotrophs
K_{UAP}	Half-saturation constant for UAP
kWhr	Kilowatt-hours
LMH	Filtration flux, L/(m^2 h)
log *D*	Effective octanol–water partitioning coefficient
M_{BOD}	Daily load of BOD5
MF	Microfiltration
MLSS	Mixed liquor suspended solids, mg/L
MLVSS	Mixed liquor volatile suspended solids, mg/L

M_{NO_x}	Daily load of nitrous oxide
M_O	Daily oxygen demand
M_{XAnox}	Amount of anoxic biomass
$n_{\mu OHO,Ax}$	Reduction factor for anoxic growth of X_{OHO}
N_T	Nitrogen removal efficiency
NaOCl	Sodium hypochlorite
NaOH	Caustic soda lye
NEB	Net environmental benefit
NF	Nanofiltration
$NH_3 - N$	Ammonia nitrogen, mg/L
NH_4-N	Ammonium Nitrogen
NP	Net profit
O_3/H_2O_2	Ozonation and Hydrogen Peroxide
OA_m	Mass percentage of oxygen in air
OA_v	Volume percentage of oxygen in air
OTR	Oxygen transfer rate
P	Phosphorous
P_{atm}	Atmospheric pressure
P_T	Phosphorous removal efficiency
PE	Polyethylene
pH	potential of Hydrogen
PI_{NO}	Potential impact of the null option
PI_{TP}	Potential impact of the treatment process
PI_{TW}	Potential impact of the treated water
PO_4^{3-}	Phosphate
Q	Feed flow rate, m^3/d
Q	Inflow, inflow
Q_{design}	Design flow
Q_{peak}	Peak inflow, peak inflow
Q_R	Recycled flow rate, m^3/d
q_{BAP}	Maximum specific BAP utilization rate
q_{UAP}	Maximum specific UAP utilization rate
Q_A	Airflow
r	Returned sludge ratio ($=Q_R/Q$)
R_i	Recycle rate, index i
R_{tot}	Total filtration resistance
R_P	Hydraulic resistance of the single pore
r_s	Substrate utilization rate
r_{UAP}	Specific UAP utilization rate

Rc	cake layer resistance, 1/m
RE	Heavy metal removal efficiency, %
$r_{X,s}$	Specific nitrogen removal rate
S_F	Fermentable organic matter
S	Substrate concentration, mg/L
S_0	Feed substrate concentration, mg/L
S_{Ac}	Fermentation product (considered to be acetate)
S_{Alk}	Alkalinity (HCO_3^-)
$S_{B,N}$	Soluble biodegradable organic N
S_B	Soluble biodegradable organics
S_{N2}	Dissolved nitrogen gas
S_{NHx}	Ammonium and ammonia nitrogen (NH_4 + NH_3)
S_{NO2}	Nitrite and acid nitrous (NO_2 + HNO_2)
S_{NOx}	Nitrate and nitrite (considered NO_3 only for stoichiometry)
S_{O2}	Dissolved oxygen
S_{PO4}	Soluble inorganic phosphorus
S_U	Soluble undegradable organics
S_{BAP}	BAP concentration
S_{EBAP},	EBAP endogeny-associated BAP
S_{GBAP}	GBAP growth-associated BAP
S_{SMP}	Total soluble microbial products
S_{UAP}	UAP concentration
SADm	Membrane area-specific aeration demand
SEDAm	Membrane area-specific energy demand for aeration
SF	Safety factor
$sF_{SP,Bio}$	Specific biological sludge production
SOTE	specific oxygen transfer efficiency
SRT	Solids retention time, d
SS	Suspended solids concentration, mg/L
T	Temperature
TiO_2/UV	TiO_2-photocatalyzed UV oxidation
TMP	Transmembrane pressure, kPa
TN	Total nitrogen
TP	Total phosphorous
TSS	Total suspended solids
Tsust	Sustainable flux operating time
V	permeate volume per unit area, m^3/m^2
V	volume of a reactor, m^3
Vbio	Bioreactor volume

VMnet	Net volume of membrane compartment
VSS	Volatile suspended solids, mg/L
Vtot	Total volume of membrane bioreactor
X	MLSS concentration, mg/L
X_{ANO}	Autotrophic nitrifying organisms (NH_4 to NO_3-)
X_{MeOH}	Metal hydroxide compounds
X_{MeP}	Metal phosphate compounds
$X_{OHO,Stor}$	Storage compound in OHOs
X_{OHO}	Ordinary heterotrophic organisms
$X_{PAO,PHA}$	Storage compound in PAOs
$X_{PAO,PP}$	Stored polyphosphates in PAOs
X_{PAO}	Phosphorus accumulating organisms
X_{TSS}	Total suspended solids
$X_{U,E}$	Particulate undegradable endogenous products
$X_{U,Inf}$	Particulate undegradable organics from the influent
X_U	Particulate undegradable organics
X_a	Active biomass
X_{EPS}	EPS concentration
Xbio	Solids concentration in bioreactor
$XC_{B,N}$	Particulate biodegradable organic N
XC_B	Particulate biodegradable organics
XDN	Biomass concentration denitrification basin
XM	Solids concentration in membrane tank
XN	Biomass concentration nitrification basin
XRi	Solids concentration in reactor
Y	Biomass yield, mass of cells formed per mass of substrate consumed, gVSS/gBOD
Y_{ANO}	Yield for X_{ANO} growth
Y_{EBAP}	Yield coefficient for growth on EBAP
Y_{GBAP}	Yield coefficient for growth on GBAP
Y_{OHO}	Yield for X_{OHO} growth
Y_{PAO}	Yield for X_{PAO} growth
$Y_{SB,STO}$	Yield for $X_{OHO,Stor}$ formation per SB
$Y_{Stor\_OHO,Ax}$	Yield for X_{OHO} growth per $X_{OHO,Stor}$ (Anoxic)
$Y_{Stor\_OHO,Ox}$	Yield for X_{OHO} growth per $X_{OHO,Stor}$ (Aerobic)
Y_{UAP}	Yield coefficient for growth on UAP
yi	Immersion depth

About the editors

This book on membrane biological reactors (MBR) has been edited by a team of outstanding academics who have eminently complementary expertise, strongly focusing on water and wastewater treatment and reuse. In addition to a strong international research network, the editors have an extensive collaborative network with the water industry and have led several prestigious international projects in this field. To date, the editors remain highly active in research and publication encompassing development of next generation MBRs.

Dr. Faisal I. Hai is an associate professor and an academic program leader at the School of Civil, Mining and Environmental Engineering of the University of Wollongong (UOW), Australia. Dr. Hai is the director of the Strategic Water Infrastructure Laboratory (SWIL) which has a longstanding track record of collaboration with leading MBR research groups worldwide. He is also the coordinator of the Water and Bioresource Technologies group of the Simulation, Modelling, Analysis, Research and Teaching (SMART) Research Facility - a major multidisciplinary strategic research initiative at UOW. Dr. Hai is the Environmental Engineering cluster representative to the Faculty Research Committee. Environmental Engineering at UOW has received the highest rating in the 2018 Excellence in Research for Australia (ERA). With nearly two decades of research experience gained in Japan and Australia in the field of MBR, the main breakthroughs achieved by Dr. Hai in this particular field include: development of fouling-resistant compact membrane modules for MBR, formulation of strategies for bio-augmentation of special degrading microbes, resulting in improved biodegradation of recalcitrant pollutants by MBR and coupling of physicochemical processes with aerobic/anaerobic MBR to enhance overall pollutant removal. A recipient of Japan Society for Promotion of Science fellowship (2007) and UOW Vice Chancellor's fellowship (2010), Dr. Hai was awarded with the outstanding lecturer award by the Japanese Society on Water Environment in 2010 and the UOW Outstanding Contribution to Teaching and Learning (OCTAL) award in 2015. He is an editor of Water Science and Technology journal (IWA publishing).

Professor Kazuo Yamamoto is a recognised authority on MBR technology. The first descriptions of the MBR technology date back the 1960s. However, the breakthrough for this technology came in 1989 when Prof Yamamoto's research group reported the development of the submerged MBR system. After completing an illustrious career in research and administration at the University of Tokyo, Prof Yamamoto remains research active through his affiliations with University of Tokyo, Chuo University and Kindai University, Japan. A few of the landmark contributions in this specific field from his research group include: invention of the submerged MBR, ultra-low pressure nanofiltration for water treatment, introduction of the high retention (e.g., nanofiltration) MBR for advanced wastewater treatment and the simultaneous water/energy production concept via MBR. The applied research conducted by Prof. Yamamoto has been in the forefront to shape the MBR technology into a reliable wastewater reuse technology worldwide. His contribution to the field has been recognized by several prestigious awards: Membrane Technology Award (2009, IWA) and Sidney Loeb Award (2008, European Desalination Society) to name a few.

Professor Chung-Hak Lee is an emeritus professor at the School of Chemical and Biological Engineering, Seoul National University, Korea. Through around four decades of distinguished research activities on membrane application to water environment, Professor Lee's group (Water Environment Membrane Technology Group, Seoul National University, Korea) made significant contribution to the development of the next generation membrane-based innovative water treatment technologies. Prof. Lee's research has opened up a new horizon to the analysis and control of membrane fouling in various membrane processes: elucidating the key foulants (inorganic precipitates) in anaerobic MBR, experimental determination of porosity of biocake on membrane in MBR and control of biofouling from the viewpoints of molecular biology (enzymatic and bacterial quorum quenching). Prof. Lee has served in a number of international scientific bodies and editorial boards of journal: he served as an associate editor of the journal *Water Research* and the chair of Membrane Technology Specialist Group, IWA (UK).

Preface

It is a great pleasure to present to you the second edition of this book following the kind acceptance of the first edition by the colleagues at the academia and industry. The first edition has been one of the best sellers from IWA publishing in recent years.

The MBR market continues to experience a massive growth, and best practice is constantly changing. This up-to-date second edition comprehensively covers the salient features and emerging issues associated with MBR technology. It provides thorough coverage, starting from biological aspects and fundamentals of membranes, via modeling and design concepts, to practitioners' perspective and good application examples. Particularly, the book presents the current status of the technology including market drivers/restraints and development trends. Coverage of process fundamentals (both biological and membrane components) has been improved and two new chapters have been added on reuse applications and practical design problems. Newly presented are developments and hybrid processes for specialised wastewaters, energy efficiency and sustainability, process automation, case studies and worked out design examples. The new edition is a valuable reference for academics and professionals and suitable for undergraduate and postgraduate teaching in Environmental Engineering, Chemical Engineering and Biotechnology and acts as a valuable information base for state of the art research.

The content of the book has been designed to include enough basic information for teaching purposes and practical information for the practitioners, while sufficient depth and breadth of information has been presented for use in research. In this edition, most chapters have gone through moderate to significant revisions. Some chapter numbers in the second edition have been rearranged to better fit the newly introduced content. With two new chapters (6 and 15), there are now 15 chapters. Chapter 1 systematically introduces the membrane biological reactors, a brief history of the MBR technology development and the drivers/constraints as well as the current status of the MBR market. The membrane and biological fundamentals are well covered in Chapter 2. The design, operation and maintenance

Editor(s): Faisal I. Hai, Kazuo Yamamoto and Chung-Hak Lee
doi: 10.2166/9781780409177_xxxi

issues are covered in Chapter 3 and 4, with Chapter 7 particularly covering the potential impact of hazardous events on MBR operation. Chapter 5, 6 (new) and 13 focus on important aspects of advanced wastewater treatment and reuse, namely, nutrient, pathogen and emerging contaminants removal. Chapter 8 evaluates the economic and environmental implications of MBR using Cost Benefit Analysis and Life Cycle Assessment, while Chapter 9 provides a state of the art review of the MBR modeling studies for system design and process optimization, with some discussion on models for evaluating operational and control strategies. Chapter 10 presents a candid overview of the other forms of MBRs besides the common biosolids separation MBR, namely, gas-diffusion, extractive, biocatalytic, and electrochemical MBR for the treatment of, or resource recovery from, wastewater. Chapter 11 provides a timely comprehensive review of the anaerobic MBR technology. Chapter 12 introduces the near future of the MBR technology—from its integration with conventional water reclamation technologies to the development of novel membranes and novel MBR configurations. The commercial MBR membrane products and their application to wastewater treatment are described in Chapter 14. Chapter 15 (new) includes mathematical/design problems to illustrate the concepts communicated in the previous chapters – this is likely to make the book even more attractive to students and practicing engineers.

Some overlaps between the chapters were inevitable to maintain continuity; however, effort was made to minimize repetitions while ensuring a degree of coherence throughout the book via cross-referencing between sections and chapters. A detailed table of contents and a list of symbols, nomenclature and abbreviations are placed at the beginning of each chapter for easy reference. We have tried our best to ensure that the information presented is as accurate and complete as possible. The authors cannot be held liable for the inaccuracies and omissions that may still exist despite the best efforts. The editors, of course, welcome any constructive criticism from readers.

This preface would remain incomplete without offering the most grateful and sincere thanks to the contributors of the aforementioned chapters—over 50 authors from Australia, Canada, China, Germany, Italy, Japan, Korea, Singapore, Spain, Sweden, Thailand, UK and USA, both from academia and industry. We gratefully acknowledge the advice and encouragement of many top researchers in the field, including, in alphabetical order, Professor Duu-Jong Lee (National Taiwan University, Taiwan), Professor Garry Amy (King Abdullah University of Science and Technology, Saudi Arabia), Professor Ismail Koyuncu (Istanbul Technical University, Turkey), Professor Roger Ben Aim (Institut De La Filtration Et Des Techniques Separatives, IFTS, France), Professor Simon Judd (Cranfield University, UK – and the author of 'The MBR Book'), Professor Saravanamuth Vigneswaran (UTS, Australia), Professor Shinichiro Ohgaki (Japan Water Research Center, Japan), Prof Thomas Wintgens (University of Applied Sciences and Arts, Northwestern, Switzerland), Professor Tony Fane (UNSW, Australia), Professor Tor Ove Leiknes (King Abdullah University of Science and Technology, Saudi Arabia) and Professor Xia Huang (Tsinghua University, China). Last but not least, special thanks go to the wonderful editorial and marketing team of IWA publishing, particularly the book commissioning editor Mark Hammond and production editor Niall Cunniffe.

Faisal I. Hai, University of Wollongong, Australia
Kazuo Yamamoto, University of Tokyo, Japan
Chung-Hak Lee, Seoul National University, Korea

Dedication

The editors dedicate this humble attempt to their loving families.

Faisal Hai dedicates this work to his late father Md. Abdul Hai, who was a great admirer of his work and a constant source of inspiration:

"Abba, I vividly remember how excited and proud you were when the first edition of this book came out. Wish I could send the first copy of this edition too to you!"

Chapter 1

Introduction to membrane biological reactors

Faisal I. Hai[1], Kazuo Yamamoto[2,3,4] and Chung-Hak Lee[5]*

[1]Strategic Water Infrastructure Laboratory, School of Civil, Mining and Environmental Engineering, The University of Wollongong, Australia
[2]Environmental Risk Management and Quality Control Technology Laboratory, Environmental Science Center, University of Tokyo, Japan
[3]Research and Development Initiative, Chuo University, Japan
[4]Kindai University, Japan
[5]Water Environment Membrane Technology Laboratory, Seoul National University, Korea
**Corresponding author: faisal@uow.edu.au*

ABSTRACT

This chapter delineates the rationale of combining the membrane and bioreactor technologies and systematically introduces the membrane biological reactors that are the focus of this book. This is followed by a brief account of the history of the various uses of membranes in conjunction with bioreactors and how the initial formats led to the development of the present day commercialized ones. The relative advantages of membrane biological reactors over the conventional biological processes are presented. The current status of the membrane bioreactor (MBR) market is described and the drivers propelling its growth along with the constraints are highlighted. The trends in world-wide MBR research is succinctly presented in order to assess whether academic research has so far been able to maintain close relationship with the specific practical requirements in the industry.

1.1 MEMBRANE BIOLOGICAL REACTORS – DEFINITION AND APPLICATION

Membrane biological reactors refer to the technologies based on the combination of membranes and biological reactors for the treatment or resource recovery from wastewater. Combining membrane

 Membrane Biological Reactors: Theory, Modeling, Design, Management and Applications to Wastewater Reuse
Editor(s): Faisal I. Hai, Kazuo Yamamoto and Chung-Hak Lee
doi: 10.2166/9781780409177_0001

separation with biochemical conversion has led to a range of innovative environmental biotechnology applications, namely, biosolids separation, gas diffusion, extractive, biocatalytic, and electrochemical membrane biological reactors (Hai & Yamamoto, 2011; Hai *et al.*, 2013; Judd, 2011). In biosolids separation membrane biological reactors, membranes reject the solid materials developed by the biological process to provide a clarified and disinfected product. Gas-diffusion membrane biological reactors facilitate efficient delivery of a gaseous oxidizing or reducing agents such as oxygen, hydrogen, or methane to microbial biofilms treating wastewater (Hwang *et al.*, 2009). Extractive membrane biological reactors have been devised for the transfer of degradable organic pollutants from hostile industrial wastewaters, via a nonporous silicone membrane, to a nutrient medium for subsequent biodegradation (Livingston, 1994). Biocatalytic membrane biological reactors utilize enzymes or immobilized microbial cells for degradation of persistent xenobiotics or for synthesis of fine chemicals (Hai *et al.*, 2013). Electrochemical membrane biological reactors make it possible to utilize organic compounds in wastewater for production of energy or chemicals (Logan & Rabaey, 2012).

Biosolids separation is, however, the most widely studied type and has found full-scale applications in many countries. Recent comprehensive reviews point to the vast majority of research on biosolid separation type applications, paralleling the commercial success in this field. In line with the current trend of research and commercial application, this book will focus on the biosolids separation membrane biological reactors, which, henceforth, will be denoted as membrane bioreactor (MBR) according to the common trend. However, Chapter 10 focuses on the remainder of the aforementioned types of membrane biological reactors that are currently in the research and development stage, but can potentially contribute to more efficient removal of pollutants and recovery of resources from wastewater. Discussion on other forms of membrane biological reactors such as that for waste gas treatment (Reij *et al.*, 1998), is beyond the scope of this book.

Solid–liquid separation by membranes in MBRs combines clarification and filtration of a conventional activated sludge (CAS) process into a simplified, single-step process. Membranes are seldom used by themselves to filter untreated wastewater, since fouling prevents the establishment of a steady state and because water recovery is too low (Fuchs *et al.*, 2005; Schrader *et al.*, 2005). However, when used in conjunction with the biological process, it converts dissolved organic matter into suspended biomass, reducing membrane fouling and allowing recovery to be increased (Gallucci *et al.*, 2011; Hai & Yamamoto, 2011). In addition, the membrane filtration process introduced into bioreactors not only replaces the settling unit for solid–liquid separation but also forms an absolute barrier to solids and bacteria and retain them in the process tank, giving rise to several advantages (see section 1.3) over the CAS process.

1.2 HISTORICAL DEVELOPMENT OF BIOSOLIDS SEPARATION IN MBRS

The period between the 1960s and the 1980s is often regarded as being the golden age of membrane science (Judd, 2011). The crucial breakthrough was the development of the asymmetric cellulose acetate membrane for reverse osmosis by Loeb and Souriarajan in 1963 (Loeb & Sourirajan, 1964). This along with some other early-stage developments, as listed in Table 1.1, paved the way for the development of the present-day membranes for MBRs and still has an impact on academic research and industrial applications.

Different combinations of membrane solid–liquid separators in biological treatment processes have been studied since the 1960s (Visvanathan *et al.*, 2000). The trends that led to the development of today's MBR are depicted in Table 1.2 and Figure 1.1. The progress of membrane manufacturing technology and its applications led to the eventual replacement of tertiary treatment steps by microfiltration or ultrafiltration (Figure 1.1a). Parallel to this development, microfiltration or ultrafiltration was used for solid–liquid separation in the biological treatment process, thereby eliminating the need of the sedimentation step (Figure 1.1b). The original process, which combined the use of an activated sludge bioreactor with a crossflow membrane filtration loop, was introduced by Dorr-Olivier Inc. (Smith *et al.*, 1969) (Table 1.2). Due to the poor economics of the first-generation MBRs (e.g. high cost of membranes and high-energy consumption due to high-pressure recirculation of sludge through the membrane), apart from a few examples such as installations at the basement level of sky scrapers in Tokyo, Japan for the wastewater reuse for toilet flushing, they usually found applications only in niche areas with special needs such as isolated trailer parks or ski resorts for example (Hai & Yamamoto, 2011).

Table 1.1 Historical milestones leading to the development of present day porous membranes for MBR at a glance.

Milestone	Selected References
Fick's phenomenological laws of diffusion	Fick (1855)
van't Hoff's (1887, 1888) osmotic pressure equation	van't Hoff (1887, 1888)
Use of bovine heart membranes (1–50 nm) to separate soluble *Acacia* by Schmidt – arguably the first documented ultrafiltration (UF) experiment	Schmidt (1856)
Graham's pioneering work in gas separation using both porous and dense membranes	Graham (1861, 1866)
First synthetic UF membranes preparation; introducing membrane bubble points test; proposing the term 'ultrafilter'	Bechhold (1907)
'Dry inversion' method to produce porous collodion membrane in an industrial scale (leading to the establishment of the world's first commercial microporous membrane supplier, Sartorius Werke GmbH in 1925)	Zsigmondy and Bachmann (1918, 1922)
Introduction of 'vapour-induced phase separation' formation method leading to establishment of Millipore Corporation in 1954	Goetz and Tsuneishi (1951)
Development of the higher-flux, asymmetric cellulose acetate membrane by 'wet phase inversion' or 'non-solvent-induced phase separation' (NIPS)	Loeb and Sourirajan (1964)
General applicability of new kinds of UF membranes prepared by using various polymers on an industrial scale (collaboration between Amicon Inc. collaborated with Dorr-Oliver Inc.)	Michaels (1963)
Commercialization of thermally induced phase separation (TIPS) microfiltration (MF) membranes (greater flux than NIPS membranes)	Castro (1981)
Radiation track etching method of membrane development (limited application in the manufacture of flat membrane due to its poor permeability and high cost)	Fleischer *et al.* (1969)
Development of the less expensive melt extrusion and cold-stretching method by Celanese Corp. in 1974	Druin *et al.* (1974)
Development of a hollow-fibre (HF) polyethylene (PE) MF membrane by melt extrusion and cold-stretching method in 1977 by Mitsubishi Rayon Corp.	Kamada *et al.* (1977)

The breakthrough for the MBR came in 1989 by submerging the membranes in the bioreactor itself and withdrawing the treated water by suction through the membrane (Yamamoto *et al.*, 1989) (Figure 1.1c). Eventually two broad trends have emerged today, namely submerged or immersed MBRs (iMBR) and sidestream MBRs (sMBR). Submerged technologies tend to be more cost effective for larger scale lower strength applications, and sidestream technologies are favoured for smaller scale higher strength

Table 1.2 Milestones at the early stage of development of the commercial MBRs from the CAS processes at a glance.

Milestone	Selected References
Sidestream	
Early bench-scale reports of solids separation by ultrafiltration for concentrated activated sludge	Grethlein (1978), Smith *et al.* (1969)
The first commercial sMBRs developed by Dorr-Oliver Inc., combining UF with a CAS process, was introduced in 1969 for application to ship-board sewage treatment	Bailey *et al.* (1971), Bemberis *et al.* (1971)
In 1974, Thetford Systems (later became part of Zenon Environmental which is now GE Water & Process Technologies) manufactured an on-site treatment and recycling system called 'Cycle-Let' incorporating an aerobic–anoxic activated sludge process with tubular ultrafiltration in a two-pump feed and bleed loop	Benedek and Cote, (2003), Gaines *et al.* (2005)
In 1974, Dorr-Oliver Inc. system marketed in Japan under license to Sanki Engineering, with some success up until the early 1990s.	Mishra *et al.* (1994)
Development of systems combining anaerobic digester and UF membrane in South Africa in late 1970s	Botha *et al.* (1992), Grethlein (1978)
Commercialization of Dorr-Oliver Inc. anaerobic sMBRs for high-strength wastewater treatment in early 1980s	Sutton *et al.* (1983)
Submerged	
In Japan (1989), pioneering work by Yamamoto *et al.* led to development of a submerged hollow-fibre UF MBR process	Yamamoto *et al.* (1989)
Development of a FS microfiltration submerged MBR by Kubota in Japan.	Ishida *et al.* (1994)
In Japan, Mitsubishi Rayon introduced its SUR MBR membrane module, based on its Sterapore product in 1993	Mitsuhashi and Saito (1993)
Zenon Environmental receives patent for their ZenoGem submerged HF UF MBR process	Behmann (1996), Tonelli and Canning (1993)

applications. The sMBR envelope has been extended in recent years by the development of the air lift concept, which bridges the gap between submerged and crossflow sMBR, and may have the potential to challenge submerged systems in larger scale applications (Pearce, 2008) (Figure 1.1e). It is worth noting that anaerobic MBRs (AnMBR) have also been developed to combine advantages such as requirement of no energy for aeration, low sludge production and energy (biogas) production (Liao *et al.*, 2006). Chapter 11 focuses on AnMBRs. As for the development of MBR products in the late 1990s onwards, it is interesting to note that, whereas the first half of the 1990s saw the launch of only three major iMBR products, namely, Zenon, Kubota and Mitsubishi Rayon, from just two countries (the United States and Japan), the first five years of the following decade saw the launch of at least 10 products originating from seven countries (Judd, 2011), reflecting the steadily expanding nature of the global MBR market. Chapter 14 provides detailed information regarding the commercial MBR formats and the key suppliers.

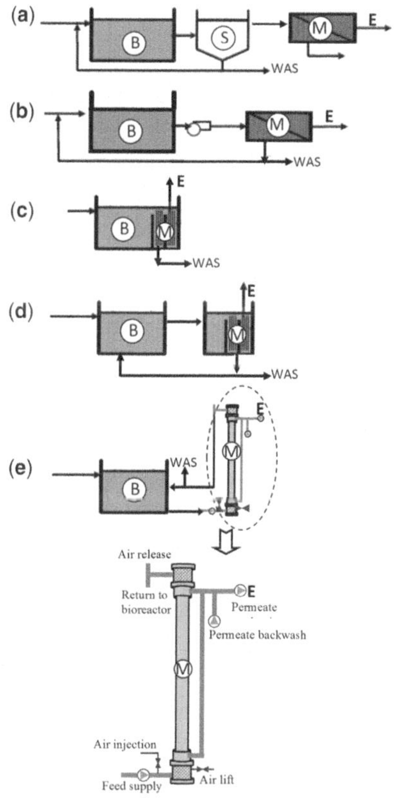

Figure 1.1 Evolution of membrane use in conjunction with bioreactor. (a) CAS followed by tertiary filtration (MF/UF), (b) sMBR, (c) Submerged MBR (integrated), (d) Submerged MBR (separated), (e) Airlift sMBR. (B = bioreactor, S = settling tank, M = membrane, E = effluent (permeate), WAS = waste-activated sludge).

1.3 PROCESS COMPARISON WITH CAS PROCESS

In this section, some important basic characteristics of CAS process and MBR are compared and the relative advantages of MBRs are highlighted.

1.3.1 Treatment efficiency

The biological treatment in MBR is performed according to the principles known from activated sludge treatment. However, colloidal and some soluble compounds, after being adsorbed on suspended solids (SS), can be retained in MBR and may be better biodegraded. Some studies have ascribed the better removal of soluble COD in MBR to the fact that the effluent is particle free (Cote *et al.*, 1997; Engelhardt *et al.*, 1998; Judd, 2008). Due to the complete retention of sludge by membrane, compared to CAS, microbial communities in MBR may better acclimatize to the fluctuations in wastewater characteristics and bioreactor operating conditions. Thus, the quality of the treated effluent from MBR may be better than that from CAS process (Radjenović *et al.*, 2008).

MBR produces quality effluent suitable for reuse applications or as a high-quality feed water source for reverse osmosis treatment. Indicative output quality include SS < 1 mg/L, turbidity <0.2 NTU and up to 6 log removal of virus (depending on the membrane nominal pore size and type of virus) (Hai *et al.*, 2014). In addition, it provides a barrier to certain chlorine-resistant pathogens such as *Cryptosporidium* and *Giardia*. In comparison to the CAS process that typically achieves 95%, COD removal can be increased to 96–99% in MBRs (Berube *et al.*, 2010). Total nitrogen removal through the inclusion of an anoxic zone is possible in MBR systems. Enhanced biological phosphorus removal (EBPR) can be implemented, which requires an additional anaerobic process step. Some characteristics of MBR technology render EBPR in combination with post-denitrification an attractive alternative that achieves very low nutrient effluent concentrations (Drews *et al.*, 2005). Detailed information on performance of full-scale MBRs for the removal of bulk organics, nutrients and pathogens is available in Chapters 5 and 6.

1.3.2 Sludge production and handling

Sludge yield in MBR is slightly higher due to complete retention of particles and colloids by the membrane. Compared to the CAS process, MBRs can be operated at comparatively longer solids retention times (SRT) with higher mixed liquor suspended solids (MLSS) concentration (low F/M ratio) to reduce the sludge production. However, this is associated with inefficient mixing, raising aeration demand for the biological metabolism and membrane cleaning, which largely offsets the economic advantage of less sludge production (Lesjean *et al.*, 2011). It should, however, be noted that the primary sludge production in case of MBR may be lower. Compared to primary settling, fine screen may be better suited for MBR due to better removal of fibrous materials that can affect membrane fouling. Screened water was observed to contain 30% more solids than settled water (Jimenez *et al.*, 2010). Waste-activated sludge (WAS) handling and disposal for MBRs are not significantly different from CAS systems (WEF, 2012). The presence of higher levels of filamentous bacteria and colloidal material in MBR sludge can lead to less efficient centrifugation and floatation (Merlo *et al.*, 2007). However, arguably MBR sludge requires similar flocculant dosage to reach the highest filterability and lowest specific filtration resistance (Chu *et al.*, 2007; Galil & Jacob, 2009; Khongnakorn & Wisniewski, 2007). Because MBR systems can be designed to work with high MLSS concentration and long SRT, sludge dewatering is required in reduced frequencies. Sludge dewatering can take place directly from the aeration tank, requiring no clarifier/sludge holding tank.

1.3.3 Space requirements

One of the advantages of MBR is its compactness, because large sedimentation tanks are not needed. A further reduction in foot print is made possible by the high MLSS concentration that can be applied in an MBR. Brescia wastewater treatment plant (WWTP) in Italy, which was the world's largest MBR retrofit of an existing conventional plant in 2002, gives a full-scale example of a ratio of two when comparing the area needed by CAS and MBR (Brepols *et al.*, 2008). The area required for large scale ($\geq$10,000 m^3/d) MBRs may be in the range of 0.5–1.2 m^2/m^3 d, while the area required for CAS-based wastewater treatment plants (treatment capacity = 10,000–50,000 m^3/d) with and without tertiary treatment has been reported to range from 0.85 to 1.2 and 1.2 to 1.6 m^2/m^3 d, respectively (Xiao *et al.*, 2018).

1.3.4 Wastewater treatment cost

It is difficult to make a general economic comparison between MBR and CAS systems. First of all, the reference system should not simply be an activated sludge system, but a system that produces an effluent of the same quality. It should be furthermore noted that although the equipment and energy costs of MBR are higher than conventional options, total water costs can be competitive due to the smaller footprint and lower installation costs (Lesjean *et al.*, 2004, 2011; Pearce, 2008). Xiao *et al.* (2018) performed statistical analysis of the capital cost of large scale ($\geq$10,000 m^3/d) MBRs in China. They reported that capital cost of MBR ranges from approximately 380–800 USD/m^3 d, and the average capital cost of MBR is about 600 USD/m^3 d. On the other hand, the average capital cost of all CAS-based municipal wastewater in China is 380 USD/m^3 d. However, the addition of tertiary treatment processes (e.g. coagulation and adsorption) with the CAS process for achieving effluent quality comparable to that achieved by MBR, the capital cost of both technologies becomes almost the same (Xiao *et al.*, 2018). Notably, Young *et al.* (2013, 2014) compared the capital and operational expenditures of MBR and CAS, and demonstrated the cost benefits of MBR over CAS technology. In addition, the analysis of net present value of all costs over the full life revealed that MBR is a cost-effective technology as compared to CAS for situations where land price is high and effective nutrient removal is required (Young *et al.*, 2013, 2014).

1.3.5 Comparative energy usage

MBR provides an equivalent treatment level to membrane filtration after CAS (CAS-UF/MF), but at the expense of higher aeration demand since MBRs are generally operated at higher MLSS concentrations and longer SRT (and this is one of the reasons of the excellent biodegradation efficiency). CAS/UF-MF is estimated to add 0.1–0.2 kWhr/m^3 to the energy, equivalent to a total energy use for CAS-UF/MF of 0.35–0.5 kWhr/m^3 in a new facility (Lesjean *et al.*, 2004). Experience of large-scale commercial MBRs shows an energy usage of around 1.0 kWhr/m^3, though smaller scale facilities typically operate at 1.2–1.5 kWhr/m^3 or higher (Judd, 2011). However, in comparison to these values, energy consumption of around 1.9 kWhr/m^3 was reported in 2003 (Zhang *et al.*, 2003) and up to 2.5 kWhr/m^3 in 1999 (Ueda & Hata, 1999). This proves that gradual improvement in MBR design is occurring.

1.3.6 Relative advantages of MBR

The prime advantages of MBRs are the reuse standard treated water quality, the smaller footprint of the plant, the potential of less sludge production, and the flexibility of operation (Visvanathan *et al.*, 2000; WEF, 2012). As noted earlier, MBRs may provide an equivalent treatment level to membrane filtration after CAS (CAS-UF/MF), but in a single step, compact unit. The MBR system is particularly attractive when applied in situations where long SRTs are necessary and physical retention and subsequent hydrolysis are critical to achieving biological degradation of pollutants (Chen *et al.*, 2003).

MBRs eliminate several process difficulties of CAS. For example, because all the MBR process units can be tightly closed, odour dispersion can be prevented much efficiently. The absence of a clarifier, which also acts as a natural selector for settling organisms, enables sensitive, slow-growing species (e.g. nitrifying bacteria) to develop and persist in the system (Cicek *et al.*, 2001; Rosenberger & Kraume, 2002). MBRs can operate independent of the problems associated with settling, which is usually the most troublesome part of wastewater treatment by CAS processes. The potential for operating MBRs at comparatively higher SRTs without having the obstacle of settling allows high biomass (MLSS) concentrations in the bioreactor. Consequently, higher strength wastewater can be treated and lower biomass yields are realized (Muller *et al.*, 1995). This also results in more compact systems than conventional processes, significantly reducing plant footprint making it desirable for water recycling applications (Konopka *et al.*, 1996). The higher biomass concentrations also increases shock tolerance, which is particularly important where feed is highly variable (Xing *et al.*, 2000). The increased endogenous (autolytic) metabolism of the biomass (Liu & Tay, 2001) under long SRT allows development of predatory and grazing communities, with the accompanying trophic-level energy losses (Ghyoot & Verstraete, 2000). These factors, in addition to resulting in lower overall sludge production, may lead to higher mineralization efficiency than those of a CAS process. High molecular weight soluble compounds, which are not readily biodegradable in conventional systems, are retained in the MBR (Cicek *et al.*, 2003). Thus, their residence time is prolonged and the possibility of oxidation is improved. The system is also able to handle fluctuations in nutrient concentrations due to extensive biological acclimation and retention of decaying biomass (Cicek *et al.*, 1999).

1.4 FACTORS INFLUENCING PERFORMANCE/DESIGN CONSIDERATIONS

This section sheds light on some important design considerations of MBR. More detailed information has been furnished in Chapters 2 and 3.

1.4.1 Pre-treatment

Because of membrane sensitivity to the presence of foreign bodies, fine pre-screening of the feed (and sometimes of the mixed liquors) must occur. The type of sieve installed is very important regarding the total screening of hair, and fibres that may become embedded or wrapped around the hollow fibres (HFs) (Frechen *et al.*, 2006; Schier *et al.*, 2009). It is noteworthy that the investment in the pretreatment is of little use if the bioreactor is uncovered in which case different sorts of debris can easily enter the bioreactor.

1.4.2 Membrane maintenance

Membrane fouling occurs due to undesirable deposition and accumulation of microorganisms, colloids, solutes, and cell debris within pores or on membrane surfaces (Meng *et al.*, 2009). Membranes need to be replaced when it is irreversibly fouled. Irreversible fouling could be either due to the entrapment of colloidal/sludge particles in membrane channels/pores or due to the collapse of membrane pores. Clogging increases the selectivity of the membrane but reduces the permeability of the membrane. By contrast, breach in membrane active layer increases the permeability but affects the selectivity of membranes (Branch *et al.*, 2016; Meng *et al.*, 2009). Backwashing, relaxation and chemical cleaning of the membranes are the backbone of stable filtration performance (WEF, 2012). The end of the membrane lifetime may result from mechanical failure of the membrane or the module construction or irrecoverable loss of filtration performance. Faulty installation is one obvious reason of membrane failure. Other reasons associated with regular operation include frequent and/or extended contact between membrane and cleaning solution causing delamination of the membrane, scoring and cleaving of the membrane resulting from the presence of abrasive or sharp-edged materials in the influent, and operating stress and strain occurring in the system due to fibre movement and membrane backwashing (Childress *et al.*, 2005; Doyen *et al.*, 2010; Gijsbertsen-Abrahamse *et al.*, 2006). The life span of the membranes used is of key concern in determining the viability of MBR technology. Following a thorough literature survey, Judd (2017) noted that increasing the replacement frequency from five to 10 years can reduce the operating cost of a HF-iMBR by approximately 20%, and this reduction could be more than 20% for flat-sheet MBR. In a study conducted by Cote *et al.* (2012), membrane life was estimated by analysing the data from all the MBR plants that have been installed in North America since 1996. They found that the average life of the membrane is eight years, and the membrane life is even longer for the latest membrane models (Cote *et al.*, 2012).

1.4.3 Membrane selection and applied flux

An MBR membrane needs to be mechanically robust, chemically resistant to high chemical concentrations used in cleaning, and nonbiodegradable (Pearce, 2008) (see Chapter 14). Process flux in treating a wastewater feed is important since it will directly affect capital cost due to its effect on membrane area and footprint. Process flux also influences operating costs due to its effect on evolution of fouling and on chemical and air use (for cleaning).

1.4.4 SRT and MLSS concentration

While early MBRs were operated at SRTs as high as 100 d with MLSS concentrations of up to 30 g/L to obtain a highly compact process and to minimize sludge production or eliminate it all together, the recent trend is to apply a shorter SRT (around 10–20 d), resulting in more manageable MLSS levels (10–15 g/L).

1.5 MUNICIPAL VS INDUSTRIAL MBRs

1.5.1 Membrane types and materials

Membrane modules can be divided into three categories, namely, flat sheet (FS), HF and tabular/multi-tube (MT) membranes. FS and HF membranes can be employed in iMBR, while MT membranes are only used in sMBR (Judd, 2010b; Meng *et al.*, 2009). Ceramic and polymeric materials are commonly used for the manufacturing of UF/MF membranes. Ceramic membranes are suitable for high-strength wastewater due to the resistance against chemical agents (Ciora & Liu, 2003; Hofs *et al.*, 2011; Jin *et al.*, 2010). Since ceramic membranes are expensive (Hofs *et al.*, 2011), different polymeric materials, including polyvinylidene difluoride (PVDF), polyethylsulphone (PES), polyethylene (PE), polypropylene (PP) and polytetrafluoroethylene (PTFE) have been used for the synthesis of UF or MF membrane (Ahmed & Lan, 2012; Lin *et al.*, 2012; Ruel *et al.*, 2011). Indeed, a survey of full-scale municipal and industrial MBRs indicate that polymeric membranes have been preferred in MBRs (Figure 1.2). Among different polymeric membranes, PVDF, PE and PES membranes have been most commonly used in both municipal and industrial iMBRs (Judd, 2010b, 2014; Lin *et al.*, 2012).

The type of membrane modules (i.e. FS and HF) and materials (i.e. ceramic or polymeric) do not directly affect the biological performance of MBRs. For instance, Torres *et al.* (2008) operated parallel MBRs integrated with PVDF and PE membranes for the treatment of refinery wastewater, and achieved comparable chemical oxygen demand (COD) removal. Membrane modules and materials are selected based on factors such as cost and type of wastewater. Because compared to polymeric membranes, inorganic membranes show better thermal, hydraulic and chemical resistance, the latter may be more suitable for industrial wastewater. A few studies have investigated the performance of inorganic membranes such as aluminium, titanium dioxide and zirconium in lab-scale MBR (Meng *et al.*, 2009; Zhang *et al.*, 2005, 2006).

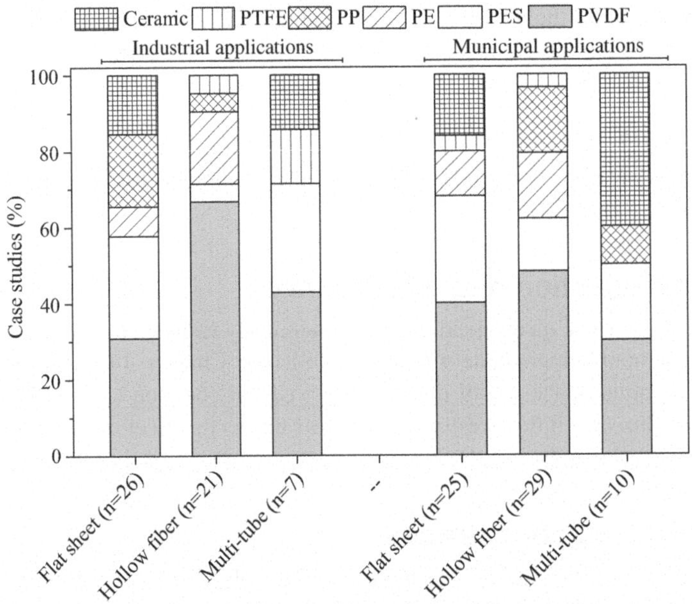

Figure 1.2 Materials used in different configuration of membranes for the treatment of industrial and municipal wastewater. 'n' indicates number of case studies; PTFE: polytetrafluoroethylene; PVDF: polyvinylidene difluoride; PE: polyethylenel; (PES): polyethylsulphone; and PP: polypropylene. Information Source (Judd, 2010b, 2014; Lin *et al.*, 2012).

1.5.2 Permeate flux

Permeate flux in MBRs is governed by a number of factors including MBR configuration, average/peak flows and type of wastewater streams (Judd, 2016). A wide variation in permeate flux applied for MBRs is noted in Table 1.3. The reported permeate flux varies from 8.5 to 25 and 10 to 45 L/m^2 h (LMH) for

Table 1.3 Permeate flux studied in different configurations of municipal and industrial MBRs. Information source: (Judd, 2010b, 2104).

Wastewater Stream	MBR Configuration	Permeate Flux (*L*/m^2 h)	
		Minimum – Maximum	Average ± SD
Food	iMBR	9–25	14 ± 3.7 (*n* = 13)
Food	Air-lift sMBR	30–45	40 ± 1.8 (*n* = 3)
Food	Pressurized sMBR	115–190	140 ± 2.1 (*n* = 5)
Textile	iMBR	20–30	28 ± 5.2 (*n* = 3)
Leachate	Pressurized sMBR	55–90	80 ± 2.3 (*n* = 4)
Municipal	iMBR	8.5 – 25^a/10 – 45^b	19 ± 2.4^a/25 ± 2.5^b (*n* = 38)
Municipal	Air-lift sMBR	NA	50 (*n* = 1)

[a]Based on average daily flow;
[b]Based on PDF.
SD: Standard deviation; n: no. of studies; NA: not available

municipal iMBRs at average and peak flow rates, respectively (Table 1.3). On the other hand, for the treatment of food industry wastewater, permeate flux ranged from 9 to 25, 30 to 45 and 115 to 190 LMH in iMBR, air-lift sMBR and pressurized sMBR, respectively (Table 1.3). Wastewater characteristics can significantly influence the hydraulic performance of MBRs. For example, permeate flux (55–90 LMH) achieved for the treatment of leachate in pressurized sMBR was significantly lower than those reported (115–190 LMH) for pressurized sMBR treating food industry wastewater (Judd, 2010b, 2014). In general, iMBR, that is the most common configuration, are operated at a permeate flux of 15–25 LMH (Xiao *et al.*, 2018).

1.5.3 Treatment performance

Activated sludge consumes the organic matter in wastewater and use the energy for cell growth and cell maintenance. This results in the removal of bulk organics. In addition, biosorption and accumulation in bacterial cells as well as sorption on the cake layer formed on membrane surface also contribute to the removal of organics in MBR (Carucci *et al.*, 2001; Radjenović *et al.*, 2008). More than 2000 full-scale MBRs have been installed or under construction, and most of them are for municipal wastewater treatment (Judd, 2016, 2010b). MBR has been reported to consistently achieve 90–95, 80–99 and 70–99% COD, ammonium nitrogen (NH_4-N) and total phosphorous (TP) removal, respectively (Giménez *et al.*, 2011; Gurung *et al.*, 2016, 2017; Khan *et al.*, 2017; Krzeminski *et al.*, 2012; Pagotto *et al.*, 2014; Phan *et al.*, 2015; Shen *et al.*, 2012). On the other hand, COD removal by MBR from industrial wastewater may not be as effective as reported for municipal MBRs. This is because industrial wastewater may contain resistant organic pollutants that are not easily biodegradable. For example, wastewater originating from paper and pulp contains di-terpene alcohols and chlorinated and acidic resins, and their removal by activated sludge is challenging. Similarly, high concentration of salts (1000–2000 mg/L) and heavy metals (e.g. chromium) in textile industry wastewater can inhibit the activated sludge in MBR (Lin *et al.*, 2012). Following a thorough survey of 51 reference sites, Judd (2014) reported that MBR can achieve 90–95% COD removal during the treatment of food and beverage wastewater. On the other hand, MBRs achieved COD removals ranging between 50 and 95% during the treatment of wastewater from textile, paper and pulp, pharmaceutical and petroleum industries (Judd, 2014). MBRs were reported to achieve effective COD removal only from young leacheates. COD removal ranged in between 40 and 80% for old leachates (Ahmed & Lan, 2012; Klimiuk & Kulikowska, 2006).

1.5.4 Capital cost

The capital cost of aMBR plant includes the costs of civil work, piping and plumbing, and equipment. In general, primary treatment, MBR plant (i.e. biological treatment and membrane), tertiary treatment, and sludge treatment units account for 10–20, 55–85, 0–30, and 5–10% of the total plant construction cost, respectively (Xiao *et al.*, 2014). Xiao *et al.* (2018) compared the capital cost of municipal and industrial MBRs in China: the average capital cost of industrial MBRs was approximately 50% more than municipal MBRs. The capital cost of industrial MBRs varies between 600 and 1200 USD/m^3 d, while 300–380 USD/m^3 d is required for the construction of municipal MBRs. It is worth noting that the unit capital and operating costs of municipal and industrial MBR reduces significantly with the increase in their treatment capacities (Judd, 2016; Santos *et al.*, 2010; Xiao *et al.*, 2014; Young *et al.*, 2013, 2014). For instance, Xiao *et al.* (2018) demonstrated that the unit capital cost of municipal MBR could reduce by almost 40% if the treatment capacity increases from 10,000 to 100,000 m^3/d.

1.6 MARKET DRIVERS/RESTRAINTS AND DEVELOPMENT TREND

1.6.1 Current status and typical drivers

The future of MBRs as a technology providing consistent and high-quality product water at a reduced footprint is highly promising, driven by increasing freshwater scarcity and the simultaneous drive to reuse wastewater (Judd, 2011). As summarized in Table 1.4, there are a number of drivers for MBR installation.

Table 1.4 MBR market dynamics – drivers, restraints, and way forward.

Drivers	Scarcity of Water Enhances Demand for High-Quality Effluent Required for Water Reuse Applications
	• Stringent environmental regulations • Legislation promoting the use of MBR technology. For example; o Common ministerial decision, 145116/2011 (Greece) o Decreto Ministeriale (Italuy) o State of Vermont statute (USA) o K.D.P. 269/2005 (Cyprus) o Baltic Sea Action Plan (nine countries from Baltic sea area) • Economy of space compared to other treatment equipment • MBR modules are easy to automate • Demand for large municipal applications and medium industrial onsite packaged solutions leads to the growth of various MBR technologies • Decrease in membrane prices help cope with the current financial constraints
Restraints	• High capital costs • Competition with other conventional water reclamation and recycling technologies hinders the penetration of MBR technology • Sophisticated system requires trained workers • Compared to CAS processes, MBR technology is new, and membrane equipment systems are unique to each vendor; therefore, preselection or prepurchase is required to ensure that membrane systems are designed adequately.
Opportunity	• High demand of MBRs in developing regions for municipal and industrial applications
Way forward	• Innovative solutions/ technology upgradation: compact design with higher flux and treatment quality with lower energy consumption is expected.

Legislation exerts arguably the greatest influence on the global MBR market. Legislation determines the discharge water quality standards, and, therefore, influences the choice of water and wastewater treatment technology for meeting statutory requirements. However, even without legislation, local water resourcing problems alone can provide sufficient motivation for recycling in water stressed countries (Judd, 2011; Lesjean *et al.*, 2011).

Membrane systems have modular designs allowing MBR technology to be successfully applied on a wide range of scales, ranging from package facilities treating the wastewater from individual buildings up to major centralized WWTPs (WEF, 2012). MBR technology is particularly well suited to small treatment systems because MBRs are highly automated and provide reliable treatment. Moreover, MBRs can be easily operated remotely and require minimal daily operator intervention. MBRs are easy to operate as long as the operators have been properly trained, pay strict attention to the proper operation, corrective maintenance and preventive maintenance tasks. MBR technology has become more competitive due to reduced module price with longer warranty, increased energy efficiency and improved design and operation practices (Lesjean *et al.*, 2011).

In the past, due to its perceived novelty and the unavailability of extensive reference data needed to support investment decisions, there has been some reluctance to invest in the MBR process in some areas. The maturing of the immersed technology (now 25 years old) and the much wider knowledge base of the process, in particular the key aspects of energy optimization and process failure risk, has promoted greater confidence in the technology. The fact that MBRs are increasingly becoming the technology of choice for water and wastewater applications is evident by the substantial expansion of the MBR market over recent years. Confidence in the process appears to be increasing as the number and size of reference installations grow, with plants over 100 MLD peak daily flow (PDF) capacity now installed (Oppenheimer *et al.*, 2012; WEF, 2012). There are drivers behind the use of MBRs on a larger scale. MBR technology may offer a cost-competitive treatment solution for site-constrained facilities that must expand or upgrade using small-footprint treatment processes that can provide a high-level nutrient reduction and/or high-quality effluent for water reuse. Analysing the data regarding the top largest MBRs in the world from two surveys conducted in 2009 and 2011, Judd (2011) anticipated that there will be nine MBRs of 100 MLD or more PDF capacity by April 2012, compared with just the one in October 2009, demonstrating again the growing confidence in the process. The trend continues to date. Figure 1.3 shows the treatment capacity of the tentatively top largest MBR plants in the world for the last decade.

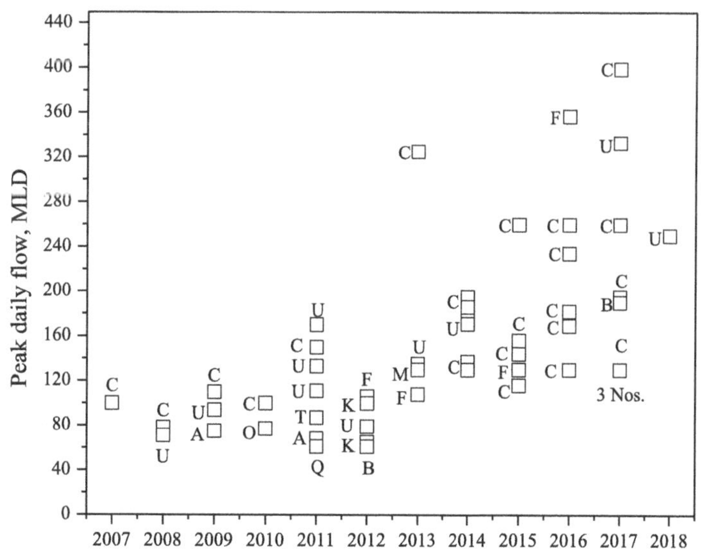

Figure 1.3 Treatment capacity of largest MBRs in the world (A = Australia, B = Brazil, C = China, F = France, K = Korea, T = Turkmenistan, U = USA). Date source: (Judd, 2018).

Whereas in the past, MBRs may have been disregarded in favour of conventional treatment plants, it is now the case that for applications where footprint is limited and a high product water quality is demanded, and for reuse in particular, the MBR is the technology of choice. The stringent environmental regulations, the scarcity of available water combined with the economy of space and ease of automation that MBRs provide are driving the growth in its demand, usage, and range of applications (Judd, 2011; Oppenheimer *et al.*, 2012).

1.6.2 Challenges

Notwithstanding the drivers propelling the MBR market, higher capital and operating costs (energy demand) as compared to the CAS processes, and relative process complexity requiring trained workers, generally impede the more widespread adoption of the technology as the preferred process over competing technologies for municipal wastewater treatment (Judd, 2010a).

1.6.2.1 Public perception of wastewater reclamation

A key aspect governing the uptake of MBRs is the acceptance of the technology by the various stakeholders, including the public, the politicians and decision makers within the procuring organization. The most contentious perception issue directly impacting on the uptake of MBR technology is wastewater reuse. Despite the very persuasive technical capability, negative public perception often impedes even the indirect potable reuse (IPR) of municipal wastewater (Hurlimann & Dolnicar, 2010), let alone direct reuse, in that only one such 'toilet to tap' plant currently exists in the world – the New Goreangab Reclamation Plant at Windhoek in Namibia, commissioned in 2002.

1.6.2.2 Capital and operating costs

Compared to the CAS systems, MBRs typically require a higher level of pretreatment to protect membranes and avoid excessive fouling, regardless of whether primary clarification is eliminated or not (WEF, 2012). Again, MBRs have a limited peak flow capacity, requiring multiple process tanks and extra membrane trains. These additional unit processes require added capital and operational costs. Furthermore, membranes are expensive and create a high initial capital investment; there is also a long-term cost for membrane replacement. Though membrane prices have decreased significantly over the past few years due to the continuous research and development efforts to reduce costs, inflation and rising material costs are still likely to influence MBR prices. Therefore, the operational costs of MBRs may be higher than CAS systems.

1.6.2.3 Technical challenges

Operationally, membrane fouling and clogging is considered the Achilles' heel of the MBR technology (Hai & Yamamoto, 2011). In a 2012 MBR Survey (Judd, 2012), the top four technical challenges facing MBRs were identified as screening/pretreatment (16%), membrane/aerator clogging (11%), overloading/under-design (13%), and membrane fouling (15%). In addition, the issues such as energy demand, operator knowledge, sludge quality and uneven aeration between them attracted 35% of nominations. There were also a number of other issues identified as being important, most significantly: membrane integrity, hydraulic and organic (and thus F/M ratio) fluctuations, out of spec feed water. It was pointed out that many of the issues are inter-related. For example, 'uneven aeration' and 'clogged aerators', and 'clogging' and/or 'fouling' and 'poor sludge quality'. Surveys conducted in subsequent years identified similar issues. This is further discussed in Section 1.8.

1.6.2.4 Competition with other water reclamation techniques

The MBR technology is often in strong competition with tertiary filtration (membrane filtration, sand filtration or even microsieves) for water reuse applications, or also reverses osmosis pre-treatment. For large plants, tertiary treatments are often more competitive than MBR technology (Lesjean *et al.*, 2004) due to lower total energy demand, greater membrane flux and longer membrane life span. Very large tertiary membrane treatment plants have been commissioned in the recent years (e.g. the Sulaibiya WWTP plant in Kuwait with 375,600 m^3/d), and such systems can also achieve advanced phosphorus removal (20–50 mg TP/L) with total specific energy requirement below 0.6 kWh/m^3 for the treatment steps activated sludge plus tertiary treatment (Gnirss & Dittrich, 2000).

1.6.2.5 Energy-intensive decentralized MBRs

Despite the great expectations raised at the early stage of development (Fane, 2004) regarding the application of decentralised and semi-central municipal applications (4 p.e up to 5,000 p.e.), with the current commercial solutions, MBR systems are not cost-effective for most decentralised or semi-central applications (<5,000 p.e.), with the exception of a market niche for household applications (<50 p.e.) and water reuse (Lesjean *et al.*, 2011). The viability of systems for the range 50–5,000 p.e. is impaired by unnecessary treatment quality for this scale of application, and significant specific energy requirements due to low-energy efficiency of the equipment (Stüber *et al.*, 2010).

1.6.2.6 Environmental impact

Although some studies have demonstrated competitive life cycle costs (LCC) to other technologies achieving equivalent degree of treatment for medium-size MBR plants (Brepols *et al.*, 2010; Lesjean *et al.*, 2011), in general, in LCAs conducted to date, MBRs have not been rated as the best available technology for water reuse due to the high electricity consumption (Lesjean *et al.*, 2011). This calls for radical technological breakthroughs. However, as more toxic compounds need to be removed from wastewater, there might be a point when the advantages of avoiding the adverse impacts due to implementation associated with the release of pollutants outweigh the environmental burdens imposed by the very treatment technique. Under such circumstances, MBR could emerge as the best option. For example, according to Santos and Judd (2010), MBRs remove on average 10% more heavy metals from water than the CAS process.

1.6.2.7 Manpower requirement and staff training

Additional manpower requirement for scheduled and unscheduled membrane cleaning and membrane replacement still impairs the competitiveness of MBR systems (Lesjean *et al.*, 2011). Fail-safe operation of an MBR requires a skilled workforce. The principles and procedures routinely used to design and operate the biological component of clarifier-based CAS systems are generally applicable to MBRs; however, the specific conditions and constraints generally incorporated into MBRs must be recognized. This basic knowledge must be supplemented by a fundamental understanding of particle removal (i.e. microfiltration and ultrafiltration) membranes to successfully implement MBRs (WEF, 2012).

1.6.2.8 Non-standard design and performance-evaluation method

Numerous manufacturers of MF/UF membrane systems for MBRs exist, each with their own proprietary technologies. The differences between proprietary systems can often be overwhelming and present significantly varying design considerations with diverse geometries, module capacities, and operational modes (De Wilde *et al.*, 2008; Lesjean & Huisjes, 2008). Although this situation promotes a competitive market, it is detrimental for the acceptance of the technology as 'state-of-the-art' process, and raises concern with potential clients or end users. From the point of view of the MBR operators, the possibility to interchange filtration modules of different companies/suppliers would facilitate the replacement of the modules at the end of their life, and would reduce the risk of a supplier withdrawing from the market or releasing a new series of product. In addition, the stakeholders in the industry employ various methods of membrane characterization and performance evaluation. This creates confusion among the users and prohibits fair comparison. It is interesting to note that, most of the newcomers in the market are developing their systems so that they can easily replace the products of the two main suppliers (Zenon-GE and Kubota). A standardization process driven by the end users could accelerate this evolution and contribute to the market development (Lesjean & Huisjes, 2008). Pearce (2008) also pointed out that, although the dimensions of the relatively newer Puron products are not identical to Zenon's ZW 500d or MRE's SADF, the elements are similar, and cassettes made from the elements could be used interchangeably. This begins to introduce retrofit possibilities into what has up to now been a fragmented market with no standardization.

1.6.3 The way forward

The rising demand in developing regions can counter the above-mentioned restraints. It is also noteworthy that comparison of process technologies conducted in the past have not always taken account of the complete process envelope including, for example, sludge processing and management costs and the embedded carbon associated with the more extensive infrastructure (i.e. the concrete tanks) of the larger conventional plant (Judd, 2011; WEF, 2012). Indeed, provided that a long membrane life can be assumed, the costs of installing and running MBRs can be comparable with those of conventional treatment plants on a whole-life basis, with the added benefit of improved effluent quality. System integrators and engineering companies will have a key role to play in optimizing the overall system including the membrane performance, pretreatment requirements, the biological system design and proper management of peak flows. MBR membrane manufacturers and systems designers continue to reduce the system costs to make the process more competitive, and the difference in energy demand between MBRs and conventional processes is steadily decreasing, such that it is likely that parity in carbon footprint between conventional and MBR wastewater treatment processes is attainable. Advanced control tools to facilitate the filtration operation while optimizing the energy requirement and reducing the operational risk through the automatic detection of irreversible fouling will further increase the acceptance of the MBR technology (de la Torre *et al.*, 2010; Huyskens *et al.*, 2008; Manic *et al.*, 2008).

1.7 MBR MARKET

1.7.1 Global market overview

The global MBR market was valued at \$838.2 million in 2011, and recorded an annual growth rate of 22.4%. The compound annual growth rate (CAGR) from 2011 to 2018 is estimated to be around 15.2% (Frost & Sullivan, 2013). Another estimate stipulated that the global MBR operation was expected to grow from 3.879 million m^3/d (3879 MLD) in 2011 to 12.344 million m^3/d (12344 MLD) by 2017, at an estimated CAGR of 20.8% for the period 2012–2017 (Research & Markets, 2012). An updated estimate shows that the MBR market is expected to reach $5590 million by 2026, expanding at a CAGR of 15.46% from 2018 to 2026 (Research & Markets, 2018).

1.7.2 Regional key drivers and constraints and market trend

Notwithstanding generally the high global growth rates, regional implementation of the MBR technology varies significantly (Research & Markets, 2012). While Europe and North America have been the traditional markets for MBR sales, it is Asia Pacific and the Middle East that are anticipated to be the leading markets in the upcoming years with anticipated growth rates of between 16.8 and 23.5% (Frost & Sullivan, 2013). Municipalities, which are important large buyers of MBR technology, will be the driving forces in Asia Pacific while many companies will look to industrial clients in Europe and North America for sales.

1.7.2.1 Asia Pacific

Water stress in Australia, South East Asia, and India has caused increasing local government interest in wastewater reclamation and recycling. The trend of industrial wastewater recycling and reuse to achieve the goal of zero liquid discharge has a positive impact on the MBR market. China experienced the highest growth in the world in 2011. Over the next five to six years, the industry is forecast to grow further, driven by a steady increase in demand. Japan accounted for only around 2.0% of the regional market revenue in 2011 due to the low addressable market base and the highly developed community sewage treatment systems. South Korea is also a low growth market. The wastewater reclamation rate in South Korea is already high; therefore, the market is moving on to the development of small-to-medium community or residential MBR plants. Taiwan is another low-growth region with the MBR market predominantly comprising small-scale industrial wastewater reclamation projects. The Indian MBR market offers great potential in both the municipal and industrial sectors and is expected to witness high growth for the next five to 10 years. Southeast Asia has relatively medium potential for MBR technology. The growth potential of Singapore is limited as its market is relatively mature. Of the remaining Southeast Asian regions, Malaysia and Indonesia offer high market potential, while Vietnam and Thailand are yet to gain user awareness and the cost of the MBR systems has to attain acceptability in order to penetrate these markets (Frost & Sullivan, 2013; Research & Markets, 2012).

1.7.2.1.1 Australia

Following its emergence elsewhere in the world, the investigation of MBR technology in Australia commenced in the late 1990s, while the first full-scale Australian MBR (at Picnic Bay on Magnetic Island near Townsville, North Queensland) has been operational since 2002. Although the main drivers for MBR technology are similar to those worldwide, water scarcity is a key factor in Australia which is the world's second driest continent (second only to Antarctica). Between 2006 and 2008, revised legislation and guidelines were published in Australia covering water recycling (NWI, 2011). Other specific drivers include space constraints and discharge limits (ecologically sustainable development). Relatively high population densities in coastal areas and legislative or other barriers to approvals for new wastewater treatment sites (e.g. pumping costs to more remote locations in relatively flat coastal zones) are significant driver in terms of overall plant footprint considerations. On the other hand, the large MBR projects in North Queensland (e.g. Townsville, Cairns) have been driven partly by environmental concerns over the Great Barrier Reef to cap or reduce nutrient loads discharged to the marine environment due to human activity, including wastewater and agricultural run-off. MBRs were found to be best suited to producing high-quality low-nutrient effluent suitable for water recycling (e.g. land irrigation of golf courses, sports fields, public open spaces and toilet flushing), and, hence, either zero or limited marine discharges (Judd, 2011).

A significant challenge to the application of MBRs in Australia is handling of wet weather flows in municipal applications. Whilst wet weather events are infrequent, due to high local rainfall intensities, it is common for design peak wet weather flow rates to be >3–11 (typically 5) times the average dry weather flow rates. This either requires installation of additional membrane modules (which need to be kept idle during dry weather) or some alternative strategy for handling wet weather flow.

Industrial MBR applications have largely resulted from recycling initiatives and space constraints. An interesting application is the so-called Gippsland Water Factory in the state of Victoria. This project was driven by a number of different factors, including the need to produce a very high treated water quality for reuse and concerns over activated sludge settleability when treating pulp and paper effluent in conventional biological treatment processes that depend on sedimentation for secondary clarification. As of mid-2010 there were at least 44 full-scale applications of MBR in Australia, either operating or under construction, excluding smaller 'on-site' systems at household or cluster housing scale and mine sites (Judd, 2011).

1.7.2.1.2 China

With the modest beginning in the early 1990s with lab-scale research and pilot-scale demonstration projects (Wang *et al.*, 2008), the current average annual growth rate of MBR market in China is more than 50% – considerably higher than the average annual global growth rate of 11.5–12.5% (Frost & Sullivan, 2013). The earliest practical applications were generally <100 m^3/day capacity in the early 2000s. However, since 2006, the number of large-scale MBRs has been substantially increasing. Of the numerous large-scale MBR plants in China, particularly noteworthy are the MBRs of Beijing Miyun wastewater treatment plant and Wenyu River water treatment plant. The Beijing Miyun MBR plant was the first >10 MLD MBR plant in China, and the Beijing Wenyu River MBR plant the first 100 MLD plant worldwide. Between 2003 and the end of 2009, there were more than 100 MBRs installed in China providing total wastewater treatment capacity of close to 1200 MLD, based on MBRs having an installed capacity greater than 0.1 MLD.

In 2008, the MBR market in China exceeded 1.6 billion CNY (over $US230 million). Today, China has become one of the most MBR-active countries. Municipal wastewater applications account for about 60% of installed capacity and industrial wastewater plants about 30%, and the rest are for polluted river water treatment (e.g. Wenyu River plant) and other applications (Judd, 2011).

Of the many factors influencing the MBR market in China, water scarcity is the most important. Water shortage is a significant problem in China, particularly in the north-eastern and north-western areas. This problem is further exacerbated by water pollution. For example, in some sensitive drainage basins such as at Tai Lake and Dian Lake, eutrophication is a serious problem. Increasingly stringent discharge regulations in those areas have made MBR an option of choice. On the other hand, although China covers a large area, some large cities do not have sufficient available land for the construction of municipal wastewater treatment plants. The small footprint incurred by the MBR technology is especially attractive for these areas. In addition, a significant decrease in the MBR investment costs as well as increased maturing and acceptance of the MBR technology, especially relating to domestic companies, has continued to sustain the high level of growth of the Chinese MBR market (Figure 1.3). Initially most large-scale MBRs treating municipal wastewater were centred in Beijing and Jiangsu provinces, both of which are more developed than most of the other provinces (Judd,

2011; Wang *et al.*, 2008). However, in the last five years or so, large MBR plants have been built in other provinces too.

1.7.2.1.3 India

The centralized sewage treatment systems and sewerage lines of most cities and metropolitan areas in India were constructed several decades ago and have not undergone modernization or expansion to meet the challenges of a rapidly growing population. The state and Central Pollution Control Boards (CPCBs) require that that all medium and large establishments, including housing societies, hospitals, hotels, educational institutions, factories and shopping malls have their own wastewater treatment systems in-house, and have enforced standards for point-of-use discharge. Because the point-of-use discharge standards prescribed by the CPCBs do not necessitate the use of MBR technology, unlike many other regions of the world, legislation is not the key driver for the growth of MBRs in India. The most significant driver for MBRs in India is probably the shortage of clean water. In view of the dwindling fresh water resources and the brackish groundwater and uncertain rainwater, water reuse has become increasingly important, making MBRs more attractive (Frost & Sullivan, 2011).

Historically, India has been late in adopting the latest water treatment technologies (Judd, 2011). It was not before a decade after the introduction of RO plants (since late 1980s) and UF- and MF-based treatment plants (since late 1990s), respectively, that the technologies were adopted on a large scale in India. The same trend has been observed in case of the MBRs: they have only been effectively marketed since around 2007 and the next five years saw growth of the MBR market in India. All the major global players in the MBR industry now have their presence in India. Current major MBR applications include on-site reclamation of municipal sewage at public facilities (e.g. Cubbon Park, Bangalore) and industrial effluent recycling (e.g. food, textile and petrochemical industries).

1.7.2.1.4 Japan

In Japan, several domestic wastewater management systems, controlled by different legislation and national ministries, co-exist. The status of MBR installations in respect of individual systems has been summarized in Table 1.5. Japan has played an important role in pioneering trials on iMBR systems as well as the development of a variety of commercial membrane systems (Hai & Yamamoto, 2011). However, although the first full-scale commercial MBR plants in Japan were installed for in-building wastewater recycling in the early 1980s, the application of MBRs to municipal WWTPs did not initiate before 1999 and 2005 for small-scale rural sewerage systems and larger-scale urban sewerage systems, respectively. Possible reasons for this include: (i) relatively lax effluent regulations (e.g. nutrient removal is required only for limited areas), (ii) little requirement for water reuse, since for most of the area water scarcity is not an issue, and (iii) the somewhat conservative nature of policy-makers and engineers. Nevertheless, in March 2009, the total number of MBR installations in Japan was at least 3870. MBRs have also been used for many types of industrial process wastewater treatment, with around 1610 installations as in March 2009, predominantly for the food industry (33%), stock farms (18%) and medical industry (14%) (Judd, 2011).

Notable membrane-related national R&D initiatives in Japan, promoting MBR market, include: publication of the 'Guideline for Membrane Technology' by MLIT (2009); launch of 'Advance of Japan Ultimate Membrane bioreactor Technology Project' (A-JUMP), which, based on data and experience derived from two demonstration MBR plants, fed into a revision of the above guideline (2010); R&D projects commenced by the New Energy and Industrial Technology Development Organization (NEDO) for the development of low-energy MBRs.

Table 1.5 Role of MBR technology amongst the co-existing systems in domestic wastewater management in Japan.

System	Description
Urban sewerage systems	• Typically for urban areas, • Managed by the Ministry of Land, Infrastructure and Transport (MLIT), • >2000 plants (~75% of the domestic population), MBR technology has been installed in only ~20 plants (from 2005–), • Standardized process configuration and design approach in accordance with the JS MBR Design Recommendations prepared by the Japan Sewage Works Agency (JS) in 2003 for <3 MLD plants.
Rural sewerage systems	• Small-scale sewerage systems (typically agricultural), • Supervised by the Ministry of Agriculture, Forestry and Fisheries (MAFF). • >5000 plants (~3% of the domestic population), MBR technology has been installed in only ~50 plants (from 1999–), • Any one of three types of submerged MBR configurations developed by The Japan Association of Rural Resource Recycling Solutions (JARUS) is used.
Johkasou systems	• On-site systems treating wastewater from individual houses/public facilities (e.g. schools), commercial buildings, restaurants and offices, • Serves ~9% of the domestic population, approximately 1950 MBR installations (from 1985–).
Night soil treatment systems	• Alternative system in some rural areas for toilet wastewater (night soil), often together with excess sludge collected from *Johkasous*, • Around 1000 plants, MBR technology has been installed in ~210 plants (from 1988–), • MBRs made it possible to treat the high-strength wastewater treatment without dilution.
In-building wastewater recycling systems	• Obligatory on-site wastewater treatment and recycling (e.g. toilet flushing) systems in large buildings, • Traditionally external MBR systems have been used since the early 1980s, increased use of submerged MBR systems from the late 1990s, • ~75 installations (~11 MLD) in total.

1.7.2.1.5 Malaysia

Despite various research projects, in Malaysia, full-scale MBR plants were nonexistent until very recently. This is due to a number of factors including the perception that newer technologies such as MBRs are very expensive to install and operate; the significant capital investment, often beyond the capability of privately operated wastewater treatment operators; lack of perceived importance of environmental issues and the general lack of awareness of the technology among policymakers (Judd, 2011). In recent years, the Malaysian government has strengthened its water quality regulations – particularly for industrial wastewater. Accordingly, the processing industries have been required to take prompt measures (Frost & Sullivan, 2013). For example, Kubota Corporation supplied an MBR plant for the treatment of the effluent and the recovery of biogas from the palm oil-producing process in Sarawak, Malaysia in 2013. The biogas recovery plant (AnMBR) and effluent treatment plant (MBR) have a plant capacity of maximum 0.1 MLD. Pentair X-Flow was awarded a contract to supply Airlift™ MBR skids to the four

MLD wastewater project 'Proposed Wastewater Treatment Plant for Livestock Farming Area' in Samarahan, Malaysia. The main objective of this project, initiated by the Natural Resources and Environment Board (NREB), is to design a WWTP that focuses on recycling and environmental protection in the agricultural sector. The result is a plant which substantially reduces the water consumption by reusing four MLD wastewater for washing and cleaning purposes.

1.7.2.1.6 Singapore

Significant strategic investment in water technology and management has helped Singapore to overcome the problems such as water shortage and flooding and pollution in its rivers in the 1960s and 1970s, and today, Singapore has a robust, diversified and sustainable water supply from four different sources known as the Four National Taps (water from local catchment areas, imported water from Malaysia, reclaimed water known as NEWater and desalinated water) (Frost & Sullivan, 2013).

Because land is scarce in Singapore, MBR systems are particularly attractive due to their small footprint. The 'MBR technology study programme' initiated in 2002 by the national water agency, Public Utilities Board (PUB), generated valuable information on the design and operation of the MBR systems under a tropical environment. For example, the reliability of the MBR technology and its effectiveness in producing better feed water quality for the production of NEWater through MBR-RO was demonstrated. Important PUB demonstration plants include Bedok Water Reclamation Plant (WRP), Ulu Pandan WRP and Jurong WRP. MBR technology is also gaining acceptance in the private industrial sector in Singapore for complex industrial wastewater treatment and reuse (Judd, 2011).

1.7.2.2 Europe

Overall, the MBR market in Europe has attained maturity, with established municipal wastewater reclamation projects in the member countries. However, Eastern Europe is expected to be the hub of growth in the future. With more non-EU countries becoming EU members, compliance with water and wastewater treatment standards and reclamation requirements is expected to drive the market (Frost & Sullivan, 2013; Lesjean & Huisjes, 2008; Research & Markets, 2012).

1.7.2.2.1 Germany

The first full-scale municipal MBR plant in Germany was of a capacity of ~500 population equivalent (p.e.) and went into operation in 1999. Today there are at least 15 municipal MBRs, with capacities from 700 to 80,000 p.e. Although water scarcity is uncommon, maintaining a high water quality (nutrient and pathogen removal) according to the highest standard of the EU Directive (91/271/EWG, 1998) is one of the major goals of the national environmental policy, and has been the key drive behind the establishment of MBR market in Germany (Lesjean & Huisjes, 2008). An additional factor has been the government support for capacity building and new technologies in the water sector. Numerous initiatives and institutions have promoted membrane technology, and MBR technology in particular, e.g. DEBRANE (2010), an R&D network for promotion of membrane technology; BMBF (2009), a federal ministry for education and research; and SIMAS (2010), a training institute for membrane applications in wastewater treatment.

A large number of decentralized MBRs (package-plant MBRs with sizes from 4 to 40 p.e.) have been established in environmentally sensitive areas. For example, in specific geological situations where pathogens in the effluent of WWTPs might contaminate the aquifer via rivers that percolate entirely (Monheim or Glessen), or where the receiving rivers are used as bathing waters (Hutthurm). Strict discharge standards and site restrictions, the drive for water reuse, and the opportunities for retrofitting to existing WWTPs can be seen as major drivers for MBR technology implementation in the industrial sector in Germany. Between 50 and 60 full-scale installations were in operation in 2007 (Lesjean & Huisjes, 2008) for the treatment of landfill leachate, and also for wastewater treatment in industries such as pharmaceutical and food and beverage industries. Maritime use on board large ships is another field of application (Judd, 2011).

1.7.2.2.2 Italy

Italy was one of the first European regions to embrace MBR, and some of its oldest plants have already reached a 10-year life (Lesjean *et al.*, 2009). In 2005, the size of the Italian MBR market was estimated as \$12.1 million with a compound annual growth rate (CAGR) of 11.5% until 2011, which when compared to a CAGR of 7.9% (until 2013) of the total European market, establishes Italy as one of the key regions for MBR implementation in Europe. The main drivers in the Italian municipal market include European regulations for nutrient removal in sensitive areas and the need for retrofitting the ageing WWTPs. Coastal areas and mountain regions are particularly increasingly favouring MBRs due to its reduced footprint with reusable effluent quality and little impact on bathing waters. For the industrial sector, implementation is driven by increasingly stringent requirements concerning direct discharge into water bodies (Lesjean & Huisjes, 2008).

The industrial market mainly comprises food and beverage industries, along with the oil and gas, tannery, tanker washing and laundry industries. The market is characterized by medium- and small-sized municipal and industrial MBR systems with a few exceptions at larger scale. It is anticipated that most MBR plants in the near future will be in the range of up to 3 MLD in capacity, with perhaps one project per year in between 3 MLD and 10 MLD (Judd, 2011).

1.7.2.2.3 The Netherlands

The Netherlands has been internationally recognized for the country's commitment to MBR technology. Supported by the Dutch water authorities and the Stichting Toegepast Onderzoek Waterbeheer (STOWA), the Dutch Foundation for Applied Water Research, and the timely funding at the turn of the millennium saw the establishment of a series of pioneering research programme and demonstration MBR plants, namely, Beverwijk, Varsseveld, Heenvliet and Ootmarsum (Judd, 2011). The main driver for the Netherlands MBR programme was water quality improvement and footprint reduction, both of which have been verified in the last five years of operation of Varsseveld and the other demonstration plants. The adoption of the MBR technology has been more widespread in case of industrial wastewater (e.g., food, leachate, waste handling, and tanker washing industries) where the drivers are based on cost savings, space requirements, water reuse, discharge quality and robustness of operation (Lesjean *et al.*, 2009).

1.7.2.2.4 Spain

With the first industrial plants installed in 2000, and the first municipal plant installed in 2003, Spain embraced MBRs slightly later than other European regions; however, the market developed much faster than in the neighbouring countries. The Spanish MBR market, with an estimated value of \$13 million (Frost & Sullivan, 2005), was predicted to follow a CAGR of 16.8% until 2011 (Lesjean *et al.*, 2009). Indeed, the cumulative number of MBR plants in Spain grew from 47 at the end of 2005 up to 111 at the end of 2008. However, the impact of the global recession on the MBR market appears to have been more profound in Spain such that no new municipal projects appear to have been implemented since 2009.

Medium-sized municipal and small-sized industrial installations govern the MBR plants in Spain, although large-scale municipal plants with a capacity of 10–50 MLD are also present. Major drivers for MBR technology in the Spanish municipal market is wastewater reuse, especially in the regions around the major economic centres, and the European regulations for nutrient removal in sensitive areas. For the industrial sector (e.g. food and beverage, wineries and fruit processing plants, leachate, cosmetics and pharmaceutical sectors), reuse, limited footprint and rising pollution discharge costs remain the major incentives for installing MBRs.

1.7.2.2.5 UK

Europe's first full-scale municipal MBR plant Porlock WWTP was commissioned in England in 1998. The MBR market in the United Kingdom and Ireland is mature, with over 10 years operational experience in full-scale plants in both the municipal and industrial sectors. In the municipal sector, the main legislative drivers for the application of MBRs in the United Kingdom and Ireland relate to EU legislations, namely, the Urban Wastewater Treatment Directive, the Water Resources Act, the Bathing Water Directive and the Shellfish Directive. The drivers in the industrial sector are the Integrated Pollution Prevention and Control (IPPC) regulations and the associated schemes such as the Best Available Technologies (BAT) and the Enhanced Capital Allowance (ECA) (ECA, 2013).

A steady number of opportunities for the future municipal MBRs are anticipated at sites where space is limited, the effluent standards are being tightened and the existing assets are at the end of their operational life. The increasing costs of potable water supply and wastewater disposal and the reduction of the investment payback periods to acceptable timeframes due to the ECA allowances are expected to increase the MBR market in the industrial sector (Lesjean *et al.*, 2009).

Despite the significant reductions made in capital cost and energy demand, these factors are still considered the main impediments to the United Kingdom and Ireland MBR market. However, thanks to the MBR membrane manufacturers and systems designers, with the continuous reduction of the system costs, the difference in energy demand between MBRs and conventional processes is steadily decreasing, such that parity in carbon footprint between conventional and MBR wastewater treatment processes appears attainable (Frost & Sullivan, 2013).

1.7.2.3 Americas (North America and Latin America)

Growth in the mature North American market has been slow but steady. It is one of the most dynamic markets, with the United States dominating the market share in this region, while Canada accounting for less than 10% of the regional market revenue. A market shift from the municipal sector to the industrial sector is anticipated in North America. Market opportunities comprise newly built MBR treatment plants, upgrading wastewater reclamation plants, and replacements in the existing plants. Latin America is an evolving market, currently heavily influenced by the industrial sector (Frost & Sullivan, 2013).

1.7.2.3.1 USA

In the United States, the total number of large municipal MBR installations (with capacity greater than or equal to 1 MGD, or 3.8 MLD) grew more than fivefold from 13 to 68 in the five years between 2004 and 2009 (Frost & Sullivan, 2013; Research & Markets, 2012). Improved water quality and reliability requirements and footprint limitation have been the key drivers behind selecting the MBR process in the United States. A higher number of MBR installations have been observed in the states that have developed regulations or guidelines that strongly encourage water reuse for groundwater recharge and IPR as a strategy for conserving water resources. Since these applications require high-quality effluent, low in organic matter and microbial contaminants, they promote the use of membrane technologies such as micro/ultrafiltration, MBR and RO. The primary barriers to the use of recycled water in the United States include numerous federal, state and regional regulations, capital investment costs and public perception. However, recent approval of projects such as the City of San Diego's advanced water treatment demonstration plant has shown increased public acceptance of IPR (CSD, 2006).

1.7.2.4 Middle East and Africa

The MBR market in the Middle East and Africa witnessed rapid growth over the last three years. The market is expected to witness a CAGR of 16.8%; most of the revenue will come from the GCC region, with UAE leading the implementation of projects during the last three years. Other countries that have been at the forefront of implementing MBR systems are Saudi Arabia, Qatar, and Oman, with Egypt and Algeria pitching in. The municipal sector witnessed the commissioning of some of the biggest MBR treatment plants in the region, such as the Oman's Al-ansab facility. The most dominant consumer of Middle Eastern systems is the commercial segment, which has driven the demand for MBR systems, primarily due to urbanization and real estate sector development in the region. The key industrial end-user sectors in the region are oil and gas, refineries, petrochemicals, and metals and minerals (Frost & Sullivan, 2013; Research & Markets, 2012).

In Africa, the MBR market is driven by the chronic water shortages prevalent in some regions. North Africa has been slow to adopt advanced wastewater treatment systems, especially in the municipal sector. The demand for MBR systems in North Africa is expected to be driven by the small packaged wastewater treatment systems in the industrial sector in the next few years.

1.7.2.4.1 South Africa

The South African MBR market is in its incipient stages, with only a few MBR plants greater than 1 MLD in capacity. For example, Illovo Sugar MBR plant (at Sezela, commissioned in 2005, capacity 1.2 MLD), Zandvliet plant (in the Western Cape, commissioned in 2009, capacity 18 MLD), Malmesbury plant (in the Western Cape, commissioned in 2012, capacity 20 MLD), and a few upcoming plants including Belville plant (in the Western Cape, went out to tender in 2010, capacity 40 MLD and Coega Industrial Zone (in the Eastern Cape, commissioned in 2011, capacity 100 MLD). However, in the past five years or so several of the international MBR vendors have established a presence in South Africa, and there has been a flurry of pilot-plant trials driven by a looming water shortage crisis, and the need to meet discharge standards, footprint limitations and waste minimization (Frost & Sullivan, 2013). Some of the larger industries are actively involved in MBR pilot trials that are likely to lead to full-scale applications in the very near future. A possible barrier to implementation is the driver being limited to meeting discharge standards, which can be obtained by existing chemical treatment processes, rather than water reuse. A few MBR units can be identified in the neighbouring countries to South Africa, namely, Botswana, Mozambique and Harare.

South Africa also offers great potential, as Veolia Water has won the contract for its largest MBR project worth $24.0 million, in 2011. Indian water technology provider Shivsu Canadian Clear International Ltd. is very active in Kenya, Tanzania, and Ethiopia with orders worth $6.0 million in 2011 (Judd, 2011).

1.8 WORLDWIDE RESEARCH TREND

Three stages may be identified in the worldwide MBR research (van Nieuwenhuizen *et al.*, 2008; Yang *et al.*, 2006): (1) An entry-level stage spanning from 1966 to 1980, during which lab-scale research was mainly conducted. Membranes of the time had low flux and short life span due to the undeveloped membrane manufacturing technology. (2) A slow-to-moderate growth period from 1980 to 1995, when MBR technology was well investigated especially in Japan, Canada and the United States. During this stage, new membrane material development, MBR configuration design and MBR operation were critically studied. The immersed (submerged) MBR concept was put forward by Japanese researchers in 1989. (3) The rapid development stage starting from 1995 till now, when MBR technology underwent a rapid development featured by deep understanding of the technology in research communities and the installation of full-scale MBRs.

In a comprehensive review, Yang *et al.* (2006) grouped the available worldwide publications regarding MBR into six main research areas (1) literature and critical reviews; (2) fundamental aspect; (3) municipal and domestic wastewater treatment; (4) industrial wastewater and landfill leachate treatment; (5) drinking water treatment; (6) others, which includes gas removal, sludge treatment, hydrogen production and gas diffusion. The fundamental research category was based on studies that exclusively looked at membrane fouling, operation and design parameters, sludge properties, microbiological characteristics, cost and modelling. Studies, which focused on applied research and general reactor performance, were categorized by influent (feed) type (groups 3–6). Membrane fouling, which has been widely considered to be one of the major limitations to faster commercialization of MBRs, has been investigated from various perspectives including the causes, characteristics, mechanisms of fouling and methods to prevent or reduce membrane fouling. More than one third of studies in the fundamental aspects group were found to deal with issues related to membrane fouling.

Another review of the research conducted in MBRs for wastewater treatment over the past two decades, based on Scopus, confirmed the aforementioned trends (Santos *et al.*, 2010). In this study, within the 1450 publications between 1990 and 2009, identified based on selected keywords, a search of papers in key subject areas revealed the subject of fouling to be the most prevalent. Along with the topic of micropollutants or trace organic contaminants, papers concerning fouling were also the ones growing the most rapidly in number – around 36% growth p.a.

Much of the published information on MBRs to date, however, has mainly focused on bench- or pilot-scale studies, performance results of treating a specific type of wastewater, and short-term operations. Regardless of the source of wastewater, whether it is municipal or industrial, few publications involved full-scale studies for long-term operational periods. It has been argued that academic research is addressing only some of the crucial issues and often failing to provide practical outcomes (Lesjean *et al.*, 2004, 2011). For instance, while many publications on fouling are being contributed and some cost studies are conducted, there is a dearth of articles adequately addressing membrane life span, pretreatment and scale-up issues. Although it is estimated that one third of research investigations and publications related to MBR are dedicated to understanding and controlling the membrane fouling mechanism, and this has resulted in progress in knowledge on fouling, it is observed that many results are contradictory and the practical outcomes are limited compared with the level of investment in research. Many factors, such as polysaccharide and protein concentration, capillary suction time or time to filter, were claimed to be directly correlated to irreversible fouling but they actually provide indirect indication of the suspension filterability while informing on the state of flocculation of the biomass (de la Torre *et al.*, 2010).

In addition to the widely studied membrane surface fouling, membrane 'clogging' is an important technical issue faced by the end users during the operation of MBR. Clogging is closely related to inefficient pretreatment, and can take different forms: 'Sludging' is the accumulation sludge solids within membrane channels and depends on process design (membrane module and aerator, pre-treatment), while 'Ragging' (or 'braiding') refers to the blocking of membrane channels with particles agglomerated as long rag-like particles. Clogging is likely as significant as membrane fouling but only a handful of studies are available in peer-reviewed literature (Gabarrón *et al.*, 2013; Stefanski *et al.*, 2011). Zsirai *et al.* (2012) investigated the clogging/ragging phenomenon in 12 full-scale iMBRs. They reported a wide variation in the degree of membrane clogging, and pointed out that the underlying reasons were still unclear. Depending on the properties of the activated sludge, sludge is retained within the membrane channels, consequently forming agglomerate with time. Notably, in this case, regular chemically enhanced backwashing was observed to be ineffective for flux recovery, and flux recovery was only achieved by declogging by physical means (Zsirai *et al.*, 2012). Significant R&D efforts are required to develop an effective solution. Otherwise, end users must either modify the properties of the sludge or physically clean the membrane regularly for preventing this kind of fouling. Approaches such as air-jet along with periodic backwashing (Chang & Judd, 2002) and vibrating membranes (Chatzikonstantinou *et al.*, 2016) are worth considering for clogging control in MBRs.

Judd (2016) conducted a survey in 2015 through the 'MBR site' (http://www.thembrsite.com) to assess the technical issues/challenges associated with the operation of MBR. The survey mainly asked two questions: (i) in your experience, what are the main technical issues or limitations that prevent MBRs working as they should? and (ii) how will MBR technology develop in the future? A total of 186 industry representatives provided their responses to this online survey. Based on the responses selected by the respondents from the given options for question 1, the greatest challenges are the clogging of membrane channels/aerators and membrane fouling. More than 10% respondents in this survey indicated that operator knowledge is also an important technical issue. Similar surveys were also conducted via the MBR site (http://www.thembrsite.com) in 2010 and 2012, and their comparison with the current survey showed that screening/pre-treatment, membrane fouling, clogging and membrane chemical cleaning remain the main technical issues of MBR to date (Judd, 2016; Santos *et al.*, 2010). Question 2 which did not have options to select from, got a wide range of responses. Based on the analysis of keywords, Judd (2016) reported that the five most frequently mentioned terms were 'energy/power', 'reuse/recycle', 'cost', 'industrial applications' and 'fouling'. Since MBR stakeholders such as suppliers, contractors and end users participated in this survey, its significance cannot be ignored.

Future research in MBRs is likely to focus on better understanding of peak flow management, pretreatment, membrane channel clogging and chemical cleaning (Chapters 3 and 4), water-reuse issues such as removal of nutrients, pathogens and emerging contaminants (Chapters 5, 6, and 13), impact of hazard incidents on MBR performance (Chapter 7), environmental impact and cost–benefit analysis and modelling (Chapters 8 and 9), energy production (e.g. AnMBRs, Chapter 11), innovative MBR designs for wastewater reuse and resource recovery (Chapter 12), and, efficient membrane system development and overall reduction in energy demand (Chapter 14). Together with these, a renewed interest in other forms of membrane biological reactors, namely, gas-diffusion, extractive, biocatalytic, and electrochemical MBR (Chapter 10), is anticipated. Yamamoto (2009) suggested that in addition to the alleviation of the technology bottlenecks, a radical shift from the conventional concept of 'organic wastewater treatment' to 'water/biomass production' is necessary. This can be materialized by developing next-generation MBRs where the membrane acts as a separator of water and biomass, and biomass is utilized for energy production.

1.9 SUMMARY AND FUTURE OUTLOOK

Membrane separation in MBRs combines clarification and filtration of a CAS process into a simplified, single-step process capable of reliably and consistently producing water of high quality – substantially clarified, disinfected and almost fully nitrified. The MBR technology is now the preferred option for industrial and often municipal wastewater treatment, especially when advanced treatment is required and/or a compact system is desired. With the sustained annual growth of the market since the introduction of the immersed MBR technology, it has demonstrated its potential to be one of the key available technologies for future wastewater treatment.

In most countries, continued growth of the MBR market is predicted, although growth rates vary markedly between countries and/or regions. Currently, the most rapid growth is observed in China. The MBR market is now characterized with many different MBR membrane products and MBR technology suppliers, although three main suppliers still dominate. For any region, a number of different factors, namely, legislation, water scarcity, return on investment, environmental considerations and public and political perception/engagement influence the market. The commercial significance of the technology is manifested in, along with the exponential growth of the global market, a decreased gestation time between the commercialization of an MBR membrane product and its large-scale implementation, gradually decreasing capital and operating costs, and increasing public acceptance of water reuse.

Notwithstanding the benefits and drivers, there is still much room for improvement in order to fully utilize the potential of this technology. The more widespread adoption of the technology as the preferred process over competing technologies will essentially depend on further improvement to make it competitive with other technological options: the challenges will centre on energy saving/production, ease of operation, simplified membrane cleaning and replacement strategies, and peak flow management. It is likely that a combination of technical advances and the demand for ever improved water quality can sustain, or even increase, the growth in the MBR market to the stage where it becomes the automatic first choice for wastewater treatment and reuse. The international adventure on research and development of the MBR technology continues.

REFERENCES

Ahmed F. N. and Lan C. Q. (2012). Treatment of landfill leachate using membrane bioreactors: a review. *Desalination*, **287**, 41–54.

Bailey J. R., Bemberis I., Hubbard P. J., Leonard F. B. and Presti J. B. (1971). Phase I Final Report – Shipboard Sewage Treatment System, National Technical Information Service (NTIS), San Diego, California.

Bechhold H. (1907). Kolloidstudien mit der Filtrationsmethode. *Zeitschrift für Elektrochemie und angewandte physikalische Chemie*, **13**(32), 527–533.

Behmann H. (1996). Membrane bioreactor system with in-line gas micronizer. US Patent 5151187 A. United States Patent and Trademark Office, Alexandria, VA 22314, USA.

Bemberis I., Hubbard P. J. and Leonard F. B. (1971). Membrane Sewage Treatment Systems – Potential for Complete Wastewater Treatment. American Society for Agricultural Engineers Winter Meeting, Chicago, Illinois, 1–28.

Benedek A. and Cote P. (2003). Long Term Experience with Hollow Fibre Membrane Bioreactors. BAH03-180, 2003. International Desalination Association, Topsfield, Massachusetts.

Berube P., Isabel C. E. and Andrea I. S. (2010). Membrane bioreactors: theory and applications to wastewater reuse, In: Sustainability Science and Engineering, Isabel C. Escobar, Andrea I. Schäfer (eds), Elsevier, Amsterdam, The Netherlands, Chapter 9, 255–292.

Botha G. R., Sanderson R. D. and Buckley C. A. (1992). Brief historical review of membrane development and membrane applications in wastewater treatment in southern Africa. *Water Science and Technology*, **25**(10), 1–4.

Branch A., James G., Trinh T., Truong Ta M., Leslie G. and Le-Clech P. (2016). Membrane Ageing and Replacement–Impact on Pathogen Removal in Full-Scale MBR. Proceedings of the American Water Works Association Annual Water Quality Technology Conference & Exposition, 2016, 13–17.

Brepols C., Dorgeloh E., Frechen F. B., Fuchs W., Haider S., Joss A., de Korte K., Ruiken C., Schier W., van der Roest H., Wett M. and Wozniak Th. (2008). Upgrading and retrofitting of municipal wastewater treatment plants by means of membrane bioreactor (MBR) technology. *Desalination*, **231**(1–3), 20–26.

Brepols C., Schäfer H. and Engelhardt N. (2010). Considerations on the design and financial feasibility of full-scale membrane bioreactors for municipal applications. *Water Science and Technology*, **61**(10), 2461–2468.

Carucci A, Dionisi D, Majone M, Rolle E. and Smurra P. (2001). Aerobic storage by activated sludge on real wastewater. *Water Research*, **35**(16), 3833–3844.

Castro A. J. (1981). Methods for making microporous products. US Pat: 4247498. United States Patent and Trademark Office, Alexandria, VA 22314, USA.

Chang I.-S. and Judd S. J. (2002). Air sparging of a submerged MBR for municipal wastewater treatment. *Process Biochemistry*, **37**(8), 915–920.

Chatzikonstantinou K, Tzamtzis N., Pappa A. and Liodakis S. (2016). Membrane fouling control using high-frequency power vibration, in an SMBR pilot system – preliminary studies. *Desalination and Water Treatment*, **57**(25), 11550–11560.

Chen T. K., Chen J. N., Ni C. H., Lin G. T. and Chang C. Y. (2003). Application of a membrane bioreactor system for opto-electronic industrial wastewater treatment – a pilot study. *Water Science and Technology*, **48**, 195–202.

Childress A. E., Le-Clech P., Daugherty J. L., Chen C. and Leslie G. L. (2005). Mechanical analysis of hollow fiber membrane integrity in water reuse applications. *Desalination*, **180**(1–3), 5–14.

Chu C. P., Wu C. M., Wu Y. S., Lin C. C. and Chung Y. J. (2007). Structural analysis and dewatering characteristics of waste sludge from WWTP MBR. *Separation Science and Technology*, **42**(16), 3713–3726.

Cicek N., Franco J., Suidan M. and Urbain V. (1999). Effect of phosphorus on operation and characteristics of MBR. *Journal of Environmental Engineering*, **125**(8), 738–746.

Cicek N., Macomber J., Davel J., Suidan M. T., Audic J. and Genestet P. (2001). Effect of solids retention time on the performance and biological characteristics of a membrane bioreactor. *Water Science and Technology*, **43**, 43–50.

Cicek N., Suidan M. T., Ginestet P. and Audic J. M. (2003). Impact of soluble organic compounds on permeate flux in an aerobic membrane bioreactor. *Environmental Technology*, **24**(2), 249–256.

Ciora R. J. and Liu P. K. T. (2003). Ceramic membranes for environmental related applications. *Fluid/Particle Separation Journal*, **15**(1), 51–60.

Cote P., Buisson H., Pound C. and Arakaki G. (1997). Immersed membrane activated sludge for the reuse of municipal wastewater. *Desalination*, **113**(2–3), 189–196.

Cote P., Alam Z. and Penny J. (2012). Hollow fiber membrane life in membrane bioreactors (MBR). *Desalination*, **288**, 145–151.

CSD. (2006). Water reuse study, Final Draft Report, City of San Diego.

de la Torre T., Iversen V., Meng F., Stüber J., Drews A., Lesjean B. and Kraume M. (2010). Searching for a universal fouling indicator for membrane bioreactors. *Desalination and Water Treatment*, **18**(1–3), 264–269.

De Wilde W., Richard M., Lesjean B. and Tazi-Pain A. (2008). Towards standardisation of the MBR technology? *Desalination*, **231**(1–3), 156–165.

Doyen W., Mues W., Molenberghs B. and Cobben B. (2010). Spacer fabric supported flat-sheet membranes: a new era of flat-sheet membrane technology. *Desalination*, **250**(3), 1078–1082.

Drews A., Evenblij H. and Rosenberger S. (2005). Potential and drawbacks of microbiology–membrane interaction in membrane bioreactors. *Environmental Progress*, **24**(4), 426–433.

Druin M. L., Loft J. T. and Plovan S. G. (1974). Novel open-celled microporous film. US patent 3801404. United States Patent and Trademark Office, Alexandria, VA 22314, USA.

ECA. (2013). https://etl.beis.gov.uk/ (accessed April, 2019).

Engelhardt N., Firk W. and Warnken W. (1998). Integration of membrane filtration into the activated sludge process in municipal wastewater treatment. *Water Science and Technology*, **38**(4–5), 429–436.

Fane A. G. (2004). Exploration of the potential of membrane technology for sustainable decentralized sanitation, 23–26 April 2003, Villa Serbelloni, Bellagin, Italy. *Journal of Membrane Science*, **228**(2), 127–128.

Fick A. (1855). Ueber diffusion [On Diffusion]. *Annalen der Physik* [*Annals of Physics*], **170**(1), 59–86.

Fleischer R. L., Price R. B. and Walker R. M. (1969). Nuclear tracks in solids. *Scientific American*, **220**(6), 30–39.

Frechen F. B., Schier W. and Wett M. (2006). Pre-treatment of municipal MBR applications in Germany – current status and treatment efficiency *Water Practice & Technology*, **33**(12), 1–2. doi:10.2166/wpt.2006.057.

Frost and Sullivan. (2005). Global Membrane Bioreactor (MBR) Market Report. Frost & Sullivan, San Antonio, TX 78229-5616, USA.

Frost and Sullivan. (2011). Snapshot of Membrane Bioreactor Market in India, Report # 9835-00-2E. Frost & Sullivan, San Antonio, TX 78229–5616, USA.

Frost and Sullivan. (2013). Global Membrane Bioreactor (MBR) Market. Report # M7E2-15. Frost & Sullivan, San Antonio, TX 78229–5616, USA.

Fuchs W., Braun R. and Theiss M. (2005). Influence of Various Wastewater Parameters on the Fouling Capacity During Membrane Filtration. Proceedings of the International Congress on Membranes and Membrane Processes (ICOM), 7–10 June 2005, Seoul, Korea.

Gabarrón S, Gómez M, Monclús H, Rodríguez-Roda I. and Comas J. (2013). Ragging phenomenon characterisation and impact in a full-scale MBR. *Water Science and Technology*, **67**(4), 810–816.

Gaines F. R., Marshall J. and Halloran D. (2005). Wastewater Reuse from Onsite Systems. Proceedings of the 78th Annual Water Environment Federation Technical Exhibition and Conference, October 29–November 2 2005, Washington D.C., USA. Water Environment Federation, Alexandria, Virginia.

Galil N. I. and Jacob L. (2009). Comparative characterization of biosolids from a membrane bioreactor and from a sequencing batch reactor. *Environmental Engineering Science*, **26**(5), 1001–1008.

Gallucci F., Basile A. and Hai F. I. (2011). Introduction – a review of membrane reactors. In: Membranes for Membrane Reactors: Preparation, Optimization and Selection, A. Basile and F. Gallucci (eds.), John Wiley & Sons, Ltd, West Sussex, United Kingdom, 1–61.

Ghyoot W. and Verstraete W. (2000). Reduced sludge production in a two-stage membrane-assisted bioreactor. *Water Research*, **34**(1), 205–215.

Gijsbertsen-Abrahamse A. J., Cornelissen E. R. and Hofman J. A. M. H. (2006). Fiber failure frequency and causes of hollow fiber integrity loss. *Desalination*, **194**(1–3), 251–258.

Giménez J. B., Robles A., Carretero L., Durán F., Ruano M. V., Gatti M. N., Ribes J., Ferrer J. and Seco A. (2011). Experimental study of the anaerobic urban wastewater treatment in a submerged hollow-fibre membrane bioreactor at pilot scale. *Bioresource Technology*, **102**(19), 8799–8806.

Gnirss R. and Dittrich J. (2000). Microfiltration of municipal wastewater for disinfection and advanced phosphorus removal: results from trials with different small-scale pilot plants. *Water Environment Research*, **72**(5), 602–609.

Goetz A. and Tsuneishi N. (1951). Application of molecular filter membranes to the bacteriological analysis of water. *Journal of American Water Works Association*, **43**(12), 943–969.

Graham T. (1861). On liquid transpiration in relation to chemical composition. *Philosophical Transactions of the Royal Society of London*, **151**, 373–386.

Graham T. (1866). LV. On the absorption and dialytic separation of gases by colloid septa. *Philosophical Magazine Series 4*, **32**(218), 401–420.

Grethlein H. E. (1978). Anaerobic digestion and membrane separation of domestic wastewater. *Journal of Water Pollution Control Federation*, **50**, 754–763.

Gurung K., Ncibi M. C., Fontmorin J.-M., Särkkä H. and Sillanpää M. (2016). Incorporating submerged MBR in conventional activated sludge process for municipal wastewater treatment: a feasibility and performance assessment. *Journal of Membrane Science and Technology*, **58**(3), 6.

Gurung K., Ncibi M. C. and Sillanpää M. (2017). Assessing membrane fouling and the performance of pilot-scale membrane bioreactor (MBR) to treat real municipal wastewater during winter season in Nordic regions. *Science of The Total Environment*, **579**, 1289–1297.

Hai F. I. and Yamamoto K. (2011). Membrane biological reactors. In: Treatise on Water Science, W. Peter (ed.), Elsevier, Oxford, 571–613.

Hai F. I., Nghiem L. D. and Modin O. (2013). Biocatalytic membrane reactors: removal of recalcitrant and emerging pollutants. In: Handbook of Membrane Reactors, Vol. **2**, A. Basile (ed.), Woodhead Publishing, Cambridge.

Hai F. I., Riley T., Shawkat S., Magram S. and Yamamoto K. (2014). Removal of pathogens by membrane bioreactors: a review of the mechanisms, influencing factors and reduction in chemical disinfectant dosing. *Water*, **6**(12), 3603.

Hofs B., Ogier J., Vries D., Beerendonk E. F. and Cornelissen E. R. (2011). Comparison of ceramic and polymeric membrane permeability and fouling using surface water. *Separation and Purification Technology*, **79**(3), 365–374.

Hurlimann A. and Dolnicar S. (2010). When public opposition defeats alternative water projects – The case of Toowoomba Australia. *Water Research*, **44**(1), 287–297.

Huyskens C., Brauns E., Van Hoof E. and De Wever H. (2008). A new method for the evaluation of the reversible and irreversible fouling propensity of MBR mixed liquor. *Journal of Membrane Science*, **323**(1), 185–192.

Hwang J. H., Cicek N. and Oleszkiewicz J. A. (2009). Membrane biofilm reactors for nitrogen removal: state-of-the-art and research needs. *Water Science & Technology*, **60**(11), 2739–2747.

Ishida H., Izumi K., Moro M., Shimizu K., Soeda Y. and Yamada Y. (1994). Filtration membrane cartridge. US Pat: EP 0602560 A1. United States Patent and Trademark Office, Alexandria, VA 22314, USA.

Jimenez J., Grelier P., Meinhold J. and Tazi-Pain A. (2010). Biological modelling of MBR and impact of primary sedimentation. *Desalination*, **250**(2), 562–567.

Jin L., Ong S. L. and Ng H. Y. (2010). Comparison of fouling characteristics in different pore-sized submerged ceramic membrane bioreactors. *Water Research*, **44**(20), 5907–5918.

Judd S. (2008). The status of membrane bioreactor technology. *Trends in Biotechnology*, **26**(2), 109–116.

Judd S. (2010a). MBRs – a global picture. The MBR site: https://www.thembrsite.com/features/membrane-bioreactors-a-global-picture-june-2010/ (accessed 26 November 2018).

Judd S. (2010b). The MBR Book: Principles and Applications of Membrane Bioreactors for Water and Wastewater Treatment. Butterworth-Heinemann, Oxford (United Kingdom).

Judd S. (2011). Chapter 1 – introduction. In: The MBR Book, 2nd edn, S. Judd and C. Judd (ed.), Butterworth-Heinemann, Oxford, 1–54.

Judd S. (2012). The 2012 MBR Survey – the results. The MBR site: https://www.thembrsite.com/features/the-2012-membrane-bioreactor-survey-the-results-december-2011/ (accessed 26 November 2018).

Judd S. (2014). Industrial MBRs: Membrane Bioreactors for Industrial Wastewater Treatment. Judd and Judd Ltd, United Kingdom.

Judd S. (2016). The status of industrial and municipal effluent treatment with membrane bioreactor technology. *Chemical Engineering Journal*, **305**, 37–45.

Judd S. (2017). Membrane technology costs and me. *Water Research*, **122**, 1–9.

Judd S. (2018). Largest MBR plants (over 100 MLD) – Worldwide. The MBR site: https://www.thembrsite.com/largest-membrane-bioreactor-plants-worldwide/ (accessed 26 November 2018).

Kamada K., Minami S. and Yoshida K. (1977). Porous polypropylene hollow filaments and method making the same. US patent 4055696 A. United States Patent and Trademark Office, Alexandria, VA 22314, USA.

Khan S. J., Hasnain G., Fareed H. and Ben A. R. (2017). Evaluation of treatment performance of a full-scale membrane bioreactor (MBR) plant from unsteady to steady state condition. *Journal of Water Process Engineering*. https://www.sciencedirect.com/science/article/pii/S2214714416306444 (DOI: 10.1016/j.jwpe.2017.03.004).

Khongnakorn W. and Wisniewski C. (2007). Production of sludge in a submerged membrane bioreactor and dewatering aspects. *International Journal of Chemical Reactor Engineering*, **5**. DOI: 10.2202/1542–6580.1416.

Klimiuk E. and Kulikowska D. (2006). Organics removal from landfill leachate and activated sludge production in SBR reactors. *Waste Management*, **26**(10), 1140–1147.

Konopka A., Zakharova T., Oliver L., Camp D. and Turco R. F. (1996). Biodegradation of organic wastes containing surfactants in a biomass recycle reactor. *Applied and Environmental Microbiology*, **62**(9), 3292–3297.

Krzeminski P., Iglesias-Obelleiro A., Madebo G., Garrido J. M., van der Graaf J. H. J. M. and van Lier J. B. (2012). Impact of temperature on raw wastewater composition and activated sludge filterability in full-scale MBR systems for municipal sewage treatment. *Journal of Membrane Science*, **423–424**, 348–361.

Lesjean B. and Huisjes E. H. (2008). Survey of the European MBR market: trends and perspectives. *Desalination*, **231**(1–3), 71–81.

Lesjean B., Rosenberger S., Schrotter J. -C. and Recherche A. (2004). Membrane-aided biological wastewater treatment – an overview of applied systems. *Membrane Technology*, **2004**(8), 5–10.

Lesjean B., Ferre V., Vonghia E. and Moeslang H. (2009). Market and design considerations of the 37 larger MBR plants in Europe. *Desalination and Water Treatment*, **6**(1–3), 227–233.

Lesjean B., Tazi-Pain A., Thaure D., Moeslang H. and Buisson H. (2011). Ten persistent myths and the realities of membrane bioreactor technology for municipal applications. *Water Science & Technology*, **63**(1), 32–39.

Liao B.-Q., Kraemer J. T. and Bagley D. M. (2006). Anaerobic membrane bioreactors: applications and research directions. *Critical Reviews in Environmental Science and Technology*, **36**(6), 489–530.

Lin H., Gao W., Meng F., Liao B.-Q., Leung K.-T., Zhao L., Chen J. and Hong H. (2012). Membrane bioreactors for industrial wastewater treatment: a critical review. *Critical Reviews in Environmental Science and Technology*, **42**(7), 677–740.

Liu Y. and Tay J.-H. (2001). Strategy for minimization of excess sludge production from the activated sludge process. *Biotechnology Advances*, **19**(2), 97–107.

Livingston A. G. (1994). Extractive membrane bioreactors: a new process technology for detoxifying chemical industry wastewaters. *Journal of Chemical Technology and Biotechnology*, **60**(2), 117–124.

Loeb S. and Sourirajan S. (1964). High flow porous membranes for separating water from saline solutions. US patent 3133132. United States Patent and Trademark Office, Alexandria, VA 22314, USA.

Logan B. E. and Rabaey K. (2012). Conversion of wastes into bioelectricity and chemicals by using microbial electrochemical technologies. *Science*, **337**(6095), 686–690.

Manic G., David N., Lesjean B., Tazi Pain A. and Buisson H. (2008). Generic Online Diagnosis Method for the Filtration Performances of Membrane Bioreactors. 2008 IWA World Water Congress, 7–12 September 2008, Vienna, Austria.

Meng F., Chae S.-R., Drews A., Kraume M., Shin H. -S. and Yang F. (2009). Recent advances in membrane bioreactors (MBRs): membrane fouling and membrane material. *Water Research*, **43**(6), 1489–1512.

Merlo R. P., Trussell R. S., Hermanowicz S. W. and Jenkins D. (2007). Effects of sludge properties on the thickening and dewatering of waste activated sludge. *Water Environment Research*, **79**(12), 2412–2419.

Michaels A. S. (1963). Ultrafiltration. In: Progress in Separation & Purification, E. S. Perry (ed.), Wiley-Interscience, New York, 297–334.

Mishra P. N., Sutton P. M. and Mourato D. (1994). Innovative Liquid Phase Treatment Systems: Combining Membrane Technology and Biological Processes. Proceedings of the 67th Annual Water Environment Federation Technical Exposition and Conference, 15–19 October 1994, Chicago, Illinois, USA. Water Environment Federation: Alexandria, Virginia.

Mitsuhashi T. and Saito M. (1993). Modular hollow-fiber filter unit. US patent 5225079 A.

Muller E. B., Stouthamer A. H., van Verseveld H. W. and Eikelboom D. H. (1995). Aerobic domestic waste water treatment in a pilot plant with complete sludge retention by cross flow filtration. *Water Research*, **29**(4), 1179 1189.

NWI (2011). The National Water Initiative – securing Australia's water future: 2011 assessment. Australian Government http://www.nwc.gov.au/reform (accessed 1 June 2013).

Oppenheimer J., Rittmann B., DeCarolis J., Hirani Z. and Kiser A. (2012). Investigation of membrane bioreactor effluent water quality and technology, Report Number WRF-06-007; WateReuse Research Foundation, Alexandria, Virginia.

Pagotto R., Rossetto R., Gasperi R. L. P., Andrade J. P., Trovati J., Vallero M. V. G., Okumura A. and Arntsen B. (2014). SANASA Capivari II-the first full-scale municipal membrane bioreactor in Latin America. *Water Science and Technology*, **70**(2), 272.

Pearce G. (2008). Introduction to membranes – MBRs: manufacturers' comparison: part 1. *Filtration and Separation*, **March 2008**, 28–31.

Phan H. V., Hai F. I., McDonald J. A., Khan S. J., Zhang R., Price W. E., Broeckmann A. and Nghiem L. D. (2015). Nutrient and trace organic contaminant removal from wastewater of a resort town: comparison between a pilot and a full scale membrane bioreactor. *International Biodeterioration & Biodegradation*, **102**, 40–48.

Radjenović J., Matošić M., Mijatović I., Petrović M. and Barceló D. (2008). Membrane bioreactor (MBR) as an advanced wastewater treatment technology. In: Emerging Contaminants from Industrial and Municipal Waste, D. Barceló and M. Petrovic (ed.), Springer, 37–101.

Reij M. W., Keurentjes J. T. F. and Hartmans S. (1998). Membrane bioreactors for waste gas treatment. *Journal of Biotechnology*, **59**(3), 155–167.

Research and Markets. (2012). Membrane Bioreactor Systems Market – by Types, Configuration & Applications – Trends & Forecasts to 2017.

Research and Markets. (2018). Global membrane bioreactor systems market size, market share, application analysis, regional outlook, growth trends, key players, competitive strategies and forecasts, 2018 to 2026. Report ID: 4564636.

Rosenberger S. and Kraume M. (2002). Filterability of activated sludge in membrane bioreactors. *Desalination*, **146**(1–3), 373–379.

Ruel S. M., Choubert J. M., Esperanza M., Miège C., Madrigal P. N., Budzinski H., Le Ménach K., Lazarova V. and Coquery M. (2011). On-site evaluation of the removal of 100 micro-pollutants through advanced wastewater treatment processes for reuse applications. *Water Science and Technology*, **63**(11), 2486–2497.

Santos A. and Judd S. (2010). The fate of metals in wastewater treated by the activated sludge process and membrane bioreactors: a brief review. *Journal of Environmental Monitoring*, **12**(1), 110.

Santos A., Ma W. and Judd S. J. (2010). Membrane bioreactors: two decades of research and implementation. *Desalination*, **273**(1), 148–154.

Schier W., Frechen F. B. and Fischer St. (2009). Efficiency of mechanical pre-treatment on European MBR plants. *Desalination*, **236**(1–3), 85–93.

Schmidt W. (1856). Versuche über Filtrationsgeschwindigkeit verschiedener Flüssigkeiten durch thierische Membran. [Experiments on filtration rate of different fluids through membrane]. *Annalen der Physik* [Annals of Physics], **175**(11), 337–388.

Schrader G. A., Zwijnenburg A. and Wessling M. (2005). The effect of WWTP effluent zeta-potential on direct nanofiltration performance. *Journal of Membrane Science*, **266**(1–2), 80–93.

Shen Y.-x., Xiao K., Liang P., Sun J.-y., Sai S.-j. and Huang X. (2012). Characterization of soluble microbial products in 10 large-scale membrane bioreactors for municipal wastewater treatment in China. *Journal of Membrane Science*, **415**, 336–345.

Smith C. V., Gregorio D. O. and Talcott R. M. (1969). The Use of Ultrafiltration Membranes for Activated Sludge Separation. Proceedings of the 24th annual Purdue Industrial Waste Conference, 1969, Lafayette, IN, USA. 1300–1310.

Stefanski M., Kennedy S. and Judd S. (2011). The determination and origin of fibre clogging in membrane bioreactors. *Journal of Membrane Science*, **375**(1–2), 198–203.

Stüber J., Schallehn F., Lüdicke C., Lesjean B., Gnirss R. and Kraume M. (2010). Economical Evaluation of Decentralised MBRs Achieving Varying Effluent Qualities. IWA World Water Congress, 20–24 September 2010, Montreal, Canada.

Sutton P. M., Li R. R. and Dorr-oliver's K. S. R. (1983). Fixed Film Suspended Growth Anaerobic Systems for Industrial Wastewater Treatment and Energy Recovery. Proceedings of the 37th Annual Purdue Industrial Waste Conference, 11–13 May 1983, Lafayette, IN, USA, 667–675.

Tonelli F. A. and Canning R. P. (1993). Membrane bioreactor system for treating synthetic metalworking fluids and oil based products. US Pat: 5204001. United States Patent and Trademark Office, Alexandria, VA 22314, USA.

Torres A. P. R., Santiago V. M. J. and Borges C. P. (2008). Performance evaluation of submerged membrane bioreactor pilot units for refinery wastewater treatment. *Environmental Progress*, **27**(2), 189–194.

Ueda T. and Hata K. (1999). Domestic wastewater treatment by a submerged membrane bioreactor with gravitational filtration. *Water Research*, **33**(12), 2888–2892.

Van't Hoff J. H. (1887). The role of osmotic pressure in the analogy between solutions and gases. *Zeitschrift für Physikalische Chemie (International Journal of Research in Physical Chemistry and Chemical Physics)*, **1**, 481–508.

van't Hoff J. H. (1888). XII. The function of osmotic pressure in the analogy between solutions and gases. *Philosophical Magazine Series 5*, **26**(159), 81–105.

van Nieuwenhuizen A. F., Evenblij H., Uijterlinde C. A. and Schuting F. L. (2008). Review on the state of science on membrane bioreactors for municipal wastewater treatment. *Water Science & Technology*, **57**(7), 979–86.

Visvanathan C., Ben A. R. and Parameshwaran K. (2000). Membrane separation bioreactors for wastewater treatment. *Critical Reviews in Environmental Science and Technology*, **30**(1), 1–48.

Wang Z., Wu Z., Mai S., Yang C., Wang X., An Y. and Zhou Z. (2008). Research and applications of membrane bioreactors in China: progress and prospect. *Separation and Purification Technology*, **62**(2), 249–263.

WEF (2012). Membrane Bioreactors, WEF Manual of Practice No 36. McGraw-Hill, New York.

Xiao K., Xu Y., Liang S., Lei T., Sun J., Wen X., Zhang H., Chen C. and Huang X. (2014). Engineering application of membrane bioreactor for wastewater treatment in China: current state and future prospect. *Frontiers of Environmental Science & Engineering*, **8**(6), 805–819.

Xiao K., Liang S., Wang X., Chen C. and Huang X. (2018). Current state and challenges of full-scale membrane bioreactor applications: a critical review. *Bioresource Technology*, **47**(3), 681–682.

Xing C. H., Tardieu E., Qian Y. and Wen X. H. (2000). Ultrafiltration membrane bioreactor for urban wastewater reclamation. *Journal of Membrane Science*, **177**(12), 73–82.

Yamamoto K. (2009). Submerged MBR technology: An Unfinished International Adventure of 20 Years. Final MBR-Network Workshop, 31 March–1 April 2009, Berlin, Germany, 83–84.

Yamamoto K., Hiasa M., Mahmood T. and Matsuo T. (1989). Direct solid-liquid separation using hollow fiber membrane in an activated sludge aeration tank. *Water Science and Technology*, **21**(4–5), 43–54.

Yang W., Cicek N. and Ilg J. (2006). State-of-the-art of membrane bioreactors: worldwide research and commercial applications in North America. *Journal of Membrane Science*, **270**(12), 201–211.

Young T., Smoot S., Peeters J. and Côté P. (2013). When does building an MBR make sense? How variations of local construction and operating cost parameters impact overall project economics. *Proceedings of the Water Environment Federation*, **2013**(8), 6354–6365.

Young T., Smoot S., Peeters J. and Côté P. (2014). Cost-effectiveness of membrane bioreactors treatment system for low-level phosphorus reduction from municipal wastewater. *Water Practice and Technology*, **9**(3), 316–323.

Zhang S., van Houten R., Eikelboom D. H., Doddema H., Jiang Z., Fan Y. and Wang J. (2003). Sewage treatment by a low energy membrane bioreactor. *Bioresource Technology*, **90**(2), 185–192.

Zhang S., Qu Y., Liu Y., Yang F., Zhang X., Furukawa K. and Yamada Y. (2005). Experimental study of domestic sewage treatment with a metal membrane bioreactor. *Desalination*, **177**(1–3), 83–93.

Zhang S., Yang F., Liu Y., Zhang X., Yamada Y. and Furukawa K. (2006). Performance of a metallic membrane bioreactor treating simulated distillery wastewater at temperatures of 30 to 45 C. *Desalination*, **194**(1–3), 146–155.

Zsigmondy R. and Bachmann W. (1918). Über neue Filter [New filter]. *Zeitschrift für anorganische und allgemeine Chemie* [*Journal of Inorganic and General Chemistry*], **103**(1), 119–128.

Zsigmondy R. and Bachmann W. (1922). Filter and method of producing the same. US patent 1421341. United States Patent and Trademark Office, Alexandria, VA 22314, USA.

Zsirai T., Buzatu P., Aerts P. and Judd S. (2012). Efficacy of relaxation, backflushing, chemical cleaning and clogging removal for an immersed hollow fibre membrane bioreactor. *Water Research*, **46**(14), 4499–4507.

Chapter 2

Process fundamentals: from conventional biological wastewater treatment to MBR

Veeriah Jegatheesan[1] *and Chettiyappan Visvanathan*[2]
[1]*School of Engineering, RMIT University, Melbourne, VIC 3000, Australia*
[2]*School of Environment, Resources and Development, Asian Institute of Technology, PO Box 4, Klong Luang, Pathumthani 12120, Thailand*

ABSTRACT

Understanding of microbial processes and kinetics of degradation of various organic and inorganic micro- and macro-pollutants helped in developing the conventional activated sludge processes in the late 19th century. Research and development in wastewater treatment have further helped the innovation of advanced processes such as membrane bioreactors (MBRs). This chapter illustrates the fundamentals of the MBR technology, which has recently emerged as a membrane-based biological process capable of advanced treatment of wastewater, producing water that is reusable for various purposes.

2.1 INTRODUCTION

Microbes that are present in water environment have the ability to degrade organic and inorganic compounds; this ability has been put into use since the late 19th century when trickling filters and septic tanks were introduced for wastewater treatment. Since then extensive studies have been made to understand (i) aerobic, anaerobic and anoxic environments that are conducive to different microbial cultures, (ii) the degradation pathways of pollutants by various microbes and (iii) the process kinetics associated with the degradation of different pollutants. This in turn has helped to design treatment systems that are capable of treating wastewater containing various pollutants. Advances in technology further enhanced the efficiency of treatment systems and one such advanced technology is membrane bioreactor (MBR). An MBR combines biological treatment of wastewater with membrane filtration, and hence it is important to know the basics of both biological processes and membrane filtration. This

Editor(s): Faisal I. Hai, Kazuo Yamamoto and Chung-Hak Lee
doi: 10.2166/9781780409177_0035

chapter first outlines the basics of both biological processes and membrane filtration and then explains how those two components can be seamlessly combined into an integrated process as in an MBR.

2.2 NEED FOR BIOLOGICAL TREATMENT

From the early 1800s the world's population started to increase at a faster rate which inevitably led to the increased generation of waste in the form of the solids, liquids and gases. Since then, these wastes have contributed significantly to the ill health of the population and have forced scientists, microbiologists and engineers to find ways of treating them. The industrial development that occurred alongside the increase in population obviously caused further deterioration of the environment through industrial emissions. The waste to be treated was composed of organic and inorganic substances, some of which are toxic even in trace levels when exposed to living organisms. Pathogenic microorganisms that are also present in the waste would cause harm when they came into contact with living organisms. This chapter considers the liquid waste that is being produced from domestic uses and industrial activities. The typical constituents of wastewater generated from different sources are given in Table 2.1.

Table 2.1 Constituents of wastewater generated by different sources and their typical concentrations (Coombes *et al.*, 2000; Hammer, 1986; Metcalf & Eddy, 2003).

	Domestic Wastewater	Industrial Wastewater	Storm Water
Suspended solids (mg/L)	100–240	70–12,000	150
BOD (mg/L)	120–200	800–18,000	12
COD (mg/L)	240	200–4,800	90
pH	6.0–8.5	3.5–8.0	6.0–9.0
Total nitrogen (mg/L)	35	40–3100	3
NH_3-N (mg/L)	20	–	–
Total phosphorus (mg/L)	10	5–370	0.5
PO_4-P (mg/L)	7	–	–
E. coli (per 100 mL)	10^7		40–67,000

The naturally occurring aerobic and anaerobic processes in carbon, nitrogen, phosphorus, and sulphur cycles helped scientists to further increase their knowledge on treating all the three forms (solids, liquid and gaseous) of waste to reduce their impact on the environment as well as the health of humans and other living organisms. Understanding of microbial degradation of organic and inorganic compounds paved the way to the first biological wastewater treatment system in 1880s as trickling filters and septic tanks.

2.3 MICROBIAL COMMUNITIES, THEIR ENVIRONMENTS AND DEGRADATION PATHWAYS OF POLLUTANTS

Microbes require food to generate energy that could be used for their mobility, transport of nutrients, and so forth as well as for the synthesis of cells. The organic and inorganic compounds (in wastewater) that are in the form of electron donors and electron acceptors are used by microbes as food for this purpose. The carbon sources are used for the cell synthesis (growth) and energy for maintenance. The principal carbon sources used in the cell synthesis are either carbon dioxide or organic wastes and the source for the energy production is either light or chemical compounds (Madigan *et al.*, 2000). The classification of microbes is given in Table 2.2.

Table 2.2 Classification of microbes based on the selection of energy and carbon sources (adapted from Jegatheesan *et al.*, 2004).

Energy Source	Carbon Source	
	Carbon Dioxide (Autotrophs)	Organic Compounds (Heterotrophs)
Light	***Photoautotrophs*** Higher plants, algae, photosynthetic bacteria	***Photoheterotrophs*** Purple and green bacteria
Chemical compounds	***Chemoautotrophs*** Energy derived from reduced organic compounds, reduced form of sulphur and ferrous ion; examples are nitrogen-fixing bacteria (in soils), sulphur bacteria (deep-sea thermal vents), nitrifying bacteria and iron oxidizing bacteria (in lava beds)	***Chemoheterotrophs*** Carbon and energy usually derived from the metabolism of a single organic carbon; most bacteria, fungi and protozoa

Photoautotrophs such as the higher plants, algae and photosynthetic bacteria use carbon dioxide as carbon sources, while photoheterotrophs use the organic substances. Similarly, the chemoautotrophs produce the energy from the reduced organic compounds such as NH_3, NO_2^-, H_2; reduced forms of sulphur (H_2S, S, $S_2O_3^{2-}$), or ferrous iron and the chemoheterotrophs derive carbon and energy from the metabolism of a single organic carbon. In order to adsorb the substrates that are present in the waste, first microbial cells secrete extracellular enzymes called *hydrolases*. These enzymes hydrolyse the substrates that are present in the form of complex molecules (such as cellulose and other polysaccharides, proteins, etc.) to simple molecules (such as the sugars, amino acids, fatty acids, etc.) in wastewater. These simple molecules then dialyse through the cell walls. Once they enter the cells, these simple molecules are oxidized to produce energy. The intracellular enzymes called *desmolases* or respiratory enzymes, act as catalysts for oxidation–reduction reactions (Jegatheesan *et al.*, 2004).

The substrates such as carbohydrates, proteins, fatty acids, methanol, and so on present in waste, can act as electron donors in oxidation–reduction reactions to produce energy as well as to provide carbon for cell synthesis. The chemoheterotrophic microorganisms utilizing these substrates require electron acceptors such as oxygen, nitrate (NO_3^-), sulphate (SO_4^{2-}) or carbon dioxide (CO_2) for oxidation (Figure 2.1). For example, if the waste is under aerobic conditions, the specific chemoheterotrophic microorganisms will utilize oxygen (as an electron acceptor) and organic substances (as electron donors) present in the waste to produce energy. Similarly, under anoxic conditions, the electron acceptor is nitrate (NO_3^-) and under anaerobic conditions, it is either sulphate (SO_4^{2-}) or carbon dioxide (CO_2), which oxidizes the organic substances (electron donor) present in the waste. In addition, more energy is produced from oxidation–reduction reactions that occur under the aerobic conditions. For example, when acetate ion is oxidized by aerobic bacteria, the following chemical reaction will take place:

$$CH_3COO^- + 2O_{2(aq)} \rightarrow HCO_3^- + CO_{2(aq)} + H_2O_{(l)} \tag{2.1}$$

Using free energy calculations, it can be shown that at 25°C, when 1 mole of acetate ion is oxidized by aerobic bacteria, 208.17 kcal of energy will be produced. On the contrary, if the reaction is carried out by anaerobic bacteria, the following reaction will take place and 1 mole of acetate ion will produce 6.69 kcal of energy:

$$CH_3COO^- + H_2O_{(l)} \rightarrow HCO_3^- + CH_{4(g)} \tag{2.2}$$

Therefore, the aerobic microorganisms have more energy for the cell synthesis compared to the anaerobic ones. Thus, the production of the microbial mass (or sludge) is higher in aerobic and anoxic treatment processes than in anaerobic processes. These phenomena are brought into practice in biological waste treatment.

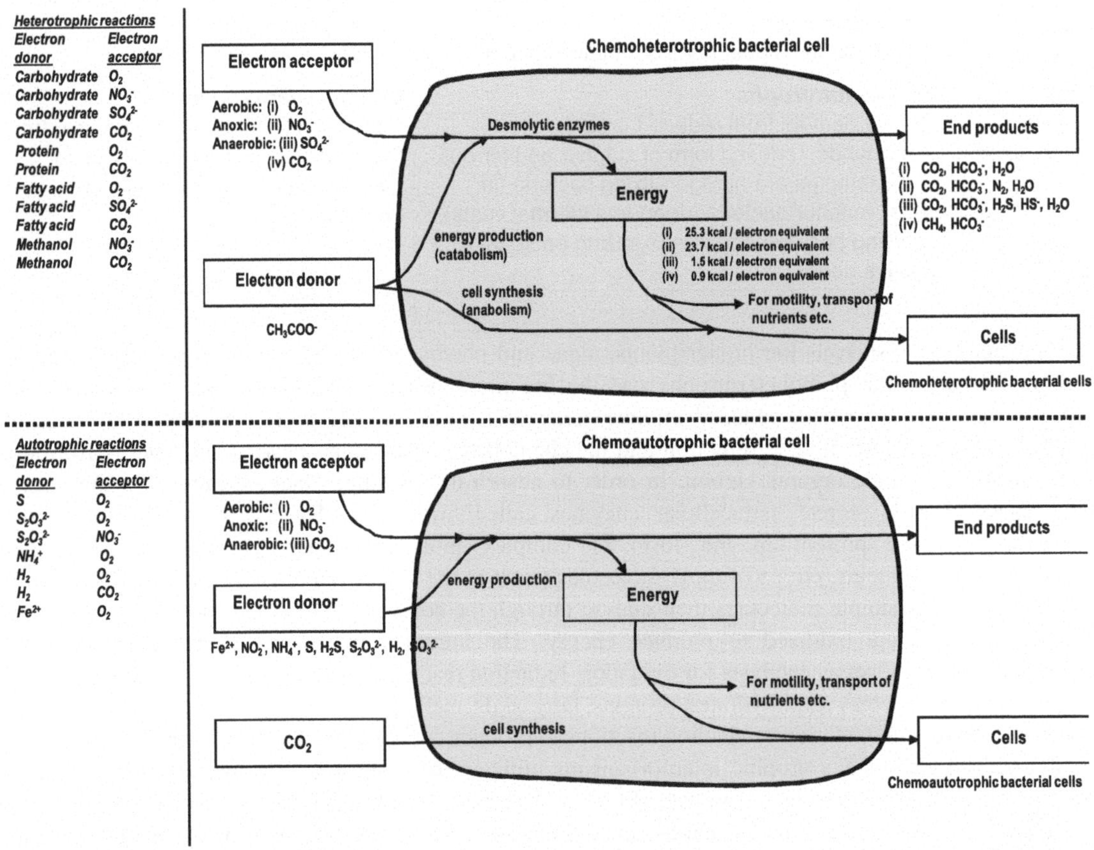

Figure 2.1 Degradation pathway of organic and inorganic compounds by chemoautotrophic and chemoheterotrophic bacteria (Jegatheesan *et al.*, 2008, reprinted with permission from Asiatech Publishers, Inc.).

2.4 BIOLOGICAL TREATMENT FUNDAMENTALS

Wastewater that reaches a treatment plant will undergo the following: (i) preliminary treatment composed of screening and grit removal, (ii) if required, primary treatment in the form of a primary sedimentation tank to reduce excessive suspended solids, (iii) secondary treatment in the form of aeration and secondary sedimentation to biologically remove organic substances (Metcalf & Eddy, 2003). Figure 2.2 shows the schematic of typical secondary wastewater treatment plant processes with design parameters and changes in quality parameters along the treatment process. Tertiary treatment is required to remove the nutrients (nitrogen and phosphorus) and to further reduce the organic substances as well as suspended solids. The typical quality of treated wastewater after each treatment process mentioned is shown in Table 2.3.

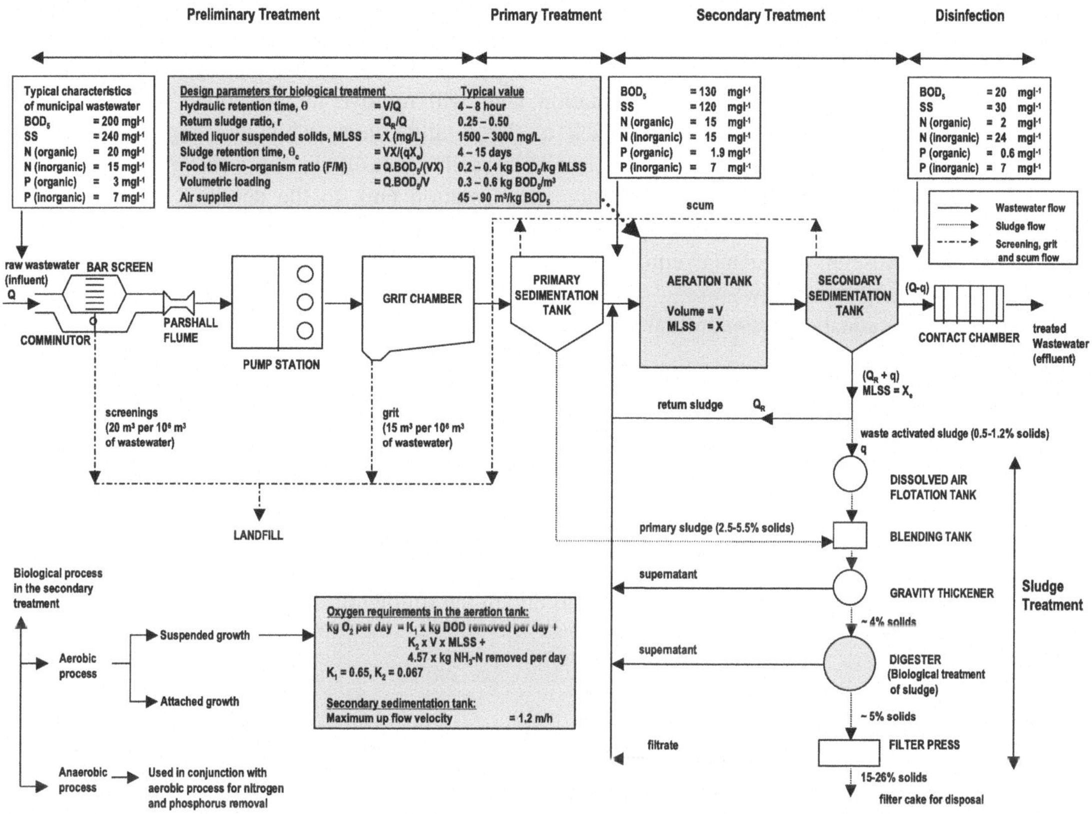

Figure 2.2 Schematic of a secondary wastewater treatment plant (Jegatheesan *et al.*, 2008, reprinted with permission from Asiatech Publishers, Inc.).

Note: The flow rates of raw wastewater into the treatment plant and the wastewater into the aeration tank are assumed to be the same although the latter will be reduced by the amount of the primary sludge flow rate.

Table 2.3 Raw and treated wastewater quality after primary, secondary and tertiary treatment (Metcalf & Eddy, 2003; Ragsdale, 2007).

Parameter	Raw Wastewater	Effluent from Primary Treatment	Effluent from Secondary Treatment	Effluent from Tertiary Treatment
BOD (mg/L)	300	200	15	2
Suspended solids (mg/L)	300	120	20	2
Total nitrogen (mg/L)	50	45	20	4
Total phosphorus (mg/L)	10	9	7	<1

2.4.1 Conventional activated sludge (CAS) process basics

The aeration tank in the secondary treatment will facilitate the microbial degradation of organic substances (which are referred to as substrate) into CO_2 and H_2O and in doing so will allow the growth of heterotrophic bacteria in the aeration tank. The microbial suspension in the aeration tank is called mixed liquor suspended solids (MLSS) and the secondary treatment process is called activated sludge process. The activated sludge will consist of rotifers, nematodes, protozoa, and bacteria. In order to maintain the concentration of MLSS in the aeration tank, a secondary sedimentation tank is provided after the aeration tank. The sedimentation tank will remove the microbial cells that leave the aeration tank and a portion will return back to the aeration tank, while the rest is sent to the sludge treatment process. Mass balances for microbial cells as well as the substrate will help to compute the return flow rate from the secondary sedimentation tank to the aeration tank as well as the volume of sludge to be wasted every day. Understanding of microbial kinetics is essential in order to develop those mass balance equations. Figure 2.3 is used to derive some important mass balance equations as given below:

The mass balance equation for microbial biomass:

$$\frac{\mu_{max} XS}{K_S + S} - k_d X = \frac{qX_e}{V} \tag{2.3}$$

The mass balance equation for substrate (or biochemical oxygen demand (BOD)):

$$\frac{\mu_{max} S}{K_S + S} - k_d X = \frac{QY}{VX}(S_o - S) \tag{2.4}$$

By combining the above two equations, we could obtain the following:

$$X = \frac{\theta_c Y(S_o - S)}{\theta(1 + k_d \theta_c)} \tag{2.5}$$

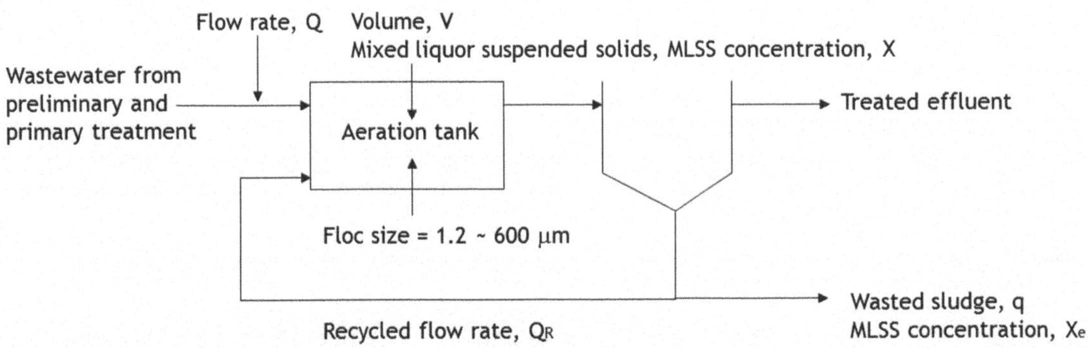

Figure 2.3 Schematic diagram of a secondary biological treatment system.

where X and X_e are the concentrations of MLSS in the aeration tank and in the sludge exiting the secondary sedimentation tank; μ_{max} is the maximum specific growth rate of the microbial culture that is present in the MLSS and utilizing the substrate; K_S is half-saturation constant, k_d is the endogenous decay rate constant of the microbial culture; $Y(=(dX/dS))$ is the specific yield, S_o and S

are the concentrations of substrate that is entering and exiting the secondary treatment, Q is the flow rate of wastewater that is entering the secondary treatment, θ $(=V/Q)$ and $\theta_c(=(VX/qX_e))$ are hydraulic and sludge retention times, respectively, and q is the amount of sludge wasted per unit time. Thus, it is easy to design a secondary treatment system when the kinetic parameters of the microbial culture that is dominant in the aeration tank are known. Further, the food to microorganism ratio $((F/M)=(QS_o/VX))$ and the recycling ratio, Q_R/Q should also be maintained at appropriate values in order for the treatment system to function consistently and efficiently. The above information on mass balance covers only the fundamental concepts and detailed expressions can be seen elsewhere (Judd & Judd, 2011; Metcalf & Eddy, 2003).

2.4.2 Nitrogen removal

The nitrogen attached to complex molecules (such as protein) that are present in wastewater should undergo the following if it is to be removed from the wastewater: (i) hydrolysis into simple molecules such NH_3, (ii) oxidation (nitrification) to NO_2^- by the chemoautotrophic bacteria *Nitrosomonas*, (iii) further oxidation (nitrification) to NO_3^- by chemoautotrophic bacteria *Nitrobactor*, and (iv) denitrification to N_2 gas by heterotrophic bacteria under anoxic conditions. Thus both aerobic and anoxic environments should prevail in order to remove nitrogen. Since denitrification is carried out by heterotrophic bacteria, organic substrates are required as electron donors during this process. Thus, denitrification will also reduce the overall oxygen demand required for the removal of organic substrates. The corresponding chemical reactions are given below (Ahmed, 2012):

Nitrification (cell synthesis neglected (Antoniou *et al.*, 1990)):

$$NH_4^+ + 1.5O_2 \rightarrow NO_2^- + 2H^+ + H_2O \tag{2.6}$$

$$NO_2^- + 0.5O_2 \rightarrow NO_3^- \tag{2.7}$$

Nitrification (cell synthesis included (McCarty *et al.*, 1969)):

$$\textit{Nitrosomonas}: 55NH_4^+ + 76O_2 + 109HCO_3^- \rightarrow C_5H_7NO_2 + 54NO_2^- + 57H_2O + 104H_2CO_3 \tag{2.8}$$

$$\textit{Nitrobacter}: 400NO_2^- + NH_4^+ + 4H_2CO_3 + HCO_3^- + 195O_2 \rightarrow C_5H_7NO_2 + 400NO_3^- + 3H_2O \tag{2.9}$$

$$\textit{Overall}: NH_4^+ + 1.83O_2 + 1.98HCO_3^- \rightarrow 0.021C_5H_7NO_2 + 1.04H_2O + 0.98NO_3^- + 1.88H_2CO_3 \tag{2.10}$$

From equations 2.8 and 2.9, the yield of *Nitrosomonas* and *Nitrobacter* can be estimated to be 0.15 g cells/g $NH_4^+ - N$ and 0.02 g cells/g $NO_2^- - N$, respectively. Similarly, the oxygen consumption by *Nitrosomonas* and *Nitrobacter* can be estimated to be 3.22 mg O_2/mg $NH_4^+ - N$ and 1.11 mg O_2/mg $NO_2^- - N$, respectively. Thus, a total of 4.18 mg of O_2 is required to nitrify 1 mg of $NH_4^+ - N$. Further, the alkalinity destruction during nitrification with and without cell synthesis is considered to be 7.07 and 7.14 mg as $CaCO_3$/mg $NH_4^+ - N$ oxidized, respectively.

Denitrification: (when methanol is used as H^+ donor (McCarty *et al.*, 1969)):

$$NO_3^- + 0.33CH_3OH \rightarrow NO_2^- + 0.33H_2O + 0.33H_2CO_3 \tag{2.11}$$

$$NO_2^- + 0.5CH_3OH + 0.5H_2CO_3 \rightarrow 0.5N_2 + H_2O + HCO_3^- \tag{2.12}$$

Thus, the overall dissimilation of NO_3^- can be given by

$$NO_3^- + 0.83CH_3OH + 0.17H_2CO_3 \rightarrow 0.5N_2 + 1.33H_2O + HCO_3^- \tag{2.13}$$

Cell synthesis during denitrification when methanol is used as carbon source can be expressed by the following reaction:

$$3NO_3^- + 14CH_3OH + 4H_2CO_3 \rightarrow 3C_5H_7NO_2 + 20H_2O + 3HCO_3^- \quad (2.14)$$

Therefore, the overall nitrate removal can be given by

$$NO_3^- + 1.08CH_3OH + 0.24H_2CO_3 \rightarrow 0.056C_5H_7NO_2 + 0.47N_2 + 1.68H_2O + HCO_3^- \quad (2.15)$$

During denitrification, 3.57 mg alkalinity as $CaCO_3$ will be produced for every milligram of $NO_3^- - N$ removed. Both autotrophic and heterotrophic bacteria are capable of denitrification and Table 2.4 provides additional details of their requirements.

Table 2.4 Types of denitrifying bacteria and corresponding hydrogen donors.

Type of Microbes	Species	H_2 Donor
Heterotrophic	*Pseudomonas, Micrococcus, Spirillum, Achromobacter, Alkaligenes*	Organic substances
Autotrophic	*Hydrogenomonas denitrificans, Thiobacillus denitrificans*	Hydrogen gas, sulphides

In order to optimize this process, the denitrifying compartment will be placed in front of the nitrifying compartment so that the organic substrate entering the biological treatment will be first utilized for denitrification (Hammer, 1986). Further, denitrification is an alkalinity producing process (1 meq of alkalinity produced per mole of NO_3^- consumed) while nitrification is an alkalinity consuming process (2 meq of alkalinity consumed per mole of NH_4^+ reacted). Thus, denitrification followed by nitrification will maintain a stable pH throughout the treatment system as well. However, recycling of MLSS from the nitrification compartment to the denitrification compartment is necessary to supply enough NO_3^- for denitrification to occur.

2.4.3 Phosphorus removal

Biological removal of phosphorus depends on increasing the phosphorus content in the sludge. Introduction of an anaerobic zone at the front of the aeration tank is required to do so. This introduces a metabolic function in the microorganisms to release phosphorus in the anaerobic phase and to uptake phosphorus drastically by poly-phosphate accumulating organisms (PAOs) in the aerobic phase (Seviour *et al.*, 2003). PAOs accumulate carbon sources as poly-β-hydroxylalkanoates (PHA) whose composition depends on the carbon source assimilated. Acetate assimilation will produce poly-β-hydroxybutyrate (PHB) and propionate assimilation will produce poly-β-hydroxyvalerate (PHV). The phosphorus content of PAOs is 0.30 g P/g VSS as compared to 0.02 g P/g VSS of other bacteria. Figure 2.4 shows the metabolic pathways associated with phosphorus removal in anaerobic and aerobic phases where accumulation of PHB is assumed. Phosphorus removal can be affected by the following factors: organic loading, hydraulic retention time of anaerobic and aerobic phases, concentration of cations Mg^{2+} and K^+, concentration of anions NO_2^- and NO_3^-, pH, temperature and sludge retention time. Usually, 1 mg/L of phosphorus in the treated effluent is achievable through biological removal process. However, biological instability could occasionally lead to higher phosphorus concentrations in the effluent. Combination of chemical and biological phosphorus removal can alleviate this situation.

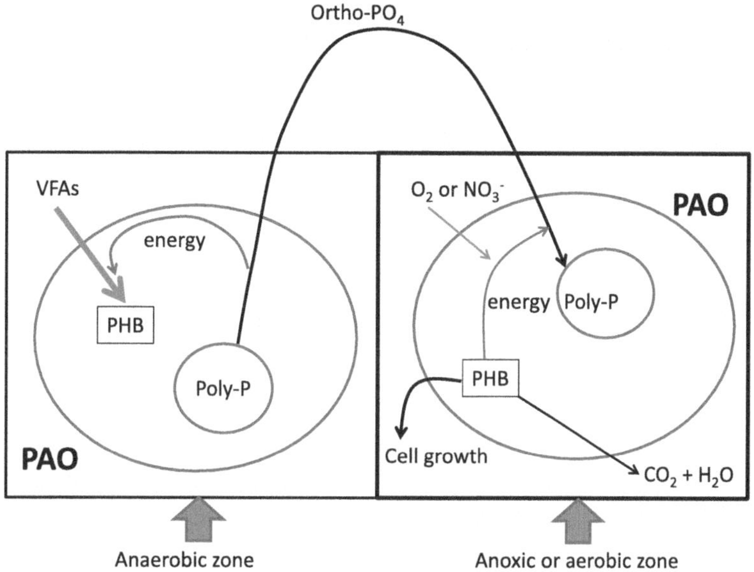

Figure 2.4 Metabolic pathways associated with phosphorus removal.

2.4.4 Combined biological nutrient removal (BNR)

Combining the processes discussed in Sections 2.4.1 through 2.4.3 will give an advanced biological treatment system that is capable of removing organic substances as well nutrients (e.g. nitrogen and phosphorous). Figure 2.5 shows such a system called the three-stage Bardenpho process or A^2/O

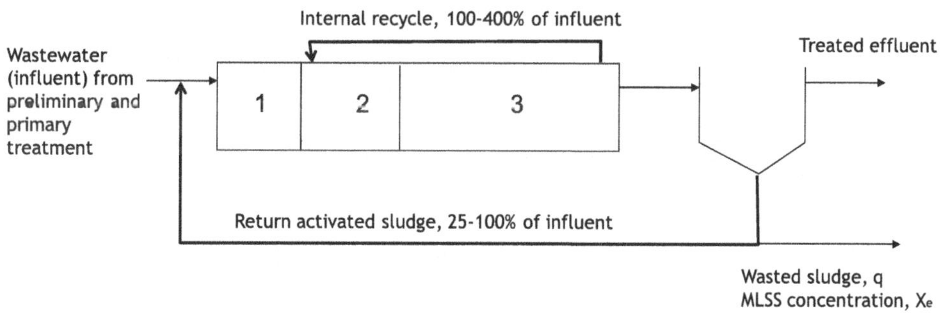

1. **Anaerobic zone (HRT, θ = 0.5 to 1.5 hours)**
   Heterotrophs produce volatile fatty acids as substrates for PAO and break down organic compounds
2. **Anoxic zone (HRT, θ = 0.5 to 1 hours)**
   Heterotrophs denitrify NO_3^- to N_2 gas and break down organic compounds
3. **Aerobic zone (HRT, θ = 4 to 8 hours; SRT, θ_c = 5 to 25 days)**
   Heterotrophs break down organic compounds
   Autotrophic nitrifiers convert ammonia to NO_3^-
   Heterotrophic PAO uptake excess phosphorus

Figure 2.5 Schematic diagram of three-stage Bardenpho process (also known as A^2/O process) to remove both nitrogen and phosphorus along with biodegradable organic substances from wastewater.

process (Jeyanayagam, 2005). The activated sludge of a BNR system will consist of fermenters and hydrolysers (6%), Nitrifiers (7%), denitrifiers (29%), PAO (13%), filamentous bacteria (29%) and others such as iron and sulphate reducers (27%) (Nielsen *et al.*, 2010). Table 2.5 provides details of the kinetic parameters of different microbial communities that are present in different compartments of the system. This information can be used to design the treatment system with the aid of equations 2.3 through 2.5. Further, temperature effects on kinetic constants such as the specific growth rate and the half-saturation constant of nitrifiers need to be incorporated in design computations through Arrhenius equations (Metcalf & Eddy, 2003). In the A^2/O process, the MLSS concentration is maintained at 3000–4000 mg/L and this process produces good settling sludge and saves energy (Metcalf & Eddy, 2003). Operating this system is relatively easy as well.

Table 2.5 Kinetic parameters of different bacterial communities (Jegatheesan *et al.*, 2008; Judd & Judd, 2011; Metcalf & Eddy, 2003).

	μ_{max} (d^{-1})	K_S (mg/L)	Y (mg VSS/mg Substrate)	k_d (d^{-1})
Heterotrophs	3.0–13.2 (6.0)	5–40 (20)	0.3–0.5 (0.4)	0.06–0.20 (0.12)
Nitrosomonas	0.6–0.8	0.3–0.7	0.15–0.20	0.03–0.06
Nitrobacter	0.6–1.0	0.8–1.2	0.5–1.5	0.03–0.06
PAOs	0.67–1.6 (1.0)	1–10 (4)	0.46–0.69 (0.625)	0.1–0.3 (0.2)

Note: Typical values are shown within parentheses.

2.4.5 Operational requirements

2.4.5.1 Oxygen

Oxygen is required to oxidize biodegradable organic compounds and ammonia-nitrogen in the wastewater as well as for the respiration of microorganisms. Denitrification in the anoxic zone reduces oxygen demand as 1 kg of NO_3^- reduced will be equivalent to 2.86 kg of O_2 that is required to oxidize biodegradable organic compounds. Thus a simple formula as given below can be used to compute the oxygen requirement in a BNR system (WEF, 1998; Metcalf & Eddy, 2003):

$$\text{kg } O_2 \text{ required/d} = K_1 \times \text{kg BOD removed per day} + 4.57 \times \text{TKN removed} + K_2 \times \text{Volume of the aeration tank} \times \text{MLSS} \tag{2.16}$$

where K_1 (carbonaceous oxygen demand) = 0.65 and K_2 (respiration oxygen demand) = 0.067/d. Then estimation of standard oxygen transfer into mixed liquor should be computed to size aerators. Here TKN refers to total Kjeldahl nitrogen which is the sum of the concentrations of total organic nitrogen and total ammonia nitrogen present in the wastewater.

2.4.5.2 Sludge management

Disposal facilities of wasted sludge from a wastewater treatment plant usually represent 40–60% of the construction cost of the wastewater treatment plant and account for as much as 50% of the operating costs. Disproportionate shares of operating difficulties are also caused by disposal facilities (Metcalf & Eddy, 2003).

The volume of sludge produced per day in a primary sedimentation tank (PST) can be computed by its efficiency of removing the incoming suspended solids as given by the following equations. Mass of the suspended solids, M_p removed (in kg/d) by a PST can be given by

$$M_p = \eta \times SS \times Q \tag{2.17}$$

where η is the removal efficiency of suspended solids by the PST, SS is the concentration of incoming suspended solids to the PST and Q is the inflow. Similarly, the volume of the sludge, V_p generated (in m^3/d) by the PST, can be given by

$$V_p = \frac{M_p}{[\rho_{psl} \times (P_p/100)]} \tag{2.18}$$

where ρ_{psl} is the density of the sludge and P_p is the percentage of solids present in the sludge. Further, the volume of the solids produced in the aeration tank due to the growth of microbial biomass can be computed using the utilization of BOD by the microbes and their subsequent growth. Thus, the mass of the solids, M_s produced (in kg/d) in the aeration tank can be computed by the following equation:

$$M_s = Y \times \text{Reduction in BOD concentration in the aeration tank} \times Q' \tag{2.19}$$

where Y is the biomass conversion factor and Q' is the inflow from the PST to the aeration tank. The corresponding volume of the secondary sludge, V_s (in m^3/d) produced can be given by

$$V_s = \frac{M_s}{[\rho_{ssl} \times (P_s/100)]} \tag{2.20}$$

where ρ_{ssl} is the density of the sludge and P_s is the percentage of solids present in the sludge. Therefore, the total mass of the solids, M_T and volume, V_T of sludge that reaches a thickener will be equal to $(M_p + M_s)$ and $(V_p + V_s)$, respectively. Two types of digesters are available to achieve the above-mentioned purposes: (i) anaerobic digester (AD) which is commonly used to deal with wastewater containing primary sludge, which would generally contain large amounts of readily available organic substances and (ii) aerobic digester which is used to treat biological sludge in the absence of primary sludge.

There are two types of anaerobic digesters available for dewatering: (i) standard-rate AD and (ii) high-rate AD. A standard-rate AD will receive sludge intermittently and the supernatant will be returned to the aeration tank. Accumulated digested sludge will be stored at the bottom of the digester and will be disposed at desired time intervals. Thus, the volume, V of a standard-rate AD can be computed by the following equation:

$$V = t_1 \times \frac{(V_1 + V_2)}{2} + t_2 \times V_2 \tag{2.21}$$

where V_1 is the raw sludge loading rate (m^3/d), V_2 is the digested sludge accumulation rate (m^3/d), t_1 is the period of digestion (d) and t_2 is the period of storage of digested sludge (d). Generally, t_1 and t_2 will be 30 and up to 90 d, respectively. Assuming negligible solids escape with the thickener supernatant, the mass of the solids in the thickened sludge that is entering the digester will be M_T. If these solids contain γ % organic fraction and of which δ % is converted into liquids and gaseous end products after digestion, then the digested sludge accumulation rate, V_2 can be computed by

$$V_2 = \frac{\{M_T \times (\gamma/100) \times (1 - \delta/100) + M_T \times (1 - \gamma/100)\}}{[\rho_{dsl} \times (P_d/100)]} \tag{2.22}$$

where ρ_{dsl} is the density of the digested sludge and P_d is the percentage of solids present in the digested sludge.

A high-rate AD has two stages where in the first stage, solid contents will be removed but dewatering will not occur. Therefore the volume, V^{1st} of the first-stage reactor can be computed by

$$V^{1st} = V_1 \times t_1' \tag{2.23}$$

where V_1 is the raw sludge loading rate (m^3/d) to the first stage reactor that is similar to the one used in the standard-rate AD and t_1' is the digestion time in the first-stage reactor which is generally around 10 d. In the second stage, dewatering of the sludge would occur and the volume, V^{2nd} of the second-stage reactor will be computed using the equation

$$V^{2nd} = \frac{(V_1 + V_2)}{2} \times t_1'' + V_2 \times t_2 \tag{2.24}$$

where V_1 is the sludge loading rate (m^3/d) to the second-stage reactor, which will be same as the raw sludge loading rate (m^3/d) to the first-stage reactor, V_2 is the digested sludge accumulation rate (m^3/d) (this will be similar to the value obtained for the standard-rate AD), t_1'' is the period of digestion (d) and t_2 is the period of storage of digested sludge (d). Generally, t_1'' and t_2 will be 3 and up to 90 d, respectively. High-rate AD reactors have smaller total volume compared to that of single-stage AD. However, when a treatment plant processes less than 4,000 m^3/d of wastewater, standard-rate AD is usually preferred due to economic reasons and ease of operation.

Digested sludge will need further dewatering through either vacuum filtration or pressure filtration. Vacuum filtration is a reliable but energy-intensive process. Pressure filtration can be carried out either by belt filter press or plate-and-frame filter press and both of them are less energy intensive. The dewatered sludge can be used as fertilizers or soil conditioners; otherwise, the sludge can be either sent to landfill or incinerated.

Discussion until this point emphasizes that maintaining different microbial communities to remove various chemical constituents present in wastewater is a challenging task; thus, creating a system that could form an ecology within which diverse microbial communities can co-exist would improve the efficiency of a wastewater system enormously. Additionally, to maintain a small food to microorganisms ratio in the aeration tank, either the volume of the aeration tank or the concentration of the microbes (MLSS) present in the aeration tank need to be increased. Increasing the aeration tank volume may not be feasible always and increasing the concentration of microbes will require larger sedimentation tank to settle them subsequently. Thus, the search to create a system that could satisfy the above requirements led to the introduction of membrane bioreactors (MBRs).

2.5 MEMBRANE FUNDAMENTALS

In wastewater treatment, membranes are used mainly as a filtration barrier to remove suspended, colloidal, or dissolved particles. This section briefly reviews membrane process fundamentals to provide an understanding of commonly used terms to define the operations of membrane. Section 2.6 includes an overview of the issues associated with the incorporation of the membrane process into MBRs, namely, membrane process configurations, system components for MBRs, and the fundamentals of membrane operation and maintenance (O&M). The concepts introduced in this chapter are discussed in detail in Chapter 3 (Design, operation and maintenance of MBR) and Chapter 14 (Commercial MBR products and selected case studies).

2.5.1 Membrane performance parameters

The most important process performance matrices are flux (LMH), trans-membrane pressure TMP (kPa, psi) and permeability (LMH/kPa). The flux (normally denoted J) is the quantity of material passing through a unit area of membrane per unit time.

$$J(m^3/m^2\,h^-1) = \frac{\text{Flow } (m^3/h)}{\text{Total membrane area } (m^2)} \tag{2.25}$$

Thus flux takes SI units of $m^3/(m^2\,s)$, or simply m/s, and is occasionally referred to as the permeate or filtration velocity. For the convenience of more accessible numbers, other non-SI units, namely, litres/$m^2\,h^{-1}$ (or LMH) and m/d are also used. In fact, the most usual units for MBRs are LMH and m^3/m^2 d. The Imperial unit of gallons per square foot per day (GFD) is still used in the United States. MBRs generally operate at fluxes between 10 and 150 LMH (Judd & Judd, 2011).

Flux can further be calculated as

(1) Instantaneous flux
(2) Average (gross) flux
(3) Net (production) flux

and they are influenced by the following parameters:

(a) Membrane resistance
(b) Operational driving force per unit membrane area
(c) Hydrodynamic conditions at the membrane–liquid interface and
(d) Fouling and subsequent cleaning of the membrane surface.

Transmembrane pressure or TMP can be calculated by pressure differential across the membrane and the permeability can be calculated dividing instantaneous flux by TMP.

Transmembrane pressure

$$\text{TMP(Pa)} = \left(\frac{1}{2}\right) \times [\text{inlet pressure (Pa)} + \text{outlet pressure (Pa)}] - \text{permeate pressure (Pa)} \tag{2.26}$$

$$\text{Permeability } (m^3/m^2\,h^{-1}\,Pa^{-1}) = \frac{\text{Flux } (m^3/m^2\,h^{-1})}{\text{TMP (Pa)}} \tag{2.27}$$

Manufacturers can offer charts showing relationship between membrane permeability and transmembrane pressure/flux to calibrate the system. However, field data may vary from the standard conditions due to difference in testing conditions and the effect of the following conditions should be taken into account while performing calibrations:

(a) Operational conditions (e.g. shock loading, temperature and pH variation, solids loading, etc.)
(b) Dead volumes (unutilized membrane areas due to hindered circulation) due to packing density and membrane configurations
(c) Effluent during backwash and membrane cleaning
(d) Deterioration of membrane with time
(e) Physical damages.

2.5.2 Membrane classifications

Membranes can be classified depending on various categories like porosity, driving force, material used (organic/inorganic), texture, characteristics of membrane (hydrophobic/hydrophilic), and so on.

However in this section discussions are confined to pore size which is relevant to the size of the solute to be removed. Figure 2.6 illustrates the basic structure of membranes with pores and their relative selectivity of particles according to their pore size.

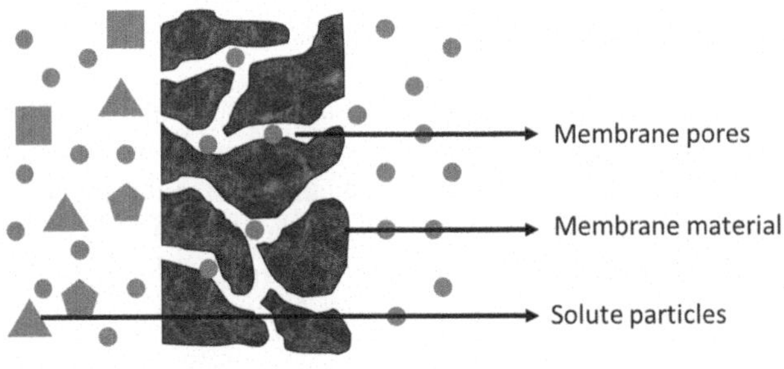

Figure 2.6 Pores in membrane structure.

According to the pore size, membranes can be classified into:

(1) Microfiltration (MF)
(2) Ultrafiltration (UF)
(3) Nanofiltration (NF)
(4) Reverse osmosis (RO)

As discussed in Chapter 1, MF and UF membranes are used in MBRs. MF and UF membranes operate on size-exclusion mechanism, unlike the NF and RO processes, in which other mechanisms are involved. For background purposes, a brief summary of the other membrane categories (i.e. NF and RO) are also provided here.

A broad classification of membranes defined by their pore size as compared with that of common particles is illustrated in Figure 2.7. Another common classification is to divide membranes into two groups, namely, those which can remove dissolved particles (solutes) and those which can remove suspended or colloidal particulates.

Membranes that can remove soluble species include NF and RO systems. NF membranes are generally called as membrane softeners due to their capability to remove synthetic organic compounds and multivalent ions like calcium and magnesium. But NF membranes are less effective in removing monovalent ions. To remove such single charged species RO membranes are used where diffusion and sieving mechanism through the membrane are responsible for the removal. Basic principle for both NF and RO membranes is applying a pressure greater than that of the natural osmotic pressure of the solution that contains solutes that are to be removed. This pressure may vary from less than 700 kPa (100 psi) for fresh water in NF applications to more than 7000 kPa (1000 psi) for sea water in RO applications (Allgeier *et al.*, 2005). As RO and NF membrane separation is mainly due to diffusion rather than size exclusion, they are generally characterized by molecular weight cutoff (MWCO) rather than nominal pore size. The MWCO of NF and RO typically range from 200 to 1000 Da and 100 Da, respectively. Characterization of RO and NF membranes can also be carried out by their percentage rejection of monovalent and divalent ion as salt rejection, respectively. Suspended and colloidal particles can be easily removed using MF and UF using physical filtration characteristics of the membrane; MF and NF membranes are often referred to as *membrane filtration*.

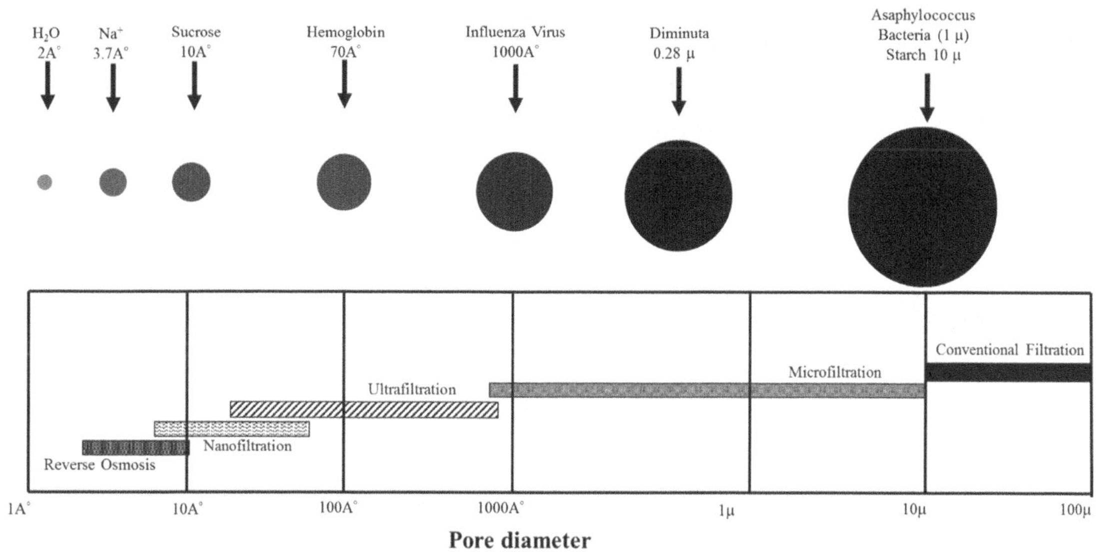

Figure 2.7 Membrane size-exclusion spectrum.

2.5.3 Membrane interactions with suspended impurities

In wastewater, the organic and inorganic impurities are present in dissolved, colloidal or suspended forms. These impurities are present in different sizes and they are retained by a membrane based on its pore size. For example, in MF, most of the suspended impurities are retained on the membrane surface. This eventually leads to reduced pore size, which facilitates retention of colloidal impurities by MF membranes. Both colloidal and suspended impurities are naturally retained by UF membranes.

The retention of suspended impurities on the membrane surface during filtration triggers build up of another layer of particles. During process operations, the retention of suspended impurities on the membrane surface increases with time. This may result in the formation of filter cakes (due to particles and colloids) or gel layers (due to macro molecules) which consequently increase the resistance of the membrane. Organic membrane materials may be deterioriated by the microbial mass entrapped in the filter cake. Both MF and UF membranes can be operated either using a pressure-driven or a vacuum-driven process in which the filtrate is 'pushed' (pressure-driven) or 'pulled' (vacuum-driven) through the membrane, which generates transmembrane pressure. Generally, TMP for MF and UF are less than NF and RO but it can vary according to the particle density, size, fouling potential, cleaning efficiency, and so forth. The TMP typically ranges from 20 to 100 kPa (3–14.5 psi). Typically, MF and UF membranes are associated with MBRs because of their physical properties for particle exclusion, while NF and RO are used in wastewater treatment as a final purification step. The pore-size distribution of a membrane provides information about the range of pore sizes present in the membrane. Though the separation efficiency of a membrane cannot be solely predicted based on the pore-size distribution, pore size and its distribution provide a fair quantitative estimation of the separation characteristics of the membrane. Surface porosity of UF and MF membranes typically ranges from 0.1 to 1% and 5 to 70%, respectively. In ideal cases, an MF or UF membrane is thought to possess same pore size across the membrane. However, most polymeric membranes (MF and UF) do not possess exactly the same pore sizes throughout the membrane surface (Figure 2.8). Pore-size distribution of MF and UF membranes can be measured by scanning electron microscopy, transmission electron microscopy, bubble point method, solute rejection method, mercury intrusion porosimetry, thermophotometry, and the adsorption method. Large pore size reduces the external surface fouling potential in the system, which is a tradeoff between selection of UF or MF. Therefore this decision must be made early while designing the system. UF membranes are mainly used for MBRs as these membranes strike a balance among effluent quality, energy requirement and reduced internal membrane clogging.

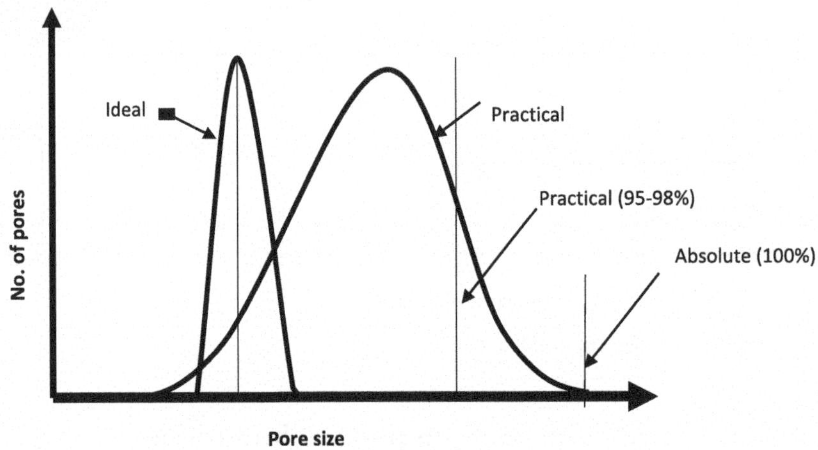

Figure 2.8 Membrane pore-size distribution.

Pathogen removal is an added advantage of MF and UF systems. Removal of all bacteria and some viruses leads to reduced downstream disinfection and brings down the cost of plant construction, operation and maintenance. The MF membrane pores provide a physical barrier for particulates, bacteria, and protozoan cysts, but not viruses. Traditionally, MF membranes with a nominal pore size of 0.1 μm with a range of 0.1–0.2 μm have been used for MBR process. But in certain industrial cases larger pore sizes (0.4 μm) for MF are being used for reducing potential fouling problems that are generally associated with MBR. UF membranes are adept in removing organic macromolecules such as particulates, bacteria, protozoan cysts, and some viruses and, therefore they can be defined with molecular weight cut off (MWCO) (Allgeier *et al.*, 2005). UF membranes can handle molecules with a range between 10,000 and 500,000 Da. The pore size of UF membranes are in the order of 0.01–0.05 μm with a nominal pore size of 0.01 μm. Due to the presence of a cake layer on the membrane, MF membranes, in addition to bacteria, can remove some viruses, which would not be possible otherwise.

2.5.4 Membrane materials, system configurations, and operating modes

The majority of membranes are manufactured from polymeric materials. Historically, ceramic membranes have been considered too expensive and, thus, have not had a significant effect on the municipal wastewater market. However, because of successful demonstrations of their high integrity, chemical resistance, and low operating costs, there have been recent market advances of ceramic membranes (WEF, 2012). Selection of membrane material entails both manufacturing process and performance considerations. Further details regarding the characteristics and suitability of these membrane materials are available in Chapter 14.

Hydrophobicity or hydrophilicity of a membrane is a property of the membrane that determines its wettability based on the attractive or repulsive surface reaction to water. Hydrophobic polymeric membranes have low affinity for water while hydrophilic membranes show high affinity for water or are 'water loving'. The surface chemistry of hydrophilic membranes allows the membrane material to be wetted, forming a water film or coating on the surface, thus permitting the passage of water. Contact angle is one of the common ways to measure the wettability of a surface or material. To limit contact with water, hydrophobic particles cluster to reduce surface area and as well lower their surface tension in hydrophilic environments. Generally, if the water contact angle is smaller than 90°, the solid surface is considered hydrophilic and if the water contact angle is larger than 90°, the solid surface is considered hydrophobic. Hydrophobic membranes made of, for example, polypropylene (PP), polyvinylidene fluoride (PVDF) and polytetrafluoroethylene (PTFE, Teflon) are commercially available in capillary forms for MF purposes due to high void volume, well-controlled porosity, high thermal and chemical stability, and low cost. However, foulants such as colloids, starch, protein, clay, oil, paraffin, and surfactants in water exhibit hydrophobicity. To avoid applying high pressure to get permeate from hydrophobic membranes, and also to reduce membrane fouling due to hydrophobic interactions, hydrophobic membrane surface can be treated to make hydrophilic without changing its bulk properties.

High surface area membranes are desirable in MBR application. High membrane surface areas allow operation under lower flux for the same volume of treated wastewater, which in turn helps reducing membrane fouling. The total effective membrane surface area involved in a membrane filtration process may be obtained by multiplying the surface area contained within each membrane element by the number of elements in the membrane system. The total available surface area for a membrane can be increased or decreased by the module configuration design. However, tradeoffs associated with each configuration in terms of cost, energy input and cleaning process must be carefully considered during membrane selection for an MBR system. Three main membrane-element types are presented in Section 2.6.3.2 based on the type of membrane used (i.e. hollow-fibre, flat-sheet, and tubular membrane-element types). Further discussions in this line can be seen in Chapter 14.

Pressure-driven membrane processes generally can be operated in either cross flow or dead-end mode as illustrated in Figure 2.9. Cross flow being a single passage of the feed over the membrane surface while only a fraction is being generated as permeate on the other side of the membrane and dead end being no retentate stream on the feed side of the membrane. Filtration always leads to an increase in resistance to the flow due to solid accumulation on the surface of the membrane. In the case of dead-end mode, the resistance increases with the thickness of the cake layer formed on the membrane surface. In crossflow mode this deposition continues until the adhesive forces binding the cake to the membrane are balanced by the scouring forces of the fluid. The relationship between cake formation and flux is clearly shown in Figure 2.10. All other things being equal, a crossflow filtration process would be expected to attain steady-state conditions determined by the degree of concentration polarization (CP). In practice, only pseudo-steady-state (or stabilized) conditions are attained due to the unavoidable deposition or adsorption of fouling material.

2.5.5 Membrane resistance

In membrane filtration applications, membrane resistance is an important functional property of a membrane that has an inverse relation with the permeate flux when driving force/applied pressure is held constant. The membrane resistance of clean membrane changes over time during process application of the membrane. The change in membrane resistance however may also result from the deteriorating conditions or process conditions of the fluid. The membrane resistance specified by the membrane manufacturer is referred to as the clean membrane resistance. Pure water permeate flux of a membrane will remain constant at a constant pressure under clean membrane resistance conditions. However, the permeation resistance of the membrane may be influenced by internal or external fouling of the membrane. The change in membrane resistance may significantly alter the performance of the membrane. Theoretically, the total resistance of a membrane measured at a given time is the sum of the intrinsic membrane resistance and other secondary resistances (Figure 2.11).

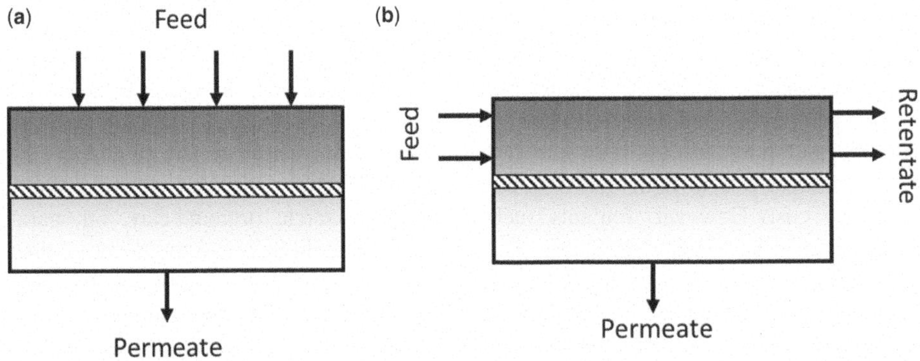

Figure 2.9 Two basic operational modes for membranes. (a) Dead End mode, (b) Cross-flow mode.

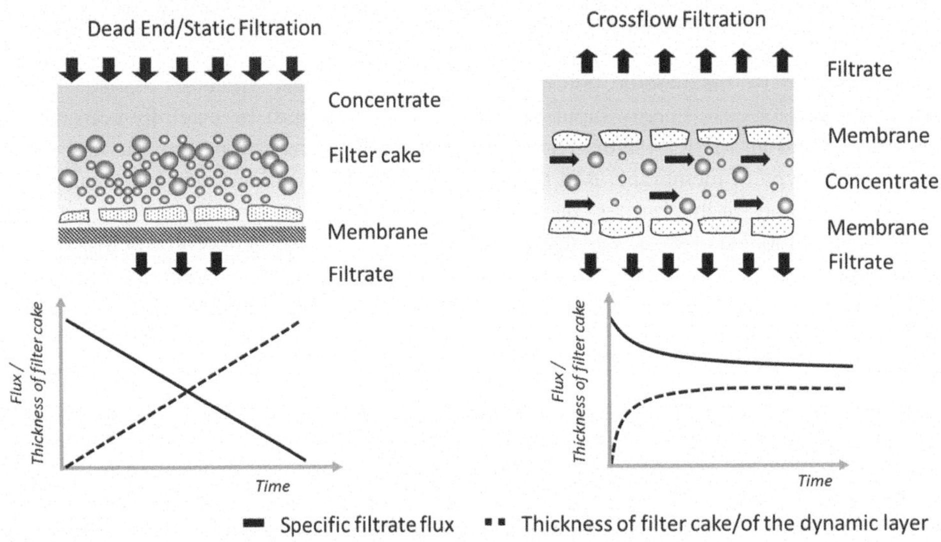

Figure 2.10 Relationship between flux and filter cake in dead-end and crossflow modes.

Thus, total resistance of a membrane measured at a given time is essentially the sum of the primary clean membrane resistance and secondary resistances due to the fouling condition of the membrane and or process conditions of the fluid. The possible resistances encountered in a typical MF and UF application are clean membrane resistance, pore blocking resistance, adsorption resistance, gel-layer formation resistance, and resistance due to CP.

$$J = \frac{\Delta P}{\mu \times R_t} \tag{2.28}$$

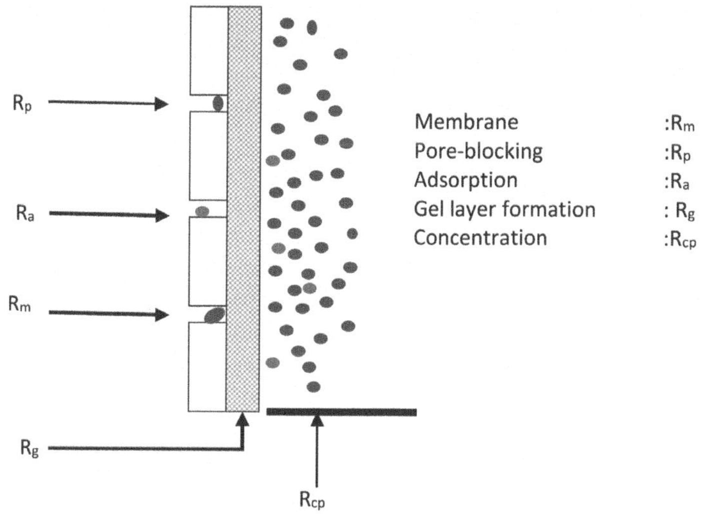

Figure 2.11 Membrane resistance and permeation behaviour.

where J is the permeate flux ($m^3/m^2\ h^{-1}\ bar^{-1}$), ΔP is the driving force in bars, μ is the viscosity of the fluid (for MF, UF, $\mu \approx 0$) and R_t (m^{-1}) is the total resistance. Typically, in MF and UF applications, membrane resistance may change with time according to the type of membrane fouling. In cases where filter cake layer develops over the membrane surface and blocks the pores of the membrane, the total membrane resistance increases.

2.6 FUNDAMENTALS OF MBR

2.6.1 History of MBR technology

The first patent for MBR concept was issued in the United States in 1969 to Dorr-Oliver Inc. Later on in 1970s and 1980s, small MBR systems were commercialized and used in the United States and Japan. The MBR process was very energy intensive at that time (see Chapter 1). In 1989, Yamamoto *et al.* (1989) published a famous paper as the 'proof of concept' of the submerged MBR process which revolutionized the outlook towards membrane-based wastewater treatment. In recent years, there has been an exponential growth of MBR applications in both municipal and industrial wastewater treatment applications. Due to various technological advancements, reduced capital cost and energy optimization, MBR can compete with the conventional systems when the situation presents itself for a need to have higher quality effluent.

2.6.2 Differences between CAS and MBR processes

Water reuse is becoming increasingly popular all over the world as societies require advanced wastewater treatment processes to achieve the standards required for reuse water. The lower the $F/M(=QS_o/VX)$ ratio, the higher the BOD removal efficiency by the microorganisms that consume the BOD. For given Q and S_o, the two ways to achieve lower F/M ratio are (i) to increase V, which is not always economical and (ii) to increase X, which is possible if the MLSS could be concentrated in the biological reactor (Ben Aim & Semmens, 2002). By extracting the treated effluent through a membrane, the MLSS concentration can be increased in the biological reactor. An MBR utilizes this concept. Another difference between the CAS and the MBR is that while the former would retain only the microorganisms that would have better settlement in the sedimentation tank, the latter one would retain microorganisms that would even have poor settling properties. Therefore, the microbial environment would be completely different in an MBR compared to that of a conventional ASP system.

Another parameter that affects the performance of a biological reactor is the SRT($=VX/qX_e$). In a CAS process, the values of X and X_e will be around 3,000 and 10,000 mg/L, respectively. However, in an MBR both X and X_e will be equal and in the range from 15,000 to 20,000 mg/L. Thus, the SRT of MBR can be computed by V/q. Therefore, the SRT of an MBR could be three times more than the CAS (the SRT of a conventional CAS is around 4–15 d). Due to longer SRT, the MBR will retain even slow-growing microorganisms such as nitrifiers, microorganisms that grow on synthetic chemicals and so on, which would usually be washed out in a CAS system. Therefore, an MBR is suitable for performing nitrification as well as treating industrial wastewater that contain synthetic chemicals in addition to BOD removal. Longer SRT and higher MLSS cause stress to the microorganisms in an MBR which requires more energy for cell maintenance and therefore leave less energy for cell production. This leads to lower sludge production in an MBR compared to that of a CAS system.

Thus, MBR is increasingly becoming the first choice to treat domestic, aquaculture and industrial effluents for the purpose of reuse. As mentioned in the above two paragraphs, the advantages of an MBR compared to the CAS process that is used to treat wastewater effluents are (i) production of high-quality-treated effluent, (ii) higher biomass concentration and therefore lower F/M (food to microorganisms) ratio and (iii) reduced cost of sludge handling due to the application of longer SRT and potentially lower sludge growth and (iv) low investment cost due to smaller footprint. However, the disadvantages of an MBR are the disintegration of microorganisms and excretion of soluble microbial products (SMP) that lead to frequent fouling of membranes. Thus, appropriate membranes should be selected for the use in MBRs. Further differences between the CAS and MBR processes are presented in Table 2.6.

Table 2.6 Differences between CAS and MBR systems (WEF, 2012).

	CAS	MBR
Fine screen	Not required	Required (1–3 mm)
Reactor volume	Larger than MBR for a given sludge inventory or SRT	Smaller than CAS for a given sludge inventory
Typical MLSS	1,500–3,000 mg/L	4,000–15,000 mg/L
RAS flow (% of influent flow)	50 –100	300–500
Wastage location	Secondary clarifiers	MBR tanks

2.6.3 Design of MBR systems

The design of an efficient MBR system pivots on the design of a suitable membrane system. In addition, the effect of changes in the (biochemical) process parameters (such as organic and hydraulic loading, sludge age, etc.) due to change in the solid–liquid separation system needs to be considered in the reactor design and other unit processes.

2.6.3.1 MBR arrangements

As mentioned in Chapter 1 and detailed in Chapter 14, several MBR variants exist in the market these days, but they are all originally variants of two membrane configurations: submerged or immersed MBR and side-stream MBR. In view of the market-dominance of the submerged MBRs, this section discusses the two common formats of submerged MBRs, namely, internal submerged MBR and external submerged MBR. An external submerged MBR system allows improved chemical cleaning and lowers fouling conditions. This results in better control over clogging and foaming; and biological reactors are optimized independently with no fluctuations, dead zones and short circuits resulting in better effluent quality. The major challenge is that, this system has a larger footprint and higher operational cost in comparison to internal submerged MBR.

(a) **Internal submerged MBR**
This type of membrane configuration is very common due to its compatibility with the activated sludge process, as the membrane module can be directly immersed in the reactor vessel as shown in Figure 2.12. Using negative pressure, permeate is sucked using a vacuum pump and the biomass remains in the vessel. Due to their compactness, low-energy requirements and ease of sludge wasting directly from the reactor, sMBR have become very popular but they are suitable for wastewater with good filterability and require more membrane area for effective treatment.

(b) **External submerged MBR**
In this type of MBRs, membrane modules are located outside the reactor basin as shown in Figure 2.13. In this system, the mixed liquor from the reactor is pumped into the external membrane module. External MBRs are also commercially used in industries as these require less membrane area compared to submerged MBRs and work better for high-strength wastewater with poor filterability. However, these MBRs consume more energy as more pumps are required for pumping and recirculating the sludge; they also need additional space and manifolds for active treatment.

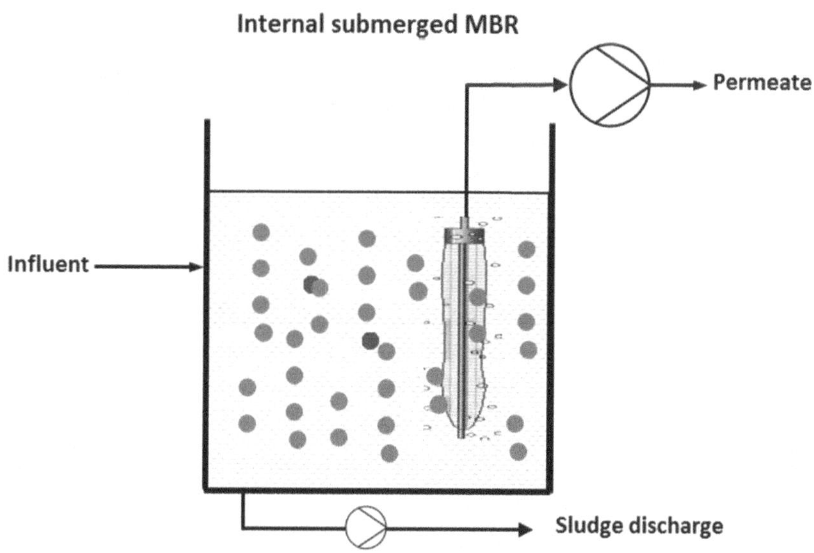

Figure 2.12 Schematics of internal submerged MBR.

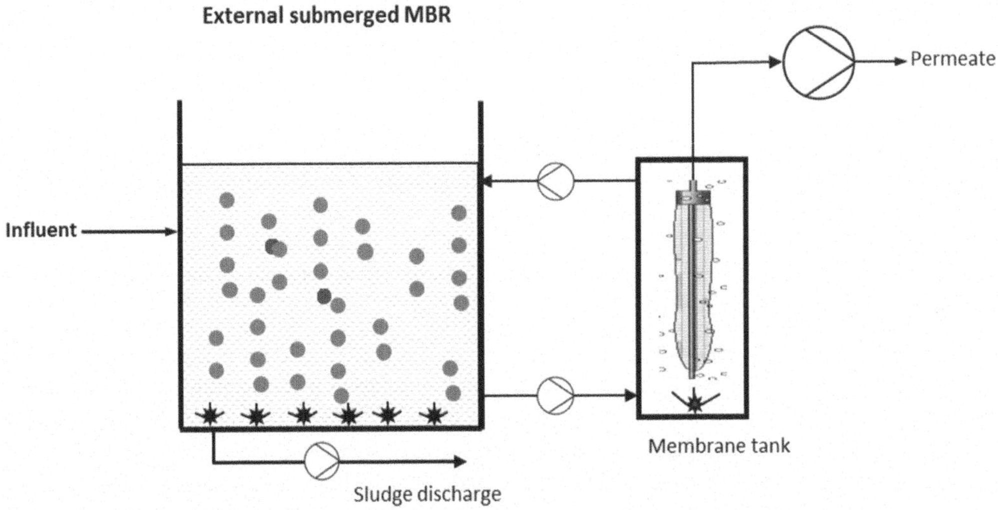

Figure 2.13 Schematics of external submerged MBR.

2.6.3.2 *Element types used in MBR system*

Hollow fibre and tubular membranes are commonly used in MBRs, but flat-sheet membranes are also commercially available. These membranes are operated in crossflow mode in which permeate is obtained by dead-end filtration. The membrane modules are generally assembled in compact modules containing several individual units as shown in Figure 2.14. While packing membranes inside the modules, proper care should be taken; if membranes are packed too densely then the mixed liquor circulation on the

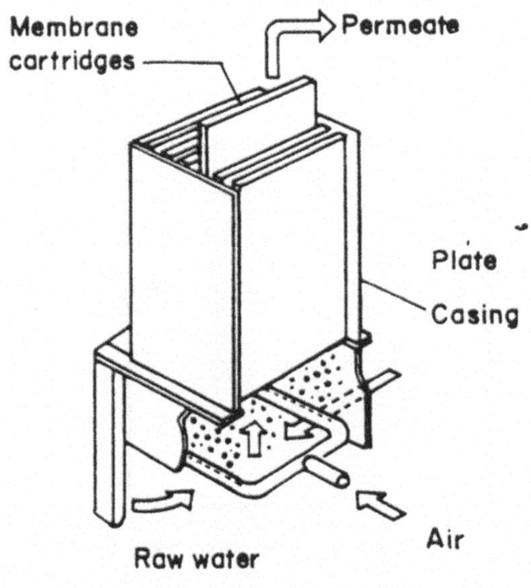

Figure 2.14 Flat-sheet membrane arranged in compact modules.

surface of the membranes will be lessened and high dead volume zones of low flow circulation will be created. For all submerged MBRs, while installing inside the reactor vessel the orientation can be vertical or horizontal with frames and other holding devices depending on the particular design capacity or functionality. Table 2.7 shows a comparison between three commonly available membrane elements. Chapter 13 provides more details on the commercial membrane elements available in the market.

Table 2.7 Comparison between flat sheet, hollow fibre and tubular membranes.

Criteria	Flat Sheet	Hollow Fibre	Tubular Membrane
Operational flux	Moderate to high	High	High
Packing density	Moderate	Moderate	Low
Clogging tendency	Moderate	Moderate	Low
Cleaning	Poor (no or only soft backflush)	Moderate (possibility of backflush)	Good
Energy Usage	Moderate to high	Moderate	High
Capital cost	Moderate to high	Moderate	High

2.6.3.3 Membrane unit life span in MBR

Membrane life span varies depending on configuration, membrane material, operating conditions and maintenance. Membrane manufacturers provide information on the physical/chemical and design properties of membranes. Utilizing the membrane, in accordance with this information may prolong its life span. In general, internal submerged membranes last 5 years and external submerged membranes for about 7 years, after which physical deterioration and irreversible fouling start, which damage the membrane permanently (Cote *et al.*, 2012).

2.6.4 Process overview

In addition to an activated sludge process, components which are generally found in an MBR are a fine screen; membrane elements, themselves typically submerged system in a separate filtration tank and a membrane scour blower which aerates the membrane to keep it clean. In most systems there will be a permeate pump with a backpulse system or cleaning in place (CIP) system to keep the membrane system clean. A typical membrane system can be observed in Figure 2.13. Generally, membranes with a nominal pore size of 0.01–0.1 μm are used; these can be classified accordingly as UF or MF and configured in a variety of arrangements such as tubular or flat sheets or hollow fibres (Figure 2.15).

Elevated MLSS allows higher organic loading in the range of 0.8–1.5 kg $COD/m^3\,d^{-1}$ for MBR systems as compared to 0.4–0.8 kg $COD/m^3\,d^{-1}$ for conventional systems. Higher organic loading allows reduced reactor size but increases the chance of membrane fouling. Therefore, solid loading should be limited to within 1.5 kg $COD/m^3\,d^{-1}$.

Hydraulic retention time (HRT) is used to calculate the volume of the reactor. For industrial wastewater, higher HRT is required compared to domestic wastewater to provide sufficient time for completion of the biochemical reaction. Studies show that for mixed industrial sewage a HRT of 15–24 h may be required. In some cases, HRT can be as long as 2–3 d. Biomass concentration increases with the decrease of HRT (higher loading). However, very short HRT tends to produce more extra-cellular polymeric substances (EPS) that are known to aggravate membrane fouling. Table 2.8 illustrates the design criteria that have evolved over three generations of MBR.

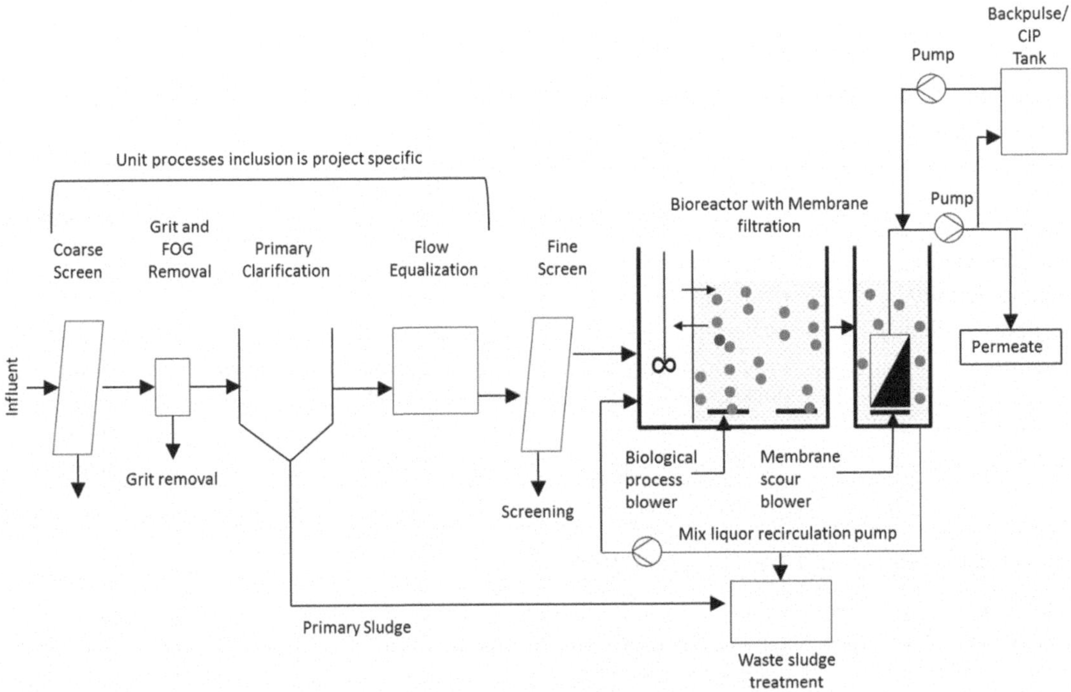

Figure 2.15 Typical components of an MBR system (modified from WEF (2012)).

Table 2.8 Design criteria of MBR over three generations (Crawford *et al.*, 2001).

	First Generation	Second Generation	Third Generation
SRT (d)	50+	20+	10–15
MLSS (mg/L)	>20,000	20,000	10,000
NH_3 removal	Yes	Yes	Yes
Total N removal	No	Yes	Yes
P removal	No	Yes	Yes

2.6.5 Biology in MBR

A simple replacement of sedimentation tank with a membrane module radically changes the microbiology of the system, which leads to process design reconsideration. Higher forms of microorganisms such as protozoa and rotifers play crucial roles in consuming suspended organic matter and controlling sludge concentration by scavenging bacteria. A simple replacement of sedimentation tank with a membrane module radically changes the microbiology of the system, which leads to process design reconsideration. Higher forms of microorganisms such as protozoa and rotifers play crucial roles in consuming suspended organic matter and controlling sludge concentration by scavenging bacteria. Foaming in MBR plants has been observed even in the absence of actinomycetes, which are generally the cause of foaming in activated sludge processes. The degree of foaming is related to EPS concentration in an MBR process. Sludge foaming is characterized by higher hydrophobicity and low membrane permeability.

MBR applications require aeration for effective biodegradation and for scouring to reduce membrane fouling. Aeration is considered as one of the major cost factors in MBR applications. Aeration for biodegradation depends on the wastewater strength, i.e. influent nutrient (Kjeldahl nitrogen), COD and or BOD levels to be oxidized. In addition, the aeration requirement of an MBR system depends on the membrane type. Aeration to clean membranes is normally achieved through bubble aerators positioned beneath the membrane units. Air supply is required to maintain a good level of turbulence in the MLSS to keep the biosolids from settling and to prevent membrane fouling. The reduction in fouling is achieved by the abrasion properties of the supplied air against the membrane surface.

The aeration requirements for aerobic MBRs is often considered in terms of the oxygen required (MO_2) for organic pollutant oxidation by heterotrophic bacteria and autotrophic nitrifying bacteria. The oxygen requirement can be estimated using Equation 2.29 according to Martin *et al.* (2011).

$$MO_2 = Q(COD_{IN} - COD_{EFF}) + 4.33Q(TN_{IN} - TN_{EFF}) - 2.83QNO_{3,EFF} \tag{2.29}$$

where COD_{IN}, COD_{EFF} are influent and effluent COD concentrations, respectively (mg COD/L) and Q is the influent flow (L/d). The oxygen requirement for nitrification is accounted for by the difference in the total nitrogen in influent, TN_{IN} and the total nitrogen in effluent TN_{EFF} (mg NO_3 - N/L). $NO_{3,EFF}$ (mg NO_3 - N/L) is the oxygen savings due to denitrification. Further, modification of Equation 2.29 to include air requirements for scouring in the case of a submerged aerated MBR modelled by Martin *et al.* (2011) is shown as

$$Q_{air,bio} = \frac{MO_2 - MO_{2,MEM}}{0.21 \cdot 1000 \cdot \alpha \cdot \beta \cdot \gamma \cdot OTE} \tag{2.30}$$

where

$$MO_{2,MEM} = 0.21 \cdot 1000 \cdot \alpha \cdot \beta \cdot \gamma \cdot OTE_{MEM} \cdot Q_{air,MEM}$$

where β and γ are oxygen transfer efficiency factors, taken as 0.95 and 0.89, respectively. Oxygen transfer efficiencies (OTEs) of 0.02 and 0.05 were assumed for membrane coarse bubble aeration and biological fine bubble diffusers. α is the oxygen transfer efficiency relating to the mixed liquid suspended solids (MLSS) concentration, according to the equation

$$\alpha = e^{-0.082MLSS} \tag{2.31}$$

2.6.6 Operation of the membrane system in MBR

Fouling is the major drawback of any membrane-based system which affects operation and effective functioning of the MBR system. Due to the application of negative pressure on the permeate side, a deposition of solid layer develops on the membrane. The more the cake layer builds up, the more the hydraulic resistance builds up, thus reducing the flux. To maintain flux in a partially fouled membrane the operating TMP can be increased. But eventually membrane cleaning is inevitable for restoration of original flux. Furthermore, pore blocking of the membrane requires chemical cleaning. Over a period of time the flux is reduced to a point where the membrane cannot be used any further and needs to be replaced. Chapter 4 provides a detailed account of the causes and remedies of membrane fouling in MBR. However, the classification of fouling and the fouling control strategies are briefly presented here.

2.6.6.1 Fouling classification

As noted earlier, membrane fouling can occur on membrane surface and within membrane pores. Fouling, therefore, can be classified as:

(a) Physically reversible: The foulants are loosely attached to the membrane and can be cleaned easily.
(b) Chemically reversible: These are stronger foulants and require a chemical treatment to clean the membrane.
(c) Irreversible: Fouling occurs over a period of time and as the damage is permanent no cleaning is possible.

Fouling can also be described by the mechanisms that cause it to occur:

(a) Cake fouling: Happens at the membrane surface and larger particles play a role. The mechanism can be observed in Figure 2.16.
(b) Pore blocking: Membrane pores are blocked by smaller particles and may include inorganic precipitation. The mechanism can be observed in Figure 2.17.

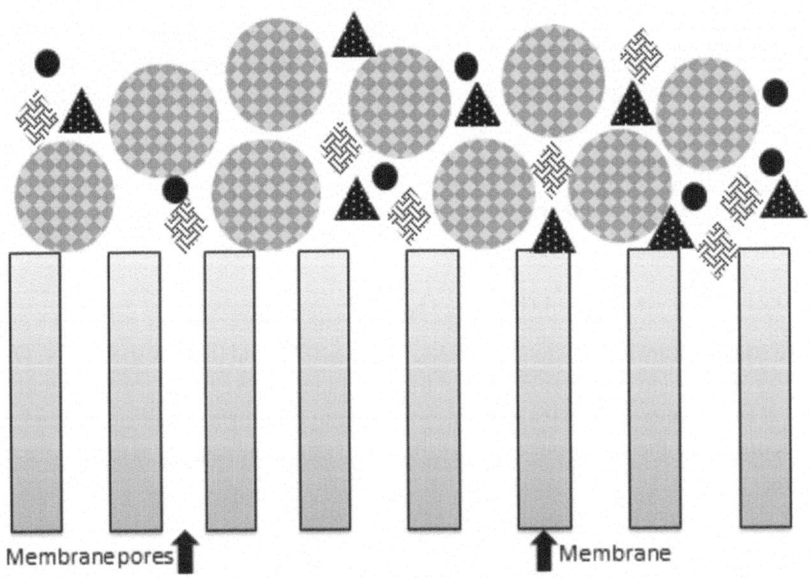

Figure 2.16 Mechanism of cake fouling on membrane surfaces.

MBRs can be operated in either constant TMP or constant flux mode. In constant TMP mode, the TMP is kept constant as the flux drops as the cake layer develops and finally, when the flux drops to a predetermined level, the process is stopped and membrane is backwashed before any further operation. Conversely, in constant flux mode, the flux is maintained at a constant level by increasing the TMP as fouling develops. When the TMP reaches a predetermined set point, the operation is stopped and the membrane is taken out for cleaning by backwashing or chemical treatment. In real practice, backwashing is done before the quality and quantity of permeate is reduced.

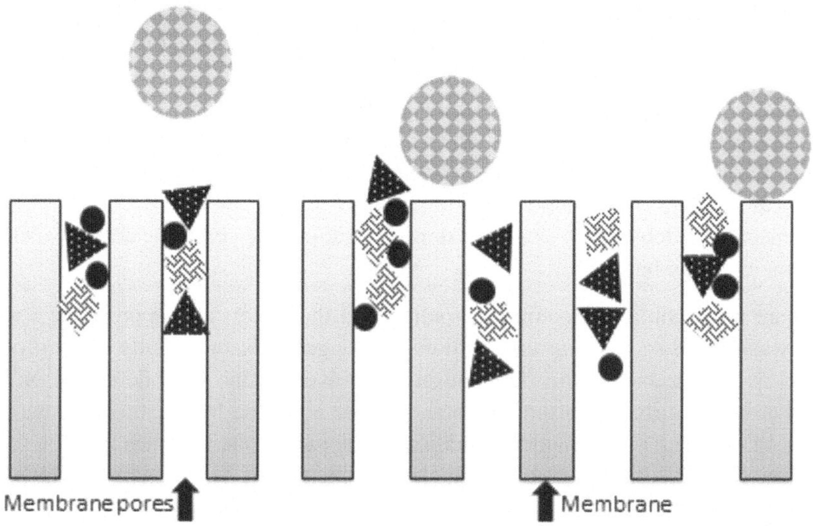

Figure 2.17 Mechanism of pore blocking inside the membrane.

2.6.6.2 Fouling control

There are several ways to control the fouling of membranes. The major causes of membrane fouling and its monitoring, characterization and mitigation techniques are presented in detail in Chapter 3. However, some of the common methods of fouling control in MBR are presented below:

(1) Air scour
Pushing air bubbles on the surface of submerged membrane to clean it. Air produces a shear at the membrane surface and this can reduce the risk of cake fouling. Too much or too little scouring has a negative effect as excessive fouling can cause membrane breakage and floc erosion thus increasing the EPS in the MLSS; and insufficient scouring results in localized dewatering on membrane surface.

(2) Relaxation
Introduction of a pause on the throughput of the membrane with maintained air scour. This results in a concentration gradient on the surface of the membrane that drives the cake formation to the surrounding mix liquor.

(3) Absorbent/coagulant addition
Chemicals are used for fouling control like alum, powder activated carbon to reduce the SMP concentration and EPS.

(4) Backwashing/back flushing
Permeate is generally pushed back in the reverse direction for a short duration to remove the surface fouling on the membrane.

(5) Chemical cleaning
This method is used to remove tough foulants which cannot be removed by physical cleaning. Organic foulant removal is done by sodium hypochlorite and inorganic foulants can be cleaned using oxalic/citric/other acids or caustic soda depending on the fouling criteria. There are different types of chemical cleaning:

(a) Maintenance cleaning
Reverse flow of cleaning solution back through the membrane for a short duration (0.5–1 h) with the mix liquor inside the reactor tank.
(b) Recovery cleaning (intensive cleaning/CIP)
The cleaning solution flows in either directions both in-situ and ex-situ. Frequency of this operation is rather less, 1–2 times per year, for about 2 or more hours.

(6) Mechanical cleaning (manual cleaning)
Removing cake or debris such as rags or dewatered solids are generally done by using water spray, or in some cases by hand.

Various factors are responsible for membrane fouling and the fastest and appropriate way to treat it is membrane backwashing. Backwashing can be done by tangentially or radially forcing pulsating air or water (permeate) across the membrane. This pulsating air loosens the cake deposited on the surface of the membrane. Further disturbance to the membrane can be achieved by creating turbulence close to the membrane using air scouring or mechanical vibration of the membrane modules.

To reduce deep pore scaling, membranes are cleaned chemically in addition to regular backwashing during operation. CIP using chemical agents is not advisable as it has been observed that chorine-based compounds e.g. sodium hypochlorite can unfavourably affect the microbiological activities. It is always advisable to use chemical cleaning ex-situ followed by a rinse with clean water to remove any residual chemicals. An external membrane generally has a cleaning frequency of 1–2 months while submerged membranes can be cleaned at an interval of 2–6 months. Wherever in-situ cleaning is unavoidable, the chemical dosage should be moderated to avoid damage to microbial activity.

2.6.6.3 Membrane integrity

Due to the use of MF and UF membranes in drinking water process much attention has been given to the integrity of these membranes as a barrier to *Giardia, Cryptosporidium*, and some viruses. The same criteria do not apply to wastewater treatment applications but it plays an important role in membrane life, process performance and maintenance of the overall system. *Membrane integrity* can be defined as veracity of individual membrane elements.

The overall failures in membrane integrity can be broadly divided into two categories

Table 2.9 Manufacturing and operational issues in membrane integrity.

Manufacturing Defects	Operational Problems
Macroporous voids can potentially be a source of weakness and can result in breakage.	Scratches and piercing from debris during operation caused by screen failure or permeate system contamination
Mishandling in module manufacture during module assembly resulting in breech in physical structure of membrane	Contact of large particles with the membrane surface due to improper pretreatment
Slits caused by crushing membrane surface during manufacturing thus creating a potential for fatigue failure during normal conditions	Strong chemicals causing embrittlement and pore collapse
Problems with the epoxy potting procedures giving weakness near the tube sheet causing embrittlement	Slits caused by collapse from over-pressurization

(1) Manufacturing defects
(2) Operational problems.

The major issues in both categories are discussed in Table 2.9.

The following additional care must be taken while designing MBR system for membrane integrity:

(1) Coatings and/or lining of pipes, and other equipment may flake off and damage the membranes.
(2) Proper selection of cleaning agents with appropriate dosages must be taken into consideration for improving membrane life.
(3) Finally, proper control on pump valves and other equipment during backwashing and air scouring can help prevent water hammer.

Designing an MBR system with these design considerations will improve both membrane integrity and performance of the overall system.

2.6.7 Energy utilization in MBR

Much of the energy requirements in MBR system are due to aeration. Though the energy consumption of an MBR may be influenced by other factors such as the MBR system design, volume of wastewater being treated and other operational strategies, membrane aeration may exceed more than 50% of the energy requirements of an MBR system (Krzeminski *et al.*, 2012). However, some studies reported aeration for biological pollutant oxidation to represent as much as 88–93% of the total energy demand in MBR systems (Martin *et al.*, 2011). The mixing and recirculation of sludge only accounts for 12% of the energy demand (Krause & Dickerson, 2011). The energy demand for a typical MBR process is shown in Figure 2.18. The internal submerged MBR typically uses less energy than the external submerged MBRs. However, in some packaged units the energy requirement for external MBRs has been shown to be reduced significantly.

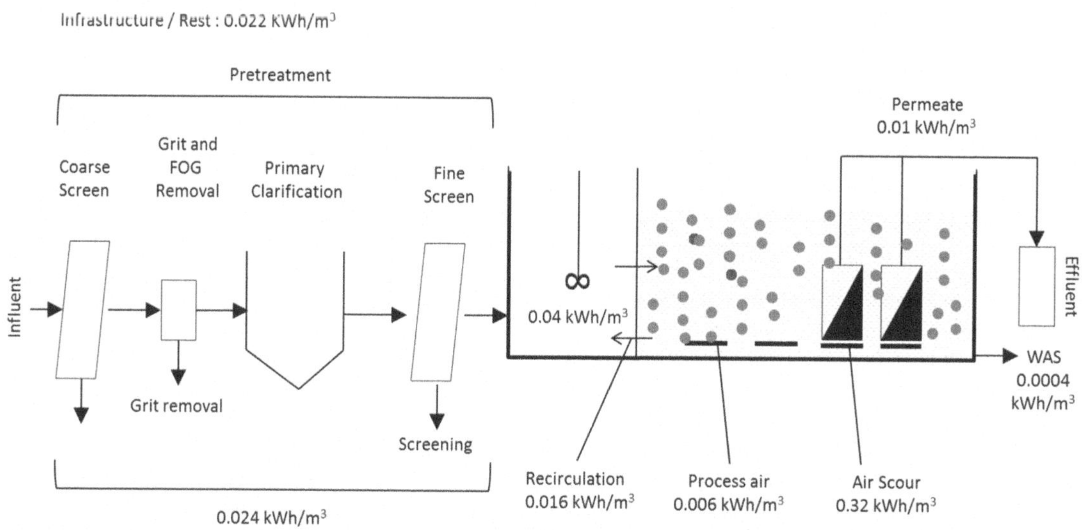

Figure 2.18 Energy demand for an MBR process (Krause & Dickerson, 2011).

In terms of energy demand, a biological wastewater treatment system consumes 0.4–0.8 kWh/m^3. From 1960 to 1980, at the early stage of development, the average energy consumption was around 3–6 kWh/m^3. This high consumption of energy was an obstacle against the commercial success of the technology. However, with the introduction of immersed MBR, and subsequent optimization of aeration system, energy consumption further came down to 1 kWh/m^3 by the turn of the millennia, and currently energy consumption as low as *ca.* 0.5 kWh/m^3 is often reported, which is comparable to that required in conventional wastewater treatment plants (Krzeminski *et al.*, 2012).

2.7 SUMMARY AND FUTURE OUTLOOK

Building on the preliminary discussion furnished in Chapter 1, this chapter reviews the fundamentals of biological treatment and membrane filtration, and provides an overview of the issues associated with the incorporation of the membrane process in MBRs. The concepts introduced in this chapter are discussed in detail in Chapter 3 (Design, operation and maintenance) and Chapter 14 (Commercial MBR products). The commercial success of the MBR technology is faced with technical challenges, namely, pretreatment requirement, membrane fouling, and limitations in management of wet weather flow (peak flow demand management). The energy consumption is also a key challenge in an MBR system due to the inevitable high aeration demand to mitigate membrane fouling. Further, with over 20 years of commercial application, MBR can be considered a mature technology; however, the MBR systems are yet to be standardized and each vendor has a unique design, so pre-selection is an important aspect of cost-effective design for a particular treatment system. Worldwide research and development initiatives are ongoing in this line.

REFERENCES

Ahmed Z. (2012). Microbial communities in nutrient-removing membrane bioreactors: a review. *Journal of Environmental Science and Technology*, **5**(1), 16–28.

Allgeier S., Alspach B. and Vickers J. (2005). Draft Membrane Filtration Guidance Manual. U.S. Environmental Protection Agency, Cincinnati, Ohio.

Antoniou P., Hamilton J., Koopman B., Jain R., Holloway B., Lyberatos G. and Svoronos S. A. (1990). Effect of temperature and pH on the effective maximum specific growth rate of nitrifying bacteria. *Water Research*, **24** (1), 97–101.

Ben Aim R. and Semmens M. J. (2002). Membrane bioreactors for wastewater treatment and reuse: a success story. *Water Science and Technology*, **47**(1), 1–5.

Coombes P. J., Kuczera G., Kalma J. D. and Dunstan R. H. (2000). Rainwater quality from roofs, tanks and hotwater systems at Figtree place. Proceedings of the Third International Hydrology and Water Resource Symposium, pp. 1042–1047.

Cote P., Alam Z. and Penny J. (2012). Hollow fiber membrane life in membrane bioreactors (MBR). *Desalination*, **288**, 145–151.

Crawford G., Fleischer E., Lozier J. and Daigger G. (2001). The evolution of membrane system designs for wastewater treatment. Proceedings of International Conference on Membrane Technology for Wastewater Reclamation and Reuse, 9–13 September 2001, Tel-Aviv, Israel.

Hammer H. J. (1986). Water and Wastewater Technology. John Wiley & Sons, New York.

Jegatheesan V., Visvanathan C. and Ben Aim R. (2004). Advances in biological wastewater treatment. In: Concise Encyclopedia of Bioresource Technology, A. Pandey (ed.), The Haworth Press, Inc., New York.

Jegatheesan V., Shu L., Visvanathan C. and Thanh B. X. (2008). Aerobic environmental processes. In: Advances in Fermentation Technology, A. Pandey et al. (eds.), Asiatech Publishers, Inc., Delhi, India, pp. 607–639.

Jeyanayagam S. (2005). True confessions of the biological nutrient removal process. *Florida Water Resources Journal*, **January**, 37–46.

Judd S. and Judd C. (2011). The MBR Book. Principles and applications of membrane bioreactors for water and wastewater treatment, 2nd edn. Elsevier, London.

Krause S. and Dickerson D. (2011). Energy efficient MBR process. In: *Membrane Bioreactor Technology*, Water Treatment Academy, Bangkok, Thailand, Vol. **1**, 148–158.

Krzeminski P., van der Graaf J. H. and van Lier J. B. (2012). Specific energy consumption of membrane bioreactor (MBR) for sewage treatment. *Water Science and Technology*, **65**(2), 380–392.

Madigan M. T., Martinko J. M. and Parker J. (2000). *Biology of Microorganisms*, Prentice-Hall, NJ.

Martin I., Pidou M., Soares A., Judd S. and Jefferson B. (2011). Modelling the energy demands of aerobic and anaerobic membrane bioreactor for wastewater treatment. *Environmental Technology*, **32**(9), 932–932.

McCarty P. L., Beck L. and St. Amant P. (1969). Biological denitrification of wastewater by addition of organic materials. Proceedings of 24th Industrial Waste Conference, Purdue Univ., LaKyatte, Ind., pp. 1271–1285.

Metcalf & Eddy. (2003). Wastewater Engineering: Treatment and Reuse, McGraw-Hill, New York.

Nielsen P. H., Mielczarek A. T., Kragelund C., Nielsen J. L., Saunders A. M., Kong Y., Hansen A. A. and Vollertsen J. (2010). A conceptual ecosystem model of microbial communities in enhanced biological phosphorus removal plants. *Water Research*, **44**(17), 5070–5088.

Ragsdale D. (2007). Advanced Wastewater Treatment to Achieve Low Concentration of Phosphorus, Report EPA-910-R-07-002, USEPA, Washington, USA.

Seviour R. J., Mino T. and Onuki M. (2003). The microbiology of biological phosphorus removal in activated sludge systems. *FEMS Microbiology Reviews*, **27**, 99–127.

WEF (1998). Design of Wastewater Treatment Plants, 4th ed., Manual of Practice No. 8, Water Environment Federation, Alexandria, VA.

WEF (2012). Membrane Bio Reactors, WEF Manual of Practice No 36, McGraw-Hill, New York.

Yamamoto K., Hiasa M., Mahmood T. and Matsuo T. (1989). Direct solid-liquid separation using hollow fiber membrane in an activated sludge aeration tank. *Water Science and Technology*, **21**(4), 43–54.

Chapter 3

Membrane bioreactors design, operation, maintenance

Christoph Brepols

Erftverband, Am Erftverband 6, D-50126, Bergheim, Germany

ABSTRACT

Key aspects of design, operation and maintenance of full-scale membrane bioreactors (MBR) are portrayed in this chapter. Qualitative and quantitative data from case studies, guidelines and research literature is presented with special attention to practical experiences from full-scale operations. A course of the action in MBR project development is proposed, from proper identification of influent data and consideration of alternative process layouts to basic bioreactor and membrane unit design and cost calculations. This is complemented by hands-on information on plant management, process reliability, automation and operation.

3.1 INTRODUCTION

Today wastewater reuse has become an important issue in rural regions as well as the world's megacities as a way to uncouple human water consumption from dwindling natural resources. Technological designs of reuse applications are embedded into complex socio-economic, environmental and legislative frameworks. Reports on policies, strategies, research and technologies of reuse schemes around the world enlist a multitude of factors that require consideration in applying wastewater reuse for various purposes and on different technological levels (UNEP/WHO/HABITAT/WSSCC, 2004; WHO, 2006; WWAP, 2017). This is also echoed in the complexity of possible technological solutions where treatment processes of distinct quality levels are employed to construct a variety of treatment schemes (Kazner *et al.*, 2012; Melin and Wintgens, 2006). In this context membrane bioreactor (MBR) technology with micro- and ultrafiltration (UF) membranes combines highly efficient biological wastewater treatment, filtration and basic disinfection to a single process.

Editor(s): Faisal I. Hai, Kazuo Yamamoto and Chung-Hak Lee
doi: 10.2166/9781780409177_0067

This chapter provides hands-on information on MBR design, operation and maintenance. It starts with a concise overview on technical reuse concepts employing MBR (Section 3.2) followed by a brief description of selected case studies in different fields of application (Section 3.3).

Essential information on MBR design will be presented in Section 3.4. Qualitative and practical information is complemented by quantitative tables and equations, which provide basic knowledge for drafting an MBR scheme. A general overview on the fundamentals of MBR can be found in Chapter 2 of the book, while this chapter discusses various features of MBR design in greater detail. In addition, more thorough information on investment and life-cycle cost is presented in Chapter 8, while an overview on commercial membrane products and more case studies are provided in Chapter 14 of the book. Additional considerations on MBR operation and management are given in Section 3.5 with highlights on operational procedures and process reliability.

3.2 TECHNICAL CONCEPTS

MBRs provide an excellent effluent quality which is suitable for many reuse purposes without any further treatment and thus hold a distinctive position among the available treatment technologies employed in wastewater reuse (Bérubé, 2010; Hoinkis *et al.*, 2012; Melin *et al.*, 2006; Wisniewski, 2007). MBRs are considered one of the most promising technologies for integrated, decentralized wastewater management for resource recovery in rural and peri-urban areas (Capodaglio, 2017). Even where the MBR effluent was not initially intended for reuse, it has been considered later on (Zanetti *et al.*, 2010). Furthermore, MBRs have the potential to ease operation of subsequent treatment steps. Suspended solids and also pathogens are already removed by MBR to a great extent, thus loadings to eventual supplementary treatment such as disinfection, nano-filtration, reverse osmosis (RO) or activated carbon filtration can be lowered significantly. Table 3.1 gives a general overview of technologies which can be arranged to form multi-barrier systems, which are essential for technically reliable reuse schemes. As can be seen, multiple options exist for pre-treatment of MBR influent as well as post-treatment of MBR effluent. As a matter of fact, not all of the treatment steps are obligatory to design, build and operate an MBR. Their selection depends mainly on the reuse purpose.

Table 3.1 Possible process units of multi-barrier reuse schemes.

Process unit	Available Technologies
Collection	Mixed sewer system, separate sewer system, road haulage, separate collection of domestic and industrial water, sewer mining
Storage/equalization	Storm water retention, influent buffering, septic tanks
Primary/mechanical treatment	Coarse screen, fine screen, sieve, grit chamber, grease trap, floatation, primary settling
Secondary/biological treatment	
Natural systems	Constructed wetland
Fixed biomass	Trickling filter, biofilm reactor, moving-bed reactor
Suspended biomass	CAS systems, combined anaerobic/aerobic treatment, **MBR**, powdered activated carbon dosage
Tertiary treatment/filtration	Sand filtration, cloth filter, membrane filtration, **MBR**
Quaternary treatment	Membrane filtration (UF and NF), RO, UV-disinfection, chlorination, advanced oxidation, activated carbon filtration
Storage/buffering/blending	Natural reservoir, designed storage tank, ambient buffering

3.3 REFERENCE DATA ON DESIGN AND OPERATION

MBRs for wastewater treatment have been built in increasing numbers during the past decade. MBRs have replaced, supplemented or retrofitted existing conventional waste water treatment plants (WWTPs) or they have been built as green-field installations. Stand-alone MBRs as well as MBRs in parallel or in sequence with conventional activated sludge (CAS) treatment units can be found.

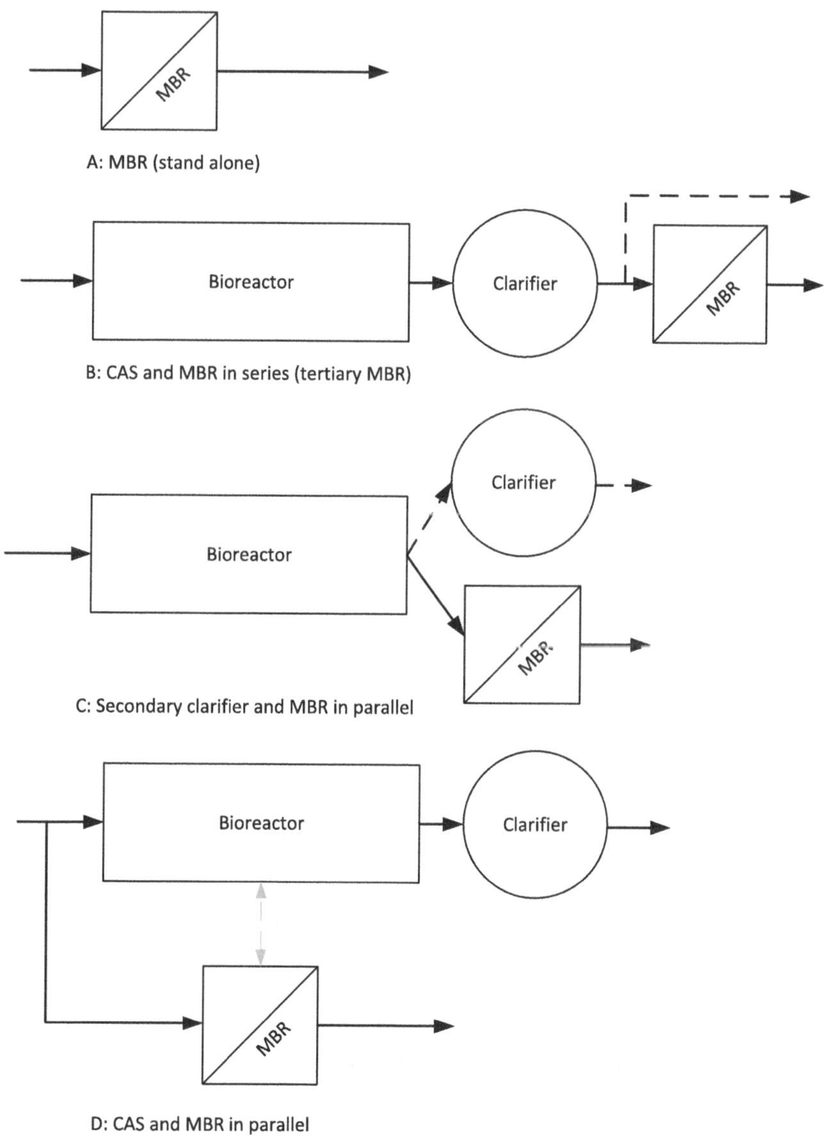

Figure 3.1 Process flow schemes for combination of MBR and CAS treatment (Ch. Brepols *et al.*, 2008; Lesjean & Leiknes, 2009).

Figure 3.1 gives an overview of possible combinations between CAS and MBR systems. The four different process schemes presented here have been applied in practical applications (Ch. Brepols *et al.*, 2008): Type A is widely used at small-scale urban reuse applications as described for example in Section 3.3.1. However, case studies representing this process configuration can be found at all scales and kinds of applications (Brepols, 2010; Judd, 2011). Often these are niche applications to tackle water scarcity and/or protect highly sensitive environments, ranging in scale from single households (Abegglen *et al.*, 2009) to schools (Busse, 2018) and office buildings, small communities and holiday resorts (Birks, 2008; Boehler *et al.*, 2007; Verrecht *et al.*, 2012), and even ships and Antarctic research stations (Pyper, 2013). An example of type B is presented in Section 3.3.1, where the MBR is used to treat a part of the secondary effluent to fulfil the quality requirements for the intended reuse. Examples of type C and D have been in use as the so-called hybrid MBRs at the WWTPs in Heenvliet and Ootmarsum in the Netherlands (Krzeminski *et al.*, 2012; Mulder, 2009; STOWA, 2009) or as a so-called dual configuration at Schilde WWTP in Belgium (De Wilde *et al.*, 2007; Fenu *et al.*, 2010; Lesjean & Leiknes, 2009). The WWTPs at Heenvliet, Ootmarsum and Schilde were designed to treat regular dry weather flow from mixed sewer systems at the MBR line while excess flow during storm weather has been channelled to the old, existing CAS treatment line. These configurations have a potential of reducing investment cost and energy consumption by lowering the maximum inflow to the MBR line (Fenu *et al.*, 2010). However, it has to be acknowledged that only a part of the wastewater treated at the MBR line might fulfil the requirements for an intended reuse. Another example of a parallel MBR configuration that is described in more detail in Section 3.3.1 is the Ulu Pandan MBR, Singapore (PUB, 2012).

The following examples have been selected based on various modes (e.g. municipal/industrial/ground water treatment) and scales of application at different sites; however, this is not intended to be an exhaustive list.

3.3.1 Municipal/urban applications

3.3.1.1 Ulu Pandan MBR, Singapore

Key information:

Location/Operator:	Singapore, Public Utilities Board (PUB)
Date of commissioning:	2006
Wastewater source:	Settled sewage of mainly domestic origin, 10% industrial sewage
Pre-treatment:	Intermediate 2 mm mesh sieve between bioreactor and filtration units
Post-treatment:	RO, UV disinfection, storage
Reuse purpose:	Non-potable industrial use, blending with raw water for potable reuse

During recent decades, Singapore's Public Utilities Board (PUB) has evaluated MBR systems under tropical conditions to become part of the national water reuse programme (Lay *et al.*, 2017). The MBR at Ulu Pandan water reclamation plant is a part of Singapore's NEWater scheme. In 2006 a side-stream MBR with five separate filtration units and upstream denitrification in two retrofitted bioreactors of the original CAS plant was commissioned (see Table 3.2), while several parallel CAS lines remained operative. In 2012 PUB issued plans to retrofit more CAS lines with membrane filtration instead of secondary clarifiers (see Figure 3.2) (Judd, 2011; PUB, 2012; Tao *et al.*, 2009).

Table 3.2 Key technical data, Ulu Pandan MBR (2006 unit), Singapore.

Data	Values
Average inflow Q	23,000 m^3/d
Peak inflow Q_{peak}	23,000 m^3/d
Influent load COD/TN/TP	6,700/1,100/150 kg/d
Bioreactor volume	n.a.
F:M ratio	0.1 kg BOD_5/(kg MLVSS d)
HRT	>3 h
SRT	>10 d
Bulk temperature, average	30°C
Effluent concentration COD/TN/TP	n.a./6.3/3.3 mg/L
Membrane type	GE Zenon ZW 500C
Membrane tank configuration/volume	Separate/n.a.
Installed membrane surface	37,920 m^2
Energy consumption, average	0.435 kWh/m^3

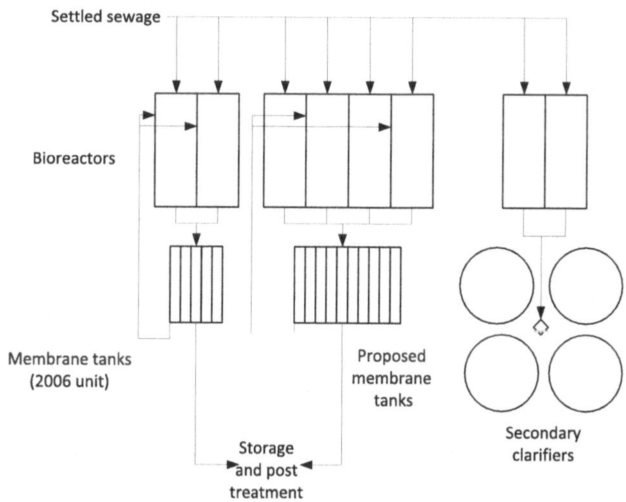

Figure 3.2 Process scheme of Ulu Pandan MBR, Singapore.

3.3.1.2 Shingu Chuo sewage treatment plant (Aqua Shingu), Fukuoka prefecture, Japan

Key information:

Location/Operator:	Town of Shingu, Fukuoka/Fukuoka Prefecture (Water and Sewer System Division)
Date of commissioning:	2010
Wastewater source:	Domestic, ca. 17,000 population equivalents
Pre-treatment:	Mesh screens, equalization basin
Post-treatment:	Ozone treatment, sterilization, depending on reuse
Reuse purpose:	Toilet water, irrigation, landscape

For this plant, wastewater is collected in a separate sewer system. The water treatment facilities have been built underground in high-rise buildings. After ozonation, one part of the treated water is fed to an artificial stream in a surrounding public park area. Another part of the water is sterilized and supplied to a neighbouring train station, park and major commercial facilities for flushing toilets. The remainder of the treated water is discharged into the Mutagawa River without further post-treatment. The WWTP also serves as a centre for scientific exchange and public information (Table 3.3).

Table 3.3 Key technical data, Aqua Shingu MBR, Japan.

Data	Values
Average inflow Q	6,060 m^3/d
Peak inflow Q_{peak}	9,090 m^3/d
SRT	25 d
Membrane type	Kubota FS
Membrane tank configuration/volume	Integrated/n.a.
Installed membrane surface	n.a.

3.3.1.3 Cloudcroft, New Mexico, USA

Key information:

Location:	Cloudcroft, New Mexico
Date of commissioning:	2007
Wastewater source:	Domestic, ca. 1,000 population equivalents
Pre-treatment:	Tertiary treatment, aerated equalization basin
Post-treatment:	RO, advanced oxidation, storage
Reuse purpose:	Irrigation, storage for potable water supply

The original WWTP was converted into an MBR process (see Figure 3.3). The MBR effluent is disinfected with chloramines and pumped uphill to a water storage tank, where it stays for a 30-day natural treatment by diffusion and sunlight. From there, some of the water is diverted for irrigation, while 440 m^3/d flow downhill to the water treatment plant. The water treatment plant contains an RO system. Permeate from the RO system is disinfected with peroxide and ultraviolet light, before it is discharged to a covered reservoir. From there, the water flows into another covered reservoir, where it is blended with spring and ground waters. The purified wastewater constitutes up to 50% of the drinking water supply. Finally, the blended water passes through UF membranes and granular activated carbon, than it is disinfected using sodium hypochlorite before entering the distribution system and residents' homes (Freeman & Levesque, 2009).

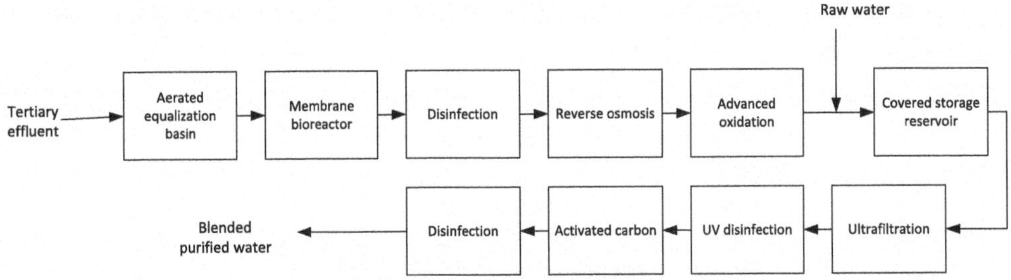

Figure 3.3 Process scheme Cloudcroft, New Mexico, USA (Leverenz *et al.*, 2011).

3.3.1.4 Beddington zero energy development MBR, Great Britain

Key information:

Location/Operator:	Beddington, Great Britain/Thames Water
Date of commissioning:	2008
Wastewater source:	Domestic
Pre-treatment:	Pumping stations, septic tanks, 3 mm screen
Post-treatment:	Activated carbon filtration, chlorination
Reuse purpose:	Toilet flushing and irrigation

The MBR produces reclaimed water at Beddington Zero Energy Development (BedZED), a prestigious sustainable development scheme in south London. The original water reclamation plant at the BedZed housing project did not produce water of sufficient quality and was finally discontinued in 2005 because of high electricity consumption and labour-intensive monitoring. The company which installed and operated it had been dissolved, making it more difficult to operate the plant affordably and reliably (Schoon, 2016). A regional water utility took over responsibility for the water reclamation and finally commissioned a small-scale MBR in 2008 (Birks, 2008; Verrecht *et al.*, 2010a). Due to legislative constraints water reclamation could only be provided at a local level with chemical disinfection as a prerequisite. The MBR is equipped with upstream denitrification and two immersed hollow fibre (HF) membrane cassettes integrated to the nitrification basin. The membrane filtration is operated intermittently, while the nitrification basin buffers the inflow (see also Figure 3.4 and Table 3.4).

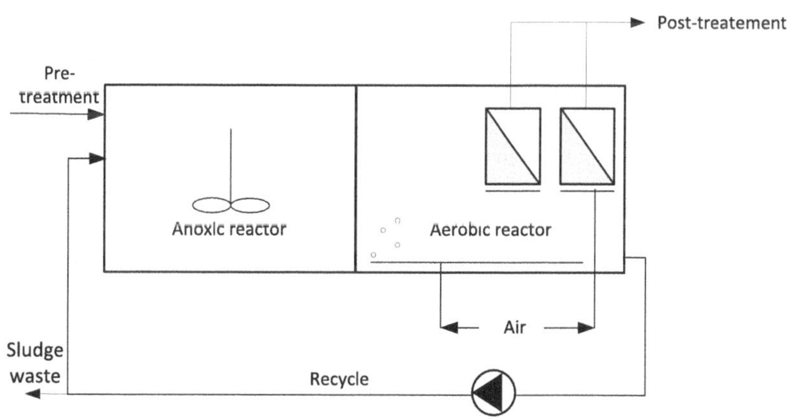

Figure 3.4 Process scheme of BedZED MBR, Great Britain (Verrecht *et al.*, 2010a).

3.3.1.5 Colegio Johannes Gutenberg, school building, Lima/Peru

Key information:

Location/Operator:	Lima, Peru/n.a.
Date of commissioning:	2017
Wastewater source:	Domestic
Pre-treatment:	aerated septic tanks with screen
Post-treatment:	none
Reuse purpose:	Toilet flushing and irrigation

Table 3.4 Key technical data, BedZED MBR, Great Britain.

Data	Values
Average inflow Q	25 m^3/d
Influent load COD/TN	12/2 kg/d
Bioreactor volumes anoxic/aerobic	10/13 m^3
Bulk temperature	14–27°C
MLSS concentration	8,000 mg/L
HRT	22 h
SRT	35 d
DO set point	1.3 mg/L
Effluent concentration NH_4-N/TN	$<0.5/<20$ mg/L
Membrane type	GE Zenon ZW 500C
Membrane tank configuration/volume	Integrated
Installed membrane surface	139 m^2
Flux during filtration	11–28 L/(m^2 h)
Membrane aeration	42 N m^3/h
Energy consumption, average	3.1 kWh/m^3

In April 2017, two package plant MBRs were commissioned at Colegio Johannes Gutenberg school building where 1,700 pupils are educated. The school is situated in a favela district of the City of Lima/Peru. The reclaimed water is used to irrigate the school garden and for toilet flushing (Busse, 2018, n.d.; Machuca, 2017). The plant can be seen as an example of many decentralized MBRs worldwide which can be found at single houses, hotels, small communities and tourist resorts as well as in urban areas. They are commercially available as packaged plants or can be individually designed depending on their capacity and the whereabouts of their installation (Judd, 2011). A main focus of these plants is high reliability and easy maintenance while providing high-quality treatment (Table 3.5).

Table 3.5 Key technical data, Colegio Johanes Gutenberg MBR, Peru.

Data	Values
Average inflow Q	2 × 20 m^3/d
Membrane type	Busse MF Type C100/200
Membrane tank configuration/volume	Integrated
Footprint	6 × 12 m

3.3.2 Industrial applications

3.3.2.1 COOPERL Lamballe, abattoir wastewater, France

Key information:

Location/Operator: Lamballe, France/COOPERL
Date of commissioning: First line 2002, second line 2006

Wastewater source:	Meat production
Pre-treatment:	Floatation, buffer tank
Post-treatment:	RO
Reuse purpose:	Cleaning, boiler feed water

One third of the treated wastewater from this plant is reused via RO units. The energy consumption in Table 3.6 includes the floatation and the RO units (Brockmann, 2008).

Table 3.6 Key technical data, Cooperl Lamballe MBR, France.

Data	Values, Line 1	Values, Line 2
Average inflow *Q*	9,500 m^3/week	6,600 m^3/week
Influent load COD	18,000 kg/week	12,500 kg/week
Effluent concentration COD/N/P	$<30/<3/0.1$ mg/L	
Membrane type	GE Zenon ZW	KMS Puron
Membrane tank configuration/volume	Separate	Separate
Energy consumption, average	3.6 kWh/m^3	

3.3.2.2 Aquapolo ambiental, S.A., Brazil

Key information:

Location/Operator:	Sao Paolo, Brazil/Aquapolo ambiental S.A.
Datc of commissioning:	2012
Wastewater source:	Municipal/industrial
Pre-treatment:	Biological treatment, secondary clarifier, 2 mm screen
Post-treatment:	RO of side stream, to meet conductivity limits for industrial re-use

According to the operating company the Aquapolo installation is the largest provider of recycled water for industrial use in Latin America ("Aquapolo," 2016). The process is described as a polishing or tertiary MBR process (TMBR). The installation is situated at an existing CAS plant site. Several critical discharge parameters of the CAS operation seriously compromised the ability to reuse the existing effluent, including oxygen demand (COD and BOD_5), suspended solids, ammonium, total phosphorus and conductivity (see Table 3.7). The influent to the MBR is secondary effluent. An additional screen is used to shield the MBR from eventual floating debris. A part of the TMBR effluent is further treated by RO to meet conductivity levels required for intended industrial water reuse (Kullmann *et al.*, 2011).

3.3.2.3 Ujams MBR, Windhoek, Namibia

Key information:

Location/Operator:	Windhoek, Namibia/BuildOwnOperateTransfer
Date of commissioning:	2014
Wastewater source:	Industrial
Pre-treatment:	Fine sieving
Post-treatment:	UV disinfection

Table 3.7 Key technical data, Aquapolo MBR, Brazil (Kullmann *et al.*, 2011).

Data	Values
Average inflow Q	56,160 m^3/d
Peak inflow Q_{peak}	86,400 m^3/d
Influent concentrations COD/TN/TP	50–120/>20/>5 mg/L
F:M ratio	kg BOD/(kg MLVSS d)
HRT	3 h
SRT	12–20 d
Bulk temperature, average	15–30°C
Effluent concentration COD/N/P	20/1.0/0.5 mg/L
Membrane type	Puron KMS
Membrane tank configuration/volume	Separate/n.a.
Installed membrane surface	94.500 m^2

Namibia is known to be one of the most arid countries in the world and thus the capital of Windhoek looks back on decades of water reuse. Lahnsteiner *et al.* (2018) have reported that Windhoek is one of two direct potable reuse cases in southern Africa where domestic and industrial wastewater are strictly separated. While the domestic wastewater is treated in multi-barrier systems which do not employ MBRs, a major portion of the industrial effluents are centrally treated at Ujams MBR. The used water comes mainly from an abattoir, a brewery and a tannery. After a period of pilot testing, the full-scale plant went into operation in October 2014. Currently, the reclaimed water is mainly reused for the augmentation of the ephemeral Klein Windhoek river. However, due to the ongoing water crisis in southern Africa, industries have become more interested in reusing the water (Lahnsteiner *et al.*, 2018).

Data	Values
Average inflow *Q*	5,200 m^3/d
Influent concentrations COD/TN	3,314 (peaks > 10,000)/96 (peaks > 300) mg/L
Hydraulic retention time (HRT)	h
Sludge retention time (SRT)	d
Bulk temperature, average	°C
Effluent concentration COD/NH_4-N/NO_3-N	68/<0.1/6.2
Membrane type	ZeeWeed 500C
Membrane tank configuration/volume	Separate/n.a.
Installed membrane surface	m^2

3.3.3 Groundwater replenishment

3.3.3.1 Glessen MBR, Germany

Key information:

Location/Operator: Bergheim, Germany/Erftverband
Date of commissioning: 2008

Wastewater source:	Domestic
Pre-treatment:	Storm water retention, pumping stations, 6 mm screen, sand and grit chamber, 1.5 mm screen
Post-treatment:	Granular activated carbon filtration (planned)
Reuse purpose:	Environmental discharge, groundwater recharge

The Glessen MBR (9,000 P.E.) consists of a retrofitted bioreactor with surface aerators and intermittent denitrification and a four-line separate membrane filtration (see also Figure 3.5 and Table 3.8). The WWTP discharges into a small stream in a groundwater protection area. A few kilometres downstream, the water flows into a wetland draining away in a porous geological gravel formation. This wetland

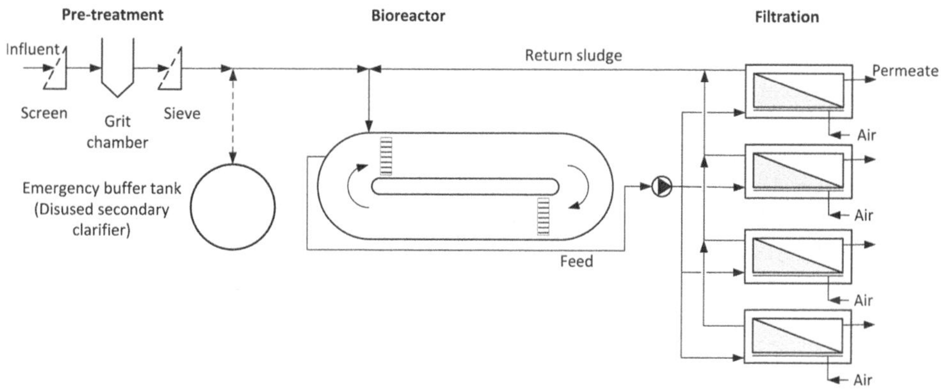

Figure 3.5 Process flow scheme of Glessen MBR, Germany.

Table 3.8 Key technical data, Glessen MBR, Germany.

Data	Values
Average inflow Q	2,400 m^3/d
Peak inflow Q_{peak}	268 m^3/d
Influent load COD/TN	1,008/99 kg/d
Bioreactor volume	1,400 m^3
Bulk temperature	5–25°C
MLSS concentration in bioreactor	8.0 kg/m^3
HRT	17 h
SRT	25 d
DO set point	1.5 mg/L
Effluent concentration NH_4-N/TP	$<1.5/<0.6$ mg/L
Membrane type	GE Zenon ZW 500C
Membrane tank configuration/volume	Separate (4 lines)
Installed membrane surface	12,096 m^2
Flux during filtration	17–27 $L/(m^2\ h)$
Energy consumption entire plant, average	1.1 kWh/m^3

feeds directly into the underground drinking water reserve of the nearby city of Cologne. The water is withdrawn again at a neighbouring water works. For decades, the wetland has been used to supplement the groundwater body which is influenced by open-cast lignite mining in the region. However, there was increasing concern that pathogens from the existing CAS could contaminate the drinking water. As a consequence, the existing WWTPs were retrofitted as UF MBR. Additionally, all stormwater sewer overflows have been channelled to newly built retention soil filters, which reduce hydraulic stress and pollution discharge to the river (Brepols, 2010). Currently, plans for a post-treatment to remove micropollutants by granular activated carbon filtration have been considered (Brepols *et al.*, 2017a).

3.4 MBR DESIGN

3.4.1 Design workflow

Dimensioning of membrane filtration, bioreactor volume and aeration capacity are at the core of MBR design procedure. Several authors (Brepols, 2010; Brepols *et al.*, 2005; Judd, 2011; Pinnekamp & Friedrich, 2006; Verrecht, 2010) have proposed design workflows and calculations for key parts of entire MBR plants. Several institutional recommendations for the design of MBRs have also been published (Brockmann *et al.*, 2007; DWA, 2014; Sewage Technical Meeting on Membrane Technology, 2011; Water Environment Federation, 2012).

Various interactions between biological treatment and other MBR process steps require recurring calculations of size, layout and properties of these process steps in order to find an optimized solution. Plant footprint, effluent quality, process stability, energy consumption or life-cycle cost are among possible targets of optimization. Ideally, these target functions can be defined ahead of an MBR planning project, in practice these are also often reconsidered in the design process.

Planning methods grow in complexity and sophistication from first preliminary designs to final design. A first draft might be sketched based on a few heuristic design rules while verification of a final design might require dynamic modelling of the biological treatment process and the use of elaborate cost models. At least the treatment of specific wastewaters of unusual composition might demand pilot plant tests to assess membrane and bioreactor performance. Details on particular steps within the design workflow-scheme of Figure 3.6 will be discussed in the following sections.

3.4.2 General plant layout

Numerous options can be considered to fulfil the project's purpose depending on site-specific constraints, reuse applications and demand of reclaimed water. Even more so, different types of membranes and commercial membrane products can be combined with biological treatment in various ways (see Table 3.9).

As can be concluded from several reports (Buer & Cumin, 2010; Lesjean & Huisjes, 2008; Zheng *et al.*, 2010) aerobic submerged or immersed MBRs (iMBR) with microfiltration (MF) or UF currently play the most prominent role among the processes which are applied for wastewater treatment on a large scale. The following sections therefore focus on the application of these iMBRs.

3.4.2.1 Integrated or separate filtration units

Figure 3.7 shows the basic configurations of iMBRs which are distinguished by the position of membrane modules relative to the bioreactor.

The advantages of an integrated or inside configuration of the membrane filters on the one hand and a separate or outside configuration on the other have been discussed by De Wever *et al.* (2009). The following aspects can be considered:

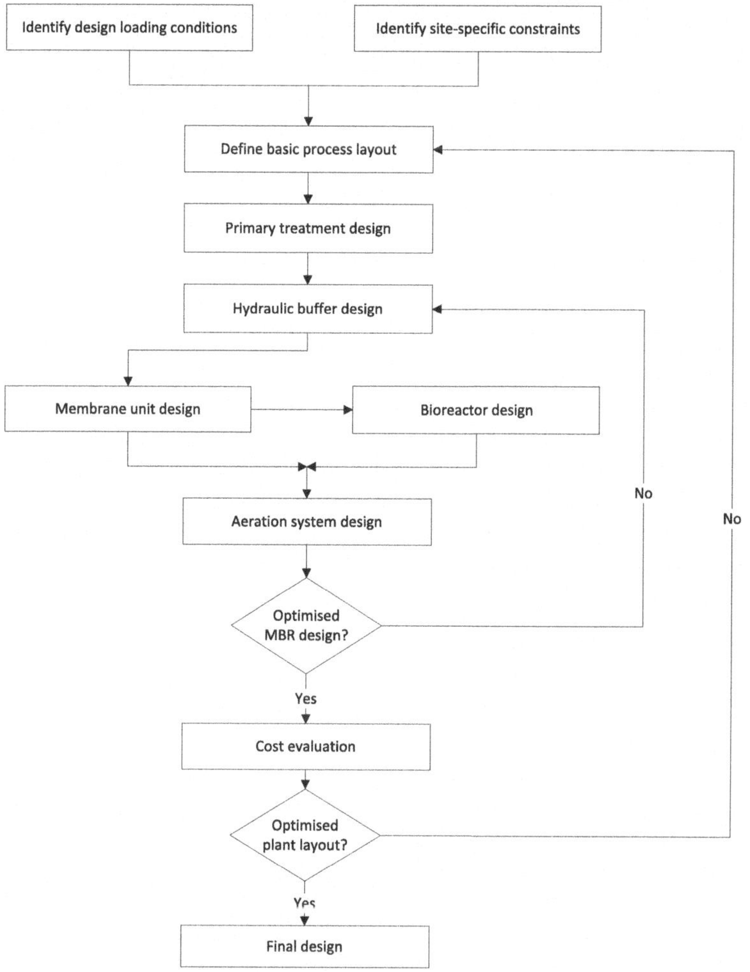

Figure 3.6 Proposed MBR design workflow.

Table 3.9 Possible combinations of membrane filtration MF and bioreactors.

Membrane Type	HF	FS	Tubular Membrane
Module type	Submerged (immersed)		Side-stream
Liquid flow on membrane	Aerated flow		Pump feed/aerated flow
Relative position of membrane unit to bioreactor	Integrated or separate filtration tanks		External

Size and footprint requirements of the plant. The inside configuration allows for smaller bioreactor volumes because of resulting higher MLSS in the bioreactors. No additional filtration tanks have to be built. The footprint can thus be reduced.

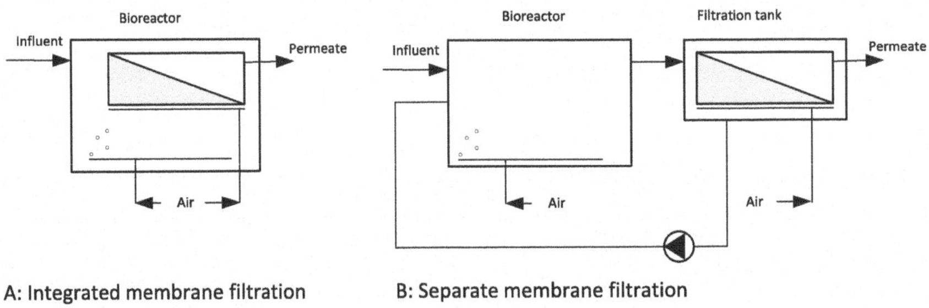

Figure 3.7 Basic process configuration of iMBR.

Energy demand. With an inside configuration at least the energy for sludge recycling can be saved. The cross flow aeration of the membranes can be used for the aeration of the bioreactor as well. Especially for small MBRs there is often no additional aeration required for the biological treatment step.

Nitrogen removal. With the outside configuration return sludge and nitrate recycle can be held separately. Oxygen carryover to the denitrification zone is thus reduced. Filtration and biological treatment can be optimized separately. A slightly increased overall volume with additional biological treatment inside the filtration tanks makes the process configuration less vulnerable to shock loadings.

3.4.2.2 Standalone, parallel or hybrid configurations

An MBR as full stream installation is designed to treat the entire inflow from the sewer catchment. Examples of this MBR configuration exist at all scales. Parallel MBRs or hybrid configurations have been built to supplement the capacity of existing WWTPs. Both operate in parallel to conventional treatment lines. While side stream MBRs can be seen as entirely separate installations at the same site, hybrid configurations share at least a part of the sludge cycle with a CAS treatment line (Krzeminski *et al.*, 2012). Hybrid and parallel MBRs are often used to treat a constant inflow in the MBR, while the gross of the inflow variations is treated at the conventional line. It can be considered that constant flow might lead to an optimized hydraulic MBR layout, because the design does not need to account for high inflow variations. Furthermore, permeate production might be adjusted to meet the demand of reclaimed water, in case the incoming wastewater flow constantly exceeds reuse demand.

3.4.2.3 Number of treatment and filter lines

Similar to large CAS, large MBRs are designed with multiple treatment lines. But even for small MBRs it can be recommended to have at least two separate filtration lines. Multi-line filtration units provide operational safety in case of cleaning, maintenance, repair or an unforeseeable breakdown of one filter line (US EPA, 2007). Installing a higher number of separate filtration lines bears the potential for improved plant operations and reduced energy consumption (Veltmann *et al.*, 2011) especially when dealing with considerable inflow variations.

As experiences at Nordkanal MBR have shown, scale-up of filtration lines might have practical limits (each one of the eight filtration lines at Nordkanal MBR contains approximately 10,000 m^2 of membrane surface area) while numbering-up results into multiple constructions and equipment (Brepols, 2010). The right balance between scale-up und numbering-up has to be found to balance advantages and disadvantages of both concepts. Scaling-up might reduce the complexity of the process layout and can be a way to avoid cost of installing multi-lane equipment. Numbering up improves operational flexibility and increases plant availability in case of operational disruptions.

3.4.3 Wastewater composition, volume and temperature

Identifying key design conditions requires a profound examination of basic influent data consisting of physical, chemical and biological characteristics. Concerning the key wastewater constituents there are no general differences compared to the design of conventional WWTPs. However, besides the inflow of sewage from external sources, eventual internal recycle or waste streams from subsequent treatment steps within the reuse process chain require particular consideration.

3.4.3.1 Wastewater components

Typical and well-known physical, chemical, biochemical and biological wastewater characteristics such as chemical and biological oxygen demand (COD and BOD_5), nitrogen (N) and phosphorous (P) components, suspended solids, pH and alkalinity are key constituents for MBR design. On a case-by-case basis other constituents like pathogens, priority pollutants, refractory organics, heavy metals or dissolved inorganics can be of interest.

Actual loading and possible variations of key constituents are best evaluated with the help of a detailed investigation of the catchment and dedicated sampling and analytical campaigns. Comparable standard values to assess the load are often documented in guidelines (ATV DVWK, 2000; WEF, 2009) or textbooks (Henze *et al.*, 2001; Tchobanoglous *et al.*, 2003). In case of strong seasonal variations, which for example, can be found in tourist areas or when treating wastewater from agricultural or industrial production campaigns, design conditions have to be considered for each season separately.

It has to be acknowledged that membranes and especially organic membrane materials may be vulnerable to various substances such as solvents, coating agents and petrochemical products among others. The sewer catchment should be carefully screened for potential sources of these substances. Even if no contributors can be identified, precautions should be taken to avoid risks from accidental spills to the sewer system. As an example: laundries, paper mills or textile industries may eventually discharge high loads of fibrous materials to the sewer that has to be removed during the pre-treatment. Chemical or petro-chemical industries or food processing may also contribute unwanted ingredients such as solvents or coating agents to the sewer. Abrasive materials may cause mechanical membrane damage. Undesirable substances may run-off to the sewer after traffic accidents.

Besides these external sources, internal waste streams, coming from subsequent treatment steps within the overall process chain, require particular consideration. For example: Recycling concentrate streams from subsequent NF or RO treatment will lead to accumulation of retained substances which may eventually yield inhibiting effects on the biological treatment or impair the MBR process. Sludge liquor from on-site sludge handling and dewatering can be a source of additional nitrogen loads but also of polymeric substances used in sludge conditioning. Residual chemicals from membrane cleaning might also have effects on biological treatment.

3.4.3.2 Hydraulic inflow profile

Average dry and wet weather inflow conditions may be sufficient to determine the average hydraulic residence time in bioreactor design. In addition to that, membrane layout has to also consider duration, frequency and height of maximum peak flow conditions. As an example, Kochaba *et al.* (2008) have examined the sensitivity of flow variations on the design of a large-scale plant, while Wedi and Joss (2008) have proposed a case- sensitive design approach. For this purpose, the inflow hydrograph can be classified by its peak hour, daily, weekly and monthly average inflow to the MBR. Figure 3.8 gives an example of the inflow hydrograph of a large MBR treating combined sewer influent.

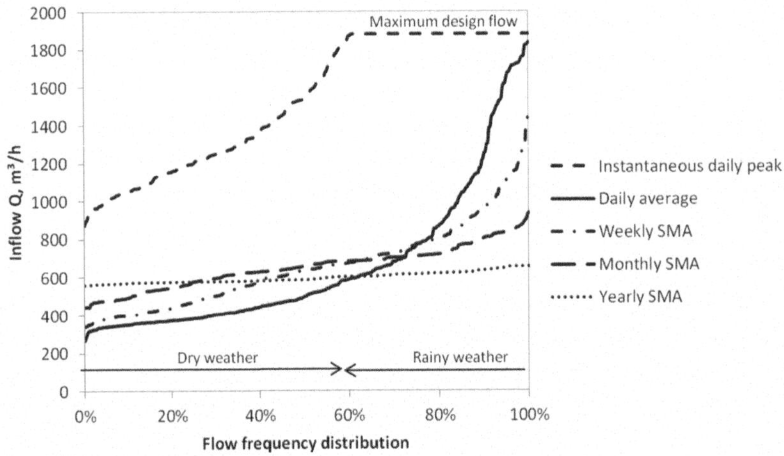

Figure 3.8 Inflow profile of an MBR (80,000 population equivalents), peak inflow, daily average and simple moving average (SMA) at distinctive intervals.

3.4.3.3 Temperature

Wastewater temperature may change rapidly depending on the location of the plant, the local weather and climate, run-off, sun-radiation or mixtures of industrial and municipal sewage; and such temperature variations affect treatment performance. Wastewater temperature thus is a key design parameter for both bioreactor and membrane unit design. For example, nitrification might eventually be suspended at wastewater temperatures below 10°C while full organic removal is still achieved. If the temperature declines further, full hydraulic performance of membrane units are still required to treat the inflow entirely. Generally inflow from sewer systems cannot be reduced deliberately and any overspills of untreated wastewater have to be avoided. Even for MBRs treating mixed sewer effluent in mild climate temperatures below 6°C have been observed during snow-melt at least on single days of the year (Brepols, 2010). As a result design temperatures for the biological process and the filtration units can be different and should be considered independently.

Maximum wastewater temperatures may also be of interest as there are maximum tolerable temperatures for certain biological processes as well as for many commercial membrane materials. It is known, that the growth rate of nitrifying bacteria stagnates between 30 and 35°C and then between 35 and 40°C rapidly deteriorates towards zero (Henze *et al.*, 2001). According to many manufacturer guidelines, organic membrane materials should not be exposed to temperatures above approximately 35°C in the long run, while ceramic membranes can be suitable for operation at higher temperatures.

3.4.4 Process units: inflow equalization

Influent buffering can be a way to optimize MBR layout by equalizing inflow variations and reducing peak inflow. There is thus a trade-off between the size of the influent buffer, the required MBR reactor volume and the required membrane surface area (Verrecht *et al.*, 2010b). Municipal combined sewers often already inherit means of influent buffering in the form of storm water-retention tanks.

Retrofitting existing WWTPs with MBR technology often offers an opportunity to convert unused reactors or secondary clarifiers into buffering tanks at relatively low expenses. At least, these can be of use to buffer influent or store biomass temporarily under extraordinary operating conditions (Brepols, 2010).

It has to be considered, that prolonged storage of raw or only primarily treated sewage might lead to unwanted effects, such as unintended anaerobic decomposition and odour emissions, which may practically limit the size and storage time of influent buffer tanks.

A report of Kochaba *et al.* (2008) reveals the practical importance of an analysis of hourly flows and membrane peak design flow rates. By making use of existing buffer volume inside the bioreactors, a reduced membrane peak design flow can be obtained, which allows for a corresponding reduction in membrane area.

3.4.5 Process units: mechanical pre-treatment

The risk of severe membrane damage is decisively reduced by high-quality removal of debris upstream of the MBR. Experiences of operators (Brepols, 2010; Judd, 2011) and institutional recommendations (US EPA, 2007), have highlighted the importance of sufficient pre-treatment.

In practice, multi-step primary treatment is often provided in larger installations, although not in most small-to-medium-sized installations. However, MBR systems require smaller than 1–3-mm-cutoff fine sieves upstream of the MBR unit. Alternative pre-treatment configurations include two stages of screening, sieves located downstream of a primary settling or intermediate sieves between bioreactor and filtration units (Ch. Brepols *et al.*, 2008; Frechen *et al.*, 2008; Schier *et al.*, 2009a, 2012) (see Figure 3.9).

Studies conducted in Germany (Schier *et al.*, 2009b) and Great Britain (Stefanski *et al.*, 2011) have found that sieve and screen debris largely consists of paper-pulp and fine cotton-fibres although composition and amount of materials may vary widely. Based on such reports as well as anecdotal evidence it can be assumed that the amount and composition of materials are strongly affected by consumer habits in the sewer catchment area. From country to county or even region to region, diverse personal and cultural sanitation and hygiene practices can be found, (UNEP/WHO/HABITAT/WSSCC, 2004; Webster, 2007), together with different technological sanitation standards and attitudes towards waste disposal

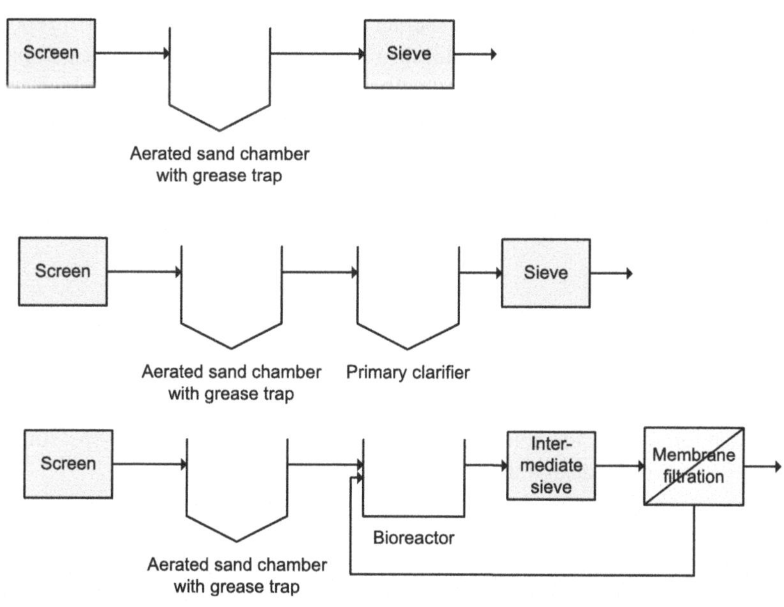

Figure 3.9 Possible pre-treatment configurations (Frechen *et al.*, 2008; Schier *et al.*, 2009b).

these may influence the actual composition of sewage. For example, a report by WWF (2013) highlights that the per capita consumption of toilet paper in Europe is four times higher than the worldwide average, while another source of debris which can be related to consumer habits is the improper disposal of solid waste into household toilets (Berkley, 2007).

The amount of screened-out debris increases exponentially with the decrease of the screen width (Seyfried *et al.*, 1985). The role of sieve geometries on the removal of fibrous materials was highlighted by Frechen *et al.* (2008). Two-dimensional sieve geometries like mesh or punched holes are favourable over one-dimensional grid of slot geometries to remove fibrous materials.

The survey by Frechen *et al.* (2008) has also shown that from a number of 19 European MBR, a great majority of 14 use two-stage screening equipment with a 2–6 mm cut-off at the first stage and 0.5–3.0 mm cut-off at the second stage. The remainder employs only one stage of screening with fine sieves. A cut-off of less than 1.0 mm is used independently of type of membranes (HF of FS). The two screening steps are regularly separated by an aerated sand and grit chamber to remove mineral components and fat from the sewage. The latter may otherwise cause operational problems with the fine sieving by blocking the sieve pores especially at low wastewater temperatures (Brepols, 2010; Schier *et al.*, 2012). Interactions between sieves and sand-grit chambers need to be considered (Schier *et al.*, 2012).

To date primary clarifiers have been seldom used in combination with MBRs. A few cases exist where CAS plants have been retrofitted as MBRs that already employed primary clarification. Still sieves have been installed in those plants as a protection against floating debris (e.g. Wädenswil MBR, which was the first large-scale MBR in Switzerland commissioned in 2005). Currently, the Nordkanal MBR is retrofitted with a primary clarifier which will lower energy consumption for the biological treatment step by reducing the COD load to the MBR. Energy-rich primary sludge will be further used to produce biogas at the newly built anaerobic digestion and co-generation (CHP) facility on site. The project was planned to be completed in 2018 (Brepols *et al.*, 2017a; Schäfer, 2017).

Unfortunately, sieves have been a major source of operational disruptions (see Section 7.5.2). Determining the actual content of sieving debris, the removal efficiency and the resulting appropriate hydraulic capacity of fine sieves is challenging (Schier *et al.*, 2012). Brepols (2010) reported a series of replacements of sieves at large scale before finding a sufficient mesh-width as a compromise between removal efficiency and hydraulic throughput. Today many suppliers offer sieves which are dedicated for operation on MBRs and membrane suppliers regularly give mandatory recommendations on the proper mesh width of sieve installations for their type of membrane filtration.

In case high amounts of falling leaves, twigs or other debris might contaminate the biomass via an open bioreactor surface covering the bioreactor with nets or intermediate screens may ease the problem (De Wilde *et al.*, 2005).

3.4.6 Process units: biological treatment

Practical evidence has shown that biological treatment efficiency at MBRs is generally comparable to other applications of suspended activated sludge (Brepols, 2010; Judd, 2011; Lesjean *et al.*, 2011). Observed variations in biological removal efficiencies and sludge production are often site-specific and thus should have no general relevance for the design at full scale for most applications.

Compared to organics removal, nutrient removal is more sensitive to the design parameters. However, the use of similar biological design parameters such as food-to-microorganisms ratio (F/M ratio) and SRT will lead to a similar range of removal efficiencies for CAS and MBR systems. Bioreactor design can thus be carried out using well-known empirical and heuristical design rules as well as static models or dynamic models such as the activated sludge models (ASM) (Henze *et al.*, 2000) or benchmark simulation models

(BSM) (Alex *et al.*, 2008; Maere *et al.*, 2011). Several guidelines (ATV DVWK, 2000; US EPA, 2008) and textbooks (Henze *et al.*, 2001; Tchobanoglous *et al.*, 2003), which are intended for CAS design are also applicable for MBR bioreactor design as long as some key aspects are provided due consideration.

Differences to CAS design mainly arise from higher bioreactor MLSS concentrations and consequently reduced HRT. Furthermore, higher return sludge rates and eventually different sludge cycles can be found at MBR plants. These shall be highlighted in the following subsections. Equations, tables and figures given here are used to illustrate the proposed course of action in drafting a first preliminary design for municipal wastewater of usual composition. More detailed information on biological process design can be found in established references (ATV DVWK, 2000; Henze *et al.*, 2001; Tchobanoglous *et al.*, 2003; US EPA, 2008; WEF, 2009).

In recent years many municipal MBRs have been designed with an SRT of approximately 25 d to achieve simultaneous aerobic biomass stabilization (Brepols, 2010; Judd, 2011). Table 3.10 shows a collation of key design values from existing full-scale immersed MBRs and separate or side-stream MBRs (sMBR).

Design at shorter SRTs may require separate sludge stabilization. If not compensated by a comparably low MLSS concentration, a reduced SRT will result into small bioreactors with a short HRT. This might bear an increasing risk of short circuits or incomplete removal of shock loadings and therefore reactor design requires careful consideration. Additionally, equipment such as aeration devices which needs to be fitted into the bioreactor may play a role in determining the reactor footprint and volume.

Table 3.10 Typical design values for municipal and industrial MBRs (Brepols, 2010; Judd, 2011).

Parameter	Symbol	Unit	Municipal	Industrial
MLSS concentration in bioreactor	X_{bio}	g/L	5–14 (iMBR)	<15 (iMBR) <30 (sMBR)
SRT	θ_x	d	10–35	5–>50
Hydraulic retention time	HRT	h	>3.6	–

3.4.6.1 MLSS concentration, mass balances and bioreactor volume

Excessively high MLSS concentrations in the membrane tank may lead to increased operational problems with fouling, stagnant biomass, clogging and sludging of membrane modules. From practical experience at a small municipal MBR Brepols (2010) has reported severe operational problems, when a maximum biomass concentration in the membrane tanks (X_M) of 18 gMLSS/L was exceeded over a long period. The maximum tolerable MLSS concentration at the membrane filters is therefore a decisive quantity in the whole reactor design (see Section 3.4.7).

Large MBRs often employ compartmentalized or cascaded bioreactors with various biomass cycles for return sludge and nitrate recycle. MLSS concentrations in each upstream bioreactor compartment can thus differ significantly from that in downstream reactors or the filtration tank. Still the overall amount of biomass in the plant must be sufficient to achieve the desired biological treatment quality. Prior to determining the actual reactor and compartment volumes, it is therefore suggested to determine the total amount of biomass (M_X, kg) as the product of required SRT (θ_X, d) and sludge production (F_{SP}, kg/d):

$$M_X = \theta_X \cdot F_{SP} \tag{3.1}$$

Then MLSS concentrations for each reactor compartment can be determined: Starting with the design MLSS concentration at the membrane filtration unit (X_M) the MLSS concentration in the upstream bioreactor (X_{bio}) can be calculated. Sludge recycle rates between filtration, bioreactors or different bioreactor compartments determine the MLSS concentration in each reactor compartments and thus the overall reactor volume of the installation. Figure 3.10 exemplifies MLSS concentration balances of different process configurations. With the known recycle rate between reactor compartments (R_i) and the MLSS concentration ($X_{\text{R},i}$) in the downstream reactor compartment the volume for each reactor compartment (i out of n) can be calculated subsequently (Brepols *et al.*, 2005).

$$X_{\text{R},i} = \frac{R_i}{1+R_i} \cdot X_{\text{R},i+1} \tag{3.2}$$

$$V_{\text{R},i} = \frac{M_{X,i}}{X_{\text{R},i}} \tag{3.3}$$

The overall volume of the MBR (V_{to}) including the net volume of membrane units ($V_{\text{M,net}}$) is then

$$V_{\text{tot}} = \sum_{i=1}^{n} \frac{M_{X,i}}{X_{\text{R},i}} \tag{3.4}$$

For example, the bioreactor volume (V_{bio}, m^3) of an MBR plant with pre-denitrification and separate membrane filtration tanks can be calculated as

$$V_{\text{bio}} = \frac{M_{X,\text{anox}}}{X_{\text{DN}}} + \frac{M_{X,\text{aer}} - V_{M,\text{net}} \cdot X_{\text{M}}}{X_{\text{N}}} \tag{3.5}$$

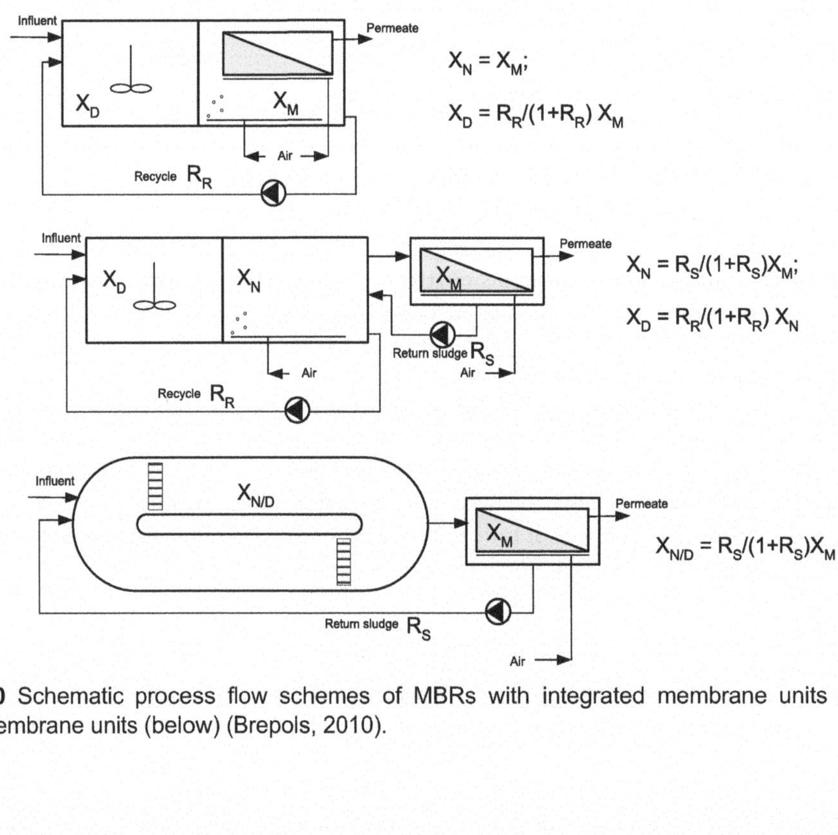

Figure 3.10 Schematic process flow schemes of MBRs with integrated membrane units (above) and separate membrane units (below) (Brepols, 2010).

Here (X_N) and (X_{DN}) represent the biomass concentration in the aerobic nitrification and the anoxic denitrification basin and ($M_{X,\text{aer}}$) and ($M_{X,\text{anox}}$) are the respective total amount of biomass operated under aerobic and anoxic conditions.

3.4.6.2 Plants for nitrification and denitrification

Reuse applications might require a high degree of nitrogen removal. Activated sludge plants for nitrification and denitrification can be designed based on a SRT (θ_x) which is required to achieve the target effluent quality (Figure 3.11).

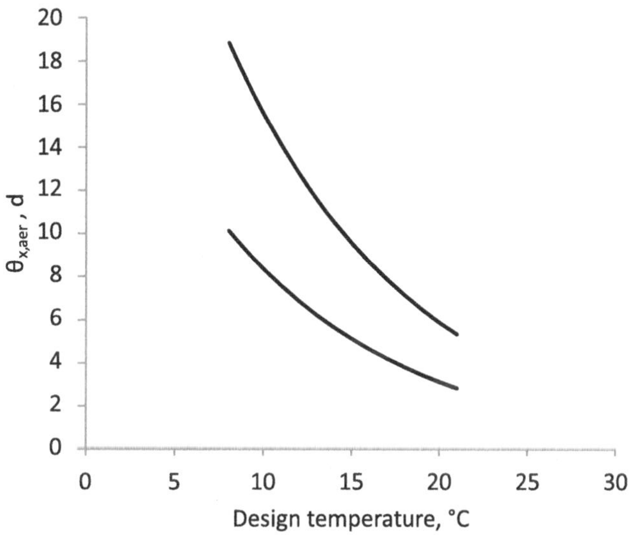

Figure 3.11 Aerobic SRT required for full nitrification (DWA, 2016).

The required aerobic SRT ($\theta_{x,\text{aer}}$) is decisive for the nitrification process. It can be calculated based on the design value of maximum net growth rate of the nitrifying bacteria ($\mu_{\text{A,max}}$) at the required design temperature and a process factor PF, which accounts for the attained ammonia effluent concentration, load and inflow variations or possible inhibitory effects. Figure 3.12 shows a range of resulting SRTs. According to DWA (2016), the process factor ranges between 1.5 and 2.8 while the safety factor SF is given at 1.6 and $\mu_{\text{A,max}}$ is 0.47 d^{-1} at a temperature T of 15°C.

$$\theta_{x,\text{aer}} = PF \cdot \frac{1}{\mu_{A,\max}} \cdot SF \cdot 1{,}103^{(15-T)} \tag{3.6}$$

To achieve denitrification, an additional portion of biomass is operated under anoxic conditions. The required portion of anoxic biomass ($M_{X,\text{anox}}$, kgMLSS) is proportional to the total amount of removed nitrous oxide ($M_{\text{NO}x}$, in kgNO_X/d), which can be calculated based on mass balances of the amount of removed nitrogen (M_N). Other process conditions that have to be evaluated when determining the nitrogen removal rate ($r_{X,s}$, kgNO_X/kgMLSS/d) include: temperature, nitrate, substrate and oxygen concentrations in bioreactor influent, recycle rates and efficiency of the chosen process (Henze *et al.*,

2001). The resulting value of $r_{X,s}$ is usually determined iteratively with the help of tabulated values or computer-aided design.

$$M_{X,\text{anox}} = \frac{M_{NOx}}{r_{X,s}} \tag{3.7}$$

As an example, a range of resulting $M_{X,\text{anox}}/M_{X,\text{aer}}$ values for municipal wastewater of usual composition are shown in Figure 3.12 for a pre-denitrification process and intermittent or simultaneous nitrification. The biomass ratio is dependent on calculated mass balances for nitrogen, oxygen and the amount of available carbonaceous substrate as a fraction of COD_{tot} (DWA, 2016). The concentration of nitrogen which has to be denitrified $NO_{x,DN}$ is balanced against the concentration of COD_{DN} which is available for denitrification.

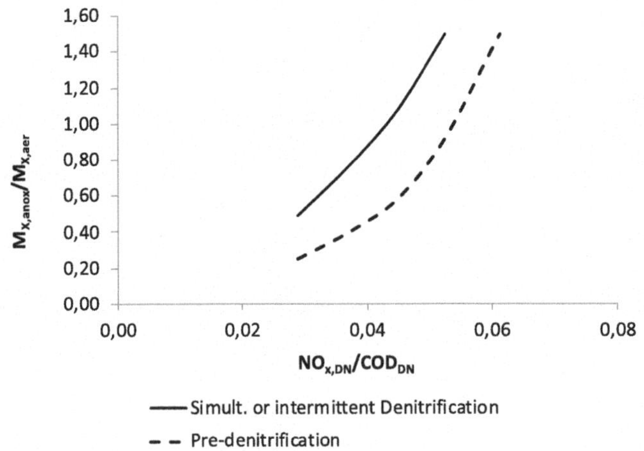

Figure 3.12 Anoxic biomass ratio required for denitrification (DWA, 2016).

The overall required minimum SRT (θ_x) can hence be calculated as

$$\theta_x = \left(1 + \frac{M_{X,\text{anox}}}{M_{X,\text{aer}}}\right) \cdot \theta_{X,\text{aer}} \tag{3.8}$$

The required recycle rate for pre-denitrification can be calculated from the ratio of denitrified and effluent NO_X concentrations.

$$R_{\text{DN}} = \frac{NO_{X,\text{DN}}}{NO_{X,\text{eff}}} \tag{3.9}$$

At least for large MBRs such preliminary designs should be verified with the help of dynamic simulations of the activated sludge process, to gain more insight into process dynamics and limitations of biological treatment efficiency. Otherwise, safety margins might be applied to avoid insufficiently low HRTs or to prevent adverse effects from oxygen carryover to denitrification reactors. Chapter 9 provides a comprehensive overview of activated sludge modelling of MBRs.

3.4.6.3 Activated sludge production

The SRT is further required to determine the excess activated sludge production (F_{SP}) from which finally the overall amount of MLSS can be calculated.

$$M_X = M_{X,\text{aer}} + M_{X,\text{anox}} = \theta_X \cdot F_{SP} \tag{3.10}$$

Sludge production is the sum of biological growth ($F_{SP,\text{bio}}$), including the inflow of inert matter, and sludge from an eventual chemical phosphorous precipitation ($F_{SP,P}$). Figure 3.13 shows an example for the estimated specific biological sludge production ($sF_{SP,\text{bio}}$) based on SRT, and inflow concentrations of COD and TSS for usual municipal wastewater. Detailed calculation based on the fractionation of influent COD and TSS can be made according to DWA (2016) or other references (Henze *et al.*, 2008; Tchobanoglous *et al.*, 2003).

$$F_{SP} = F_{SP,\text{bio}} + F_{SP,P} \tag{3.11}$$

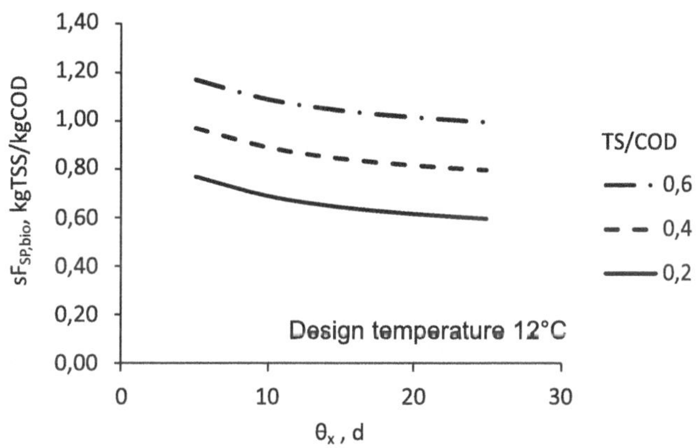

Figure 3.13 Example of specific biomass production at a wastewater temperature of 12°C for municipal wastewater (DWA, 2016).

Past experiences from full-scale MBRs have shown no significant differences in sludge production compared to CAS at similar SRTs (Brepols, 2010; Judd, 2011). Accordingly, Lesjean *et al.* (2011) have reported that, at least for municipal applications and common SRTs, reduced sludge production by MBRs is a misconception. In fact, the actual sludge extraction can even be slightly higher because the small portion of biomass that does not settle in the secondary clarifier, and is thus discharged with the CAS effluent, is retained by the membrane in MBR (Lesjean *et al.*, 2011).

3.4.7 Process units: membrane unit design

Primarily, membrane filtration at an MBR plant can be seen as a means of phase separation between treated wastewater and biomass. The design and layout of the filtration step thus influences the design of the biological treatment step. It can be recommended to carry out the design of the membrane filtration unit prior to the design of the biological treatment step because the MLSS concentration in the membrane unit and the required recycle rates directly influence the bioreactor design (Brepols *et al.*, 2010; DWA, 2011; Judd, 2011).

Table 3.9 shows a range of typical design values for municipal and industrial applications. It has to be acknowledged that these values should only be used as rough estimates when drafting a preliminary design (Table 3.11).

Table 3.11 Typical design values for iMBRs (Brepols, 2010; Brepols *et al.*, 2017a; Drensla & Janot, 2017; Judd, 2011; Pinnekamp & Friedrich, 2006).

Parameter	Symbol	Unit	Value (Range)
Flux	$J_{net,design}$	$L/(m^2\ h)$	15–40
Membrane specific footprint	F_S	m^2/m^2	70–165
Immersion depth of membrane filter modules	y_i	m	2.5–5.0
Membrane area specific aeration demand	SAD_m	$N\ m^3/(m^2\ h)$	0.3–1.5
Specific energy demand for membrane aeration	$SEDA_m$	kWh/m^2	0.3–0.6

3.4.7.1 Design flux and required membrane area

Variations in temperature, loading or biomass properties affect membrane filter performance. However, at most applications and under most circumstances it is not acceptable to reduce plant inflow to compensate for an involuntary loss in membrane performance. On the other hand, installation and maintenance of excess membrane surface as a precaution is costly. As a consequence, the determination of the critical design conditions and proper dimensioning of the required membrane surface area need to be carried out with care and prudence. Identifying a sustainable design flux (J'_{design}) is a key to facilitate long-term safe and stable membrane operation. Typical and critical operating conditions should be identified and considered to calculate the required membrane surface (Pinnekamp, 2006; Wedi & Joss, 2008).

The basic membrane process properties can be described as follows. According to Darcy's law the membrane flux (J') can be described by pressure difference as the driving force (Δp), the filtration resistance (R_{tot}) and the temperature-related viscosity of the water (η_T).

$$J' = \frac{Q}{A} = \frac{\Delta p}{\eta_T\, R_{tot}} \tag{3.12}$$

Although dynamic viscosity of mixed liquor in a bioreactor as well as supernatant can be somewhat different from the clear water values, it is often assumed that membrane permeability is directly proportional to the dynamic viscosity of clear water which is a function of temperature that can be approximated by the following equation (Judd, 2011). Judd (2011) cites typical values for (η_{20}) between 1.015 and 1.040 with 1.024 being the ASCE standard temperature correction factor of viscosity η at 20°C.

$$\eta_T = 1.024\ \eta_{20}^{(20-T)} \tag{3.13}$$

Total filtration resistance (R_{tot}) can be understood as a summation of all resistances in a row and includes intrinsic membrane resistance as well as cake layer and fouling resistance (Wintgens *et al.*, 2003). It is dependent on membrane properties but also on biomass properties and operational modes. Filtration resistance is not constant over time. With on-going operation, a build-up of filtration resistances occurs, which is also influenced by a superposition of temperature related or seasonal effects (Lyko *et al.*, 2008; Wintgens *et al.*, 2003).

Actually, membrane performance is influenced by a combination of a number of factors including: (a) operating conditions such as temperature, SRT and biomass composition, (b) wastewater properties, and (c) membrane cleaning regimes as well as membrane materials and module construction (Bütehorn, 2010; Judd, 2011; Lyko *et al.*, 2008; Pan *et al.*, 2010; Van den Broeck *et al.*, 2012; Yan *et al.*, 2012). In practice, these factors can be subject to seasonal or long-term variations. Furthermore, ageing of materials and equipment contribute to membrane performance decay. Safety measures should be taken in order to prevent critical situations or in the worst case an irreversible deterioration of membrane performance (Brepols, 2010).

For example, membrane performance can deteriorate over time at unsustainably high fluxes. On a large pilot MBR operated on municipal wastewater, Guglielmi *et al.* (2007) examined the long-term behaviour of flux at different flow rates, to determine for how long a certain flux can be sustained without a significant (and unsustainable) increase in TMP. As illustrated by Figure 3.14, the sustainable time span (t_{sust}) decreases with an increasing flux. In this case study, chemical cleanings were only carried out between the test trials. The authors have defined t_{sust} as the last point at which a correlation factor for an exponential fitting of the TMP curve is more than 95%. The observations were made with a FS membrane. Guglielmi *et al.* (2007) also verified the results on a large pilot-scale MBR equipped with a HF UF membrane at imposed average fluxes between 10 and 34 L/(m^2 h).

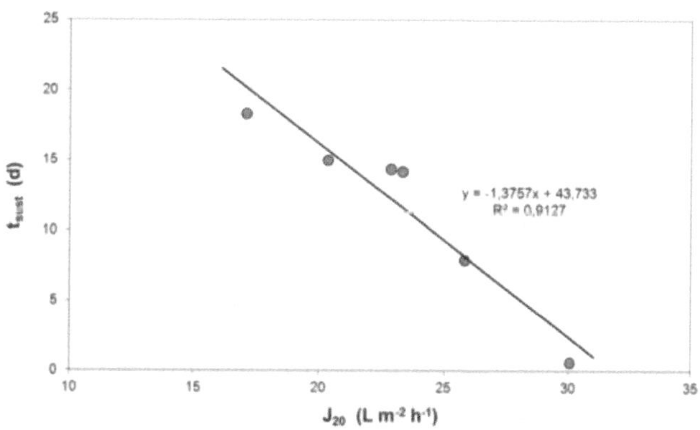

Figure 3.14 Values of t_{sust} at different sub-critical fluxes ($SAD_m = 0.35$ N m^3/(m^2 h); $X_M = 7.7$ gMLSS/L; SRT = 20 d for a municipal MBR pilot (Guglielmi *et al.*, 2007).

Numerical relationships between peak-flow duration, design temperature and design flux are in most cases proprietary knowledge of membrane suppliers or need to be investigated via pilot trials which unfortunately carry a risk of underestimating later scale-up effects (Kang *et al.*, 2008). It has to be acknowledged that unsubstantiated high design flux or lack of prudence in determining peak flow conditions may lead to sustained operational problems or a significantly reduced membrane service life. This is the reason why many full-scale MBRs have been designed with rather conservative estimations of the design flux (Judd, 2011; Verrecht *et al.*, 2010b). Figure 3.15 exemplifies a relation between flow conditions, their relative occurrence and design flux.

A case-sensitive design for a typical range of critical operating conditions is highly recommendable. Membrane surface area (A_M, m^2) is then calculated based on design flow (Q_{design}, L/h) and respective design net flux ($J_{net,design}$, L/(m^2 h)) for sets of process conditions determined by maximum inflow, average flow for distinctive periods and the respective water temperatures. This procedure was originally suggested by Wedi and Joss (2008). Although the context illustrated by Figure 3.15 applies generally, in practice detailed calculations in many cases require proprietary knowledge of membrane suppliers on performance characteristics of their actual type of membrane filters.

$$A_{\mathrm{M}} = \frac{Q_{\mathrm{design}}}{J'_{\mathrm{net,design}}} \tag{3.14}$$

Furthermore, scenarios for downtime of filter lines due to maintenance, cleaning and inspections, or even eventual replacement of filters or relevant parts should be carefully considered in determining the required membrane surface as well as number and size of individual filtration lines and compartments.

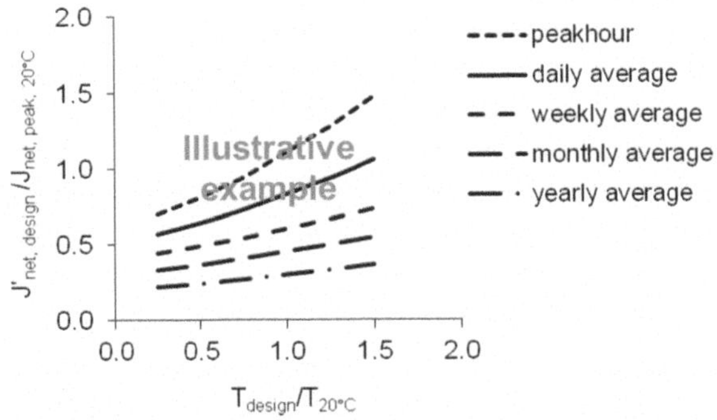

Figure 3.15 Membrane design flux, operating temperature and flow conditions.

3.4.7.2 Filtration compartment

The volume of the filtration compartment of an iMBR (V_{M}) is determined by the required membrane surface, the packing density of the membrane modules, characterized as specific footprint (F_{S}, $\mathrm{m}^2/\mathrm{m}^2$) and their required immersion depth (y_i, m). In addition to the mere membrane modules, filtration compartments may also be equipped with bearing constructions and inflow, outflow, permeate and aeration pipes. In practice, an additional volume (V_{inst}) is required to install ancillary equipment but also to facilitate proper biomass circulation around the membrane modules. The volume of the membrane compartment thus can be determined as shown below.

$$V_{\mathrm{M}} > \frac{A_{\mathrm{M}}}{F_{\mathrm{S}}} \cdot y_i + V_{\mathrm{inst}} \tag{3.15}$$

On the basis of suggestions of Brepols (2010) some main objectives in filtration tanks design are:

- Occasional but regular visual inspections of submerged filtration units are inevitable to detect failures like broken or clogged aerators or braiding and clogging of modules. This can be either facilitated by having a crane to remove the filtration units from the basin or a basin that can be easily emptied and accessed by an operator.
- Inflow and outflow of the membrane compartments should be designed to facilitate a complete distribution of biomass along the membrane modules, avoid stagnant zones in the reactor and avoid biomass accumulation on the membrane filters and sedimentation inside the tank.
- Often the geometry of the tanks can only harbour a single type of commercially available membrane modules. This reduces the flexibility in later membrane replacement. Later replacements could eventually be anticipated by constructions that allow easy adjustments to suit other products.

- When chemical membrane cleaning is carried out inside the filtration tanks, the material of the tanks should be sufficiently resistant to chemical corrosion.

However, filtration compartment design requires thorough consideration. For example, the filtration compartment may also serve as the plants bioreactor. In this case, its volume might be substantially larger than the volume required to harbour the filtration modules. The reactor is then eventually equipped with mixers, recycle pumps or fine-bubble aerators. Operation may result in complex and alternating flow patterns inside the reactor which potentially interfere with the intended flow patterns on the membrane modules. Unintended deposition of solids within the modules, physical wear of membrane modules and frames can be a result leading to increased fouling (Brepols, 2010). Besides that, physical wear can also be a reason for reduced membrane life time (Ayala *et al.*, 2011). Additional equipment like shrouds, baffles or plates may be a way to protect membrane modules against undesirable currents in the bioreactor. But then again these have to be designed to avoid stagnant biomass and accumulation of debris at the membrane modules (Brepols, 2010).

3.4.7.3 Fouling control and mitigation

Membrane fouling can be seen as a consequence of interactions between biological processes and membrane filtration. As the year-long research on membrane fouling has revealed, it is a highly complex phenomenon (Drews, 2010; Judd, 2011), which is scrutinized in more detail in Chapter 4. From a practical viewpoint some aspects shall be highlighted in the following:

Not every decline in membrane performance which can be observed at full-scale MBR installations is genuine membrane fouling. Sections 3.5.1 and 3.5.2 give some insight into potential sources of membrane failure.

In order to reduce and mitigate membrane fouling, in practice it is often vaguely recommended to aim at a 'good quality' of the biomass. It can be supposed that a sufficient HRT, a medium-to-long SRT, a high degree of nitrogen removal and a diverse microbial population are key constituents of such a good quality biomass (Al-Halbouni *et al.*, 2008; Lyko *et al.*, 2008; Van den Broeck *et al.*, 2012; Wagner, 2005; Wang *et al.*, 2008; Zhang *et al.*, 2006). Deflocculating biomass by high mechanical shear through pumps, mixers or aerators should be prevented. Extensive biomass foaming or bulking should best be reduced by ways of appropriate process design (Pan *et al.*, 2010; Prendl and Kroiß, 1998). The use of anti-foaming agents should be approved by the membrane supplier, because their ingredients might be incompatible to the membranes materials and thus eventually yield adverse effects on membrane performance (Brepols, 2010).

Hydrodynamics at the membrane modules must be seen as another key factor in fouling mitigation even if not all effects are yet fully understood (Böhm *et al.*, 2012). Numerous hydrodynamic studies have been carried out to gain more insight into the flow patterns of MBRs and membrane modules mostly at lab- or pilot scale or by using computational fluid dynamics modelling (Böhm *et al.*, 2012; Brannock *et al.*, 2009; Ratkovich *et al.*, 2012; Saalbach and Hunze, 2008; Tacke *et al.*, 2008; Wang *et al.*, 2008). Unfortunately, there seems to be a significant difference in lab-scale observations and full-scale operation. Several studies (Bütehorn *et al.*, 2012; Kang *et al.*, 2008; Liu *et al.*, 2010) revealed difficulties in transferring experiences from membrane module design at lab- or pilot-scale to full-scale applications.

3.4.7.4 Membrane cleaning

Membrane cleaning is an intrinsic and important part of the MBR process. Reliable and sufficient cleaning is a key to maintain stable membrane performance. Not only does the decisive remedial measure against genuine membrane fouling but it also cures adverse effects from unintended clogging, sludging or braiding of the membrane modules. Designing the membrane cleaning equipment, therefore, requires particular attention.

Suitable cleaning methods and concentrations of cleaning agents depend on the type of membrane modules, the installation and the nature of the wastewater/biomass. Membrane suppliers regularly give recommendations for the cleaning procedures which are often tied in with warranty limitations in case certain cleaning intervals or concentration limits are violated.

However it remains in the responsibility and the core interest of the plant designer and operator to achieve stable operation through appropriate precautionary cleaning (Drensla, 2011). Providing equipment and strategies for remedial cleaning in case of operating disruptions can be recommended (C. Brepols *et al.*, 2008). Different types of cleaning equipment for HF membranes have been described by Brepols (2010).

Cleaning methods can be categorized into mechanical and chemical cleaning methods, although practical cleaning procedures regularly are combinations of elements from both categories. Following cleaning processes are regularly employed.

Mechanical cleaning

- Cross-flow (continuous)
- Air scouring (continuous or periodic)
- Backwashing (periodic)
- Relaxation (periodic)

Air scouring, backwashing and relaxation are carried out automatically continuously or periodically.

Chemical cleaning

- CIP, cleaning-in-place also known as maintenance cleaning or chemically enhanced backwash at intervals depending on process properties (e.g. often once or twice a week at municipal MBRs)
- COP, cleaning out-of-place or out-of-process also known as intensive cleaning, here intervals may range from once in several months to once in several years.

CIP can be used to prevent short-term decline of permeability due to cake-layer formation and reversible fouling. On larger time scales of several months or even years, permeability may decline to unsustainable levels by heavier forms of reversible fouling, clogging or sludging of membrane modules. Intensive COP is then carried out either as a preventive cleaning or as a relieve measure in terms of a filtration failure. It includes all of the above-mentioned methods, which are then often carried out at higher concentrations of cleaning agents, longer durations and higher water temperatures to increase the cleaning effect. If necessary, the intensive cleaning can incorporate also a manual mechanical cleaning to remove sludge and braids from the filtration modules. Besides aiming at high cleaning efficiency, a superior strategy should be developed with a view towards fulfilling the following objectives (C. Brepols *et al.*, 2008):

- minimizing emission of cleaning chemicals and by-products in the effluent,
- being compatible with the whole treatment process,
- reducing the risk of damaging the membranes,
- minimizing plant downtime during cleaning,
- lowering labour input,
- reducing chemical cost,
- avoiding any hazards to plant operators.

Diverse chemical cleaning agents are used in sequence or in alternating order to achieve different cleaning effects (see Table 3.10). Storage, handling and dosing of these chemicals has to be designed in accordance with substance properties and with respect to safety guidelines.

Cleaning equipment can either be designed for in-line dosage of cleaning agents into the permeate system during backwash or with separate tanks for preparing the cleaning agents prior to the backwash. In any case the unintentional mixing of acidic and alkaline cleaning agents has to be avoided. In most cases permeate is used for backwashing. The volume of the permeate storage needs to be designed with respect to the permeate consumption for backwashing, cleaning and other operational purposes (Brepols, 2010).

3.4.8 Process units: aeration

Aeration design is usually evaluated for different loading conditions and operating temperatures in order to find the determining values for the oxygen demand and supply. The oxygen demand (M_O, kg/d) of the process can be calculated based on the following empirical equation, including removal of organic substances, endogenous respiration, nitrification and denitrification. The following equation is adapted from DWA (2016) and relies mainly on a mass balance for the amount of COD which is degraded biologically (referred to as $M_{COD,bio}$).

$$M_O = M_{COD,bio} + 4.3M_{NOx} - 2.9O(M_{NOx} - M_{NOx,eff}) \tag{3.16}$$

On the supply side, two aspects need consideration, the bioreactor aeration and the oxygen carryover from the membrane air scouring. It has been widely reported that, compared to the CAS process, the main difference in MBR aeration design comes from elevated MLSS concentrations in MBRs, which influence the oxygen transfer coefficient α. Various authors (see Table 7.11) have determined correlations between MLSS concentration and α-coefficient described by the following equation.

$$\alpha = a \cdot e^{-\omega \cdot MLSS} \tag{3.17}$$

Figure 3.16 illustrates observed values of the α-coefficient for the fine bubble aeration at a small municipal MBR (design capacity of 3.000 P.E.) at different MLSS concentrations. The error bar indicates the minimum and maximum values. It has to be acknowledged the in practice α-coefficients show considerable variations even at constant MLSS concentrations. Oxygen transfer is subject to a large number of influences (Cornel *et al.*, 2003; Germain *et al.*, 2007; Henkel *et al.*, 2011) and MLSS concentration is thus not much more than an auxiliary quantity to get an estimate for the design α-coefficient.

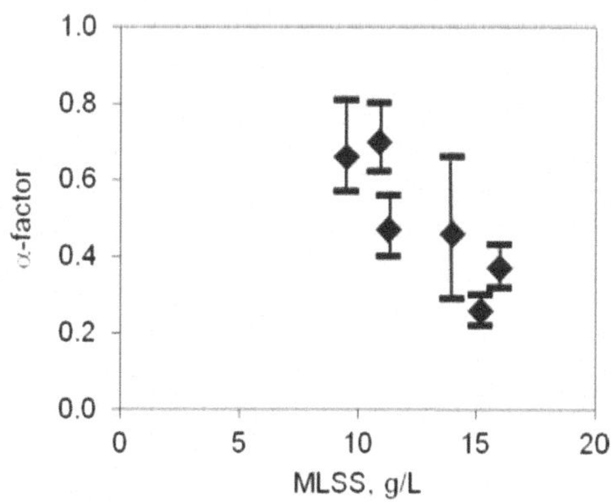

Figure 3.16 Observed α factor at different MLSS concentrations for fine bubble aeration at Rödingen MBR, Germany (Erftverband, 2004).

Furthermore, air scouring of the membrane module leads to high oxygen concentrations in the filtration compartments. This oxygen load is eventually fed back to the bioreactor with the sludge recycle and is partially lost with the effluent. Figure 3.17 exemplifies how oxygen carryover increases proportionally to the return sludge rate. Values in the figure were obtained during a measurement campaign at an MBR plant with HF modules installed in separate filtration tanks (Brepols, 2010).

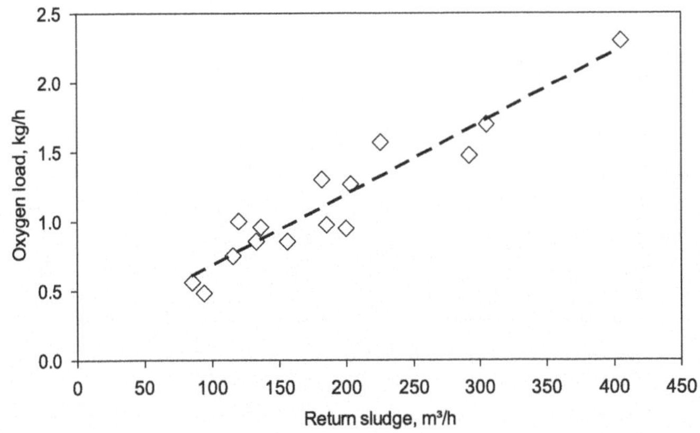

Figure 3.17 Example of the recycled oxygen load at Rödingen MBR, Germany (Brepols, 2010).

The oxygen transfer rate (OTR) of the aeration system has to be designed to meet the oxygen demand of the biological treatment. In the design of the aeration system, following values have to be calculated: (1) the oxygen supply by coarse bubble membrane air scouring and (2) the resulting oxygen carryover to the bioreactor and (3) the required additional bioreactor aeration. Iterative calculations are required to determine the actual layout.

Based on literature, an aeration energy model for MBRs, which is intended for application in ASM and BSM simulations of the wastewater treatment process was developed (Maere *et al.*, 2011). The model will be described below.

The oxygen transfer rate (OTR) is the product of airflow (Q_A), the density of air (ρ_A), the mass percentage of oxygen ($O_{A,m}$) in air and the actual oxygen transfer efficiency (AOTE).

$$\mathrm{OTR} = Q_A \cdot \rho_A \cdot O_{A,m} \cdot \mathrm{AOTE} \tag{3.19}$$

The AOTE is calculated from the standard oxygen transfer efficiency with the help of several process-related parameters, which include operational and design values like aerator depth y, dissolved oxygen concentration C, and the alpha factor which is dependent on the TSS concentration. The equations have to be solved iteratively.

$$\mathrm{AOTE} = \mathrm{SOTE} \cdot y \cdot \frac{(\beta \cdot C^{*,\mathrm{av}}(T) - C)}{C^{*}(20)} \cdot \varphi^{T-20} \cdot \alpha \cdot F \tag{3.20}$$

$$\alpha = e^{-\omega \cdot \mathrm{TSS}} \tag{3.21}$$

$$C^{*,\mathrm{av}}(T) = \frac{1}{2} \cdot C^{*}(T) \cdot \left(\frac{P_d}{P_{\mathrm{atm}}} + \frac{O_{\mathrm{out}}}{O_{A,v}}\right) \tag{3.22}$$

$$P_d = P_{\mathrm{atm}} + \rho_{\mathrm{sludge}} \cdot g \cdot h \tag{3.23}$$

$$O_{\mathrm{out}} = \frac{100 \cdot O_{A,v} \cdot (1 - \mathrm{AOTE})}{(1 - O_{A,v} \cdot \mathrm{AOTE})} \tag{3.24}$$

The model parameters provided in Tables 3.12 and 3.13 can be considered average values or at least within the range of reported literature values. They might be of use for the design of the bioreactor aeration and the assessment of the contribution of the membrane aeration to the biological process as well. It has to be acknowledged that the actual values for any given MBR system can be different from those in the table and therefore have to be assessed separately. Because of their different properties, coarse bubble aeration in membrane compartements and fine bubble aeration in bioreactors need to be calculated separately (Table 3.14).

Table 3.12 Overview of different commercially available cleaning agents and their cleaning effects.

Category	Chemical	Main Effect
Acidic	Citric acid, HCl, HNO_3	Solubilization, chelation
Alkaline	NaOH	Hydrolysis, solubilization
Oxidative	NaOCl, H_2O_2, peroxyacetic acid	Oxidation, disinfection
Detergents	Surfactants, chelating agents, enzymes	Emulsification, dispersion, chelation, surface conditioning

Table 3.13 Correlations between MLSS concentration and α-coefficient.

Reference	*a*	$-\omega$
Muller *et al.* (1995)	1.507	−0.045
Günder (2001)	1.000	−0.083
Krampe & Krauth (2003)	1.000	−0.088
Germain *et al.* (2007)	4.318	−0.233
Krause (2005)	1.000	−0.056

Table 3.14 Oxygen transfer design parameters (Maere *et al.*, 2011).

Parameter		Unit	Value
β	Salinity-surface tension correction factor	–	0.95
F	Correction factor for fouling of air diffusors	–	0.9^a–0.7^b(1.0 for clean)
g	Gravitational acceleration	m/s^2	9.81
$O_{A,m}$	Mass percentage of oxygen in air	%	23.2
$O_{A,v}$	Volume percentage of oxygen in air	%	21
P_{atm}	Atmospheric pressure	Pa	101,325
ρ_A	Density of air at standard conditions	g/m^3	1,200
ρ_{sludge}	Density of sludge	kg/m^3	1,000
SOTE	Standard oxygen transfer efficiency	%/m	2^a–6^b
T	Temperature of the mixed liquor	°C	5–20
y_i	Aerator immersion depth	m	2.5–5
φ	Temperature correction factor for oxygen transfer	–	1.024
ω	α-factor exponent coefficient	–	0.05^a–0.083^b

[a]Coarse bubble.
[b]Fine bubble.

3.4.9 Process units: automation

Due to the sophistication of the process and the complexity of filtration routines, MBR operation relies on automation more than conventional WWTPs (Ferrero *et al.*, 2012). Figures 3.18 and 3.19 present classifications of the automation of the filtration and the biological process provided by Ferrero *et al.* (2012).

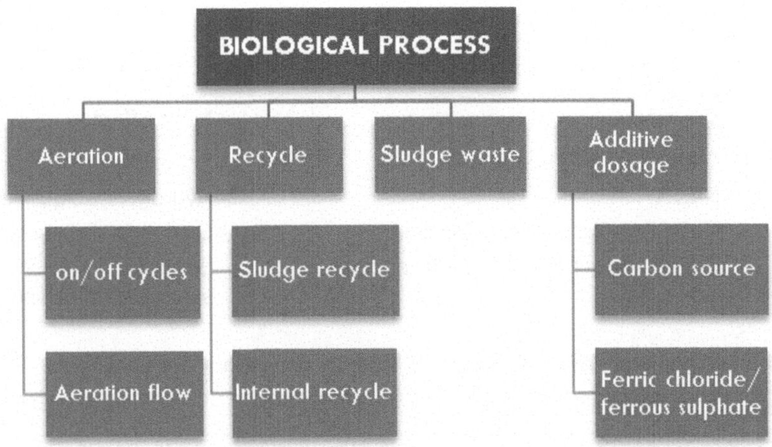

Figure 3.18 Classification of control actions for biological process control (Ferrero *et al.*, 2012).

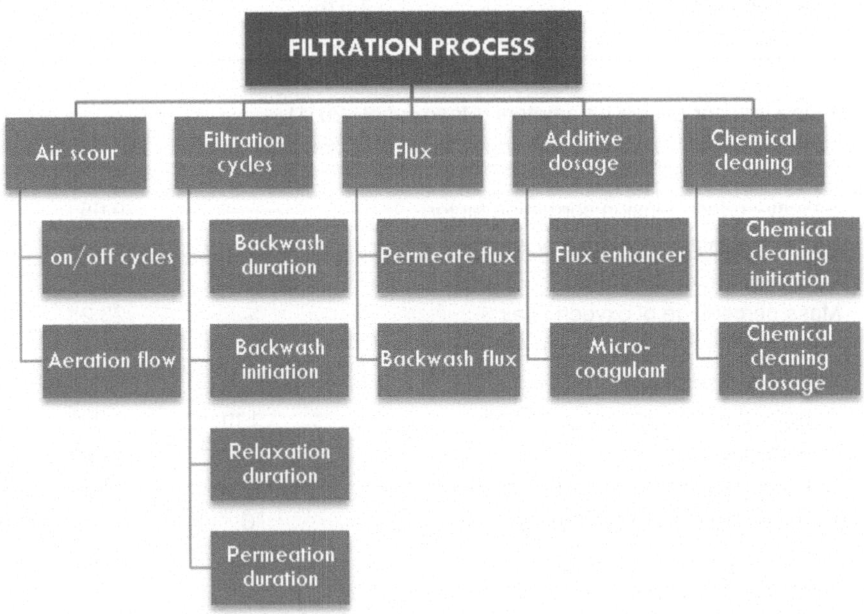

Figure 3.19 Classification of control actions for filtration process control (Ferrero *et al.*, 2012).

Several authors (Fenu *et al.*, 2010; Lazarova *et al.*, 2012; Veltmann *et al.*, 2011) have described the importance of optimized automation for the energy-efficient operation of MBRs. Maere *et al.* (2011) have described the use of BSM models to develop and test automation routines already during early stages of the design process. Still the knowledge on application, operation, maintenance and control of MBRs has not reached a degree of outright maturity. Therefore, automation concepts and programming of Supervisory Control and Data Acquisition (SCADA) systems and Programmable Logic Controllers (PLC) systems should allow for later optimizations of automation routines.

3.4.10 Energy consumption

Energy audits and estimations have shown that the dominant energy consumer is cross-flow aeration, which is used for cake layer and fouling control on the membrane surface (Akhondi *et al.*, 2014; Fenu *et al.*, 2010; Gil *et al.*, 2010; Mannina & Cosenza, 2013). However, the specific energy consumption for the filtration process has gradually decreased over the years by new technological development. In the 1960s, the early MBRs relied on pressure-driven cross-flow filtration. High flux rates, high transmembrane pressure and cross flow resulted in high specific energy consumption. For side-stream UF membranes, values around 3.5 kWh/m^3 have been reported (Tchobanoglous *et al.*, 2003). A decrease in production cost for membrane filters and new technical developments allowed the implementation of the immersed membranes, which have been applied for municipal wastewater treatment on a large scale since the early 1990s. Direct integration into bioreactors required larger membrane surfaces, which resulted in lower flux rates and thus lower transmembrane pressures. Air-sparging as an alternative cleaning mechanism, compared to conventional cross-flow filtration, turned out to be less energy intensive. Early full-scale MBRs relied on constant high air-flow rates regardless of the actual membrane flux, while later on intermittent cross-flow aeration was introduced, which immediately reduced the average aeration rate to almost 50% (Erftverband, 2004). Specific energy consumption for MBRs thus dropped down to values around 1.0 kWh/m^3 (Judd, 2011; Lesjean *et al.*, 2004). Developments in module constructions, membrane materials, air-sparging and operational regimes further lowered the energy consumption (Buer & Cumin, 2010; Judd, 2011; Lesjean & Leiknes, 2009; Miyoshi *et al.*, 2018). Many commercially available membrane filters nowadays use different aeration rates under different operational conditions such as high vs low inflow or filtration vs relaxation period. Control of optimized operational set points and automated flow distribution to parallel membrane units have been used to reduce energy consumption of MBRs as a whole (Brepols, 2012; Drensla & Janot, 2017). Specific values of <0.2 kWh/m^3 have already been reported for full-scale applications (Iglesias *et al.*, 2017).

Many factors contribute to the energy consumption of an MBR which can be influenced during the planning and building phase. Number and redundancy of membrane tanks and membrane filters, biological processes, biomass treatment, pumping and internal recycling of water offer trade-offs between operational comfort and energy consumption. Even at existing MBRs, a large number of measures can be taken to improve the energy consumption (Drensla & Janot, 2016, 2017; PUB, 2011; Sun *et al.*, 2016). For example, MLSS concentrations, bioprocess aeration and sludge recycle influence the energy demand of the biological process and the MBR as a whole (see Figure 3.20). To calculate and optimize MBR energy consumption, models have been applied (Maere *et al.*, 2011; Martin *et al.*, 2011; Sun *et al.*, 2016; Verrecht *et al.*, 2008, 2010a). A broader overview of practical measures to reduce energy consumption at existing installations is given in Section 3.5.

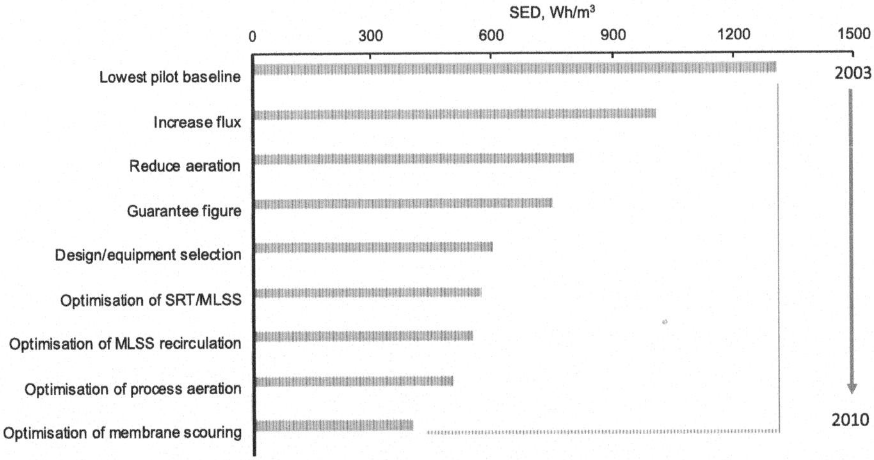

Figure 3.20 Step-by-step specific energy demand reductions for MBR from the start of the project (1300 Wh/m^3) to the optimized operation (370 Wh/m^3), adapted from PUB (2011).

Biological treatment in MBRs is often defined by simultaneous aerobic biomass stabilization, high SRTs and high MLSS concentrations. Only a few MBRs exists that use separate anaerobic biomass stabilization. These installations often resulted from conventional WWTP retrofits using MBR (e.g. Rietliau WWTP in Wädenswil, CH; Traverse City WWTP in Michigan, USA). A recent case study (Brepols *et al.*, 2017b) shows that separate anaerobic biomass treatment can also be used to retrofit an existing MBR. Here, energy consumption for the biological treatment step is reduced while electricity is produced on-site from digester gas using a combined heat and power (CHP) unit.

Energy consumption of MBR processes is not only an issue of cost but also of environmental footprint. For example, Ioannou-Ttofa *et al.* (2016) examined the carbon-footprint and environmental impact of an MBR based on a life-cycle approach. Results indicate that the majority of the environmental impacts of the MBR pilot unit can be attributed to indirect emissions, tracing back to electricity generation. The second main contributor to the total environmental footprint is identified to be the membrane units. Nonetheless, due to their high life expectancy, they have only a low contribution to the total environmental footprint. However, it should be highlighted that the total emissions of this unit operated for the treatment of 1 m^3 of urban wastewater corresponds to approximately 1% of the daily GHG emissions per capita, demonstrating thus the systems' environmental sustainability. The environmental impact and the overall sustainability of the MBR system are highly depended on the different origins of the electricity consumed. Energy production from renewable resources thus significantly lowers the environmental footprint of an MBR and/or material production (Ioannou-Ttofa *et al.*, 2016).

3.4.11 Cost evaluations

Life-cycle cost of MBR plants are mainly determined by three factors: investment cost, operating cost (personnel, energy, waste disposal) and cost for membrane replacement. Interactions between wastewater pre-treatment, biological treatment and membrane filtration result in a number of trade-offs within the cost of a particular plant design and later operation. Attempts to determine life-cycle cost of MBR plants and to develop elaborate cost models in order to assess the influence of several design options have been reported in a number of recent studies (Fletcher *et al.*, 2007; Friedler & Hadari, 2006; Krzeminski *et al.*, 2012; Maere *et al.*, 2011; Verrecht *et al.*, 2010b).

For the situation in Germany, Brepols *et al.* (2010) have shown that investment cost off full-scale MBR are similar or even lower than for CAS sharing comparable standards in equipment and treatment efficiency. Energy cost for CAS with tertiary treatment can also be in a range similar to that of MBRs at least for small installations. For reuse applications in Spain, Iglesias *et al.* (2017) reported that CAPEX of the MBR technology was found to be around 10% lower than that of a reuse installation based on conventional extended aeration with tertiary membrane polishing. MBR costs were, however, 30–34% higher than those associated with classical tertiary clarification by coagulation followed by sand media filtration and disinfection at capacities below 10,000 m^3/d. The cost benefits of MBRs increase with increasing plant size due to the decreased labour requirement, though the energy demand is higher than for conventional processes (Iglesias *et al.*, 2017).

It has often been assumed that short service-life of the filters increases life-time expenditure significantly. A few authors (Ayala *et al.*, 2011; De Wilde *et al.*, 2007) have tried to estimate the membrane life-time based on a projection of permeability decline or an assessment of fouling and module integrity; however, a wide range of values have been reported. Actual operational experiences today draw a different picture. Anecdotal evidence has it that some of the earlier full-scale MBRs have been operating with their original set of membrane modules for 10–15 years. For example, the membrane modules at Nordkanal MBR (Brepols, 2010; Brepols *et al.*, 2017a; Drensla & Janot, 2017) have been in continuous operation since 2004. Until 2018 less than 15% of the membrane modules have been replaced because of age-related defects while filtration performance of the remaining membranes is still sufficient. For HF membrane applications in North America also life times of over 10 years have been reported (Cote *et al.*, 2012). Although specific cost for membrane filters have declined remarkably during the last decade, membrane filters are still the single most expensive part of equipment. Membrane lifetime and cost for membrane replacement remain a key factor in the life-cycle costs, which can be influenced by good operational and design practice. More detailed information on life-cycle costs will be given in Chapter 8.

However, not only capital expenditure (CAPEX) and operational expenditure (OPEX), but also social, environmental, administrative and technical aspects should be considered in a decision-making process (see Table 3.15) (Bertanza *et al.*, 2017; Sun *et al.*, 2015; Zaerpour, 2014). To support the decision process, formalized multi-criteria decision methods have been explored in wastewater treatment and reuse (Benedetti *et al.*, 2010; Garrido-Baserba *et al.*, 2016).

Table 3.15 Evaluation criteria and methods.

CAPEX	OPEX	Socio-environmental Factors	Technical Factors
Pretreatment; MBR design; Membrane cost; Membrane replacement; Residuals treatment	Energy demand; Maintenance; Membrane cleaning; Residuals treatment and disposal; Water revenue	Effluent quality; Legislation; Footprint; Public acceptance; Noise pollution; Landscape integration; GHG-emissions	Space; General experience; Reliability; Variability; Robustness; Operator qualification
Methods for evaluation: Net Present Value (NPV) Return on invest (ROI) Effluent quality index (EQI) Weighed decision matrix Multi-criteria-decision-making method			

Table 3.16 Risk matrix for the reliability of operation of an iMBR (Lazarova, 2008).

Occurrence/Severity	Very Unlikely	Remote	Occasional	Probable	Frequent
Very high (catastrophic)	–	–	–	–	–
Critical	–	Failure of grease and oil removal	Membrane damage or heavy clogging	3 mm screen clogging	–
Major	Failure of PLC	Failure of chemical dosage (polymer, $FeCl_3$)	Fault of recirculation	Fault of membrane blowers and aeration	Clogging of 0.75 mm screen
Moderate	Failure of SCADA and HMI	Fault of inlet pumps	Failure of membrane auxiliary equipment	Fault of blowers and diffusers of aeration basins	Fault of PLC-link (modems, etc.)
Minor	–	Failure of buffer tank equipment	Failure of odour treatment	Failure of sludge treatment	–

3.4.12 Alternative MBR concepts

Although iMBR plays a dominant role in the application of MBRs, alternative MBR concepts exist and they also operate at industrial scale. These are discussed in Chapters 12 and 14. According to the properties of the process, design values and workflow can be different from what is described here, especially when different biological processes are employed. While moving-bed or biofilm bioreactors (MBBR-MBR) (Canziani *et al.*, 2006; Leiknes & Odegaard, 2006; Sun *et al.*, 2010; Zhang *et al.*, 2009) or pressure-driven cross-flow filtration systems (Canziani *et al.*, 2006) employ aerobic wastewater treatment processes their biomass properties and the related filtration performance can be entirely different. These processes can be used in treating special wastewater or wastewater components (Ahmed & Lan, 2012; Sun *et al.*, 2010). Furthermore, anaerobic MBRs constitute an entirely different process (Martin *et al.*, 2011; Pillay *et al.*, 2007; Wang *et al.*, 2018; Yan *et al.*, 2012; You *et al.*, 2010) that will be discussed in Chapter 11.

3.5 OPERATION AND PLANT MANAGEMENT

Although a higher level of technical sophistication is associated with MBRs, the technical, administrative and managerial practices required in maintaining and operating MBRs are based on well-known practices from CAS plant operations. Some noteworthy MBR specifics are mentioned below.

3.5.1 Membrane cleaning and maintenance

From the practical experience of operators, MBR maintenance policies rely on following key measures:

Membrane filtration

- Regular de-aeration of permeate collectors and modules
- Regular preventive maintenance cleaning (once per week) and intensive cleaning of filters (once every one to two years)

- Integrity testing of membrane modules (approximately once per year)
- Maintenance and repair upon failure of modules, cassettes, aeration hoses, permeate collectors
- Replacement of malfunctioning membrane modules

Peripheral membrane equipment

- Preventive maintenance of critical equipment subject to high wear (e.g. actuators for membrane aeration)
- Repair upon failure
- Replacement upon failure

Regular and efficient membrane cleaning is the backbone of sustainable operation (C. Brepols *et al.*, 2008; Stuber *et al.*, 2009; Zsirai *et al.*, 2012). Improper cleaning can lead to a critical loss of permeability within a relatively short time span of several weeks or months. In practice membrane units have been designed conservatively and membranes are operated at low fluxes (see Section 3.4.7) while then ineffective energy consumption is the reverse of the medal.

Cleaning intervals, cleaning agents and recipes depend on the type of membrane and their installation, as well as the treatment process and properties of the wastewater. It has to be acknowledged that a considerable effort is required to adjust and optimize chemical cleaning (C. Brepols *et al.*, 2008; Drensla, 2011, 2015). Proper dilution and mixing of the cleaning agents as well as the proper distribution along the membrane modules need to be scrutinized. Partial cleaning of membranes may soon lead to loss of actively operating membrane surface.

Air and gas may enter permeate collector pipes primarily by degassing of permeate under vacuum conditions on the permeate side but also through small fissures and leakages in joints and welding. As a result, air bubbles may become entrapped inside modules and piping and air cushions inside the piping build up which then impede filtration, backwashing and the proper distribution of cleaning agents. Regular de-aeration or ventilation of the permeate system is thus highly recommended, at least prior to every chemical cleaning (Drensla, 2011).

In the long run, residuals and biofilm also accumulate inside the membrane lumen, permeate collectors and pipes. These residuals cause unintentional and premature reactions of the cleaning agents. Cleaning results and efficiency should thus be evaluated regularly and the efficiency of cleaning procedures needs to be reconsidered time and again (Drensla, 2011; Lyko *et al.*, 2008).

Inspections of module integrity should also be carried out on a regular basis. For example, they can be carried out in combination with intensive cleaning, when modules are removed from the process or the filtration compartments are drained.

3.5.2 Process reliability

Generally, it is assumed, that membrane fouling can be controlled by appropriate design and operation. In many cases system failures are thus not primarily caused by membranes themselves but by their peripheral equipment. Based on full-scale data Lazarova (2008) has developed a matrix of system failures vs their severity as shown in Table 3.13.

Here, malfunctions of mechanical pre-treatment, SCADA and PLC systems or the human–machine interface (HMI) have been identified as causes of severe system breakdowns. Similar findings were reported by Judd (2011) based on a survey conducted among MBR operators. It can be seen that malfunctions of pre-treatment and sieves cause incomplete removal of debris, high wear and tear and unintended and regular breakdowns or overflows frequently cause major-to-critical disturbances to be the consequences. Therefore, failures on mechanical pre-treatment have a high potential to inflict subsequent membrane filtration. Even minor breakdowns of peripheral equipment may lead to cascades of aggravating malfunctions of the membrane filtration itself. For a number of potential initial failures, Judd (2011) and Brepols (2010) have formed lists of subsequent malfunctions and adequate countermeasures. The fishbone diagram in Figure 3.21 exemplifies causes that might deteriorate membrane filtration performance.

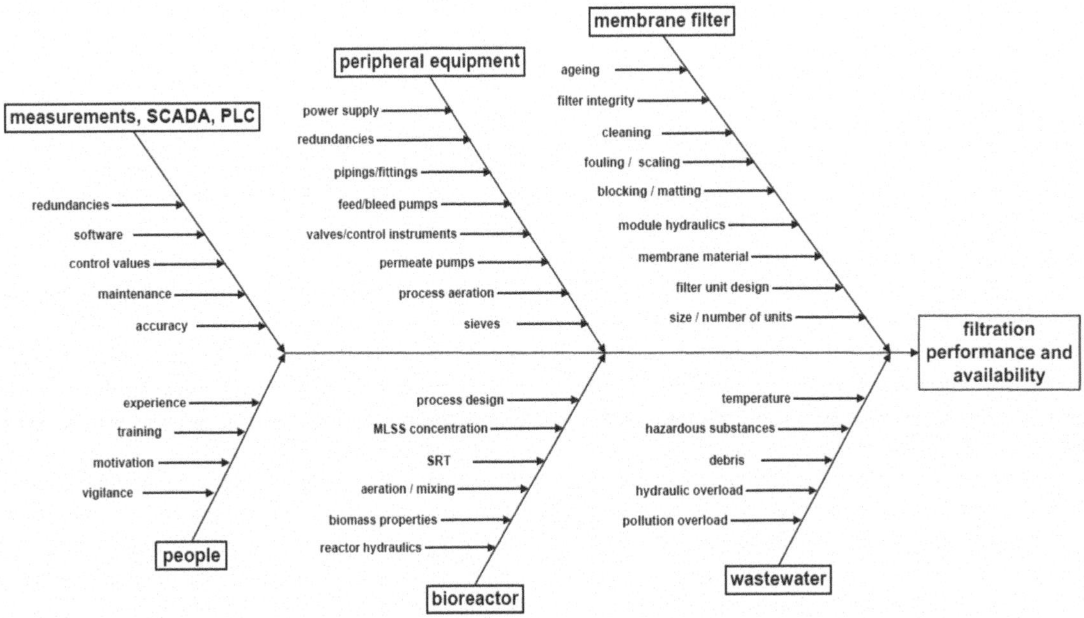

Figure 3.21 Cause–effect diagram for the decrease of membrane filter performance at a municipal iMBR.

3.5.3 Residuals and sludge management

Biomass production of full-scale MBRs is equal or at least slightly higher to that of comparable CAS installations (Lesjean *et al.*, 2011). Dewatering capability of excess biomass derived from full-scale MBR plants shows no significant differences to that from CAS operated at similar SRT and aeration conditions (Brepols, 2010).

Unlike CAS large amounts of residuals may come from wastewater pre-treatment especially debris from fine sieves. Sieving debris consist of pulp from toilet papers and similar sanitary products which are discarded with sewage. In case large amounts of these materials need to be handled, it has been shown that mixing these matters with excess biomass and treating it in combination can be an efficient form of disposal (Brepols, 2010).

Residuals from the chemical membrane cleaning often remain in the process after backwashing and flushing the membranes in situ. Residuals from intensive cleaning contain higher concentrations of cleaning agents. When contamination of the plant effluent with reaction by-products from the cleaning is undesirable, the cleaning residuals are temporarily stored on site, reused, disposed separately or returned to the plant influent (Brepols, 2010).

3.5.4 Energy improvements in existing MBRs

It can be seen that full-scale applications do not operate membrane filters with their optimal efficiency, as inflow variations result in operation at low flux rates or long periods of stand-by operation. As a consequence, ideal energy consumption values often cannot be achieved in the first place. Still there can be a huge potential to reduce the energy consumption of existing MBRs (Drensla & Janot, 2016; PUB, 2011; Sun *et al.*, 2016). For example, Drensla and Janot (2016) reported a significant cutdown of energy consumption at Nordkanal MBR (80,000 population equivalents) between 2010 and 2015: The MBR is in operation since 2004. Initially energy consumption remained stable at its original level. After several years of experience and more confidence in the operational stability of the filtration units, attempts were taken to divert from original operational set-points and improve control strategies of the biological treatment as well as the filtration process. A bundle of measures was then tested and implemented over several years.

- Increased instantaneous flux from 25 LMH to 30–40 LMH, increased duration of filtration intervals from 600 to 900 s, reduced duration of backwash, and relaxation instead of backwashing were measures that increased permeate yield while reducing cross-flow aeration by changing to a new air-cycling mode.
- Operation of bioreactors and filtration further improved by actively channelling inflow to dedicated treatment lines. As a result, active filtration lines operate at their optimal set-point while remaining lines rest in an energy-reduced stand-by mode.
- Bioreactor operation also contributed to energy savings. MLSS concentration could be lowered from its original design value of 12 gMLSS/L to less than 8 gMLSS/L without hampering effluent quality. Increased aeration efficiency by a higher α-factor and reduced oxygen demand for endogenous biomass respiration were observed.
- Finally, replacements of inefficient equipment like blowers, additional measurements and control units, and a re-programming of much of the existing SCADA system were undertaken.

As a result, total energy consumption of the WWTP including mechanical pre-treatment, biological treatment and on-site sludge processing was lowered from 1.0 to 0.65 kWh/m^3. Energy consumption of the filtration process decreased from 0.42 to 0.22 kWh/m^3 (Drensla & Janot, 2016, 2017). It is anticipated that energy consumption for the biological treatment can be further reduced. Nordkanal MBR will be refurbished with primary clarification and anaerobic sludge treatment. Energy demand of the bioprocess will thus be lowered further while more than 30% of the required electrical energy can be produced by an on-site CHP. Construction works started in February 2017 (Brepols *et al.*, 2017b).

3.5.5 Personnel and qualification

Failure safe and energy-efficient operation of MBRs requires skilled and motivated operators. Complexity and sophistication can easily overstrain operators that are not accustomed to the process. MBR operation thus requires theoretical and practical training and close supervision to increase operators knowledge and experience with special properties and control procedures of MBR systems to help them assess systems performance properly (Brepols, 2010). Crucial process factors need to be identified and appropriate actions need to be taken. MBR operation relies on automation. The number of automated parts of equipment is significantly higher than at many CAS. Basic performance data such as duration of filtration cycles, TMP, flux and permeability can only be observed indirectly via SCADA systems. At the same time, high-level automation and sophisticated and closed-control loops bear a risk of lowering the personnel's attention towards the entire process. Small deviations from good operating conditions may thus go unnoticed until problems aggravate critically. Complex automation routines such as chemical membrane cleaning for example can be a source of various malfunctions and should thus be surveyed by experienced operators.

3.6 SUMMARY

MBRs serve as an intermediate or final treatment step in applications for wastewater reuse. In many cases, MBRs already produce an effluent that is fit for high-quality reuse applications without further treatment. Generally, the design of MBRs integrated into reuse schemes is not different from that in non-reuse applications.

Appropriate mechanical pre-treatment is an intrinsic feature of good plant design and operations. It is important for successful and long-term operation.

MBR design requires careful consideration of wastewater composition, temperature and inflow variations, including eventual feed-back streams from subsequent or parallel treatment steps. Membrane units are designed with respect to wastewater temperature and frequency and duration of peak inflow conditions. Influent buffering thus can be used to adjust natural inflow to a more favourable inflow regime and opens opportunities to optimize CAPEX and OPEX for membrane filters. Membrane filter compartments should avoid zones with stagnant biomass and accumulating debris. Proper de-aeration of membrane modules and permeate pipes during operation is important to maintain process stability during filtration but also for proper and even distribution of cleaning agents during inevitable cleaning in place.

Scale-up from lab-scale or small-scale pilots to large installations with many membrane modules in a row requires prudence: membrane module hydraulics and membrane performance, distribution of flows, biomass, permeate and cleaning agents, aeration efficiency as well as wear and tear of actuators and equipment might be subject to scale-effects which can be difficult to predict.

Bioprocess design reflects influent loading and effluent requirements but also other environmental and technical constraints. Flow distribution, mixing, potential oxygen carryover from immersed membrane air scouring and effects of elevated biomass concentration on bioreactor aeration need to be considered during reactor design. A high-quality biological treatment is also a pre-requisite for safe and reliable filter operation.

MBRs are employed in ever-increasing numbers because of their obvious benefits such as high effluent quality, compactness and robustness. Nonetheless there are still perceived disadvantages which have limited a more widespread application of MBR technology. Energy consumption, complexity of process operation and control, membrane cost and reliability are still among the key arguments. Huge improvements have been achieved. Energy consumption of MBRs has been lowered significantly not only by new filter and module designs but also by improvements in bioprocess and reactor design, control and automation and optimizations of operational set-points for membrane filters and bioprocess alike. Engineers and scientists have contributed their experience to a good-practice of well-suited plant design, reliable automation, appropriate membrane cleaning and maintenance. Last but not least, skilled and motivated operators hold a key to efficient and successful MBR operation in the long term.

REFERENCES

Abegglen C., Joss A., McArdell C. S., Fink G., Schlüsener M. P., Ternes T. A. and Siegrist H. (2009). The fate of selected micropollutants in a single-house MBR. *Water Research*, **43**, 2036–2046. https://doi.org/10.1016/j.watres.2009.02.005

Ahmed F. N. and Lan C. Q. (2012). Treatment of landfill leachate using membrane bioreactors: a review. *Desalination*, **287**, 41–54. https://doi.org/10.1016/j.desal.2011.12.012

Akhondi E., Wicaksana F. and Fane A. G. (2014). Evaluation of fouling deposition, fouling reversibility and energy consumption of submerged hollow fiber membrane systems with periodic backwash. *Journal of Membrane Science*, **452**, 319–331. http://dx.doi.org/10.1016/j.memsci.2013.10.031

Alex J., Benedetti L., Copp J., Gernaey K. V., Jeppsson U., Nopens I., Pons M. N., Rieger L., Rosen C. and Steyer J. P. (2008). Benchmark Simulation Model no. 1 (BSM1). Report by the IWA Taskgroup on Benchmarking of Control Strategies for WWTPs.

Al-Halbouni D., Traber J., Lyko S., Wintgens T., Melin T., Tacke D., Janot A., Dott W. and Hollender J. (2008). Correlation of EPS content in activated sludge at different sludge retention times with membrane fouling phenomena. *Water Research*, **42**(6), 1475–88. https://doi.org/10.1016/j.watres.2007.10.026

Aquapolo [WWW Document]. (2016). http://www.aquapolo.com.br/ (accessed 04 December 2018).

ATV DVWK (Ed.) (2000). Bemessung von einstufigen Belebungsanlagen, ATV DVWK. GFA, Verl. für Abwasser, Abfall und Gewässerschutz, Hennef.

Ayala D. F., Ferre V. and Judd S. J. (2011). Membrane life estimation in full-scale immersed membrane bioreactors. *Journal of Membrane Science*, **378**, 95–100. https://doi.org/10.1016/j.memsci.2011.03.013

Benedetti L., De Baets B., Nopens I. and Vanrolleghem P. A. (2010). Multi-criteria analysis of wastewater treatment plant design and control scenarios under uncertainty. *Environmental Modelling & Software*, Thematic Issue on Modelling and Automation of Water and Wastewater Treatment Processes **25**, 616–621. https://doi.org/10.1016/j.envsoft.2009.06.003

Berkley C. (2007). Sewage Related Litter: Flushing Toilets Onto Beaches, Research Report, Environmental Campaigns, Wigan.

Bertanza G., Canato M., Laera G., Vaccari M., Svanström M. and Heimersson S. (2017). A comparison between two full-scale MBR and CAS municipal wastewater treatment plants: techno-economic-environmental assessment. *Environmental Science and Pollution Research*, **24**, 17383–17393. https://doi.org/10.1007/s11356-017-9409-3

Bérubé P. (2010). Membrane bioreactors: theory and applications to wastewater reuse, In: Sustainable Water for the Future: Water Recycling versus Desalination, I. C. Escobar and A. I. Schäfer (eds.), Chapter 9 Elsevier, Amsterdam, pp. 255–292.

Birks R. (2008). Community scale non-potable reuse in London using an MBR. *Water Practice & Technology*, **3**(2), https://doi.org/10.2166/wpt.2008.047

Boehler M., Joss A., Buetzer S., Holzapfel M., Mooser H. and Siegrist H. (2007). Treatment of toilet wastewater for reuse in a membrane bioreactor. *Water Science and Technology*, **56**(5), 63. https://doi.org/10.2166/wst.2007.557

Böhm L., Drews A., Prieske H., Bérubé P.R. and Kraume M. (2012). The importance of fluid Dynamics for MBR fouling mitigation. *Bioresource Technology*, **122**(10), 50–61. https://doi.org/10.1016/j.biortech.2012.05.069

Brannock M.W.D., De Wever H., Wang Y. and Leslie G. (2009). Computational fluid dynamics simulations of MBRs: Inside submerged versus outside submerged membranes. *Desalination*, **236**(1–3), 244–251. https://doi.org/10.1016/j.desal.2007.10.073

Brepols C. (2010). Operating Large Scale Membrane Bioreactors for Municipal Wastewater Treatment. IWA Publishing, London.

Brepols C. (2012). Performance improvement of full scale membrane bioreactors. In: Water-Energy Interactions in Water Reuse, V. Lazarova, K. Choo and P. Cornel (eds.), IWA Publishing, 127–137.

Brepols C., Schäfer H. and Engelhardt N. (2005). Recommendations on the process integration of submerged membrane filters in municipal membrane aeration plants (Hinweise zur verfahrenstechnischen Integration getauchter Membranfilter in kommunalen Membranbelebungsanlagen), KA.

Brepols C., Drensla K., Janot A., Trimborn M. and Engelhardt N. (2008). Strategies for chemical cleaning in large scale membrane bioreactors. *Water Science and Technology*, **57**(3), 457. https://doi.org/10.2166/wst.2008.112

Brepols Ch., Dorgeloh E., Frechen F.-B., Fuchs W., Haider S., Joss A., de Korte K., Ruiken C., Schier W., van der Roest H., Wett M. and Wozniak T. (2008). Upgrading and retrofitting of municipal wastewater treatment plants by means of membrane bioreactor (MBR) technology. *Desalination*, **231**, 20–26. https://doi.org/10.1016/j.desal.2007.11.035

Brepols C., Schäfer H. and Engelhardt N. (2010). Considerations on the design and financial feasibility of full-scale membrane bioreactors for municipal applications. *Water Science and Technology*, **61**(10), 2461. https://doi.org/10.2166/wst.2010.179

Brepols C., Drensla K., Janot A., Beyerle L. and Schäfer H. (2017a). Future perspectives for MBR applications at the Erftverband. In: Frontiers in Wastewater Treatment and Modelling FICWTM 2017. G. Mannina (ed.), Springer, 149–152.

Brepols C., Drensla K., Janot A., Beyerle L. and Schäfer H. (2017b). Future perspectives for MBR applications in municipal wastewater treatment (Zukunftsperspektiven für Membranbelebungsanlagen in der kommunalen Abwasserreinigung). Wassertechnologie in Der Wasseraufbereitung Und Abwasserbehandlung. Presented at the 12. Aachener Tagung Wassertechnologie, Aachen, 24–25, October 2017.

Brockmann M. (2008). Use of membrane processes for cycle closure using the example of the slaughterhouse (Einsatz von Membranverfahren zur Kreislaufschließung Am Beispiel des Schlachthofes) COOPERL Lamballe. GWF Special 149 (2008) Nr. 14, 55–58.

Brockmann M., Brüß U., Cornel P., Debus O., Kiefer M., Krause S., Kunz P.M. *et al.* (2007). Treatment of Industrial Wastewater and Process Water with Membrane Processes and Membrane Bioreactor Technology. Edited by Abwasser und Abfall Deutsche Vereinigung für Wasserwirtschaft. DWA-Topics. Hennef: DWA, November 2007.

Buer T. and Cumin J. (2010). MBR module design and operation. *Desalination*, **250**, 1073–1077. https://doi.org/10.1016/j.desal.2009.09.111

Busse R.-P. (2018). Membrane technology for Peru and Namibia (Membrantechik für Peru und Namibia). WWT 21.

Busse (Ed.), n.d. Referenz Abwasserrecycling: Colegio Johannes Gutenberg, Lima/Peru. http://www.busse-is.de/de/download/V-U-RB-DE_ReferenzAbwasserrecyclingGutenbergSchuleLima.pdf

Bütehorn S. (2010). Experimental and Numerical Investigation of the Hydrodynamics of Microfiltration Processes Using a Multi-Scale Approach. Dissertation, RWTH Aachen, Shaker Verlag, Aachen.

Bütehorn S., Brannock M., Le-Clech P., Leslie G., Volmering D., Vossenkaul K., Wintgens T., Wessling M. and Melin T. (2012). Limitations for transferring lab-scale microfiltration results to large-scale membrane bioreactor (MBR) processes. *Separation and Purification Technology*, **95**, 202–215.

Canziani R., Emondi V., Garavaglia M., Malpei F., Pasinetti E. and Buttiglieri G. (2006). Effect of oxygen concentration on biological nitrification and microbial kinetics in a cross-flow membrane bioreactor (MBR) and moving-bed biofilm reactor (MBBR) treating old landfill leachate. *Journal of Membrane Science*, **286**, 202–212. https://doi.org/10.1016/j.memsci.2006.09.044

Capodaglio A. (2017). Integrated, decentralized wastewater management for resource recovery in rural and peri-urban areas. *Resources*, **6**, 22. https://doi.org/10.3390/resources6020022

Cornel P., Wagner M. and Krause S. (2003). Investigation of oxygen transfer rates in full scale membrane bioreactors. *Water Sci. Technol.*, **47**, 313–319.

Cote P., Alam Z. and Penny J. (2012). Hollow fiber membrane life in membrane bioreactors (MBR). *Desalination*, **288**, 145–151. https://doi.org/10.1016/j.desal.2011.12.026

De Wever H., Brepols C. and Lesjean B. (2009). Decision Tree for Full-Scale Submerged MBR Configurations. Book of Proceedings. Presented at the Final MBR-Network Workshop "Salient outcomes of the European R&D projects on MBR Technology", Berlin (Germany), 31 March–1 April 2009.

De Wilde W., Thoeye C. and De Gueldre G. (2005). Operational Experiences and Optimisations Two Years after Start-Up of the First Full-Scale MBR for Domestic Wastewater Treatment in the Benelux. Presented at the 6. Aachener Tagung Siedlungswasserwirtschaft und Verfahrenstechnik, Aachen, 26–27, October 2005.

De Wilde W., Thoeye C., De Gueldre G. (2007). Membrane Life Expectancy Assessment after 3 Years of MBR. Conference Proceedings. Presented at the 4th IWA Conference on membranes for water and wastewater treatment, IWA (Ed.), 15–17 May 2007, Harrogate (United Kingdom).

Drensla K. (2011). Fouling Mechanisms and Cleaning Strategies in Membrane Bioreactors for Municipal Wastewater Treatment, Conference Proceedings. Presented at the IWA-Memebrane Technology Conference, Aachen, 4–7, October 2011.

Drensla K. (2015). Chemical cleaning of submerged low-pressure hollow-fiber membranes on large-scale municipal MBR plants (Chemische Reinigung von getauchten Niederdruck-Hohlfasermembranen auf großtechnischen kommunalen MBR-Anlagen) (Kassel University Press, (Reihe Wasser – Abwasser – Umwelt 37)). Universität Kassel, Kassel.

Drensla K. and Janot A. (2016). Funding focus on energy-efficient wastewater plants – Erftverband Bergheim "Energetic optimization of the membrane wastewater treatment plant using the example of the Nordkanal group sewage treatment plant" (Förderschwerpunkt Energieeffiziente Abwasseranlagen – Erftverband Bergheim "Energetische Optimierung der Membrankläranlage am Beispiel des Gruppenklärwerks Nordkanal").

Drensla K. and Janot A. (2017). New characteristics of a membrane plant as a balance of progressive operation and technical development. (Neue Kenndaten einer Membrananlage als Bilanz des progressiven Betriebs und technischer Entwicklung). Wassertechnologie in Der Wasseraufbereitung Und Abwasserbehandlung. Presented at the 12. Aachener Tagung Wassertechnologie, Aachen.

Drews A. (2010). Membrane fouling in membrane bioreactors–characterisation, contradictions, cause and cures. *Journal of Membrane Science*, **363**(1–2), 1–28. https://doi.org/10.1016/j.memsci.2010.06.046

DWA (2011). Leaflet DWA-M 227 Membrane bioreactor process. Draft February 2011 (Merkblatt DWA-M 227 Membranbelebungsverfahren. Entwurf Februar 2011), DWA Regelwerk, Hennef.

DWA (2014). Membrane bioreactor process (MBR process) (leaflet No. DWA-M 277). (Membran-Bioreaktor-Verfahren (MBR-Verfahren) (Merkblatt No. DWA-M 277)). DWA-Regelwerk, DWA, Hennef.

DWA (Ed.) (2016). Design of single-stage activated sludge plants, June 2016, DWA worksheet (Bemessung von einstufigen Belebungsanlagen, Juni 2016, DWA-Arbeitsblatt). Deutsche Vereinigung für Wasserwirtschaft, Abwasser und Abfall, Hennef.

Erftverband. (2004). Optimisation of an activated sludge plant with membrane filtration (waste water treatment plant Rödingen) (Optimierung einer Belebungsanlage mit Membranfiltration (Kläranlage Rödingen)), Technical Report No. 1–3, Erftverband, Bergheim/Erft.

Fenu A., Roels J., Wambecq T., De Gussem K., Thoeye C., De Gueldre G. and Van De Steene B. (2010). Energy audit of a full scale MBR system. *Desalination*, **262**, 121–128. https://doi.org/10.1016/j.desal.2010.05.057

Ferrero G., Rodriguez-Roda I. and Comas J. (2012). Automatic control systems for submerged membrane bioreactors: a state-of-the-art review. *Water Research*, **46**, 3421–3433. https://doi.org/doi:10.1016/j.watres.2010.03.055

Fletcher H., Mackley T. and Judd S. (2007). The cost of a package plant membrane bioreactor. *Water Research*, **41**, 2627–2635. https://doi.org/10.1016/j.watres.2007.02.038

Frechen F. B., Schier W. and Linden C. (2008). Pre-treatment of municipal MBR applications. *Desalination*, **231**, 108–114. https://doi.org/10.1016/j.desal.2007.10.025

Freeman S. and Levesque S. (2009). It is no secret where this small town's water comes from, in: A new frontier [WWW Document]. American City and Country, Penton Media. http://americancityandcounty.com/water/treatment/cloudcroft-wastewater-reuse-system-200902 (accessed 05 September 2012).

Friedler E. and Hadari M. (2006). Economic feasibility of on-site greywater reuse in multi-storey buildings. *Desalination*, **190**, 221–234. https://doi.org/10.1016/j.desal.2005.10.007

Garrido-Baserba M., Reif R., Molinos-Senante M., Larrea L., Castillo A., Verdaguer M. and Poch M. (2016). Application of a multi-criteria decision model to select of design choices for WWTPs. *Clean Technologies and Environmental Policy*, **18**, 1097–1109. https://doi.org/10.1007/s10098-016-1099-x

Germain E., Nelles F., Drews A., Pearce P., Kraume M., Reid E., Judd S. J. and Stephenson T. (2007). Biomass effects on oxygen transfer in membrane bioreactors. *Water Research*, **41**, 1038–1044. https://doi.org/10.1016/j.watres.2006.10.020

Gil J. A., Túa L., Rueda A., Montaño B., Rodríguez M. and Prats D. (2010). Monitoring and analysis of the energy cost of an MBR. *Desalination*, **250**, 997–1001. https://doi.org/10.1016/j.desal.2009.09.089

Guglielmi G., Saroj D., Chiarani D. and Andreottola G. (2007). Sub-critical fouling in a membrane bioreactor for municipal wastewater treatment: experimental investigation and mathematical modelling. *Water Research*, **41**, 3903–3914. https://doi.org/10.1016/j.watres.2007.05.047

Günder B. (2001). The Membrane-Coupled Activated Sludge Process in Municipal Wastewater Treatment. Technomic Publishing, Lancaster.

Henkel J., Cornel P. and Wagner M. (2011). Oxygen transfer in activated sludge–new insights and potentials for cost saving. *Water Science and Technology*, **63**, 3034–3038.

Henze M., Gujer W., Mino T. and van Loosdrecht M. (2000). Activated Sludge Models ASM1, ASM2, ASM2d and ASM3. IWA Publishing, London.

Henze M., Harremoes P., Jansen J. la C. and Arvin E. (2001). Wastewater Treatment Biological and Chemical Processes, 3rd edn, Springer, Berlin.

Henze M., Loosdrecht M. C. V. and Ekama G. A. (2008). Biological Wastewater Treatment: Principles, Modeling and Design. IWA Publishing, London.

Hoinkis J., Deowan S. A., Panten V., Figoli A., Huang R. R. and Drioli E. (2012). Membrane bioreactor (MBR) technology – a promising approach for industrial water reuse. *Procedia Engineering*, **33**, 234–241. https://doi.org/10.1016/j.proeng.2012.01.1199

Iglesias R., Simón P., Moragas L., Arce A. and Rodriguez-Roda I. (2017). Cost comparison of full-scale water reclamation technologies with an emphasis on membrane bioreactors. *Water Science and Technology*, **75**(11), 2562–2570. https://doi.org/10.2166/wst.2017.132

Ioannou-Ttofa L., Foteinis S., Chatzisymeon E. and Fatta-Kassinos D. (2016). The environmental footprint of a membrane bioreactor treatment process through life cycle analysis. *Science of the Total Environment*, **568**, 306–318. https://doi.org/10.1016/j.scitotenv.2016.06.032

Judd S. (2011). The MBR Book, Second Edition: Principles and Applications of Membrane Bioreactors for Water and Wastewater Treatment, 2nd edn, Butterworth-Heinemann, Oxford.

Kang C.-W., Hua J., Lou J., Liu W. and Jordan E. (2008). Bridging the gap between membrane bio-reactor (MBR) pilot and plant studies. *Journal of Membrane Science*, **325**, 861–871. https://doi.org/10.1016/j.memsci.2008.09.016

Kazner C., Wintgens T. and Dillon P. J. (2012). Water Reclamation Technologies for Safe Managed Aquifer Recharge. IWA Publishing, London, New York.

Kochaba T., Pellegrin M.-L., Kinnear D., Grissom M. and McNamara B. (2008). Peak Interest: Sensitivity to Design Flows in Evaluating Large-Plant MBR Retrofits. Presented at the WEFTEC, Orlando, Florida.

Krampe J. and Krauth K. (2003). Oxygen transfer into activated sludge with high MLSS concentrations. *WsT*, **47**(11), pp. 297–303.

Krause S. (2005). Untersuchungen zum Energiebedarf von Membranbelebungsanlagen. Dissertation. Verein zur Förderung des Institutes WAR der Technischen Universität Darmstadt, Darmstadt.

Krzeminski P., Langhorst W., Schyns P., de Vente D., Van den Broeck R., Smets I. Y., Van Impe J. F.M., van der Graaf J. H.J. M. and van Lier J. B. (2012). The optimal MBR configuration: hybrid versus stand-alone – Comparison between three full-scale MBRs treating municipal wastewater. *Desalination*, **284**, 341–348. https://doi.org/10.1016/j.desal.2011.10.038

Kullmann C., Lawrence D. and Costa E. (2011). An MBR System for Aquapolo Ambiental S.A., Water Today Magazine. Water Today Pvt. Ltd., Chennai, India.

Lahnsteiner J., van Rensburg P. and Esterhuizen J. (2018). Direct potable reuse – a feasible water management option. *Journal of Water Reuse and Desalination*, **8**, 14–28. https://doi.org/10.2166/wrd.2017.172

Lay W. C.L., Lim C., Lee Y., Kwok B. H., Tao G., Lee K. S., Chua S. C., Wah Y. L., Ghani Y. A. and Seah H. (2017). From R&D to application: membrane bioreactor technology for water reclamation. *Water Practice and Technology*, **12**, 12–24. https://doi.org/10.2166/wpt.2017.008

Lazarova V. (2008). Reliability of operation and failure management of membrane wastewater treatment. *Water Practice & Technology*, **3**. https://doi.org/10.2166/wpt.2008.044

Lazarova V., Choo K. and Cornel P. (2012). Water-Energy Interactions in Water Reuse. IWA Publishing, London.

Leiknes T. and Odegaard H. (2006). The development of a biofilm membrane bioreactor. *Desalination*, **202**, 135–143.

Lesjean B. and Huisjes E. H. (2008). Survey of the European MBR market: trends and perspectives. *Desalination*, **231**, 71–81. https://doi.org/10.1016/j.desal.2007.10.022

Lesjean B. and Leiknes T. (Eds.) (2009). Salient outcomes of the European R&D projects on MBR Technology. Final MBR-Network Workshop, 31 March-1 April, Berlin (Germany) Berlin 2009 Book of proceedings. MBR-Network, Berlin.

Lesjean B., Rosenberger S., Schrotter J.-C. and Recherche A. (2004). Membrane-aided biological wastewater treatment – an overview of applied systems. *Membrane Technology*, **2004**, 5–10. https://doi.org/10.1016/S0958-2118(04)00200-9

Lesjean B., Tazi-Pain A., Thaure D., Moeslang H. and Buisson H. (2011). Ten persistent myths and the realities of membrane bioreactor technology for municipal applications. *Water Science and Technology*, **63**(1), 32. https://doi.org/10.2166/wst.2011.005

Leverenz H. L., Tchobanoglous G. and Asano T. (2011). Direct potable reuse: a future imperative. *Journal of Water Reuse and Desalination*, **01.1**, 2–10.

Liu N., Zhang Q., Chin G.-L., Ong E.-H., Lou J., Kang C.-W., Liu W. and Jordan E. (2010). Experimental investigation of hydrodynamic behavior in a real membrane bio-reactor unit. *Journal of Membrane Science*, **353**(1–2), 122–134. https://doi.org/10.1016/j.memsci.2010.02.042

Lyko S., Wintgens T., Al-Halbouni D., Baumgarten S., Tacke D., Drensla K., Janot A., Dott W., Pinnekamp J. and Melin T. (2008). Long-term monitoring of a full-scale municipal membrane bioreactor—Characterisation of foulants and operational performance. *Journal of Membrane Science*, **317**, 78–87. https://doi.org/10.1016/j.memsci.2007.07.008

Machuca M. V. (2017). Next Inauguration: MODERN TECHNOLOGY WITH WASTEWATER TREATMENT PLANT. (Próxima Inauguración: MODERNA TECNOLOGÍA CON PLANTA DE TRATAMIENTO DE AGUAS RESIDUALES). Asociación Cultural Johannes Gutenberg/Boletin/Actividades.

Maere T., Verrecht B., Moerenhout S., Judd S. and Nopens I. (2011). BSM-MBR: a benchmark simulation model to compare control and operational strategies for membrane bioreactors. *Water Research*, **45**, 2181–2190. https://doi.org/10.1016/j.watres.2011.01.006

Mannina G. and Cosenza A. (2013). The fouling phenomenon in membrane bioreactors: assessment of different strategies for energy saving. *Journal of Membrane Science*, **444**, 332–344. https://doi.org/10.1016/j.memsci.2013.05.047

Martin I., Pidou M., Soares A., Judd S. and Jefferson B. (2011). Modelling the energy demands of aerobic and anaerobic membrane bioreactors for wastewater treatment. *Environmental Technology*, **32**, 921–932. https://doi.org/10.1080/09593330.2011.565806

Melin T. and Wintgens T. (Eds.) (2006). Integrated Concepts for Reuse of Upgraded Wastewater, Final Project Report (AQUAREC).

Melin T., Jefferson B., Bixio D., Thoeye C., De Wilde W., De Koning J., van der Graaf J. and Wintgens T. (2006). Membrane bioreactor technology for wastewater treatment and reuse. *Desalination*, **187**, 271–282. https://doi.org/10.1016/j.desal.2005.04.086

Miyoshi T., Nguyen T. P., Tsumuraya T., Tanaka H., Morita T., Itokawa H. and Hashimoto T. (2018). Energy reduction of a submerged membrane bioreactor using a polytetrafluoroethylene (PTFE) hollow-fiber membrane. *Frontiers of Environmental Science & Engineering*, **12**. https://doi.org/10.1007/s11783-018-1018-y

Mulder J. W. (2009). Operational Experiences with the Hybrid MBR Heenvliet, a Smart Way of Retrofitting, Final MBR-Network Workshop "Salient Outcomes of the European R&D Projects on MBR Technology", B. Lesjean and T. Leiknes (eds.), 31 March–1 April, Berlin (Germany) Berlin 2009 Book of Proceedings. MBR-Network, Berlin.

Muller E. B., Stouthamer A. H., van Verseveld H. W. and Eikelboom D. H. (1995). Aerobic domestic waste water treatment in a pilot plant with complete sludge retention by cross-flow filtration. *Water Research*, **29**, 1179–1189. https://doi.org/10.1016/0043-1354(94)00267-B

Pan J. R., Su Y.-C., Huang C. and Lee H.-C. (2010). Effect of sludge characteristics on membrane fouling in membrane bioreactors. *Journal of Membrane Science*, **349**, 287–294. https://doi.org/10.1016/j.memsci.2009.11.055

Pillay S., Foxon K. M. and Buckley C. A. (2007). An Anaerobic Baffled Reactor/Membrane Bioreactor (ABR/MBR) for On-Site Sanitation in Low Income Areas. Presented at the IWA MTC, Harrogate, United Kingdom.

Pinnekamp J. (2006). Membrantechnik für die Abwasserreinigung. FiW-Verl., Aachen.

Pinnekamp J. and Friedrich H. (Eds.) (2006). Siedlungswasser- und Siedlungsabfallwirtschaft Nordrhein-Westfalen Bd. 2. Membrane Technology for Waste Water Treatment. FiW-Verl., Aachen.

Prendl L. and Kroiß H. (1998). Bulking sludge prevention by an aerobic selector. *Water Science and Technology*, **38**(8–9), 19–27. https://doi.org/10.1016/S0273-1223(98)00673-8

PUB (2011). Innovation in Water Singapore, Singapore.

PUB (2012). NEWater [WWW Document]. http://www.pub.gov.sg/water/newater/visitors/Pages/default.aspx

Pyper W. (2013). Grand wastewater designs. *Australian Antarctic Magazine*, **30**.

Ratkovich N., Chan C.C.V., Bentzen T.R. and Rasmussen M.R. (2012). Experimental and CFD simulation studies of wall shear stress for different impeller configurations and MBR activated sludge." *Water Science & Technology*, **65** (11), 2061. https://doi.org/10.2166/wst.2012.106

Saalbach J. and Hunze M. (2008). Flow structures in MBR-tanks. *Water Science & Technology*, **57**(5), 699. https://doi.org/10.2166/wst.2008.122

Schäfer H. (2017). Membrane Bioreactor Nordkanal - Expansion by sewage sludge digestion and deammonification (Membranbelebungsanlage Nordkanal – Erweiterung um eine Klärschlammfaulung und Deammonifikation).

Schier W., Frechen F.-B. and Fischer S. (2009a). Efficiency of mechanical pre-treatment on European MBR plants. *Desalination*, **236**, 85–93. https://doi.org/10.1016/j.desal.2007.10.054

Schier W., Exler H., Ohme M. and Frechen F.-B. (2009b). Optimization of the mechanical pre-treatement stage of municipal wastewater treatment plants (final report research project). (Optimierung der mechanischen Vorreinigungsstufe kommunaler Abwasserbehandlungsanlagen (Abschlußbericht Forschungsvorhaben)). Universität Kassel, Kassel.

Schier W., Drensla K., Janot A., Exler H., Engelhardt N. and Frechen F.-B. (2012). Mechanical pre-treatment (MPT) – revitalised by MBR process. *Water Science and Technology*, **66**(12), 2524. https://doi.org/10.2166/wst.2012.416

Schoon N. (2016). The-BedZED-Story [WWW Document]. BedZED – Bioregional. https://www.bioregional.com/wp-content/uploads/2016/04/The-BedZED-Story.pdf (accessed 04 December 2018).

Sewage Technical Meeting on Membrane Technology. (Ed.) (2011). Guidelines for Introducing Membrane Technology in Sewage Works, 2nd edn. Japan, Ministry of Land, Infrastructure, Transport and Tourism website. www.mlit.go.jp/common/000213616.pdf.

Seyfried C. F., Lohse M., Schüßler H. and Bebendorf G. (1985). Comparison of the cleaning performance of rakes, sieves and screen rakes as well as their influence on the further cleaning stages (Vergleich der Reinigungsleistung von Rechen, Sieben und Siebrechen sowie deren Einfluß auf die weiteren Reinigungsstufen), Veröffentlichungen des Institutes für Siedlungswasserwirtschaft und Abfalltechnik der Universität Hannover. Hannover.

Stefanski M., Kennedy S. and Judd S. (2011). The determination and origin of fibre clogging in membrane bioreactors. *Journal of Membrane Science*, **375**, 198–203. https://doi.org/10.1016/j.memsci.2011.03.047

STOWA (2009). Ervaringen met hybride MBR Ootmarsum (Experiences with hybrid MBR Ootmarsum). STOWA, Utrecht.

Stuber J., Ludicke C., Lesjean B., Gnirss R. and Kraume M. (2009). Operation of MBR membrane modules used in a decentralised wastewater treatment plant: field study and comparison of different cleaning strategies. *Desalination and Water Treatment*, 181–188. https://doi.org/10.5004/dwt.2009.769

Sun C., Leiknes T., Weitzenböck J. and Thorstensen B. (2010). Development of a biofilm-MBR for shipboard wastewater treatment: the effect of process configuration. *Desalination*, **250**, 745–750. https://doi.org/10.1016/j.desal.2008.11.034

Sun J., Xiao K., Yan X., Liang P., Shen Y., Zhu N. and Huang X. (2015). Membrane bioreactor vs. oxidation ditch: full-scale long-term performance related with mixed liquor seasonal characteristics. *Process Biochemistry*, **50**, 2224–2233. https://doi.org/10.1016/j.procbio.2015.09.010

Sun J., Liang P., Yan X., Zuo K., Xiao K., Xia J., Qiu Y., Wu Q., Wu S., Huang X., Qi M. and Wen X. (2016). Reducing aeration energy consumption in a large-scale membrane bioreactor: process simulation and engineering application. *Water Research*, **93**, 205–213. https://doi.org/10.1016/j.watres.2016.02.026

Tacke D., Pinnekamp J., Prieske H. and Kraume M. (2008). Membrane bioreactor aeration: Investigation of the velocity flow pattern. *Water Science & Technology*, **57**(4), 559. https://doi.org/10.2166/wst.2008.123

Tao G., Kekre K. A., Oo M. H., Viswanath B., Lew C. H., Kan L. M. and Seah H. (2009). Large Scale Membrane Bioreactor Plant Design (Retrofit) and Optimisation. Abstract Book. Presented at the 5th IWA Specialised membrane Technology Conference for Water and Wastewater Treatement, Beijing, P.R. China, pp. 111–112, 1–3, September 2009.

Tchobanoglous G., Burton F. L. and Stensel H. D. (2003). Wastewater Engineering: Treatment and Reuse, 4th edn, McGraw-Hill, Boston.

UNEP/WHO/HABITAT/WSSCC (2004). Global Water Supply and Sanitation Assessment 2000 Report. UNEP/GPA Coordination Office, The Hague, The Netherlands.

US EPA (Ed.) (2007). Wastewater Management Fact Sheet Membrane Bioreactors.

US EPA (Ed.) (2008). Municipal Nutrient Removal Technologies Reference Document Technical Report Volume 1. United States Environmental Protection Agency, Washington, D.C.

Van den Broeck R., Van Dierdonck J., Nijskens P., Dotremont C., Krzeminski P., van der Graaf J. H.J. M., van Lier J. B., Van Impe J. F.M. and Smets I. Y. (2012). The influence of solids retention time on activated sludge bioflocculation and membrane fouling in a membrane bioreactor (MBR). *Journal of Membrane Science*, **401–402**, 48–55. https://doi.org/10.1016/j.memsci.2012.01.028

Veltmann K., Palmowski L. M. and Pinnekamp J. (2011). Modular operation of membrane bioreactors for higher hydraulic capacity utilisation. *Water Science and Technology*, **63**(6), 1241. https://doi.org/10.2166/wst.2011.366

Verrecht B. (2010). Optimisation of a Hollow Fibre Membrane Bioreactor for Water Reuse PhD thesis, Cranfield University, Cranfield.

Verrecht B., Judd S., Guglielmi G., Brepols C. and Mulder J. W. (2008). An aeration energy model for an immersed membrane bioreactor. *Water Research*, **42**, 4761–4770. https://doi.org/10.1016/j.watres.2008.09.013

Verrecht B., Maere T., Benedetti L., Nopens I. and Judd S. (2010a). Model-based energy optimisation of a small-scale decentralised membrane bioreactor for urban reuse. *Water Research*, **44**, 4047–4056. https://doi.org/10.1016/j.watres.2010.05.015

Verrecht B., Maere T., Nopens I., Brepols C. and Judd S. (2010b). The cost of a large-scale hollow fibre MBR. *Water Research*, **44**, 5274–5283. https://doi.org/10.1016/j.watres.2010.06.054

Verrecht B., James C., Germain E., Birks R., Barugh A., Pearce P. and Judd S. (2012). Economical evaluation and operating experiences of a small-scale MBR for nonpotable reuse. *Journal of Environmental Engineering*, **138**, 594–600. https://doi.org/10.1061/(ASCE)EE.1943-7870.0000505

Wagner M. (2005). Microbial ecology of activated sludge. *Microbiology Today*, **32**, 28–31.

Wang K. M., Cingolani D., Eusebi A. L., Soares A., Jefferson B. and McAdam E. J. (2018). Identification of gas sparging regimes for granular anaerobic membrane bioreactor to enable energy neutral municipal wastewater treatment. *Journal of Membrane Science*, **555**, 125–133. https://doi.org/10.1016/j.memsci.2018.03.032

Wang Z., Wu Z., Yin X. and Tian L. (2008). Membrane fouling in a submerged membrane bioreactor (MBR) under sub-critical flux operation: Membrane foulant and gel layer characterization. *Journal of Membrane Science*, **325**(1), 238–244. https://doi.org/10.1016/j.memsci.2008.07.035

Water Environment Federation (Ed.) (2012). Membrane Bioreactors, WEF Manual of Practice. WEF Press; McGraw-Hill [distributor], Alexandria, VA, New York.

Webster J. (2007). Culture's influence: towards understanding stakeholder interactions in rural water, sanitation and hygiene promotion projects.

Wedi D. and Joss A. (2008). Dimensioning of membrane bioreactors for municipal wastewater treatment. *Water Science and Technology*, **57**(6), 829.

WEF (2009). Design of Municipal Wastewater Treatment Plants, MOP 8, 5th edn, WEF Press, Alexandria, Virginia.

WHO (Ed.) (2006). Guidelines for the Safe Use of Wastewater, Excreta and Greywater, Volumes 1–4, Policy and Regulatory Aspects – Wastewater Use in Agriculture – Wastewater and Excreta Use in Aquaculture – Excreta and Greywater Use in Agriculture. World Health Organization, WHO Press, Geneva, Switzerland.

Wintgens T., Rosen J., Melin T., Brepols C., Drensla K. and Engelhardt N. (2003). Modelling of a membrane bioreactor system for municipal wastewater treatment. *Journal of Membrane Science*, **216**, 55–65. https://doi.org/10.1016/S0376-7388(03)00046-2

Wisniewski C. (2007). Membrane bioreactor for water reuse. *Desalination*, **203**, 15–19. https://doi.org/10.1016/j.desal.2006.05.002

WWAP (Ed.) (2017). The United Nations World Water Development Report 2017. Wastewater: The Untapped Resource. UNESCO Publishing, Paris.

WWF3 (2013). Hygienepapier [WWW Document]. Hygienepapier – WWF Deutschland. http://www.wwf.de/themen-projekte/waelder/papierverbrauch/hygienepapier/

Yan X., Bilad M.R., Gerards R., Vriens L., Piasecka A. and Vankelecom I.F.J. (2012). Comparison of MBR performance and membrane cleaning in a single-stage activated sludge system and a two-stage anaerobic/aerobic (A/A) system for treating synthetic molasses wastewater. *Journal of Membrane Science*, **394–395**, 49–56. https://doi.org/10.1016/j.memsci.2011.12.014

You S.-J., Damodar R. A. and Hou S.-C. (2010). Degradation of reactive black 5 dye using anaerobic/aerobic membrane bioreactor (MBR) and photochemical membrane reactor. *Journal of Hazardous Materials*, **177**, 1112–1118. https://doi.org/10.1016/j.jhazmat.2010.01.036

Zaerpour M. (2014). Design, Cost & Benefit Analysis of a Membrane Bioreactor. Politcnico di Milano.

Zanetti F., De Luca G. and Sacchetti R. (2010). Performance of a full-scale membrane bioreactor system in treating municipal wastewater for reuse purposes. *Bioresource Technology*, **101**, 3768–3771. https://doi.org/10.1016/j.biortech.2009.12.091

Zhang J., Chua H.C., Zhou J. and Fane A.G. (2006). Factors affecting the membrane performance in submerged membrane bioreactors. *Journal of Membrane Science*, **284**(1–2), 54–66. https://doi.org/10.1016/j.memsci.2006.06.022

Zhang Y., Zhou J., Zhang J. and Yuan S. (2009). An innovative membrane bioreactor and packed-bed biofilm reactor combined system for shortcut nitrification-denitrification. *Journal of Environmental Sciences*, **21**, 568–574. https://doi.org/10.1016/S1001-0742(08)62309-8

Zheng X., Zhou Y., Chen S., Zheng H. and Zhou C. (2010). Survey of MBR market: trends and perspectives in China. *Desalination*, **250**, 609–612. https://doi.org/10.1016/j.desal.2009.09.034

Zsirai T., Buzatu P., Aerts P. and Judd S. (2012). Efficacy of relaxation, backflushing, chemical cleaning and clogging removal for an immersed hollow fibre membrane bioreactor. *Water Research*, **46**(14), 4499–4507. https://doi.org/10.1016/j.watres.2012.05.004

Chapter 4

Monitoring, characterization and control of membrane biofouling in MBR[†]

Katsuki Kimura[1], Fangang Meng[2], In-Soung Chang[3] and Chung-Hak Lee[4]*

[1]Prof. Graduate School of Engineering, Hokkaido University, Japan (kkatsu@eng.hokudai.ac.jp)
[2]Prof. School of Environmental Science and Engineering, Sun Yat-sen University, China (fgmeng80@126.com)
[3]Prof. Department of Environmental Engineering, Hoseo University, Korea (cis@hoseo.edu)
[4]Prof. School of Chemical and Biological Engineering, Seoul National University, Korea (leech@snu.ac.kr)
**Corresponding author*

ABSTRACT

In the operation of a membrane biological reactor (MBR), a membrane is continuously in contact with mixed liquor suspension comprising microbial cells and various soluble inorganic and organic constituents. As such, deterioration of membrane permeability (i.e., membrane fouling) occurs inevitably. Membrane fouling increases both capital and operation costs of MBRs and it is reported that more than 50% of operating cost in the operation of full scale MBRs is related to membrane fouling. In order to address the fouling problem, a broad spectrum of approaches has been adopted: monitoring of fouling, analysis and characterization of foulants/fouling layer, and methods for fouling control. In this chapter, principles, classifications, advantages and limitations of these techniques are described.

[*]Section 4.3 has been reprinted (with significant modifications in figures and tables) from Meng FG, Liao BQ, Liang SA, Yang FL, Zhang HM, and Song LF, 2010 Morphological visualization, componential characterization and microbiological identification of membrane fouling in membrane bioreactors (MBRs). ***Journal of Membrane Science***, 361: 1–14., with permission from Elsevier.

 Membrane Biological Reactors: Theory, Modeling, Design, Management and Applications to Wastewater Reuse
Editor(s): Faisal I. Hai, Kazuo Yamamoto and Chung-Hak Lee
doi: 10.2166/9781780409177_0115

4.1 INTRODUCTION

Membrane fouling can be defined as the undesirable deposition and accumulation of microorganisms, colloids, solutes, and cell debris within pores or on membrane surface (Meng *et al.* 2009). It results from the interaction between the membrane material and the components of the activated sludge liquor, which include biological flocs formed by a large range of living microorganisms along with soluble and colloidal compounds. Thus it is not surprising that the fouling behavior in MBRs is more complicated than that in most membrane applications. The suspended biomass has no fixed composition and varies both with feed water composition and MBR operating conditions employed. Accordingly, although many investigations of membrane fouling have been published, the diverse range of operating conditions and feed water matrices employed, and the limited information reported in most studies on the biomass composition in suspension or on the membrane, has made it difficult to establish any generic behavior pertaining to membrane fouling in MBRs (Le-Clech *et al.* 2006; Meng *et al.* 2009; Hai & Yamamoto, 2011). In this chapter, the principles and classifications of the conventional and emerging membrane fouling monitoring and characterizing techniques and fouling control methods are critically described and their advantages and limitations are highlighted (Figure 4.1). It is worth-noting that two major types of membrane fouling in MBRs can be identified depending on their reversibility: reversible (mostly cake layer) fouling and irreversible fouling. The former type of fouling can be removed by physical cleaning such as backwashing or air-souring, whereas the removal of the latter requires chemical cleaning.

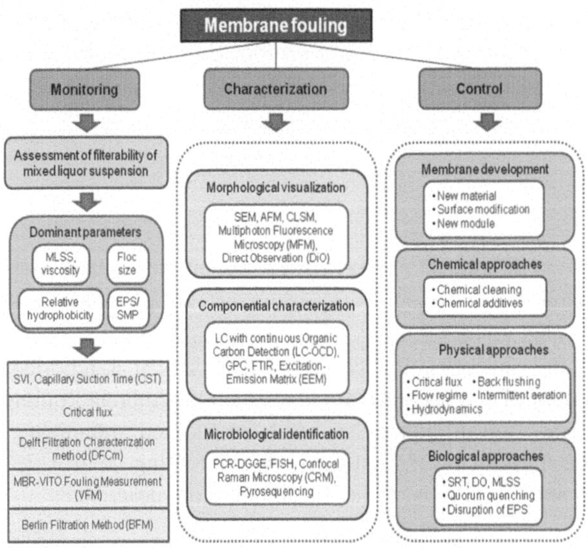

Figure 4.1 Schematic illustration showing the contents of Chapter 4.

4.2 MONITORING

4.2.1 Importance of monitoring

As experienced in numerous MBRs operated at different scales, filterability of mixed liquor suspensions in MBRs varies significantly depending on various factors: there are clearly suspensions with high filterability and low filterability. Operators can take immediate and appropriate actions (e.g., preparation of chemical cleaning or reduction of flux) to deal with the problem if sufficient information on filterability of mixed liquor suspension is available. Continuous monitoring of filterability is thus important, but no universal method has yet been established.

In this section, methods that have been used for the assessment of filterability of mixed liquor suspensions in MBRs are firstly described. Many researchers have attempted to identify properties/parameters that govern the filterability of mixed liquor in MBRs. A variety of properties/parameters have been claimed as main factors in filterability of mixed liquor suspension. These are also summarized in this section.

4.2.2 Methods used for assessment of filterability of mixed liquor

4.2.2.1 Simple methods used in conventional activated sludge (CAS)

Settleability and dewaterbility of sludge in conventional activated sludge (CAS) have been assessed by sludge volume index (SVI) and capillary suction time (CST). These methods are readily available and have been used for assessment of the filterability of a mixed liquor suspension in the MBR process. Results were, however, not always consistent: conflicting conclusions are found in the literature regarding the applicability and usefulness of conventional methods: (Chae *et al.* 2006; Fan *et al.* 2006; Wu *et al.* 2007; Lyko *et al.* 2008; Pan *et al.* 2010). This led to attempts to develop tailored methods for assessment of the filterability of a mixed liquor suspension as described below.

4.2.2.2 Critical flux measurements

The concept of critical flux was proposed by Field *et al.* (1995). They defined a flux below which no fouling would be observed. The critical flux concept has been well accepted in the field of MBRs. For the determination of critical flux for a specific suspension, the flux-stepping protocol developed by Le-Clech *et al.* (2003) has been used with modifications. However, no standardized protocol for critical flux measurement is available. As a result, comparison of different datasets is difficult. It is also known that fouling can occur even when the operational flux is set below the critical flux (Guglielmi *et al.* 2007), a phenomenon that is often explained by irreversible fouling.

4.2.2.3 Use of stirred dead-end cells

Lab-scale stirred dead-end cells have been used for assessment of the filterability of mixed liquor suspensions collected from MBRs (e.g., Shen *et al.* 2012). Although ease of operation and availability of the apparatus are advantages of this method, there are concerns about the fact that filtration is carried out at a constant TMP, which is considerably higher than those seen in MBRs, in this assessment. As a result, flux in an early stage of the assessment becomes very high, leading to rapid membrane fouling and different fouling characteristics.

4.2.2.4 Recently proposed methods

Based on the context described above, new methods for assessment of filterability of a mixed liquor suspension have recently been proposed with the objective of standardization of protocols and comparison among different datasets. Three methods proposed by European groups are introduced in this section. These methods use unique apparatuses to mimic hydrodynamic conditions in MBR processes.

(1) ***Delft Filtration Characterization method (DFCm)***
Evenblij *et al.* (2005) developed a well-defined protocol for quantification of the filterability of mixed liquor suspension. The filtration apparatus consists of a single tubular membrane

module that is used in cross-flow mode (cross flow velocity: 1 m/s). Effects of backwash and chemical cleaning are also assessed in this unit. Figure 4.2 shows a schematic overview of the filtration unit used in the DFCm. Constant flux (normally 80 L/m^2h) is maintained by a peristaltic pump.

(2) ***MBR-VITO Fouling Measurement (VFM)***
This is an *in situ* method proposed by Huyskens *et al.* (2008). In their method, a tubular membrane is submerged in the biological reaction tank of an MBR with its own aerator (Figure 4.3). Crossflow conditions are created by injection of air bubbles in the lumen of the membrane tube, generating a gas-liquid two-phase crossflow. Inside-out filtration is carried out by the vacuum pump. Constant-pressure filtration is carried out at 10 kPa. A cyclical measurement protocol was developed for quick assessment of the mixed liquor's reversible and irreversible fouling propensity. Like the DFCm, the MBR-VFM is operated in hydrodynamic conditions that lead to accelerated fouling.

(3) ***Berlin Filtration Method (BFM)***
A small flat-sheet membrane module immerged in the biological tank of an MBR is used in the BFM proposed by a group at the Technical University of Berlin (de la Torre *et al.* 2010). Like the MBR-VFM, the small membrane module has its own aerator at the bottom. Figure 4.4 shows a schematic of the test cell used in the BFM. Filterability is expressed in terms of critical flux in this method.

In situ methods such as the VFM or the BFM would be superior because they do not involve potential changes of biomass properties caused by sampling or storage Huyskens *et al.* (2008).

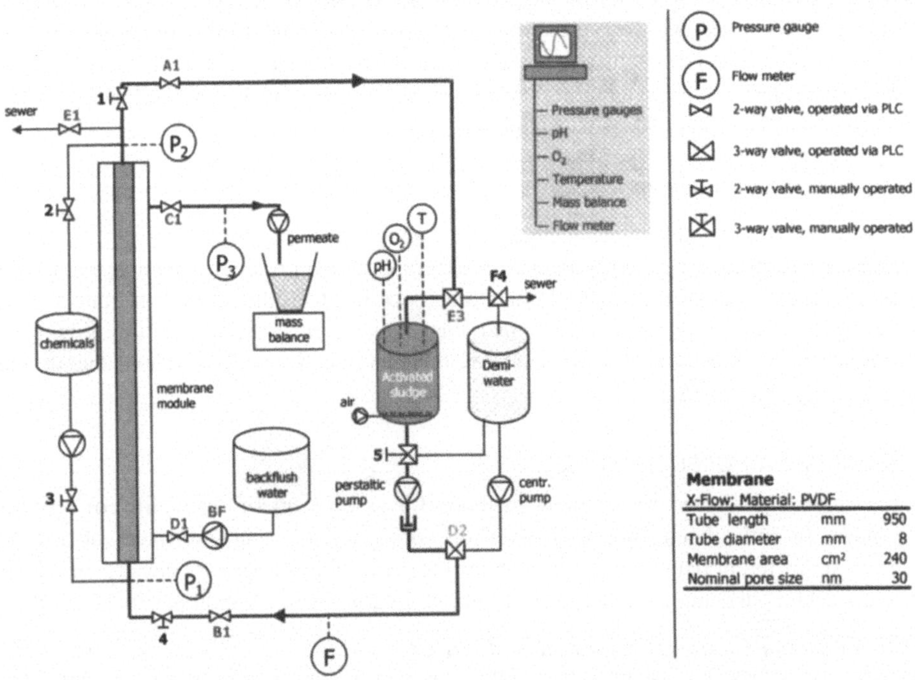

Figure 4.2 Schematic overview of the filtration unit used in the DFCm (Evenblij *et al.* 2005).

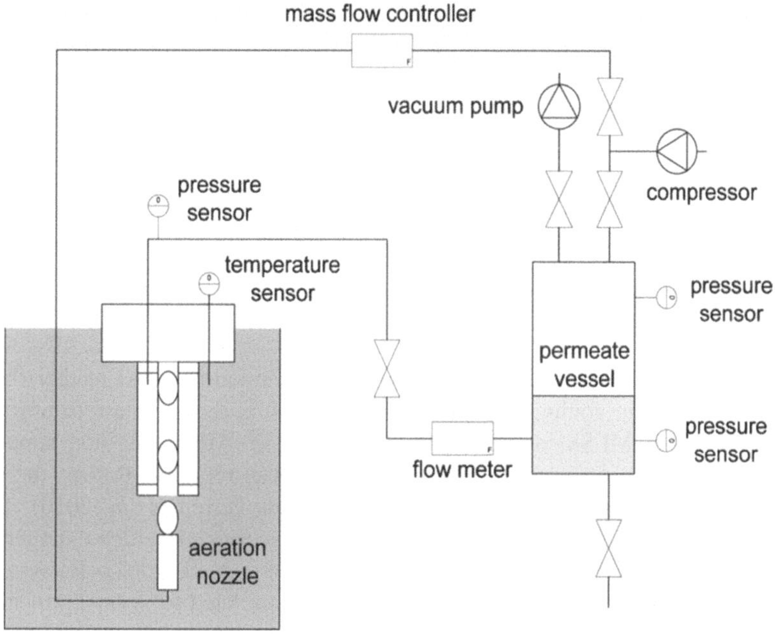

Figure 4.3 Schematic overview of the MBR-VFM set-up (Huyskens *et al.* 2008).

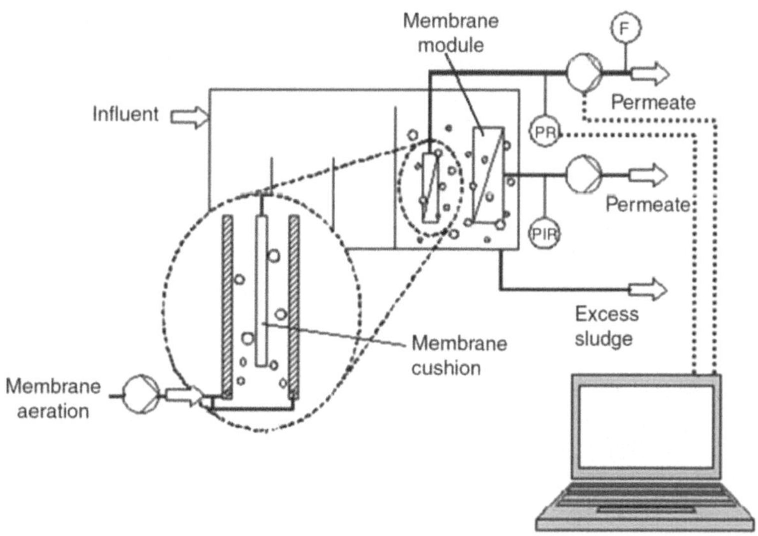

Figure 4.4 Scheme of the BFM test cell (de la Torre *et al.* 2010).

4.2.3 Identification of dominant parameters in filterability of mixed liquor

Many efforts have been made to identify the dominant factor(s) in the filterability of a mixed liquor suspension, assessed by one of the methods described above, since identification of the dominant factor(s) may enable quick selection of appropriate measures for effective control of membrane fouling in MBRs. At present, however, there is no general consensus regarding the dominant factor: for virtually all of the parameters described below, conflicting results have been reported. Considering the complexity in membrane fouling phenomena in MBRs, which is affected by biomass properties, hydrodynamics, MBR design and operational characteristics including selection of membranes, raw wastewater characteristics and temperature variations, conflicting results are not surprising. Dominant factors would be different in each MBR depending on their specific conditions. Thus, in the following section, major parameters claimed as dominant factors in the literature are summarized without intention of conclusion.

4.2.3.1 MLSS and viscosity

In early studies on membrane fouling in MBRs, biomass concentration in the reactor (i.e., MLSS) was often considered the dominant fouling parameter. Although it is the simplest parameter to measure, direct correlation between MLSS and membrane fouling in MBRs is not supported now by engineers/researchers, since many contradicting results have been reported. Rather, the existence of an optimal MLSS concentration range has been suggested (Lousada-Ferreira *et al.* 2010). The relationship between MLSS and viscosity of sludge is well established: viscosity of sludge exponentially increases with elevation of MLSS concentration (Itonaga *et al.* 2004; Wu *et al.* 2007). However, evolution of membrane fouling in pilot-scale MBRs could not be directly correlated to viscosity of the mixed liquor suspension (Kimura *et al.* 2005).

4.2.3.2 Relative hydrophobicity (RH)

The relative hydrophobicity of activated sludge can be measured as adherence to hydrocarbons (Chang *et al.* 1998). It is thought that less hydrophobicity brought about floc deterioration in sludge and caused severe membrane fouling (Van den Broeck *et al.* 2010). The opposite result, a positive correlation between RH of activated sludge and membrane fouling, was reported by Meng *et al.* (2006). When RH is used for assessment of the filterability of sludge, it seems to be necessary to take membrane hydrophobicity into consideration as well.

4.2.3.3 Particle size

There might be a linkage between filterability of a suspension and submicron-sized particles: worse filterability was linked to a larger number of submicron-sized particles. Several recent studies have shown the importance of submicron-sized particles (Jiang *et al.* 2007; Ivanovic *et al.* 2008; Van den Broeck *et al.* 2011). Submicron-sized particles might be released from the deflocculation of larger flocs and/or derived from microbial activities. Submicron-sized particle would plug/narrow micropores of membranes with sizes close to the dimensions of those particles, leading to evolution of irreversible fouling. Also, such particles are likely to be deposited on the membrane surface since their back transport velocity seem to be very low in mixed liquor suspensions in MBRs. Jiang *et al.* (2007) estimated that particles of 0.1 μm in diameter had the lowest back transport velocity. Once those particles have been deposited on the membrane surface, the cake layer formed would be much less permeable than those formed by flocs with the size of tens of micrometers, which are dominant in mixed liquor suspensions.

Larger particles of organic substances classified as biopolymer clusters (BPC) have also attracted attention. It was reported that BPC ranging from 2.5 to 60 μm in size played a crucial role in sludge cake deposition and membrane fouling in MBRs (Wang *et al.* 2008; Sun *et al.* 2008).

4.2.3.4 EPS/SMP

Microorganisms excrete organic polymers in their metabolic/decay activities. These polymers are called extracellular polymeric substances (EPS) or soluble microbial products (SMP) and are thought to play important roles in membrane fouling in MBRs. Thus, many attempts have been made to correlate EPS/SMP and filterability of a mixed liquor suspension. The terminologies EPS/SMP should be carefully defined in their use depending on whether they are bound to flocs (EPS) or are freely present in the bulk solution (SMP). Polysaccharides and proteins are the major components of EPS/SMP and are often measured in studies on membrane fouling in MBRs. The phenol-sulfuric acid method developed by Dubois (1956) and the Lowry method (Lowry *et al.* 1951) have usually been used for quantification of polysaccharides and proteins, respectively, although they are not necessarily standardized methods. In addition, extraction of EPS is not standardized, which certainly impacts on the results. Although various methods for extraction of EPS from sludge are available, the method using an ion-exchange resin developed by Frølund *et al.* (1996) seems to have been mostly used in recent studies.

EPS alter the surface properties of flocs and are thought to affect the deposition of flocs on the membrane surface. Thus, when control of reversible fouling (cake resistance) is attempted, monitoring of EPS makes good sense. The soluble fraction in a mixed liquor suspension has been increasingly focused in studies on membrane fouling in MBRs as a correlation between membrane fouling in MBRs and concentrations of SMP in the reactors has been reported (e.g., Rosenberger *et al.* 2006). It should be pointed out that in conventional colorimetric methods used for analysis of EPS/SMP, the total amount of polysaccharides and proteins is measured. However, membrane fouling in MBRs seems to be eventually caused by specific fractions of EPS/SMP in MBRs (Kimura *et al.* 2009). Based on this stance, it may be reasonable not to observe a clear correlation between EPS/SMP measured by conventional methods and membrane fouling/filterability of mixed liquor suspensions in MBRs. The number of reports in which a clear relationship between membrane fouling and EPS/SMP is denied has been increasing (Kimura *et al.* 2005; Drews *et al.* 2008; Lyko *et al.* 2008).

4.2.4 Problems to be addressed in monitoring of the filterability of mixed liquor

Despite more than a decade of worldwide research on the issue of fouling in MBR, some emerging results tend to be contradicting which can partly be attributed to the large variety of non-standardised fouling (or filterability) monitoring methods used (Drews, 2010). It should also be noted that the current understanding of properties of foulants is still poor. Although there is a consensus that polysaccharides and/or proteins originating from EPS/SMP play an important role in membrane fouling in MBRs, the specific fractions of polysaccharides/proteins causing membrane fouling in MBRs have not been identified yet. Thus, specific monitoring of 'important' polysaccharides/proteins is not possible. Fractionations of polysaccharides/proteins should be intensively investigated in the future. Some trials for this topic have been initiated (Miyoshi *et al.* 2010, 2011).

Filterability of a mixed liquor suspension is eventually reflected in permeability decline (TMP increase) of the plant. The influence of the 'history' of the membrane used in the plant needs to be considered in the assessment of plant performance. In monitoring plant performance, a sudden TMP jump following a slow and gradual increase of TMP is often observed (Hwang *et al.* 2008). This sudden increase in TMP may be caused by corresponding changes in filterability of the suspension, which is a major motivation for monitoring filterability. However, the sudden jump in TMP may be explained by the 'local-critical flux' concept proposed by Ognier *et al.* (2004). As a result of evolution of membrane fouling in the plant, the number of open and 'working' pores of membranes is reduced and actual flux in the remaining open pores increases to maintain the overall flow rate of the plant. When the actual flux exceeds the critical flux defined above, rapid membrane fouling could occur in the plant. Thus, membrane history affects the fouling process in MBRs, but it is difficult to relate this factor to the filterability of a mixed liquor suspension assessed by the currently available methods.

To make matters worse, due to distinct and inherent differences between lab and full-scale operation, prediction of fouling rates encountered in full-scale MBRs by lab-scale filtration tests is very difficult (Kraume *et al.* 2009). Membrane fouling in MBRs is such a complex phenomenon that it might be reasonable to assume that a single parameter cannot explain it. Use of multiple-component analysis might be appropriate for assessing filterability. Van der Broeck *et al.* (2011) reported that a combination of parameters was more reliable for predicting filterability.

4.3 CHARACTERIZATION OF MEMBRANE FOULANTS IN MBRs

So far, membrane fouling in MBRs is still not fully understood due to the complex nature of membrane foulants. The challenge to the study of membrane fouling behavior and membrane fouling mechanisms calls for the use and/or development of novel approaches for clearer understanding of membrane foulants such as structure and configuration of fouling layer, physicochemical and biological nature of membrane foulants, and deposition behavior of microorganisms on membranes. As such, MBR fouling can be characterized by the following: (i) visualization of cake morphology, (ii) analysis of chemical composition, and (iii) identification of microbial community structure. Indeed, a number of characterization approaches have been used or mentioned in the published papers more or less in order to open the Black Box of MBR fouling (i.e., get more detailed knowledge in regard to characterisitics of fouling layer and mechanisms underlying fouling) and to explain or verify the effectiveness of some implemented fouling control strategies. However, the basic principles, benefits and especially limitations of the characterization techniques are not fully paid enough attention to. In addition, there is a need to know what extent of the fouling behavior can be explained by a given characterization technique. It is also of particular interest to point out the possible issues of each technique when being used for fouling analysis.

4.3.1 Approaches to morphological visualization

4.3.1.1 Scanning electron microscopy (SEM)

SEM is one of the most versatile instruments available for examination and analysis of the structure of fouling layer at nano/micro-meter scale. SEM provides not only high resolution images of the structure of fouling layer, but also valuable additional clues regarding the origins of foulants (e.g., *Bacillus*, *Coccus*, filamentous bacteria). It can also depict the size and shape of foulants clearly. Additionally, some glue-like substances (i.e., biopolymers) embedding the microbes can be observed, which maintain the bacteria aggregate with each other, or bound to the membranes tightly. As such, SEM is possible to provide a micro-scale observation of membrane fouling behavior, particularly for the analysis of surface morphology. Nevertheless, the pretreatment protocols of SEM samples including sample dehydration and gold coating remain as the key issue limiting the wider application of SEM for membrane fouling study. They would inevitably destroy the original structure of fouling layer composed of microorganisms and biopolymers (Meng *et al.* 2005).

One of the most important developments in electron microscopy technique is environmental scanning electron microscopy (ESEM). It is able to characterize the surface of almost any specimen, wet or dry, since the high-vacuum environment is no longer needed and thereby makes direct observation of membrane fouling possible. As such, attempts have been made to use ESEM for membrane fouling observation (Le-Clech *et al.* 2007). However, the ESEM visualization is difficult to obtain clear images when operated under wet mode. The ESEM images are usually not resolved well due to the fact that the microorganisms and extracellular polymeric substances (EPS) are not electron dense (Priester *et al.* 2007).

4.3.1.2 Atomic force microscopy (AFM)

AFM has been thought as an extremely powerful tool for membrane fouling study both in MBRs and in conventional membrane filtration processes (Bowen *et al.* 1999; Lee *et al.* 2001; Tansel *et al.* 2008; Yamamura *et al.* 2008). It can provide three-dimensional images with resolution at around atomic level and quantitative information of cake morphology with the aid of the associated analysis software. Unlike SEM or ESEM, no special sample pretreatment is required for AFM test. The AFM technique is composed of a cantilever with a sharp probe at one end (Hilal *et al.* 2006). When the probe is close to the sample, forces (e.g., Van der Waals forces, capillary forces, chemical bonding, and electrostatic forces) between the probe and the sample lead to a deflection of the cantilever. According to this principle, the three-dimensional image, surface roughness, and adhesion force can be obtained. Lee *et al.* (2001) reported that the fouling layer formed with suspended growth microorganisms (87 nm) had higher roughness than that formed with attached growth microorganisms (34 nm) in an MBR. Low roughness usually corresponds to a compact structure (Yu *et al.* 2006), therefore, a high roughness of fouling layer would be helpful to obtain better filtration performance. Another advantage of AFM technique is the measurement of the force profile (i.e., force as a function of separation distance) and the adhesive force between two interacting surfaces (Brant *et al.* 2002), which are of high importance for the understanding of membrane fouling behavior. According to the adhesive force measured by AFM, it enables one to compare the fouling potential of membranes to be used quickly and to implement more effective strategies for fouling control. As an example, AFM measurements can characterize membrane-to-foulant, foulant-to-foulant, and foulant-to-cleaning agent interactions at a molecular level (Mi & Elimelech 2010; Li *et al.* 2004), consequently it can explain the mechanism underlying membrane fouling and membrane cleaning. To date, the measurement of force by AFM is limited to membrane fouling caused by colloids (e.g., natural organic matter, NOM) and microbes (Li *et al.* 2004; Lee *et al.* 2006), and there is little investigation for MBR fouling.

4.3.1.3 Confocal laser scanning microscopy (CLSM)

During CLSM observation, the laser beam is focused onto the specimen by the microscope objective, which then excites fluorescence (see Figure 4.5a). The fluorescence light, because of its longer wavelength, passes through the dichroic mirror and comes to a focus at the plane of the detector pinhole (Ferrando *et al.* 2005). Only the fluorescence emission from the in-focus spot is able to pass through the pinhole detector and then, form a CLSM image. In addition, serious optical images in the *xy* plane are taken at a number of depths or thicknesses, and subsequently can be reconstructed to a 3D image (Hughes *et al.* 2006b). One of the main advantages of CLSM over most of conventional wide-field techniques (i.e., SEM, ESEM) is that it can differentiate between membrane foulants (Ferrando *et al.* 2005). An appropriate choice of fluorophores will allow the identification of membrane foulants in MBRs (see Figure 4.5b). Two-dimensional (2D) and three-dimensional (3D) CLSM images are often taken to characterize the morphology of the fouling layer, and to identify the components of fouling layer. Another notable advantage of CLSM is that there are considerable fluorescence probes able to target almost all the biopolymers in fouling layer or sludge aggregates (e.g., *refs*. (Chu *et al.* 2005; Chen *et al.* 2006; Ng *et al.* 2006; Hwang *et al.* 2007; Chen *et al.* 2007b; Tian *et al.* 2008b; Tian *et al.* 2008a)). For example, Hwang *et al.* (2007) stained the cell, polysaccharide and protein in the fouling layer using the following probes: SYBR Green I, ConA conjugated with tetramethyl rhodamine isothiocyanate (TRITC-ConA), and benzoxanthene yellow. Therefore, the multiple staining protocols make it possible to know how the membranes are fouled during the operation of MBRs, in what way the foulants interact with membranes, and how the foulants interact with each other.

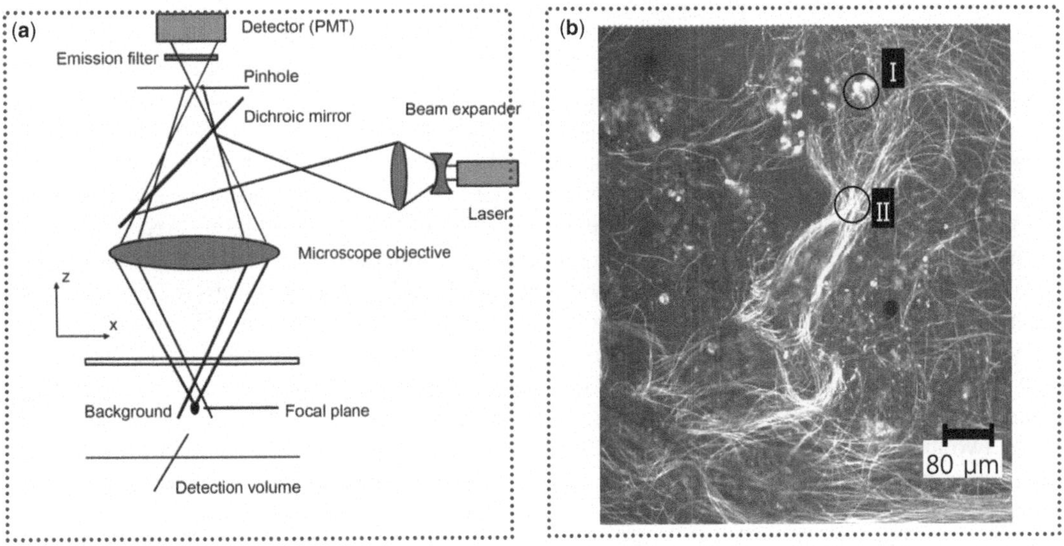

Figure 4.5 (a) Schematic illustration of the operating principle of CLSM (Ferrando *et al.* 2005), and (b) a CLSM image of fouling layer (I, II indicate the deposited free bacteria and filamentous bacteria on membrane, respectively).

Of particular interest of CLSM is the optical sectioning and quantitative analysis of fouling layer, which can be achieved by combining the 3D CLSM image and image analysis software such as Image Structure Analyzer (Beyenal *et al.* 2004; Staudt *et al.* 2004). By using the optical section of CLSM, the porosity can be obtained with the help of image analysis software. In general, the porosity of the fouling layer is low near the membrane surface because of the accumulation of biopolymers, for example 17-polysaccharide (Chen *et al.* 2006). In recent years, an architecture concept has been originally developed by Lee's group at Seoul National University to describe structural characteristics of fouling layer (Jin *et al.* 2006; Yun *et al.* 2006; Lee *et al.* 2007; Park *et al.* 2007; Hwang *et al.* 2007; Lee *et al.* 2008; Hwang *et al.* 2008). Park *et al.* (2007) found that the porosity of a fouling layer measured by CLSM was comparable to that evaluated by the Carman-Kozeny equation, and the porosity of fouling layer, which was formed by 0.1 μm of fluorescent polystyrene latex, is in the range of 0.75–0.95 depending on operating condition (e.g., suction pressure). With the help of 3D CLSM image and image analysis, the spatial distribution of porosity could be determined. Chen *et al.* (2006) however, found that near the membrane surface, the porosity of the fouling layer was low (<0.28), and it increased with increasing cake thickness and then became steady. Additionally, Lee *et al.* (2007) found that the porosity of the fouling layer changed greatly along the length of the hollow fibers, and the lowest porosity was observed at the potted ends of membrane fibers which can be easily compressed by suction pressure. This finding is of great help for the design of membrane module and the implementation of aeration.

Compared with SEM or ESES, a drawback of CLSM is the low magnification. The resolution of CLSM is therefore substantially lower than that (around nanometer scale) of SEM. Application of CLSM is also constrained by effectiveness of laser penetration, usually in the range of a few hundred micrometers dependent on the transparency of the specimen, self-shading, diffraction, and so on. Meanwhile, poor penetration of certain fluorescent dyes may cause a rapid decrease in fluorescence as a function of depth, especially for thick fouling layers. The available depth of the specimen CLSM can be effectively detected is dependent upon physical and chemical properties of the samples and the preparation protocol used for the samples. To avoid the limitations of CLSM, a more straightforward and powerful approach, multiphoton fluorescence microscopy (MFM), was used to visualize membrane fouling, which also enable optical sectioning and thus, can perform *in situ* three-dimensional characterization (Hughes *et al.* 2006a, b, 2007). Unlike CLSM, fluorescence excitation in a MFM is performed by two or more photons of light. The main advantages of MFM over CLSM are the reduced photo-bleaching and deeper optical sectioning.

4.3.1.4 Direct observation (DiO)

DiO was developed to visualize the deposition behavior of particles during membrane filtration (Li *et al.* 1998, 2000, 2003; Mores & Davis 2001, 2002; Neal *et al.* 2003; Subramani *et al.* 2009; Zhang *et al.* 2006b). The DiO technique is mainly composed of a membrane filtration cell equipped with a membrane and a microscope. Depending on the membrane used, the DiO technique can be classified as two groups (see Figure 4.6): direct observation through the membrane (DOTM, e.g., Li *et al.* (Li *et al.* 1998; Li *et al.* 2000) at The University of New South Wales Sydney) and direct visual observation on the membrane (DVO, e.g., Mores *et al.* (Mores & Davis 2001; Mores & Davis 2002) at University of Colorado). For DOTM, the membrane used is transparent in wet mode, allowing the microscope to capture the image of the foulants on the other side of the membrane. The DVO, however, allows for the study of commercially used membranes such as hollow fiber membranes. The restriction of DOTM is that only the fouling layer on the surface of the membrane can be observed clearly. On the contrary, the DVO is good for observing the top layer.

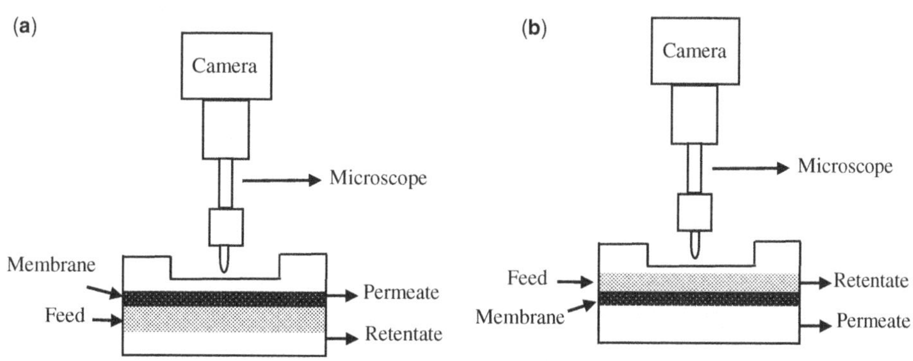

Figure 4.6 Illustration showing the apparatus of DOTM (a) and DVO (b) (Li *et al.* 2000; Mores & Davis 2001).

The DiO was used to observe the deposition of yeast, latex beads and submicron bacteria on membranes (Li *et al.* 1998, 2000, 2003; Mores & Davis 2001; Subramani *et al.* 2009). Li *et al.* (1998) noticed that particle deposition took place gradually during membrane filtration of yeast or latex beads, especially when the imposed flux was higher than the critical flux. In addition, the particle deposition on membranes relies on particle size and cross-flow velocity. Attempts also have been performed to study MBR fouling by using DiO technique. For example, evidence showed that parts of deposited sludge flocs were found to roll and slide across the membrane surface, and subsequently leave the membranes (Zhang *et al.* 2006a). It suggests that the DiO is a powerful approach that can be applied to record the formation and detachment of membrane foulants. The results of Le-Clech *et al.* (2007) provided further information about the dynamic behavior of foulant formation and detachment. They determined the moving velocity of large foulants (around 60 μm in diameter) as 0.02 mm/s, as bulk velocity was controlled at 3 mm/s. The low removal velocity of foulants was attributed to the hindering force induced by the gel on membranes. During the course of backwashing, parts of deposited microbes were removed from the membrane surface. Therefore, the use of DiO technique makes a visual track of the whole fouling development process and foulant removal process possible.

To advance the performance of DiO technique, new generations have been developed (e.g., replacement of flat sheet membranes with hollow fiber membranes (Jankhah & Berube 2009; Marselina *et al.* 2008, 2009), which allowed real time visualization of membrane fouling history possible. For example, change of cake height and velocity profiles of foulants on membranes during filtration can be observed by DiO technique and, cake expansion and erosion during backwashing of the fouled membranes are also possible to be monitored non-invasively (Marselina *et al.* 2009). Of all new generations of DiO technique, it is worth to mention the development of a microfluidic membrane chip which was designed for non-invasive *in situ* monitoring of membrane fouling via the use of synthetic embedded channel membranes (Ngene *et al.* 2010). By observing the side view of the channel membranes, it is possible to monitor the cake height and particle trajectory or velocity.

4.3.2 Approaches to componential characterization

4.3.2.1 Gel permeation chromatography (GPC)

As mentioned earlier, monitoring different fractions of EPS/SMP may be more relevant than the total amount of protein and polysaccharide in EPS/SMP (Meng *et al.* 2009). To understand the molecular weight distribution or size of EPS/SMP compounds, characterization approaches of every sort have been developed, among which one of the most powerful approaches is gel permeation chromatography (GPC) (Akram & Stuckey 2008; Ni *et al.* 2009; Tian *et al.* 2008b, 2009; Wang & Wu 2009). GPC can analyze samples on the basis of size exclusion, so it is also called size exclusion chromatography (SEC), in which gels are used as stationary phase. The EPS/SMP compounds of different size can be separated when eluting through the chromatography columns. The organic compounds (i.e., SMP/EPS) in membrane foulants are expected to be heterogeneous (e.g., large-size biopolymers, small-size acids) and thus exhibit a broad molecular weight distribution. Besides molecular weight distribution, size number average molecular weight (M_n), weight average molecular weight (M_w), size average molecular weight (M_z) and polydispersity index (*PI*) can also be determined by GPC (Omoike & Chorover 2004). For a GPC technique, the signals of eluting samples are monitored continuously with a detector such as UV absorption, differential refractometer (DRI) or refractive index (RI). However, UV detector is mainly sensitive to the UV active compounds like humic substances, proteins and nucleic acids, and it therefore cannot detect polysaccharides effectively. The DRI or RI detector is based on a measure of the decrease of the speed of light or waves through the samples. Thus, a conjunctive use of UV detector and RI detector can provide complementary information about EPS/SMP.

To get more detailed information about EPS/SMP, liquid chromatography with continuous organic carbon detection (LC-OCD) was applied, which can detect organic matter in water and wastewater in a single step. And, the measurement of biopolymer concentration in samples is also possible. As a piece of state-of-the-art equipment, it has been popularly used to analyze the fouling-causing substances in membrane processes (Haberkamp *et al.* 2007). After being injected into a chromatography column, the components of the sample are separated according to their size: (1) biopolymers (i.e., mainly polysaccharides and macromolecular proteins; $t_{elution} = 35–55$ min), (2) small-size proteins and humic substances ($t_{elution} = 55–64$ min), (3) low molecular weight acids and neutrals ($t_{elution} > 64$ min). Zheng *et al.* (2009a) also found that the LC-OCD chromatography of feedwater and permeate differed significantly in the range of large-size substances identified as biopolymer peak during UF membrane filtration of WWTP effluents, suggesting the UF membranes can retain biopolymers in feedwater completely. Both bound EPS and SMP contain biopolymers, and the biopolymer peak of bound EPS is usually more evident (Al-Halbouni *et al.* 2008), implying that the large molecular size substances in sludge supernatant might be originated from bound EPS (e.g., hydrolysis of bound EPS due to biomass metabolism and release of bound EPS from sludge flocs due to shear stress induced by aeration). In addition, the peaks of the chromatogram can be integrated to obtain the equivalent amount of organic compounds (Halle *et al.* 2009; Zheng *et al.* 2009b). Since MBR fouling is mostly determined by biopolymers, it is of high significance to calculate the gross concentration of biopolymers in the sludge supernatant. In fact, the biopolymers only account for a small proportion in the total DOC (e.g., 6–25% reported in *ref.* (Rosenberger *et al.* 2005)).

GPC and LC-OCD only allows for the analysis of solubilised samples, thus the membrane foulants have to be removed from the fouled membranes. The main drawback of GPC and LC-OCD technique is that the elution of samples through the columns depends on size of compounds as well as the adhesive/repulsive force between samples and column packing. For example, an adhesive force will delay the eluting time of samples, resulting in an underestimation of molecular weight. On the contrary, a repulsive force can lead to an overestimation of molecular weight. As such, the measured molecular weight of components rely on the column packing, samples themselves, and the standard compounds used for calibration.

4.3.2.2 Spectroscopic techniques for organic matter characterization

As a complex mixture, the organic matter (i.e., EPS/SMP) in membrane foulants exhibits strong optical characteristics such as UV absorption (UV_{254}, UV_{260}, and UV_{280}), fluorescence absorption and infrared absorption. Therefore, the optical sensitive compounds can be detected by techniques such as three-dimensional excitation-emission matrix (EEM) fluorescence spectroscopy, Fourier transform infrared spectroscopy (FTIR), and so on. EEM fluorescence spectroscopy can monitor EPS/SMP non-destructively. One important advantage of this technique is the fluorescence characteristics can be entirely acquired by changing excitation wavelength and emission wavelength simultaneously (Sheng & Yu, 2006). Owing to the high sensitivity and the good performance, EEM technique has been widely used to identify and characterize NOM and EPS substances (Kimura *et al.* 2004; Li *et al.* 2008; Lu *et al.* 2008; Nam & Amy 2008; Sheng & Yu 2006; Tsai *et al.* 2008). Kimura *et al.* (2009) compared effects of solid retention times (SRTs) on membrane foulants by means of EEM, and found that relative dominance of the protein peak (Ex/Em = 270/320 nm) differs in the order of MBR-A (SRT = 15 d) > MBR-C (SRT = 102 d) > MBR-B (SRT = 51 d) and humic substances were found to play more significant roles in MBR-B and MBR-C than in MBR-A. A more recent investigation performed by Wang *et al.* (2009) reported that Peak A and Peak B (proteins) are much stronger than Peak C (humic substances) in membrane foulants, indicating that the membrane foulants are dominated by proteins. Unfortunately, the polysaccharides in membrane foulants are not detectable by EEM (Her *et al.* 2003). Another significant challenge for the use of EEM fluorescence is the impacts of ambient conditions (e.g. metal ions, pH, and temperature) which can alter the fluorescence intensity (Henderson *et al.* 2009).

Unlike EEM fluorescence spectroscopy, FTIR can help us know the functional groups contained in organic matter. In a typical FTIR spectrum of membrane foulants, the broad peak present at 1000–1200 cm^{-1} is due to symmetric and asymmetric C=O stretch in polysaccharides or polysaccharide-like substances. The peak at around 1400 cm^{-1} is attributed to symmetrical stretches of –COO-associated with amino acids (Badireddy *et al.* 2008; Omoike & Chorover 2004). There are two peaks at 1650 and 1540 cm^{-1} in the spectrum which are unique to protein secondary structure, called amides I and II. In some cases, however, the 1600–1650 cm^{-1} band is assigned to quinines and aromatic rings or humic-like substances (Kimura *et al.* 2005; Richard *et al.* 2007). By comparing the presence and intensity of these IR peaks, the composition of membrane foulants is easily discerned, for example, a stronger peak intensity indicates the occurrence of more severe membrane fouling (Kim & Jang, 2006). One benefit of FTIR technique is that it can map the distribution of protein/polysaccharide on a region of interest if a FTIR-microscope is available. FTIR-microscope study is proved to be a powerful methodology for quantitative mapping of proteins on membranes (Delaunay *et al.* 2006; Xu & Fu, 2004). So far, there is no such study in MBRs. Similarly, the structural compositions, such as proteins (peaks at 175 and 55 ppm), aromatic carbon (110–165 ppm), and polysaccharides (peaks at 75 and 105 ppm), can be detected using ^{13}C-NMR (Kimura *et al.* 2005).

By now, the analytical methods on EPS/SMP in membrane foulants are limited to the identification of protein-like substances, polysaccharide-like substances and humic substances. The word 'like' indicates that the chemical nature of EPS/SMP compounds is not well-known. As a matter of fact, protein-like substances, polysaccharide-like substances and humic substances are of similar characteristics, for example, both proteins and humic substances are UV and fluorescence active. There are still a number of opening issues that need to be clarified. For example, how to understand the chemical nature of organic matter in detail? Which components or fractions of organic matter are more important to MBR fouling? To address these critical problems, it calls for further research on EPS/SMP such as the study of protein secondary structure through fingerprint analysis of EEM and FTIR data (Omoike & Chorover, 2004; Ramesh *et al.* 2006, 2007).

4.3.3 Approaches to microbiological identification

In MBRs, the unwanted adhesion and proliferation of microbes on membranes, also called membrane biofouling, is a complex phenomenon. Once microbial adhesion occurs, the membrane surface exhibits a more hydrophobic nature, enabling other microbes to deposit more readily. Nonetheless, the adhesion tendency of each microbial species differs depending on the shape, size and surface properties of microbes. After the initial deposition of microbes on membranes, temporal and spatial shift of microbial community structure takes place in the course of MBR operation due to bacteria proliferation and death. The microbial species are of different tendencies to metabolize and proliferate in oligotrophic and anoxic conditions of the fouling layer or biofilm. Moreover, the microbes can self-organize and coordinate themselves by transferring quorum sensing (QS) molecule signals, also known as autoinducers, to neighboring microbes, enabling more microbes of the same species congregate (Kim *et al.* 2009; Yeon *et al.* 2009a, b). To get more insights into fouling behaviors of microbes, the microbial community structures and microbial colonization on the membranes should be analyzed by using polymerase chain reaction denaturing gradient gel electrophoresis (PCR-DGGE) and Fluorescence *In situ* Hybridization (FISH).

By using FISH, the *Gammaproteobacteria* was found to be prevalent at the earlier stage of NF and UF filtration of drinking water source, whereas *Alphaproteobacteria* and *Betaproteobacteria* were the dominant species of mature fouling layer (Horsch *et al.* 2005). In contrast to this study, Miura *et al.* (2007) reported that the *Betaproteobacteria*, probably was the predominant species of the fouling layer in MBRs. In fact, the MBR in each study are subject to different operating conditions such as aeration rate, imposed flux, SRT, substrate, physical/chemical cleaning, and so on, which result in the contradictions of reported results. Taking Figure 4.7 as an example, we can see that when the aeration rate shifted from 3500 L/h to 5000 L/h, the relative abundance of *Betaproteobacteria* dramatically increased from around 30% to 70%, indicating that, compared with other microorganisms, the *Betaproteobacteria* microbes adhere to membranes more tightly, becoming the dominant microbes on membranes when operated under higher aeration stress. But, the *Betaproteobacteria* microbes can be effectively removed by chemical cleaning. More recently, the high-throughput pyrosequencing was applied for the identification of membrane fouling in MBRs (Kim *et al.* 2012; Lim *et al.* 2012; Ma *et al.* 2013). Lim *et al.* found that *Enterobacter* and *Dyella* were closely associated with the initial and late biofouling stages, respectively. In fact, the pyrosequencing analysis is of high help for the study of spatial and temporal changes of microbial community structures.

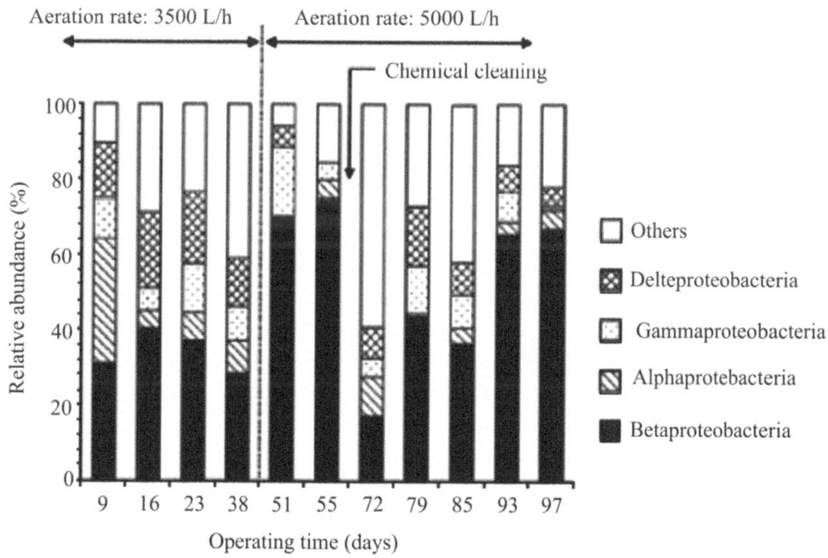

Figure 4.7 Relative abundance of major phylogenetic groups in fouling layer (Miura *et al.* 2007).

But, the procedure of FISH and PCR-DGGE analysis is complex and time-consuming. Particularly, knowledge of microbiology is a prerequisite for the use of these analytical methods. Indeed, FISH analysis is similar to the CLSM observation of protein and polysaccharide. The only difference is that FISH analysis needs more gene probes to target most of the microbial community involved in the fouling layer. However, PCR-DGGE is a destructive approach, which requires decomposition process such as destruction of the fouling layer, DNA extraction, and DNA amplification. More to the point, most of current MBR fouling study is focused on SMP/EPS, the microbe attachment/detachment and the microbial ecology on membranes are not fully illuminated. Therefore, there is a need to develop or apply a non-invasive and easy-to-operate method to identify microorganisms in the fouling layer. Confocal Raman microscopy (CRM) has been demonstrated to be a non-invasive approach for microbial analysis in biofilm (Patzold *et al.* 2006, 2008), which is able to map the spatial distribution of microbes in biofilm without destroying it. It indicates that an online analysis of microbial communities in fouling layer is possible by using CRM. Currently, however, there is a lack of a spectral database of microorganisms (Patzold *et al.* 2006). Additionally, Raman microscopy seems to be a powerful tool for recording the distribution of EPS/SMP matrix in biofilm (Ivleva *et al.* 2009; Wagner *et al.* 2009).

4.3.4 Summary of approaches to characterization

Over decades of research, MBR fouling is partially illuminated at all levels, from macro- to micro-view, by what we use to analyze and characterize it. As summarized in Table 4.1, based on the review of the literature and our knowledge, the main conclusions of this section can be drawn as follows:

- SEM, AFM, CLSM and DiO techniques are powerful techniques for autopsying structure of fouling layer (i.e., cake density, thickness, porosity), though each has its own strengths. SEM allows for the monitoring of fouling layer with high resolution, but needs a rigid dehydration process. AFM is possible to obtain three-dimensional images with resolution at around atomic level and the adhesive force between membranes and foulants. CLSM can not only prove the three-dimensional images of fouling layer, but differentiate microbes, proteins and polysaccharides as well when fluorescence probes are used. The problem of photobleaching should be borne in mind during CLSM analysis. One promising feature of DiO technique is the on-line observation of the development history of fouling layer during membrane filtration and the detachment behavior of foulants during membrane cleaning.
- GPC or LC-OCD are popularly used to determine the SMP/EPS and soluble substances in foulants in terms of molecular weight. The LC-OCD seems to more reliable due to the fact that the eluted samples are detected by an on-line instantaneous TOC measurement. FTIR and EEM are capable of detecting the compounds in foulants with IR and fluorescence adsorption. A fingerprint analysis of FTIR and EEM spectra is expected to provide useful information about membrane fouling mechanisms.
- AFM is of particular interest to measure the membrane-to-microbe interactions, or to detect the adhesion propensity of microbes. But, the development of novel AFM probes is a prerequisite of AFM measurements. FISH is a nondestructive approach that can detect microbe species on membranes. FISH analysis needs probes to target the microbes to be detected in CLSM observation. The findings of PCR-DGGE analysis are expected to be informative and reliable though a complex procedure of sample preparation is indispensable.
- A number of techniques are proven to be multi-purpose. For example, the CLSM can perform both morphological visualization (e.g., 2-D and 3-D images of fouling layer) and componential characterization (e.g., identifying the proteins, polysaccharides and microbes). Likewise, the use of AFM enables morphological visualization, as well as adhesive force measurement.

Table 4.1 Capabilities, strengths and limitations of current techniques for membrane fouling characterization in MBRs.

	Approach	Capabilities	Strengths	Limitations	References
Morphology visualization	SEM	2D imaging of fouling layer	Taking image with high resolution	Dehydration and precoating of samples is required	Le-Clech *et al.* (2007); Meng *et al.* (2005)
	AFM	3D imaging of fouling layer and; offering force profile between foulants and membranes	Dehydration and precoating of samples are not required and; It provides more information about fouling layer (e.g., surface roughness, force adhesion)	It strongly depends on the probe used	Lee *et al.* (2001); Lee and Elimelech (2006); Li and Elimelech (2004)
	CLSM	3D imaging of fouling layer, identifying of deposited bacteria and; EPS measurement of cake thickness and cake porosity	Dehydration of samples is not required; Foulants differentiation using fluorescence probes and; optical sectioning and quantitative analysis of fouling layer.	It needs fluorescence staining; Lower resolution compared to SEM and; Fluorescence photobleaching	Chen *et al.* (2006)
	MFM	*In situ* monitoring of fouling history	*In situ* three-dimensional characterization and; Reduced photobleaching and deeper optical sectioning	The filtration tests need to be stopped temporarily to obtain images with higher quality	Hughes *et al.* (2006a, b, 2007)
	DiO	*In situ* monitoring of fouling history	Monitoring of the development of cake height and; Observation of foulant movement on membranes	A special membrane filtration cell needs to be fabricated	Li *et al.* (1998, 2000); Mores and Davis (2001, 2002)
EPS/SMP characterization	GPC	Measuring the molecular size of EPS/SMP	Offering molecular weight distribution of EPS/SMP and; Determining the key fouling-causing substances	The data given by GPC is strongly depends on the detector used (UV detector can only monitor the UV active matter)	Akram and Stuckey (2008); Ni *et al.* (2009); Tian *et al.* (2008b, 2009); Wang and Wu (2009)

(*Continued*)

Table 4.1 Capabilities, strengths and limitations of current techniques for membrane fouling characterization in MBRs (*Continued*).

	Approach	Capabilities	Strengths	Limitations	References
	LC-OCD	Measuring the molecular size of EPS/SMP	Monitoring of the molecular weight distribution of EPS/SMP by continuous organic carbon detection and; Possible to give biopolymer concentration	It is much more expensive than conventional GPC	Haberkamp *et al.* (2007)
	FTIR	Characterizing functional groups in foulants	Differentiating chemical composition of organic compounds in membrane foulants via fingerprint analysis	Sample dehydration is needed	Badireddy *et al.* (2008); Omoike and Chorover (2004)
	EEM	Characterizing organic substances in foulants	Monitoring EPS/SMP non-destructively and; Proteins and humic substances can be characterized	Polysaccharides cannot be monitored	Meng *et al.* (2011)
Microbiological identification	PCR-DGGE	Detecting the predominant microorganisms	Fingerprint analysis of microbial community structures and; Semi-quantitative analysis of microorganisms	It is complex and time-consuming and; knowledge of microbiology is a prerequisite	Huang *et al.* (2008)
	FISH	Measuring the relative abundance of microorganisms	Quantitative analysis of microorganisms in fouling layer	It needs fluorescence staining	Horsch *et al.* (2005)
	CRM	Detecting microorganisms by Raman spectrum	It is a non-invasive approach for microbial analysis	It is lack of a spectral database of microorganisms	Patzold *et al.* (2006, 2008)
	Pyrosequencing	Detecting the predominant microorganisms	Analysis of a great amount of samples	Missing of the information of some bacteria	Kim *et al.* (2012); Lim *et al.* (2012); Ma *et al.* (2013)

4.4 BIOFOULING CONTROL

As the control of membrane fouling (particularly biofouling) in MBR has been recognized as a key factor for reducing the operating and maintenance costs through energy saving, numerous methods of fouling control have been studied, reported and implemented in pilot and/or industrial scale MBRs (Drews, 2010; Wu *et al.* 2012).

Various approaches to fouling control would be categorized as following: Membrane development, Chemical, Physical, and Biological Approaches, some of which will be described in more detail in the following sections (Figure 4.8).

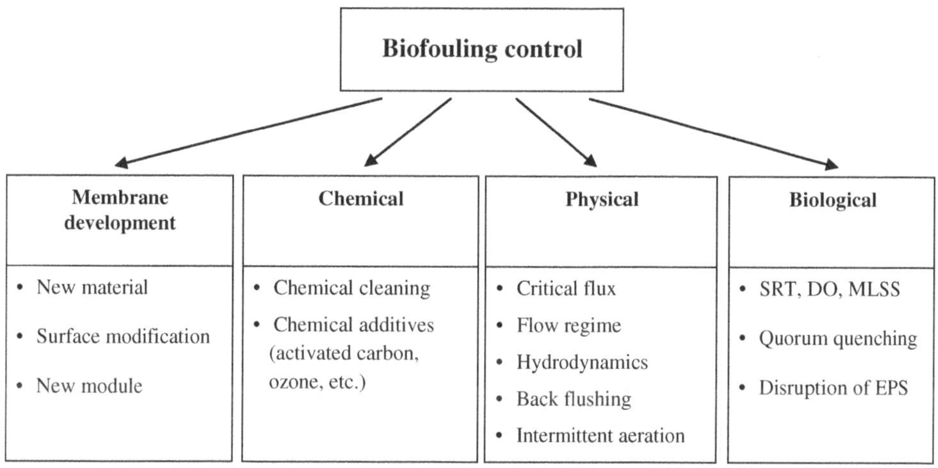

Figure 4.8 Various approaches to fouling control in MBR for wastewater treatment.

4.4.1 Membrane development

Over a long period, physico-chemical or biological modifications of membranes or membrane materials have been conducted to inhibit membrane biofouling. Recent interests focus on the application of nano-or bio-materials to modify the membrane properties. Chae *et al.* (2009) investigated the effect of fullerene C_{60} on bio-fouling of MF membranes with a model microorganism *Escherichia coli* K12. They found C_{60} inhibited microbial respiratory activity and/or attachment to membrane surface. Celik *et al.* (2011) synthesized a blended membrane with MWCNT and polyethersulfone (PES) by phase inversion. The blended membrane showed higher water flux and slower fouling rate than intact PES membrane when surface water was filtrated. They explained it as an increased hydrophilicity of the blended membrane. Kim *et al.* (2011a) directly immobilized a quorum quenching enzyme (acylase) onto a conventional NF membrane. They observed that a great anti-biofouling feature of the enzyme immobilized membrane by suppressing EPS secretion and thus biofilm maturation. A patterned surface membrane was also reported to be effective against biofouling. Won *et al.* (2012) introduced a patterned morphology (pyramid-, prism- and embossing-pattern) on the membrane surface in the preparation of MF membrane using a lithographic method. Deposition of microbial cells on the patterned membrane was significantly reduced compared to that on the flat membrane in the membrane bioreactor (MBR) for wastewater treatment. This was attributed to hydraulic resistance of the apex of the patterned surface, which induced local turbulence.

Innovative changes of module configurations have also been implemented to mitigate membrane fouling. Hai *et al.* (2008) developed a hybrid module obtained by winding a rigid spacer on the surface of a hollow fiber module. This type of spacer-filled module played the critical role in avoidance of intrusion of sludge while simultaneously allowing the washout of the small amount of sludge trapped within it. Vortex generating membrane module creates Kármán vortices at the membrane interface, without in contact with the membranes, which maintains the foulants in suspension to be carried away by the bulk flow. Kang *et al.* (2011) used the flat UF membranes equipped with a rotating vortex generator (FMX, BKT Ltd.) for the filtration of anaerobic digestion sludge, which allows the membrane system to cope with high solids concentration feeds up to 5%. Jie *et al.* (2012) designed a helical membrane to achieve enhanced flux without increasing aeration intensity. Figure 4.9 shows (a) the support spacer and (b) helical membrane. They reported that the helical membrane in laboratory scale MBR produced rotational flows near the membrane surface and enhanced the shearing rate and permeate flux.

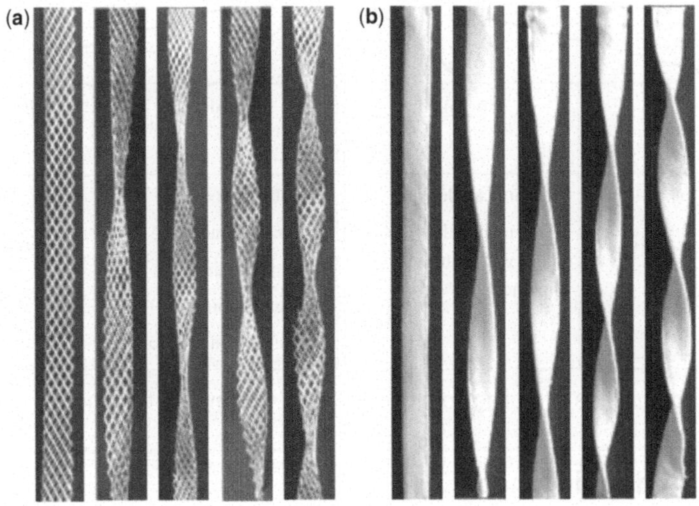

Figure 4.9 The illustration of (a) the support spacer and (b) final helical membranes (adapted from Jie *et al.* 2012).

Hollow fiber bundle property (packing density and fiber diameter) is important to design a module because those are directly related to permeate productivity. Pollet *et al.* (2008) showed that hollow fiber module with the higher fiber diameter and the lower packing density led to a better performance in terms of fouling. But their finding has a limitation for applying submerged membrane bioreactor for wastewater treatment because they used synthetic clay particles and an air lifted sidestream flow regime.

Although many studies have been made to develop low fouling membranes using the various methods described above, applications of those technologies to real MBRs for water treatment have not yet been reported, indicating that further research & development works are still needed.

4.4.2 Chemical approaches

4.4.2.1 Chemical cleaning

Although chemical cleaning has several drawbacks such as inconveniences in transportation, storage, preparation, and uses as well as production of secondary contaminants, it has been used as a primary tool to restore membrane permeability (Le-Clech, 2006). Reversible fouling caused by the deposition of sludge flocs can be prevented by sub-critical flux operation or physical cleaning such as backwashing, air scouring, and so on. The irreversible fouling, however, caused by adsorption and/or chemical bonding (or interaction) between membrane and foulants cannot be managed by sub-critical flux operation or other physical cleanings. Thus, periodical chemical cleanings are required in full scale MBR plants. Chemical cleaning using various reagents, such as oxidants, acids, alkalis, and chelating agents is carried out in two ways: off-line cleaning and cleaning-in-place (CIP). In off-line cleaning, membrane modules are taken out of bioreactor and immersed in a separate tank containing cleaning reagent or membrane modules are immersed directly in membrane tank full of cleaning agent after draining off activated sludge. But in CIP, chemical agents are directly injected into the membrane in a reverse direction to normal filtration while membrane modules are still submerged in the bioreactor.

Compared with off-line cleaning, CIP is much simpler and cheaper (Wei *et al.* 2011a). The periodical CIP cleaning, often called 'maintenance cleanings', normally employs a combination of NaOCl (primarily for removing organic polymers through oxidation) and mineral and/or organic acids (for dislodging scales and metal oxides through solubilisation) (Judd, 2010). Occasionally CIP takes the form of a chemically enhanced backflushing, where chemicals are added to the backflushing water to enhance foulants removal (Zsirai *et al.* 2012).

Among the many cleaning chemicals, sodium hypochlorite (NaOCl) remains a popular choice for both the off-line cleaning and the CIP because of its availability, reasonable price, and its disinfection ability for the biofouled membrane. Sodium hypochlorite concentration for cleanings in MBR ranges typically from 1000 to 3000 mg/L. If the membranes have been continuously exposed to NaOCl for cleaningthey can be damaged. Wang *et al.* (2010) studied the impact of hypochlorite cleaning on the characteristics of polyvinylidene fluoride (PVDF) membrane which is the most frequently used membrane material in MBR plants worldwide. Just like the basic kinetic approach of disinfection, they used a standardized value to evaluate the impact of hypochlorite exposure on membrane, $C \cdot t$, where C is a hypochlorite concentration and t is the exposure time. They found the sodium hypochlorite cleaning did not damage the chemical structure of PVDF membrane, but had impacts on its surface properties. They concluded that PVDF membranes could stand normal chemical cleaning conditions over several years.

4.4.2.2 Chemical additives

The addition of chemicals to MBR has been envisaged to remove or change the size of foulants (colloids or EPS) in the mixed liquor. Addition of powdered activated carbon (PAC) to MBR was tried to reduce membrane fouling (Kim *et al.* 1998). They reported that the membrane permeability of PAC-added-MBR was greatly enhanced compared with that of normal MBR. It is because the PAC led to the decrease of not only the compressibility of sludge flocs but the content of extracellular polymeric substances (EPSs) inside the microbial floc. This resulted in the increase of the porosity of biocakes and thus enhanced the membrane flux. Satyawali and Balakrishnan (2009) summarized the effects, causes and mechanism occurring in an MBR after PAC addition. Figure 4.10 shows the effects of PAC addition and the underlying reasons for the associated benefits. The PAC addition is expected not only to mitigate membrane fouling, but also to enhance the biodegradation of recalcitrants or slowly biodegradable compounds, which is particularly relevant to MBR treating industrial wastewater.

Charged polymers have been reported to be effective in biofouling control. For example, Nalco Company has developed a cationic polymer-based flocculant with the trade mark of 'MPE50 (Membrane Performance Enhancer)' to enhance membrane flux in MBR. With the MPE dose of 200 ~ 400 PPM, the average membrane flux was reported to increase by 50–150% depending on temperature and feed characteristics in the long term operation of a pilot- and full-scale MBRs (Yoon *et al.* 2005, 2006). The flux increase with the addition of MPE was attributed to the increase in porosities as well as the the decrease in biovolumes along the length of submerged hollow fiber membranes, and to the decrease in the concentration of soluble EPS (Lee *et al.* 2007; Hwang *et al.* 2007).

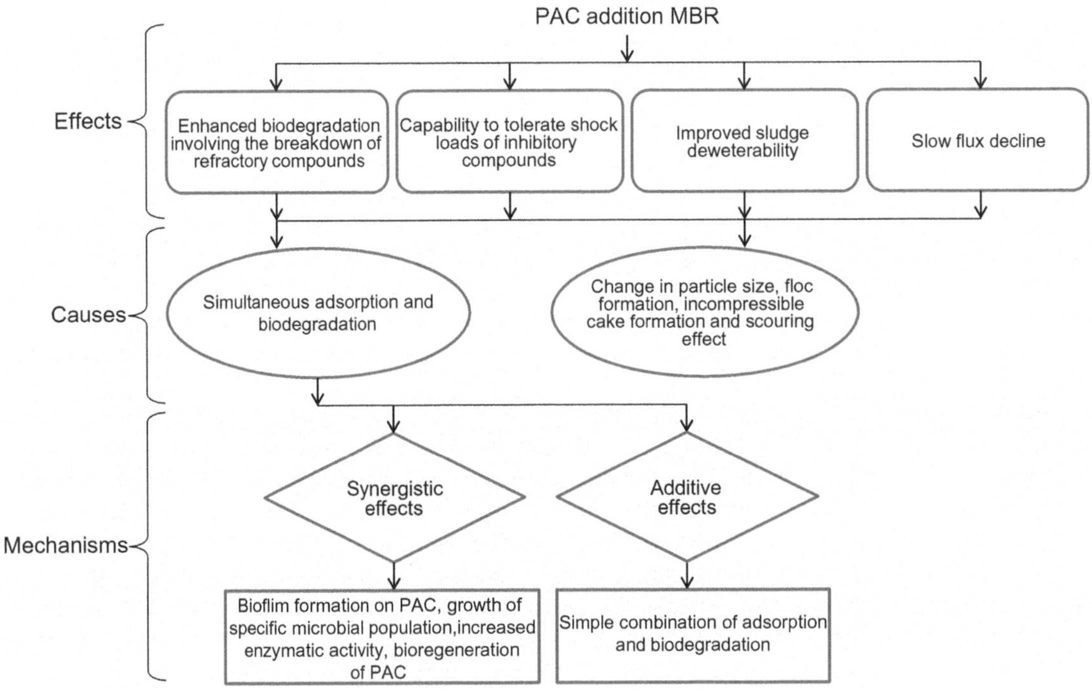

Figure 4.10 Schematic illustrating the effects, causes and mechanism occurring in an MBR after PAC addition (Adapted from Satyawali *et al.* 2009).

Several studies have introduced ozone to MBR for the purpose of controlling membrane fouling as well as excess sludge disposal (Chu *et al.* 2009). Huang and Wu (2008) studied the effect of ozonation on the membrane filterability in MBR in laboratory scale. They found the membrane filterability of the mixed liquor was improved by ozonation with a dosage less than 0.7 g O_3/kg MLSS. During long-term operation of MBR, ozonation did not affect COD and ammonia removal, which means ozonation did not influence the microbial activity within the ozone dosage applied.

Hwang *et al.* (2010) combined a turbulent jet flow ozone contactor (TJC) with MBR to establish an MBR with no excess sludge by destruction of excess sludge with TJC followed by its recycle to MBR. The TJC-MBR gave rise to higher membrane permeability despite the recycle of excess sludge because the recycle of disintegrated sludge increased the average floc size in the mixed liquor, thereby increased the porosity of biocake.

4.4.3 Physical (hydrodynamic, mechanical) approaches

4.4.3.1 Aeration

An excessive and extensive aeration to a MBR tank is a most widely practiced way to vibrate the submerged membrane mechanically and remove sludge foulants on the membrane. Unfortunately, it requires huge amounts of energy (Judd, 2008). For example, 35% of total energy needed for the Grasse WWTP in France was consumed for fine bubble aeration of activated sludge and 29% was due to the coarse aeration to prevent membrane fouling (Barllion, 2011). Janot *et al.* (2011) reported that the air scouring was responsible for 49% of the energy consumption of the Nordkanal MBR plant in Germany. Many attempts have been made to overcome this energy problem but the excessive coarse aeration is still most frequently practiced in MBR plants because it is easily installed and operated and oxygen should be supplied for microorganisms in MBR tank regardless.

Air cycling (air scour system in LEAPmbr, GE), one of the commercial examples seeking to reduce energy cost for aeration in MBR plants, involves delivering air to the membrane modules or cassettes in 20 or 40 second intervals of air-on and -off for the purpose of scouring the membrane fibers (Janot *et al.* 2011).

The MemPulse™ MBR system (Siemens Water Technologies, 2013) uses a mechanical device that supplies irregular pulses of air to the MBR module. The system is supplied with a continuous air supply that is accumulated in the base of the membrane module. It periodically releases irregular pulses of air to the MBR module, creating plug flow, providing shear stress to the membrane surface and refresh the membrane surface to prevent fouling.

One of wise strategies for the effective use of air supplied to membrane module is to produce slug (or plug) flow, which enhances mass transfer near membrane surface. Enhancement of membrane performance by two phase flow (liquid and gas flows) in MF and UF was extensively reviewed by Cui *et al.* (2003). Chang and Judd (2002) reported that a sort of slug flow in tubular membrane enhanced the flux significantly (~43%) in MBR treating sewage. Zhang *et al.* (2011) also reported that slug bubbles showed better antifouling performance in flat sheet MBR than the free bubbling. They pointed out that slug bubbling is effective in preventing the irreversible membrane fouling at the initial stage and dramatically reducing reversible fouling for long term MBR operation. Braak *et al.* (2011) reviewed aeration and hydrodynamics in submerged MBR and addressed that one of the most efficient strategies to limit fouling is the use of a gas/liquid two-phase flow to enhance the mass transfer. However, the multiphase aeration still remains incompletely understood due to the complexity of flows and of the nature of activated sludge.

Another issues related to aeration in membrane bioreactor are bubble size and distribution. Phattaranawik (2007) showed that fouling was delayed and aeration demand was lowered by using a bubble size transformer enabling the coalescence of fine bubbles into coarse bubbles. The bubble diameter could be increased by approximately 14 times and the coarse bubbles obtained were well distributed over the membrane surfaces. On the other hand, Lee *et al.* (2009) found uneven distribution of TMPs over positions on the membrane length due to the pressure drop along the lumen of hollow fibers. To overcome the maldistribution of TMP buildup, they suggested positioning a sparger in the vicinity of the suction point.

4.4.3.2 Backflushing

Backflushing (or backwashing) has been considered as a primary tool for the control of reversible fouling in most membrane filtration processes including MBR. Backflushing is typically carried out with permeate, pure water and/or cleaning chemicals. Frequency and pressure of backflushing are usually guided by membrane manufacturers. But periodic backflushing can cause membrane disintegration, so that the life span of a membrane should be considered first for backflushing.

Intermittant suction can provide an alternative method for the suppression of fouling (Yamamoto *et al.* 1989; Chiemchaisri *et al.* 1993; Liu *et al.* 2000). Temporary cessation of suction, for example, relaxation, creates back transport of permeate which then helps to dislodge the cake layer on membrane surface. This technique is also called cyclic filtration. However, an optimal time interval for on- and off- filtration largely depends on experimental data.

4.4.3.3 Moving media

Moving media can also mitigate membrane fouling by abrasion in MBR. Adding soft sponge balls (Ngo *et al.* 2008), hard plastic media such as polyethylene (Leiknes *et al.* 2007; Yang *et al.* 2012) or polypropylene (Siembida *et al.* 2010) and granular activated carbon (Johir *et al.* 2011) into MBR can reduce membrane fouling. Such carriers work as abrasive particles, that is, they move toward biocakes on the membrane surface and then, take them off by mechanical scouring, leading to enhanced membrane permeability. Commercially, Microdyne-Nadir (2013) developed a mechanical scouring system in MBR. BIO-CEL®-MCP process is supported by the cross-flow aeration and the use of inert organic granulate which is added directly into the MBR tank. The airflow draws the granulate up between the membrane sheets and the membrane surface is continually cleaned through the direct contact of the granulate with the sludge on the membrane surface.

4.4.3.4 Critical flux

After the critical flux concept was introduced by Field *et al.* (1995), sub-critical flux operation in membrane processes has been practiced commonly for the purpose of retardation of severe membrane fouling. Typical value of critical flux in MBR plants ranges from 10 to 30 $L/(h \cdot m^2)$ depending on membranes property (Bottino *et al.* 2009), flow regime (side-stream or submerged type, Judd, 2010), aeration rate (Guglielmi *et al.* 2007; Howell *et al.* 2004), module configuration (Lebegue *et al.* 2008) and microbial characteristics (Gui *et al.* 2002; Yoon & Collins, 2006; Wu *et al.* 2008).

4.4.3.5 Electrical control

Like electrophoresis for bioassay and electro-static precipitation (ESP) for dust collection, application of electricity to improve performance of pressure driven membrane filtration processes has been practiced for a long time. Recently the application of electricity to MBR for fouling control has attracted attention more and more. The electrical application is categorized into three: (i) induction of electric field, (ii) *in-situ* electro-coagulation, (iii) high voltage impulse.

Typical suspended and colloidal particles including activated sludge flocs have negative charges in aqueous solution. A DC electric field makes it possible to migrate these charged particles from membrane surface to electrodes, facilitating back transport of particles from membrane surface to bulk solution (Huotari *et al.* 1999). Liu *et al.* (2012) reported that minute electric field (0.036 to 0.073 V/cm) improved microbial growth and activity, and thus reduced EPS production. As a result, it enhanced the permeate flux in submerged MBR significantly. Akamatsu *et al.* (2010) proposed 'on and off' electric field MBR system. They found that an intermittent electric field ranging from 4 to 6 V/cm made it possible to keep the permeate flux 3.5 times as high as that with no electric field. They found no gas bubble formation from electrolysis of water, which means that the higher flux was not owing to the gas scouring which can remove biocakes from the membrane surface but to the electric field. Chen *et al.* (2007a) also reported that flux of the hollow fiber submerged MBR increased by about 10% with electric field strength from 15 to 20 V/cm, then kept constant after 20 V/cm.

The other type of application of electricity to fouling control in MBR is electro-coagulation (EC), which has recently been the subject of a growing research interest, based on its potential ability to treat various kinds of wastewater. Basically EC based on *in situ* formation of cations at anode such as aluminum and ferric ion which could be acted as a coagulant to reduce the electrical double layer of the negatively charged colloidal matters. Membrane filterability of the submerged membrane electro-bioreactor (SMEBR) proposed by Bani-Melhem and Lelktorowicz (2010) was enhanced by reducing the fouling rate up to 16% without any backflushing. They explained the cause of the improvement in filterability was closely associated with a decrease in zeta potential of the mixed liquor flocs from −31 to −15 mV.

Bacterial viability should be considered primarily as a crucial parameter for this type of EC-MBR because electric current in bioreactor may influence microbial activity. The advantages of membrane filterability should overcome or at least compensate for the disadvantages caused by possible damages of biomass which could lead to the decline of biodegradation. Wei *et al.* (2011b) studied the effect of electrical intensity on microbial viability and found that the bacterial activity was not significantly affected (less than 10% of death percentage) when the applied electric current density was less than 6.2 A/m^2 for 4 hr. But the percentage of live cells dropped by 15% and 29% at current density of 12.3 and 24.7 A/m^2, respectively, indicating that there is a critical point of current density for affecting microbial activity. They observed the key role of pH for cell viability. In general, the hydroxyl ion (OH^-) is generated at cathode due to the electrolysis of water, thus solution pH increases due to the electrochemical reactions. This alkaline environment could be harmful for bacteria. As the current density increased, the solution pH increased significantly, which is partially responsible for the decline of cell viability.

Due to the huge amounts of sludge production by either metal hydroxide or metal phosphorous, MLSS concentration in EC-MBR is higher than the normal MBR plants. Bani-Melhem and Lelktorowicz (2011) reported that the MLSS concentration increased from 3500 to 5000 mg/L within 30 days of operation. They indicated that the ratio of MLVSS to MLSS is about 70%, which is relatively lower than that of normal activated sludge suspension, 85–95%. The low ratio of MLVSS/MLSS is obviously due to the inorganic portion of the sludge originated from the precipitated sludge. Therefore, strategy for disposal of the produced 'heavy' sludge and cost consideration should be taken into account at initial stage of design for the EC-MBR.

Recently high voltage impulse (HVI) technique has been introduced in order to control membrane fouling. HVI has been known as pulsed electric fields (PEF) in food industry for decades. HVI has been used for non-thermal sterilization of foods. In principle, bacterial cell membranes are damaged by HVI and thus, torn and holed (electroporation). After all, the microorganisms are killed by the HVI induction. Kim *et al.* (2011b) studied HVI disinfection of *E. coli*. They suggested kinetic information and proposed a possibility of membrane bio-fouling control by HVI. Lee *et al.* (2010) reported that activated sludge flocs were solubilized by HVI induction ranging from 10 to 20 kV/cm of electric field and several hundred nanoseconds of pulse width, which led to mitigation of bio-fouling by means of removing the solubilized bio-cake on the membrane surface. Since the HVI application to MBR is still at its initial stage of research, further studies on design parameters and engineering data are needed.

4.4.4 Biological approaches

Recently, some biological control strategies against biofouling have been reported, including (i) inhibition of quorum sensing (i.e., quorum quenching), (ii) nitric oxide-induced biofilm dispersal, (iii) enzymatic disruption of extracellular polysaccharides, proteins, and DNA, and (iv) disruption of biofilm by bacteriophage (Xiong *et al.* 2010). Although they have not yet arrived at the mature stage, they would be new research niches that merit further study. They are promising alternatives to conventional techniques and have great potential in alleviating membrane biofouling.

4.4.4.1 Control of membrane biofouling by inhibiting quorum sensing

Microorganisms are known to coordinate their communal behaviors by quorum sensing, for example, biofilm formation, swarming motility, production of extracellular polymeric substances, and so on. (Gonzalez *et al.* 2006). The QS-coordinated process is achieved by producing, releasing, and detecting small signal molecules known as autoinducers (AIs). Increasing bacterial density gives rise to an accumulation of AIs. Once the critical AI concentrations are achieved, the regulator proteins are triggered and further induce target DNA sequences, leading to transcription of quorum sensing-regulated genes, followed by changes of bacterial social behaviors. It has been known that the cell–cell communication essential for biofilm formation is closely regulated by AIs (Figure 4.11). AIs which have been identified so far are as follows: oligopeptides, *N*-acylhomoserine lactones (AHL), and autoinducer-2 (AI-2) synthesized by LuxS (Figure 4.12). Oligopeptides and AHL are merely involved in cellular communication of Gram-positive and Gram-negative bacteria, respectively, whereas AI-2 is universal for interspecies communication of both Gram-positive and Gram-negative bacteria (Xavier *et al.* 2003). Recently, at least two additional QS systems have been identified in Gram-negative bacteria. These include autoinducer 3 (AI-3), which is associated with virulence regulation in EHEC O157:H7 (Chiemchaisri *et al.* 1993) and the *Pseudomonas* quinolone signal (PQS), which is associated with *Pseudomonas aeruginosa* (Chae *et al.* 2009). It has been believed that AIs-mediated QS systems play a prominent role in the regulation of microbial attachment and subsequent biofilm formation.

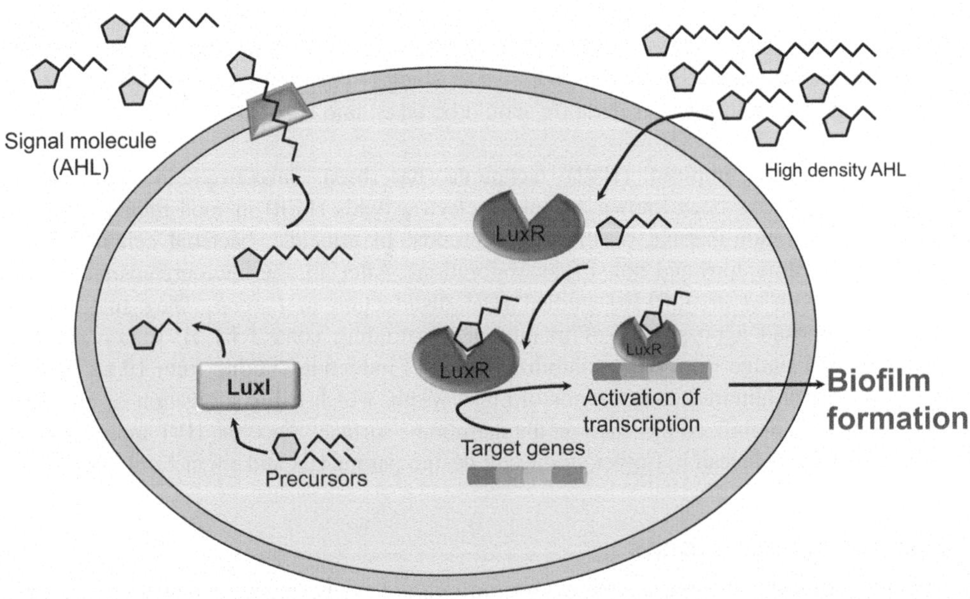

Figure 4.11 Quorum sensing mechanism in Gram-negative bacteria (AHL) (Boyen *et al.* 2009).

Since 2009, the concept of bacterial quorum sensing has been introduced to an MBR by Lee's group (Yeon *et al.* 2009a, b; Oh *et al.* 2012; Kim *et al.* 2013; Cheong *et al.* 2013). They showed that membrane biofouling due to a mixed culture of Gram-negative and Gram-positive bacteria could be efficiently removed by the addition of AHLs inhibitors (Yeon *et al.* 2009a). In detail, it was experimentally observed that quorum sensing is closely associated with the formation of a biofouling layer on the immersed membrane surface in a submerged MBR for wastewater treatment. Furthermore, a magnetic enzyme carrier (MEC) was prepared by immobilizing the quorum quenching enzyme (acylase) on magnetic particles to overcome the technical limitations of free enzyme (acylase). When MEC was applied to MBR in a continuous operation, it enhanced the membrane permeability to a large extent compared with a conventional MBR with no enzyme (Yeon *et al.* 2009b). On the other hand, Kim *et al.* (2011a) attempted to immobilize the AHL-acylase directly onto the NF membrane surface and observed that this quorum quenching membrane exhibits a great anti-biofouling feature by suppressing EPS secretion and thus biofilm formation.

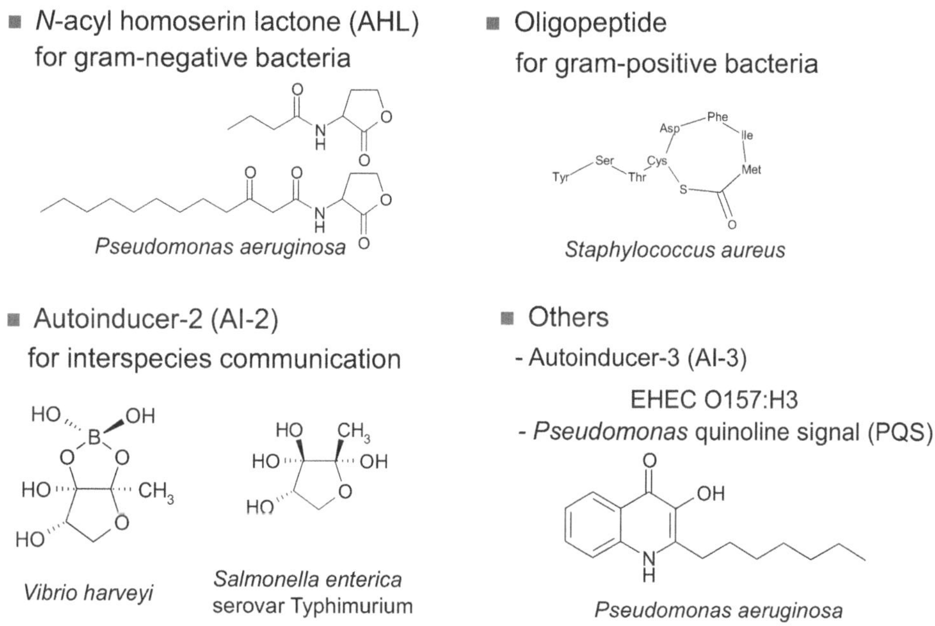

Figure 4.12 Representative signal molecules involved in bacterial quorum sensing.

Enzymatic disruption of signal molecules would have several inherent drawbacks that, to some extent, may limit its large-scale application. The enzyme is unstable in the environment, for example, the activity of the enzyme is highly sensitive to the pH, temperature, and salt concentration. In 2012, an application of bacterial quorum quenching to biofouling control in MBR was reported to overcome the limitations of enzymatic quorum quenching (Oh *et al.* 2012). They isolated quorum quenching bacteria (*Rhodococcus* sp.) producing AHL-lactonase *Rhodococcus* sp. from a real MBR plant for wastewater treatment and then encapsulated them into a microporous hollow fiber membranes (Figure 4.13). They have shown that interspecies quorum quenching could be a novel promising technique for the control of biofouling in MBR (Figure 4.14). Bacterial quorum quenching can be more economic than enzymatic quorum quenching because the former has longer life span and does not need an enzyme purification process.

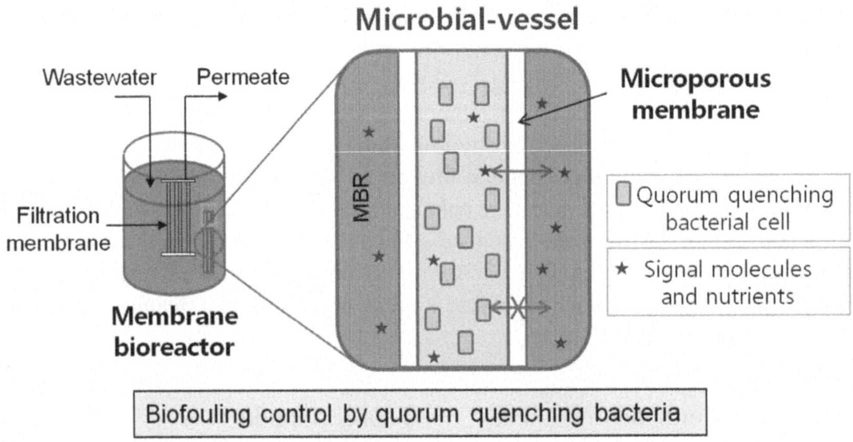

Figure 4.13 Biofouling control in MBR by quorum quenching bacteria encapsulated in microporous membrane (Adapted from Oh *et al.* 2012).

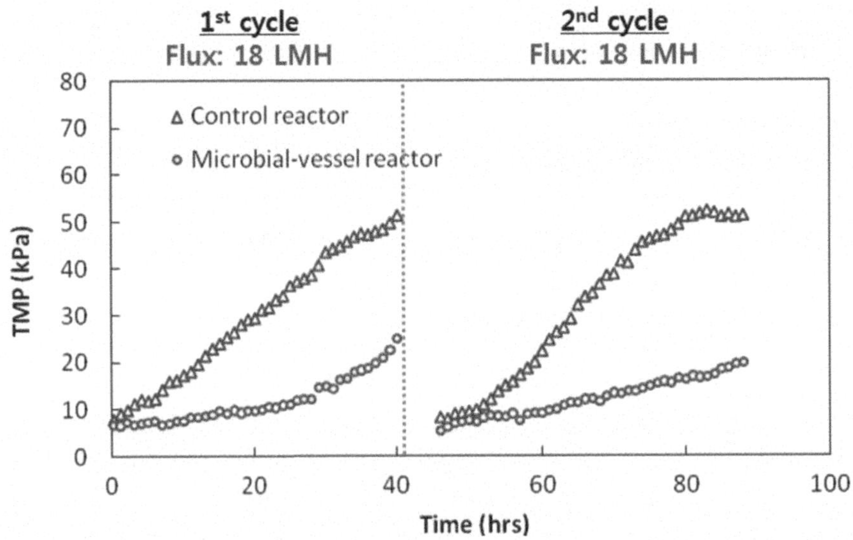

Figure 4.14 Transmembrane pressure (TMP) profile in control MBR and the MBR containing quorum quenching bacteria encapsulated microporous membrane (Adapted from Oh *et al.* 2012).

Furthermore, Kim *et al.* (2013) applied free-moving beads entrapped with the *Rhodococcus* sp. to the inhibition of biofouling in a MBR (Figure 4.15). In MBRs provided with CEBs, the time to reach the same transmembrane pressure (TMP) of 70 kPa was ten times longer than without CEBs (Figure 4.16). The mitigation of biofouling by cell entrapping beads (CEBs) was attributed to not only the biological (quorum quenching) effects but also the physical (friction) effects which occur as a result of the collisions between the moving CEBs and membranes.

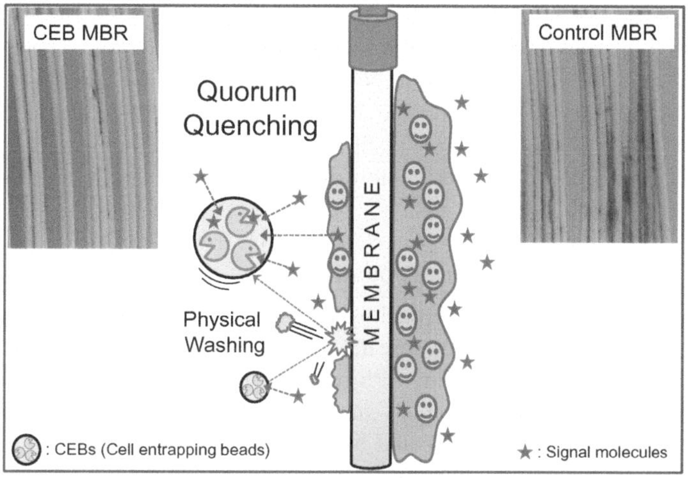

Figure 4.15 Biofouling control in MBR by CEBs (Adapted from Kim *et al.* 2013).

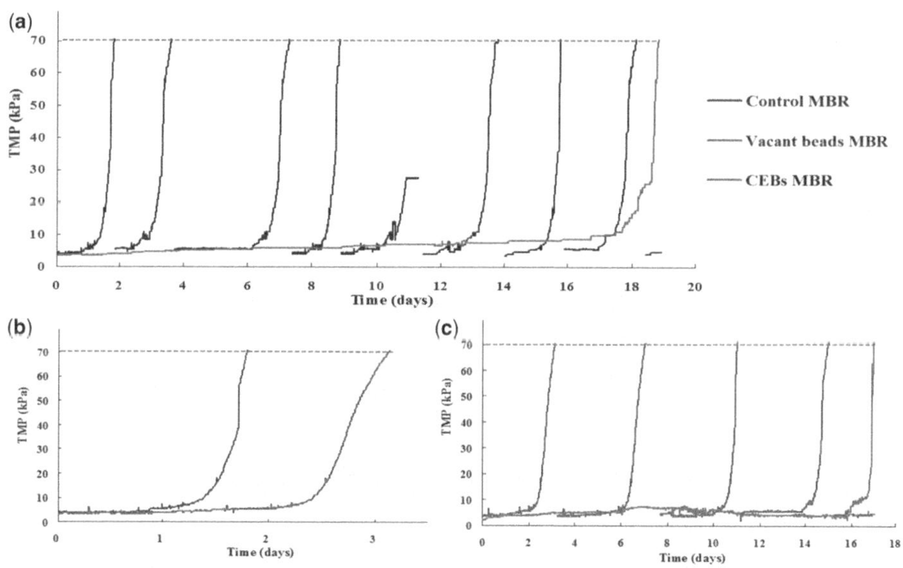

Figure 4.16 Comparison of TMP between (a) control and CEBs MBRs, (b) control and vacant beads MBRs and (c) vacant beads and CEBs MBRs under the same operating conditions (adapted from Kim *et al.* 2013).

4.4.4.2 Control of membrane biofouling by nitric oxide

Nitric oxide (NO), as a biologically ubiquitous gas molecule, can antagonize the useless cell proliferation (Sarti *et al.* 2002). This specific molecule has been identified as an important messenger molecule that regulates biofilm dispersal. For example, addition of NO at low, nontoxic concentrations would result in the dispersal of *P. aeruginosa* biofilm (Barraud *et al.* 2006). Similar to the QS inhibitors, nitric oxide attenuates the bacterial infection by the signaling mechanisms rather than the toxic effect.

Barraud *et al.* (2009) reported that pretreatment of biofilm grown on the RO membrane by 100 nM of SNP which served as the donor of nitric oxide led to a two-fold increase in the biofilm removal efficiency in contrast with the control. Nitric oxide would have great application potential in controlling microbial attachment and membrane biofouling. However, it should be pointed out that the NO-based method for biofouling control is still at the initial developmental stage and stays on in vitro, thus more effort is needed to further explore its potential application at industrial scale (Xiong *et al.* 2010).

4.4.4.3 Control of membrane biofouling by enzymatic disruption of EPS

Extracellular polymeric substances (EPS) secreted by bacteria can facilitate the attachment of bacteria to a solid surface such as membranes. EPS can decrease the susceptibility of bacteria to antibiotics and thus act as a shelter to protect the bacteria in the biofilm–EPS matrix (Figure 4.17). EPS have been believed to be irreversible foulants of membrane fouling, which cannot be efficiently removed by traditional physical or chemical cleaning methods. However, EPS could be hydrolyzed by some specific enzymes, implying a novel means to control EPS-mediated microbial attachment and membrane biofouling (Xiong *et al.* 2010). EPS are mainly composed of proteins, polysaccharides and extracellular DNA. Therefore the enzymes such as protease, polysaccharases, and DNase, which degrade those substances, can provide a feasible and effective mean for controlling membrane biofouling. Poele and van der Graaf used protease to remove biofouling on ultrafiltration (UF) membranes for wastewater treatment (Poele *et al.* 2005). Compared to the traditional cleaning method by alkaline, enzymatic cleaning by protease exhibited a much higher efficiency in removing biofouling, leading to a high-efficiency recovery of the permeate flux. Moreover, enzymatic cleaning of the fouled inorganic UF membranes by whey proteins was also tested by Arguello *et al.* (Arguello *et al.* 2002, 2003), and results showed that over 90% of removal efficiency would be achievable.

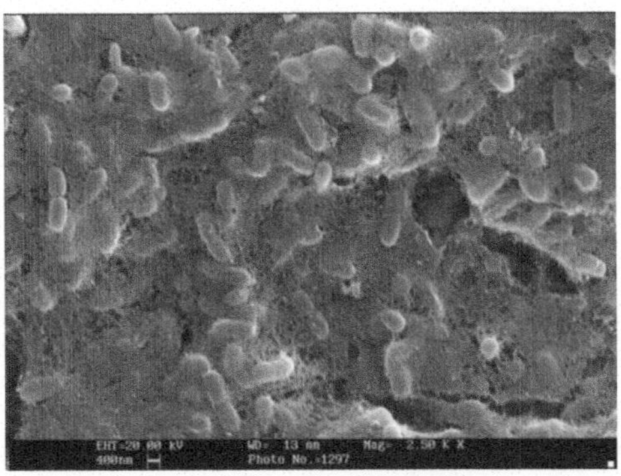

Figure 4.17 EPS matrix observed in aerobic granular biofilm (Adapted from Xiong *et al.* 2010). EPS can act as a shelter to protect the bacteria in the biofilm from antibiotics.

4.4.4.4 Control of membrane biofouling by bacteriophage

Bacteriophage or phage can infect the host bacteria by the rapid replication of virions to cause lysis of the host cells or by incorporation into the host cell's genome (Xiong *et al.* 2010). One important application of phage is to inhibit or disrupt biofilm development on solid surfaces such as membranes. Goldman *et al.* (2009) employed bacteriophage to control ultrafiltration membrane biofouling. Results showed that the addition of phages could reduce microbial attachment to membrane surfaces by 40% on average, and the performance of the MBR treating the effluents from the sewage treatment plant was improved significantly in terms of membrane permeability.

Although the study by Goldman *et al.* (2009) sheds lights on the potential application of bacteriophage in mitigating membrane biofouling without the use of antimicrobial agents, the specific-parasite characteristics of bacteriophage would eventually pose a challenge on its application in large-scale wastewater treatment, which needs to be taken into serious account in future investigations.

4.5 CONCLUSION AND FUTURE OUTLOOK

Three approaches to research and development encompassing membrane fouling mitigation can be identified: (i) monitoring, (ii) characterization, and (iii) control of membrane fouling/foulants.

Development of monitoring techniques to predict the filterability of a mixed liquor suspension is important, because it reflects the potential of membrane permeability decline in the operation of MBR. Despite more than a decade of worldwide research on the issue of membrane fouling in MBR, often contradictory results are reported, which can partly be attributed to the large variety of non-standardised fouling monitoring/characterization methods used. To make matters worse, due to inherent differences between lab and full-scale operation, prediction of fouling rates encountered in full-scale MBRs by lab-scale filtration tests is very difficult. Use of multiple-component analysis method might be appropriate for assessing filterability to explain the complex phenomenon of membrane fouling in MBR.

A number of methods of fouling control, including antifouling membrane development and physical and chemical fouling mitigation approaches, have been reported and implemented in MBRs to reduce the operating and maintenance costs. An emerging strategy is to control the physiology of microorganisms in MBR. Although further study is needed, such biological approaches are promising alternatives to conventional techniques and have great potentials in alleviating membrane biofouling.

It must be conceded that, various techniques developed so far for monitoring, characterization, and control of membrane foulants/fouling in MBR have merits but require further development. This necessitates continuous R&D efforts.

REFERENCES

Akamatsu K., Lu W., Sugawara T. and Nakao S. (2010). Development of a novel fouling suppression system in membrane bioreactors using an intermittent electric field. *Water Research*, **44**(3), 825–830.

Akram A. and Stuckey D. C. (2008). Flux and performance improvement in a submerged anaerobic membrane bioreactor (SAMBR) using powdered activated carbon (PAC). *Process Biochemistry*, **43**(1), 93–102.

Al-Halbouni D., Traber J., Lyko S., Wintgens T., Melin T., Tacke D., Janot A., Dott W. and Hollender J. (2008). Correlation of EPS content in activated sludge at different sludge retention times with membrane fouling phenomena. *Water Research*, **42**(6–7), 1475–1488.

Arguello M. A., Alvarez S., Riera F. A. and Alvarez R. (2002). Enzymatic cleaning of inorganic ultrafiltration membranes fouled by whey proteins. *Journal of Agricultural and Food Chemistry*, **50**(7), 1951–1958.

Arguello M. A., Alvarez S., Riera F. A. and Alvarez R. (2003). Enzymatic cleaning of inorganic ultrafiltration membranes used for whey protein fractionation. *Journal of Membrane Science*, **216**(1–2), 121–134.

Badireddy A. R., Korpol B. R., Chellam S., Gassman P. L., Engelhard M. H., Lea A. S. and Rosso K. M. (2008). Spectroscopic characterization of extracellular polymeric substances from Escherichia coli and Serratia marcescens: suppression using sub-inhibitory concentrations of bismuth thiols. *Biomacromolecules*, **9**(11), 3079–3089.

Bani-Melhem K. and Elektorowicz M. (2010). Development of a novel submerged membrane electro- bioreactor (SMEBR): performance for fouling reduction. *Environmental Science and Technology*, **44**(9), 3298–3304.

Bani-Melhem K. and Elektorowicz M. (2011). Performance of the submerged membrane electro-bioreactor (SMEBR) with iron electrodes for wastewater treatment and fouling reduction. *Journal of Membrane Science*, **379**(1–2), 434–439.

Barllion R. S. M. L. V. (2011). Full Scale Assessment of Energy Consumption in MBRs. Paper Presented at the Proceedings of 6th IWA Specialist Conference on Membrane Technology for Water & Wastewater Treatment, 4–7th October, Aachen, Germany.

Barraud N., Hassett D. J., Hwang S. H., Rice S. A., Kjelleberg S. and Webb J. S. (2006). Involvement of nitric oxide in biofilm dispersal of Pseudomonas Aeruginosa. *Journal of Bacteriology*, **188**(21), 7344–7353.

Barraud N., Storey M. V., Moore Z. P., Webb J. S., Rice S. A. and Kjelleberg S. (2009). Nitric oxide-mediated dispersal in single- and multi-species biofilms of clinically and industrially relevant microorganisms. *Microbial Biotechnology*, **2**(3), 370–378.

Beyenal H., Donovan C., Lewandowski Z. and Harkin G. (2004). Three-dimensional biofilm structure quantification. *Journal of Microbiological Methods*, **59**, 395–413.

Bottino A., Capannelli G., Comite A. and Mangano R. (2009). Critical flux in submerged membrane bioreactors for municipal wastewater treatment. *Desalination*, **245**(1–3), 748–753.

Bowen W. R., Hilal N., Lovitt R. W. and Wright C. J. (1999). Characterisation of membrane surfaces: direct measurement of biological adhesion using an atomic force microscope. *Journal of Membrane Science*, **154**(2), 205–212.

Boyen F., Eeckhaut V., Van Immerseel F., Pasmans F., Ducatelle R. and Haesebrouck F. (2009). Quorum sensing in veterinary pathogens: Mechanisms, clinical importance and future perspectives. *Veterinary Microbiology*, **135**(3–4), 187–195.

Braak E., Alliet M., Schetrite S. and Albasi C. (2011). Aeration and hydrodynamics in submerged membrane bioreactors. *Journal of Membrane Science*, **379**(1–2), 1–18.

Brant J. A. and Childress A. E. (2002). Membrane-colloid interactions: comparison of extended DLVO predictions with AFM force measurements. *Environmental Engineering Science*, **19**(6), 413–427.

Celik E., Park H., Choi H. and Choi H. (2011). Carbon nanotube blended polyethersulfone membranes for fouling control in water treatment. *Water Research*, **45**(1), 274–282.

Chae S., Ahn Y., Kang S. and Shin H. (2006). Mitigated membrane fouling in a vertical submerged membrane bioreactor (VSMBR). *Journal of Membrane Science*, **280**(1–2), 572–581.

Chae S. R., Wang S. Y., Hendren Z. D., Wiesner M. R., Watanabe Y. and Gunsch C. K. (2009). Effects of fullerene nanoparticles on Escherichia coli K12 respiratory activity in aqueous suspension and potential use for membrane biofouling control. *Journal of Membrane Science*, **329**(1–2), 68–74.

Chang I. S. and Judd S. J. (2002). Air sparging of a submerged MBR for municipal wastewater treatment. *Process Biochemistry*, **37**(8), 915–920.

Chang I. S. and Lee C. H. (1998). Membrane filtration characteristics in membrane-coupled activated sludge system – the effect of physiological states of activated sludge on membrane fouling. *Desalination*, **120**(3), 221–233.

Chen M. Y., Lee D. J., Yang Z., Peng X. F. and Lai J. Y. (2006). Fluorecent staining for study of extracellular polymeric substances in membrane biofouling layers. *Environmental Science & Technology*, **40**(21), 6642–6646.

Chen J. P., Yang C. Z., Zhou J. H. and Wang X. Y. (2007a). Study of the influence of the electric field on membrane flux of a new type of membrane bioreactor. *Chemical Engineering Journal*, **128**(2–3), 177–180.

Chen M. Y., Lee D. J. and Tay J. H. (2007b). Distribution of extracellular polymeric substances in aerobic granules. *Applied Microbiology Biotechnology*, **73**, 1463–1469.

Cheong W. S., Lee C. H., Moon Y. H., Oh H. S., Kim S. R., Lee S. H., Lee C. H. and Lee J. K. (2013). Isolation and Identification of indigenous quorum quenching bacteria, Pseudomonas sp. 1A1, for biofouling control in MBR. *Industrial & Engineering Chemistry Research*, **52**(31), 10554–10560, doi: 10.1021/ie303146f.

Chiemchaisri C. and Yamamoto K. (1993). Biological Nitrogen Removal under Low-Temperature in a Membrane Separation Bioreactor. *Water Science and Technology*, **28**(10), 325–333.

Chu H. P. and Li X. Y. (2005). Membrane fouling in a membrane bioreactor (MBR): Sludge cake formation and fouling characteristics. *Biotechnology and Bioengineering*, **90**(3), 323–331.

Chu L., Yan S., Xing X.-H., Sun X. and Jurcik B. (2009). Progress and perspectives of sludge ozonation as a powerful pretreatment method for minimization of excess sludge production. *Water Research*, **43**, 1811–1822.

Cui Z. F., Chang S. and Fane A. G. (2003). The use of gas bubbling to enhance membrane processes. *Journal of Membrane Science*, **221**(1–2), 1–35.

de la Torre T., Mottschall M., Lesjean B., Drews A., Iheanaetu A. and Kraume M. (2010). Filterability assessment in membrane bioreactors using an in-situ filtration test cell. *Water Science and Technology*, **61**(11), 2809–2816.

Delaunay D., Rabiller-Baudry M. and Paugam L. (2006). Mapping of protein fouling by FTIR-ATR as experimental tool to study fluid velocity profile in spiral membrane. *Desalination*, **200**(1–3), 205–207.

Drews A. (2010). Membrane fouling in membrane bioreactors – characterisation, contradictions, cause and cures. *Journal of Membrane Science*, **363**, 1–28.

Drews A., Vocks M., Bracklow U., Iversen V. and Kraume M. (2008). Does fouling in MBRs depend on SMP?. *Desalination*, **231**(1–3), 141–149.

Dubois M., Gilles K., Hamilton J., Rebers P. and Smith F. (1956). Colorimetric method for determination of sugars and related substances. *Analytical Chemistry*, **28**(3), 350–356.

Evenblij H., Geilvoet S., van der Graaf J. and van der Roest H. (2005). Filtration characterisation for assessing MBR performance: Three cases compared. *Desalination*, **178**(1–3), 115–124.

Fan F., Zhou H. and Husain H. (2006). Identification of wastewater sludge characteristics to predict critical flux for membrane bioreactor processes. *Water Research*, **40**(2), 205–212.

Ferrando M., Rozek A., Zator M., Lopez F. and Guell C. (2005). An approach to membrane fouling characterization by confocal scanning laser microscopy. *Journal of Membrane Science*, **250**(1–2), 283–293.

Field R. W., Wu D., Howell J. A. and Gupta B. B. (1995). Critical flux concept for microfiltration fouling. *Journal of Membrane Science*, **100**(3), 259–272.

Frolund B., Palmgren R., Keiding K. and Nielsen P. (1996). Extraction of extracellular polymers from activated sludge using a cation exchange resin. *Water Research*, **30**(8), 1749–1758.

Goldman G., Starosvetsky J. and Armon R. (2009). Inhibition of biofilm formation on UF membrane by use of specific bacteriophages. *Journal of Membrane Science*, **342**(1–2), 145–152.

Gonzalez J. E. and Keshavan N. D. (2006). Messing with bacterial quorum sensing. *Microbiology and Molecular Biology Reviews*, **70**(4), 859–875.

Guglielmi G., Chiarani D., Judd S. and Andreottola G. (2007). Flux criticality and sustainability in a hollow fibre submerged membrane bioreactor for municipal wastewater treatment. *Journal of Membrane Science*, **289**(1–2), 241–248.

Gui P., Huang X., Chen Y. and Quia Y. (2002). Effect of operational parameters on sludge accumulation on membrane surfaces in a submerged membrane bioreactor. *Desalintion*, **151**(2), 185–194.

Haberkamp J., Ruhl A. S., Ernst M. and Jekel M. (2007). Impact of coagulation and adsorption on DOC fractions of secondary effluent and resulting fouling behaviour in ultrafiltration. *Water Research*, **41**, 3794–3802.

Hai F. I. and Yamamoto K. (2011). 4.16 – membrane biological reactors. In: Treatise on Water Science, W. Peter (ed.), Elsevier, Oxford, pp. 571–613.

Hai F. I., Yamamoto K., Fukushi K. and Nakajima F. (2008). Fouling resistant compact hollow-fiber module with spacer for submerged membrane bioreactor treating high strength industrial wastewater. *Journal of Membrane Science*, **317**(1–2), 34–42.

Halle C., Huck P. M., Peldszus S., Haberkamp J. and Jekel M. (2009). Assessing the performance of biological filtration as pretreatment to low pressure membranes for drinking water. *Environmental Science and Technology*, **43**(10), 3878–3884.

Henderson R. K., Baker A., Murphy K. R., Hambly A., Stuetz R. M. and Khan S. J. (2009). Fluorescence as a potential monitoring tool for recycled water systems: a review. *Water Research*, **43**, 863–881.

Her N., Amy G., McKnight D., Sohn J. and Yoon Y. M. (2003). Characterization of DOM as a function of MW by fluorescence EEM and HPLC-SEC using UVA, DOC, and fluorescence detection. *Water Research*, **37**(17), 4295–4303.

Hilal N., Bowen W. R., Alkhatib L. and Ogunbiyi O. (2006). A review of atomic force microscopy applied to cell interactions with membranes. *Chemical Engineering Research and Design*, **84**(A4), 282–292.

Horsch P., Gorenflo A., Fuder C., Deleage A. and Frimmel F. H. (2005). Biofouling of ultra- and nanofiltration membranes for drinking water treatment characterized by fluorescence in situ hybridization (FISH). *Desalination*, **172**(1), 41–52.

Howell J. A., Chua H. C. and Arnot T. C. (2004). In situ manipulation of critical flux in a submerged membrane bioreactor using variable aeration rates, and effects of membrane history. *Journal of Membrane Science*, **242**(1–2), 13–19.

Huang X. and Wu J. L. (2008). Improvement of membrane filterability of the mixed liquor in a membrane bioreactor by ozonation. *Journal of Membrane Science*, **318**(1–2), 210–216.

Huang L. N., De Wever H. and Diels L. (2008). Diverse and distinct bacterial communities induced biofilm fouling in membrane bioreactors operated under different conditions. *Environmental Science & Technology*, **42**(22), 8360–8366.

Hughes D., Tirlapur U. K., Field R. and Cui Z. F. (2006a). In situ 3D characterization of membrane fouling by yeast suspensions using two-photon femtosecond near infrared non-linear optical imaging. *Journal of Membrane Science*, **280**(1–2), 124–133.

Hughes D. J., Cui Z. F., Field R. W. and Tirlapur U. K. (2006b). In situ three-dimentional characterizaiton of membrane fouling by protein suspensions using multiphoton microscopy. *Langmuir*, **22**, 6266–6272.

Hughes D. J., Cui Z. F., Field R. W. and Tirlapur U. K. (2007). Membrane fouling by cell-protein mixtures: In situ characterisation using multi-photon microscopy. *Biotechnology and Bioengineering*, **96**(6), 1083–1091.

Huotari H. M., Tragardh G. and Huisman I. H. (1999). Crossflow membrane filtration enhanced by an external DC electric field: A review. *Chemical Engineering Research & Design*, **77**(A5), 461–468.

Huyskens C., Brauns E., Van Hoof E. and De Wever H. (2008). A new method for the evaluation of the reversible and irreversible fouling propensity of MBR mixed liquor. *Journal of Membrane Science*, **323**(1), 185–192.

Hwang B. K., Lee W. N., Park P. K., Lee C. H. and Chang I. S. (2007). Effect of membrane fouling reducer on cake structure and membrane permeability in membrane bioreactor. *Journal of Membrane Science*, **288**(1–2), 149–156.

Hwang B. K., Lee W. N., Yeon K. M., Park P. K., Lee C. H., Chang I. S., Drews A. and Kraume M. (2008). Correlating TMP increases with microbial characteristics in the bio-cake on the membrane surface in a membrane bioreactor. *Environmental Science & Technology*, **42**(11), 3963–3968.

Hwang B. K., Kim J. H., Ahn C. H., Lee C. H., Song J. Y. and Ra Y. H. (2010). Effect of disintegrated sludge recycling on membrane permeability in a membrane bioreactor combined with a turbulent jet flow ozone contactor. *Water Research*, **44**(6), 1833–1840.

Itonaga T., Kimura K. and Watanabe Y. (2004). Influence of suspension viscosity and colloidal particles on permeability of membrane used in membrane bioreactor (MBR). *Water Science and Technology*, **50**(12), 301–309.

Ivanovic I., Leiknes T. and Odegaard H. (2008). Fouling control by reduction of submicron particles in a BF-MBR with an integrated flocculation zone in the membrane reactor. *Separation Science and Technology*, **43**(7), 1871–1883.

Ivleva N. P., Wagner M., Horn H., Niessner R. and Haisch C. (2009). Towards a nondestructive chemical characterization of biofilm matrix by Raman microscopy. *Analytical and Bioanalytical Chemistry*, **393**(1), 197–206.

Jankhah S. and Berube P. R. (2009). Investigation of Fouling Mechanism in Submerged Membrane Systems Through Direct Observation of Foulant Behavior. Paper Presented at the 5th IWA Specialised Membrane Technology Conference for Water and Wastewater Treatment, Beijing.

Janot D. K. E. (2011). Reducing the Energy Consumption of a Large-scale Membrane Bioreactor. Proceedings of 6th IWA Specialist Conference on Membrane Technology for Water & Wastewater Treatment. 4–7th October. Aachen, Germany.

Janot N. A., Cumin J., Marschall M., Turák T. P., Vizvardi K. and Koops H. (2011). Reducing the Cost of MBR: The Continuous Optimization of GE's ZeeWeed Technology. Paper Presented at the Proceedings of 6th IWA Specialist Conference on Membrane Technology for Water & Wastewater Treatment, 4–7th October, Aachen, Germany.

Jiang T., Kennedy M., Yoo C., Nopens I., van der Meer W., Futselaar H., Schippers J. and Vanrolleghem P. (2007). Controlling submicron particle deposition in a side-stream membrane bioreactor: A theoretical hydrodynamic modelling approach incorporating energy consumption. [Article]. *Journal of Membrane Science*, **297**(1–2), 141–151.

Jie L., Liu L. F., Yang F. L., Liu F. X. and Liu Z. J. (2012). The configuration and application of helical membrane modules in MBR. *Journal of Membrane Science*, **392**, 112–121.

Jin Y.-L., Lee W. N., Lee C. H., Chang I. S., Huang X. and Swaminathan T. (2006). Effect of DO concentration on biofilm structure and membrane filterability in submerged membrane bioreactor. *Water Research*, **40**(15), 2829–2836.

Johir M. A. H., Aryal R., Vigneswaran S., Kandasamy J. and Grasmick A. (2011). Influence of supporting media in suspension on membrane fouling reduction in submerged membrane bioreactor (SMBR). *Journal of Membrane Science*, **374**(1–2), 121–128.

Judd S. (2008). The status of membrane bioreactor technology. *Trends in Biotechnology*, **26**, 109–116.

Judd S. J. (2010). The MBR Book, Principles and Applications of Membrane Bioreactors in Water and Wastewater Treatment, 2nd edn, Elsevier, Oxford, UK.

Kang S. J., Olmstead K., Schraa O., Rhu D. H., Em Y. J., Kim J. K. and Min J. H. (2011). Activated Anaerobic Digestion with a Membrane Filtration System. Paper Presented at the Proceedings of 84th Annual Conference and Exhibition of Water Environment Federation (WEFTECH), 15–19th October, Los Angeles, USA.

Kim I. S. and Jang N. (2006). The effect of calcium on the membrane biofouling in the membrane bioreactor (MBR). *Water Research*, **40**(14), 2756–2764.

Kim H. W., Oh H. S., Kim S. R., Lee K. B., Yeon K. M., Lee C. H., Kim S. and Lee J. K. (2012). Microbial population dynamics and proteomics in membrane bioreactors with enzymatic quorum quenching. *Applied Microbiology and Biotechnology*, **97**(10), 4665–4675, doi: 10.1007/s00253012-4272-0.

Kim J. H., Choi D. C., Yeon K. M., Kim S. R. and Lee C. H. (2011a). Enzyme-immobilized nanofiltration membrane to mitigate biofouling based on quorum quenching. *Environmental Science & Technology*, **45**(4), 1601–1607.

Kim J. Y., Lee J. H., Chang I. S., Lee J. H. and Yi C. W. (2011b). High voltage impulse electric fields: Disinfection kinetics and its effect on membrane bio-fouling. *Desalination*, **283**, 111–116.

Kim J. S., Lee C. H. and Chun H. D. (1998). Comparison of ultrafiltration characteristics between activated sludge and BAC sludge. *Water Research*, **32**(11), 3443–3451.

Kim S., Lee S., Hong S., Oh Y., Seoul M., Kweon J. and Kim T. (2009). Biofouling of reverse osmosis membranes: microbial quorum sensing and fouling propensity. *Desalination*, **247**(1–3), 303–315.

Kim S. R., Oh H. S., Jo S. J., Yeon K. M., Lee C. H., Lim D. J., Lee C. H. and Lee J. K. (2013). Biofouling control with bead-entrapped quorum quenching bacteria in MBR: physical and biological effects. *Environmental Science & Technology*, **47**(2), 836–842.

Kimura K., Hane Y., Watanabe Y., Amy G. and Ohkuma N. (2004). Irreversible membrane fouling during ultrafiltration of surface water. *Water Research*, **38**(14–15), 3431–3441.

Kimura K., Yamato N., Yamamura H. and Watanabe Y. (2005). Membrane fouling in pilot-scale membrane Bioreactors (MBRs) treating municipal wastewater. *Environmental Science & Technology*, **39**(16), 6293–6299.

Kimura K., Naruse T. and Watanabe Y. (2009). Changes in characteristics of soluble microbial products in membrane bioreactors associated with different solid retention times: relation to membrane fouling. *Water Research*, **43**(4), 1033–1039.

Kraume M., Wedi D., Schaller J., Iversen V. and Drews A. (2009). Fouling in MBR: what use are lab investigations for full scale operation? *Desalination*, **236**(1–3), 94–103.

Lebegue J., Heran M. and Grasmik A. (2008). Membrane bioreactor: distribution of critical flux throughout an immersed HF bundke. *Desalination*, **231**(1–3), 245–252.

Le Clech P., Jefferson B., Chang I. S. and Judd S. J. (2003). Critical flux determination by the flux-step method in a submerged membrane bioreactor. *Journal of Membrane Science*, **227**(1–2), 81–93.

Le-Clech P., Chen V. and Fane T. A. G. (2006). Fouling in membrane bioreactors used in wastewater treatment. *Journal of Membrane Science*, **284**, 17–53.

Le-Clech P., Marselina Y., Ye Y., Stuetz R. M. and Chen V. (2007). Visualisation of polysaccharide fouling on microporous membrane using different characterisation techniques. *Journal of Membrane Science*, **290**(1–2), 36–45.

Lee C. H., Park P. K., Lee W. N., Hwang B. K., Hong S. H., Yeon K. M., Oh H. S. and Chang I. S. (2008). Correlation of biofouling with the bio-cake architecture in an MBR. *Desalination*, **231**(1–3), 115–123.

Lee J., Ahn W. Y. and Lee C. H. (2001). Comparison of the filtration characteristics between attached and suspended growth microorganisms in submerged membrane bioreactor. *Water Research*, **35**(10), 2435–2445.

Lee J. S., Yang S. H. and Chang I. S. (2010). Solubilization of the Activated Sludge Using the High Voltage Impulse Techniques in MBR. Paper Presented at the Proceedings of 6th IWA Specialist Conference on Membrane Technology for Water & Wastewater Treatment, 4–7th October, Aachen, Germany.

Lee S. and Elimelech M. (2006). Relating organic fouling of reverse osmosis membranes to intermolecular adhesion forces. *Environmental Science & Technology*, **40**(3), 980–987.

Lee W. N., Chang I. S., Hwang B. K., Park P. K., Lee C. H. and Huang X. (2007). Changes in biofilm architecture with addition of membrane fouling reducer in a membrane bioreactor. *Process Biochemistry*, **42**(4), 655–661.

Lee W. N., Cheong W. S., Yeon K. M., Hwang B. K. and Lee C. H. (2009). Correlation between local TMP distribution, bio-cake porocity on the membrane in a submerged MBR. *Journal of Membrane Science*, **332**(1–2), 50–55.

Leiknes T. and Odegaard H. (2007). The development of a biofilm membrane bioreactor. *Desalination*, **202**(1–3), 135–143.

Li H., Fane A. G., Coster H. G. L. and Vigneswaran S. (1998). Direct observation of particle deposition on the membrane surface during crossflow microfiltration. *Journal of Membrane Science*, **149**(1), 83–97.

Li H., Fane A. G., Coster H. G. L. and Vigneswaran S. (2000). An assessment of depolarisation models of crossflow microfiltration by direct observation through the membrane. *Journal of Membrane Science*, **172**(1–2), 135–147.

Li H., Fane A. G., Coster H. G. L. and Vigneswaran S. (2003). Observation of deposition and removal behaviour of submicron bacteria on the membrane surface during crossflow microfiltration. *Journal of Membrane Science*, **217**(1–2), 29–41.

Li Q. L. and Elimelech M. (2004). Organic fouling and chemical cleaning of nanofiltration membranes: Measurements and mechanisms. *Environmental Science & Technology*, **38**(17), 4683–4693.

Li W.-H., Sheng G. P., Liu X. W. and Yu H. Q. (2008). Characterizing the extracellular and intracellular fluorescent products of activated sludge in a sequencing batch reactor. *Water Research*, **42**(12), 3173–3181.

Lim S., Kim S., Yeon K. M., Sang B. I., Chun J. and Lee C. H. (2012). Correlation between microbial community structure and biofouling in a laboratory scale membrane bioreactor with synthetic wastewater. *Desalination*, **287**, 209–215.

Liu L. F., Liu J. D., Gao B. and Yang F. L. (2012). Minute electric field reduced membrane fouling and improved performance of membrane bioreactor. *Separation and Purification Technology*, **86**, 106–112.

Liu R., Huang X., Wang C. W., Chen L. J. and Qian Y. (2000). Study on hydraulic characteristics in a submerged membrane bioreactor process. *Process Biochemistry*, **36**(3), 249–254.

Lousada-Ferreira M., Geilvoet S., Moreau A., Atasoy E., Krzeminski P., van Nieuwenhuijzen A. and van der Graaf J. (2010). MLSS concentration: Still a poorly understood parameter in MBR filterability. *Desalination*, **250**(2), 618–622.

Lowry O. H., Rosebrough N. J., Farr A. L. and Randall R. J. (1951). Protein measurement with the folin phenol reagent. *Journal of Biological Chemistry*, **193**(1), 265–275.

Lu F., Chang C. H., Lee D. J., He P. J., Shao L. M. and Su A. (2008). Dissolved organic matter with multi-peak fluorophores in landfill leachate. *Chemosphere*, **74**, 575–582.

Lyko S., Wintgens T., Al-Halbouni D., Baumgarten S., Tacke D., Drensla K., Janot A., Dott W., Pinnekamp J. and Melin T. (2008). Long-term monitoring of a full-scale municipal membrane bioreactor – characterisation of foulants and operational performance. *Journal of Membrane Science*, **317**(1–2), 78–87.

Ma J., Wang Z., Yang Y., Mei X. and Wu Z. (2013). Correlating microbial community structure and composition with aeration intensity in submerged membrane bioreactors by 454 high-throughput pyrosequencing. *Water Research*, **47**(2), 859–869.

Marselina Y., Le-Clech P., Stuetz R. and Chen V. (2008). Detailed characterisation of fouling deposition and removal on a hollow fibre membrane by direct observation technique. *Desalination*, **231**(1–3), 3–11.

Marselina Y., Lifia, Le-Clech P., Stuetz R. M. and Chen V. (2009). Characterisation of membrane fouling deposition and removal by direct observation technique. *Journal of Membrane Science*, **341**, 163–171.

Meng F., Zhang H., Li Y., Zhang X. and Yang F. (2005). Application of fractal permeation model to investigate membrane fouling in membrane bioreactor. *Journal of Membrane Science*, **262**(1–2), 107–116.

Meng F. G., Zhang H. M., Yang F. L., Zhang S. T., Li Y. S. and Zhang X. W. (2006). Identification of activated sludge properties affecting membrane fouling in submerged membrane bioreactors. *Separation and Purification Technology*, **51**(1), 95–103.

Meng F., Chae S.-R., Drews A., Kraume M., Shin H.-S. and Yang F. (2009). Recent advances in membrane bioreactors (MBRs): Membrane fouling and membrane material. *Water Research*, **43**(6), 1489–1512.

Meng F., Zhou Z., Ni B.-J., Zheng X., Huang G., Jia X., Li S., Xiong Y. and Kraume M. (2011). Characterization of the size-fractionated biomacromolecules: tracking their role and fate in a membrane bioreactor. *Water Research*, **45**(15), 4661–4671.

Mi B. and Elimelech M. (2010). Organic fouling of forward osmosis membranes: Fouling reversibility and cleaning without chemical reagents. *Journal of Membrane Science*, **348**(1–2), 337–345.

Microdyne-Nadir (2013). BIO-CEL® submerged membrane. http://www.microdyn-nadir.de/cms/produkt_details.php?page=4&id=5&lang=en (accessed 22 July 2013).

Miura Y., Watanbe Y. and Okabe S. (2007). Membrane biofouling in pilot-scale membrane bioreactors (MBRs) treating municipal wastewater: Impact of biofilm formation. *Environmental Science & Technology*, **41**(2), 632–638.

Miyoshi T., Tanaka I., Tsuyuhara T., Watanabe E., Aizawa T., Kimura K. and Watanabe Y. (2010). Fouling potentials of polysaccharides in membrane bioreactors (MBRs) assessed by lectin affinity chromatography. *Water Science and Technology*, **61**(7), 1787–1792.

Miyoshi T., Aizawa T., Kimura K. and Watanabe Y. (2011). Characteristics of proteins involved in membrane fouling in membrane bioreactors (MBRs) treating municipal wastewater: the application of metaproteomic analyses. *Desalination and Water Treatment*, **34**(1–3), 150–155.

Mores W. D. and Davis R. H. (2001). Direct visual observation of yeast deposition and removal during microfiltration. *Journal of Membrane Science*, **189**(2), 217–230.

Mores W. D. and Davis R. H. (2002). Direct observation of membrane cleaning via rapid backpulsing. *Desalination*, **146**(1–3), 135–140.

Nam S. N. and Amy G. (2008). Differentiation of wastewater effluent organic matter (EfOM) from natural organic matter (NOM) using multiple analytical techniques. *Water Science and Technology*, **57**, 1009–1015.

Neal P. R., Li H., Fane A. G. and Wiley D. E. (2003). The effect of filament orientation on critical flux and particle deposition in spacer-filled channels. *Journal of Membrane Science*, **214**(2), 165–178.

Ng H. Y., Tan T. W. and Ong S. L. (2006). Membrane fouling of submerged membrane bioreactors: Impact of mean cell residence time and the contributing factors. *Environmental Science and Technology*, **40**(8), 2706–2713.

Ngene I. S., Lammertink R. G. H., Wessling M. and van der Meer W. (2010). A microfluidic membrane chip for in situ fouling characterization. *Journal of Membrane Science*, **346**(1), 202–207.

Ngo H. H., Guo W. S. and Xing W. (2008). Evaluation of a novel sponge-submerged membrane bioreactor (SSMBR) for sustainable water reclamation. *Bioresource Technology*, **99**(7), 2429–2435.

Ni B.-J., Fang F., Xie W. M., Sun M., Sheng G. P., Li W. H. and Yu H. Q. (2009). Characterization of extracellular polymeric substances produced by mixed microorganisms in activated sludge with gel-permeating chromatography, excitation–emission matrix fluorescence spectroscopy measurement and kinetic modeling. *Water Research*, **43**(5), 1350–1358.

Ognier S., Wisniewski C. and Grasmick A. (2004). Membrane bioreactor fouling in sub-critical filtration conditions: a local critical flux concept. *Journal of Membrane Science*, **229**(1–2), 171–177.

Oh H. S., Yeon K. M., Yang C. S., Kim S. R., Lee C. H., Park S. Y., Han J. Y. and Lee J. K. (2012). Control of membrane biofouling in MBR for wastewater treatment by quorum quenching bacteria encapsulated in microporous membrane. *Environmental Science & Technology*, **46**(9), 4877–4884.

Omoike A. and Chorover J. (2004). Spectroscopic study of extracellular polymeric substances from Bacillus subtilis: Aqueous chemistry and adsorption effects. *Biomacromolecules*, **5**(4), 1219–1230.

Pan J. R., Su Y. C., Huang C. P. and Lee H. C. (2010). Effect of sludge characteristics on membrane fouling in membrane bioreactors. *Journal of Membrane Science*, **349**(1–2), 287–294.

Park P.-K., Lee C. H. and Lee S. (2007). Determination of cake porosity using image analysis in a coagulation-microfiltration system. *Journal of Membrane Science*, **293**(1–2), 66–72.

Patzold R., Keuntje M. and Ahlften A. A.-v. (2006). A new approach to non-destructive analysis of biofilms by confocal Raman microscopy. *Analytical and Bioanalytical Chemistry*, **386**(2), 286–292.

Patzold R., Keuntje M., Theophile K., Muller J., Mielcarek E., Ngezahayo A. and Ahlften A. A. V. (2008). In situ mapping of nitrifiers and anammox bacteria in microbial aggregates by means of confocal resonance Raman microscopy. *Journal of Microbiological Methods*, **72**(3), 241–248.

Phattaranawik J., Fane A. G., Pasquir A. C. S. and Bing W. (2007). Membrane bioreactor with bubble size transformer: design and fouling control. *AIChE*, **53**(1), 243–249.

Poele S. T. and van der Graaf J. (2005). Enzymatic cleaning in ultrafiltration of wastewater treatment plant effluent. *Desalination*, **179**(1–3), 73–81.

Pollet S., Guigui C. and Cabassud C. (2008). Fouling, its reversibility in relation to flow properties, module design in aerated hollow fibre modules for membrane bioreactors. *Water Science, Technology*, **57**(4), 629–636.

Priester J. H., Horst A. M., van de Werfhorst L. C., Saleta J. L., Mertes L. A. K. and Holden P. A. (2007). Enhanced visualization of microbial biofilms by staining and environmental scanning electron microscopy. *Journal of Microbiological Methods*, **68**(3), 577–587.

Ramesh A., Lee D.-J. and Hong S. G. (2006). Soluble microbial products (SMP) and soluble extracellular polymeric substances (EPS) from wastewater sludge. *Applied Microbiology and Biotechnology*, **73**, 219–225.

Ramesh A., Lee D. J. and Lai J. Y. (2007). Membrane biofouling by extracellular polymeric substances or soluble mcirobial products from membrane bioreactor sludge. *Applied Microbiology and Biotechnology*, **74**, 699–707.

Richard C., Guyot G., Rivaton A., Trubetskaya O., Trubetskoj O., Cavani L. and Ciavatta C. (2007). Spectroscopic approach for elucidation of structural peculiarities of Andisol soil humic acid fractionated by SEC-PAGE setup. *Geoderma*, **142**(1–2), 210–216.

Rosenberger S., Evenblij H., te Poele S., Wintgens T. and Laabs C. (2005). The importance of liquid phase analyses to understand fouling in membrane assisted activated sludge processes–six case studies of different European research groups. *Journal of Membrane Science*, **263**(1–2), 113–126.

Rosenberger S., Laabs C., Lesjean B., Gnirss R., Amy G., Jekel M. and Schrotter J. C. (2006). Impact of colloidal and soluble organic material on membrane performance in membrane bioreactors for municipal wastewater treatment. *Water Research*, **40**(4), 710–720.

Sarti P., Avigliano L., Gorlach A. and Brune B. (2002). Superoxide and nitric oxide–participation in cell communication. [Congresses Research Support, Non-U.S. Gov't]. *Cell Seath and Sifferentiation*, **9**(10), 1160–1162.

Satyawali Y. and Balakrishnan M. (2009). Performance enhancement with powdered activated carbon (PAC) addition in a membrane bioreactor (MBR) treating distillery effluent. *Journal of Hazardous Materials*, **170**(1), 457–465.

Shen Y. X., Xiao K., Liang P., Sun J. Y., Sai S. J. and Huang X. (2012). Characterization of soluble microbial products in 10 large-scale membrane bioreactors for municipal wastewater treatment in China. *Journal of Membrane Science*, **415**, 336–345.

Sheng G. P. and Yu H. Q. (2006). Characterization of extracellular polymeric substances of aerobic and anaerobic sludge using three-dimensional excitation and emission matrix fluorescence spectroscopy. *Water Research*, **40**(6), 1233–1239.

Siembida B., Cornel P., Krause S. and Zimmermann B. (2010). Effect of mechanical cleaning with granular material on the permeability of submerged membranes in the MBR process. *Water Research*, **44**(14), 4037–4046.

Siemens Water Technologies. (2013). MemPulse Membrane Bioreactor (MBR) System. http://www.water.siemens.com/en/products/biological_treatment/membrane_biological_reactor_systems_mbr/Pages/envirex_product_integrated_mbr.aspx. (accessed 22 July 2013).

Staudt C., Horn H., Hempel D. C. and Neu T. R. (2004). Volumetric measurements of bacterial cells and extracellular polymeric substance glococonjugates in biofilms. *Biotechnology and Bioengineering*, **88**(5), 585–592.

Subramani A., Huang X. F. and Hoek E. M. V. (2009). Direct observation of bacterial deposition onto clean and organic-fouled polyamide membranes. *Journal of Colloid and Interface Science*, **336**(1), 13–20.

Sun F. Y., Wang X. M. and Li X. Y. (2008). Visualisation and characterisation of biopolymer clusters in a submerged membrane bioreactor. *Journal of Membrane Science*, **325**(2), 691–697.

Tansel B., Sager J., Garland J., Xu S. H., Levine L. and Bisbee P. (2008). Biofouling affinity of membrane surfaces under quiescent conditions. *Desalination*, **227**(1–3), 264–273.

Tian J.-y., Liang H., Li X., You S.-j., Tian S. and Li G.-b. (2008a). Membrane coagulation bioreactor (MCBR) for drinking water treatment. *Water Research*, **42**(14), 3910–3920.

Tian J.-y., Liang H., Yang Y.-l., Tian S. and Li G.-b. (2008b). Membrane adsorption bioreactor (MABR) for treating slightly polluted surface water supplies: As compared to membrane bioreactor (MBR). *Journal of Membrane Science*, **325**(1), 262–270.

Tian J.-y., Liang H., Yang Y.-l., You S.-j. and Li G.-b. (2009). Submerged membrane bioreactor (sMBR) for the treatment of contaminated raw water. *Chemical Engineering Journal*, **148**(2–3), 296–305.

Tsai B. N., Chang C. H. and Lee D. J. (2008). Fractionation of soluble microbial products (SMP) and soluble extracellular polymeric substances (EPS) from wastewater sludge. *Environmental Technology*, **29**(10), 1127–1138.

Van den Broeck R., Van Dierdonck J., Caerts B., Bisson I., Kregersman B., Nijskens P., Dotremont C., Van Impe J. F. and Smets I. Y. (2010). The impact of deflocculation–reflocculation on fouling in membrane bioreactors. *Separation and Purification Technology*, **71**(3), 279–284.

Van den Broeck R., Krzeminski P., Van Dierdonck J., Gins G., Lousada-Ferreira M., Van Impe J. F. M., van der Graaf J. H. J. M., Smets I. Y. and van Lier J. B. (2011). Activated sludge characteristics affecting sludge filterability in municipal and industrial MBRs: unraveling correlations using multi-component regression analysis. *Journal of Membrane Science*, **378**(1–2), 330–338.

Wagner M., Ivleva N. P., Haisch C., Niessner R. and Horn H. (2009). Combined use of confocal laser scanning microscopy (CLSM) and Raman microscopy (RM): Investigations on EPS - Matrix. *Water Research*, **43**(1), 63–76.

Wang X. M. and Li X. Y. (2008). Accumulation of biopolymer clusters in a submerged membrane bioreactor and its effect on membrane fouling. *Water Research*, **42**(4–5), 855–862.

Wang P., Wang Z. W., Wu Z. C., Zhou Q. and Yang D. H. (2010). Effect of hypochlorite cleaning on the physiochemical characteristics of polyvinylidene fluoride membranes. *Chemical Engineering Journal*, **162**(3), 1050–1056.

Wei C. H., Huang X., Ben Aim R., Yamamoto K. and Amy G. (2011a). Critical flux and chemical cleaning-in-place during the long-term operation of a pilot-scale submerged membrane bioreactor for municipal wastewater treatment. *Water Research*, **45**(2), 863–871.

Wei V., Elektorowicz M. and Oleszkiewicz J. A. (2011b). Influence of electric current on bacterial viability in wastewater treatment. *Water Research*, **45**(16), 5058–5062.

Wu B. and Fane A. G. (2012). Microbial relevant fouling in membrane bioreactors: influencing factors, characterization, and fouling control. *Membranes*, **2**, 565–584.

Wu Z., Wang Z. W., Zhou Z., Yu G. P. and Gu G. W. (2007). Sludge rheological and physiological characteristics in a pilot-scale submerged membrane bioreactor. *Desalination*, **212**(1–3), 152–164.

Wu Z., Wang Z., Huang S., Mai S., Yang C., Wang X. and Zhou Z. (2008). Effects of various factors on critical flux in a submerged membrane bioreactor for municipal wastewater treatment. *Separation ana Purification Technology*, **62**(1), 56–63.

Xavier K. B. and Bassler B. L. (2003). LuxS quorum sensing: more than just a numbers game. *Current Opinion in Microbiology*, **6**(2), 191–197.

Xiong Y. H. and Liu Y. (2010). Biological control of microbial attachment: A promising alternative for mitigating membrane biofouling. *Applied Microbiology and Biotechnology*, **86**(3), 825–837.

Xu T. and Fu R. (2004). Determination of effective diffusion coefficient and interfacial mass transfer coefficient of bovine serum albumin (BSA) adsorption into porous polyethylene membrane by microscope FTIR-mapping study. *Chemical Engineering Science*, **59**(21), 4569–4574.

Yamamoto K., Hiasa M., Mahmood T. and Matsuo T. (1989). Direct solid-liquid separation using hollow fiber membrane in an activated-sludge aeration tank. *Water Science and Technology*, **21**(4–5), 43–54.

Yamamura H., Kimura K., Okajima T., Tokumoto H. and Watanabe Y. (2008). Affinity of functional groups for membrane surfaces: Implications for physically irreversible fouling. *Environmental Science and Technology, 42* (14), 5310–5315.

Yang F., Wang Y., Bick A., Gilron J., Brenner A., Gillerman L., Herzberg M. and Oron G. (2012). Performance of different configurations of hybrid growth membrane bioreactor (HG-MBR) for treatment of mixed wastewater. *Desalination*, **284**, 261–268.

Yeon K. M., Cheong W. S., Oh H. S., Lee W. N., Hwang B. K., Lee C. H., Beyenal H. and Lewandowski Z. (2009a). Quorum sensing: a new biofouling control paradigm in a membrane bioreactor for advanced wastewater treatment. *Environmental Science and Technology*, **43**(2), 380–385.

Yeon K. M., Lee C. H. and Kim J. (2009b). Magnetic enzyme carrier for effective biofouling control in the membrane bioreactor based on enzymatic quorum quenching. *Environmental Science and Technology*, **43**(19), 7403–7409.

Yoon S. H. and Collins J. H. (2006). A novel flux enhancing method for membrane bioreactor (MBR) process using polymer. *Desalination*, **191**(1–3), 52–61.

Yoon S. H., Collins J. H., Musale D., Sundararajan S., Tsai S. P., Hallsby G. A., Kong J. F., Koppes J. and Cachia P. (2005). Effect of flux enhancing polymer on the characteristics of sludge in membrane bioreactor process. *Water Science and Technology*, **51**(6–7), 151–157.

Yu S.-l., Zhao F.-b., Zhang X.-h., Jing G.-l. and Zhen X.-h. (2006). Effect of components in activated sludge liqour on membrane fouling in a submerged membrane bioreactor. *Journal of Environmental Sciences*, **18**(5), 897–902.

Yun M. A., Yeon K. M., Park J. S., Lee C. H., Chun J. and Lim D. J. (2006). Characterization of biofilm structure and its effect on membrane permeability in MBR for dye wastewater treatment. *Water Research*, **40**(1), 45–52.

Zhang J., Chua H. C., Zhou J. and Fane A. G. (2006a). Factors affecting the membrane performance in submerged membrane bioreactors. *Journal of Membrane Science*, **284**(1–2), 54–66.

Zhang Y. P., Fane A. G. and Law A. W. K. (2006b). Critical flux and particle deposition of bidisperse suspensions during crossflow microfiltration. *Journal of Membrane Science*, **282**(1–2), 189–197.

Zhang K. S., Wei P., Yao M., Field R. W. and Cui Z. F. (2011). Effect of the bubbling regimes on the performance and energy cost of flat sheet MBRs. *Desalination*, **283**, 221–226.

Zheng X., Ernst M. and Jekel M. (2009a). Identification and quantification of major organic foulants in treated domestic wastewater affecting filterability in dead-end ultrafiltration. *Water Research*, **43**(1), 238–244.

Zheng X., Mehrez R., Jekel M. and Ernst M. (2009b). Effect of slow sand filtration of treated wastewater as pre-treatment to UF. *Desalination*, **249**(2), 591–595.

Zsirai T., Buzatu P., Aerts P. and Judd S. (2012). Efficacy of relaxation, backflushing, chemical cleaning and clogging removal for an immersed hollow fibre membrane bioreactor. *Water Research*, **46**(14), 4499–4507.

Chapter 5

Nutrient removal and disinfection by MBR

So-Ryong Chae[1], Yongtae Ahn[2], Yuhoon Hwang[3], Duksoo Jang[3], Fangang Meng[4], Jeffrey Shi[1], Sang-Hyup Lee[5,6] and Hang-Sik Shin[3]*

[1]*Department of Chemical and Environmental Engineering, University of Cincinnati, Cincinnati, OH 45221-0012, USA*
[2]*Department of Energy Engineering, Gyeongnam National University of Science and Technology, Jinju 52725, Republic of Korea*
[3]*Department of Environmental Engineering, Seoul National University of Science and Technology, Seoul 01811, Republic of Korea*
[4]*SYSU-HKUST Research Center for Innovative Environmental Technology (SHRCIET), School of Environmental Science and Engineering, Sun Yat-sen University, Guangzhou 510275, PR China*
[5]*Center for Water Resource Cycle Research, Korea Institute of Science and Technology, Seoul 136-791, Republic of Korea*
[6]*Graduate School of Convergence Green Technology & Policy, Korea University, Seoul 136-701, Republic of Korea*
**Corresponding author (e-mail: chaesg@ucmail.uc.edu)*

ABSTRACT

Over the past century, human activity has dramatically changed the global nutrient cycles (e.g., eutrophication due to the high nitrogen flux from municipal wastewater), ultimately having a detrimental effect on human health and the Earth's system dynamics. On the other hand, the global depletion of fresh drinking water supplies has seen the introduction of a variety of wastewater recycling strategies. During the last two decades, membrane bioreactors (MBR) have been widely used in the recycling of municipal

 Membrane Biological Reactors: Theory, Modeling, Design, Management and Applications to Wastewater Reuse.
Editor(s): Faisal I. Hai, Kazuo Yamamoto and Chung-Hak Lee
doi: 10.2166/9781780409177_0155

and industrial wastewaters. In this chapter, we reviewed recent progress towards the development of MBRs for the enhanced removal and recovery of nutrients from wastewaters, the microbial community for nutrient removal in MBRs, and the challenges associated with the disinfection of MBR effluents. MBRs are a promising strategic option for wastewater recycling and reuse but the elimination/recovery of nutrients and elimination of pathogenic virus in the MBR effluents requires careful consideration.

5.1 INTRODUCTION

By 2025, the number of people living in areas with water shortages is expected to increase from 1 billion to 2.4 billion (WHO & UNICEF, 2000). Moreover, climate change and global warming have had a negative effect on global water supply systems, which primarily depend on surface water. Therefore, the global water supply has been challenged in recent years by increasing water stress, in terms of both water scarcity and quality deterioration (Chang *et al.* 2007). As a result, alternative water resources are required in order to maintain a sustainable water supply.

Moreover, over the past century, human activity has dramatically changed the global nutrient cycles, resulting in detrimental effects on human health and the Earth's ecosystem dynamics (Schlesinger, 2009). In particular, eutrophication has been reported as an emerging issue in both natural and anthropogenic sources due to the high nutrient flux from municipal wastewater (Inglett *et al.* 2011). Eutrophication is a major cause of critical environmental problems, including the reductions in light penetration and increases in harmful algal bloom. In nutrient-sensitive estuaries, wastewater treatment plants (WWTPs) are required to implement more advanced treatment methods in order to meet the increasingly stringent effluent guidelines for total nitrogen (TN) and total phosphorus (TP) outputs into the water system.

The continued depletion of fresh drinking water supplies throughout the world has increased the need for a variety of water treatment and recycling strategies. Wastewater recycling is often cited as a viable alternative to seawater desalination (Meng *et al.* 2009). During the last two decades, membrane bioreactors (MBRs) have been widely used in the recycling of municipal and industrial wastewaters (Judd, 2006; Wang *et al.* 2008; Yang *et al.* 2006; Lesjean & Huisjes, 2008; Judd, 2008). According to the literature, MBR processes that incorporate coupled nitrification/denitrification have the potential to remove TN to a level of approximately 1.5–4 mg/L (Lazarova *et al.* 2012). However, the understanding of microbial community structures in MBRs and the post-treatment of pathogens and phosphorus in MBR effluents are still challengeable.

More than 2500 MBRs are in operation worldwide (Meng *et al.* 2012). Although most of these have a small treatment capacity, they are used to treat a wide variety of wastewaters from communities, municipalities, and industries, as well as aboard seagoing vessels. Municipal wastewater treatment is the largest application of MBRs, accounting for about 80% of all systems based on treatment capacity. However, small-scale MBR plants for smaller communities, housing developments, tourist resorts, hotels, schools, shopping centers and so on, are also in demand (Meng *et al.* 2012).

The MBRs can remove pathogens to the very high degree. However, significant problems arise if the membrane integrity fails, as the pathogen removal efficiency can deteriorate dramatically. In addition, the regrowth of pathogens has been reported in the distribution system irrespective of the treatment technology used (Jjemba *et al.* 2010). Thus, the post-treatment of MBR effluents as well as effective effluent monitoring to identify the integrity of the process is necessary in order to ensure environmental and public health protection.

This chapter details the recent progress towards the development of MBRs for the enhanced removal and recovery of nutrients from wastewaters, the microbial community for nutrient removal in MBRs, and the challenges associated with the disinfection of MBR effluents are reviewed here.

5.2 REUSE AND RECYCLING OF RECLAIMED WASTEWATER

The deliberate use of reclaimed water or wastewater must comply with the appropriate rules associated with the purposes of reuse. Recent estimates indicate that approximately 7 to 8% of wastewater is reused in the United States (USEPA, 2012). Several countries, including Australia and Singapore, have established goals for reuse, expressed in terms of the percentage of municipal wastewater effluent that is treated to a higher quality and beneficially reused. Australia currently reuses approximately 8% of its treated wastewater with a goal of reusing 30% by 2015. Saudi Arabia currently reuses 16% with a goal to increase reuse to 65% by 2016. Singapore reuses 30% and has long-term planning in place to diversify its raw water supplies and reduce dependence on supplies from outside sources (i.e., Malaysia). Israel has attained the highest national percentage by beneficially reusing 70% of the generated domestic wastewater (USEPA, 2012).

Water reuse standards or guidelines vary with the type of applications, the regional context, and the overall risk perception. However, the common guideline is ensuring public health and safety. The treatment of urban wastewater is typically designed to meet water quality objectives based on suspended solids, organic content, biological indicators, nutrient levels, and chlorine residual in some cases. Additional water quality parameters for irrigation include salinity, sodium adsorption ratio (SAR), boron concentration, heavy metals content, and phytotoxic compounds. The use of reclaimed municipal wastewater for industrial purposes may require effluent limits for dissolved solids, ammonia, disinfection byproducts and other specific inorganic and organic constituents. The following seven categories of use are described in detail considering the limitations and guidelines placed on the reuse (Table 5.1) (USEPA, 2012).

Table 5.1 Guidelines of water quality for reuse.

Reuse Category and Description	Treatment	Reclaimed Water Quality
Urban Reuse		
Unrestricted: The use of reclaimed water in nonpotable applications in municipal settings where public access is not restricted.	• Secondary • Filtration • Disinfection	• pH = 6.0–9.0 • ≤10 mg/L BOD • ≤2 NTU • No detectable fecal coliform/100 mL • 1 mg/L Cl_2 residual (min.)
Restricted: The use of reclaimed water in nonpotable applications in municipal settings where public access is controlled or restricted by physical or institutional barriers, such as fencing, advisory signage, or temporal access restriction.	• Secondary • Disinfection	• pH = 6.0–9.0 • ≤30 mg/L BOD • ≤30 mg/L TSS • ≤200 fecal coliform/100 mL • 1 mg/L Cl_2 residual (min.)
Agricultural Reuse		
Food Crops: The use of reclaimed water for surface or spray irrigation of food crops which are intended for human consumption, consumed raw.	• Secondary • Filtration • Disinfection	• pH = 6.0–9.0 • ≤10 mg/L BOD • ≤2 NTU • No detectable fecal coliform/100 mL • 1 mg/L Cl_2 residual (min.)
Processed Food Crops: The use of reclaimed water for surface irrigation of	• Secondary • Disinfection	• pH = 6.0–9.0 • ≤30 mg/L BOD • ≤30 mg/L TSS

(Continued)

Table 5.1 Guidelines of water quality for reuse (*Continued*).

Reuse Category and Description	Treatment	Reclaimed Water Quality
food crops which are intended for human consumption, commercially processed.		• ≤200 fecal coli/100 mL • 1 mg/L Cl_2 residual (min.)
Non-Food Crops: The use of reclaimed water for irrigation of crops which are not consumed by humans, including fodder, fiber, and seed crops, or to irrigate pasture land, commercial nurseries, and sod farms.	***Impoundments***	
Unrestricted: The use of reclaimed water in an impoundment in which no limitations are imposed on body-contact.	• Secondary • Filtration • Disinfection	• pH = 6.0–9.0 • ≤10 mg/L BOD • ≤2 NTU • No detectable fecal coliform/100 mL • 1 mg/l Cl_2 residual (min.)
Restricted: The use of reclaimed water in an impoundment where body-contact is restricted.	• Secondary • Disinfection	• ≤30 mg/L BOD • ≤30 mg/L TSS • ≤200 fecal coliform/100 mL • 1 mg/L Cl_2 residual (min.)
Environmental Reuse		
The use of reclaimed water to create wetlands, enhance natural wetlands, or sustain stream flows.	• Variable • Secondary and disinfection (min.)	Variable, but not to exceed: • ≤30 mg/L BOD • ≤30 mg/L TSS • ≤200 fecal coliform/100 mL • 1 mg/L Cl_2 residual (min.)
Industrial Reuse		
Once-through Cooling	• Secondary	• pH = 6.0–9.0 • ≤30 mg/L BOD • ≤30 mg/L TSS • ≤200 fecal coliform/100 mL • 1 mg/L Cl_2 residual (min.)
Recirculating Cooling Towers	• Secondary • Disinfection (chemical coagulation and filtration may be needed)	Variable, depends on recirculation ratio: • pH = 6.0–9.0 • ≤30 mg/L BOD • ≤30 mg/L TSS • ≤200 fecal coliform/100 mL • 1 mg/L Cl_2 residual (min.)
Groundwater Recharge – Nonpotable Reuse		
The use of reclaimed water to recharge aquifers which are not used as a potable drinking water source.	• Site specific and use dependent • Primary (min.) for spreading • Secondary (min.) for injection	• Site specific and use dependent

(*Continued*)

Table 5.1 Guidelines of water quality for reuse (*Continued*).

Reuse Category and Description	Treatment	Reclaimed Water Quality
Indirect Potable Reuse		
Groundwater Recharge by Spreading into Potable Aquifers	• Secondary • Filtration • Disinfection • Soil aquifer treatment	Includes, but not limited to, the following: • No detectable total coliform/100 mL • 1 mg/L Cl_2 residual (min.) • pH = 6.5–8.5 • ≤2 NTU • ≤2 mg/L TOC of wastewater origin • Meet drinking water standards after percolation through vadose zone
Groundwater Recharge by Injection into Potable Aquifers	• Secondary • Filtration • Disinfection • Advanced wastewater treatment	Includes, but not limited to, the following: • No detectable total coliform/100 mL • 1 mg/L Cl_2 residual (min.) • pH = 6.5–8.5 • ≤2 NTU • ≤2 mg/L TOC of wastewater origin • Meet drinking water standards
Augmentation of Surface Water Supply Reservoirs	• Secondary • Filtration • Disinfection • Advanced wastewater treatment	Includes, but not limited to, the following: • No detectable total coliform/100 mL • 1 mg/L Cl_2 residual (min.) • pH = 6.5–8.5 • ≤2 NTU • ≤2 mg/L TOC of wastewater origin • Meet drinking water standards

Source: USEPA (2012).

5.2.1 Urban reuse

While there are several major categories of reuse, urban reuse is one of the highest volume uses in the United States. Applications such as recreational field and golf course irrigation, landscape irrigation, and other applications, including fire protection and toilet flushing, are important components of the reclaimed water portfolio of many urban reuse programs. Urban reuse is often divided into applications that are either accessible to the public or have restricted access, in settings where public access is controlled or restricted by physical or institutional barriers, such as fences or temporal access restriction. When used to irrigate residential areas, golf courses, public school yards, and parks, reclaimed water receives treatment and high-level disinfection and is not considered a threat to public health (USEPA, 2012).

5.2.2 Agricultural reuse

Because agricultural reuse is one of the most significant uses of reclaimed water globally, it is critical to understand the factors that determine success or failure of a farming operation dependent upon reclaimed water for irrigation. The same concerns for chemical constituents are applicable to all sources of irrigation water, and reclaimed water is no exception. Thus, when considering the use of reclaimed water in agriculture, it is important to identify the key constituents of concern for agricultural irrigation. Salinity is a key parameter in determining the suitability of the water to be used for irrigation. Using total dissolved solids (TDS) as a measure of salinity, no detrimental effects are usually noticed below 500 mg/L, TDS between 500 and 1000 mg/L in irrigation water can affect sensitive plants; at concentrations above 1000 to 2000 mg/L, TDS levels can affect many crops, so careful management practices should be followed (USEPA, 2012).

5.2.3 Impoundments

Uses of reclaimed water for maintenance of impoundments range from water hazards on golf courses to full-scale development of water-based recreational impoundments involving incidental contact (fishing and boating) and full body contact (swimming and wading). Regulation of impoundments that are maintained using reclaimed water is typically according to the potential for contact for that use. In impoundments where body contact is prohibited, such as a manmade facility that is created for storage, landscaping, or for aesthetic purposes only, less stringent requirements may apply (USEPA, 2012).

5.2.4 Environmental reuse

Environmental reuse primarily includes the use of reclaimed water to support wetlands and to supplement stream and river flows. Aquifer recharge also may be considered environmental reuse. Development has altered the landscape, including changing the timing and quantities of stormwater and surface water flows and lowering of the groundwater tables, which affect environmental systems that have adapted and depend on these for their existence. Reclaimed water could be used to mitigate some of these impacts. Application of reclaimed water serves to restore and enhance wetlands that have been hydrologically altered. As with impoundments, water quality requirements for river or stream augmentation will be based on the designated use of the water course and the aim to enhance an acceptable appearance. In addition, there should be an emphasis on creating a product that can promote native aquatic life. The quality of the reclaimed water discharged to the receiving water body is critical to evaluating its benefits to the stream (USEPA, 2012).

5.2.5 Industrial reuse

Traditionally, pulp and paper facilities, textile facilities, and other facilities using reclaimed water for cooling tower purposes, have been the primary industrial users of reclaimed water. Any contamination of the cooling water through process in-leakage, atmospheric deposition, or treatment chemicals will also impact the water quality. While reclaimed water generally has very low concentrations of microorganisms due to the high level of treatment, one of the major issues with reclaimed water use in cooling towers relates to occurrence of biological growth when nutrients are present. Scaling can also be a problem in cooling towers. The primary constituents resulting in scale potential from reclaimed water are calcium, magnesium, sulfate, alkalinity, phosphate, silica, and fluoride. Constituents with the potential to form scale must be evaluated and controlled by chemical treatment and/or by adjusting the cycles of concentration. Therefore, reclaimed water quality must be evaluated, along with the scaling potential to establish the use of specific scale inhibitors (USEPA, 2012).

The food and beverage manufacturing industry was initially reluctant to use – and publicize the use of – reclaimed water because of public perception concerns. As knowledge of water reuse principles has increased, so has the reuse of highly-treated process waters that meet water quality criteria and address public health concerns. In many cases, not only is reuse of water at a manufacturing site "green," but it also can reduce operating costs and an industry's water footprint and, in some cases, provide better water quality than the public water supply (USEPA, 2012).

The use of reclaimed water in high-technology manufacturing, such as the semiconductor industry, is a relatively new practice. Within the semiconductor industry, there are two major processes that use water: microchip manufacturing, which has rarely utilized reclaimed water, and the manufacture of circuit boards. In circuit board manufacturing, water is used primarily for rinse operations; similar to production of boiler feed water, reclaimed water for circuit board manufacturing requires extensive treatment. While only circuit board manufacturing uses reclaimed water in the actual production process, both semiconductor and circuit board manufacturing facilities do use reclaimed water for cooling water and site irrigation (USEPA, 2012).

5.2.6 Groundwater recharge – nonpotable reuse

Groundwater recharge to aquifers not used for potable water has been practiced for many years, but has often been viewed as a disposal method for treated wastewater effluent. In addition to providing a method of treated effluent disposal, groundwater recharge of reclaimed water can provide a number of other benefits including (1) recovery of treated water for subsequent reuse or discharge, (2) recharge of adjacent surface streams, and (3) seasonal storage of treated water beneath the site with seasonal recovery for agriculture. In many cases, groundwater can be recharged in a manner that also utilizes the soil or aquifer system where reclaimed water is applied as an additional treatment step to improve the reclaimed water quality (USEPA, 2012).

5.2.7 Potable reuse

The use of reclaimed water to augment potable water supplies has significant potential for helping to meet future needs, but planned potable water reuse only accounts for a small fraction of the volume of water currently being reused. Planned indirect potable reuse involves a proactive decision by a utility to discharge or encourage discharge of reclaimed water into surface water or groundwater supplies for the specific purpose of augmenting the yield of the supply. The unplanned reuse of wastewater effluent as a water supply is common, with some drinking water treatment plants using waters from which a large fraction originated as wastewater effluent from upstream communities, especially under low-flow conditions. To date, no regulations or criteria have been developed or proposed specifically for direct potable reuse of reclaimed water in the United States. Past regulatory evaluations of this practice generally have been deemed unacceptable due to a lack of definitive information related to public health protection. Considering that unplanned reuse is already widely practiced, direct potable reuse may be a reasonable option based on significant advances in treatment technology and monitoring methodology in the last decade and health effects data from indirect potable reuse projects and direct potable reuse demonstration facilities (USEPA, 2012).

5.3 ADVANCED DESIGNS OF MBRs FOR NUTRIENT REMOVAL

The biological nutrient removal (BNR) processes has been widely used for nutrient removal. Biological nitrification is an oxic process of conversion of ammonia to nitrite (NO_2-N) and then to nitrate (NO_3-N). Following nitrification, these compounds can be removed from wastewater by reducing nitrate to nitrogen gas (N_2) by denitrification under anoxic conditions. A key for efficient nitrification is to keep a high concentration of nitrifying bacteria in biological reactors. The increase in solids residence time

(SRT) could reduce the wash-out of nitrifying bacteria from the bioreactors, improving the nitrification capability of activated sludge. To achieve denitrification, usually an anoxic tank is added upstream from the aerated tank. It can be also possible by introducing an intermittent aeration mode. Phosphorus removal can be achieved by wasting sludge after the aerobic period when the biomass accumulate a high level of polyphosphates (poly-P) (Chung *et al.* 2006). The P removal could be improved by adding chemicals. The chemical addition to removal P to very low level (0.025 mg P/L) did not affect other biological processes (i.e., organic and nitrogen removal) but it might increase sludge production (Liu *et al.* 2011).

Nutrient removal using the CAS processes has been widely studied. Recently, it has been reported that the modified CAS processes such as anaerobic-anoxic-oxic (A2/O) process and the University of Cape Town (UCT) process increase the TN removal efficiency (Table 5.2). In particular, Puig *et al.* (2010) split an anoxic zone in the UCT process into two parts and introduced return sludge to the first part (upstream), and used the second part (downstream) for denitrification of the nitrate recycled. By this change, nitrate concentration in the anaerobic zone dramatically reduced and the overall nutrient removal efficiency significantly increased to 80–90% (Puig *et al.* 2010). The activated sludge can be separated by gravitational settling. However, treatment efficiency of the CAS processes is usually limited by *the difficulties in separating the suspended solids (SS)* (Xing *et al.* 2000).

Table 5.2 Comparison of nitrogen and phosphorus removal between the MBR and CAS processes.

Process	Configuration (Capacity, m^3/d)	Operating Conditions	Raw Wastewater Characteristics (mg/L)	TN Removal (%)	TP Removal (%)	Reference
CAS	CAS (34,800)	HRT = 3–75 h SRT = 4–9 d	COD: 220 TN: 26, TP: 3.9	46	87	Wan *et al.* (2011)
	A2/O (400,000)	HRT = 15.3 h SRT = 20–25 d	COD: 476 TN: 57, TP: 5.9	80	96	Fan *et al.* (2011)
	Modified UCT (10,140)	HRT = 24 h SRT = 30 d	COD: 401 TN: 13.5, TP: 3.2	90	76	Puig *et al.* (2010)
MBR	Submerged (6520)	HRT = 3.5–5 h SRT = 14–21 d	COD: 220 TN: 26, TP: 3.9	70	92	Wan *et al.* (2011)
	Submerged (2400)	HRT = 16.6 h SRT = 40 d	COD: 447 NH_4-N: 26 TP: 7.3	99	93	Silva *et al.* (2012)
	Submerged (55,800)	HRT = 4.1 h SRT = 27 d	COD: 578 NH_4-N: 48 TP: 7.5	98	96	Silva *et al.* (2012)
	Submerged (8300)	HRT = 3.6 h SRT = 21 d	COD: 175 NH_4-N: 17 TP: 9.3	>95	89	Silva *et al.* (2012)

The MBR process offers many benefits over the CAS process, which includes small space and reactor requirements, better effluent quality, disinfection, increased volumetric loading, and less sludge production (Fan *et al.* 1996; Adam *et al.* 2002; Lesjean *et al.* 2002; Patel *et al.* 2005; Ramphao *et al.* 2005; Kraume *et al.* 2005). Additional advantages would include the control of SRT, operational reliability and stability, easy automatic control and compactness of the whole system.

5.3.1 Design of MBRs for removal of organic matter and nitrogen

In MBRs, membrane tanks have much higher dissolved oxygen (DO) concentrations than the CAS process. Returning biomass with high DO concentrations from membrane tank to anoxic zone leads to a decrease in the denitrification efficiency and nitrogen removal performance of the systems. To reduce DO concentrations in the return activated sludge (RAS) to the anoxic zone, a deoxygenation zone is typically introduced prior to the anoxic zone or internal mixed liquor recycle (IMLR) from the aerobic to the anoxic zone.

In the A2/O-MBRs, the following three configurations can be used: (1) RAS flow to the aerobic zone and IMLR from aerobic to anoxic zone, (2) RAS flow to a deoxygenation zone and IMLR from aerobic to anoxic zone, and (3) RAS flow to a deoxygenation zone and no IMLR (WEF, 2012; Phan *et al.* 2015).

To achieve more complete nitrogen removal, a four-stage Bardenpho process, which consists of a series of two anoxic and two aerobic zones with recycling of mixed liquor from the first aerobic zone to the first anoxic zone, can be coupled with membrane. In this system, RAS containing high DO concentrations can be returned to the first aerobic zone with use of IMLR from the aerobic zone to the first anoxic zone, or RAS is returned to a deoxygenation zone prior to the first anoxic zone with or without the IMLR (WEF, 2012).

5.3.2 Design of MBRs for simultaneous removal of nitrogen and phosphorus

The simultaneous removal of nitrogen and phosphorus requires a process that has anaerobic zone(s), anoxic zone(s), and aerobic zones(s). The anaerobic zones are needed to give a competitive advantage to polyphosphate accumulating organisms (PAOs), anoxic zones for denitrification, and aerobic zones for nitrification and phosphorus uptake. It is important to restrict DO and oxidized nitrogen entries to the anaerobic zones because biological phosphorus removal is highly dependent on a lack of nitrite and nitrate in the zone. This can be achieved in numerous ways.

The examples of three-stage CAS process coupled with membrane include (1) RAS from membrane tank to deoxygenation zone between anaerobic and anoxic zones, first IMLR from aerobic zone to anoxic zone, and second IMLR from anoxic zone to anaerobic zone, (2) RAS from membrane tank to aerobic zone, first IMLR from aerobic zone to anoxic zone, and second IMLR from anoxic to anaerobic zone, and 3) RAS from membrane tank to deoxygenation zone between anaerobic and anoxic zone, IMLR from anoxic to anaerobic zone.

Also five-stage MBR processes can be configured in numerous ways. (1) Three-recycle configuration: RAS to deoxygenation zone, IMLR from aerobic zone to the first anoxic zone, IMLR from first anoxic to the anaerobic zone. (2) Three-recycle configuration: RAS to aerobic zone, IMLR from aerobic zone to the first anoxic zone, and IMLR from first anoxic to the anaerobic zone. (3) Two-recycling configuration: RAS to deoxygenation zone, IMLR from first anoxic to the anaerobic zone.

Unlike the five-stage CAS process, the five-stage MBR configuration does not require a small reaeration zone because reaeration can be achieved in the membrane tank (WEF, 2012). Moreover, when chemical addition is required for additional phosphorus removal, chemicals can be directly injected upstream of membrane tanks. Then the phosphorus precipitates can be removed by MF or UF membranes.

By introducing anoxic/aerobic sequencing or simultaneous nitrification/denitrification in the MBRs, more than 90% of nutrients in feed wastewater can be removed (Table 5.2). For example, a sequencing anoxic/anaerobic membrane bioreactor (SAM) was developed to enhance biological phosphorus removal. The SAM system was composed of an anoxic/anaerobic zone and aerobic zone where the flat-sheet microfiltration (MF) membrane module was immersed. The confirmed phosphorus removal efficiency rate was about 93% (Ahn *et al.* 2003).

Kimura *et al.* (2008) determined that simultaneous nitrification/denitrification in a single and small reaction tank was possible by inserting baffles into a normal submerged MBR. Aerobic and anoxic conditions are alternatively created in the outer zone at a constant interval, and improvement in nitrogen removal is therefore expected in comparison to that of normal submerged MBRs, which are principally aerobic zone (Kimura *et al.* 2008). Also, Zhang *et al.* (2009) developed a modified UCT-MBR, which is an adaptation of the UCT process with a final membrane filtration unit instead of the conventional secondary clarifier (Zhang *et al.* 2009). The anoxic and anaerobic membrane bioreactor process was composed of a continuous aerated MBR and an alternating anaerobic and anoxic zone containing two separated bioreactors. The results showed that nitrogen and phosphorus removal efficiency attained high treatment levels (Yuan *et al.* 2008). Recently, Wei *et al.* (2012) developed a MBR coupled with electrodes to reduce membrane fouling and improve filtrate flow. The electrically enhanced MBR (EMBR) treating sewage achieved >99% of ammonium-N, >95% of dissolved COD, and >90% ortho-P removal at an operating temperature of 10°C (Wei *et al.* 2012).

5.4 EFFECTS OF THE MICROBIAL COMMUNITY ON NUTRIENT REMOVAL IN MBRs

Municipal wastewater contains a complex mixture of polysaccharides, proteins, lipids, and other compounds (Raunkjær *et al.* 1994). Biological wastewater treatment is achieved using a complex consortium of microorganisms (primarily bacteria) that metabolize both soluble organic and inorganic pollutants. The integration of membranes into a BNR process is becoming increasingly popular due to its various advantages in wastewater treatment. However, the method of efficiently operating MBR systems remains a topic of contention due to the lack of sufficient information on the development of microbial community structures in MBRs. Therefore, it is important to study the microbial community structures of the MBR systems for better understanding of the reactor performance and associated bacterial activity/population and membrane fouling for optimal MBR design and operation.

The recent developments of molecular biological techniques and chemotaxonomy methods have offered opportunities for analyzing the microbial community structures in order to obtain a sounder basis for engineering their control. The microbial communities responsible for nutrient removal and biofouling in MBRs fed with synthetic or municipal wastewater have been widely studied using various molecular biological tools such as fluorescence *in situ* hybridization (FISH) (Silva *et al.* 2012), polymerase chain reaction (PCR) coupled with either denaturing gradient gel electrophoresis (DGGE) (Chen & LaPara, 2006) or temperature gradient gel electrophoresis (TGGE) (Molina-Munoz *et al.* 2009), terminal restriction fragment length polymorphism (T-RFLP) (Falk *et al.* 2009), and pyrosequencing (Lim *et al.* 2012; Phan *et al.* 2016) based on the 16S ribosomal RNA (rRNA) sequences. In particular, PCR-DGGE/TGGE, a molecular method through which deoxyribonucleic acid (DNA) fragments of the same length but different base pair sequences can be separated and bands can further be excised from the gel and sequenced in order to identify the phylogenetic affiliation of the community members (Jeon *et al.* 2003). Quinone profiles, which could possess a very high sensitivity and repeatability, can only reveal the structure at the 'phylum' level and some phylogenetically different groups of bacterium share the same quinone groups (Kurisu *et al.* 2002). Pyrosequencing can generate numerous DNA sequences in a single run, which provides information on species richness and diversity in biological wastewater treatment plants (Sanapareddy *et al.* 2009).

The effects of feed wastewater on the structure and function of the microbial communities growing in MBRs were investigated (Table 5.3). Chen and LaPara (2006) monitored the shift of microbial

Table 5.3 Microbial communities in MBRs with various carbon sources.

Characteristics of Feed Water	Operating Conditions	Removal Efficiency	Major Microorganisms	Reference
Synthetic wastewater: Glucose, 1.0 g/L $(NH_4)_2SO_4$, 0.5 g/L	HRT: 20 h SRT: 30 d	COD: 96%	• *Proteobacteria (73%)* = *α-proteobacteria (32%)* + *γ-proteobacteria (33%)* + *others* • *Actinobacteria (17%)*	Lim *et al.* (2012)
Synthetic wastewater: Gelatin, 50–500 mg/L $(NH_4)Cl$, 150 mg/L	HRT: 12 h	COD: 84–97%	• *Bacteriodetes (~50%)* • *Proteobacteria (~40%)* • *Actinobacteria (~5%)*	Chen and LaPara (2006)
Synthetic wastewater: CH_3COOH, 400 mg/L as COD $(NH_4)_2SO_4$, 0.236 g/L	HRT: 24 h SRT: 20 d	COD > 94% NH_4^+)-N > 94%	• *Proteobacteria (59%)* = *β-proteobacteria (58%)* + *γ-proteobacteria (1%)* • *Sphingobacteria (22%)* • *Flavobacteria (8%)* • *Actinobacteria (5%)*	Falk *et al.* (2009)
Synthetic wastewater: Starch, 450 mg/L as COD $(NH_4)Cl$, 150 mg/L	HRT: 12 h	COD > 93%	• *Bacteriodetes* • *Proteobacteria: β-proteobacteria, γ-proteobacteria*	LaPara *et al.* (2006)
Municipal wastewater: COD: 220 mg/L TN: 26 mg/L	HRT = 3.5–5 h SRT = 14–21 d	COD: 89% TN: 70%	• *Proteobacteria (41%)* = *α-proteobacteria (14%)* + *β-proteobacteria (27%)* • *Bacteriodetes (25%)* • Others (34%): *δ-proteobacteria, Actinobacteria, Acidobacteria, Chorobi, Firmicutes*	Wan *et al.* (2011)
Municipal wastewater: COD: 175 mg/L TN: 17 mg/L	HRT = 3.6 h SRT = 21 d	COD > 85% TN > 94%	• *Accumulibacter (11%)* • *β-proteobacteria (8%)* • *Actinobacteria (5%)* • Others: *Competibacter, α-proteobacteria, γ-proteobacteria*	Silva *et al.* (2012)
Municipal wastewater: COD: 480 mg/L TN: 53 mg/L	HRT = 8–15 h	COD > 95%	• Major: *α-proteobacteria, β-proteobacteria, γ-proteobacteria* • Minor: *Actinobacteria, Gemmatimonadetes, Δ-proteobacteria*	Molina-Munoz *et al.* (2009)

communities in bench-scale MBRs fed with gelatin wastewater. The biomass concentrations initially increased rapidly in the MBRs, but subsequently declined until an asymptote was reached. The bacterial community analysis *via* PCR-DGGE suggested that the bacterial community structure completely changed from the beginning to the end of each MBR (Chen & LaPara, 2006). Li *et al.* (2006) reported that *Nitrosomonas* sp. and *Nitrospira* sp. were dominant nitrifying bacteria responsible for ammonia and nitrite oxidation, respectively in a MBR fed with inorganic ammonium-bearing wastewater (Li *et al.* 2006). Furthermore, Miura *et al.* (2007) showed that characteristics of feed wastewater had a great impact on bacterial community structures in MBRs treating municipal wastewater (Miura *et al.* 2007). Ahmed *et al.* (2008) investigated the effects of carbon sources (i.e., acetate, propionate, glucose, and methanol) on the diversity of microbial communities. The authors found that the β-subclass of *Proteobacteria* was dominant; however, its population comparatively decreased when the substrate was glucose dominant or methanol dominant. A relatively higher and more stable enhanced biological phosphorus removal performance was observed when the methanol-dominant substrate was used concurrently with an increase in the γ-subclass of *Proteobacteria*. The population of the α-subclass of *Proteobacteria* increased slightly along with a decrease in the phosphate removal activity when the substrate was glucose-dominant (Ahmed *et al.* 2008).

However, when MBRs are operated *with real municipal wastewater*, the MBRs can support relatively complex bacterial communities compared with the MBRs fed with synthetic feed media. For example, Molina-Munoz *et al.* (2009) operated a pilot-scale MBR treating domestic wastewater over nine months. In their study, it was found that the temperature, HRT, and VSS concentration were the factors that most significantly influenced the community structure. The 16S-rRNA sequences retrieved from the sludge were mostly related to the α-, β-, and γ-*Proteobacteria* classes (Molina-Munoz *et al.* 2009). Wan *et al.* (2011) analyzed microbial communities in a full-scale MBR for over one year and found that the α- and β-*Proteobacteria* and *Bacteroidetes* dominated the gene library. The vast majority of sequences retrieved was not closely related to classified organisms or displayed relatively low levels of similarity with any known 16S rRNA gene sequences; thus, they represent organisms that constitute new taxa (Wan *et al.* 2011). Silva *et al.* (2012) analyzed microorganisms involved in enhanced biological phosphorus removal (EBPR) in eight MBRs fed with municipal wastewater using FISH technique and found that the dominant bacteria group was the β-proteobacteria (ammonia-oxidizing), followed by the γ-proteobacteria in the all MBRs. Also *Tetrasphaera*-related *Actinobacteria* (PAOs) were present within the range of 1–8% in the plants (Silva *et al.* 2012).

5.5 CASE STUDIES: REUSE AND RECYCLING OF MBR EFFLUENTS

The applications of MBR technology for water reuse and recycling are rapidly expanding (Meng *et al.* 2012). MBR systems can be designed to pull wastewater directly from the sewer at the remote points of reuse. MBRs could provide suspended solids and bacteria free effluents due to the physical barrier of the membrane (Melin *et al.* 2006). The MBRs mostly coupled with polyvinylidene fluoride (PVDF) or polyethylene (PE) membranes for various water reuse purposes (Table 5.4). The effluent of MBR can generally meet the requirement of irrigation purpose, toilet flushing, and possible industrial applications (Merz *et al.* 2007). Unfortunately, the low-pressure membranes (i.e., MF or ultrafiltration (UF)) which are used for MBR cannot completely reject humic substances, virus, and waterborne pathogens. At this point, the post-treatment of MBR effluent by more dense membrane (i.e., nanofiltration (NF) or reverse osmosis (RO)) has been recently applied for the safe reuse of wastewater (Shannon *et al.* 2008). In the MBR-RO process, the RO membrane can remove the remaining compounds present in the MBR effluent. On the other hand, MBR acts as a pretreatment to reduce fouling of RO membranes.

Table 5.4 Applications of MBRs for water reuse and recycling.

Location	Membrane Supplier	Membrane Material	Capacity (x 10^3 m^3/d)	Purpose
Running Springs (USA)	Kubota	PE (0.4 μm)	2.3	Irrigation
Yas Island WWTP (UAE)	Toray	PVDF (0.08 μm)	57	Irrigation (golf course), etc.
Basic American Foods (USA)	Zenon	PVDF (0.04 μm)	6.4	Irrigation
Rocca Imperiale (Italy)	Zenon	PVDF (0.04 μm)	0.96	Demo project for agricultural reuse
Beixiao River plant (China)	Siemens	PVDF (0.04 μm)	60~100	Irrigation in the Olympic village and recharging the lake
Bejing Miyun (China)	Mitsubishi Rayon	PVDF (0.4 μm)	45	Landscape irrigation
Glen Meadows Retirement Home (USA)	Dynatec Systems	Not available	0.34	Domestic, irrigation
The Queen Mary II MBR WWTP (USA)	Orelis Environment	Not available	1.1	Domestic
Peoria WRF (USA)	Zenon	PVDF (0.04 μm)	38	Aquifer recharge
IWVA Torreele water reuse plant (Belgium)	Zenon	PVDF (0.04 μm)	7.0	Groundwater recharge
Ulu Pandan WRP (Singapore)	Zenon	PVDF (0.04 μm)	23	Indirect portable use
Fuji photographic production plant (The Netherlands)	TORAY	PVDF (0.08 μm)	0.84	Industrial use (cooling tower water)
Dalsung (Korea)	KMS	PE (0.4 μm)	15	Industrial use (paper production)
Shenghong Printing & Dyeing Co. Ltd (China)	Memstar	PVDF (<0.1 μm)	9.7	Industrial use
Sinopec Guangzhou Co., Ltd (China)	Memstar	PVDF (<0.1 μm)	6.5	Industrial use (cooling tower)
Eemshaven (The Netherlands)	Multitube	PVDF (0.04 μm)	1.4	Industrial use (malting process)
Immingham Docks (UK)	Multitube	Not available	1.7	Industrial use (boiler feed water)
Collegno (Italy)	Zenon	PVDF (0.04 μm)	0.96	Industrial use
Drews Meerane (Germany)	Zenon	PVDF (0.04 μm)	0.86	Industrial use (textile industry)
Kanes Food (UK)	Pentair	PVDF (<0.1 μm)	0.815	Industrial use (food industry)

Source: Moeslang and Brockmann (2010); Melin *et al.* (2006); Judd (2011).

The water reuse project in Singapore is the one of example for this application of MBR-RO for indirect portable use. Singapore established a water master plan in 1972 that produces NEWater from wastewater. The public utilities board (PUB) in Singapore reported that NEWater meets 30% of Singapore's total water demand (PUB, 2011). PUB has constructed a 23,000 m^3/d MBR demonstration plant at the Ulu Pandan water reclamation plant. The demonstration plant has been in operation since 2006 and has had stable operation at a membrane flux of 25.3 L/m^2/h with an energy consumption between 0.5–0.6 kWh/m^3 (PUB, 2011). Most NEWater supplied to the industrial sector is used as non-portable reuse, but indirect portable reuse has also been attempted by mixing and blending NEWater with raw water in the reservoirs prior to undergoing conventional treatment at the water treatment plant.

As shown in Figure 5.1, a dual MF/UF-RO membrane process plus UV irradiation has been successfully applied to produce high grade NEWater. Recently, the MBR technology coupled with RO has been tested as an alternative economical option to convert the conventional two-step process (activated sludge process and membrane filtration) into a single process. The MBR technology coupled with RO produces better quality NEWater compared with the conventional activated sludge followed by membrane filtration process (Qin *et al.* 2006). In addition, as this process eliminates the final effluent tanks as well as the pre-treatment step (MF or UF) to RO in the NEWater production, it can potentially reduce the production costs of NEWater by approximately 20% (Qin *et al.* 2006).

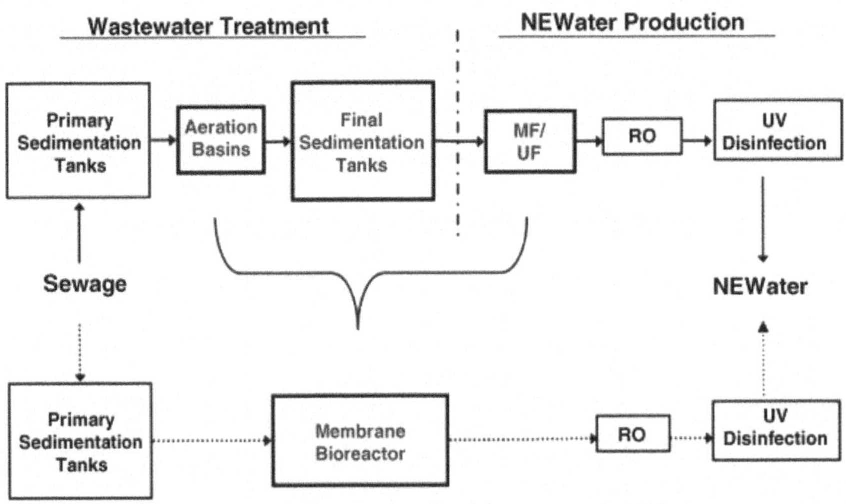

Figure 5.1 Schematic diagram of the NEWater plants with MBR and RO (Qin *et al.* 2006).

Also, the MBR process has been widely applied for water recycling in industrial fields. The MBR (14,000 m^3/d) has been applied to a bio-energy facility (Roseisle, Scotland). Through the combination of an anaerobic treatment and aerobic MBR, the wastewater was treated effectively for reuse. The anaerobic reactor removes approximately 80% of the soluble COD; the MBR then eliminates any residual biodegradable COD and performs a full nitrification and denitrification. The MBR effluent is free from particles, and the microbiological quality is also excellent (total coliform <100 cfu/100 mL).

5.6 NUTRIENT RECOVERY FROM MBR EFFLUENTS

Wastewaters have traditionally been viewed as environmental and human health concerns, and thus resources have been expended to manage them. Energy and chemicals are typically expended in order to produce water that is of satisfactory quality for discharge or release to the environment. However, there is a growing interest in either reducing the inputs required to treat wastewaters or in generating products of value from the wastewater itself (i.e., treating it as a resource). This increased interest is evidenced by the recent activities related to making wastewater treatment facilities more sustainable. The increasing market value of wastewater components such as ammonia and phosphorus are also acting as drivers for resource recovery from wastewater.

As the human population and demand for food grow, a worldwide depletion of phosphorus resources is occurring. It is estimated that the world may already have passed the 'peak phosphorus' level indicating that demand has already outpaced the known reserves of rock phosphate. However, phosphorus is present in both organic and inorganic forms in municipal wastewaters. The typical generation of phosphorus in wastewater is 2.7–4.5 g/capita/day (Metcalf & Eddy, 2004). The majority of phosphorus in municipal wastewaters is present as phosphate (PO_4^{3-}) (Melcer *et al.* 2003), which is soluble. At the same time, the production and use of nitrogen fertilizers from natural gas has significant greenhouse implications. These looming constraints make the recovery and reuse of these nutrients an economic and environmental necessity. Municipal wastewater contains reduced nitrogen in inorganic (NH_4) and organic (proteins) forms. Typically, 5–12 g/capita/day of NH_4-N and 4–10 g/capita/day of organic nitrogen are generated in municipal wastewaters (Metcalf & Eddy, 2004).

The recovery of water as a nutrient resource from wastewater is known as water reuse and it is widely used in agriculture. Irrigation with nutrient-bearing wastewaters is an example of the simultaneous reuse of nutrients and water. Reclaimed municipal wastewater is becoming an increasingly important source of water for agricultural use in water-poor countries such as Israel (25% of the total agricultural water in 2000 and projected to be 46% in 2020) (Dobrowolski *et al.* 2008). However, reclaimed water containing nutrients can only be provided near the water reclamation facility and this may not align with the demand. Accordingly, the extraction of nutrients from wastewater has been intensively studied.

The most widely applied technology for nitrogen and phosphorus recovery from the MBR effluent is the ion exchange process. The ion-exchange technology offers a number of advantages including the ability to handle shock loadings and operate over a wider range of temperatures (Johir *et al.* 2011; Rittmann *et al.* 2011). For example, polymeric ion exchangers are also reported to have high macro-porosity. An ion exchange process with macro-porous anion-exchange resins was observed to effectively remove the nutrients from the effluent of the MBR. The retention of PO_4^{3-} and NO_3^- by two ion-exchange columns in series was 85% and 95%, respectively (Johir *et al.* 2011). Also, a selective ion exchange resin was applied in order to remove nitrate from the water and it was found that the removal efficiency increased with a larger resin dose (Samatya *et al.* 2006). Another media for capturing TN and TP is the oyster-zeolite packed bed adsorbent. A combination of the oyster-zeolite media with membrane filtration could achieve up to 90% and 53% removal efficiencies of TN and TP, respectively (Jung *et al.* 2006).

The use of an innovative zirconium-based adsorbent for P removal from biologically treated municipal wastewater has also been evaluated in a pilot-scale study (Ebie *et al.* 2008; Lee *et al.* 2007). Experimental results showed that the zirconium sulphate (ZS) with meso-porous structure had a very high sorption capacity for the phosphorus (Watanabe & Kimura, 2006). The amount of phosphate ions exchanged into the solid was as great as 3.4 mM/g-ZS (Lee *et al.* 2007). Also, the hydrotalcite ($Mg_6Al_2(CO_3)(OH)_{16}\cdot 4(H_2O)$) compounds (HTALs) and trimesic acid (benzene-1,3,5-tricarboxylic acid; TMA) coated basic alumina have been studied as selective adsorbents for P removal. The nutrient in the highly concentrated brine stream from an ion exchange process can be separated using struvite ($MgNH_4PO_4$) formation and precipitation with the addition of magnesium (Kuzawa *et al.* 2006). Lime was also used as a precipitant to recover the phosphorus in the supernatant of aerobic basin of an A2/O-MBR system. In this study, the solubilized phosphorous was recovered in the form of calcium phosphate before it enters into main stream. So, the possibility of phosphorus increase in the effluent due to sludge reduction practices has been eliminated (Banu *et al.* 2009; Watanabe & Kimura, 2006).

Recently, algal photo-bioreactors for the simultaneous removal of nutrients as a post-treatment process and the production of a crop that is of value have gained significant interest (Olguin, 2012; Xiao *et al.* 2012). Algae are photosynthetic autotrophic organisms that use light as an energy source and carbon dioxide as a carbon source. As with other plant-based crops, algae require nitrogen and phosphorus as macronutrients for growth. Selected strains of algae, growing under specific conditions, have been found to have a substantial lipid content that can be converted into biofuels such as biodiesel (Olguin, 2012). In addition, the algal biomass that is generated may be processed in order to produce animal feed (Chae *et al.* 2006) or to be used as a feedstock for energy generation (Xiao *et al.* 2012).

5.7 CHALLENGES ASSOCIATED WITH PATHOGEN REMOVAL BY MBRs

Pathogenic contamination of water causes disease outbreaks and contributes to background disease rates around the world, which most seriously impacts the developing world. Although a significant proportion of disease is caused by classical water-related pathogens, such as typhoid and cholera, newly recognized pathogens and new strains of established pathogens are being discovered; these present important additional challenges to both water and public health sectors (WHO, 2003).

The microbial pathogens that can be potentially present in wastewater can be divided into three groups. These groups are bacteria, viruses, and the pathogenic protozoa (Bitton, 2010). Bacteria are the most common of the microbial pathogens found in wastewater, ranging from approximately 0.2 to 10 µm in length (Figure 5.2). One group of intestinal bacteria, the coliform bacteria, has historically been used as an indication that an environment has been contaminated by human sewage. Second, the viruses that are potentially the most hazardous of the pathogens found in wastewater range in size from approximately 20 to 200 nm. The commonly detected pathogenic viruses are enteroviruses, hepatitis A, adenoviruses, rotaviruses, reoviruses, astroviruses, and calciviruses (Sima *et al.* 2011; Gantzer *et al.* 1998). Another microorganism of concern in wastewater is the protozoa. The most commonly detected protozoa are *Entamoeba histolytica*, *Giardia lamblia*, and *Cryptosporidum parvum*, which range in size from 10 µm to 1 mm (Rose & Clark, 1986). The pathogenic bacteria and their concentrations in domestic wastewater are listed in Table 5.5.

Generally, pathogen removal using the wastewater treatment process including MBR is expressed using a log removal value (LRV), which can be calculated using the following equation:

$$\text{LRV} = \log\left(\frac{C_o}{C}\right)$$

where C_o is the pathogen number in the influent and C is the pathogen number in the effluent. For a given pathogen, 2 LRVs reflect 99% removal, whilst 5 LRVs reflect 99.999% removal.

A significant advantage of MBRs over conventional activated sludge systems is that MBRs are capable of almost complete retention of bacteria and pathogens from wastewater. The microorganisms can be removed using MBR through several different mechanisms. Physical sieving is the most dominant pathogen removal mechanism with membrane filtration technology (Huang *et al.* 2012). This occurs when the membrane pore size is smaller than the pathogen size. For example, the MF membrane can remove algae and protozoa effectively; however, most viruses cannot be removed. It has been reported that the removal of bacteriophage is closely related to the membrane pore size (Madaeni, 1999). The LRV of somatic coliphage exhibited a significantly positive correlation ($R = 0.991$) with the ratio of phage size to membrane pore size (Wu *et al.* 2010). The other possible mechanism is sorption to the membrane or cake layer.

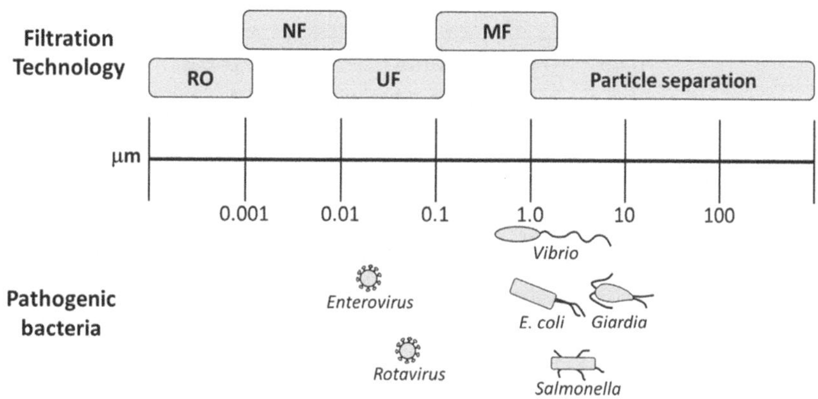

Figure 5.2 Membrane filtration spectrum for rejection of pathogenic microorganisms.

Table 5.5 Pathogenic bacteria in municipal wastewater.

Organism	Disease	Symptoms	Reported Population (per 100 mL)	Reference
Salmonella enteritidis	Salmonellosis	Diarrhea, abdominal cramps	150–4600	Koivunen *et al.* (2003)
Vibrio cholera O1	Cholera	Diarrhea	48	Oragui *et al.* (1993)
Giardia lamblia	Giardiasis	Diarrhea, nausea, abdominal pain	$22–4.3 \times 10^2$	Bertrand *et al.* (2004)
E. coli	Gastroenteritis, urinary tract infections	Diarrhea, cramps, dysentery	1.3×10^8	Lee *et al.* (2006)
Shigella	Shigellosis	Diarrhea, nausea	$1 \times 10^{5.8}$	Hench *et al.* (2003)

Some pathogens may adsorb to the membrane surface directly or to the cake layer on top of membrane surface. Thus, some pathogens with a larger size than the membrane pore size are effectively removed in the MBR with MF/UF membranes (Lv *et al.* 2006). The contributions of the suspended biomass and attached biofilm to the pathogen removal are dependent on the inter-related parameters including the concentration of mixed liquor suspended solids (MLSS), the sludge retention time (SRT), and the food-to-microorganism (F/M) ratio (Shang *et al.* 2005). This phenomenon increases the difficulty of comparing results from different studies, as most researchers do not generally consider the cake layer in the results presented. The size exclusion capability of the MF and UF membranes demonstrates the potential for concurrent pathogen removal. The MBR can effect up to approximately 7 LRVs, depending on the treatment processes and wastewater characteristics. The removal efficiencies of representative pathogens and indicators using the MBR process are summarized in Table 5.6.

Table 5.6 Comparison of pathogen log removal efficiencies with various MBR processes.

Influent	Membrane Pore Size (μm)	Target Pathogen	LRV	Reference
Sewage	0.45	Calicivirus	3.3–6.8	Sima *et al.* (2011)
Sewage	N.A.	Human adenovirus (HAdV)	4.1	Simmons and Xagoraraki (2011)
Sewage	N.A.	Enteric virus (EV)	3.6	Simmons and Xagoraraki (2011)
Sewage	~0.4	*E. coli*	>6.11*	Francy *et al.* (2012)
Sewage	0.04	Human adenovirus (HAdV)	5.0 ± 0.6	Kuo *et al.* (2010)
Sewage	0.4	Coliphage	1.7–2.6	Wu *et al.* (2010)
Settled sewage	0.4	Fecal coliforms	6.86	Ueda and Horan (2000)
Animal waste	0.03	*E. coli*	6.9	Wong *et al.* (2009)
Animal waste	0.03	Enterococci	7.3	Wong *et al.* (2009)
Sewage	0.4	Coliphage	0.8–3.1	Ottoson *et al.* (2006)
Hospital wastewater	0.4	*E. coli*	~2	Wen *et al.* (2004)

*Median value of data from four different wastewater treatment plants.

According to the literature, MBRs coupled with tight UF or NF membranes (pore size = 1–30 nm) are able to reject virus by sieving mechanisms (Hirani *et al.* 2010; Francy *et al.* 2012; Hai *et al.* 2014). Moreover, the adsorption of virus to sludge also plays an important role in virus removal in MBRs because most pathogenic bacteria have a tendency to attach to solids, and the MBR provides a better solid separation than that of the CAS process, which relies on settling (Simmons & Xagoraraki, 2011; Ueda & Horan, 2000). Sometimes, it is necessary to distinguish between the microbial removal that occurred within the activated sludge itself, and the removal that was attributable solely to the membrane (which includes the biofilm that accumulates on the membrane). One study examined the virus removal mechanism using the MBR process with a more detailed categorization method (Lv *et al.* 2006). For example, the cake layer, gel layer, and membrane contributed 6.3, 3.1, and 1.7 LRVs to phage removal, respectively.

The MBR can remove pathogens to the very high degree that can also be achieved through chemical oxidative disinfection, and it is without the associated problems and costs of storing and using corrosive agents. However, significant problems arise if the membrane integrity fails as the pathogen removal efficiency can deteriorate dramatically. Additionally, the regrowth of pathogens, such as *Aeromonas, Mycobacterium*, and *Legionella*, has been reported in the distribution system irrespective of the treatment technology used, including CAS processes and MBRs (Jjemba *et al.* 2010; Thayanukul *et al.* 2013). Thus, the post-treatment of MBR effluents as well as effective effluent monitoring to identify the integrity of the process is necessary in order to ensure environmental and public health protection.

5.8 POST-TREATMENTS FOR DISINFECTION OF THE MBR EFFLUENTS

MBRs provide a higher quality effluent than the CAS processes (Cicek *et al.* 1999; De Wever *et al.* 2007; González *et al.* 2007); however, the MBR effluent sometimes does not meet the disinfection requirements (Chae *et al.* 2007). The favorable microbiological quality of an MBR effluent is a significant factor in their frequent selection for water reuse, particularly considering the distribution and storage components of a full-scale system, which can be prone to regrowth of microorganisms and contamination from various sources. Thus, when MBR effluents are discharged into receiving waters, the reduction of bacterial numbers to minimize health hazards is a very desirable goal.

Disinfection is the treatment of the effluent that destroys all pathogens. Another term that is sometimes used in describing the destruction of microorganisms is sterilization. While disinfection indicates the destruction of all disease-causing microorganisms, there is no attempt to obtain sterilization in wastewater treatment. However, the disinfection procedures applied to wastewaters will result in a substantial reduction of all microbes so that the bacterial numbers are reduced to a safe level.

In general, disinfection can be achieved using any method that destroys pathogens, and there are a variety of physical or chemical methods that are capable of destroying microorganisms (Lazarova *et al.* 1999; Watts *et al.* 1995; Kitis, 2004; Dalrymple *et al.* 2010). The physical methods might include membrane filtration and irradiation with gamma-rays or UV rays (Bounty *et al.* 2012; Farooq *et al.* 1993; Madaeni, 1999). The chemical methods might include the use of chlorine, ozone, and ions of heavy metals (Xu *et al.* 2002; Abad *et al.* 1994).

As the UF membranes are an efficient barrier for bacteria and viruses, the MBR process with UF membranes provides a considerable level of physical disinfection. For example, the CAS system treating the hospital wastewater decreased fecal coliforms for 1.4 LRVs, whereas the MBR decreased for 3.6 LRVs (Pauwels *et al.* 2006). Moreover, the disinfection of MBR effluent requires less dose and frequency of chemical supply compared to those for CAS effluent. For example, the sand filtration process decreased fecal coliforms from 3.4×10^5 to 1.4×10^5 cfu/mL, while only 3.5×10^2 cfu/mL was detected from the effluent of MBR with UF membrane. Thus, the averaged chlorine demands for MBR effluent decreased to 3.5 mg/L, which was 56% less than that required for sand filtration effluent (8 mg/L) (Friedler *et al.* 2006).

5.8.1 Chlorination

Chlorine disinfection is one of the most widely used methods and is very effective for the deactivation of pathogenic microorganisms. Chlorine is used in the form of gas, hypochlorites, and other compounds. The different forms of chlorine that are currently used at wastewater treatment plants are gaseous chlorine, sodium hypochlorite solution, calcium hypochlorite, and bromium chloride; the most commonly used is gaseous chlorine (Cl_2). The chlorine gas hydrolyzes in water according to the following reactions (Hammer, 2001):

$$Cl_2 + H_2O \longleftrightarrow HOCl + Cl^- + H^+$$
$$HOCl \longleftrightarrow ClO^- + H^+$$

The amount of hypochlorous acid and hypochlorite formed is a function of the pH and temperature: the more acidic a solution is, the greater the concentration of hypochlorous acid and the more alkaline a solution is, the greater the concentration of hypochlorite ions. The effectiveness of chlorination depends on the dose, the chlorine demand of the wastewater, the chlorine residual, the amount of time the wastewater is in contact with the chlorine, the fecal coliform count in the wastewater, and other wastewater characteristics. The required degree of disinfection can be achieved by varying the dose and contact time for any chlorine disinfection system. The chlorine dose usually ranges from 5 to 20 mg/L (Lazarova *et al.* 1999). For optimum performance, the chlorination system must be designed so that the wastewater flows turbulently, similar to a whirlpool, throughout the contact chamber, ensuring complete mixing. This mixing allows the chlorine to have maximum contact with the wastewater and ensures that there are no dead areas in the tank (Hammer, 2001).

However, because chlorine is toxic to fish even at low concentrations, as well as the possibility that potentially harmful disinfection by-products (DBPs) may be formed, has made chlorination less favored as a disinfectant of choice in wastewater treatment (Bougeard *et al.* 2010; Richardson, 2003). A number of alternative methods are available for disinfection. These include ozone, chlorine dioxide, and ultraviolet (UV) radiation. Each disinfection technology has unique strengths and limitations, and no single disinfection method is right for all circumstances; in fact, a multi-step process may be appropriate to meet the overall treatment goals.

5.8.2 Ultraviolet irradiation

UV irradiation causes microbial damage through the deactivation of cells. In very simple terms, this occurs as a result of the nature of the energy absorption by DNA molecules, which causes cell deactivation. Unlike chemical disinfectants, UV does not "kill" an organism; rather, it damages the nucleic acids, or genetic material of a cell (DNA), such that the cell is unable to replicate. Compared with chlorine disinfection, a UV dose response is not affected by UV absorbance, temperature, and pH (USEPA, 2006). The major disadvantage of using UV radiation is that it leaves no residual disinfectant in the treated water. However, the microbial response to UV light can vary significantly between microorganisms (USEPA, 2006).

5.8.3 Ozonation

Ozone (O_3) is a highly reactive chemical with a high oxidation-reduction potential of 2.07 V (i.e., 1.36 V for chlorine). Its use in aqueous conditions usually causes the simultaneous production of secondary oxidants, such as hydroxyl radical whose oxidation power is much greater than molecular ozone (Wang *et al.* 2003). Ozone will also raise the dissolved oxygen level of the water. However, ozone must be generated and has only recently begun to compete favorably with chlorination in terms of economic costs. When bromide ions are present in raw water, they are oxidized to bromine and this leads to the formation of brominated THMs and other brominated DBPs, as in the case of chlorination.

5.8.4 Other post-treatments for MBR effluents

There are a number of different advanced oxidation processes (AOPs) that can be applied in MBR effluent treatment, including O_3/H_2O_2, UV/H_2O_2, UV/TiO_2, and UV/O_3. All of these AOPs have an item in common: the generation of hydroxyl radical, which is extremely reactive, and it possesses a very high oxidation potential. For example, the hydroxyl radical is produced when water-containing H_2O_2 is exposed to UV light as in the following reaction:

$$H_2O_2 + UV \leftrightarrow HO^\bullet + HO^\bullet$$

The oxidation potential of common oxidants used in water treatment systems is summarized in Table 5.7.

Table 5.7 Comparison of various oxidants potential.

Oxidizing Agent	Electrochemical Oxidation Potential (EOP)	EOP Relative to Chlorine
Hydroxyl radical	2.80 eV	2.06
Ozone	2.07 eV	1.52
Hydrogen Peroxide	1.77 eV	1.30
Chlorine Dioxide	1.50 eV	1.10
Chlorine	1.36 eV	1.00
Oxygen	1.23 eV	0.90

Source: USEPA (1998).

5.8.5 Applications of AOPs for MBR effluents

AOPs are powerful treatment barriers that can be effectively implemented in the post-treatment of MBR effluents for the destruction of pathogens, taste and odor compounds, endocrine disrupting compounds (EDCs), pharmaceuticals and personal care products (PPCPs). Thus, high quality, reusable effluents can be produced by combining the AOPs with MBR. For example, complete removal of a viral indicator *via* UV irradiation was reported in treating MBR effluents of grey water (Friedler & Gilboa, 2010). Furthermore, there are several studies that have reported the combination of MBR with various AOPs in order to maximize the reuse potential of MBR effluent by removing some pollutants of emerging concern (Table 5.8). Chapter 6 reviews the trace organic contaminant removal by MBRs.

Table 5.8 Performances of AOP processes treating MBR effluents.

Processes	Wastewater	Target Compound	Removal Efficiency (%)	Reference
UV/H_2O_2	Industrial wastewater	Nalidixic acid	75 (0)**	Laera *et al.* (2012)
Ozonation	Municipal wastewater	Pharmaceuticals	90–99	Huber *et al.* (2003)
O_3/H_2O_2	Municipal wastewater	Diethyl phthalate	70–80 (22)**	Park *et al.* (2010)
Ozonation	Textile wastewater	Color (SAC*)	93	Brik *et al.* (2004)
TiO_2/UV	Submarine wastewater	Anionic surfactant	95 (43)**	Chen *et al.* (2009)

SAC*: spectra absorption coefficient.
( )**: removal via MBR alone.

5.9 SUMMARY AND FUTURE OUTLOOK

MBRs with varying capacities, from a few thousand m^3/d to over 100,000 m^3/d, have been widely used for the treatment and recycling of municipal and industrial wastewaters. According to the literature, MBRs generally surpass the CAS processes in the removal of SS, COD, and nutrients from wastewater, which is promising for reuse or recycling of the effluent; MBRs allow longer SRT than the CAS processes, which results in the reduction of sludge production; from the microbial community analysis, the higher microbial activities for nutrient removal were correlated with more diversity of bacterial populations in MBRs; and MBRs could achieve 2–6 LRVs for bacteria using UF membranes and even better with

AOPs as a post-treatment. However, the elimination of nutrients and pathogenic virus in MBR effluents remains a key challenge. Therefore, it requires further innovations to develop efficient formats for advanced treatment. Moreover, research and development on cost effective recovery of resources (i.e., carbon, nitrogen, and phosphorus) from wastewater and MBR effluents is highly recommended.

REFERENCES

Abad F. X., Pintó R. M., Diez J. M. and Bosch A. (1994). Disinfection of human enteric viruses in water by copper and silver in combination with low levels of chlorine. *Applied and Environmental Microbiology*, **60**, 2377–2383.

Adam C., Gnirss R., Lesjean B., Buisson H. and Kraume M. (2002). Enhanced biological phosphorus removal in membrane bioreactors. *Water Science and Technology*, **46**(4–5), 281–286.

Ahmed Z., Lim B. R., Cho J., Song K. G., Kim K. P. and Ahn K. H. (2008). Biological nitrogen and phosphorus removal and changes in microbial community structure in a membrane bioreactor: effect of different carbon sources. *Water Research*, **42**, 198–210.

Ahn K.-H., Song K.-G., Choa E., Cho J., Yun H., Lee S. and Me J. (2003). Enhanced biological phosphorus and nitrogen removal using a sequencing anoxic/anaerobic membrane bioreactor (SAM) process. *Desalination*, **157**, 345–352.

Banu J. R., Uan D. K. and Yeom I. T. (2009). Nutrient removal in an A2O-MBR reactor with sludge reduction. *Bioresource Technology*, **100**, 3820–3824.

Bertrand I., Gantzer C., Chesnot T. and Schwartzbrod J. (2004). Improved specificity for Giardia lamblia cyst quantification in wastewater by development of a real-time PCR method. *Journal of Microbiological Methods*, **57**, 41–53.

Bitton G. (2010). Wastewater Microbiology. John Wiley & Sons, Inc, Hoboken, New Jersey, USA.

Bougeard C. M. M., Goslan E. H., Jefferson B. and Parsons S. A. (2010). Comparison of the disinfection by-product formation potential of treated waters exposed to chlorine and monochloramine. *Water Research*, **44**, 729–740.

Bounty S., Rodriguez R. A. and Linden K. G. (2012). Inactivation of adenovirus using low-dose UV/H_2O_2 advanced oxidation. *Water Research*, **46**, 6273–6278.

Brik M., Chamam B., Schoberl P., Braun R. and Fuchs W. (2004). Effect of ozone, chlorine and hydrogen peroxide on the elimination of colour in treated textile wastewater by MBR. *Water Science and Technology*, **49**(4), 299–303.

Chae S. R., Hwang E. J. and Shin H. S. (2006). Single cell protein production of Euglena gracilis and carbon dioxide fixation in an innovative photo-bioreactor. *Bioresource Technology*, **97**, 322–329.

Chae S. R., Kang S. T., Lee S. M., Lee E. S., Oh S. E., Watanabe Y. and Shin H. S. (2007). High reuse potential of effluent from an innovative vertical submerged membrane bioreactor treating municipal wastewater. *Desalination*, **202**, 83–89.

Chang H., Franczyk J., Im E. S., Kwon W. T., Bae D. H. and Jung I. W. (2007). Vulnerability of Korean water resources to climate change and population growth. *Water Science and Technology*, **56**(4), 57–62.

Chen R. and Lapara T. M. (2006). Aerobic biological treatment of low-strength synthetic wastewater in membrane-coupled bioreactors: the structure and function of bacterial enrichment cultures as the net growth rate approaches zero. *Microbial Ecology*, **51**, 99–108.

Chen Z., Zhou A., Ren N., Tian Y. and Hu D. (2009). Pollutants removal and simulation model of combined membrane process for wastewater treatment and reuse in submarine cabin for long voyage. *Journal of Environmental Sciences-China*, **21**, 1503–1512.

Chung J., Kim Y., Lee D.-J., Shim H. and Kim J.-O. (2006). Characteristics of denitrifying phosphate accumulating organisms in an anaerobic-intermittently aerobic process. *Environmental Engineering Science*, **23**, 981–993.

Cicek N., Franco J. P., Suidan M. T., Urbain V. and Manem J. (1999). Characterization and comparison of a membrane bioreactor and a conventional activated-sludge system in the treatment of wastewater containing high-molecular-weight compounds. *Water Environment Research*, **71**, 64–70.

Dalrymple O. K., Stefanakos E., Trotz M. A. and Goswami D. Y. (2010). A review of the mechanisms and modeling of photocatalytic disinfection. *Applied Catalysis B: Environmental*, **98**, 27–38.

De Wever H., Weiss S., Reemtsma T., Vereecken J., Müller J., Knepper T., Rörden O., Gonzalez S., Barcelo D. and Dolores Hernando M. (2007). Comparison of sulfonated and other micropollutants removal in membrane bioreactor and conventional wastewater treatment. *Water Research*, **41**, 935–945.

Dobrowolski J., O'neill M., Duriancik L. and Throwe J. (2008). Opportunities and Challenges in Agricultural Water Reuse: Final Report. USDA-CSREES.

Ebie Y., Kondo T., Kadoya N., Mouri M., Maruyama O., Noritake S., Inamori Y. and Xu K. (2008). Recovery oriented phosphorus adsorption process in decentralized advanced Johkasou. *Water Science and Technology*, **57**(12), 1977–81.

Falk M. W., Song K. G., Matiasek M. G. and Wuertz S. (2009). Microbial community dynamics in replicate membrane bioreactors – Natural reproducible fluctuations. *Water Research*, **43**, 842–852.

Fan X.-J., Urbain V., Qian Y. and Manem J. (1996). Nitrification and mass balance with a membrane bioreactor for municipal wastewater treatment. *Water Science and Technology*, **34**(1-2), 129–136.

Fan Z., Wu S., Chang H. and Hu J. (2011). Behaviors of glucocorticoids, androgens and progestogens in a municipal sewage treatment plant: comparison to estrogens. *Environmental Science and Technology*, **45**, 2725–2733.

Farooq S., Kurucz C. N., Waite T. D. and Cooper W. J. (1993). Disinfection of wastewaters: high-energy electron vs gamma irradiation. *Water Research*, **27**, 1177–1184.

Francy D. S., Stelzer E. A., Bushon R. N., Brady A. M. G., Williston A. G., Riddell K. R., Borchardt M. A., Spencer S. K. and Gellner T. M. (2012). Comparative effectiveness of membrane bioreactors, conventional secondary treatment, and chlorine and UV disinfection to remove microorganisms from municipal wastewaters. *Water Research*, **46**, 4164–4178.

Friedler E. and Gilboa Y. (2010). Performance of UV disinfection and the microbial quality of greywater effluent along a reuse system for toilet flushing. *Science of the Total Environment*, **408**, 2109–2117.

Friedler E., Kovalio R. and Ben-Zvi A. (2006). Comparative study of the microbial quality of greywater treated by three on-site treatment systems. *Environmental Technology*, **27**, 653–663.

Gantzer C., Maul A., Audic J. M. and Schwartzbrod L. (1998). Detection of infectious enteroviruses, enterovirus genomes, somatic coliphages, and Bacteroides fragilis phages in treated wastewater. *Applied and Environmental Microbiology*, **64**, 4307–4312.

González S., Petrovic M. and Barceló D. (2007). Removal of a broad range of surfactants from municipal wastewater–Comparison between membrane bioreactor and conventional activated sludge treatment. *Chemosphere*, **67**, 335–343.

Hai F. I., Riley T., Shawkat S., Magram S. F. and Yamamoto K. (2014). Removal of pathogens by membrane bioreactors: A review of the mechanisms, influencing factors and reduction in chemical disinfectant dosing. *Water (Switzerland)*, **6**, 3603–3630.

Hammer M. J. (2001). Water and Wastewater Technology. Prentice Hall, New Jersey, USA.

Hench K. R., Bissonnette G. K., Sexstone A. J., Coleman J. G., Garbutt K. and Skousen J. G. (2003). Fate of physical, chemical, and microbial contaminants in domestic wastewater following treatment by small constructed wetlands. *Water Research*, **37**, 921–927.

Hirani Z. M., Decarolis J. F., Adham S. S. and Jacangelo J. G. (2010). Peak flux performance and microbial removal by selected membrane bioreactor systems. *Water Research*, **44**, 2431–2440.

Huang H. O., Young T. A., Schwab K. J. and Jacangelo J. G. (2012). Mechanisms of virus removal from secondary wastewater effluent by low pressure membrane filtration. *Journal of Membrane Science*, **409**, 1–8.

Huber M. M., Canonica S., Park G. Y. and Von Gunten U. (2003). Oxidation of pharmaceuticals during ozonation and advanced oxidation processes. *Environmental Science and Technology*, **37**, 1016–1024.

Inglett P. W., Rivera-Monroy V. H. and Wozniak J. R. (2011). Biogeochemistry of nitrogen across the everglades landscape. *Critical Reviews in Environmental Science and Technology*, **41**, 187–216.

Jeon C. O., Lee D. S. and Park J. M. (2003). Microbial communities in activated sludge performing enhanced biological phosphorus removal in a sequencing batch reactor. *Water Research*, **37**, 2195–2205.

Jjemba P. K., Weinrich L. A., Cheng W., Giraldo E. and Lechevallier M. W. (2010). Regrowth of potential opportunistic pathogens and algae in reclaimed-water distribution systems. *Applied and Environmental Microbiology*, **76**, 4169–4178.

Johir M. A. H., George J., Vigneswaran S., Kandasamy J. and Grasmick A. (2011). Removal and recovery of nutrients by ion exchange from high rate membrane bio-reactor (MBR) effluent. *Desalination*, **275**, 197–202.

Judd S. (2006). The MBR Book. Elsevier Ltd., UK.

Judd S. (2008). The status of membrane bioreactor technology. *Trends in Biotechnology*, **26**, 109–116.

Judd S. (2011). The MBR Book: Principles and Applications of Membrane Bioreactors for Water and Wastewater Treatment. Elsevier Ltd., UK.

Jung Y.-J., Koh H.-W., Shin W.-T. and Sung N.-C. (2006). A novel approach to an advanced tertiary wastewater treatment: combination of a membrane bioreactor and an oyster-zeolite column. *Desalination*, **190**, 243–255.

Kimura K., Nishisako R., Miyoshi T., Shimada R. and Watanabe Y. (2008). Baffled membrane bioreactor (BMBR) for efficient nutrient removal from municipal wastewater. *Water Research*, **42**, 625–632.

Kitis M. (2004). Disinfection of wastewater with peracetic acid: a review. *Environment International*, **30**, 47–55.

Koivunen J., Siitonen A. and Heinonen-Tanski H. (2003). Elimination of enteric bacteria in biological–chemical wastewater treatment and tertiary filtration units. *Water Research*, **37**, 690–698.

Kraume M., Bracklow U., Vocks M. and Drews A. (2005). Nutrients removal in MBRs for municipal wastewater treatment. *Water Science and Technology*, **51**(6–7), 391–402.

Kuo D. H., Simmons F. J., Blair S., Hart E., Rose J. B. and Xagoraraki I. (2010). Assessment of human adenovirus removal in a full-scale membrane bioreactor treating municipal wastewater. *Water Research*, **44**, 1520–1530.

Kurisu F., Satoh H., Mino T. and Matsuo T. (2002). Microbial community analysis of thermophilic contact oxidation process by using ribosomal RNA approaches and the quinone profile method. *Water Research*, **36**, 429–438.

Kuzawa K., Jung Y. J., Kiso Y., Yamada T., Nagai M. and Lee T. G. (2006). Phosphate removal and recovery with a synthetic hydrotalcite as an adsorbent. *Chemosphere*, **62**, 45–52.

Laera G., Cassano D., Lopez A., Pinto A., Pollice A., Ricco G. and Mascolo G. (2012). Removal of organics and degradation products from industrial wastewater by a membrane bioreactor integrated with ozone or UV/H_2O_2 treatment. *Environmental Science and Technology*, **46**, 1010–1018.

Lapara T. M., Klatt C. G. and Chen R. (2006). Adaptations in bacterial catabolic enzyme activity and community structure in membrane-coupled bioreactors fed simple synthetic wastewater. *Journal of Biotechnology*, **121**(3), 368–380.

Lazarova V., Savoye P., Janex M. L., Blatchley E. R. and Pommepuy M. (1999). Advanced wastewater disinfection technologies: state of the art and perspectives. *Water Science and Technology*, **40**(4-5), 203–213.

Lazarova V., Choo K.-H. and Cornel P. (2012). Water-Energy Interactions in Water Reuse. IWA Publishing, London.

Lee D.-Y., Shannon K. and Beaudette L. A. (2006). Detection of bacterial pathogens in municipal wastewater using an oligonucleotide microarray and real-time quantitative PCR. *Journal of Microbiological Methods*, **65**, 453–467.

Lee S. H., Lee B. C., Lee K. W., Choi Y. S., Park K. Y. and Iwamoto M. (2007). Phosphorus recovery by mesoporous structure material from wastewater. *Water Science and Technology*, **55**(1-2), 169–176.

Lesjean B. and Huisjes E. H. (2008). Survey of the European MBR market: trends and perspectives. *Desalination*, **231**, 71–81.

Lesjean B., Gnirss R. and Adam C. (2002). Process configurations adapted to membrane bioreactors for enhanced biological phosphorous and nitrogen removal. *Desalination*, **149**, 217–224.

Li H. Y., Yang M., Zhang Y., Yu T. and Kamagata Y. (2006). Nitrification performance and microbial community dynamics in a submerged membrane bioreactor with complete sludge retention. *Journal of Biotechnology*, **123**, 60–70.

Lim S., Kim S., Yeon K. M., Sang B. I., Chun J. and Lee C. H. (2012). Correlation between microbial community structure and biofouling in a laboratory scale membrane bioreactor with synthetic wastewater. *Desalination*, **287**, 209–215.

Liu W. J., Hu Z. R., Walker R. L. and Dold P. L. (2011). Enhanced nutrient removal MBR system with chemical addition for low effluent TP. *Water Science and Technology*, **64**(6), 1298–1306.

Lv W., Zheng X., Yang M., Zhang Y., Liu Y. and Liu J. X. (2006). Virus removal performance and mechanism of a submerged membrane bioreactor. *Process Biochemistry*, **41**, 299–304.

Madaeni S. S. (1999). The application of membrane technology for water disinfection. *Water Research*, **33**, 301–308.

Melcer H., Dold P. L. and Jones R. M. (2003). Methods for Wastewater Characterization in Activated Sludge Modeling. WERF Report 99-WWF-3.

Melin T., Jefferson B., Bixio D., Thoeye C., De Wilde W., De Koning J., Van Der Graaf J. and Wintgens T. (2006). Membrane bioreactor technology for wastewater treatment and reuse. *Desalination*, **187**, 271–282.

Meng F. G., Chae S. R., Drews A., Kraume M., Shin H. S. and Yang F. L. (2009). Recent advances in membrane bioreactors (MBRs): membrane fouling and membrane material. *Water Research*, **43**, 1489–1512.

Meng F. G., Chae S. R., Shin H. S., Yang F. L. and Zhou Z. B. (2012). Recent advances in membrane bioreactors: configuration development, pollutant elimination, and sludge reduction. *Environmental Engineering Science*, **29**, 139–160.

Merz C., Scheumann R., El Hamouri B. and Kraume M. (2007). Membrane bioreactor technology for the treatment of greywater from a sports and leisure club. *Desalination*, **215**, 37–43.

Metcalf L. and Eddy H. P. (2004). Wastewater Engineering: Treatment and Reuse. McGraw-Hill, New York, USA.

Miura Y., Hiraiwa M. N., Ito T., Itonaga T., Watanabe Y. and Okabe S. (2007). Bacterial community structures in MBRs treating municipal wastewater: relationship between community stability and reactor performance. *Water Research*, **41**, 627–637.

Moeslang H. and Brockmann M. (2010). Membrane Bioreactors Keytechnology for Water Reuse. The 11th International Workshop on Information Security Applications, of Conference, Jeju Island, Korea.

Molina-Munoz M., Poyatos J. M., Sanchez-Peinado M., Hontoria E., Gonzalez-Lopez J. and Rodelas B. (2009). Microbial community structure and dynamics in a pilot-scale submerged membrane bioreactor aerobically treating domestic wastewater under real operation conditions. *Science of the Total Environment*, **407**, 3994–4003.

Olguin E. J. (2012). Dual purpose microalgae-bacteria-based systems that treat wastewater and produce biodiesel and chemical products within a Biorefinery. *Biotechnology Advances*, **30**, 1031–1046.

Oragui J. I., Arridge H., Mara D. D., Pearson H. W. and Silva S. A. (1993). Vibrio cholerae O1 (El Tor) removal in waste stabilization ponds in northeast Brazil. *Water Research*, **27**, 727–728.

Ottoson J., Hansen A., Bjorlenius B., Norder H. and Stenstrom T. A. (2006). Removal of viruses, parasitic protozoa and microbial indicators in conventional and membrane processes in a wastewater pilot plant. *Water Research*, **40**, 1449–1457.

Park J. H., Park C. G., Lee J. W. and Ko K. B. (2010). Degradation of diethyl phthalate in treated effluents from an MBR via advanced oxidation processes: effects of nitrate on oxidation and a pilot-scale AOP operation. *Environmental Technology*, **31**, 15–27.

Patel J., Nakhla G. and Margaritis A, (2005). Optimization of biological nutrient removal in a membrane bioreactor system. *Journal of Environmental Engineering*, **131**, 1021–1029.

Pauwels B., Ngwa F. F., Deconinck S. and Verstraete W. (2006). Effluent quality of a conventional activated sludge and a membrane bioreactor system treating hospital wastewater. *Environmental Technology*, **27**, 395–402.

Phan H. V., Hai F. I., McDonald J. A., Khan S. J., van de Merwe J. P., Leusch F. D. L., Zhang R., Price W. E., Broeckmann A. and Nghiem L. D. (2015). Impact of hazardous events on the removal of nutrients and trace organic contaminants by an anoxic-aerobic membrane bioreactor receiving real wastewater. *Bioresource Technology*, **192**, 192–201.

Phan H. V., Hai F. I., Zhang R., Kang J., Price W. E. and Nghiem L. D. (2016). Bacterial community dynamics in an anoxic-aerobic membrane bioreactor – impact on nutrient and trace organic contaminant removal. *International Biodeterioration and Biodegradation*, **109**, 61–72.

PUB (2011). Innovation in Water Singapore. An R&D publication of PUB Singapore, Public Utilities Board (PUB), Singapore.

Puig S., Van Loosdrecht M. C. M., Flameling A. G., Colprim J. and Meijer S. C. F. (2010). The effect of primary sedimentation on full-scale WWTP nutrient removal performance. *Water Research*, **44**, 3375–3384.

Qin H. J., Kekre K. A., Tao G. H., Oo M. H., Wai M. N., Lee T. C., Viswanath B. and Seah H. (2006). New option of MBR-RO process for production of NEWater from domestic sewage. *Journal of Membrane Science*, **272**, 70–77.

Ramphao M., Wentzel M. C., Merritt R., Ekama G. A., Young T. and Buckley C. A. (2005). Impact of membrane solid-liquid separation on design of biological nutrient removal activated sludge systems. *Biotechnology and Bioengineering*, **89**, 630–646.

Raunkjær K., Hvitved-Jacobsen T. and Nielsen P. H. (1994). Measurement of pools of protein, carbohydrate and lipid in domestic wastewater. *Water Research*, **28**, 251–262.

Richardson S. D. (2003). Disinfection by-products and other emerging contaminants in drinking water. *TrAC Trends in Analytical Chemistry*, **22**, 666–684.

Rittmann B. E., Mayer B., Westerhoff P. and Edwards M. (2011). Capturing the lost phosphorus. *Chemosphere*, **84**, 846–853.

Rose J. B. and Clark C. S. (1986). Microbial aspects of wastewater reuse for irrigation. *Critical Reviews in Environmental Control*, **16**, 231–256.

Samatya S., Kabay N., Yuksel U., Arda M. and Yuksel M. (2006). Removal of nitrate from aqueous solution by nitrate selective ion exchange resins. *Reactive and Functional Polymers*, **66**, 1206–1214.

Sanapareddy N., Hamp T. J., Gonzalez L. C., Hilger H. A., Fodor A. A. and Clinton S. M. (2009). Molecular diversity of a North Carolina wastewater treatment plant as revealed by pyrosequencing. *Applied and Environmental Microbiology*, **75**, 1688–1696.

Schlesinger W. H. (2009). On the fate of anthropogenic nitrogen. *Proceedings of the National Academy of Sciences of the United States of America*, **106**, 203–208.

Shang C., Wong H. M. and Chen G. H. (2005). Bacteriophage MS-2 removal by submerged membrane bioreactor. *Water Research*, **39**, 4211–4219.

Shannon M. A., Bohn P. W., Elimelech M., Georgiadis J. G., Marinas B. J. and Mayes A. M. (2008). Science and technology for water purification in the coming decades. *Nature*, **452**, 301–310.

Silva A. F., Carvalho G., Oehmen A., Lousada-Ferreira M. and Nieuwenhuijzen A. V. (2012). Microbial population analysis of nutrient removal-related organisms in membrane bioreactors. *Applied Microbiology and Biotechnology*, **93**, 2171–2180.

Sima L. C., Schaeffer J., Le Saux J. C., Parnaudeau S., Elimelech M. and Le Guyader F. S. (2011). Calicivirus removal in a membrane bioreactor wastewater treatment plant. *Applied and Environmental Microbiology*, **77**, 5170–5177.

Simmons F. J. and Xagoraraki I. (2011). Release of infectious human enteric viruses by full-scale wastewater utilities. *Water Research*, **45**, 3590–3598.

Thayanukul P., Kurisu F., Kasuga I. and Furumai H. (2013). Evaluation of microbial regrowth potential by assimilable organic carbon in various reclaimed water and distribution systems. *Water Research*, **47**, 225–232.

Ueda T. and Horan N. J. (2000). Fate of indigenous bacteriophage in a membrane bioreactor. *Water Research*, **34**, 2151–2159.

USEPA (1998). Advanced Photochemical Oxidation Processes. EPA-625-R98-004. USEPA (2006). Ultraviolet Disinfection Guidance Manual. EPA 815-R-06-007.

USEPA (2012). Guidelines for Water Reuse. United States Environmental Protection Agency, Washington, D.C., USA.

Wan C. Y., De Weyer H., Diels L., Thoeye C., Liang J. B. and Huang L. N. (2011). Biodiversity and population dynamics of microorganisms in a full-scale membrane bioreactor for municipal wastewater treatment *Water Research*, **45**, 1129–1138.

Wang F., Smith D. W. and El-Din M. G. (2003). Application of advanced oxidation methods for landfill leachate treatment – a review. *Journal of Environmental Engineering and Science*, **2**, 413–427.

Wang Z., Wu Z., Mai S., Yang C., Wang X., An Y. and Zhou Z. (2008). Research and applications of membrane bioreactors in China: progress and prospect. *Separation and Purification Technology*, **62**, 249–263.

Watanabe Y. and Kimura K. (2006). Hybrid membrane bioreactor for water recycling and phosphorus recovery. *Water Science and Technology*, **53**(7), 17–24.

Watts R. J., Kong S., Orr M. P., Miller G. C. and Henry B. E. (1995). Photocatalytic inactivation of coliform bacteria and viruses in secondary wastewater effluent. *Water Research*, **29**, 95–100.

WEF (2012). Membrane Bioreactors. WEF Manual of Practice No. 36. Water Environment Federation, McGraw-Hill, New York.

Wei V., Elektorowicz M. and Oleszkiewicz J. A. (2012). Electrically enhanced MBR system for total nutrient removal in remote northern applications. *Water Science and Technology*, **65**(11), 737–742.

Wen X., Ding H., Huang X. and Liu R. (2004). Treatment of hospital wastewater using a submerged membrane bioreactor. *Process Biochemistry*, **39**, 1427–1431.

WHO (2003). Emerging Issues in Water and Infectious Disease. World Health Organization, Geneva, Switzerland.

WHO and UNICEF (2000). Global Water Supply and Sanitation Assessment 2000 Report, Geneva, UN.

Wong K., Xagoraraki I., Wallace J., Bickert W., Srinivasan S. and Rose J. B. (2009). Removal of viruses and indicators by anaerobic membrane bioreactor treating animal waste. *Journal of Environmental Quality*, **38**, 1694–1699.

Wu J. L., Li H. T. and Huang X. (2010). Indigenous somatic coliphage removal from a real municipal wastewater by a submerged membrane bioreactor. *Water Research*, **44**, 1853–1862.

Xiao L., Young E. B., Berges J. A. and He Z. (2012). Integrated photo-bioelectrochemical system for contaminants removal and bioenergy production. *Environmental Science and Technology*, **46**, 11459–66.

Xing C. H., Tardieu E., Qian Y. and Wen X. H. (2000). Ultrafiltration membrane bioreactor for urban wastewater reclamation. *Journal of Membrane Science*, **177**, 73–82.

Xu P., Janex M.-L., Savoye P., Cockx A. and Lazarova V. (2002). Wastewater disinfection by ozone: main parameters for process design. *Water Research*, **36**, 1043–1055.

Yang W., Cicek N. and Ilg J. (2006). State-of-the-art of membrane bioreactors: worldwide research and commercial applications in North America. *Journal of Membrane Science*, **270**, 201–211.

Yuan L.-M., Zhang C.-Y., Zhang Y.-Q., Ding Y. and Xi D.-L. (2008). Biological nutrient removal using an alternating of anoxic and anaerobic membrane bioreactor (AAAM) process. *Desalination*, **221**, 566–575.

Zhang H., Wang X., Xiao J., Yang F. and Zhang J. (2009). Enhanced biological nutrient removal using MUCT–MBR system. *Bioresource Technology*, **100**, 1048–1054.

Chapter 6

Wastewater reuse applications and MBR

Muhammad B. Asif[1] *and Faisal I. Hai*[1]*

[1]*Strategic Water Infrastructure Laboratory, School of Civil, Mining and Environmental Engineering, University of Wollongong, Wollongong, NSW 2522, Australia*

**Corresponding author: faisal@uow.edu.au*

ABSTRACT

Less than 3% of water on the earth is categorized as freshwater. Rapid urbanization, population increase and climate change have created significant burden on the limited reserve of freshwater. Hence, water reuse is critical to secure the future water demand for domestic, industrial and agricultural activities. Treated effluent from the conventional wastewater treatment processes are suitable for only limited types of wastewater reuse applications such as restricted irrigation. Ineffective removal of nutrients, pathogens and emerging trace organic contaminants is harmful for the environment and public health. Wastewater discharge and water reuse regulations/guidelines adopted by many countries around the globe have been made more stringent to protect the existing freshwater resources and to foster safe water reuse. Membrane bioreactors (MBR) are attractive options to consider for applications requiring high-quality effluent, particularly in situations with stringent suspended solids, nutrient and microbiological discharge limits, or when wastewater reuse is intended. The MBR market is growing rapidly and is estimated to reach $3 billion by 2019. In this chapter, wastewater reuse applications and guidelines are critically reviewed. Features of MBR technology in the context of wastewater reuse are analysed and discussed. In addition, a summary of selected wastewater reuse projects involving MBR is presented, which shows that multiple barrier treatment systems consisting of MBR and post-treatment steps (e.g. disinfection) are favoured.

Editor(s): Faisal I. Hai, Kazuo Yamamoto and Chung-Hak Lee
doi: 10.2166/9781780409177_0183

6.1 INTRODUCTION

Due to the rapid population increase and urbanization, water demand for domestic, industrial and agricultural activities is increasing at an alarming rate. Less than 3% of water on the earth is categorized as freshwater, most of it frozen in glaciers, ice and snow, and less than 1% present as fresh groundwater and soil moisture. Less than 0.01% of it is present as surface water in lakes, swamps and rivers. Furthermore, the global water resources are unevenly distributed over different regions (Figure 6.1). The situation is exacerbated by erratic rainfall patterns due to climate change, which impose new challenges to the already water-stressed areas (Zimmerman *et al.*, 2008).

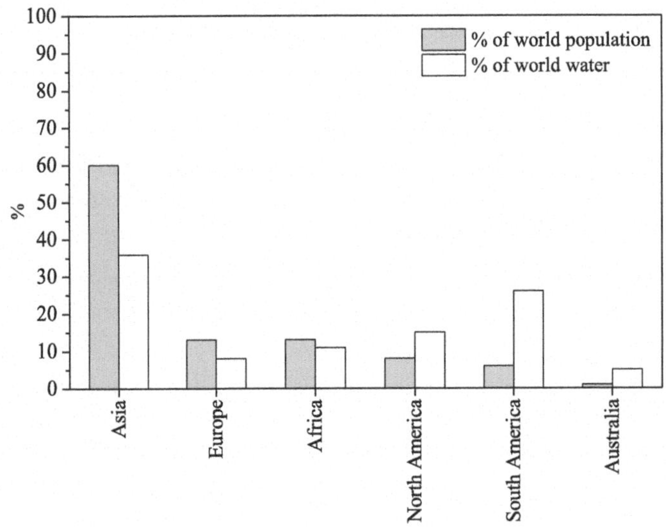

Figure 6.1 Distribution of population and freshwater in different regions of the world. Modified after Zimmerman *et al.* (2008).

The concept of water reuse dates back to the 1920s (Asano *et al.*, 1996; Chen *et al.*, 2013; Wilcox *et al.*, 2016). Given the ever-increasing water demand and uncertain freshwater supply, the importance of water reuse cannot be overstated (Marlow *et al.*, 2013; Vo *et al.*, 2014). In the last decade, wastewater reclamation and reuse have received particular attention to meet water demand during long-term droughts and to improve and strengthen water supply portfolio. A wealth of literature is available on different aspects of wastewater reclamation and reuse including the performance of different treatment technologies (De Koning *et al.*, 2008; Sano *et al.*, 2016; Venkatesan *et al.*, 2011); economics of water reclamation (Listowski *et al.*, 2013; Molinos-Senante *et al.*, 2011); environmental impacts and public safety (Qu *et al.*, 2016; Rose, 2007; Sano *et al.*, 2016); and social acceptance (Harris-Lovett *et al.*, 2015; Ross *et al.*, 2014).

Wastewater reclamation includes the treatment of municipal and industrial wastewater, or storm water to a level that it can be safely reused. Wastewater reclamation infrastructure consists of sewerage system, treatment plants and distribution networks. Wastewater reclamation allows the use of treated wastewater as an alternative water source for diverse end-user applications (Chen *et al.*, 2013; Zhang & Shen, 2017). However, there are technical, financial and management challenges to its implementation.

Currently, treated wastewater is mainly reused for agricultural activities (Lautze *et al.*, 2014; Norton-Brandão *et al.*, 2013). It is also reused for activities such as landscape irrigation, industrial activities, recreational activities and groundwater recharge (Garcia-Cuerva *et al.*, 2016). Due to the lack of adequate treatment facilities, up to 80% of the total wastewater generated has been reported to be reused in either raw or partially treated form for agriculture in some countries such as Ghana, Bolivia and Pakistan (Ensink *et al.*, 2004; Keraita *et al.*, 2003; Landa-Cansigno *et al.*, 2013; Lautze *et al.*, 2014; Zabalaga *et al.*, 2007). Water reuse without adequate treatment must be discouraged because of its adverse impacts on environment and human health (Courault *et al.*, 2017; Vo *et al.*, 2014). Conventional biological wastewater treatment techniques such as conventional activated sludge (CAS) processes can achieve partial removal of suspended solids, bulk organics and pathogens, which makes the treated wastewater suitable for only restricted irrigation. However, these treatment processes are not suitable for effective removal of nutrients and trace organic contaminants (TrOCs), thereby restricting their reuse in limited types of applications (Carr *et al.*, 2011; Vo *et al.*, 2014). Therefore, tertiary and advanced wastewater treatment techniques such as membrane technology and activated carbon adsorption are employed to comply with water reuse guidelines. An overview of treatment levels required for different wastewater reuse applications is provided in Figure 6.2.

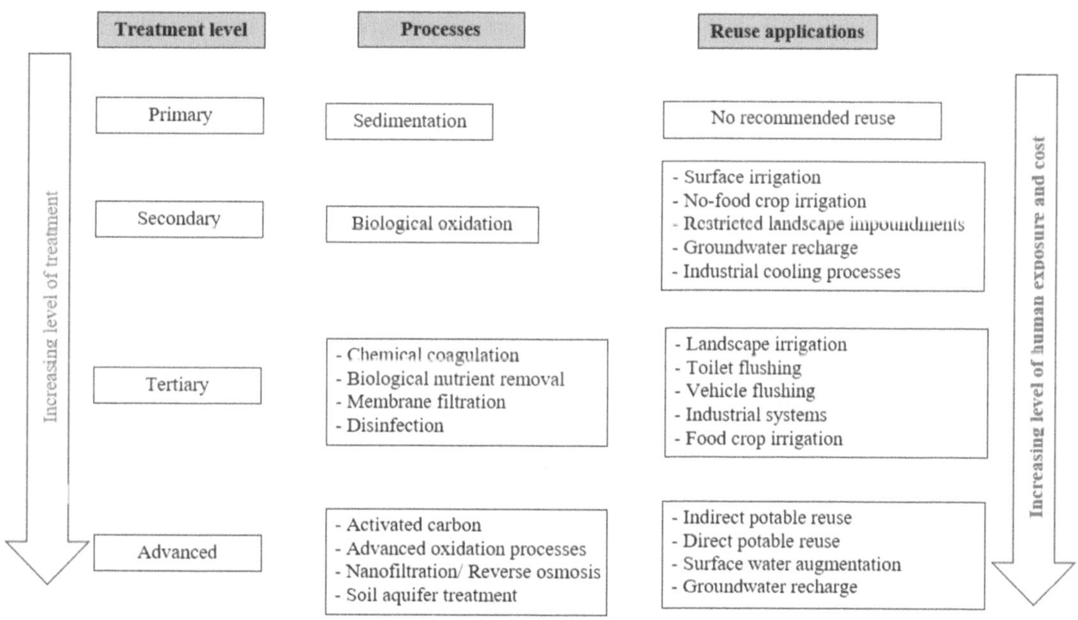

Figure 6.2 Level of treatments effective for different wastewater reuse applications. Modified after Vo *et al.* (2014).

Membrane bioreactors (MBR) are a combination of the CAS process and ultrafiltration or microfiltration membranes. It is better equipped than conventional wastewater treatment processes to produce high-quality, suspended solids-free effluent suitable for water reuse applications (Judd, 2016b; Xiao *et al.*, 2018). Notably, full-scale MBRs have been installed in more than 200 countries, and the MBR market is expected to reach $3 billion by 2019 (Judd, 2016b). As noted in previous chapters, due to the effective retention of activated sludge achieved by membranes, MBR can be operated at higher mixed liquor suspended solids (MLSS) concentration and longer solids retention time (SRT). This allows MBR to produce high-quality effluent using a compact footprint. In addition, as compared to the CAS process, MBR may achieve better and consistent removal of bulk organics, nutrients and pathogens (see Chapter 5) as well as TrOCs (see Chapter 13) (Asif *et al.*, 2018; Sano *et al.*, 2016; Xiao *et al.*, 2018). Despite its proven track record for effective wastewater treatment, reuse of MBR effluent for different applications has not been systematically reviewed. In this chapter, the current wastewater reuse practices and trends, and the suitability of MBR technology for this purpose is critically analysed. In addition, performance of MBR for the removal of bulk organics, nutrients, pathogens and TrOCs is reviewed and compared with water reuse guidelines. Finally, a succinct overview of selected wastewater reuse projects involving MBR is provided.

6.2 OVERVIEW OF CURRENT WASTEWATER REUSE PRACTICES AND TRENDS

Reclaimed wastewater can be divided into different categories based on the intended mode of water reuse. The categories of reuse as defined by the United States Environmental Protection Agency (USEPA) include urban reuse, agriculture reuse, environmental reuse, industrial reuse, groundwater reuse and indirect potable reuse. Wastewater reuse categories are comprehensively described in Chapter 5 of this book.

At the current rate of water consumption and population increase, global water demand has been estimated to increase to 6900 billion m^3/year by 2030 (2030 Water Resources Group, 2009). Therefore, it is the need of the hour for the governments to look for alternate sources of water. Globally, several countries have made policies and have set water reuse targets (Table 6.1). This is favourable for the growth of wastewater reuse market. Among the countries that have set water reuse targets, Mexico and Saudi Arabia have set targets to reuse 100% of municipal wastewater by 2030 (Al-Saud, 2013; MENR, 2010). As discussed in Chapter 14, China has taken an intensive wastewater reuse programme.

Table 6.1 Recorded and targeted wastewater reuse in different countries.

Country	Recorded Wastewater Reuse Practice		Targeted Wastewater Reuse Practice	
	%	Year	%	Year
Cyprus[a]	100	2007	–	–
Australia[b]	17	2009–2010	30	2015
Saudi Arabia[c]	30 (municipal)	2010	100 (municipal)	2030
	10 (industrial)	2010	80 (industrial)	2030
Mexico[d]	41	2009	100	2030
China[e]	10–15 (northern cities)	2011	20–25 (northern cities)	2025
	5–10 (southern cities)	2011	10–15 (southern cities)	2025

[a]Mediterranean Wastewater Reuse Report (Report, 2007).
[b]Progress against the national target of 30% of Australia's wastewater being recycled by 2015 (DSEWPaC, 2012).
[c]National Water Strategy – the road map for sustainability, efficiency and security of water future in the K.S.A. (Al-Saud, 2013).
[d]Statistics on water in Mexico. 2010 ed. (MENR, 2010).
[e]An Overview of Reclaimed Water Reuse in China (Yi *et al.*, 2011).
'–' not applicable.

Interestingly, the United States and the European countries have not officially set water reuse targets, and they are only reusing 2–3% of the total treated wastewater. Nevertheless, the European Union has carried out a project for promoting wastewater reuse practices in member countries (Futran, 2013; Vo *et al.*, 2014). A few states of the United States such as California, Florida and Washington have adopted regulations for mandatory connections to the reclaimed water network, if available (USEPA, 2012), thus creating opportunities for investment in future water reuse projects. The financial models of wastewater reuse projects are evolving to attract investments (Szyplinska, 2012). These projects are also eligible for subsidies (USEPA, 2012). For example, water recycling projects in Australia have received approximately $800 million in funding/subsidies from Water Smart Australia programme initiated by the Australian Government (Radcliffe, 2010).

Global average daily wastewater generation rate is almost 700 million m^3. Of the total wastewater, only 32 million m^3 (~4% of total wastewater) was reclaimed in 2010, and it was estimated to increase by 70% (i.e. 55 million m^3) in 2015 (GWI, 2010; Szyplinska, 2012; Yi *et al.*, 2011). Currently, irrigation for agriculture is the major consumer of freshwater and reclaimed wastewater (Figure 6.3). According to a World Bank report, wastewater, mostly in raw form, is being used for the irrigation of almost 1/10th of crops grown globally, which is unsafe for human health (Scheierling *et al.*, 2010). Therefore, development and strict reinforcement of wastewater reuse guidelines are required. An additional concern is the presence of emerging pollutants such as TrOCs in wastewater that may enter the food chain *via* bioaccumulation. Thus, the guidelines and standards for reclaimed water reuse needs to be reviewed and updated to tackle the emerging contaminants. An overview of the water quality guidelines for irrigation is presented in Table 6.2.

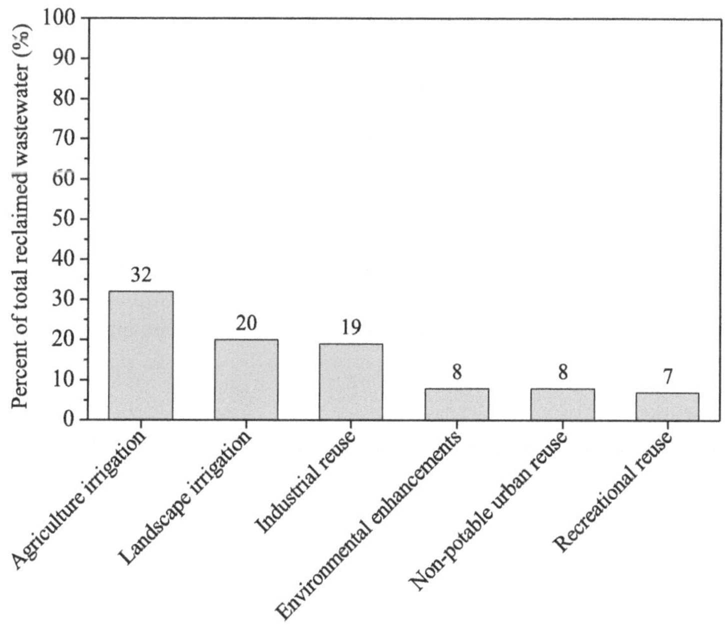

Figure 6.3 Global wastewater reuse following tertiary treatment and share of different water reuse categories. Data source: GWI (2010).

Table 6.2 Water quality guidelines of water reuse for irrigation.

Parameters	USEPA (2012)[a]		WHO (2006)[b]		Australia (2000)[c]		China (2007)[d]	
	R	UR	R	UR	R	UR	R	UR
pH	6–9	6–9	–	–	6.5–8.5	6.5–8.5	6.5–8.5	6.5–8.5
Turbidity (NTU)	–	2	–	–	–	2	–	–
SS (mg/L)	–	30	–	–	–	–	80–100	60
BOD (mg/L)	10	10	–	–	–	–	60–100	40
DO (mg/L)	–	–	–	–	–	–	$\geq$0.5	$\geq$0.5
Total coliform (CFU/100 mL)	–	–	–	–	10^2–10^4	10	–	–
Faecal coliform (CFU/100 mL)	2×10^2	Absent	–	–	–	–	4×10^4	2×10^4
E. coli (CFU/100 mL)	–	–	–	10^3	10^2–10^4	–	–	–
Nematode eggs (nos./L)	–	–	$\leq$1	$\leq$1	–	–	–	–

SS: suspended solids; BOD: biochemical oxygen demand; DO: dissolved oxygen; R: restricted; UR: unrestricted; USEPA: United States Environmental Protection Agency; WHO: World Health Organization.
[a]USEPA (2012).
[b]WHO (2006).
[c]NWQMS (2000).
[d]Yi *et al.* (2011).
'–': not applicable.

The focus of wastewater reuse is expanding from predominantly agricultural reuse to direct and indirect potable reuse (USEPA, 2012). The market for advanced wastewater reuse (i.e. direct and indirect potable) is expected to grow rapidly in the United States, Australia, Mexico, China, Spain and Saudi Arabia (Vo *et al.*, 2014). Water quality guidelines for advanced water reuse applications are presented in Table 6.3. Energy- and water-intensive industries such as textile, petroleum and beverage industries have also initiated water reuse projects to save water and energy costs. The average growth rate of water reuse projects including desalination in industrial sectors was estimated to be 11.4% from 2012 to 2017, and the total market is expected to reach \$12 billion by 2025 (GWI, 2012). This will result in the strong growth of supporting technologies, particularly of MBRs that are estimated to generate \$3 billion revenue by 2019 (GWI, 2012; Judd, 2016b). The MBR market development trend has been discussed in Chapter 1.

6.3 PERFORMANCE OF MBR TECHNOLOGY FOR POLLUTANT REMOVAL

The basic principle and the extent of wastewater treatment by MBR have been discussed in depth in the preceding chapters. To maintain continuity, some salient features of MBR treatment efficiency is discussed here.

6.3.1 Removal of bulk organics and nutrients

Microorganisms use organic matter as substrate to produce energy for cell growth and cell maintenance. This process results in the removal of bulk organics. In addition, biosorption, storage and accumulation

Table 6.3 Guidelines for different wastewater reuse categories. Modified after USEPA (2012).

Category	Guideline Values						
	pH	BOD mg/L	TOC mg/L	TSS mg/L	Turbidity NTU	Faecal Coliform CFU/100 mL	Residual Chlorine mg/L
Urban reuse							
Restricted	6–9	≤30	–	≤30	–	≤200	1
Unrestricted	6–9	≤10	–	–	≤2	0	1
Agriculture reuse							
Food crops	6–9	≤10	–	–	≤2	0	1
Processed food crops and non-food crops	6–9	≤30	–	≤30	–	≤200	1
Impoundments							
Restricted	6–9	≤30	–	≤30	–	≤200	1
Unrestricted	6–9	≤10	–	–	≤2	0	1
Environmental reuse	6–9	≤30	–	≤30	–	≤200	1
Industrial reuse	6–9	≤30	–	≤30	–	≤200	1
Groundwater recharge or non-potable reuse	Site-specific and intended use dependent						
Indirect potable reuse							
Groundwater recharge by spreading in to potable aquifer[1]	6.5–8.5	–	≤2	–	≤2	0	1
Groundwater recharge by injection in to potable aquifer[2]	6.5–8.5	–	≤2	–	≤2	0	1
Augmentation of surface water supply reservoirs[2]	6.5–8.5	–	≤2	–	≤2	0	1

BOD: biochemical oxygen demand; TOC: total organic carbon; and TSS: total suspended solids.
[1]Meets drinking water standards after percolation through vadose zone.
[2]Meets drinking water standards.
'–' not available.

also contribute to the removal of organics by the microorganisms in MBR (Carucci *et al.*, 2001; Radjenović *et al.*, 2008). For municipal wastewater treatment and reuse, MBR has emerged as a suitable technology. A wealth of studies on the comparative performance of CAS and MBR processes are available (Judd, 2010, 2016a), which show that MBRs are better placed to provide more stable chemical oxygen demand (COD) and nutrient removal (Cirja *et al.*, 2008; Gao *et al.*, 2004; Ng & Hermanowicz, 2005). MBRs achieve complete retention of biomass and particulate organics. Moreover, following adsorption on suspended solids, some soluble and colloidal organics get retained by MBR, leading to enhanced/stable COD removal (Côté *et al.*, 1997; Radjenović *et al.*, 2008). With complete retention, microbial communities in MBR have better opportunities to acclimatize to challenging wastewater, and provide improved degradation of those pollutants that are not readily biodegradable (Radjenović *et al.*, 2008).

The MBR supplier review presented in Chapter 14 shows that the four companies with the longest established commercial MBR products, i.e. Kubota, Mitsubishi, Suez (Zenon) and Econity (formerly Korea Membrane Separations) have more than 14,000 MBR plants among them. The total installed capacity of MBR of all suppliers for any plant size is thought to approach 25,000 MLD (see Chapter 14).

A survey of selected pilot- and full-scale MBR studies reveal that, in addition to complete removal of TSS, MBRs can achieve 87–97, 80–99 and 70–99% COD (Figure 6.4), ammonium-nitrogen (NH_4-N) and total phosphorous (TP) removal, respectively (Giménez *et al.*, 2011; Gurung *et al.*, 2016, 2017; Khan *et al.*, 2017; Krzeminski *et al.*, 2012a; Pagotto *et al.*, 2014; Phan *et al.*, 2015; Shen *et al.*, 2012). Thus, the effluent quality of MBR can convincingly meet the wastewater reuse guidelines (presented in Table 6.3) in terms of bulk organics removal.

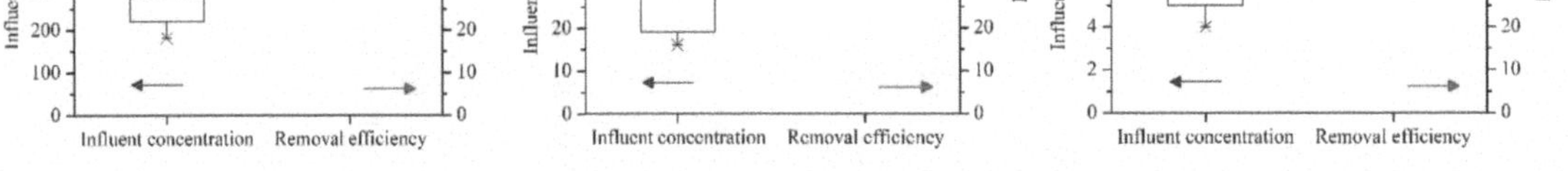

Figure 6.4 Performance of pilot- and full-scale MBR for the removal of COD, NH_4-N and TP from municipal wastewater. Data source: Giménez *et al.* (2011); Gurung *et al.* (2016, 2017); Khan *et al.* (2017); Krzeminski *et al.* (2012a); Pagotto *et al.* (2014); Phan *et al.* (2015) and Shen *et al.* (2012).

6.3.2 Removal of bacteria and viruses

For wastewater reuse, effective removal of bacteria and virus from wastewater is an important consideration. The extent and mechanism of removal of bacteria and virus by MBR have been investigated extensively in the last decade (Chaudhry *et al.*, 2015; De-Luca *et al.*, 2013; Hai *et al.*, 2014; Harb & Hong, 2017; Ma *et al.*, 2015; Purnell *et al.*, 2016; Sano *et al.*, 2016; Simmons *et al.*, 2011; Xagoraraki *et al.*, 2014). Performance of MBR technology for pathogen removal and the associated removal mechanisms are discussed in Chapter 5. This section provides a succinct overview of the guidelines established by the international organizations (e.g. WHO and USEPA) for the design of an effective wastewater treatment and reclamation process in the context of MBR technology.

According to the WHO guidelines for the safe use of wastewater, a disability-adjusted life year (DALY) loss per person per year of 10^{-6} is an acceptable additional disease burden (WHO, 2006). This is equivalent to rotavirus disease and infection risks of 10^{-3} and 10^{-4} per person per year (Mara, 2008). Although the values of acceptable/tolerable additional burden of disease can be used to estimate the acceptable concentration of pathogens (bacteria or virus) in reclaimed wastewater, WHO recommends setting wastewater treatment-specific pathogen removal targets. The removal targets are generally presented in the form of log removal (LR), which is the logarithm of the ratio of influent and effluent pathogen concentration. According to WHO guidelines, overall LR of 2–3 for virus is required for restricted irrigation, while 6–7 LR of pathogens is recommended for unrestricted irrigation activities. WHO recommends that each country establishes its own standards based on its social, economic and epidemiological requirements (WHO, 2006).

The USEPA guidelines for wastewater reuse do not set a tolerable limit for virus because virus detection is time-consuming and labour-intensive. Nevertheless, USEPA recommends achieving effective virus removal by multiple barrier treatments, i.e. primary treatment, secondary treatment and disinfection process (USEPA, 2012). In response to the recommendations of USEPA, the state of California has set targets for the removal of virus and other pathogens. For groundwater recharge using reclaimed wastewater, a 12 LR of viruses and 10 LR of both *Cryptosporidium* and *Giardia* is required. In addition, the recommended LR values for direct potable reuse in the state of Texas (USA) are 5.5, 6 and 8 LR of *Cryptosporidium*, *Giardia* and viruses, respectively (Sano *et al.*, 2016).

The guidelines for minimum virus and other pathogen reduction in Australia are based on a study of quantitative microbial risk assessment (QMRA). This study was carried out using rotavirus due to its prevalence in municipal wastewater to assess the minimum virus reduction level during wastewater treatment (NRMMC, 2006). According to this study, depending on the exposure duration to humans, a LR of 2.1–6.5 is required for the reuse of reclaimed wastewater. Similar to USEPA guidelines, the Australian national guidelines also recommend multiple barrier treatment systems. Notably, more stringent guidelines for virus reduction have been adopted by the state of Queensland (Australia) as compared to those provided by the WHO and USEPA. According to the State of Queensland guidelines, virus and bacteria/protozoa reduction of 8 and 6 LR, respectively, is required for achieving the highest quality of reclaimed water. Importantly, 9.5 LR of viruses and 8 LR of bacteria or protozoa are required in the State of Queensland (Australia) if the reclaimed water is intended for the augmentation of water supplies (Sano *et al.*, 2016). Although many European countries such as France, Greece, Italy and Spain have water reuse guidelines, only France has established regulations for virus reduction, which requires a 2–4 LR of F-specific bacteriophage for water reuse (Raso, 2013).

The performance of pilot- and full-scale MBR for the removal of pathogenic bacteria and viruses is pathogen-specific (Figure 6.5). MBRs can achieve above 5 LR of *Escherichia coli* and enterococci, while reported virus removal ranges from 4 to 9 LR (Chaudhry *et al.*, 2015; De-Luca *et al.*, 2013; Purnell *et al.*, 2016; Simmons *et al.*, 2011; Xagoraraki *et al.*, 2014). This is significantly higher than that achieved by the CAS process. However, to comply with the multiple barrier treatment concept, similar to CAS, for MBR too a post-disinfection step (e.g. chlorination or UV irradiation) is recommended. Nevertheless, MBR-treated wastewater is likely to warrant smaller dosage of disinfectants.

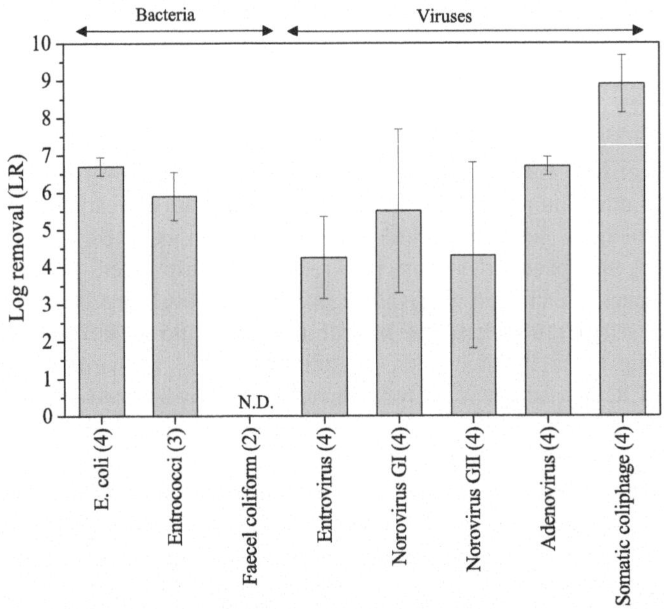

Figure 6.5 Performance of MBR for the removal of pathogenic bacteria and virus from municipal wastewater. The numbers within parentheses on the x-axis represent number of data points. Data source: Chaudhry *et al.* (2015); De-Luca *et al.* (2013); Purnell *et al.* (2016); Simmons *et al.* (2011); and Xagoraraki *et al.* (2014). '–' means not detected in MBR permeate.

6.3.3 Removal of emerging TrOCss

In addition to the removal of bulk organics and nutrients, effective removal of emerging contaminants such as TrOCs is vital for safe wastewater reuse applications. This is because even at very low concentrations, i.e. at μg/L or lower, TrOCs such as pharmaceuticals, personal care products, industrial chemicals, steroid hormones and pesticides are hazardous to aquatic lives and, potentially, to humans (Hai *et al.*, 2018; Lapworth *et al.*, 2012; Tran *et al.*, 2014). For example, at a level of approximately 1 ng/L, steroid hormones can produce endocrine disrupting effects in aquatic organisms (Schwarzenbach *et al.*, 2006). Given the documented side effects of some pharmaceuticals on the human body when taken medicinally, it is possible that pharmaceutical residues in wastewater-impacted water lead to detrimental health impacts (Hai *et al.*, 2018; Hoeger *et al.*, 2005; Soderstrom *et al.*, 2006). Accordingly, TrOC removal from wastewater is an important requirement in the context of wastewater reuse.

TrOC removal by MBR has been investigated extensively to understand the performance-governing factors (Hai *et al.*, 2011; Luo *et al.*, 2014a; Phan *et al.*, 2015; Qu *et al.*, 2009; Wijekoon *et al.*, 2013), and these factors have been critically analysed and discussed in Chapter 13 of this book. Briefly, TrOCs showing significant sorption on sludge (e.g. hydrophobic TrOCs) are well-removed from the aqueous phase. The removal of the hydrophilic TrOCs depends on their biodegradability. TrOCs with simple molecular structures are reported to be effectively biodegraded, while poor/unstable removal has been reported for structurally complex TrOCs (Cirja *et al.*, 2007). More specifically, TrOCs containing electron-withdrawing functional groups (EWGs) such as carboxyl, halogen and amide are resistant, and, like CAS, their removal by MBR is poor and/or unstable (Radjenović *et al.*, 2009; Wijekoon *et al.*, 2013). Hardly biodegradable TrOCs accumulate in sludge and pose the risk of spreading to the environment when sludge is periodically withdrawn from MBR and often reused for different purposes, including as fertilizer.

The removal of both hardly biodegradable and easily biodegradable TrOCs is often reported to be similar by CAS and MBR. However, compared to the CAS process, MBR has been reported to achieve better and stable removal of certain TrOC groups including steroid hormones and also some moderately biodegraded pharmaceuticals (e.g. sulphamethoxazole, ketoprofen, gemfibrozil, diclofenac and bezafibrate) (Asif *et al.*, 2018; Radjenović *et al.*, 2008). These aspects are discussed in detail in Chapter 13. With this understanding of comparative performance of CAS and MBR, the discussion from this point onward is focused on MBR.

Most emerging TrOCs remain unregulated to date. However, environmental quality standards (EQS) for safe disposal of treated effluent as well as wastewater reuse applications have been established/proposed for TrOCs that are commonly detected in raw wastewater. A comparison of TrOC concentration in MBR-effluent according to two selected studies (Phan *et al.*, 2015; Radjenović *et al.*, 2009) with the EQS from Switzerland, European Union and Australia is presented in Table 6.4. Concentration of the selected TrOCs in MBR permeates complied with the Australian Guidelines for Water Recycling: Augmentation of Drinking Water Supplies. However, the comparison also shows that the concentration of several TrOCs in MBR permeates exceeds chronic EQS values (Table 6.4). This implies that, although better than CAS effluents, MBR permeates can be harmful to aquatic lives over a prolonged exposure due to the presence of TrOCs.

It is challenging to evaluate the risks associated with the occurrence of TrOCs in raw and treated wastewater. The risks associated with the occurrence of TrOCs in treated effluent can be quantified based on the following two methods: (i) consumption rate posing health risk (CRPHR) for human; and (ii) risk characterization ratio (RQ) for ecological impacts.

TrOCs are potential health hazards and are regulated based on their acceptable daily intake (ADI). ADI of a chemical compound is measured experimentally in which an animal species is orally exposed to different doses of a compound until a lowest observed adverse effect level (LOAEL) is determined. LOAEL is divided by an uncertainty factor (normally 100) to set an ADI value for a compound. The purpose of the uncertainty factor is to conservatively compensate the difference between animals and human. ADI, expressed in mg/kg d, is a measure of the toxicity of a compound to humans. According to Table 6.4, ADI of different TrOCs are less than 0.01 mg/kg d, and generally range between 1×10^{-5} and 1×10^{-7} mg/kg d for estrogens. CRPHR of the selected TrOCs for a person weighing 70 kg can be calculated by using the following equation:

$$\text{CRPHR} = \frac{\text{ADI} \times 70\,\text{kg}}{\text{Concentration}\ (ng/\text{L}) \times 10^{-6}} \tag{6.1}$$

where CRPHR is the consumption rate (L/day) and ADI is the acceptable daily intake (mg/kg d); concentration of a compound is given in ng/L and 10^{-6} is the conversion factor.

TrOCs can be divided into four categories based on their CRPHR, namely: (i) high risk if CRPHR$<$2 L/day; (ii) moderate risk if CRPHR is 2–20 L/day; (iii) negligible risk if CRPHR is 21–200 L/day and (iv) no risk if CRPHR $>$200 L/day (Murray *et al.*, 2010). Based on these categories, of the TrOCs listed in Table 6.4, only estriol (CRPHR=1.4 L/day) and triclosan (CRPHR=18 L/day) appear harmful to human health. There are a few limitations in CRPHR-based categorization such as: (i) ADI values of a

Table 6.4 Comparison between the concentration of the selected TrOCs in MBR permeate with the environmental quality standards.

TrOCs	Concentration (ng/L)		EQS (ng/L)		PNEC[e] (ng/L)	ADI[f] (mg/kg d)	CRPHR[g] (L/d)
	Influent[a] ($\times 10^3$)	MBR Permeate[b]	Safe Disposal[c]	Water Reuse[d]			
Atenolol	0.14–35.1	42–210	150,000 (CH)	–	6,400,000	–	–
Diclofenac	0.001–90.5	5–180	50 (CH)	1,800	500	0.067	26,055
Ibuprofen	0.004–590	5–400	300 (CH)	400,000	1	0.57	99,750
Naproxen	0.015–1.15	5–290	1,700 (CH)	220,000	300	–	–
Carbamazepine	0.035–3.8	330–600	500 (CH)	100,000	89	0.0034	397
Gemfibrozil	0.10–17.1	5–20	–	600,000	312,500	0.013	45,500
Sulphamethoxazole	<0.01–0.98	5–310	600 (CH)	35,000	13	0.13	29,355
Trimethoprim	0.06–6.80	20–210	–	70,000	20	0.0094	3,133
Caffeine	0.22–209	31–320	–	3,500	500	–	–
Triclosan	0.03–23.9	14–730	20 (CH)	350	1.5	0.00019	18
DEET	2.56–3.19	5–31	41,000 (CH)	2,500,000	800	0.000082	185
Estrone	0.01–0.17	0.05–5	3.6 (CH)	30	3.6	0.000015	210
17β-Estradiol	0.002–0.05	0.05–5	0.4 (CH)	175	0.042	0.000005	70
17α-Ethynylestradiol	0.001–0.003	0.05–5	0.037 (CH)	175	0.004	0.000005	70
Estriol	0.125–0.80	0.05–5	–	50	67	0.0000001	1.4
Bisphenol A	0.013–2.14	10–300	1,500 (CH)	–	175	0.05	11667
Atrazine	0.02–28	40–730	600 (EU)	–	0.011	0.035	3356

[a]Luo *et al.* (2014b).
[b]Phan *et al.* (2015); Radjenović *et al.* (2009).
[c]Chronic Environmental quality standards (EQS) for inland surface waters in Switzerland, CH (WPO, 2018) and European Union, EU(EU, 2013).
[d]EPHC/NHMRC/NRMMC (2008).
[e]Predicted no effect concentration (PNEC): concentration below which exposure to a substance is not expected to cause adverse effects. Automatically calculated by the ecotoxicity database AiiDA (http://aiida.tools4env.com) based on the most sensitive species.
[f]Acceptable daily intake (ADI): USEPA (2010); Blanset *et al.* (2007); Muñoz *et al.* (2010); Schriks *et al.* (2010); Snyder (2008); and Murray *et al.* (2010).
[g]Consumption rate posing health risk values calculated based on the maximum TrOC concentration in MBR permeate.
'–': Data not available.

wide range of TrOCs are not available and (ii) ADI is calculated for a single compound and, thus, may not reflect the real impact of a mixture of compounds.

Risk characterization ratio or risk quotient (RQ) is used to categorize TrOCs based on their ecological impacts. It can be calculated by dividing the concentration of a compound with the corresponding predicted no-effect concentration (PNEC) value. The compounds with $RQ < 0.01$ pose no ecological risks; the compounds with $0.01 \leq RQ < 0.1$ pose moderate ecological risks; and the compounds with $RQ \geq 1$ are highly toxic to environmental systems (Minguez *et al.*, 2016).

RQ values calculated from the range of MBR effluent-TrOC concentrations listed in Table 6.4 are shown in Figure 6.6. It can be seen that at the concentration in which these selected TrOCs were detected in MBR permeates, they are indeed a significant risk to different environmental systems. All the selected TrOCs except atenolol and gemfibrozil were detected at levels potentially harmful to the aquatic organism (Figure 6.6).

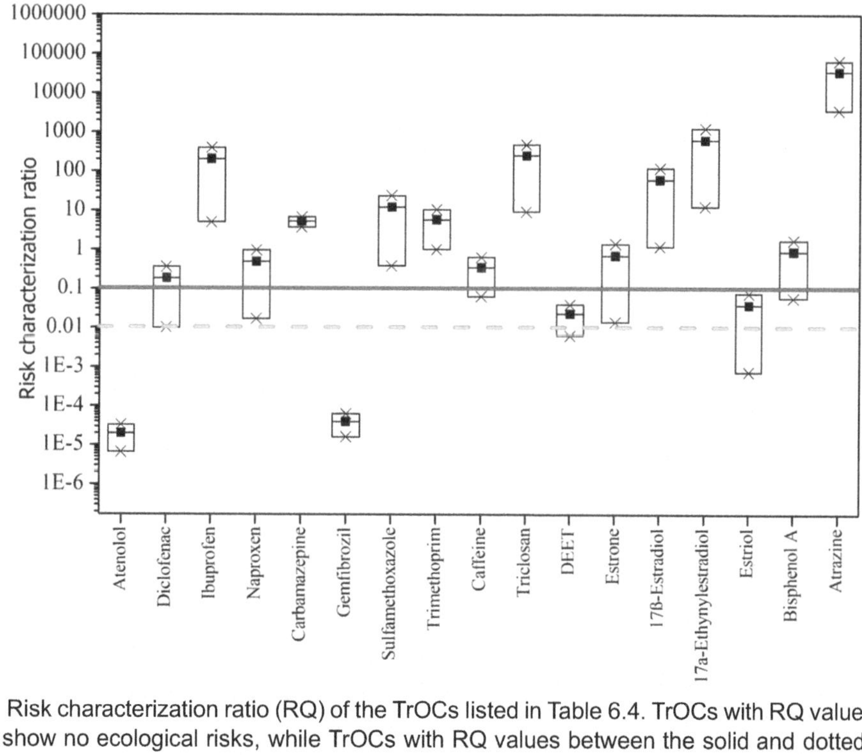

Figure 6.6 Risk characterization ratio (RQ) of the TrOCs listed in Table 6.4. TrOCs with RQ values below the dotted line show no ecological risks, while TrOCs with RQ values between the solid and dotted lines pose moderate ecological risks. TrOCs with RQ values above the solid line are of significant concern according to the concept of RQ.

On the basis of the above discussion, it is conceivable that ineffective removal of resistant TrOCs by MBR can be harmful for aquatic ecosystems. For safe disposal and water reuse applications, hybrid/integrated processes and MBRs implementing 'high retention' membranes have been investigated for effective TrOC removal (Alturki *et al.*, 2010; Asif *et al.*, 2018; Cartagena *et al.*, 2013). These innovative formats of MBRs are comprehensively discussed in Chapters 12 and 13 of this book.

6.4 WASTEWATER REUSE APPLICATION EXAMPLES INVOLVING MBR

The global capacity of full-scale MBR plants for wastewater treatment is expected to rise above 5 million m^3/d in 2019 (Abass *et al.*, 2015; Krzeminski *et al.*, 2017). Due to the decrease in membrane price over the last decade as well as its relative advantages compared to the CAS process (e.g. excellent effluent quality and small footprint), MBR market has seen a remarkable growth. However, compared to CAS, the MBR process is still costly, particularly for small and decentralised communities. It is also noteworthy that the unit capital and operating cost of an MBR is about 50% higher than that of the CAS process for the treatment of low strength municipal wastewater. This is because the costs associated with membrane replacement and chemical cleaning of membrane, and energy requirement for membrane fouling mitigation becomes proportionately more significant for low-strength wastewater and smaller plants or when high treated water quality is not required, consequently affecting the economic feasibility of MBRs (Van Haandel & Van Der Lubbe, 2012). Nevertheless, the progress in design and optimization has improved MBR process control, which continues to significantly reduce the capital and operating costs of MBR (Krzeminski *et al.*, 2017; Pretel *et al.*, 2016).

The MBR market drivers include (see Chapter 1): (i) requirement of high-quality effluent; (ii) stringent discharge and water reuse guidelines; (iii) rapidly increasing user acceptance of the technology and (iv) the opportunity of integration into the existing CAS-based treatment plants, i.e. potential of upgradation existing infrastructure (Krzeminski *et al.*, 2012b; Santos *et al.*, 2011). Historically, MBR plants were commissioned to comply with strict effluent discharge regulations. For instance, more than 40 small MBR plants were installed for the treatment of municipal wastewater at the historic places of Venice to meet the guidelines adopted by the Italian Ministry of Environment in 1999 (Krzeminski *et al.*, 2017). Henriksdal wastewater treatment plant in Stockholm was upgraded into an MBR to increase treatment capacity and to comply with EU Water Framework Directive (Malamis *et al.*, 2015). Since 2007, a steady trickle of large-scale MBR projects has occurred for Indirect Potable Reuse (IPR) and Direct Potable Reuse (DPR) schemes, particularly in North America and China, and this has massively increased the installed capacity of MBR (see Chapter 14).

Similar to other treatment technology, MBR units have been first tested at laboratory and pilot scale for the treatment and reuse of both municipal and industrial wastewater. For instance, Brik *et al.* (2006) assessed the performance of a lab-scale MBR for the treatment of wastewater from textile industry at different volumetric loading rates (i.e. 0.35–3.6 g/L d), and achieved up to 60–90 and 57–99% COD and colour removal, respectively. Since the MBR permeate did not meet the water reuse criteria, they proposed an additional polishing step, i.e. nanofiltration process for improving the quality of the treated water (Brik *et al.*, 2006). In another study by Krull and Döpkens (2004), a treatment process consisting of MBR and ozonation was developed for the treatment and recycling of textile wastewater. The MBR-ozonation process was effective, and the treated effluent met water reuse standards within the textile industry.

Carwash wastewater contain pollutants including bulk organics, nutrients, suspended solids, oil and grease. A commercial carwash station can generate up to 10,000 L/d (or 3.5 million L/year) of wastewater on an average (Rodriguez Boluarte *et al.*, 2016). Although the common practice is to dispose it into sewerage system, carwash wastewater could be reclaimed and reused in car washing processes (Zaneti *et al.*, 2011). In a study by Rodriguez Boluarte *et al.* (2016), a lab-scale anoxic−aerobic MBR was studied for the treatment of carwash wastewater. They reported that the MBR permeate met the carwash water quality standards in terms of BOD (10 mg/L), COD (50 mg/L), suspended solids (5 mg/L), turbidity (5 NTU) and pH (6.5–9.0) (Rodriguez Boluarte *et al.*, 2016; Tang *et al.*, 2007).

Greywater recycling has gained significant attention, particularly for meeting the non-potable water demand of large buildings in arid and semi-arid regions around the world (Li *et al.*, 2009; Revitt *et al.*, 2011). Merz *et al.* (2007) investigated the performance of a lab-scale MBR for the treatment of greywater originating from the showers of a leisure club. The MBR achieved above 95% COD removal, and the treated effluent complied with USEPA water recycling standards for toilet flushing. Disinfection of the treated effluent was essential for maintaining zero faecal coliform levels at all times (Merz *et al.*, 2007). In another study, a laboratory-scale MBR was tested for the treatment of greywater originating from the building of a university. The MBR permeate met the COD, TSS and coliform count-related guidelines established by the Jordan Standards and Metrology Organization for reuse of treated grey water (Bani-Melhem *et al.*, 2015). Kraume *et al.* (2010) reported long-term performance of laboratory-scale MBR for greywater treatment originating from showers. The MBR achieved TN and COD removal of 72 and 91%, respectively, which made the permeate suitable for reuse in irrigation. Similarly, greywater treatment by MBR followed by disinfection process was reported to produce effluent that met Spanish legislation for urban water reuse such as toilet flushing (Santasmasas *et al.*, 2013). In another study, during the treatment of greywater originating from a single house, MBR permeate was reported to comply with water reuse guidelines proposed by British Standards Institute, Government of Western Australia, and California State Water Resources Control Board for toilet flushing (Fountoulakis *et al.*, 2016).

In all the studies mentioned above, the fate of emerging TrOCs during the treatment of greywater has not been investigated. Donner *et al.* (2010) assessed the presence and fate of priority substances in domestic greywater treatment and reuse systems. They observed that removal of emerging TrOCs from greywater was mainly due to adsorption onto the activated sludge. They suggested that the introduction of emerging TrOCs in greywater should be controlled at the source because their adsorption onto activated sludge can make the process of sludge treatment and disposal complex and costly (Donner *et al.*, 2010; Revitt *et al.*, 2011).

If reuse of MBR-treated wastewater for potable or non-potable water is intended, additional polishing steps such as 'high retention' membrane separation processes (e.g. nanofiltration and reverse osmosis), advanced oxidation processes, disinfection and/or activated carbon adsorption may be required. Indeed, MBR coupled with post-treatment processes such as disinfection and activated carbon adsorption have been installed worldwide for a wide range of water reuse applications. An overview of the selected full-scale water reuse projects involving MBR is provided in Table 6.5.

6.5 FUTURE PERSPECTIVE AND CONCLUSION

The potential of wastewater reuse for meeting the future water demand has been recognized by many countries. However, safe wastewater reuse necessitates compliance with the water reuse guidelines (Krzeminski *et al.*, 2017; Vo *et al.*, 2014). MBRs have emerged as an attractive compact option to consider for applications requiring a high quality of treated effluent, particularly in situations with stringent suspended solids, nutrient and microbiological disposal limits, or when water reuse is required. The review of wastewater reuse projects involving MBR shows that the approach of multiple barrier system, as recommended by the World Health Organization and the USEPA, has been adopted by adding a post-treatment process such as disinfection, high retention membrane and/or adsorption. The effluent from the current MBR-based wastewater reuse projects is used for crop irrigation, landscape irrigation as well as potable reuse (mainly indirect potable reuse) applications.

Table 6.5 Selected case studies of wastewater reuse involving MBR technology.

Wastewater Type	Location	Treatment Processes	Design Capacity	Water Reuse Application	Project Funding
Process wastewater	Arizona (United States)	• Rotary drum screening • Equalization with pH adjustment • Primary sedimentation • MBR with BNR • GAC • UV disinfection • Low-pressure RO • Chlorine disinfection	2,419 m^3/d	Unrestricted industrial reuse	PepsiCo and Frito-Lay[1]
Municipal wastewater	Florida (United States)	• Packaged MBR plant • Chlorine disinfection	12,960 m^3/d	Irrigation Groundwater recharge	Marco Island Utilities[1]
Process wastewater	United States	• Ultrafiltration MBR • RO • Ozonation • UV disinfection • Chlorine residual	–	Indirect potable reuse	Coca-Cola[1]
Municipal wastewater	Massachusetts (United States)	• MBR plant • Post treatment step	100 m^3/d	Indirect potable reuse	Fay School student tuition fund[1]
Municipal wastewater	Massachusetts (United States)	• MBR • Ozonation • UV disinfection	950 m^3/d	Toilet flushing Groundwater recharge	Private and municipal funds[1]
Municipal wastewater	Sydney (Australia)	• Mechanical screening • Ultrafiltration MBR • Chlorine disinfection • RO (for water to be reused in cooling tower)	100 m^3/d	Non-potable water reuse	Cbus Property[1]

Greywater	Queensland, (Australia)	• Ultrafiltration MBR • UV disinfection • Chlorine disinfection	25 m^3/d	Non-potable water reuse	RSL care[1]
Municipal wastewater	Shoalhaven City (Australia)	• Primary sedimentation • Anoxic-aerobic-anoxic MBR • UV disinfection	743 m^3/d	Spray irrigation Onsite cleaning and washing	Shoalhaven, NSW, Australia[2]
Secondary treated municipal wastewater	Hohhot (China)	• Microfiltration MBR • Acid resins for hardness removal • Chlorine disinfection	31,000 m^3/d	Industrial reuse (cooling water)	Jinqiao power plant, Huaneng Power[1]
Municipal wastewater	Beijing (China)	• Ultrafiltration MBR • Activated carbon filter	80,000 m^3/d	Landscape irrigation Non-potable water reuse Surface water replenishment	Beijing Drainage Group[1]
Municipal wastewater	Tokyo (Japan)	• SBR (Greywater treatment due to oil and grease contents before mixing with MBR influent) • Ultrafiltration MBR • Chlorine disinfection	680 m^3/d	Toilet flushing	Project developer[1]

GAC: granular activated carbon; UV: ultraviolet; BNR: biological nitrogen removal; RO: reverse osmosis; SBR: sequencing batch reactors.
[1]USEPA (2012).
[2]Phan *et al.* (2015); Wale *et al.* (2014).
'–': not available.

REFERENCES

2030 Water Resources Group. (2009). Charting our water future: economic frameworks to inform decision-making, 2030 Water Resources Group, pp. Accessed from https://www.2030wrg.org/charting-our-water-future/ on Jan 2019.

Abass O. K., Wu X., Guo Y. and Zhang K. (2015). Membrane bioreactor in China: a critical review. *International Journal of Membrane Science and Technology*, **2**, 29–47.

Al-Saud M. (2013). National Water Strategy – The road map for sustainability, efficiency and security of water future in the K.S.A. Ministry for Water Affairs, Kingdom of Saudi Arabia.

Alturki A. A., Tadkaew N., McDonald J. A., Khan S. J., Price W. E. and Nghiem L. D. (2010). Combining MBR and NF/RO membrane filtration for the removal of trace organics in indirect potable water reuse applications. *Journal of Membrane Science*, **365**(1–2), 206–215.

Asano T., Maeda M. and Takaki M. (1996). Wastewater reclamation and reuse in Japan: overview and implementation examples. *Water Science and Technology*, **34**(11), 219–226.

Asif M. B., Ansari A. J., Chen S.-S., Nghiem L. D., Price W. E. and Hai F. I. (2018). Understanding the mechanisms of trace organic contaminant removal by high retention membrane bioreactors: a critical review. *Environmental Science and Pollution Research*, (DOI: https://doi.org/10.1007/s11356-018-3256-8).

Bani-Melhem K., Al-Qodah Z., Al-Shannag M., Qasaimeh A., Rasool Qtaishat M. and Alkasrawi M. (2015). On the performance of real grey water treatment using a submerged membrane bioreactor system. *Journal of Membrane Science*, **476**, 40–49.

Blanset D. L., Zhang J. and Robson M. G. (2007). Probabilistic estimates of lifetime daily doses from consumption of drinking water containing trace levels of *N*, *N*-diethyl-meta-toluamide (DEET), triclosan, or acetaminophen and the associated risk to human health. *Human and Ecological Risk Assessment*, **13**(3), 615–631.

Brik M., Schoeberl P., Chamam B., Braun R. and Fuchs W. (2006). Advanced treatment of textile wastewater towards reuse using a membrane bioreactor. *Process Biochemistry*, **41**(8), 1751–1757.

Carr G., Nortcliff S. and Potter R. B. (2011). Water reuse for irrigation in Jordan: plant beneficial nutrients, farmers' awareness and management strategies. *Water Science and Technology*, **63**(1), 10–15.

Cartagena P., El Kaddouri M., Cases V., Trapote A. and Prats D. (2013). Reduction of emerging micropollutants, organic matter, nutrients and salinity from real wastewater by combined MBR–NF/RO treatment. *Separation and Purification Technology*, **110**, 132–143.

Carucci A., Dionisi D., Majone M., Rolle E. and Smurra P. (2001). Aerobic storage by activated sludge on real wastewater. *Water Research*, **35**(16), 3833–3844.

Chaudhry R. M., Nelson K. L. and Drewes J. E. (2015). Mechanisms of pathogenic virus removal in a full-scale membrane bioreactor. *Environmental Science & Technology*, **49**(5), 2815–2822.

Chen Z., Ngo H. H. and Guo W. (2013). A critical review on the end uses of recycled water. *Critical Reviews in Environmental Science and Technology*, **43**(14), 1446–1516.

Cirja M., Ivashechkin P., Schäffer A. and Corvini P. F. X. (2007). Factors affecting the removal of organic micropollutants from wastewater in conventional treatment plants (CTP) and membrane bioreactors (MBR). *Reviews in Environmental Science and Bio/Technology*, **7**(1), 61–78.

Cirja M., Ivashechkin P., Schäffer A. and Corvini P. F. (2008). Factors affecting the removal of organic micropollutants from wastewater in conventional treatment plants (CTP) and membrane bioreactors (MBR). *Reviews in Environmental Science and Bio/Technology*, **7**(1), 61–78.

Côté P., Buisson H., Pound C. and Arakaki G. (1997). Immersed membrane activated sludge for the reuse of municipal wastewater. *Desalination*, **113**(2), 189–196.

Courault D., Albert I., Perelle S., Fraisse A., Renault P., Salemkour A. and Amato P. (2017). Assessment and risk modeling of airborne enteric viruses emitted from wastewater reused for irrigation. *Science of the Total Environment*, **592**, 512–526.

De-Luca G., Sacchetti R., Leoni E. and Zanetti F. (2013). Removal of indicator bacteriophages from municipal wastewater by a full-scale membrane bioreactor and a conventional activated sludge process: implications to water reuse. *Bioresource Technology*, **129**, 526–531.

De Koning J., Bixio D., Karabelas A., Salgot M. and Schäfer A. (2008). Characterisation and assessment of water treatment technologies for reuse. *Desalination*, **218**(1–3), 92–104.

Donner E., Eriksson E., Revitt D. M., Scholes L., Lützhøft H. C. H. and Ledin A. (2010). Presence and fate of priority substances in domestic greywater treatment and reuse systems. *Science of the Total Environment*, **408**(12), 2444–2451.

DSEWPaC. (2012). Progress against the National Target of 30% of Australia's Wastewater Being Recycled by 2015, Report Prepared for the Department of Sustainability, Environment, Water, Population and Communities (DSEWPaC), Marsden Jacob Associates, Melbourne. https://pitttownwater.com.au (Accessed October 2018).

Ensink J. H., Mahmood T., Van der Hoek W., Raschid-Sally L. and Amerasinghe F. P. (2004). A nationwide assessment of wastewater use in Pakistan: an obscure activity or a vitally important one? *Water Policy*, **6**(3), 197–206.

EPHC/NHMRC/NRMMC. (2008). Australian Guidelines for Water Recycling: Managing Health and Environmental Risks (Phase 2): Augmentation of Drinking Water Supplies. A publication of the Environment Protection and Heritage Council, the National Health and Medical Research Council and the Natural Resource Management Ministerial Council. ISBN: 1 921173 19 X. https://nhmrc.gov.au/ (Accessed October 2018).

EU. (2013). List of priority substances in the field of water policy. Directive 2013/39/EU of the European Parliament and of the Council of 12 August 2013 amending Directives 2000/60/EC and 2008/105/EC as regards priority substances in the field of water policy. https://eur-lex.europa.eu/eli/dir/2013/39/oj (Accessed January 2019).

Fountoulakis M. S., Markakis N., Petousi I. and Manios T. (2016). Single house on-site grey water treatment using a submerged membrane bioreactor for toilet flushing. *Science of the Total Environment*, **551–552**, 706–711.

Futran V. (2013). Tackling Water Scarcity: Israel's Wastewater Recycling as a Model for the World's Arid Lands. Global Water Forum, Canberra, Australia. Available online at: http://www.globalwaterforum.org (Accessed October 2018).

Gao M., Yang M., Li H., Wang Y. and Pan F. (2004). Nitrification and sludge characteristics in a submerged membrane bioreactor on synthetic inorganic wastewater. *Desalination*, **170**(2), 177–185.

Garcia-Cuerva L., Berglund E. Z. and Binder A. R. (2016). Public perceptions of water shortages, conservation behaviors, and support for water reuse in the US. *Resources, Conservation and Recycling*, **113**, 106–115.

Giménez J., Robles A., Carretero L., Durán F., Ruano M., Gatti M. N., Ribes J., Ferrer J. and Seco A. (2011). Experimental study of the anaerobic urban wastewater treatment in a submerged hollow-fibre membrane bioreactor at pilot scale. *Bioresource Technology*, **102**(19), 8799–8806.

Gurung K., Ncibi M. C., Fontmorin J.-M., Särkkä H. and Sillanpää M. (2016). Incorporating submerged MBR in conventional activated sludge process for municipal wastewater treatment: a feasibility and performance assessment. *Journal of Membrane Science and Technology*, **6**(3), 158.

Gurung K., Ncibi M. C. and Sillanpää M. (2017). Assessing membrane fouling and the performance of pilot-scale membrane bioreactor (MBR) to treat real municipal wastewater during winter season in Nordic regions. *Science of the Total Environment*, **579**, 1289–1297.

GWI. (2010). Municipal Water Reuse Markets 2010: Analysis, Forecasts and Inventory. Media Analytics Ltd, Oxford, United Kingdom.

GWI. (2012). Industrial Desalination and Water Reuse: Ultrapure Water, Challenging Waste Streams and Improved Efficiency. Media Analytics Ltd, Washington DC, USA.

Hai F. I., Li X., Price W. E. and Nghiem L. D. (2011). Removal of carbamazepine and sulfamethoxazole by MBR under anoxic and aerobic conditions. *Bioresource Technology*, **102**(22), 10386–10390.

Hai F. I., Riley T., Shawkat S., Magram S. and Yamamoto K. (2014). Removal of pathogens by membrane bioreactors: a review of the mechanisms, influencing factors and reduction in chemical disinfectant dosing. *Water*, **6**(12), 3603.

Hai F. I., Yang S., Asif M. B., Sencadas V., Shawkat S., Sanderson-Smith M., Gorman J., Xu Z.-Q. and Yamamoto K. (2018). Carbamazepine as a possible anthropogenic marker in water: occurrences, toxicological effects, regulations and removal by wastewater treatment technologies. *Water*, **10**(2), 107.

Harb M. and Hong P.-Y. (2017). Molecular-based detection of potentially pathogenic bacteria in membrane bioreactor (MBR) systems treating municipal wastewater: a case study. *Environmental Science and Pollution Research*, **24** (6), 5370–5380.

Harris-Lovett S. R., Binz C., Sedlak D. L., Kiparsky M. and Truffer B. (2015). Beyond user acceptance: a legitimacy framework for potable water reuse in California. *Environmental Science & Technology*, **49**(13), 7552–7561.

Hoeger B., Köllner B., Dietrich D. R. and Hitzfeld B. (2005). Water-borne diclofenac affects kidney and gill integrity and selected immune parameters in brown trout (*Salmo trutta* f. *fario*). *Aquatic Toxicology*, **75**(1), 53–64.

Judd S. (2010). The MBR Book: Principles and Applications of Membrane Bioreactors for Water and Wastewater Treatment, 2nd edn. Butterworth-Heinemann, Oxford.

Judd S. (2016a). The status of industrial and municipal effluent treatment with membrane bioreactor technology. *Chemical Engineering Journal*, **305**, 37–45.

Judd S. J. (2016b). The status of industrial and municipal effluent treatment with membrane bioreactor technology. *Chemical Engineering Journal*, **305**, 37–45.

Keraita B., Drechsel P. and Amoah P. (2003). Influence of urban wastewater on stream water quality and agriculture in and around Kumasi, Ghana. *Environment and Urbanization*, **15**(2), 171–178.

Khan S. J., Hasnain G., Fareed H. and Aim R. B. (2017). Evaluation of treatment performance of a full-scale membrane bioreactor (MBR) plant from unsteady to steady state condition. *Journal of Water Process Engineering*, (DOI: 10.1016/j.jwpe.2017.03.004).

Kraume M., Scheumann R., Baban A. and El Hamouri B. (2010). Performance of a compact submerged membrane sequencing batch reactor (SM-SBR) for greywater treatment. *Desalination*, **250**(3), 1011–1013.

Krull R. and Döpkens E. (2004). Recycling of dyehouse effluents by biological and chemical treatment. *Water Science and Technology*, **49**(4), 311–317.

Krzeminski P., Iglesias-Obelleiro A., Madebo G., Garrido J. M., van der Graaf J. H. J. M. and van Lier J. B. (2012a). Impact of temperature on raw wastewater composition and activated sludge filterability in full-scale MBR systems for municipal sewage treatment. *Journal of Membrane Science*, **423–424**, 348–361.

Krzeminski P., van der Graaf J. H. and van Lier J. B. (2012b). Specific energy consumption of membrane bioreactor (MBR) for sewage treatment. *Water Science and Technology*, **65**(2), 380–392.

Krzeminski P., Leverette L., Malamis S. and Katsou E. (2017). Membrane bioreactors – a review on recent developments in energy reduction, fouling control, novel configurations, LCA and market prospects. *Journal of Membrane Science*, **527**, 207–227.

Landa-Cansigno O., Durán-Álvarez J. and Jiménez-Cisneros B. (2013). Retention of *Escherichia coli*, *Giardia lamblia* cysts and *Ascaris lumbricoides* eggs in agricultural soils irrigated by untreated wastewater. *Journal of Environmental Management*, **128**, 22–29.

Lapworth D. J., Baran N., Stuart M. E. and Ward R. S. (2012). Emerging organic contaminants in groundwater: a review of sources, fate and occurrence. *Environmental Pollution*, **163**, 287–303.

Lautze J., Stander E., Drechsel P., da-Silva A. K. and Keraita B. (2014). Global Experiences in Water Reuse. International Water Management Institute (IWMI), Colombo, Sri Lanka. CGIAR Research Program on Water, Land and Ecosystems (WLE). p. 31. (DOI: 10.5337/2014.209).

Li F., Wichmann K. and Otterpohl R. (2009). Review of the technological approaches for grey water treatment and reuses. *Science of the Total Environment*, **407**(11), 3439–3449.

Listowski A., Ngo H. and Guo W. (2013). Establishment of an economic evaluation model for urban recycled water. *Resources, Conservation and Recycling*, **72**, 67–75.

Luo Y., Guo W., Ngo H. H., Nghiem L. D., Hai F. I., Kang J., Xia S., Zhang Z. and Price W. E. (2014a). Removal and fate of micropollutants in a sponge-based moving bed bioreactor. *Bioresource Technology*, **159**, 311–319.

Luo Y., Guo W., Ngo H. H., Nghiem L. D., Hai F. I., Zhang J., Liang S. and Wang X. C. (2014b). A review on the occurrence of micropollutants in the aquatic environment and their fate and removal during wastewater treatment. *Science of the Total Environment*, **473**, 619–641.

Ma J., Wang Z., Zang L., Huang J. and Wu Z. (2015). Occurrence and fate of potential pathogenic bacteria as revealed by pyrosequencing in a full-scale membrane bioreactor treating restaurant wastewater. *RSC Advances*, **5**(31), 24469–24478.

Malamis S., Andreadakis A., Mamais D. and Noutsopoulos C. (2015). Can strict water reuse standards be the drive for the wider implementation of MBR technology? *Desalination and Water Treatment*, **53**(12), 3303–3308.

Mara D. (2008). Quantifying health risks in wastewater irrigation. In: UNESCO Encyclopedia of Life Support Systems, W.O.K. Grabow, (ed.), EOLSS Publishers. Oxford, United Kingdom.

Marlow D. R., Moglia M., Cook S. and Beale D. J. (2013). Towards sustainable urban water management: a critical reassessment. *Water Research*, **47**(20), 7150–7161.

MENR. (2010). Statistics on water in Mexico. 2010 ed. National Water Commission of Mexico, Ministry of Environment and Natural Resources. http://www.conagua.gob.mx (Accessed October 2018).

Merz C., Scheumann R., El Hamouri B. and Kraume M. (2007). Membrane bioreactor technology for the treatment of greywater from a sports and leisure club. *Desalination*, **215**(1), 37–43.

Minguez L., Pedelucq J., Farcy E., Ballandonne C., Budzinski H. and Halm-Lemeille M.-P. (2016). Toxicities of 48 pharmaceuticals and their freshwater and marine environmental assessment in northwestern France. *Environmental Science and Pollution Research*, **23**(6), 4992–5001.

Molinos-Senante M., Hernández-Sancho F. and Sala-Garrido R. (2011). Cost–benefit analysis of water-reuse projects for environmental purposes: a case study for Spanish wastewater treatment plants. *Journal of Environmental Management*, **92**(12), 3091–3097.

Muñoz I., Bueno M. J. M., Agüera A. and Fernández-Alba A. R. (2010). Environmental and human health risk assessment of organic micro-pollutants occurring in a Spanish marine fish farm. *Environmental Pollution*, **158** (5), 1809–1816.

Murray K. E., Thomas S. M. and Bodour A. A. (2010). Prioritizing research for trace pollutants and emerging contaminants in the freshwater environment. *Environmental Pollution*, **158**(12), 3462–3471.

Ng H. Y. and Hermanowicz S. W. (2005). Membrane bioreactor operation at short solids retention times: performance and biomass characteristics. *Water Research*, **39**(6), 981–992.

Norton-Brandão D., Scherrenberg S. M. and van Lier J. B. (2013). Reclamation of used urban waters for irrigation purposes – a review of treatment technologies. *Journal of Environmental Management*, **122**, 85–98.

NRMMC. (2006). National Water Quality Management Strategy. Australian Guidelines for Water Recycling: Managing Health and Environmental Risks (Phase 1). Natural Resource Management Ministerial Council. http://www.environment.gov.au (Accessed October 2018).

NWQMS. (2000). Guidelines for Sewage Systems. Use of Reclaimed Water. National Water Quality Management Strategy, Agriculture and Resource Management Council of Australia and New Zealand.

Pagotto R., Rossetto R., Gasperi R., Andrade J., Trovati J., Vallero M., Okumura A. and Arntsen B. (2014). SANASA Capivari II – the first full-scale municipal membrane bioreactor in Latin America. *Water Science and Technology*, **70**(2), 272.

Phan H. V., Hai F. I., McDonald J. A., Khan S. J., Zhang R., Price W. E., Broeckmann A. and Nghiem L. D. (2015). Nutrient and trace organic contaminant removal from wastewater of a resort town: comparison between a pilot and a full scale membrane bioreactor. *International Biodeterioration & Biodegradation*, **102**, 40–48.

Pretel R., Robles A., Ruano M., Seco A. and Ferrer J. (2016). Economic and environmental sustainability of submerged anaerobic MBR-based (AnMBR-based) technology as compared to aerobic-based technologies for moderate-/high-loaded urban wastewater treatment. *Journal of Environmental Management*, **166**, 45–54.

Purnell S., Ebdon J., Buck A., Tupper M. and Taylor H. (2016). Removal of phages and viral pathogens in a full-scale MBR: implications for wastewater reuse and potable water. *Water Research*, **100**, 20–27.

Qu X., Zhao Y., Yu R., Li Y., Falzone C., Smith G. and Ikehata K. (2016). Health effects associated with wastewater treatment, reuse, and disposal. *Water Environment Research*, **88**(10), 1823–1855.

Qu Y.-Y., Zhou J.-T., Wang J., Xing L.-L., Jiang N., Gou M. and Salah Uddin M. (2009). Population dynamics in bioaugmented membrane bioreactor for treatment of bromoamine acid wastewater. *Bioresource Technology*, **100**(1), 244–248.

Radcliffe J. (2010). Evolution of water recycling in Australian cities since 2003. *Water Science and Technology*, **62**(4), 792–802.

Radjenović J., Matošić M., Mijatović I., Petrović M. and Barceló D. (2008). Membrane bioreactor (MBR) as an advanced wastewater treatment technology. In: Emerging Contaminants from Industrial and Municipal Waste,

D. Barceló and M. Petrovic (eds), Springer, Heidelberg, Germany, Vol. 5, pp. 37–101. (DOI: 10.1007/978-3-540-79210-9).

Radjenović J., Petrović M. and Barceló D. (2009). Fate and distribution of pharmaceuticals in wastewater and sewage sludge of the conventional activated sludge (CAS) and advanced membrane bioreactor (MBR) treatment. *Water Research*, **43**(3), 831–841.

Raso J. (2013). Updated Report on Wastewater Reuse in the European Union. EC-reference: TYPSA-reference:7452; Code: 7452-IE-ST03_WReuse_Report-Ed1.doc. http://ec.europa.eu/ (Accessed October 2018).

Report M. W. R. (2007). Mediterranean Wastewater Reuse Report, Joint Mediterranean EUWI/WFD Process EU.

Revitt D. M., Eriksson E. and Donner E. (2011). The implications of household greywater treatment and reuse for municipal wastewater flows and micropollutant loads. *Water Research*, **45**(4), 1549–1560.

Rodriguez Boluarte I. A., Andersen M., Pramanik B. K., Chang C.-Y., Bagshaw S., Farago L., Jegatheesan V. and Shu L. (2016). Reuse of car wash wastewater by chemical coagulation and membrane bioreactor treatment processes. *International Biodeterioration & Biodegradation*, **113**, 44–48.

Rose J. (2007). Water reclamation, reuse and public health. *Water Science and Technology*, **55**(1–2), 275.

Ross V. L., Fielding K. S. and Louis W. R. (2014). Social trust, risk perceptions and public acceptance of recycled water: testing a social-psychological model. *Journal of Environmental Management*, **137**, 61–68.

Sano D., Amarasiri M., Hata A., Watanabe T. and Katayama H. (2016). Risk management of viral infectious diseases in wastewater reclamation and reuse. *Environment International*, **91**, 220–229.

Santasmasas C., Rovira M., Clarens F. and Valderrama C. (2013). Grey water reclamation by decentralized MBR prototype. *Resources, Conservation and Recycling*, **72**, 102–107.

Santos A., Ma W. and Judd S. J. (2011). Membrane bioreactors: two decades of research and implementation. *Desalination*, **273**(1), 148–154.

Scheierling S. M., Bartone C., Mara D. D. and Drechsel P. (2010). Improving Wastewater Use in Agriculture An Emerging Priority. World Bank, Washington, DC.

Schriks M., Heringa M. B., van der Kooi M. M., de Voogt P. and van Wezel A. P. (2010). Toxicological relevance of emerging contaminants for drinking water quality. *Water Research*, **44**(2), 461–476.

Schwarzenbach R. P., Escher B. I., Fenner K., Hofstetter T. B., Johnson C. A., von Gunten U. and Wehrli B. (2006). The challenge of micropollutants in aquatic systems. *Science*, **313**(5790), 1072–1077.

Shen Y.-X., Xiao K., Liang P., Sun J.-Y., Sai S.-J. and Huang X. (2012). Characterization of soluble microbial products in 10 large-scale membrane bioreactors for municipal wastewater treatment in China. *Journal of Membrane Science*, **415**, 336–345.

Simmons F. J., Kuo D. H. W. and Xagoraraki I. (2011). Removal of human enteric viruses by a full-scale membrane bioreactor during municipal wastewater processing. *Water Research*, **45**(9), 2739–2750.

Snyder S. A. (2008). Occurrence, treatment, and toxicological relevance of EDCs and pharmaceuticals in water. *Ozone: Science and Engineering*, **30**(1), 65–69.

Soderstrom J., Murray L., Daly F. and Little M. (2006). Toxicology case of the month: oral hypoglycaemic overdose. *Emergency Medicine Journal*, **23**(7), 565–567.

Szyplinska P. (2012). Design–build–operate a sustainable global model. *Water World*, **26**(6). https://www.waterworld.com/articles/wwi/print/volume-26/issue-6/regulars/creative-finance/design-build-operate.html (Accessed March 2019).

Tang L., Xue-Jun T., Fu-Yi C., Zhou Q. and Yin J. (2007). Reuse of carwash wastewater with hollow fiber membrane aided by enhanced coagulation and activated carbon treatments. *Water Science and Technology*, **56**(12), 111.

Tran N. H., Urase T. and Ta T. T. (2014). A preliminary study on the occurrence of pharmaceutically active compounds in hospital wastewater and surface water in Hanoi, Vietnam. *CLEAN – Soil, Air, Water*, **42**(3), 267–275.

USEPA. (2010). Integrated Risk Information System (IRIS), Vol. **2016**, USEPA, USA.

USEPA. (2012). Guidelines for Water Reuse. United States Agency for International Development, Washington DC.

Van Haandel A. and Van Der Lubbe J. (2012). Handbook of Biological Wastewater Treatment. IWA Publishing, London, United Kingdom.

Venkatesan A. K., Ahmad S., Johnson W. and Batista J. R. (2011). Salinity reduction and energy conservation in direct and indirect potable water reuse. *Desalination*, **272**(1–3), 120–127.

Vo P. T., Ngo H. H., Guo W., Zhou J. L., Nguyen P. D., Listowski A. and Wang X. C. (2014). A mini-review on the impacts of climate change on wastewater reclamation and reuse. *Science of the Total Environment*, **494**, 9–17.

Wale R., Nayak S. and James G. (2014). Decentralised wastewater reclamation facilities and network for national trust listed village. Available online at https://www.evoqua.com/en/about/Australia/Documents/Kangaroo-Valley-Wastewater.pdf (Accessed October 2018).

WHO. (2006). Guidelines for the Safe Use of Wastewater, Excreta and Greywater. World Health Organization, France.

Wijekoon K. C., Hai F. I., Kang J., Price W. E., Guo W., Ngo H. H. and Nghiem L. D. (2013). The fate of pharmaceuticals, steroid hormones, phytoestrogens, UV-filters and pesticides during MBR treatment. *Bioresource Technology*, **144**, 247–254.

Wilcox J., Nasiri F., Bell S. and Rahaman M. S. (2016). Urban water reuse: a triple bottom line assessment framework and review. *Sustainable Cities and Society*, **27**, 448–456.

WPO. (2018). Acute and Chronic Quality Standards. From Annex 2 of "814.201 Waters Protection Ordinance (WPO)" of 28 October 1998 (Status as of 1 June 2018). https://www.admin.ch (Accessed Jan 2019).

Xagoraraki I., Yin Z. and Svambayev Z. (2014). Fate of viruses in water systems. *Journal of Environmental Engineering*, **140**(7), 04014020.

Xiao K., Liang S., Wang X., Chen C. and Huang X. (2018). Current state and challenges of full-scale membrane bioreactor applications: a critical review. *Bioresource Technology*, **271**, 473–481.

Yi L., Jiao W., Chen X. and Chen W. (2011). An overview of reclaimed water reuse in China. *Journal of Environmental Sciences*, **23**(10), 1585–1593.

Zabalaga J., Amy G. and Von Münch E. (2007). Evaluation of agricultural reuse practices and relevant guidelines for the Alba Rancho WWTP (primary and secondary facultative ponds) in Cochabamba, Bolivia. *Water Science and Technology*, **55**(1–2), 469–475.

Zaneti R., Etchepare R. and Rubio J. (2011). Car wash wastewater reclamation. Full-scale application and upcoming features. *Resources, Conservation and Recycling*, **55**(11), 953–959.

Zhang Y. and Shen Y. (2017). Wastewater irrigation: past, present, and future. *Wiley Interdisciplinary Reviews: Water*, e1234. (DOI: 10.1002/wat2.1234).

Zimmerman J. B., Mihelcic J. R. and Smith J. (2008). Global stressors on water quality and quantity. *Environmental Science and Technology*, **42**(12), 4247–4254.

Chapter 7

Impacts of hazardous events on performance of membrane bioreactors

*Trang Trinh[1], Amos Branch[1,2], Ben van den Akkerc, Pierre Le-Clech[2], Jörg E. Drewes[1,4] and Stuart J. Khan[1,4]**

[1]UNSW Water Research Centre, University of New South Wales, NSW, Australia
[2]School of Chemical Engineering, University of New South Wales, NSW, Australia
[3]SA Water Corporation, Adelaide, SA, Australia
[4]Chair of Urban Water Systems Engineering, Technical University of Munich, Garching, Germany
**Corresponding author: s.khan@unsw.edu.au.*

ABSTRACT

Membrane bioreactors (MBRs) can provide effective treatment for many wastewater contaminants including chemicals and microorganisms. Operational performances for the removal of these contaminants are most typically characterised under what are considered to be normal operating conditions. However, all MBR systems are continuously subjected to the risk of deviations in operating conditions during what have been termed 'hazardous events'. Hazardous events may include such things as sudden changes in source water composition, extreme weather events, human error and mechanical malfunctions. Depending on both the likelihood and the consequences of these events, they may ultimately define the treatment reliability and level of risk regarding meeting final water quality objectives. This chapter describes potential hazardous events and their impact on MBR operation. Expected consequences are identified, along with techniques for assessing the likelihoods for some types of hazardous events. Finally, recommendations are made for the management of hazardous events through engineered redundancy and multiple barrier treatment systems.

 Membrane Biological Reactors: Theory, Modeling, Design, Management and Applications to Wastewater Reuse
Editor(s): Faisal I. Hai, Kazuo Yamamoto and Chung-Hak Lee
doi: 10.2166/9781780409177_0207

7.1 INTRODUCTION – HAZARDOUS EVENTS IN RISK ASSESSMENT

The operational performance of any wastewater treatment system can be viewed from two distinct perspectives. The first, and most commonly considered, is the inherently variable treatment performance that may be achieved when the system is operating within a defined set of 'normal' operational conditions. The less commonly considered perspective regards the consideration of how the system may perform in the event of a disruption to normal operating conditions. In the field of risk assessment, a departure from normal operational conditions is commonly termed a 'hazardous event'.

Hazardous events that may affect the operation of wastewater treatment systems can include sudden changes in source water composition, extreme weather events, human error and mechanical malfunctions.

Since hazardous events may occur from time to time, and may have significant impacts on short-term operational performance, the characterisation of the likelihoods and consequences of these events is necessary in order to fully characterise the long-term performance of the system. Indeed, hazardous event scenarios are commonly the scenarios that present the greatest levels of risk related to final water quality. Therefore, characterisation of these events is required to properly characterise risks including those posed to the environment and to human health.

The vast majority of observed waterborne disease outbreaks in developed countries during the last few decades have been associated with hazardous events, such as unusual weather patterns, plumbing errors or treatment failures (Hrudey & Hrudey, 2007; Rizak & Hrudey, 2007). Consequently, the assessment of hazardous event scenarios has become an integral component of drinking water quality management in many countries. This approach is encapsulated within the Australian Drinking Water Guidelines (NWQMS 2011) and the World Health Organization Guidelines for Drinking Water Quality (WHO 2011).

Following this trend in drinking water management, the Australian Guidelines for Water Recycling (NRMMC & EPHC 2006) have adopted a consistent approach for the qualitative incorporation of hazardous event analysis in overall system performance assessment. In this context, potential hazardous events are identified and each is allocated a qualitative measure for both its perceived 'likelihood' (Table 7.1) and its 'consequence' or impact (Table 7.2).

Once a suitable qualitative measure of likelihood and consequences has been allocated to each identified (potential) hazardous event, a qualitative risk estimation or 'risk rating' can be applied according to the risk matrix presented in Table 7.3. The specific characterisation (e.g. low, moderate, high, and very high) of risks relating to various combinations of likelihood and consequence measures may be adapted for particular systems and applications. The example given in Table 7.3 is that used in the Australian Guidelines for Water Recycling (NRMMC & EPHC 2006) and is very similar to those presented in the Australian Drinking Water Guidelines (NWQMS 2011) and the World Health Organization Guidelines for Drinking Water Quality (WHO 2011).

Table 7.1 Qualitative measures of likelihood (NRMMC & EPHC 2006).

Level	Descriptor	Example Description
A	Rare	May occur only in exceptional circumstances. May occur once in 100 years
B	Unlikely	Could occur within 20 years or in unusual circumstances
C	Possible	Might occur or should be expected to occur within a 5- to 10-year period
D	Likely	Will probably occur within a 1- to 5-year period
E	Almost certain	Is expected to occur with a probability of multiple occurrences within a year.

Table 7.2 Qualitative measures of consequence or impact (NRMMC & EPHC 2006).

Level	Descriptor	Example Description
1	Insignificant	Insignificant impact or not detectable
2	Minor	Health – Minor impact for small population Environment – Potentially harmful to local ecosystem with local impacts contained to site
3	Moderate	Health – Minor impact for large population Environment – Potentially harmful to regional ecosystem with local impacts primarily contained to on-site.
4	Major	Health – Major impact for small population Environment – Potentially lethal to local ecosystem; predominantly local, but potential for off-site impacts
5	Catastrophic	Health – Major impacts for large population Environment – Potentially lethal to regional ecosystem or threatened species; widespread on-site and off-site impacts

Table 7.3 Qualitative risk estimation (NRMMC & EPHC 2006).

	Consequences				
Likelihood	**1-Insignificant**	**2-Minor**	**3-Moderate**	**4-Major**	**5-Catastrophic**
A Rare	Low	Low	Low	High	High
B Unlikely	Low	Low	Moderate	High	Very high
C Possible	Low	Moderate	High	Very high	Very high
D Likely	Low	Moderate	High	Very high	Very high
E Almost Certain	Low	Moderate	High	Very high	Very high

This risk assessment process provides a basis for managing risks and applying *preventive measures*. In the context of wastewater and recycled water management, preventative measures most commonly refer to actions, activities and processes used to prevent significant hazards from being present in final effluents or to reduce the hazards to acceptable levels. Risk should be assessed at two levels:

- Maximum (unmitigated) risk, which is risk in the absence of preventive measures – assessment of maximum risk is useful for identifying high-priority risks, determining where attention should be focused and preparing for emergencies.
- Residual risk, which is risk after consideration of existing and proposed preventive measures – assessment of residual risk provides an indication of the safety and sustainability of the system or the need for additional preventive measures.

The following sections are intended to provide insights to the potential impacts of hazardous events on the ongoing performance of membrane bioreactors. It is proposed that this information will be a significant value to system managers, people responsible for system performance assessment and validation, health and environmental regulators and, ultimately, to the designers and manufacturers of future, more resilient systems.

7.2 CHARACTERISATION OF POTENTIAL HAZARDOUS EVENTS AND THEIR IMPACT ON MBR OPERATION

To characterise hazardous events relevant to MBR operation, it is first necessary to describe the elements of an MBR process in relation to hazard analysis terminology. The primary hazard within the MBR process is presented by the components of the mixed liquor solution of an activated sludge (AS) system. In particular, pathogenic microorganisms within the AS constitute a human health hazard, while bulk parameters such as biochemical oxygen demand (BOD), chemical oxygen demand (COD), and total suspended solids (TSS) present environmental risks. The concentration of pathogenic microorganisms in AS has been observed to be similar to sewage for indicator species prone to biological predation, such as *Escherichia coli*. However, indicators that exhibit resistance to biological degradation and are of greater diameter than the membrane pore size, such as sulphite reducing clostridia, have been shown to accumulate within the AS (Marti *et al.*, 2011; van den Akker *et al.*, 2012). As a result the concentration factor for resistant pathogens and indicators is expected to be proportional to the MBR solid retention time (SRT).

Due to health and environmental hazards associated with the components of the mixed liquor, hazardous event scenarios are expected to include any deviation from normal MBR operation, which would lead directly, or indirectly to 'loss of containment' of the AS. Loss of containment in MBR is expected to result from membrane/module integrity failure, overflow from the bio- or membrane reactor or decrease in the treatment efficiency of the AS system. A range of threats could be defined within the various treatment steps of the MBR plant, including collection, pre-treatment, AS process, membrane and post treatment.

7.2.1 Deviation from normal operation

7.2.1.1 Collection

Collection of MBR influent may occur downstream of primary settling or pre-screening at a municipal wastewater treatment facility or following an equalisation tank in smaller decentralised systems. Nominal feed quality will be subject to diurnal, seasonal and regional variations. Shock loadings have been also widely reported to occur within the sewage collection, generally due to upstream intermittent discharge from industry, heavy rainfall event or via ingress into aged and damaged sewer mains.

Shock loads resulting from seawater ingress (Severn, 2003), unregulated upstream discharge of industrial wastes and high loadings of non-dissolved material during storm weather flow were reported to affect nominal operation of MBRs (van Bentem *et al.*, 2007). Maintenance cleaning of upstream unit operations, without appropriate isolation, can also result in shock loading of downstream processes with high concentrations of suspended solids and grease, causing clogging of pre-treatment equipment and membrane units (Lazarove *et al.*, 2008).

7.2.1.2 Pre-treatment

Effective pre-treatment is arguably one of the most important aspects of the operation of MBR, in order to preserve the integrity of downstream membranes. Pre-treatment may involve with fine screening (1–3 mm) with the possible addition of micro sieving (down to 250 μm) along with grit and grease removal. Bypass of screens due to seal and screen failure or even deliberate screen removal has been reported and can increase the likelihood of membrane damage by foreign materials (metal shavings, fibrous rag material, leaves, etc.).

Failure of fine screening caused accumulation of solids and grit in the membrane compartment leading to increased membrane cartridge damage and replacement rate; up to 50% of the inventory reported by Nishimori *et al.* (2010). Self-cleaning micro-sieve systems can also pose a source of abrasive contaminants through loss of brush fibres during operation (van Bentem *et al.*, 2010).

7.2.1.3 Activated sludge process

Threats to AS include loss of aeration and circulation due to port clogging, mechanical fault or power loss and overdose of membrane cleaning chemicals (Judd & Judd, 2011). Disturbances, particularly to influent quality, can result in foaming, leading to potential loss of containment via overflow of the aeration tanks. Simulations of hazardous events on AS in MBR have revealed decreased capacity for removal of bulk parameters such as BOD, COD and total nitrogen; however, simulations of microbial quality of the permeate was not possible with the model utilised (Friedler *et al.*, 2008).

During operation, biological treatment processes may be exposed to changing environmental conditions such as variations in the flow rate, concentration, and quality of the raw wastewater entering the process. In general, any rapidly occurring or immediate change in the chemical or physical environment might be classified as a system 'shock'.

Organic shock loads have been described in terms of quantitative shock loads and qualitative shock loads (Gaudy & Engelbrecht, 1961). *Quantitative shock load* implies a rapid increase in organic loading by rising high concentration of substrate to which the sludge is acclimated or to which it needs no acclimation. However, waste streams do not often have constant chemical composition of the organic constituents. A qualitative change in the chemical composition of the substrate (with constant TOC concentration) may constitute a serious type of system shock. This is termed a *qualitative shock load*. It implies that the composition of the carbon source has changed from that to which the sludge is normally acclimated while it does not imply that the change is toxic. For example, the substrate may change from a predominantly carbohydrate waste to a proteaceous or a fatty waste, from simple sugars to polymers, or from sucrose to lactose.

An important variation on quantitative shock loads is 'starvation shock'. Most treatment systems are designed to manage some variability in flow regimes. However, in extreme conditions, some treatment plants exhibit feed starvation periods during which no appreciable wastewater feeds the systems. This discrepancy between the conceptual design and the practical situation may lead to process upsets and unsatisfactory system performance (Beler Baykal *et al.*, 1990).

Toxic shock involves an influx of organics or inorganic constituents and radicals, which wholly or partially inhibit or damage the existing metabolic pathways or disrupt the established physiological condition of the microbial population (Gaudy & Engelbrecht, 1961). Rapid changes in pH of the waste are also considered to be in this class of shock loading although they are more easily controlled and may be of less significance than other toxicity shock loads.

Waste streams with high ammonia concentration are very commonly produced by human handling (Campos *et al.*, 2002). Sudden increase in ammonia concentration in biological treatment process can be due to increase in ammonia concentration in raw sewage or inhibition of nitrification in the biological treatment process (Hart *et al.*, 2003). Similarly, pH changes in biological treatment processes can be due to pH variation in raw sewage or due to failure of denitrification process within the biological treatment units.

Temporary interruptions to aeration of MBR systems would be expected to have a detrimental impact on the aerobic metabolic degradation of chemical contaminants and potentially lead to change within the microbial community. Loss of aeration may also lead to loss of suspension of the MLSS, potentially causing damage to MBR membranes.

7.2.1.4 Membrane filtration

Crucial threats at the membrane filtration stage regarding the containment of AS can be encompassed within the integrity failure of the membrane or the module itself (seals, gaskets, connections). Through fault tree analysis based on the top event of *Cryptosporidium* release, threats were scoped for an ultrafiltration plant (Beauchamp *et al.*, 2010) and can be equally applicable to the membrane filtration step of an MBR.

In addition to the case of membranes exhibiting manufacturing defects, solid particles and foreign bodies within the bioreactor can breach or damage the membrane. Moreover, inappropriate high dosing of cleaning chemicals, and pressure shock (due to air from integrity testing or water from hydraulic shock of a pump start up) are expected to increase the likelihood of membrane integrity failure (Beauchamp *et al.*, 2010). Integrity failure can be induced via sparks from welding in the vicinity of membranes (Ayala *et al.*, 2011) and high-pressure hosing during maintenance cleans (Le-Clech *et al.*, 2005).

Failure of the module integrity results in short circuit of the membrane by constituents present in the mixed liquor. Module weak points include seals couplings and membrane-frame/pot interface. Module failure likelihood is increased as a result of the seal being of poor quality or inevitable wear out due to an insufficient replacement regime. Coupling failure of MBR cassette has been previously attributed to the strong mechanical forces in the module header due to the air-cycling fouling mitigation system (van Bentem *et al.*, 2007).

The cleaning regime frequently imposed on membranes to remove fouling and recover hydraulic performance results in gradual changes in the physical and chemical membrane properties (especially decrease of mechanical strength) (Hajibabania *et al.*, 2012). A decrease in mechanical strength of the hollow fibre membrane is expected to significantly increase the likelihood of membrane integrity failure.

7.2.1.5 Post-treatment

MBR permeate is sometimes disinfected and/or stored shortly before discharge. The major post treatment threat can be defined as the bacterial regrowth in permeate lines or storage reservoirs, which have been reported to cause detectable levels of total coliforms in the permeate of MBRs (Zhang & Farahbakhsh, 2007).

7.3 EXPECTED CONSEQUENCES OF KEY HAZARDOUS EVENT TYPES

Few research studies have been reported to specifically examine the consequences of hazardous events to MBR performance. However, many insights can be obtained from previous studies of conventional AS systems since the biological characteristics of the two types of systems are similar. The following sections discuss the expected impacts of hazardous events on the removal of chemical and microbial constituents, with observations derived from studies on both MBR and conventional AS systems.

7.3.1 Impact on the removal of bulk organic matter and nutrients

Consequences of hazardous event conditions on conventional AS and MBR treatment performance are summarised in Table 7.4.

The results of quantitative organic shock load studies on AS systems show that reactors, which were stably operated at influent COD concentrations above 100–500 mg/L can withstand influent shock concentrations of up to 1500 mg/L COD, even when the shock durations varied from hours to weeks (Gaudy & Engelbrecht, 1961; Normand & Perdrieux, 1981; Saleh & Gaudy, 1978). However, at influent shock concentrations around 3000 mg/L COD, the change may exceed the maximum assimilation capacity of the biomass, leading to an increased deterioration of effluent quality caused by loss of

Table 7.4 Consequence of hazardous event conditions on AS and MBR treatment performance based on select studies.

Event type	System	Monitored Parameters	Consequence on Removal	Ref.
Organic shock	AS operated stably at influent COD of 100–500 mg/L	COD	Influent COD increased to $\leq$1,500 mg/L: no impact	1,2,3
			Influent COD increased to $\geq$3,000 mg/L: biomass grown rapidly, floc size decreased, filamentous forms increased and number of protozoa reduced, loss of biomass causing deterioration of effluent quality	2,4
	MBR system (hollow fibre, 0.04 μm) subjected to a shock of 5 g/L COD	pH, COD, DOC, SS, VSS, Capillary suction time (CST), fouling	COD of permeate increased sharply from 29 to 4650 mg/L after 1 h. Permeate COD reduced to 1066 mg/L, then, and finally 7 mg/L after 24, 48 and 72 h, respectively.	5
Starvation shock	AS system subjected to 10 d starvation period	Biomass characteristics	After shock 8 d, biomass concentration and respiration activity decreased sharply due to degradation of proteins polysaccharides contents in biomass	6
	AS system subjected to 21 d starvation period	pH, SS, VSS, CODd, DOC, biomass characteristics	Biomass concentration, bacteria cell size and respiration activity decreased sharply during first 4 d, disappearance of some typical microbial groups in AS. CODd and DOC in liquid phase increased sharply between days 4 and 9 due to release of organic material from death microorganisms.	5
	MBR system (hollow fibre, 0.4 μm) subjected to 5 d starvation period	COD, TOC, TSS, TKN, phosphate, biomass characteristics	After 5 d starvation, removal efficiencies of COD, TOC, TSS, TKN, phosphate reduced significantly and they recovered fully after 6 d of normal operation. Biomass concentration and activity reduced significantly and took a month to recover	8
	MBR system (hollow fibre, 0.04 μm) subjected to 6 d starvation period	pH, COD, DOC, SS, VSS, CST, fouling	After re-feeding the starved MBR, COD and DOC removal efficiencies in the reactor immediately increased to 97 and 94%, respectively. The high COD and DOC removals were stable and remained within the control	5

(Continued)

Table 7.4 Consequence of hazardous event conditions on AS and MBR treatment performance based on select studies (*Continued*).

Event type	System	Monitored Parameters	Consequence on Removal	Ref.
			range until the end of the experiment. After starvation, MLVSS concentration was 17% lower than the control range and remained at this level until the end of the experiment.	
Salinity shock	AS system subjected to NaCl up to 45 g/L	COD, biomass characteristics	COD removal and biomass settleability reduced	9
	As systems subjected to NaCl from 0 to 60 g/L	COD, biomass characteristics	NaCl $\leq$10 g/L: DOC removal slightly increased NaCl $>$10 g/L: DOC removal reduced NaCL $\geq$15 g/L: morphological changes in microbial population NaCl $\geq$30 g/L: effluent turbidity increased	10
	MBR system (hollow fibre, 0.04 μm) subjected to a shock of 20 g/L NaCl	pH, COD, DOC, SS, VSS, CST, fouling	COD removal efficiency had reduced considerably after 1 h of introducing the salinity shock. COD removal in the salinity shock experiment recovered to the control range within 24 h after introducing the shock.	5
Ammonia shock	MBR system (hollow fibre, 0.04 μm) subjected to a shock of 700 mg/L ammonia	pH, COD, DOC, SS, VSS, CST, fouling	There was a slight reduction observed in COD removal by 24 h, whereas DOC removal had considerably decreased. The COD and DOC removals then recovered back within the control range by 48 h after shock.	5
Ammonia shock	MBR system (hollow fibre, 0.04 μm) subjected to a single dose equivalent to 1,000 mg/L ammonia into the bioreactor	pH, TOC, TN, COD, TP, SS and VSS	Reduction in MLVSS from 2.6 to 2.0 g/L. Reduction in TOC removal from $>$90% to 0% at 1 and 24 h, recovering to 80% after 72 hours. Only marginal removal reduction in COD removal. Reduction in TN removal from $>$80 to 0% after 1 and 24 h and then recovering to 25% removal after 72 h.	11

1. Gaudy and Engelbrecht (1961); 2. Saleh and Gaudy (1978); 3. Normand and Perdrieux (1981); 4. Manickam and Gaudy (1985); 5. Trinh *et al.* (2016); 6. Urbain *et al.* (1993); 7. Coello Oviedo *et al.* (2003); 8. Yogalakshmi *et al.* (2007); 9. Dan *et al.* (2003); 10. Ng *et al.* (2005); 11. Phan *et al.* (2015).

biological solids (Manickam & Gaudy, 1985; Saleh & Gaudy, 1978). A 3000 mg/L COD shock load to an AS system was reported to cause a rapid growth in biomass, a noticeable change in colour of the mixed liquor, a decrease in floc size, an increase in filamentous forms and a reduction in the number of protozoa (Saleh & Gaudy, 1978). Disruption in COD removal capacity and the change in colour of an AS system were observed to be correlated with changes in the biochemical composition of the sludge (Manickam & Gaudy, 1985). In general, high organic concentration in influent wastewater is known to inhibit nitrification as it supports the growth of heterotrophic bacteria, which compete with autotrophic nitrifying bacteria for oxygen, nutrients, and space. The COD of the permeate of an MBR system was found to increase sharply from 29 to 4650 mg/L within 1 h after being subjected to a feed water shock load sufficient to achieve a reactor concentration of 5000 mg/L COD dosed as a 1:1 solution of glucose and glutamic acid. The COD concentrations of the permeate in the shocked reactor were reduced to 1066, 280 and then 7 mg/L after 24, 48 and 72 hours, respectively. The permeate pH decreased by 3.6 units within 1 h after shock loading. This was likely due to the large amount of glutamic acid added in the reactor, which exceeded the metabolic rate of the existing microorganisms, and may not occur in real systems unless a high COD and low pH shock load originated from the feedwater. The pH recovered slightly after 24 h and fully recovered to the control range (6.8 ± 0.7) within 48 h, suggesting that the microorganisms had metabolised the accumulated glutamic acid in the reactor. The organic carbon shock was observed to create foaming in the reactor which subsequently resulted in overflow of biomass out of the reactor within the first 48 h after shock (Trinh *et al.*, 2016).

Studies have shown that biomass concentrations have decreased sharply during the first four days of a starvation shock and then reduced more slowly after that (Coello Oviedo *et al.*, 2003; Urbain *et al.*, 1993). In addition, the bacteria cell size was also found to be reduced, which was described as one of the adaptive responses to starvation conditions (Coello Oviedo *et al.*, 2003; Kjelleberg *et al.*, 1987; Urbain *et al.*, 1993). These responses were related to the degradation of both proteins and polysaccharide contents of the sludge and led to a decrease in respiratory activity of the microorganisms. After 3–4 days under starvation conditions, the biomass drastically lost its ability to biodegrade exogenous nutrients reactions (Urbain *et al.*, 1993). Starvation shocks also resulted in disappearance of some of the typical microbial groups usually found in an AS, and appearance of other opportunistic microorganisms (Coello Oviedo *et al.*, 2003). A recent study reported that after recommencement of feeding of an MBR that had been starved for 6 days, COD and DOC removal efficiencies in the reactor immediately increased to 97 and 94%, respectively. This suggested a rapid microbial utilisation of the available carbon sources after a long starvation period (Li *et al.*, 2006). The high COD and DOC removals were stable and remained within the normal operational range until the end of the experiment. After starvation, MLVSS concentration was 17% lower than the control range and remained at this level until the end of the experiment (Trinh *et al.* 2016).

The removal efficiencies of COD, TOC, TSS, total Kjeldahl nitrogen (TKN) and phosphate by a MBR were reduced significantly under a feed starvation shock load of 5 days (Yogalakshmi *et al.*, 2007). In addition, a large fraction of biomass wash off and a reduction in microbial activity inside the reactor was observed. The removal of organics and nutrients was recovered back to steady-state conditions after six days of normal operation. However, it took nearly a month of continuous operation to regain the amount of biomass lost during feed starvation shock load (Yogalakshmi *et al.*, 2007).

High salt concentrations in a biological reactor have been reported to reduce organic removal efficiencies and biomass settleability (Dan *et al.*, 2003; Ng *et al.*, 2005). This is because salty conditions produce high osmotic pressure on bacterial cells, which can inhibit bacterial growth and floc formation (Dan *et al.*, 2003). Additionally, high salt concentration conditions also reduce gravity separation due to lower density difference between water and biomass (Ng *et al.*, 2005). COD removal efficiency of an MBR was found to reduce considerably after 1 h of introducing the 20 g/L salinity shock. COD removal in the salinity shock experiment recovered to the normal operating range within 24 hours after introducing the shock. This agreed with permeate conductivity data, which indicated that the majority of NaCl had been washed out of the MBR by 24 hours (Trinh *et al.* 2016).

Failure modes leading to physical membrane damage tend to be gradual rather than sudden and are easily identified by long-term changes in flux or operating pressures. Accordingly, their relevance as 'hazardous events' leading to sudden deterioration in water quality appears low. Nonetheless, there is some evidence to suggest that events such as chemical membrane cleaning and accidental exposure to excessive chlorine concentrations may physically harm some types of water treatment membranes leading to reduced performance (Beyer *et al.*, 2010; Simon *et al.*, 2009). Losses in membrane integrity may influence the removal of undissolved fractions of organic carbon and COD but would likely not significantly influence removal of dissolved components as these would likely be smaller than the nominal pore size of the membrane, typically 0.04–0.4 μm.

7.3.2 Impact on the removal of pathogens and microbial indicators

Information on the impact of hazardous events on the removal of pathogenic microorganisms by MBRs is scarce. Research has traditionally focused on studying the behaviour of microbial indicators (model pathogens) under a range of event conditions. Most of this information has been derived from lab- and pilot-scale studies, whereby key operating parameters can be easily adjusted and the MBRs challenged under controlled conditions. The impact of key operational events on the microbial removal efficiency of MBRs are summarised in Table 7.5.

What is clear from Table 7.5 is that the most important mechanisms responsible for removing microorganisms are membrane rejection, adsorption to solids and removal with waste AS and biodegradation. Generally, pathogen removal improves as membrane fouling layers develop, and thus events that lead to the removal or disturbance of fouling layers (e.g. membrane cleaning, backwashing and change in permeate flux) can adversely influence removal. The extent of membrane fouling is commonly quantified by the monitoring changes in the permeate flux or the transmembrane pressure (TMP).

MBRs are well known for their ability to remove a wide range of indicator organisms (e.g. bacteria, bacteriophages, and spores) and it is clear from the literature that each organism behaves differently. Notably, the removal of membrane fouling influences the rejection of bacteriophage more so than bacteria; simply because bacteriophages are typically (0.02–0.2 μm) smaller than the nominal pore size of membranes (0.04–0.4 μm). As a result, bacteriophage removal is typically less consistent and can be more dependent on the membrane pore size as well as to changes in operation, such as membrane TMP, permeate flux, spikes in initial feed concentrations and chemical variations from the feedwater that can influence phage-to-particle interaction (adsorption) and also predation by higher-order organisms, compared to other organisms. Therefore, bacteriophage appear to be a superior model organism for understanding the impacts of hazardous event conditions on the microbial removal efficiency of MBRs, as they challenge both size-exclusion mechanisms of the membrane and potential biological predation and adsorption within the bioreactor.

Not all bacteriophage species behave the same. Different species feature varying retention mechanisms, owing to differences in surface properties. For example, F-specific bacteriophage have been reported to have a higher tendency to adsorb to membrane surfaces and suspended biomass more so than somatic coliphage, exhibiting a more even removal pattern during maintenance cleaning events (Zhang & Farahbakhsh 2007). One study has reported that adsorption tendency to AS was highly variable when bacteriophage species with different protein content in their surface capsids were measured. The contribution to removal from

Table 7.5 Consequence of operational and event conditions on the removal of microbial indicators based on select studies.

Event Type	Membrane	Model Organisms	Consequence on LRV	Ref.
Chemical backwash	Zenon ZW-500C-SMC	Somatic coliphage F-specific coliphage	Small decrease (from 3.0 to 2.5) No significant impact	12
Formation of fouling	Memcor 0.2 μm	Indigenous somatic coliphage	Increase from 1.2 (clean) to 2.0 (fouled) No impact	13*
Increase in flux for clean membrane			Decrease from 2.2 (50 $L/m^2/h$) to 1.7 (85 $L/m^2/h$)	
Increase in flux for fouled membrane			2.3 (25 $L/m^2/h$), 2.7 (50 $L/m^2/h$) and 2.3 (85 $L/m^2/h$)	
Longer filtration/relaxation cycle	Six MBR systems	Seeded MS-2 bacteriophage	Increase from 2.9 (1 min cycles) to 3.4 (8–18 min cycles)	14
Relaxation period/air scouring			Small decrease by 0.25 LRV	
Change in pore size (0.03–0.1 μm)		Indigenous coliphage	Increase from 1.5 (0.1 μm) to 4.5 (0.03 μm) No impact, due to particle association	
Change in pore size (0.03–0.2 μm)	Nine MBR systems	Coliform bacteria Indigenous coliphage	No impact No impact	15
Increase in MLSS conc. (3–9 g/L)	Hollow fibre, 0.4 μm	Indigenous somatic coliphage	No impact on LRV, but change in biological action	16
Formation of fouling			Increase from 0.6 (clean) to 1.5 (fouled)	
Chemical backwash			Decrease by 0.5 (*attributable to biomass only)*	
Change in SRT (10–50 d)			Weak increase by 0.05 (*attributable to biomass only)*	
Change in HRT (8–13 h)			Increase from 1.5 to 1.9 (*attributable to biomass only)*	
Filtration of supernatant	Flat sheet, 0.4 μm	T-even-like indigenous bacteriophage	LRV across membrane only: 0.5	17
Operation with mixed liquor			Increase to 4	
Power failure to air scour and influent pump			Increase from 0.4 to 1.0, possibly due to increased fouling	
Clean membrane (filtration of supernatant)	Hollow fibre, 0.4 μm	MS-2 bacteriophage	0.3–0.4	18

(*Continued*)

Table 7.5 Consequence of operational and event conditions on the removal of microbial indicators based on select studies (*Continued*).

Event Type	Membrane	Model Organisms	Consequence on LRV	Ref.
High flux operation			Decrease	
Operation with mixed liquor			Increase to 1.0 (after 9 h) to 2.0 (21 d)	
Change in MLSS conc. (6–10 g/L)			No impact	
Formation of fouling	Flat sheet, 0.4 μm	Indigenous somatic coliphage	No significant impact	19
		Indigenous FRNA bacteriophage	Increase from 4.5 to 4.8	
		Bacterial indicators (*C. perfringens*, *E. coli*)	No impact	
Membrane rinsing		T4 coliphage	Decrease from 5.8 to 3.1	20
Chemical cleaning	Hollow fibre, 0.22 μm		Decrease from 5.8 to 1.7	
Chemical cleaning	Hollow fibre, 0.22 and 0.1 μm	Coliphage f2	Decrease from 3.9 to 0.8	21
Formation of fouling	KOCH membrane systems 0.04 μm	MS-2 bacteriophage	LRV of 0.9 with a clean membrane increasing to 1.5 with a fouling layer and up to 1.7 with biomass present	22
		phiX174 (somatic coliphage)	LRV of 0.4 with a clean membrane increasing to 1.1 with a fouling layer and up to 2.3 with biomass present	
		fr bacteriophage	LRV of 0.5 with a clean membrane increasing to 1.2 with a fouling layer and up to 4.2 with biomass present	
Chemically enhanced backwash (CEB)	Zenon ZW-500D Full-scale plant 0.04 μm	Adenovirus	From 4.0 immediately after CEB to 5.0, 5.1 and 5.5 at 3 hours, 1 day and 5 days after	23
		Norovirus GII	From 5.0 immediately after CEB to 5.5 after 3 hours	
		Indigenous FRNA bacteriophage	From 5.1 immediately after CEB to 5.1 and 7.0 at 1 day and 5 days after	
Fibre breakage + backflush (cut up to 4% (4 of 120 fibres))	Hollow fibre 0.04 μm pilot MBR	*C. perfringens*	No significant decrease 15 min after cut. LRV reduced from >5.0 to 1.2 immediately after backflush, recovered to >3.2 log units 5 minutes after and then 5.2 log units after 24 h	24

		Spiked FRNA bacteriophage	No significant decrease 15 min after cut. LRV reduced from 5.2 to 2.7 immediately after backflush, recovered to 5.9 log units 5 minutes after and then 5.2 log units after 24 h	
NaCl shock (20 g/L added to MBR in one dose)	Hollow fibre 0.04 µm pilot MBR	Spiked FRNA bacteriophage	LRV reduced from 5.4 the day before to 4.5 immediately after and then 3.9 the following day. recovery to >6.9 was noted after 2 days.	24
COD shock (5 g/L added to MBR in one dose)	Hollow fibre 0.04 µm pilot MBR	Spiked FRNA bacteriophage	Increased significantly from 5.4 the day before to >7.0 immediately after and then stabilised. Increase in removal was thought to be due to a significant decrease in pH due to the mixture of glutamic acid and glucose used to increase COD.	24
		Total coliforms	No significant impact to LRV overall, but the biological predation typically 0.3 log units reduced to below 0 and total coliform bacteria began to accumulate (LRV biological = −1.5) within the reactor with no recovery observed in 48 h.	
Filtration with defective membranes (5–50 µm holes)	Hollow fibre 0.04 µm pilot MBR (+ defects)	Spiked FRNA bacteriophage	Larger indicators removal (*C. perfringens* and *E. coli*) did not appear to be significantly affected for the samples taken. FRNA LRV upon filtration commencement was 3.5 significantly lower than the normal range. After 24 h FRNA LRV had increased to 4.2, within the normal range.	

12. Zhang and Farahbakhsh (2007); 13. Farahbakhsh and Smith (2004); 14. Hirani *et al.* (2010), 15. Hirani *et al.* (2012); 16. Wu *et al.* (2010); 17. Ueda and Horan (2000), 18. Shang *et al.* (2005); 19. Marti *et al.* (2011); 20. Lv *et al.* (2006); 21. Zheng and Liu (2006); 22. Chaudhry *et al.* (2015a); 23. Chaudhry *et al.* (2015b); 24. Branch *et al.* (2016).

*Direct filtration of sewage, no MBR.

adsorption alone was the most variable between different phage species and was reported to be 0.2 log units for MS-2 bacteriophage, 1.2 log for phi X174 and 3 log for bacteriophage (Chaudhry *et al.*, 2015a). Also, the removal patterns of indigenous and laboratory-grown bacteriophage strains can also differ (Hirani *et al.* 2010). Clearly, selection of the right indicator organisms (i.e. one that shares a similar fate to target pathogen) is therefore crucial when characterising the impacts of hazardous events on MBR performance.

Research characterising the removal of indicator organisms by MBRs also suggests that the suspended biomass (mixed liquor) can play a very important role in the elimination of pathogens via adsorption and predation (Table 7.5). The contribution of biomass, however, is dependent on inter-related parameters including the concentration of mixed liquor suspended solids, the sludge retention time and the food-to-mass ratio; and thus operational events that lead to changes in these parameters may influence pathogen removal. In a previous study, the half-lives of adenovirus, norovirus GII and F+ coliphage within the bioreactor of a full-scale MBR were estimated to be 1.6, 0.5 and 0.3 h, which for the design of that particular facility were equivalent to the log reduction values (LRVs) of 1.1, 1.6 and 1.7 log units, respectively (Chaudhry *et al.*, 2015b). Given the highly complex nature of the fouling layer and the wide variety of organisms investigated, the relative impact of the fouling layer on the rejection capability of the membrane has still not been clearly demonstrated. More recent research has suggested that the reversible cake layer could contribute 0.6 log units additional bacteriophage removal when compared to the clean membrane (Chaudhry *et al.*, 2015a). The role of the irrecoverable fouling layer formed over years of continuous operation is expected to be responsible for the build-up of a protective layer suitable for adsorption for viruses. For example, the irrecoverable fouling layer present on a 10-year-old MBR membrane module was reported to improve removal from 1 to 2 log units in addition to the nominally expected 1 log removal of a clean membrane (Chaudhry *et al.*, 2015a). However, Table 7.5 indicates that the various types of cleaning used in MBR maintenance generally result in lower pathogen rejection.

In previous work, LRV of indicator organisms FRNA bacteriophage, *E. coli*, total coliforms and *Clostridium perfringens* were determined on pilot MBR systems that were subjected to a range of hazardous events (Branch *et al.*, 2016). Events that resulted in significant reductions in LRV were included in Table 7.5.

Under normal conditions it was noted that the removal of indicators across the membrane and the bioreactor, measured from sewage to permeate contributed between 4 and 7 log units more than the LRV determined due to the bioreactor only (comparing sewage to mixed liquor densities). Therefore it was suspected that only events that severely compromised membrane integrity would significantly reduce overall LRV. This was largely the case, but surprisingly, the addition of NaCl at 20 g/L as a single shock load to the bioreactor resulted in a significant decrease of FRNA bacteriophage LRV from above the median of 5.4, to 4.5 and then further to 3.9 immediately and 1 day after the shock. After 2 days, when the NaCl had washed out of the system the bacteriophage LRV recovered to within the normal range >6.9. Although small increases of NaCl (0–0.02 M) have been reported to enhance membrane rejection (van Voorthuizen *et al.*, 2001), at ionic strengths above 0.1 M, the adsorption affinity of some microorganisms was reported to decrease (Stevik *et al.*, 2004). 20 g/L of NaCl is equivalent to 0.3 M and, when added to the bioreactor, may have decreased the adsorption affinity of FRNA bacteriophage, resulting in the initial decrease observed in LRV. However, the high NaCl was also thought to cause biomass stress which increased the level of fouling, indicated by trans membrane pressure, by 200–500% in two days (Branch *et al.*, 2016). Although undesirable for hydraulic productivity, the formation of a less permeable fouling layer on the membrane, may have helped mitigate any further loss of FRNA bacteriophage LRV.

Fibre breakage in combination with backflush was also investigated. Surprisingly, no significant reduction in LRV was observed when samples were taken 15 min after three out of 120 fibres from a mini-module bundle were cut. However, samples taken immediately after backflushing showed a significant loss of LRV for *C. perfringens* and FRNA bacteriophage. Nonetheless, within 5 min after the backflush, FRNA bacteriophage LRV had recovered to within the normal range. At 1 day after backflushing, rejection of all tested indicators was within the normal range. Subsequent investigation of the mini-module showed that a plug of AS had formed within the lumen of the hollow fibre membranes used. Based on the study results, the plug appeared to provide a sufficient seal to enable LRV within normal operational variability (Branch *et al.*, 2016). However, the plug was able to be dislodged by the application of only 30 kPa air backpressure, suggesting that the removal of the plug, loss of containment of microorganisms, and subsequent reforming could reoccur during operation. A defective mini-module with 5–50 μm holes in the skin was also used. LRV for none of the larger organisms, *C. perfringens*, *E. coli* or total coliforms, were significantly reduced when operating with the defective membrane. LRV for FRNA bacteriophage started at 3.5, below the normal lower limit of LRV. However, after 1 day operation, FRNA LRV had stabilised and remained above the 5th percentile of 4.1 for the remainder of the trial (Branch *et al.*, 2016). Both of these results suggest that the AS has some capacity to shield membrane defects. However, the shielding effect is difficult to continuously validate. An increase in particle count and turbidity results taken up to 3 hours after a membrane clean at a full-scale MBR showed a response that would suggest continuous shielding of defects during regular relaxation cycles (i.e. more particles in the permeate after a backflush and then a slow reduction in intensity) (Chaudhry *et al.*, 2015b). Clearly, further research is needed to investigate the impact of the potential of these transient losses of integrity on overall water quality.

7.4 ASSESSING LIKELIHOODS OF MBR HAZARDOUS EVENTS

Techniques for quantitatively assessing the likelihoods of specific hazardous events could be investigated including the use of historical data such as weather patterns and frequencies of power failures or mechanical malfunctions. An alternative approach is by the use of available mechanical reliability measures such as critical component analysis methodology (Eisenberg *et al.*, 1998, 2001; Olivieri *et al.*, 1996; Shultz & Parr, 1982).

A critical component analysis can be carried out by creating a list of all components in a facility and then categorising the components by treatment unit, component and subcomponent. Data are collected for all planned and unplanned maintenance events and then used to compute performance statistics for treatment units and for individual components in the treatment system. The performance statistics describe the expected time between failures for treatment units, the overall mean time between failures of components, and the fraction of time that a unit or component was operating, either including or excluding preventative maintenance.

This type of analysis provides a foundation from which an assessment of the inherent reliability of a treatment system may be made. For example, if it can be demonstrated that a treatment facility is operational nearly 100% of the time on a long-term basis, plant performance data may be used to evaluate the probability that the effluent will meet a specified set of criteria. Otherwise, it may be necessary to investigate if and how component failures impact treatment plant effluent quality.

The established engineering parameters mean time between failures (MTBF, a function of reliability) and mean time to repair (MTTR, a function of availability) may be used to calculate the operational availability (A_o, the probability that an item is in an operable state at any time) as shown in Equation 7.1.

Determination of operational availability from MTBF and MTTR

$$A_o = \frac{\text{MTBF}}{\text{MTBF} + \text{MTTR}} \tag{7.1}$$

Reliability of machinery can be derived through parametric models to serve as population models for failure times arising from a wide range of products and failure mechanisms. Weibull statistics provide a life distribution model, which has been useful in many engineering applications to derive failure rates (Carrasco *et al.*, 2008; Davis *et al.*, 2008; Erumban, 2008). The two-parameter Weibull distribution function has been used to derive a reliability function $R(t)$ given by the cumulative form (Equation 7.2).

Reliability function $R(t)$ from the cumulative form of the Weibull distribution

$$R(t) = \int_t^{\infty} f(x)\,\mathrm{d}x = \mathrm{e}^{-(x/\beta)^{\alpha}} \quad t \geq 0,\ a > 0,\ b > 0 \tag{7.2}$$

where α is the Weibull shape parameter, β is the scale parameter, and t is the time of operation.

The scale parameter β has the same units as t and the shape parameter α is a dimensionless quantity. When $\alpha = 1$, representing a constant failure rate, the reliability model is simplified to the form presented in Equation 7.3.

Reliability function $R(t)$ for a constant failure rate ($\alpha = 1$)

$$R(t) = \mathrm{e}^{-\lambda . t} \text{ with the failure rate } (\lambda),\ \lambda(t) = 1/\beta = 1/\text{MTBF} \tag{7.3}$$

Process reliability for an MBR system may be engineered through reliability assessments made using Weibull distribution databases for all mechanical components (Moore *et al.*, 2008). Historical MTTR for each component can be tracked and updated through corrective maintenance work orders. The MTBF and MTTR values analysed may also form part of an asset replacement strategy.

7.5 MANAGEMENT OF HAZARDOUS EVENTS THROUGH ENGINEERED REDUNDANCY AND MULTIPLE BARRIER TREATMENT SYSTEMS

It is generally not possible to guarantee the prevention of many types of hazardous events. Accordingly, systems must be designed with a degree of robustness to manage impacts to ongoing operation as well as risks to human health and the environment when hazardous events occur. Important concepts for managing hazardous events are the incorporation of multiple barriers in the design and the establishment of a monitoring program that is suitable to constantly assess proper system performance. The selection of multiple barriers and a monitoring program will depend on the context in which an MBR is employed. Meeting effluent discharge standards will require a different management approach to potential hazardous events as compared to practices where MBR effluents are used for non-potable or potable reuse applications given the higher degree of potential exposure to public health.

Multiple barriers in water treatment and reclamation are aimed at ensuring that performance goals are met by (1) expanding the variety of contaminants a process train can effectively address by providing engineered redundancy (i.e. robustness) and (2) by improving the extent of consistent performance of a unit process to attenuate a contaminant (i.e. reliability) (National Research Council, 2012).

Even when true redundancy is not provided, multiple barriers can reduce the consequences of hazardous events when they do occur. The independence of multiple barriers is a key aspect of system reliability and safety (Drewes & Khan, 2011). For example, to mitigate the risk from pathogen exposure, all MBRs usually employ a disinfection step either using a chlorine-based disinfectant or UV irradiation, in addition to the MF or UF membrane that serves as a barrier to pathogens.

The extent of system performance and water quality monitoring will depend on project-specific water quality objectives and the potential impact from hazardous events. An idealized monitoring programme would measure critical process parameters and microbial and chemical contaminants in real time in the finished product water. However, real-time monitoring comes at significant capital and maintenance expenses and needs to be balanced against the estimated likelihood of certain hazardous events.

Monitoring requirements usually become more stringent (e.g. more frequent and broader in scope) as the potential for human contact with the reclaimed water increases (e.g. non-restricted irrigation of public parks; indirect potable reuse). Monitoring programmes to assure that water quality requirements are met most commonly include effluent turbidity and residual chlorine. Operational parameters that are measured in real time include flow measurements, TMP, bioreactor tank levels, dissolved oxygen concentration of the bioreactor, as well as status of pumps and critical valves (i.e. on/off). These parameters are recorded in the Supervisory Control and Data Acquisition System (SCADA) of the treatment facility and usually linked to certain threshold levels. An exceedance of these threshold levels that might be caused by a hazardous event will result in shut-down of the system to mitigate the negative impact of that event.

7.6 CONCLUSIONS AND FUTURE OUTLOOK

The possibility or frequency of hazardous events plays a significant role in defining the overall risks to health and the environment from wastewater treatment by MBRs. Potential hazardous events are diverse and even prediction of a comprehensive suite of events that may disrupt and MBR performance is difficult. However, important examples include rapid and/or significant changes in influent water quality impacting the biological integrity and physical damage, which may impact membrane integrity. Hazardous events may lead to drastic loss of treatment performance by impeding microbial degradation processes or by impeding the retention of particulate substances by membranes.

Formalised risk assessment procedures, aimed at rating potential hazardous events in terms of their likelihood and consequences are well suited for assessing MBR system vulnerabilities. Existing risk management approaches including the multiple barrier approach and a focus on monitoring the performance of operational parameters can be effective means for managing these vulnerabilities for the protection of health and the environment.

As important as the proper assessment and management of system failures and risks may be, surprisingly little attention has been paid to this topic for MBRs. More comprehensive future risk management will benefit from focused investigation of a wider range of potential failure modes, their consequences particularly in terms of their impacts to final water quality, and statistical descriptions of their likelihood. These factors will enable informed assessment of risks and better direct efforts towards more effective risk management.

REFERENCES

Ayala D. F., Ferre V. and Judd S. J. (2011). Membrane life estimation in full-scale immersed membrane bioreactors. *Journal of Membrane Science*, **378**(1–2), 95–100.

Beauchamp N., Lence B. J. and Bouchard C. (2010). Technical hazard identification in water treatment using fault tree analysis. *Canadian Journal of Civil Engineering*, **37**(6), 897–906.

Beler Baykal B., Orhon D. and Artan N. (1990). Implications of the task group model – II. Response to intermittent loadings. *Water Research*, **24**(10), 1259–1268.

Beyer M., Lohrengel B. and Nghiem L. D. (2010). Membrane fouling and chemical cleaning in water recycling applications. *Desalination*, **250**(3), 977–981.

Branch A., Trinh T., Carvajal G., Leslie G., Coleman H. M., Stuetz R. M., Drewes J. E., Khan S. J. and Le-Clech P. (2016). Hazardous events in membrane bioreactors – Part 3: impacts on microorganism log removal efficiencies. *Journal of Membrane Science*, **497**, 514–523.

Campos J. L., Mosquera-Corral A., Sánchez M., Méndez R. and Lema J. M. (2002). Nitrification in saline wastewater with high ammonia concentration in an activated sludge unit. *Water Research*, **36**(10), 2555–2560.

Carrasco J. M. R., Ortega E. M. M. and Paula G. A. (2008). Log-modified Weibull regression models with censored data: sensitivity and residual analysis. *Computational Statistics & Data Analysis*, **52**(8), 4021–4039.

Chaudhry R. M., Holloway R. W., Cath T. Y. and Nelson K. L. (2015a). Impact of virus surface characteristics on removal mechanisms within membrane bioreactors. *Water Research*, **84**, 144–152.

Chaudhry R. M., Nelson K. L. and Drewes J. E. (2015b). Mechanisms of pathogenic virus removal in a full-scale membrane bioreactor. *Environmental Science & Technology*, **49**(5), 2815–2822.

Coello Oviedo M. D., López-Ramírez J. A., Sales Márquez D. and Quiroga Alonso J. M. (2003). Evolution of an activated sludge system under starvation conditions. *Chemical Engineering Journal*, **94**(2), 139–146.

Dan N. P., Visvanathan C. and Basu B. (2003). Comparative evaluation of yeast and bacterial treatment of high salinity wastewater based on biokinetic coefficients. *Bioresource Technology*, **87**(1), 51–56.

Davis P., De Silva D., Marlow D., Moglia M., Gould S. and Burn S. (2008). Failure prediction and optimal scheduling of replacements in asbestos cement water pipes. *Journal of Water Supply Research and Technology-Aqua*, **57**(4), 239–252.

Drewes J. E. and Khan S. J. (2011). Chapter 16: Water reuse for drinking water augmentation. In: Water Quality & Treatment: A Handbook on Drinking Water, J. K. Edzwald (ed.), 6th edn, McGraw-Hill Professional, New York, NY., 16.1–16.48.

Eisenberg D., Olivieri A., Soller J. and Gagliardo P. (1998). Reliability Analysis of an Advanced Water Treatment Facility. ASCE National Conference on Environmental Engineering American Society of Civil Engineers, 7–10, June 1998 Chicago, Illinois. https://www.abebooks.com/servlet/BookDetailsPL?bi=15298657667&searchurl=tn%3DNational%2BConference%2BEnvironmental%2BEngineering%26sortby%3D17&cm_sp=snippet-_-srp1-_-title6.

Eisenberg D., Soller J., Sakaji R. and Olivieri A. (2001). A methodology to evaluate water and wastewater treatment plant reliability. *Water Science and Technology*, **43**(10), 91–99.

Erumban A. A. (2008). Lifetimes of machinery and equipment: evidence from Dutch manufacturing. *Review of Income and Wealth*, **54**(2), 237–268.

Farahbakhsh K. and Smith D. W. (2004). Removal of coliphages in secondary effluent by microfiltration – mechanisms of removal and impact of operating parameters. *Water Research*, **38**(3), 585–592.

Friedler E., Shwartzman Z. and Ostfeld A. (2008). Assessment of the reliability of an on-site MBR system for greywater treatment and the associated aesthetic and health risks. *Water Science and Technology*, **57**(7), 1103–1110.

Gaudy A. F., Jr. and Engelbrecht R. S. (1961). Quantitative and qualitative shock loading of activated sludge systems. *Journal Water Pollution Control Federation*, **33**(8), 800–816.

Hajibabania S., Antony A., Leslie G. and Le-Clech P. (2012). Relative impact of fouling and cleaning on PVDF membrane hydraulic performances. *Separation and Purification Technology*, **90**(0), 204–212.

Hart J., Rundle H. and Stokes L. (2003). Operation of an attached growth nitrifying reactor under shock loads of ammonia and activated sludge suspended solids. *Proceedings of the Water Environment Federation*, **2003**(6), 70–81.

Hirani Z. M., DeCarolis J. F., Adham S. S. and Jacangelo J. G. (2010). Peak flux performance and microbial removal by selected membrane bioreactor systems. *Water Research*, **44**(8), 2431–2440.

Hirani Z. M., DeCarolis J. F., Lehman G., Adham S. S. and Jacangelo J. G. (2012). Occurrence and removal of microbial indicators from municipal wastewaters by nine different MBR systems. *Water Science and Technology*, **66**(4), 865–871.

Hrudey S. E. and Hrudey E. J. (2007). Published case studies of waterborne disease outbreaks – evidence of a recurrent threat. *Water Environment Research*, **79**(3), 233–245.

Judd S. and Judd C. (2011). The MBR book – Principles and Applications of Membrane Bioreactors for Water and Wastewater Treatment, 2nd edn. Elsevier, Great Britain.

Kjelleberg S., Hermansson M. and Marden P. (1987). The transient phase between growth and non-growth of heterotrophic bacteria, with emphasis on the marine-environment. *Annual Review Microbiology*, **41**, 25–49.

Lazarove V., Bonroy J.-L. and Richard J.-L. (2008). Reliability of operation and failure management of membrane wastewater treatment. *Water Practice & Technology*, **3**(2).

Le-Clech P., Fane A., Leslie G. and Childress A. (2005). MBR focus: the operators' perspective. *Filtration + Separation*, **42**(5), 20–23.

Li Z. H., Kuba T. and Kusuda T. (2006). The influence of starvation phase on the properties and the development of aerobic granules. *Enzyme and Microbial Technology*, **38**(5), 670–674.

Lv W., Zheng X., Yang M., Zhang Y., Liu Y. and Liu J. (2006). Virus removal performance and mechanism of a submerged membrane bioreactor. *Process Biochemistry*, **41**(2), 299–304.

Manickam T. S. and Gaudy A. F., Jr., (1985). Comparison of activated sludge response to quantitative, hydraulic, and combined shock for the same increases in mass loading. *Water Pollution Control Federation*, **57**(3), 241–252.

Marti E., Monclús H., Jofre J., Rodriguez-Roda I., Comas J. and Balcázar J. L. (2011). Removal of microbial indicators from municipal wastewater by a membrane bioreactor (MBR). *Bioresource Technology*, **102**(8), 5004–5009.

Moore M. R., Chapman H., Leusch F., Tan B., Drew R., Frangos J., Khan S., Leslie G. and Shaw G. (2008). Recycled Water Quality: A Guide to Determining. Monitoring and Achieving Safe Concentrations of Chemicals in Recycled Water, Adelaide, SA, Australia, National Environment Protection Council – Adelaide, pp. 176.

National Research Council (2012). Water Reuse: Potential for Expanding the Nation's Water Supply Through Reuse of Municipal Wastewater. National Academies Press, Washington, D.C.

National Resource Management Ministerial Council, Environment Protection and Heritage Council and Australian Health Ministers Conference (2006). Australian Guidelines for Water Recycling: Managing Health & Environmental Risks (Phase 1). Australian Government, Canberra, ACT, Australia.

National Water Quality Management Strategy (2011). Australian Drinking Water Guidelines. National Health and Medical Research Council. Natural Resource Management Ministerial Council - Government of Australia, Canberra.

Ng H. Y., Ong S. L. and Ng W. J. (2005). Effects of sodium chloride on the performance of a sequencing batch reactor. *Journal of Environmental Engineering*, **131**(11), 1557–1564.

Nishimori K., Tokushima M., Oketani S. and Churchouse S. (2010). Performance and quality analysis of membrane cartridges used in long-term operation. *Water Science and Technology*, **62**(3), 518–524.

Normand T. and Perdrieux S. (1981). Dynamics of activated sludge in terms of organic carbon. *Journal Water Pollution Control Federation*, **53**(5), 576–586.

Olivieri A. W., Eisenberg D. M., Cooper R. C., Tchobanoglous G. and Gagliardo P. (1996). Recycled water – A source of potable water: city of San Diego health effects study. *Water Science and Technology*, **33**(10–11), 285–296.

Phan H. V., Hai F. I., McDonald J. A., Khan S. J., van de Merwe J. P., Leusch F. D. L., Zhang R., Price W. E., Broeckmann A. and Nghiem L. D. (2015). Impact of hazardous events on the removal of nutrients and trace organic contaminants by an anoxic–aerobic membrane bioreactor receiving real wastewater. *Bioresource Technology*, **192**, 192–201.

Rizak S. and Hrudey S. E. (2007). Achieving safe drinking water – risk management based on experience and reality. *Environment Review*, **15**, 169–174.

Saleh M. M. and Gaudy A. F., Jr., (1978). Shock load response of activated sludge with constant recycle sludge concentration. *Journal Water Pollution Control Federation*, **50**(4), 764–774.

Severn R. (2003). Long term operating experience with submerged plate MBRs. *Filtration and Separation*, **40**(7), 28–31.

Shang C., Hiu M. W. and Chen G. (2005). Bacteriophage MS-2 removal by submerged membrane bioreactor. *Water Research*, **39**(17), 4211–4219.

Shultz D. and Parr V. (1982). Evaluation and Documentation of Mechanical Reliability of Conventional Wastewater Treatment Plant Components, EPA/600/2-82-044, US Environmental Protection Agency, Washington, DC.

Simon A., Nghiem L. D., Le-Clech P., Khan S. J. and Drewes J. E. (2009). Effects of membrane degradation on the removal of pharmaceutically active compounds (PhACs) by NF/RO filtration processes. *Journal of Membrane Science*, **340**(1–2), 16–25.

Stevik T. K., Aa K., Ausland G. and Hanssen J. F. (2004). Retention and removal of pathogenic bacteria in wastewater percolating through porous media: a review. *Water Research*, **38**(6), 1355–1367.

Trinh T., Branch A., Hambly A. C., Carvajal G., Coleman H. M., Stuetz R. M., Drewes J. E., Le-Clech P. and Khan S. J. (2016). Hazardous events in membrane bioreactors – Part 1: impacts on key operational and bulk water quality parameters. *Journal of Membrane Science*, **497**, 494–503.

Ueda T. and Horan N. J. (2000). Fate of indigenous bacteriophage in a membrane bioreactor. *Water Research*, **34**(7), 2151–2159.

Urbain V., Pys E., Block J. C. and Manem J. (1993). Composition and activity of activated sludge under starvation conditions. *Environmental Technology*, **14**(8), 731–740.

van Bentem A. G. N., Petri C. P., Schyns P. F. T. and Roest H. F. v. d. (2007). Membrane Bioreactors: Operation and Results of an MBR Wastewater Treatment Plant. IWA Publishing, London.

van Bentem A. G. N., Nijman N., Schyns P. F. T. and Petri C. P. (2010). MBR Varsseveld: 5 years of operational experience. *Water Practice & Technology*, **5**(1), 1–8.

van den Akker B., Trinh T., Coleman H., Stuetz R., Le-Clech P., Drewes J. E. and Khan S. (2012). Validation of a full-scale membrane bioreactor for water recycling: characterising process variability. In: Ozwater 2012 Australian Water Association (AWA), NSW, Australia.

van Voorthuizen E. M., Ashbolt N. J. and Schäfer A. I. (2001). Role of hydrophobic and electrostatic interactions for initial enteric virus retention by MF membranes. *Journal of Membrane Science*, **194**(1), 69–79.

World Health Organization (2011). Guidelines for Drinking-Water Quality, 4th edn. WHO, Geneva, Switzerland.

Wu J., Li H. and Huang X. (2010). Indigenous somatic coliphage removal from a real municipal wastewater by a submerged membrane bioreactor. *Water Research*, **44**(6), 1853–1862.

Yogalakshmi K. N., Joseph K. and Visvanathan C. (2007). Response of membrane bioreactor to feed starvation shock load. *Asian Journal of Water, Environment and Pollution*, **4**(2), 81–86.

Zhang K. and Farahbakhsh K. (2007). Removal of native coliphages and coliform bacteria from municipal wastewater by various wastewater treatment processes: implications to water reuse. *Water Research*, **41**(12), 2816–2824.

Zheng X. and Liu J. X. (2006). Mechanism investigation of virus removal in a membrane bioreactor. *Water Science and Technology*, **6**(6), 51–59.

Chapter 8

Cost benefit and environmental life cycle assessment

Gonzalo Rodriguez-García[1*], *María Molinos-Senante*[2], *Sara Gabarron*[3], *Carolina Alfonsín*[1], *Almudena Hospido*[1], *Lluis Corominas*[4], *Francesc Hernandez-Sancho*[2], *Francisco Omil*[1], *Gumersindo Feijoo*[1], *Ramón Sala-Garrido*[2], *Ignasi Rodriguez-Roda*[3,4] *and Maria Teresa Moreira*[1**]

[1]*Department of Chemical Engineering, School of Engineering, University of Santiago de Compostela, Santiago de Compostela, Spain*
[2]*Department of Applied Economics II, Faculty of Economics, University of Valencia, Valencia, Spain*
[3]*LEQUiA (Laboratory of Chemical and Environmental Engineering), University of Girona, Girona, Spain*
[4]*ICRA (Catalan Institute for Water Research), University of Girona, Girona, Spain*
***Corresponding author: maite.moreira@usc.es*

ABSTRACT

This chapter evaluates the economic and environmental implications of membrane bioreactors (MBR) using cost–benefit analysis (CBA) and life cycle assessment (LCA). A brief introduction to these methodologies and the main developments and achievements when applied to MBR is given before the presentation of two case studies. In the first case study, CBA and LCA were applied for the assessment of seven wastewater treatment plants using MBR and other membrane technologies. The results highlighted the benefits of

*Current affiliation: Department of Experimental Thermal Fluid Dynamics, Institute of Fluid Dynamics, Helmholtz-Zentrum Dresden-Rossendorf, Dresden, Germany.

 Membrane Biological Reactors: Theory, Modeling, Design, Management and Applications to Wastewater Reuse
Editor(s): Faisal I. Hai, Kazuo Yamamoto and Chung-Hak Lee
doi: 10.2166/9781780409177_0227

removal of substances responsible for eutrophication and the relevance of the scale factor regarding electricity use, one of the main environmental and economic costs. In the second case study, LCA alone was applied for the environmental assessment of four novel MBR configurations at pilot-plant scale focusing on the removal of pharmaceutical and personal care products, of which only hormones were found to be relevant in terms of toxicity terms due to their higher concentrations in the influent.

8.1 INTRODUCTION

This chapter analyzes the economic costs and environmental impacts of membrane bioreactor (MBR) design and operation. A better knowledge of them allows designers and plant operators to make membrane technologies efficient and sustainable.

As a novel technology, MBRs are usually regarded as effective yet expensive. In order to be competitive they need to overcome the decades of optimization that conventional technologies have enjoyed. For a cost–benefit analysis (CBA) operational costs need to be considered; but for projected plants, estimations are required. A brief review of current methodologies for the evaluation of MBR costs is presented as well as a tool to get a first estimation with minimum information requirements. Benefits from water reuse and from the removal of pollutants are also included, the shadow prices methodology being employed for the latter.

As a remediation technology, MBRs possess some environmental benefits. This is due to the aforementioned pollution removal from the influent. In order to achieve such removal, these reactors require chemicals and electricity and produce undesirable outputs such as sludge. In some cases, pollutants are not removed from water but transferred to other environmental compartments such as air and soil. In order to have a broader perspective of the environmental impacts associated with this technology, a holistic tool such as life cycle assessment (LCA) is required. In this chapter, the application of CBA and LCA to MBR is explained following the structure below.

Section 8.2 presents the methodology applied for the economic assessment and introduces the two tools used for that purpose: cost estimation based on plant size and removal efficiencies and shadow prices of the pollutants removed. Section 8.3 introduces the concepts of product life cycle and the different stages of a LCA study and reviews the main findings in LCA studies of wastewater treatment, focusing on membrane technologies. Section 8.4 compares four MBRs and three WwTP with other membrane technologies, namely, filtration (F), ultrafiltration (UF) and reverse osmosis (RO), according to the methodologies presented in the previous sections. Section 8.5 compares the environmental profile of four different forms of pilot-scale MBR, namely, a conventional submerged MBR, a pre-denitrification MBR, a modified University of Cape Town (UCT) MBR and an anaerobic hybrid MBR, from the context of pharmaceutical and personal care products (PPCP) removal.

8.2 COST–BENEFIT ANALYSIS

Economic feasibility assessment plays a vital role in the decision-making process enabling the selection of the most suitable option from a set of technological alternatives. In this context, one of the most popular tools is CBA (Hutton & Haller, 2004). It ensures the economic rationality of investments guaranteeing that the benefits of action outweigh the costs. In the traditional approach, only benefits and costs with market value are considered in the analysis, i.e., only internal impacts are included. However, it is interesting to highlight that pollution removal by WwTP involves significant environmental benefits that cannot be ignored from an economic perspective. Otherwise there is an underestimation of the benefits. Thus, CBA must include internal and external impacts. This approach is being increasingly recognized in the water sector (Godfrey *et al.*, 2009; Hutton & Haller, 2004; Molinos-Senante *et al.*, 2012) partially because the Water Framework Directive (EC, 2000) proposes that the measures applied to achieve water quality in rivers should consider not only financial costs but also environmental costs and related resources, all in accordance with a detailed and rigorous economic analysis.

CBA estimates the net profit (NP) by comparing benefits and costs (Equation. 8.1). If the NP is higher than zero, then the project is economically feasible, while if the NP is lower than zero, the project is not feasible in economic terms. If several alternatives are assessed, the one with the highest NP value is the most suitable from an economic perspective.

$$\mathrm{NP} = \sum B_i - \sum C_i \tag{8.1}$$

where NP is the net profit; B_i is the value of the benefit item i and C_i is the value of the cost item i.

In the case of investment projects whose duration exceeds one year, such as WwTP, both costs and benefits must be expressed in terms of present value (net present value).

8.2.1 Modeling of operational costs in WwTP and membrane technologies

Cost estimation is a basic requirement to carry out any CBA. For this purpose, cost functions are a useful tool, allowing a better understanding of the cost structure of the process. The statistical method is the most commonly used to develop cost functions in the field of wastewater treatment (Papadopoulos *et al.*, 2007; Sipala *et al.*, 2005). This methodology consists of the assessment of the relationship between the dependent variable (cost) and independent variables (e.g. flow of wastewater treated) by means of ordinary least-squares-regression analysis. Most of the contributions in this field are focused on examining how the flow treated at each facility affects the operational costs (Gonzalez-Serrano *et al.*, 2006; Nogueira *et al.*, 2007). More recently, Hernández-Sancho *et al.* (2011) included other variables such as the efficiency of pollutant removal or the construction year of the facility, developing costs functions for several wastewater treatment technologies including tertiary treatment.

Taking into account that membrane technologies are considered as tertiary treatments, the cost functions developed by Hernández-Sancho *et al.* (2011) are suitable for the empirical application presented in Section 8.4. The greatest advantage of these functions is their differentiation of secondary treatment costs from those corresponding exclusively to tertiary treatment. Moreover, the consideration of the removal efficiencies of pollutants allows the evaluation of how the performance of the facilities affects operational costs.

Costs vary widely depending on design choices and operational parameters (Verrecht *et al.*, 2010a). For example, Côté *et al.* (2004) estimated capital costs for an MBR plant treating 38,000 m^3/d, including direct, indirect and land costs, to be around 1,000 $\$/(m^3/d)$. These costs, for a 75,000 m^3/d plant were around 0.25 $\$/m^3$ (Côté *et al.*, 2005). Regarding total costs, it has been demonstrated that they are also significantly affected by economies of scale. For example, Côté *et al.* (2004) estimated that costs vary from 0.45 $\$/m^3$ for small plants (<19,000 m^3/d) to 0.20 $\$/m^3$ for large plants (<19,000 m^3/d), 33% of them due to electricity use (estimating a value of 3.46 kWh/m^3). However, more recent studies (Verrecht *et al.*, 2010a) show the use of electricity in MBR to have decreased significantly (0.7 and 1.07 kWh/m^3). Despite this reduction, the future trend in electricity costs will be a determining factor for the economic feasibility of MBR.

8.2.2 Calculation of the environmental benefits associated with WwTP: the shadow prices methodology

Externalities are costs or benefits occurring when the actions of firms or individuals have an effect on people other than themselves without economic compensation. According to this definition, the environmental benefits from wastewater treatment can be considered as positive externalities.

Since no market value is associated with these environmental benefits, their integration in CBA requires an estimation of their economic value. From the economic theory, several methodologies have been developed to quantify the economic value of environmental benefits from investment projects. In most applications related to water resources, the quantification of externalities has been made using stated preference methods (Bateman *et al.*, 2006; Birol *et al.*, 2006). In spite of their usefulness they are not free of limitations and criticisms such as instrumental biases, embedding effect, lexicographic preferences and especially high costs (Boyle *et al.*, 1993; Shabman & Stephenson, 2000). As an alternative, the concept of distance function has been suggested to estimate the economic value of undesirable outputs without market value (Färe *et al.*, 1993, 2001, 2006). A distance function generalizes the concept of conventional production functions and measures the difference between the outputs produced in the process under study and the outputs of the most efficient process. The relationship between the distance function and the revenue function allows the quantifying of the shadow prices for the undesirable outputs or pollutants generated in the production process. The empirical estimation of the distance function is made through mathematical optimization using linear programming.

Hernández-Sancho *et al.* (2010) adapted the distance function methodology to the field of wastewater treatment, adaptation for which Molinos-Senante *et al.* (2011a, b) have made successive developments. In all cases, wastewater treatment has been considered as a productive process in which a desirable output (treated water) is obtained together with a series of undesirable outputs (suspended solids, nitrogen, phosphorus and organic matter) using inputs (costs). It should be noted that pollutants extracted from wastewater are considered undesirable outputs because if discharged into the environment they would generate negative environmental impacts.

The distance function approach enables to quantify the shadow prices of any pollutant at any WwTP evaluated. Shadow prices represent the environmental damage avoided by preventing the discharge of pollutants. Assuming the current level of pollution to be optimal, marginal cost equals marginal benefit and therefore, shadow prices can be interpreted as an estimation of the environmental benefits from wastewater treatment.

The value of shadow prices obtained is always negative, indicating that, from the perspective of the production process, these prices are not associated with marketable outputs that are able to generate an income. However, from an environmental point of view they can be interpreted positively because they represent the avoided cost, and therefore an estimation of the environmental benefits from wastewater treatment.

A consideration that must be taken into account to calculate the shadow prices of pollutants is that the shadow price of the desirable output must be equal to its market price. In the wastewater treatment context, it means that it is necessary to assign a reference price for the treated water (desirable output). Therefore, in the empirical applications developed by Hernández-Sancho *et al.* (2010) and Molinos-Senante *et al.* (2010), various prices have been assigned to treated water depending on its final function.

The quantification of the environmental benefits from wastewater treatment through the distance function approach presents the following advantages:

- Authorities may use the information provided by shadow prices to set rates for using environmental services.

- The estimation of the environmental benefits in monetary terms may help society understand the benefits generated as a result of investment projects aimed to improve the quality of the effluent from wastewater treatment plants.
- This method is much cheaper than the conventional stated preference based on expensive surveying processes.

8.3 LIFE CYCLE ASSESSMENT

8.3.1 Life cycle assessment methodology

LCA is a holistic technique for assessing the environmental aspects and potential impacts associated with all the stages of a product, process or service from cradle to grave and not just those considered to be the core of the process. The origins of LCA can be traced back to the late 1960s but it was not until late 1990s that the International Standards Organization (ISO) developed a standard procedure for LCA in their 14000 series (ISO, 2006a, b). The current 14040 and 14044 ISO standards establish a general methodology but they do not detail how it should be applied for each product or process. This section describes the different stages required to perform a LCA according to the ISO standards and presents a short literature review regarding the application of LCA in WwTPs and MBR. Suggested readings for those aiming to know more about LCA methodology include Baumann and Tillman (2004), Life Cycle Initiative (2008), and European Commission-Joint Research Group-Institute for Environment and Sustainability (2010).

8.3.1.1 Goal and scope

The objectives of an LCA need to go beyond indicating which product or process is going to be assessed (ISO, 2006a, b). It is required to clearly state the reasons prompting the assessment and the targeting of the public, and especially important to indicate the expected applications of the study, which should be consistent with the goal and scope.

In order to correctly frame the study, two key questions need to be answered regarding what is going to be assessed. The first one is the definition of the functional unit (FU), to whom all further flows and impacts are going to be referred. It should reflect the function of the product or process under study, e.g. to treat wastewater, and be measurable, e.g. 1 m^3 of treated wastewater. Other functional units have been used in LCA of WwTP: 1 p.e. (Suh & Rousseaux, 2001), annual flow treated in a WwTP (Tillman *et al.*, 1998), or 1 kg PO_4^{3-} removed (Rodriguez-Garcia *et al.*, 2011).

The second question regards the selection of the system boundaries. Within the system under study, all the relevant processes required for providing the aforementioned function should be included. The boundaries can be geographical (where the study takes place); temporal (to which period of time the LCA is referred to and which stages of the life cycle are to be assessed (research and development, material extraction, component manufacturing, use and maintenance and end-of-life management)) and physical, which defines the level of detail of the LCA. There, it is possible to only include foreground sub-systems, i.e. those directly related to the product or process under study, such as the effluent discharge of a WwTP. However, it is far more common to also include background systems, which are those required for the provision of products or services for the foreground sub-system, e.g. the emissions associated with the production of the electricity that would be used in a WwTP (Corominas *et al.*, 2011).

8.3.1.2 Life cycle inventory analysis

A life cycle inventory (LCI) involves data collection and calculation procedures to quantify the environmentally relevant energy and material flows included within the product system under study

(ISO, 2006a). It usually starts by constructing a product system model, normally a flowchart, as defined in the goal and scope and is continued by the data collection of all the aforementioned flows. The sources of information commonly include direct measurements and design documents for the foreground processes and scientific literature and LCA databases such as Ecoinvent (Swiss Centre for Life Cycle Inventories, 2012) for the background processes. This is usually the most time-consuming stage (Baumann & Tillman, 2004). It is also an iterative process because as data collection proceeds more information becomes available, and the goal and scope of the study might change accordingly (ISO, 2006a).

8.3.1.3 Life cycle impact assessment

Once the inventory data are collected, it is necessary to calculate the potential impacts associated with the different flows of the system. For that purpose, the ISO standards (ISO, 2006a, b) define three mandatory stages (1–3) plus two optional ones (4 and 5):

(1) *Selection of impact categories*: Impact categories are undesired environmental effects associated with the release of the substances included in the LCI. They can be mid-point ones, those reaching a well-defined effect on the cause-and-effect chains such as global warming or eutrophication; or end-point ones, covering broader effects such as damage to ecosystem diversity or to human health.
(2) *Classification of the different emissions according to the impact category they affect*: Some substances can affect several impact categories.
(3) *Characterization*: Quantification of the impact associated with each emission. The impact of each substance is related to a reference substance for each impact mid-point categories, e.g. CO_2 for global warming.
(4) *Normalization*: Compares the potential impact associated with the system under study with those associated with a reference scenario, such as the impact generated by the whole world in 1990.
(5) *Weighting*: Aggregates all impact categories into one single score using subjective criteria such as expert opinion, monetization, or policy goals.

Ready-to-use life cycle impact assessment (LCIA) methods such as CML (Guinée *et al.*, 2002), EDIP97 (Wenzel *et al.*, 1997) or Impact 2002+ (Jolliet *et al.*, 2003), present a set of impact categories such as eutrophication, global warming, ecotoxicity and the characterization factors (CF) related to them for a large number of substances. Thus, most LCA practitioners would only need to choose which methodology suits them better and whether they require normalized and weighted results or not, without going in-depth into the previously described impact assessment stages (Baumann & Tillman, 2004).

8.3.1.4 Interpretation of results

Due to the iterative nature of LCA, findings from the LCI and the LCIA can be discussed independently to reach conclusions and recommendations consistent with the goal and scope of the study. It is also possible that results from one of the stages suggest new approaches in the other one or even define a whole new scope for the study. Interpretation usually goes beyond the mere description of results by also assessing their validity through sensibility analysis, uncertainty analysis or the analysis of alternative scenarios (Baumann & Tillman, 2004).

8.3.2 Life cycle assessment in WwTP and membrane technologies

LCA has been applied to the water cycle in general and wastewater treatment in particular since the early stages of development of the methodology (Emmerson *et al.*, 1995; Flückiger & Gubler, 1994). Reviews of these 18 years of research can be found in Larsen *et al.* (2007) and Corominas *et al.* (2011). Before focusing on the available literature concerning MBRs, the most important considerations concerning WwTP in general are given below.

Regarding the goal and scope:

- The selection of the FU is a key factor when it comes to defining which treatment alternatives are sustainable (Rodriguez-Garcia *et al.*, 2011).
- Background emissions have to be considered. Emissions related to electricity production are usually the dominant source for many impact categories (Emmerson *et al.*, 1995; Hospido *et al.*, 2012).
- Sludge treatment and disposal cannot be neglected since it can be a source of several emissions (N_2O, heavy metals) but at the same time it prevents the production of fertilizers or the generation of electricity by non-renewable sources (Hospido *et al.*, 2005; Houillon & Jolliet, 2005).

Regarding the inventory analysis:

- Differences between seasons, humid and dry, are not relevant (Hospido *et al.*, 2004).
- The quality of the influent plays a major role in the operation of the plant and not only in the quality of the effluent (Rodriguez-Garcia *et al.*, 2011).

Regarding the impact assessment:

- Eutrophication is, even after treatment, the single most relevant impact category (Larsen *et al.*, 2007). However, toxicity impacts associated with the discharged water might also become relevant as more research is conducted regarding emerging pollutants (Larsen *et al.*, 2010).
- The reduction of the impact in the eutrophication category increases the impacts in other categories (Foley, 2009; Godin *et al.*, 2012).

A large number of studies have focused on the assessment of different wastewater treatment technologies (Foley *et al.*, 2010; Muñoz *et al.*, 2009), plant configurations (Gallego *et al.*, 2008; Hospido *et al.*, 2008) and operating conditions (Clauson-Kaas *et al.*, 2006; Lundie *et al.*, 2004). In particular, membrane technologies have been included in a number of studies; typically to compare them to other water reuse technologies.

Tangsubkul *et al.* (2005) compared two advanced water recycling technologies: (a) an MBR followed by RO, (b) continuous microfiltration (CMF) with ozonation as a pretreatment and a chemical disinfection as posttreatment, and a simple one, (c) wastewater stabilization pond (WwSP). In order to assess the potential reuse of the water for agricultural purposes, a custom-made impact category, salinization potential, was included. However, the impact of pathogens, expected to play an important role when assessing water reuse technologies, was not included since its characterization factors were not developed until later (Larsen *et al.*, 2009). The combined MBR+RO configuration was considered the least appropriate technology for water reuse due to the excessive removal of ions by RO, potentially causing soil dispersion and thus making the soil unable to retain water or nutrients. Considering conventional impact categories, the MBR presented an intermediate environmental profile between the most environmentally friendly option, WwSP and the alternative with the highest impact, CMF due to its average sludge-related impact and high electricity consumption.

Subsequently, Tangsubkul *et al.* (2006) analyzed the effect that operational conditions had on the environmental profile of a microfiltration membrane. Operating at a low flux with a high maximum trans-membrane pressure resulted in the best environmental profile. The sensitivity analysis indicated that within the low flux range, the chemical cleaning frequency could affect the environmental performance of the whole process.

The study performed by Ortiz *et al.* (2007) was focused on the comparison of conventional technologies with MBR. The authors assessed the effect of stepping up of a conventional activated sludge (CAS) system to an external MBR, an internal MBR, or a UF post treatment. Both MBR configurations and UF produced a better effluent quality than CAS, but the higher electricity use resulted in a higher overall impact. The external MBR presented slightly higher impacts than the internal one due to its higher electricity use, which was found to be the key aspect when comparing these technologies. Thus, cleaner electricity mixes could make tertiary treatments more attractive from an environmental point of view. In the same topic, Memon *et al.* (2007) compared MBR and membrane chemical reactors (MCR) with other natural treatments for the recycling of graywater: reed beds and a green roof water recycling system. Twenty different scales for each technology were tested, the results being consistent among them. The construction phase was not found to be very relevant for membrane processes but MBR presented a higher overall environmental profile due to their electricity consumption.

Memon *et al.* (2007) compared MBR and membrane chemical reactors (MCR) with other natural treatments for the recycling of graywater: reed beds and a green roof water recycling system. Twenty different scales for each technology were tested, the results being consistent among them. The construction phase was not found to be very relevant for membrane processes but MBR presented a higher overall environmental profile due to their electricity consumption.

With regard to micropollutants removal, Wenzel *et al.* (2008) weighted the trade-offs associated with advanced wastewater treatment technologies for their removal. Unlike ozonation and MBR, the benefits associated with the removal of toxic compounds through sand filtration were found to be higher than the environmental losses associated with the process itself. However, no information was given regarding which impact sources were the most relevant ones. If more heavy metals than the ones appearing in the EU Water Framework (EC, 2000) were included in the analysis, MBR would produce a net environmental benefit, but it would still be behind sand filtration. The same group assessed the technical, economic and environmental performance of the technologies presented in Wenzel *et al.* (2008), including also ultraviolet disinfection (UV) and UV + advanced oxidation (Høibye *et al.*, 2008). Again, MBR appeared as the least desirable option due to its high electricity use, about 3.5 times higher than that for UV+ozonation. In environmental terms, both technologies are far behind sand filtration, considered again the most suitable alternative.

The study of Hospido *et al.* (2012) was focused on the differences between MBR configurations: (a) CAS with an internal membrane, (b) modified Lutzak-Ettinger (MLE) coupled with an external membrane, (c) modified UCT reactor coupled with an external membrane, and (d) upflow anaerobic sludge blanket (UASB) reactor followed by a hybrid aerobic reactor and an externally submerged membrane. The same items, electricity use and sludge disposal, were found to be the most relevant ones in all four technologies. Contrary to most studies on LCA in the wastewater sector until then, it was possible to appreciate from this study that, in general terms, the more advanced the technology, the lower the overall environmental impact. This was related to the removal of more pollutants (N and P), which reduced the eutrophication impact, and a more efficient use of electricity, which reduces all other impact categories assessed since, as in previous studies, electricity was found to be a major source of impact.

In another comparison between CAS and MBR, Lazarova *et al.* (2012) found their eutrophication, human toxicity, and greenhouse gases (GHG) to be comparable. MBR presented lower impacts in aquatic toxicity, even when the authors stressed that LCA did not fully cover the benefits of MBRs in terms of water sanitation: higher micropollutant removal efficiencies and a lower microbial fingerprint than river water. However, it is necessary to mention that the LCA community has already addressed the limitations of LCA in terms of capturing the benefit of MBR. See, for example, discussion related to PPCP removal in Section 8.5, or microbial influent quality in Larsen *et al.* (2009).

Kalbar *et al.* (2012) worked on the integration of different indicators. These authors conducted a multi-criteria decision analysis (MCDA) for wastewater treatment technologies for high-rise buildings in urban India. Two LCA criteria, eutrophication and global warming potential (GWP), and one life cycle costing (LCC) criteria, net present worth, were included along with other five. Compared to sequencing batch reactors (SBR) and CAS, MBR was the preferred alternative. Its lower land and workforce requirements, its higher reliability, and its better effluent quality made up for its higher capital, and operation and maintenance costs.

An emerging technology that has received some attention over the past five years is anaerobic MBR (AnMBR). In the first of a series of four publications, Pretel *et al.* (2013) evaluated the effect of temperature on submerged AnMBRs. They concluded that the increased energy production for operating at higher temperatures did not justify the additional energy and environmental costs. In a second paper, Pretel *et al.* (2015) elucidated how operational decisions influenced the technical, economic, and environmental sustainability of AnMBR. They developed 43,200 scenarios, varying solid retention time (SRT), mixed liquor suspended solids (MLSS) concentration, sludge recycling ratio, flux, and specific gas demand (SGD), as well as different alternatives for sludge and methane management. For most variables, there was a trade-off between environmental and economic performance, although minimizing SRT improved both concurrently. The two variables affecting the LCA the most were MLSS concentration and transmembrane flux. Finally, they found a number of scenarios at high temperatures where submerged AnMBRs could be net energy producers. In a narrower LCA+LCC evaluation, Pretel *et al.* (2016a) found AnMBR+CAS a more sustainable alternative for nitrogen removal than (a) CAS, (b) Aerobic MBR (AeMBR), or (c) the combination of AnMBR and AeMBR as a posttreatment. However, the latter could be a promising alternative if water is reused for fertigation. In what is to the best of our knowledge the last LCA of the authors to date, Pretel *et al.* (2016b) evaluated the effect of treating in an AnMBR urban wastewater combined with the organic fraction of municipal solid waste. While doing so enhances its environmental performance, higher benefits are obtained by simply increasing the SRT.

AnMBRs were also the target of a study by Smith *et al.* (2014), who compared their environmental and economic profile with another promising alternative in terms of energy recovery: high rate activated sludge with anaerobic digestion (HRAS+AD). They also compared these novel technologies with their more established counterparts, i.e., AeMBR+AD and CAS+AD. When treating medium strength waters, HRAS+AD was the only technology with a positive net energy balance. This was because although AnMBR recovers more energy, it also used four times as much energy. They found significant potential to reduce the energy consumption of AnMBR by improving simultaneously the SGD, gas sparging frequency, and flux. When treating high strength wastewater, AnMBR achieved a more positive net energy balance than HRAS+AD. Although both technologies recovered more energy under these conditions, HRAS also required more energy for aeration, while the AnMBR energy-demand remained almost identical. Nevertheless, AnMBR had higher environmental impacts than HRAS in all the categories evaluated. In the case of GWP, it was due to the presence of dissolved methane in the effluent, while in the other categories it was because of its higher energy demand.

In another ambitious sustainability assessment including both a techno-economic analysis and an LCA, Shoener *et al.* (2016) evaluated 150 AnMBR alternatives. These differed in their type of reactor; the configuration, type, and material of their membranes; besides the recovery and use of methane. Membrane configuration and type, as well as the inclusion of granular activated carbon (GAC), were the parameters affecting both the economic and environmental sustainability the most. A sensitive analysis conducted among the most affordable alternatives suggested that in order to improve the environmental performance of AnMBR, research should prioritize the decrease in cross-flow velocity, the removal of gas sparging, the increase in membrane life, the decrease in upflow velocity for physical media bed expansion, and the development of low-cost physical media for fouling mitigation. No design had both the lowest cost and the lowest environmental impact, although the options with lower cost (below the 10th percentile) had also comparatively lower environmental impacts. Two configurations: (a) cross-flow with multi-tube, and (b) submerged hollow fiber with GAC were less expensive than current state-of-the-art alternatives, and able to achieve energy neutral or positive treatment.

Membrane technologies are usually part of tertiary treatments, when a high-quality effluent is desired, potentially for water reuse. Holloway *et al.* (2016) compared an advanced treatment consisting of a conventional WwTP followed by low-pressure membrane filtration, RO, and ultraviolet advanced oxidation process (UV-AOP) vs a hybrid ultrafiltration osmotic MBR (UFO-MBR). The latter consisted of an MBR followed by two lines: one with a forward osmosis (FO) and an RO line to produce potable water, and another with a UF unit for non-potable water. Plant operation was the main contributor to energy use, GHG, and most of the other emissions evaluated for both alternatives. UFO-MBR showed worse environmental performance, but its difference with the advanced treatment could be reduced, and under certain circumstances surpassed, if the FO permeability was increased and energy from RO was recovered. The addition of a UV-AOP to a UFO-MBR was considered a positive improvement to increase pathogen removal, since it only slightly increased its environmental burden.

Recently, Ioannou-Ttofa *et al.* (2016) assessed the environmental impact of a pilot plant MBR treating 10 m^3/day focusing on its construction and operation. The water and sludge post-treatment impacts were excluded, so the results were not route specific (i.e. whether water was reused for irrigation, released to a body of freshwater, or released to the sea). The authors concluded that the operation of the reactor itself did not impose a direct impact on the local environment. Its environmental impact was attributed to the production of energy instead. This remained true, even with energy mixes with a larger presence of nuclear or renewables. Still, an all-solar energy mix could lead up to a 13-fold reduction of its impacts. Membrane production as well as the construction of the overall unit, had a low environmental contribution, thanks to their high life expectancy.

Following the idea of Smith *et al.* (2014) that AnMBRs can show better environmental profile when treating high strength wastewater, Becker *et al.* (2017) studied the effect of diverting food waste for its co-treatment with different wastewater treatment technologies. These were (a) HRAS followed by different sludge management alternatives: landfilling (LF), composting, AD; (b) CAS+LF (reference scenario); and (c) AnMBR. HRAS+AD and AnMBR were the most attractive based on energy balance alone. However, the benefit of energy cogeneration was offset by the effects of toxicity and ozone depletion. Methane emissions played a major role in terms of GWP for most technologies, be it as fugitive emissions in the case of LF, or as direct emissions of dissolved methane for AnMBR. A 60% recovery of the latter would be required for AnMBR to compete in terms of GWP with HRAS+AS. Co-management of food waste and wastewater could improve the environmental profile of the technologies evaluated, although 40% diversion (percentage of food waste going through co-treatment) would be required for AnMBR to achieve a positive energy balance.

Bertanza *et al.* (2017) conducted an MCDA comparing the MBR and the CAS lines of a WwTP in Brescia (Italy). Altogether, the lower costs, complexity, and energy consumption gave the conventional line a slight advantage over the MBR line, despite its better social acceptance. The overall environmental footprint was similar for both technologies: the lower P removal of the MBR resulted in higher freshwater eutrophication, which in turn was balanced by the higher marine eutrophication of the CAS line due to its lower N removal. Results were clearly dependent on local conditions, not only due to the electricity mix, or the sludge management, but also due to the perception of the stakeholders.

Nevertheless, for this case in particular, MBR would be the preferred solution only if economic aspects weighted low, and social aspects were to be considered of major relevance.

As a water constrained city, Hong Kong (China) is one of the few areas of the world where seawater is commonly used for toilet flushing. Using LCA and LLC, Lam *et al.* (2017) compared the eco-efficiency of this technology for high-rise buildings with a freshwater flushing system, and two MBR alternatives for the treatment of graywater: AeMBR, and anaerobic fluidized-bed MBR (AFMBR). Capital costs were higher for MBR treatment than for direct flushing alternatives, although operational costs made the total lowest for AFMBR. AFMBR also presented the lowest environmental impacts, making it the most promising alternative. This was true whether environmental or economic criteria were considered more important. The reasons for the superior performance of AFMBR over aerobic MBR were its lower sludge generation (less than half in dry mass) and energy use. Unlike aerobic MBR, AFMBR did not require large quantities of electricity for aeration. In addition, AFMBR can produce electricity from the methane generated.

Jeong *et al.* (2018) also examined a graywater reclamation system in the city of Atlanta, Georgia (USA), where submerged MBRs treated water from laundry, showers, faucets, and dishwashers. Their simulation included five single-family house zones (SFZ), and four multi-family apartment building zones (MFZ). The system could reduce the non-potable water demand by 26% on average, higher in SFZ than in MFZ, since in the latter less landscape irrigation occurs. Average reductions of 20% on the electricity consumption could also be achieved (17–49% in SFZ, 32–41% in MFZ). In addition, the LCA score of the reclamation system was 20–41% lower than the current centralized system, mostly due to the lower electricity consumption, with scores stabilizing at higher population densities.

In another evaluation of graywater reuse, Dominguez *et al.* (2018) assessed the treatment of laundry water from a hotel with 70 guests in Santander (Cantabria, Spain). They compared MBR to photocatalysis, and to photovoltaic solar-driven photocatalysis. For the two environmental indicators used: natural resources (NR) and environmental burdens (EB), and the three technologies under assessment, energy was the main source of impact. MBR had the highest use of NR, 46% more than the next alternative, due to the air required for treatment. It causes however, the lowest emissions to air, 28% less than the second best, but higher emissions for water (more than doubling the next competitor). The reason for the latter was the metals present in the sludge, which was incinerated before its final disposal. If sludge were to be composted before its use in land stabilization, there would be very little reduction of this or any other impact, although it could still be considered the best MBR alternative.

In summary, LCAs conducted to date highlight the high-energy demand of AeMBRs as one of their major environmental limitations. With further optimization, AnMBRs are likely to be more competitive. Overall, MBRs have proven to be an alternative for water reuse: they are suitable for graywater treatment and their high micropollutant removal could make them an attractive technology for fertigation and other high-end uses.

8.4 ECONOMIC AND ENVIRONMENTAL PROFILE OF FULL-SCALE MBR

MBRs are a novel though well-established technology present in more than 5000 WwTP around the world with a fast development market in Europe as well as other areas of the world (Lesjean & Hisjes, 2008; Lesjean *et al.*, 2009; Santos & Judd, 2010a, b). In recent years they have received substantial attention from the LCA community, mostly considering them as a tertiary treatment. However, the environmental and economic implications of MBR as the alternative for a secondary treatment are yet to be assessed. In this section, the environmental and economic profile of four full-scale MBR are assessed according to

CBA and LCA methodologies and compared with three plants where membranes are used for tertiary treatment purposes: microfiltration, UF and RO.

8.4.1 Economic profile

The economic assessment of seven full-scale WwTP using membrane technologies, four of them using MBR (WwTP 1–4) as a secondary treatment and three of them using membrane tertiary treatments (WwTP 5–7) is presented here. A brief description of the main characteristics of these plants is included in Table 8.1. All the plants were located in eastern Spain and can be considered representative of Mediterranean plants. Further information regarding WwTP 1–4 and 5–7 can be found in Gabarron *et al.* (2011) and Rodriguez-Garcia *et al.* (2011), respectively.

As a starting point for CBA, WwTP operating costs were quantified through cost functions. Since MBR is a tertiary treatment, two cost functions developed by Hernández-Sancho *et al.* (2011) were used to estimate total operational costs. The first one (Equation. 8.2) estimates the costs set-up for secondary treatment while the second cost function (Equation 8.3) provides information exclusively regarding tertiary treatment.

$$\mathrm{CS} = 2.510 * V^{0.7153} * e^{(0.007\,A+1.455\,\mathrm{BOD}+0.258\,N_{\mathrm{T}}+0.243\,P_{\mathrm{T}})} \quad R^2 = 0.7301 \tag{8.2}$$

$$\mathrm{CT} = 3.7732 * V^{0.7223} * e^{(0.6721\,\mathrm{BOD}+0.1958\,N_{\mathrm{T}}+0.7603\,P_{\mathrm{T}})} \quad R^2 = 0.9029 \tag{8.3}$$

where CS is the total operating cost for secondary treatment in €/year; CT is the total operating cost for tertiary treatment in €/year; V is the total volume of wastewater treated in m^3/year; A is the age of the WwTP in years; BOD is the biochemical oxygen demand removal efficiency; N_{T} is the nitrogen removal efficiency, and P_{T} is the phosphorous removal efficiency.

Figure 8.1(a) shows that operating costs of the evaluated WwTPs are very variable; the maximum being (0.887 €/m^3) three times higher than the minimum (0.315 €/m^3). This is mainly due to the different treatment capacity of the plants since the largest plant (WwTP 7) treats a flow of wastewater 42 times larger than the smallest one (WwTP 1). This is also the reason behind the differences between MBR and non-MBR plants since the latter are always larger than the former.

The weighted average, according to the treated volume of wastewater, is 0.3801 €/m^3, of which 46% is due to secondary treatment and the rest to tertiary treatments.

Once the WwTP operating costs are quantified, the next step in performing the CBA is determining the benefits from the wastewater treatment process. Since membrane technologies obtain a high-quality effluent which can be reused, three scenarios have been analyzed regarding the fate of the treated water: (i) no sale of treated water; (ii) sale of 50% of the treated water and (iii) sale of 100% of the treated water. With this in mind, and referencing a Spanish Environmental Ministry report (MMA, 2007), the value of 0.345 €/m^3 as a market price for treated water was allocated.

The environmental benefits, expressed in monetary units, were estimated using the shadow prices obtained by Hernández-Sancho *et al.* (2010) for the removal of different pollutants for river disposal and reuse: 0.005 and 0.010 €/kg SS removed, 0.098 and 0.140 for COD, 16.353 and 26.182 for N and 30.944 and 79.268 for P. By considering the volume of pollutants removed during the treatment process (kg/m^3) and their shadow prices (€/kg) depending on the final fate of the effluent, the environmental benefits from wastewater treatment for the three scenarios were calculated (Figure 8.1b).

Regarding the contribution of the different pollutants to the total benefit, Figure 8.1(b) shows that the greatest environmental benefit is associated with N removal, accounting for up to 65% of the total for the scenario (i) due to the high removal percentage of this pollutant. On the other hand, the contribution of the removal of SS from wastewater is almost negligible (<less than 1% of the total) and therefore not

Table 8.1 Main characteristics of the plants under study.

	Plant Description						Removal Efficiencies		
	m^3/d (d)	m^3/d (r)	pe (r)	Secondary Treat.	Tert. Treat.	Sludge Treat.	COD	N_T	P_T
WwTP 1	1100	840	5,000	MBR	–	GravT+CW	94%	82%	83%
WwTP 2	2160	1,100	8,640	MBR	–	GW+CD	98%	64%	71%
WwTP 3	3225	3,730	12,000	EA+MBR and Aux. ST	–	GravT+CD	98%	96%	82%
WwTP 4	90,000	43,890/ 7,554 (MBR)	400,000	Line 1: IFAS+ST Line 2: EA+MBR	UV	GravT+AD+CD	89%	60%	67%
WwTP 5	15,000	12,517	60,701	EA+N & P removal	F+UF	GravT+CD	93%	70%	89%
WwTP 6	30,000	28,870	117,816	AS+N removal	F+CoF+UV	GravT+MechT+ AD+LS+CD	92%	16%	65%
WwTP 7	60,000	35,613	229,154	AS+N & P removal	UF+RO	GravT+FT+AD+CD	94%	66%	65%

d, design; r, real; AD, anaerobic digestion; AS, activated sludge; CD, centrifuge dewatering; CoF, coagulation/flocculation; EA, extended aeration; F, filtration; FT, GravT, MechT, flotation, gravity and mechanical thickener; IFAS, integrated fixed film AS; LS, lime stabilization; ST, settling tank; UF, ultrafiltration; UV, ultraviolet.

shown in Figure 8.1(b). The environmental benefits accounted for organic matter (COD) ranged between 6 and 8% because its shadow price is comparatively low. Despite the high shadow price for *P*, specifically if the effluent is reused, its contribution to the total benefit is limited by the low amount of this pollutant that can be removed.

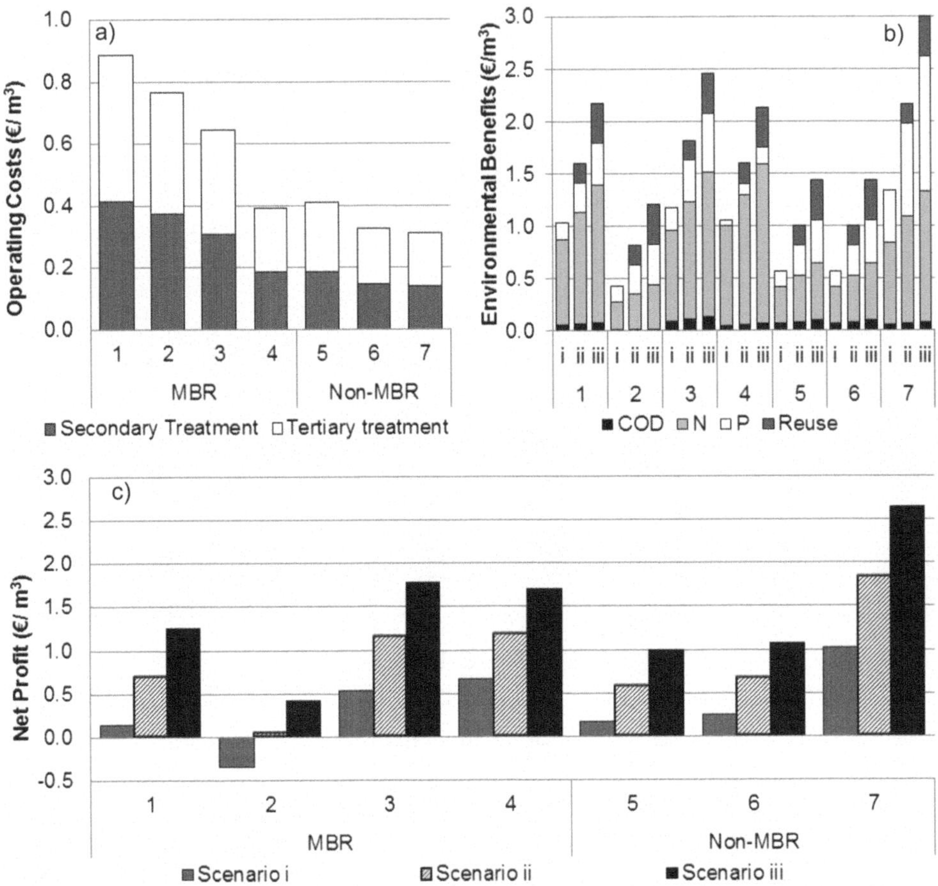

Figure 8.1 (a) Operating costs of WwTPs per m^3, (b) environmental benefits of WwTPs expressed in €/m^3, (c) net profit of WwTPs per m^3.

Figure 8.1 (b) also presents the total economic benefit for the three scenarios considered. Because the amount of pollutants removed in each plant is different, their economic benefits vary significantly. Regardless of the reuse scenario considered, WwTP 2 results in the lowest environmental benefit while WwTP 6 obtains the highest one. MBR present higher benefits than tertiary-membrane plants due to their higher pollutants removal efficiency. If all treated water is reused, the environmental benefit is 1.8 times higher than if the effluent is discharged into the river without any reuse.

According to Equation (8.1), NP is calculated considering internal and external costs and benefits. Therefore, for scenario (i), the net profit is calculated based on operating costs and environmental benefits. For scenarios (ii) and (iii), the benefits from the sale of the regenerated water were also considered.

Figure 8.1 (c) shows the average NP to be positive for the three scenarios considered, verifying that the operation of MBR technology is economically feasible if environmental benefits are included in the analysis. If treated water is reused, the operation of all plants is economically feasible since their NP is always positive. However, if treated water is not sold, one of the plants (WwTP 2) is not economically feasible. Both its low environmental benefits (the lowest) and its high costs (the second highest) are behind its negative NP. On average, MBR have a lower NP since their higher benefits do not compensate for their higher operational costs.

This CBA illustrated the importance of integrating the externalities in the assessment of the economic feasibility of wastewater treatment technologies since, in order to consider MBR operation as feasible, environmental benefits had to be taken into account.

8.4.2 Environmental profile

8.4.2.1 Goal and scope

In this section, the environmental profile of the plants presented in Table 8.1 is assessed by means of LCA methodology introduced in Section 8.3. As indicated in Section 8.3.1.1, the function of a WwTP can be defined as the treatment of a certain volume of wastewater. Using 1 m^3 of treated water as FU, the net environmental benefit (NEB) approach developed by Godin *et al.* (2012) was used here. This methodology systematically presents the potential impacts (PI) with and without treatment following Equation 8.4, for each impact category.

$$NEB = PI_{NO} - [PI_{TW} + PI_{TP}] \tag{8.4}$$

PI_{NO} being the potential impact of discharging wastewater without any treatment into the environment (null option), PI_{TW} the potential impact of the discharge of treated water into the environment and PI_{TP} the potential impacts associated with the treatment process. This approach highlights the benefits of nutrients and organic matter removal, typically considered as the main objective of a WwTP (Rodriguez-Garcia *et al.*, 2011) without neglecting the importance of removing other substances such as heavy metals.

According to Lassaux *et al.* (2007), Doka (2009) and Mahgoub *et al.* (2010), the WwTP is the most relevant stage of the water cycle in environmental terms. For that reason, the WwTP as a whole is considered within the limits of the system.

Of all the stages of the life cycle of the WwTP, Lundie *et al.* (2004) and Lassaux *et al.* (2007) found the impacts associated with the operation to be far more relevant than the ones produced by construction and disposal. The construction stage has only been found to be relevant for some technologies by Tangsubkul *et al.* (2005). In this study, construction is responsible for 17 and 31% of the (GWP) impact associated with CMF and MBR, while its impact on the other categories is almost negligible. In spite of that, the construction stage has not been considered in the present study. The reason for that is the fact that the methodology used by Tangsubkul *et al.* (2005) to estimate the impact of the construction stage, the Missing Inventory Estimation Tool (MIET) developed by Suh (2002), which correlates the environmental impacts of a process with its monetary costs, uses an input–output methodology based on data from the United States in 1998. This database is probably outdated for a fast developing technology such as membranes. Emmerson *et al.* (1995) and Zhang and Wilson (2000) indicated that the disposal stage usually contributes insignificantly to the overall impact. Such conclusion has been kept in recent reviews (Corominas *et al.*, 2011).

For that reason, this assessment considers the operation of primary, secondary, and tertiary treatments (in WwTPs 4–7) as well as the sludge treatment and its final disposal as agricultural amendment, including the avoided production of fertilizers.

8.4.2.2 Life cycle inventory analysis

The LCI presented in Table 8.2 was elaborated based on reports supplied by the managing companies and the regional authorities. They correspond to annual averages for year 2010 (WwTP 1–4) and 2008 (WwTP 5–7). Data are presented per FU (1 m^3 treated water). For the information not supplied by the aforementioned sources, the following assumptions were considered and the subsequent processes from the Ecoinvent database (Swiss Centre for Life Cycle Inventories, 2012) were used:

- *Electricity*: The Spanish Electric Production and Export mix was updated for 2010 according to REE (2009). We decided not to use the profile of 2008 for WwTP 5–7 to avoid differences related to changes in the Spanish electric profile between the years of study.
- *Chemicals*: *Acrylonitrile from Sohio process, at plant/RER* was considered for polyelectrolyte, *iron (III) chloride, 40% in H_2O, at plant/CH* for $FeCl_3$, *sodium hypochlorite, 15% in H_2O, at plant* for NaClO, *sodium hydroxide, 50% in H_2O, production mix, at plant/RER* for NaOH, *sulfuric acid, liquid, at plant/RER* for H_2SO_4, *Hydrochloric acid, 30% in H_2O, at plant/RER* for HCl, *EDTA, ethylenediaminetetraacetic acid, at plant/RER* for EDTA and *chemicals organic* as a proxy for citric acid.
- *Waste*: *Disposal, municipal solid waste, 22.9% water, to municipal incineration /CH* was used for the incineration of residues that can be considered as municipal solid waste (MSW) and *Disposal, inert waste, 5% water, to inert material landfill/CH*, for the final destination of grit. Fat treatment, consisting in the addition of cement and limestone, was described using the *Cement, unspecified, at plant/CH* and the *Limestone, milled, loose, at plant/CH* processes.
- *Transport*: A distance of 20 km was chosen for all the technosphere material inputs and outputs going to and from the WwTP. This is, for the different chemicals, from the chemical supplying company to the WwTP; for the sludge, from the WwTP to the agricultural land where it is disposed; for the municipal solid waste, from the WwTP to the incineration plant; for the sand, from the WwTP to the inert landfill and for the sand, from the WwTP to the waste treatment plant. For the chemicals, the process *Transport with van <3.5t/RER* was used while for the different sub-products and residues, the process *Transport with lorry 7.5–16 t, EURO 3/RER* was used instead. Sludge application was assimilated to *Slurry spreading, by vacuum tanker/CH*.
- *Fertilizers*: A substitutability of 50 and 70%, respectively, of the N and P present in the sludge was used to calculate the amount of avoided fertilizers as indicated by Bengtsson *et al.* (1997). The Ecoinvent processes *Ammonium sulfate, as N, at regional storehouse/RER* and *Diammonium sulfate, as P_2O_5, at regional storehouse/RER* were chosen for this purpose.
- *Direct emissions from sludge disposal*: The PO_4^{3-} leakages were calculated according to Doka (2009). The gaseous emissions of N_2O and NH_3 were considered as in Hobson (2003) and Lundin *et al.* (2000). All the heavy metals present in the sludge are supposed to remain in soil.
- *Heavy metals*: The concentration of heavy metals in the sludge was directly measured but the influent and effluent ones had to be estimated according to equations 8.5 and 8.6. Removal efficiencies were taken from Santos and Judd (2010a, b) for both MBR and non-MBR. For the latter the following enhanced removals were considered: Cu, Pb and Zn for WwTP 6 from El Samrani *et al.* (2008) due to its use of ferric coagulant and Cu, Ni and Zn for WwTP 7 from Fu and Wang (2011) due to its RO process.

$$[HM]_i(mg/m^3) = [HM]_s(mg/kg) * \frac{\text{kg sludge}}{m^3 \text{ treated water}} * \frac{1}{RE} * 100 \tag{8.5}$$

Table 8.2 Life cycle inventory of the plants under study per FU (1 m^3 treated water).

	WwTP 1	WwTP 2	WwTP 3	WwTP 4	WwTP 5	WwTP 6	WwTP 7
Inputs							
Influent							
COD (g)	646	126	954	580	750	617	623
NT (g)	61.00	25.00	55.00	96.00	30.20	41.59	55.30
PT (g)	6.00	7.00	8.70	3.00	5.78	6.37	7.52
Cd (mg)	0.38	0.09	0.84	0.19	0.66	0.54	0.21
Cr (mg)	11.46	4.02	1.11	25.95	12.65	17.76	14.91
Cu (mg)	109.69	50.05	61.91	204.06	215.07	74.43	91.48
Hg (mg)	0.12	0.13	0.08	0.16	0.06	0.09	0.24
Ni (mg)	12.48	5.25	2.05	15.91	8.30	53.82	4.53
Pb (mg)	18.72	7.51	51.58	14.91	9.22	19.74	8.85
Zn (mg)	187.56	59.42	95.20	319.81	217.07	277.30	133.93
Electricity							
From the grid (kWh)	1.16	0.69	0.7	0.43	0.50	0.13	0.69
To the grid (kWh)[1]	x	x	x	x	x	0.13	0.07
Chemical consumption							
Polyelectrolyte (g)	1.96	x	2.07	1.60	x	2.67	1.59
$FeCl_3$ (g)	3.26	x	4.29	0.03	89.93	53.30	44.37
NaClO (g)	x	0.01	6.32	0.45	x	2.24	23.19
NaOH (g)	x	x	x	x	x	1.50	0.12
H_2SO_4 (g)	x	x	x	x	x	1.25	0.26
HCl (g)	x	x	0.13	x	x	x	x
Citric acid (g)	x	x	0.34	0.07	x	x	x
EDTA (g)	x	x	0.30	x	x	x	x
Transport (kg · km)	0.10	0.00022	0.27	0.04	1.80	1.22	1.39
Outputs							
Effluent							
COD (g)	36	3	20	65	49	49	39
NT (g)	11.00	9.00	2.10	38.00	9.16	34.98	18.81
PT (g)	1.00	2.00	1.60	1.00	0.65	2.24	2.63
Cd (mg)	0.10	0.02	0.22	0.05	0.25	0.20	0.08
Cr (mg)	1.95	0.68	0.19	4.41	3.92	5.51	4.62
Cu (mg)	18.65	8.51	10.53	34.69	68.82	12.65	15.55
Hg (mg)	0.01	0.01	0.01	0.01	0.01	0.01	0.03
Ni (mg)	4.49	1.89	0.74	5.73	4.48	29.06	0.05
Pb (mg)	5.05	2.03	13.93	4.03	2.40	5.13	2.30
Zn (mg)	46.89	14.86	23.80	79.95	69.46	69.33	1.34

(*Continued*)

Table 8.2 Life cycle inventory of the plants under study per FU (1 m^3 treated water) (*Continued*).

	WwTP 1	WwTP 2	WwTP 3	WwTP 4	WwTP 5	WwTP 6	WwTP 7
Air emissions from biogas combustion (anaerobic digestion of the sludge)							
CO (mg)	x	x	x	0.71	3.92	1.81	3.17
CO_2 (mg)	x	x	x	437	2426	1117	1957
NMVOC (mg)	x	x	x	0.01	0.06	0.03	0.05
CH_4 (mg)	x	x	x	2.32	12.86	5.93	10.38
SO_2 (mg)	x	x	x	1.89	10.52	4.85	8.49
NO_2 (mg)	x	x	x	1.42	7.89	3.64	6.37
NH_3 (mg)	x	x	x	0.16	0.88	0.40	0.71
N_2O (mg)	x	x	x	0.11	0.58	0.27	0.47
Sludge, to agriculture							
To agriculture (kg WW)[2]	0.87	0.47	1.06	0.89	1.18	0.61	0.93
Transport (kg · km)	43.71	23.5	53	44.5	58.80	30.64	46.29
Application as slurry (l)	0.87	0.47	1.06	0.89	1.16	0.61	0.91
Cd (mg)	0.28	0.07	0.62	0.14	0.41	0.33	0.13
Cr (mg)	9.51	3.34	0.92	21.54	8.73	12.26	10.29
Cu (mg)	91.04	41.54	51.39	169.37	146.25	61.77	75.93
Hg (mg)	0.11	0.12	0.07	0.15	0.05	0.08	0.21
Ni (mg)	7.99	3.36	1.31	10.18	3.82	24.76	4.49
Pb (mg)	13.67	5.48	37.65	10.89	6.82	14.61	6.55
Zn (mg)	140.67	44.57	71.40	239.86	147.61	207.98	132.59
N_2O (g) to air	0.09	0.07	0.23	0.17	0.16	0.11	0.15
NH_3 (g) to air	1.75	1.41	4.43	3.30	3.06	2.10	2.82
PO_4 (g) to water	0.32	0.18	0.56	0.55	0.33	0.28	0.40
Avoided Products							
N-based fertilizer (g)	2.88	2.32	7.29	5.44	5.04	3.46	4.63
P_2O_5-based fertilizer (g)	8.92	5	16.18	15.42	6.76	5.59	8.17
Waste, to further treatment							
Grit (g)	–	–	5.13	–	13.75	14.70	5.26
MSW (g)	–	42.34	30.49	–	0.88	10.31	23.70
Fats (g)	–	x	x	–	1.29	0.36	2.04
Transport (kg · km)	–	0.85	0.71	–	0.32	0.51	0.62

[1]Some plants produce electricity by combustion of the CH_4 produced by the anaerobic digestion of their sludge that is sold to the net rather than used inside the WwTP.
[2]WW: Wet weight.
x No flow – No data available.

$$[HM]_e(mg/m^3) = [HM]_i(mg/m^3) - [HM]_s(mg/kg) * \frac{\text{kg sludge}}{m^3 \text{ treated water}} \tag{8.6}$$

where [HM] is the concentration of each heavy metal in the influent (i), effluent (e) and sludge (s) and RE is the removal efficiency expressed as a percentage.

The influents of the WwTPs under study can be considered in general terms of medium–high strength according to Tchobanoglous *et al.* (2002) and thus high differences due to influent quality are not expected. All plants present high COD removal efficiencies, even for WwTP2, where the low COD removal associated with low strength influents described in Seijas *et al.* (2003) was not observed. High N removal is achieved in MBR plants (WwTP 1–4) due to their longer SRT, which is beneficial for N assimilation during microbial growth (Judd, 2011). Non-MBR plants were designed for the removal of N. However, in WwTP 5 and WwTP 6 nutrient removal is low since water is reused for irrigation, being N and P valuable resources. Finally, the removal of P is noticeably high in all MBR plants, probably due to the low F/M ratio that is characteristic of this technology (Monclús *et al.*, 2010; Yilmaz *et al.*, 2007). Substantial P removal is also achieved by non-MBR plants due to the use of $FeCl_3$, specifically for P removal (WWPT 5 and 7) or as a coagulant (WwTP 6).

The electricity use of the MBR falls within the 0.6–1.1 kWh/m^3 indicated by Rodriguez-Roda *et al.* (2009). The effect of the economy of scale can observed since the smallest system (WwTP 1) has the highest electricity usage and the largest system (WwTP 4) has the lowest one. The other plants, WwTP 5–7, fall outside the 0.4–0.5 kWh/m^3 suggested by Günder and Krauth (1999) for CAS. Both WwTP 5 and WwTP 7 have tertiary treatments with important electricity demands as pointed out by Clauson Kaas *et al.* (2006). The combustion of the CH_4 produced by the anaerobic digestion of the sludge, generates electricity that it is usually sold to the grid. WwTP 6 is almost self-sufficient, producing practically the amount of electricity required for its operation.

Chemical consumption is a direct consequence of the plant configuration. WwTP 2–4 use several chemicals for membrane cleaning (i.e. NaClO, HCl and citric acid) while WwTP 1 did not report the use of any chemical for this purpose. polyelectrolyte is used in all but two plants for the dewatering of sludge with several degrees of intensity. However, no correlation was found between the sludge dosage and the production of sludge. Ferric coagulant ($FeCl_3$) is used in all plants but WwTP 2. Although the dosage varies significantly, MBR plants present a lower consumption than the other three WwTPs. On the one hand, this is because the complete removal of suspended solids during primary clarification results in increased membrane clogging and thus, according to Sánchez-Sánchez *et al.* (2010), Buntner *et al.* (2011) and Garrido *et al.* (2009), a large use of coagulant is not recommended in MBR configurations. On the other, all the non-MBRs have additional uses for the $FeCl_3$ aside from primary coagulation/flocculation (CoF): *P* removal (WwTP 5 and 7) and tertiary CoF (WwTP 6).

The removal of heavy metals, as indicated earlier in this section, is based on literature data; so, according to Santos and Judd (2010a, b), MBRs remove on average 10% more heavy metals from water than CAS. Interestingly, removal efficiencies from CAS processes assessed in that work were on average 4% higher than the ones presented in Doka (2009), values previously used in LCA for the estimation of the remaining fraction of heavy metals in water by Godin *et al.* (2012).

8.4.2.3 Life cycle impact assessment

8.4.2.3.1 Impact assessment methodology

According to the most recent reviews on LCA of WwTP, Larsen *et al.* (2007) and Corominas *et al.* (2011), GWP, acidification (AP) and eutrophication (EP) are the environmental indicators that have received more attention and thus they were included in this study. For these impact categories, the CML methodology v. 2.05 developed by Guinée *et al.* (2002), a widespread impact assessment methodology that has been widely used in the LCA of WwTP, was used. Although toxicity-related categories have received less attention in the environmental assessment of WwTP (Corominas *et al.*, 2011), its relevance is increasing as more toxic substances are considered (Muñoz *et al.*, 2008) and methodologies developed (Larsen *et al.*, 2009). In this case study, the toxicity-related impact categories defined by the state-of-the-art USETOX methodology developed by Rosenbaum *et al.* (2008): human toxicity (comprising both cancer and non-cancer impacts, HTP) and freshwater ecotoxicity (ETP) were considered including both recommended and interim substances.

8.4.2.3.2 General overview

The NEB approach aggregates into a single score the results that in more conventional LCAs appear as two different scenarios: with and without treatment. However, it does nothing to facilitate the comparison between impact categories or the assessment of several technologies at the same time. In order to overcome these limitations of LCA and in order to offer a quick glimpse of the five impact categories for the seven WwTPs under study, Figure 8.3(a) was constructed. It refers the impact of each WwTP to the one with the highest absolute value for each impact category. The discussion of the environmental profile for each category will be carried out in detail next. Overall discussion of the environmental profile is offered later in Section 8.4.2.4 once each category is analyzed thoroughly.

8.4.2.3.3 Eutrophication

The removal of eutrophied substances is a key aspect of WwTP according to the current European legislation (EEC, 1991; EC, 1998). If legal requirements are met, significant removals are achieved and thus, WwTP should produce an environmental benefit in this impact category. As seen in Figure 8.2(a), this is the case of all plants under study although some differences among them can be observed. Three factors can affect the results, better seen in Figure 8.2(b): (i) the removal efficiency (Table 8.1) since it increases the differences between terms PI_{NO} and PI_{TW} in Equation 8.4, (ii) the influent quality, since the higher the eutrophication potential is, the higher removal can be in absolute terms (Table 8.2), and (iii) the impact associated with the treatment process, responsible for term PI_{TP} in Equation 8.4.

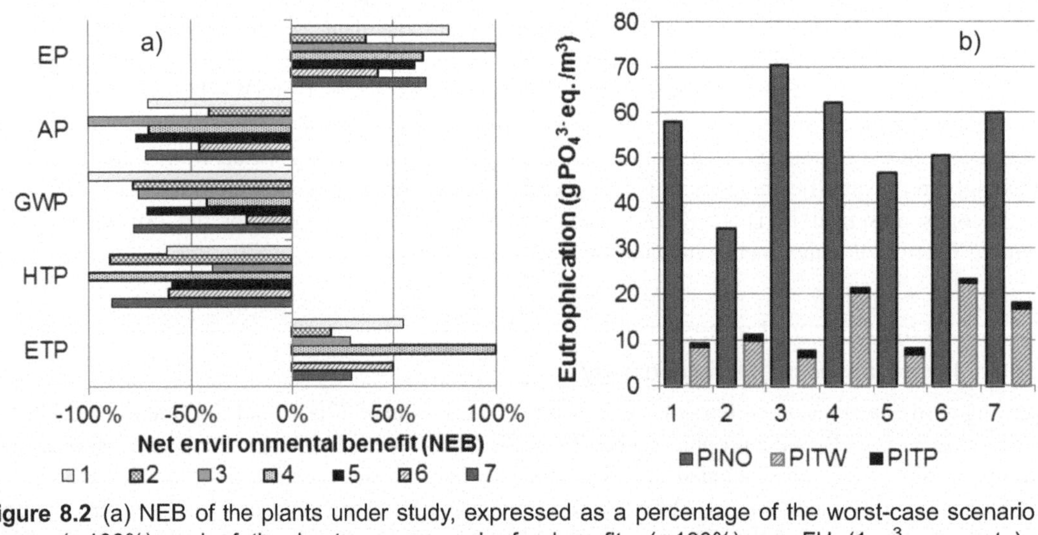

Figure 8.2 (a) NEB of the plants under study, expressed as a percentage of the worst-case scenario for losses (−100%) and of the best-case scenario for benefits (+100%) per FU (1 m^3 permeate), (b) eutrophication potential impact of the influent (PI_{NO}), the effluent (PI_{TW}) and the treatment process (PI_{TP}) of the plants under study per FU (1 m^3 treated water).

The relationship between removal efficiency and NEB has been found to be weaker than the one linking the latter with influent quality ($R^2 = 0.57$ and 0.75, respectively). This suggests that the seven plants assessed, despite their differences, constitute a relatively homogeneous group when it comes to removal efficiency and therefore, the greatest difference among them is the quality of the influent entering the plant. As seen in Figure 8.2(b), three of the four MBR (WwTP 1, 3 and 4) rank 2nd, 1st and 4th on the NEB score, but they also have the 4th, 1st and 2nd highest load in the influent. For that reason, MBRs might not necessarily be more efficient than plants with tertiary membranes. The impacts associated with the wastewater treatment process: chemicals, electricity, sludge, etc. (PI_{TP}) were found to be almost irrelevant in all plants, less than 3% of the influent impact.

8.4.2.3.4 Acidification

The potential impacts in the acidification category are only associated with the treatment process (PI_{TP}) since neither the influent nor the effluent generate any impact and thus it results in a net environmental loss for any WwTP (Figure 8.3a).

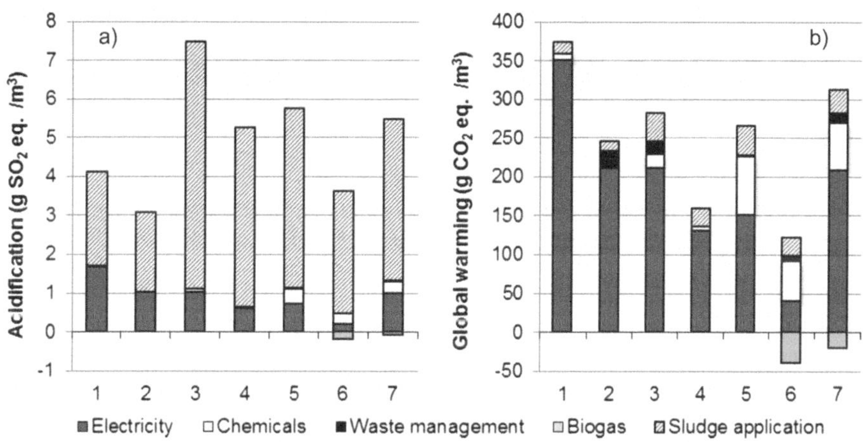

Figure 8.3 (a) Acidification and (b) global warming potential impacts of the different stages of the treatment process (PI_{TP}) of the plants under study per FU (1 m^3 treated water).

Within the treatment process (PI_{TP}), sludge application is responsible for the biggest share on all seven WwTP, from 58 to 81% (Figure 8.3a). Most of this impact is related to the release of 25% of the N present in the sludge as NH_3 as described by Hobson *et al.* (2003). Even if only 10% of the N was released as NH_3, sludge application would continue to be the most relevant source of acidification for almost all plants except WwTP 1 and 2. Due to the relevance of those emissions, plants with significant concentrations of N in their sludge, namely WwTP 3 and WwTP 4 where 8.6 and 5.6% of their sludge dry matter is N, respectively, present considerable impacts in this category.

Electricity production accounts for 5–41% of the total acidifying potential (Figure 8.3a), mainly because of the SO_2 and NO_X released during coal and gas combustion (75 and 24%, respectively). The production of chemicals is relevant for WwTPs 5–7, due to their use of larger quantities of $FeCl_3$. The manufacture of Cl_2 required for the latter is an electricity-demanding process: 1.6 kWh/kg according to the Ecoinvent Database (Swiss Centre for Life Cycle Inventories, 2012), and thus their emissions of both SO_2 and NO_X can be relevant. Finally, biogas combustion generates acidifying gases but the electricity generated compensates those emissions for WwTP 6 and 7, meaning that in these two plants the combustion of biogas results in fewer emissions per kWh generated than the Spanish electricity mix for 2010.

The overall results presented in Figure 8.4(a) and the importance of NH_3 released from sludge, a factor that has more to do with the initial concentration of N in the influent than with the place where a membrane is located, advices against considering MBR as a better or worse option than tertiary membrane plants. Regardless of that, it is possible to see the importance of the scale factor on the smallest plant (WwTP 1), where the high electricity consumption (1.16 kWh/m^3, Table 8.2) significantly affects acidification.

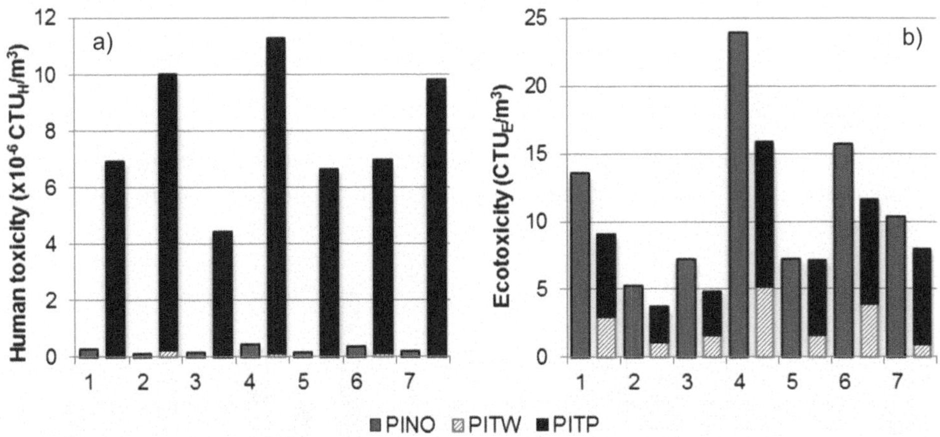

Figure 8.4 (a) Human toxicity and (b) ecotoxicity potential impacts of the influent (PI_{NO}), the effluent (PN_{TW}) and the treatment process (PI_{TP}) of the plants under study per FU (1 m^3 treated water).

8.4.2.3.5 Global warming

The IPPC methodology (Doorn *et al.*, 2006) considered untreated wastewater as a potential source of both CO_2 and CH_4. These emissions have been included in previous LCA such as Foley *et al.* (2010) and Godin *et al.* (2012). However, greenhouse gases (GHG) emissions from wastewater treatment were not included in this study due to a number of reasons: CO_2 is considered to be of biogenic origin, CH_4 is not expected to be produced in aerobic/anoxic processes and there are still important uncertainties regarding the possible emissions of N_2O during nitrification and denitrification processes (Corominas *et al.*, 2012; Ni *et al.*, 2012). Nevertheless, important advances have been done in recent years regarding GHG in WwTP (Porro *et al.*, 2011) and models specifically designed for LCA applications are being developed (Rodriguez-Garcia *et al.*, 2012).

Taking this into account, global warming results in a net environmental loss for all plants (Figure 8.2a) since only the PI_{TP} term of Equation 8.4 is present. This term is analyzed in detail in Figure 8.3(b).

Electricity production is the main source of impact (Figure 8.3b), being responsible for 48% of the impact (WwTP 6) to 94% (WwTP 1). CO_2 is the main gas generated during electricity production, which is mainly related to the combustion of gas and coal, processes that were responsible for 33% of the electricity generated in Spain in 2010 (REE, 2009). As for acidification, the production of $FeCl_3$ generates a relevant impact in those plans using this chemical intensively, WwTP 5–7 for the same reasons mentioned before. The same occurs for biogas combustion, but this time the benefits for WwTP 6 and 7 are more relevant than for acidification since the emissions of CO_2 and CH_4 from the biogas are assumed to be of biogenic origin and therefore do not count towards GWP. Sludge application has also some relevance in plants WwTP 3–7 (Figure 8.3b) due to the release of 1% of the N present in the sludge as N_2O.

Due to the larger relevance of electricity in global warming when compared to acidification, it is easier to consider that tertiary membrane plants generate a lower environmental burden than MBR and that within this group of plants, impacts are greatly affected by the scale factor. The reduction of the impact in this category for MBR goes necessarily through the optimization of electricity use but in the case of tertiary membrane plants, a reduction of the $FeCl_3$ consumption might be more effective.

8.4.2.3.6 Human toxicity

From a NEB perspective, toxicity-related impacts present a similar behavior to eutrophication. They are caused by a number of substances present in the influent that are partially removed from the effluent as it can be seen in Table 8.2. The main difference between these categories is the number of substances that can potentially impact in this category. The CML 2.05 methodology (Guinée *et al.*, 2002) includes 14 substances whose emission to water can potentially impact on eutrophication while USEtox considers thousands of potential toxics and, as indicated in Section 8.5.3.1., it is far from being a complete list. In Section 8.4.2 only heavy metals will be assessed but PPCPs will be the target pollutant in the novel MBR case study presented in Section 8.5.

The toxicity of heavy metal has been a controversial aspect for LCA practitioners for many years since most dispersion models used by ready-to-use LCIA methodologies have been developed for organic pollutants, not including particularities related with inorganic compounds such as metals (Diamond *et al.*, 2010; Jolliet *et al.*, 2006; Sonnemann, 2004). Nevertheless, a recent study comparing the toxicity of metals according to 9 different impact assessment methodologies suggested that although there is no such thing as a "best methodology", USEtox can be considered as the reference model since, among other aspects, it does not necessarily overestimate or underestimate the impact of each metal (Pizzol *et al.*, 2011).

Contrarily to eutrophication, Figure 8.2(a) shows human toxicity as an environmental loss rather than a benefit, which is caused not due to a poor removal of the toxic substances but to the large indirect impact of the treatment process (Figure 8.4a).

By far the major source of impact in the treatment process (minimum contribution of 98%) is the application of sludge: specifically, the release of heavy metals in agricultural soil. The reason why the application of sludge as soil amendment results in a net environmental loss is that, according to the USEtox methodology (Rosenbaum *et al.*, 2008), all metals but Ni have a higher toxicity when they are released to soil than when they are released to water. This is, their CF for emissions to soil 10–100 times larger than for emissions to freshwater (e.g. 4.4×10^{-2} and 1.3×10^{-3} CTU_H/kg for Zn, respectively). Metals in soil do not easily mobilize to other compartments such as freshwater, but they are transferred to crops and from there to cattle and humans faster than those emitted to water, increasing thus the probability of being ingested by humans.

The overall classification is completely dependent on the concentration of Zn in the influent and its final destination. Zn is the most relevant heavy metal; responsible for at least 98% of the impact in the influent, effluent and sludge except for the effluent of WwTP 7, where due to the 99% removal of Zn, it is responsible for 68% of the impact. As seen in Figure 8.2(a) WwTP 4 is the plant with the lowest NEB, due to its high initial concentration (Table 8.2). On the other hand, the second lowest NEB corresponds to WwTP 7, ranking 5th when it comes to Zn initial concentration, but which due to RO, transfers 99% of the total content to the sludge according to Fu and Wang (2011). This latter case exemplifies that, according to the USEtox methodology for human toxicity, when it comes to heavy metal removal, less is better. For this reason, the fact that more heavy metals are removed from MBR that from CAS pointed out by Santos and Judd (2010a, b) might constitute an environmental burden rather than a benefit.

8.4.2.3.7 Freshwater ecotoxicity

So far the USEtox methodology only includes freshwater ecotoxicity, which is expected to be most relevant when dealing with wastewater treatment technologies. Figure 8.2(a) indicates that most plants generate an environmental benefit in this category, WwTP 2 being the only exception. A more detailed analysis presented in Figure 8.4(b) shows a moderate-to-high removal of the toxicity present in the influent, from 68% for WwTP 6 to 91% for WwTP 7. This removal efficiency can practically be offset by the impact of the treatment process (WwTP 2, 5).

Zn and Cu share almost all the impact associated with the influent, from the 50–49% of WwTP 7 to the 68–26% of WwTP 6. The presence of these two metals in the sludge is also responsible for most of the impact associated with the treatment process. As for human toxicity, metals are not removed from the environment but transferred to a compartment that, in this case, is less harmful that the original one. For the USEtox methodology, metals have lower CFs for soil emissions than for freshwater ones, meaning that emissions to water have higher chances of affecting aquatic life.

As WwTP 3 is the plant with the highest concentration of Cu and Zn in the influent (Table 8.2), its partial transfer to soil makes it the one with the highest NEB (Figure 8.2a), contrary to results observed for human toxicity. The other end of the line is WwTP 5, which had the highest NEB for human toxicity due to its lower metal removal but has the lowest one regarding ecotoxicity for the same reason. It is possible to conclude that, according to the USEtox methodology, the removal of metals from the influent results in a balance between the benefits of a lower freshwater ecotoxicity potential and a higher human toxicity.

8.4.2.4 Results interpretation

As indicated in Section 8.3.1.3, weighing the different impact categories relies on subjectivity to define how important is each one of them. If all impact categories are considered equally relevant, all the NEB percentages presented in Figure 8.3(a) can be added to a single score ranging from −500 to 500 for the worst and best plants possible. Considering this approach, all plants would result in a net environmental loss since all present negative values (Table 8.3). In order to justify stricter levels of nutrient removal, Foley (2009) suggested that EP should be weighted three times more than the rest of other categories. If done so, the majority of plants would result in an environmental benefit but it would be necessary to consider EP six times more important that the rest of the categories to produce an environmental benefit.

Table 8.3 Aggregated scores, un-weighted and weighted per FU (1 m^3 treated water).

	Un-Weighted	3x EP Weighting	6x EP Weighting	6x HTP Weighting	6x EP+HTP Weighting
Scale	−500/500	−700/700	−1,000/1,000	−1,000/1,000	−1,500/1,500
WwTP 1	−99	56	290	−406	−18
WwTP 2	−152	−78	33	−600	−415
WwTP 3	−86	114	414	−284	216
WwTP 4	−48	82	276	−548	−224
WwTP 5	−146	−24	159	−445	−139
WwTP 6	−36	50	178	−339	−124
WwTP 7	−99	−10	189	−585	−253
MBR	**−96.34**	**43.49**	**253.24**	**−459.54**	**−109.96**
Non-MBR	**−93.67**	**5.26**	**175.61**	**−456.28**	**−172.38**

If weighted results are used, MBRs are less efficient than non-MBR plants. However, the moment more relevance is given to EP than to other impact categories, this tendency changes and the differences become higher the more relevance is given to eutrophication. It is necessary to remember that as indicated in Section 8.4.2.4.3, this might have more to do with the higher loads found in MBR influents than in their removal efficiencies.

Human toxicity can be considered also as a particularly relevant impact category. However, since it is an environmental burden for all plants, the higher relevance is given to it, the worse the overall results of each plant will be. Again due to the average lower concentration of Zn in the influent of the MBR, this would result in a better environmental profile than non-MBR plants if HTP was weighted more than other impact categories. If the same relevance is given to EP and HTP all plants but WwTP 3 would result in an environmental loss, but those for MBR would be lower than for other plants.

8.5 ENVIRONMENTAL PROFILE OF PILOT PLANT MBR

As mentioned in Section 8.4, MBR is an advanced technology with a well-developed market. One of the reasons why it is considered a novel technology is that a large fraction of the research associated with this technology is still carried out at pilot plant scale. MBR cannot be seen as only one technology or configuration and, therefore, we must compare different kinds of MBR and not just compare an "average" MBR against other treatment technologies. For this reason, this section of the chapter assesses four different MBR configurations, still in the early stages of their development, from an environmental perspective. Due to the scale difference between the reactors assessed here and the ones used to develop the equations for the estimation of operational costs (Equations 8.2 and 8.3) these equations were not considered suitable for pilot plant reactors and thus no economic assessment was conducted.

Another level of novelty is the focus on compounds not usually taken into account by WwTP operators such as micropollutants. The European Union is paying more attention to this kind of substances. In 2001 the EU presented a list of 23 priority pollutants (EC, 2001) including 4 metals (Cd, Hg, Ni and Pb), further expanded in 2008 to 33 substances (EC, 2008). A new addition with 15 new substances is expected in the next few years (EC, 2012) including for the first time PPCPs: two hormones, estradiol (E2) and ethinyl estradiol (EE2) and the anti-inflammatory diclofenac (DCF). Therefore, we paid special attention to the environmental aspect associated with the removal of PPCPs on three reactors (MBR 1, 2 and 4).

8.5.1 Goal and scope

The prime objective of this section is to conduct the LCA of four MBR configurations at pilot plant scale, with an additional emphasis on the removal of PPCPs.

The four reactors under study are:

- MBR 1, consisting of an activated sludge reactor (0.18 m^3) coupled with a hollow fiber membrane for the removal of organic matter and nitrification. For further information, see Reif *et al.* (2008).
- MBR 2, corresponding to a MLE configuration: anoxic tank (0.20 m^3), aerobic tank (0.20 m^3) with a hollow fiber membrane unit (0.10 m^3). This configuration removes organic matter and N using hybrid biomass (attached and suspended growth). Additional information regarding this pilot plant can be found in Expósito (2010).
- MBR 3, based on a UCT modified configuration: one anaerobic tank (0.47 m^3), two anoxic tanks (0.71 and 0.46 m^3), one aerobic tank (0.94 m^3) and the hollow fiber membrane compartment (1.40 m^3). The system removes organic matter, N and P. See Garrido *et al.* (2009) for further information.

- MBR 4, consisting of an UASB bioreactor followed by a hybrid aerobic reactor and a membrane-chamber at the end of the process. The volume of the different units is 0.12, 0.04 and 0.02 m^3, respectively. This configuration is suitable for the removal of organic matter, *N* and *P*, producing biogas during the anaerobic stage. More information regarding this system is found in Sánchez-Sánchez *et al.* (2010).

Additional information regarding the four reactors, including process diagrams and main operational parameters, is found in Hospido *et al.* (2012). As for real scale plants presented in Section 8.4, 1 m^3 of treated water was chosen as FU, called in this case 1 m^3 of permeate since all plants under study are MBR. The NEB approach developed by Godin *et al.* (2012) was also applied here.

The most relevant impacts from wastewater management are assessed when the operational phase of the whole WwTP, including sludge disposal, is within the system boundaries. Contrary to the plants presented in Section 8.4.2, the MBR assessed here are pilot-scale plants and thus two important aspects need to be considered:

- All the MBR under study but MBR 4 were located in real WwTP and fed with wastewater coming from the pretreatment while MBR 4 was fed with synthetic wastewater prepared from diluted milk. For that reason, no pretreatment was included within the system boundaries.
- The sludge generated by all MBR was not treated due to the pilot plant scale of the reactors and thus there was no information regarding the final characteristics of the sludge. Since, as pointed out by Hospido *et al.* (2005) and Houillon and Jolliet (2005), the impacts of sludge management cannot be neglected, the sludge treatment process had to be modeled based on the characteristics of the raw sludge. See Section 8.5.2 for further details.

8.5.2 Life cycle inventory analysis

The data required for the elaboration of Table 8.4 were mainly collected during the operation of the different reactors at steady-state conditions. More information regarding the frequency of the analysis and the methods followed can be found in Hospido *et al.* (2012). For those flows not directly measured, the same assumptions and Ecoinvent processes (Swiss Centre for Life Cycle Inventories, 2012) as in Section 8.4.2.2 were followed unless stated below:

- PPCPs were added to the raw wastewater to match the following theoretical concentrations (mg/m^3), noticeably higher than the ones commonly found in WwTP influents. The anti-inflammatories: ibuprofen (IBP) and naproxen (NPX), DCF, the antibiotics: sulfamethoxazole (SMX), roxithromycin (ROX), erythromycin (ERY) and trimethoprim (TRI), the antidepressants: fluoxetine (FLX) and citalopram (CTL) were added to reach 10 mg/m^3 in the influent. The antiepileptic carbamazepine (CBZ) and the tranquilizer diazepam (DZP) were added to match 20 mg/m^3. The hormones: estrone (E1), E2 and EE2 were added to reach an influent concentration of 1 mg/m^3. The fragrances: galaxolide (HHCB), tonalide (AHTN) and celestolide (ADBI) were added to reach 20 mg/m^3 in the influent of MBR 1 and 4 and 40 mg/m^3 in MBR 2.
- Phosphorus concentration was not available for the influent or the effluent of MBR 2 for the period under study. A typical concentration presented in Tchobanoglous *et al.* (2002) was used for the influent and the average removal efficiency from a later period of operation was used (96.75%).
- *Electricity*: The Spanish Electric Production and Export mix was updated for 2008 according to MITYC (2009), ONE (2008) and REE (2009). The electricity use for MBR 2 was estimated according to Brannock *et al.* (2010), Gil *et al.* (2010) and Verrecht *et al.* (2010b).

Table 8.4 Life cycle inventory of the MBRs under study per FU (1 m^3 permeate).

	MBR 1	MBR 2	MBR 3	MBR 4
Inputs				
Influent				
COD (g)	370	319	183	806
N_T (g)	20.00	19.34	18.37	12.41
P_T (g)	2.22	2.18	2.81	2.51
Cd (mg)	0.01	0.03	0.04	0.03
Cr (mg)	68.60	35.36	2.12	35.36
Cu (mg)	9.65	11.08	12.50	11.08
Hg (mg)	0.07	0.08	0.08	0.08
Ni (mg)	2.27	4.93	7.59	4.93
Pb (mg)	0.17	0.31	0.45	0.31
Zn (mg)	187.79	139.88	91.97	139.88
IBP (mg)	10.46	26.99	x	9.97
NPX (mg)	9.94	17.33	x	6.21
DCF (mg)	14.45	7.42	x	4.88
CBZ (mg)	26.05	24.64	x	27.30
DZP (mg)	24.61	33.65	x	21.36
HHCB (mg)	24.71	45.34	x	14.41
AHTN (mg)	17.15	38.83	x	11.62
ADBI (mg)	18.30	45.37	x	9.21
SMX (mg)	6.17	15.45	x	119.59
ERY (mg)	6.45	16.43	x	3.09
ROX (mg)	7.23	18.51	x	286.94
TRI (mg)	6.30	x	x	236.70
FLX (mg)	4.90	x	x	x
CTL (mg)	6.00	x	x	x
EE2 (mg)	0.62	x	x	30.66
E2 (mg)	0.18	x	x	31.32
E1 (mg)	1.06	x	x	26.41
Electricity use				
From the grid (kWh)	10.60	4.42	2.34	3.26
Chemical consumption				
Bleach (ml)	20	–	30	70
$NaHCO_3$ (g)	163.27	–	x	x
Distilled water (ml)	–	–	x	236
Tap water (ml)	540	–	x	x
Acrylonitrile (g)	0.39	–	5.23	0.06
Transport (kg · km)	3.67	–	0.70	6.12

(Continued)

Table 8.4 Life cycle inventory of the MBRs under study per FU (1 m^3 permeate) (*Continued*).

	MBR 1	MBR 2	MBR 3	MBR 4
Sludge management				
Transport (kg · km)	9.1	–	120	1.34
Agricultural application (ml)	182.0	–	2400.0	26.8
Outputs				
Effluent				
COD (g)	16.9	16.66	14.32	5.41
N_T (g)	18.21	1.96	13.29	23.57
P_T (g)	2.49	0.042	2.21	3.5
Cd (mg)	0.00	0.01	0.01	0.01
Cr (mg)	11.66	6.01	0.36	6.01
Cu (mg)	1.64	1.88	2.13	1.88
Hg (mg)	0.01	0.01	0.01	0.01
Ni (mg)	0.82	1.78	2.73	1.78
Pb (mg)	0.05	0.08	0.12	0.08
Zn (mg)	46.95	34.97	22.99	34.97
IBP (mg)	0.19	3.14	x	2.23
NPX (mg)	0.69	2.47	x	0.29
DCF (mg)	11.47	4.04	x	11.37
CBZ (mg)	15.64	29.36	x	23.49
DZP (mg)	16.37	20.31	x	20.41
HHCB (mg)	7.82	3.51	x	6.15
AHTN (mg)	5.67	1.94	x	3.92
ADBI (mg)	6.33	2.35	x	2.90
SMX (mg)	1.61	7.78	x	7.3
ERY (mg)	0.25	10.29	x	6.49
ROX (mg)	0.39	4.64	x	163.25
TRI (mg)	1.49	x	x	23.69
FLX (mg)	1.61	x	x	x
CTL (mg)	3.60	x	x	x
EE2 (mg)	0.08	x	x	5.65
E2 (mg)	0.01	x	x	0.04
E1 (mg)	0.02	x	x	1.32
Air emissions from biogas combustion (anaerobic stage of the HMBR)				
CO (mg)	X	x	x	578
CO_2 (mg)	X	x	x	357,883
NMVOC (mg)	x	x	x	8.62
CH_4 (mg)	x	x	x	1,898
SO_2 (mg)	x	x	x	1,552

(*Continued*)

Table 8.4 Life cycle inventory of the MBRs under study per FU (1 m^3 permeate) (*Continued*).

	MBR 1	MBR 2	MBR 3	MBR 4
NO_2 (mg)	x	x	x	1,164
NH_3 (mg)	x	x	x	129
N_2O (mg)	x	x	x	86
Air emissions from biogas combustion (anaerobic digestion of the sludge)				
CO (mg)	0.2	–	2.7	1.96×10^{-2}
CO_2 (mg)	127	–	1,672	1.21
NMVOC (mg)	0	–	0.04	2.92×10^{-5}
CH_4 (mg)	0.67	–	8.87	6.42×10^{-2}
SO_2 (mg)	0.55	–	7.25	5.25×10^{-2}
NO_2 (mg)	0.41	–	5.44	3.94×10^{-2}
NH_3 (mg)	0.05	–	0.6	4.38×10^{-3}
N_2O (mg)	0.03	–	0.4	2.92×10^{-3}
Sludge				
Cd to soil (mg)	0.01	0.02	0.03	0.02
Cr to soil (mg)	56.94	29.35	1.76	29.35
Cu to soil (mg)	8.01	9.19	10.38	9.19
Hg to soil (mg)	0.06	0.07	0.07	0.07
Ni to soil (mg)	1.45	3.16	4.86	3.16
Pb to soil (mg)	0.12	0.23	0.33	0.23
Zn to soil (mg)	140.84	104.91	68.98	104.91
N_2O to air (mg)	70	x	450	15
NH_3 to air (mg)	1,360	x	8,750	296
PO_4^{3-} to water (mg)	2.8	x	590	0.61
IBP (mg)	0.78	–	x	–
NPX (mg)	1.27	–	x	–
DCF (mg)	2.05	–	x	–
CBZ (mg)	ND	–	x	–
DZP (mg)	0.093	–	x	–
HHCB (mg)	1.18	–	x	–
AHTN (mg)	0.83	–	x	–
ADBI (mg)	0.69	–	x	–
SMX (mg)	0.00	–	x	–
ERY (mg)	0.003	–	x	–
ROX (mg)	0.001	–	x	–
TRI (mg)	0.05	x	x	–
FLX (mg)	0.22	x	x	x
CTL (mg)	0.13	x	x	x
EE2 (mg)	0.004	x	x	–

(*Continued*)

Table 8.4 Life cycle inventory of the MBRs under study per FU (1 m^3 permeate) (*Continued*).

	MBR 1	MBR 2	MBR 3	MBR 4
E2 (mg)	0.0016	x	x	–
E1 (mg)	0.0017	x	x	–
Avoided products				
Ammonium sulfate, as N (mg)	2,250	–	14,414	471
Diammonium sulfate, as P_2O_5 (mg)	47.8	–	12,239	4.01
Electricity (kWh)	X	x	x	0.69

x No flow – No data available.

- *Chemicals*: *Sodium hypochlorite, 15% in H_2O, at plant/RER* was considered for the production of bleach, *sodium carbonate from ammonium chloride production, at plant/GLO* for $NaHCO_3$, *water, deionized at plant/CH*, for distilled water, and *tap water, at user/RER* for tap water were considered.
- Biogas produced by MBR 4 was assumed to avoid the production of electricity (40%) and heat (50%) according to Pertl *et al.* (2010). Air emissions associated with its combustion were taken from Doka (2009).
- Sludge treatment consisted of anaerobic digestion and dewatering. For the first one, their energy requirements were assumed to be equal to the energy generated by biogas production. The electricity and polyelectrolyte requirements for sludge dewater were taken from Hospido *et al.* (2005). No information regarding sludge production was available for MBR 2 and thus, sludge-related stages were not included for this reactor.
- Direct emissions from sludge disposal. As for heavy metals, all PPCPs present in the sludge are supposed to end in soil.
- *Heavy metals*: The influent concentrations for MBR 1 and MBR 3 correspond to their actual measurements while for MBR 2 and 4 average values were used. The removal efficiencies for the MBR technologies found in Santos and Judd (2010b) were used to determine their concentration in the effluent, assuming that the fraction that does not end in the effluent, ends in the sludge instead according to Equations 8.5 and 8.6.

Table 8.4 shows influent concentrations to be very similar when it comes to *P*, the most relevant substance from an eutrophication perspective and moderately similar for *N*, the second most relevant. Both concentrations can be considered representative of low strength waters according to Tchobanoglous *et al.* (2002). COD on the other hand varies widely from low strength concentrations for MBR 3 to high strength ones for MBR 4. Contrarily, effluent values present more homogeneous values for COD but the differences are more significant for *N* and *P*. Regarding *N*, it is interesting to highlight the high removal efficiency of MBR 2 (90%) due to nitrification–denitrification. The increased concentration of N as NH_4^+ in MBR 4 is likely to be associated with the hydrolysis under anaerobic conditions of the proteins present in the milk-based feeding. The high removal of P for MBR 2 (97%) is clearly much higher than the ones found in the other reactors. A sensitivity analysis indicated that removal efficiencies of 50 and 10% would produce results more in line with the other MBR.

Concerning the presence of heavy metals in the influent, important differences can be found between MBR 1 and MBR 3. Regarding Cr for example, it is 30 times higher in the former than in the latter.

The differences on other metals are also significant, being Zn concentration in MBR 1 twice as much as the one in MBR 3. On the other hand, Ni, Pb and Cd are 2, 3 and 4 times more abundant in MBR 3 than in MBR 4 while the concentrations of Cu and Hg are similar for both reactors. The concentrations of MBR 2 and MBR 4 are somewhere in between these two since their concentrations are an average of both of them.

PPCPs concentrations in the MBR under study are 1–2 orders of magnitude higher than the concentrations usually found in WwTP in order to be easily detected and measured by current analytical methods. In general terms, MBR 2 presents 2–3 times the initial concentration of MBR 1. More relevant are the differences between MBR 1 and MBR 4 since the latter presents concentrations 1–2 orders of magnitude higher than the ones for MBR 1 for the antibiotics: SMX and ROX, and TRI and the hormones: E1, E2, and EE2. These differences can be associated with variations in the methanol solution where the PPCPs are diluted before mixing with the influent or with the pump feeding this solution.

Problems during sampling as seen in Ort *et al.* (2010) and analytical inaccuracies led to higher concentrations in the effluent than the ones in the influent for several substances: the tranquilizer CBZ for MBR 2, the antiepileptic DZP and the antibiotic ERY for MBR 4. Not taking these substances into account, the average removal efficiencies are very similar for all three reactors: MBR 1 (71%), MBR 2 (70%), and MBR 4 (68%). Although the detailed analysis of the removal mechanisms goes beyond the scope of this study, it is necessary to highlight the significantly higher removal of fragrances (HHCB, AHTN, ADBI) in MBR 2, which might be related with their higher concentration in the influent. Similar conclusions can be drawn for some of the antibiotics and hormones found in higher concentrations in MBR 4 (TRI and E2). However, this is not always the case since for the antibiotic ROX, the higher the concentration is, the lower the efficiency.

The electricity use of all the MBR is about an order of magnitude higher than the one of the full-scale plants found in Section 8.4 and the bibliography there indicated. The effect of the scale factor can be also observed within these four MBR: MBR 1, the smallest one, has the highest electricity use while MBR 3, the largest one, reports the lowest figure. The scale factor might hide the fact that MBR 3 and MBR 4 present a more complex configuration, with more recirculation, more pumping and, in theory, higher electricity consumption. The electricity use estimated for the dewatering process was found to be irrelevant compared with the one associated with the operation of the reactor (<2%) and thus not included in the inventory table.

Chemical consumption highly depends on the cleaning procedure chosen by the operator. For example, lower solid concentrations would be expected for MBR 4 than for MBR 1 since the former presents attached biomass and a separate chamber for the membrane although it has a more intense cleaning. Due to the short time the inventory is referred to (6–10 months) compared with the lifetime of a membrane, it was not possible to see the long-term effect of the different cleaning processes. Acrylonitrile consumption is proportional to amount of sludge generated, which is significantly higher for MBR 3 than for MBR 1 and MBR 4. The same can be said for air emissions associated with sludge digestion and avoided fertilizers, all affected by the amount of sludge generated, which can be seen in the sludge application row (Table 8.4).

8.5.3 Life cycle impact assessment

8.5.3.1 Impact assessment methodology

The same impact categories used in Section 8.4 and defined in Section 8.4.2.4.1 are applied here. PPCPs, despite being known as toxic substances, have suffered a lack of attention by ready-to-use LCIA methodologies. Muñoz *et al.* (2008) developed the CFs for a large number of PPCPs for the USES-LCA 1.0 and EDIP 97 methodologies. The list was further expanded for USESLCA 1.0 by the same author for Hospido *et al.* (2010) but more contemporary methodologies such as USES-LCA 2.0 or USEtox lack CFs for those substances. To overcome this problem, Alfonsin *et al.* (2012) have developed the factors (CFs) for the two methodologies used in this section.

8.5.3.2 Eutrophication

The same approach described in Section 8.4.2.3.2 is applied here to give the general overview of the four MBR presented in Figure 8.5(a). For the EP, it is possible to realize in the aforementioned figure that it constitutes an environmental benefit for all MBR. As mentioned in Section 8.4.2.3.3, the EP is affected by three different factors that can be better evidenced in Figure 8.5(b).

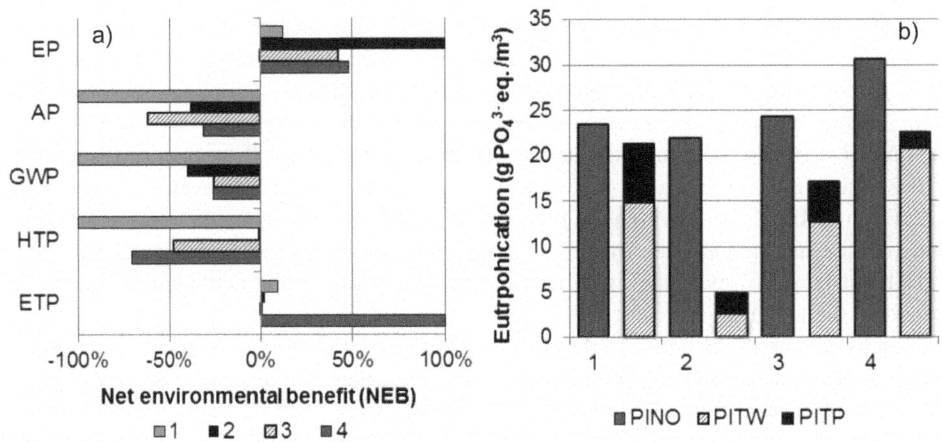

Figure 8.5 (a) NEB of the MBRs under study, expressed as a percentage of the worst-case scenario for losses (−100%) and of the best-case scenario for benefits (+100%) per FU (1 m^3 permeate), (b) eutrophication potential impact of the influent (PI_{NO}), the effluent (PN_{TW}) and the treatment process (PI_{TP}) of the MBRs under study.

Removal efficiency is the main factor affecting MBR 2 best score (Figure 8.5b). Its 54% overall efficiency is responsible for the high NEB of this reactor although, as it has been indicated before, an important part of it is associated with the high P removal. The average removal efficiencies for MBR 1, 3 and 4 are more similar (37, 38 and 32%), being the differences between their NEBs more affected by the influent quality, as it can be inferred from Figure 8.5(b). Finally, the impact associated with the treatment process (PI_{TN}) plays a relevant role in MBR 1 where the eutrophication removed through treatment is almost overcome by the one generated during the production of electricity.

8.5.3.3 Acidification

As for full-scale plants, acidification results in a net environmental loss for any MBR (Figure 8.5a). Within the treatment process electricity is responsible for most of the impact produced in this category (44–92%) due to the amounts of SO_2, HCl and NH_3 released during its generation (Figure 8.6a) and thus it would be expected the reactor with the largest electricity use, MBR 1 (Table 8.4), to be the one with the largest impact (Figure 8.6a). In spite of it, electricity production is not the only source of acidifying substances. NH_3 is also released during the application of the sludge to agriculture, being responsible for 66% of the impact associated with MBR 3. This is due to the large amount of sludge generated by this plant and the high concentration of N on it. In the rest of the MBR sludge production is much smaller and thus this emission is far less relevant (5–7%). The production of chemical used during the treatment or the generation of biogas in MBR 4 causes an even lower impact (1–3%).

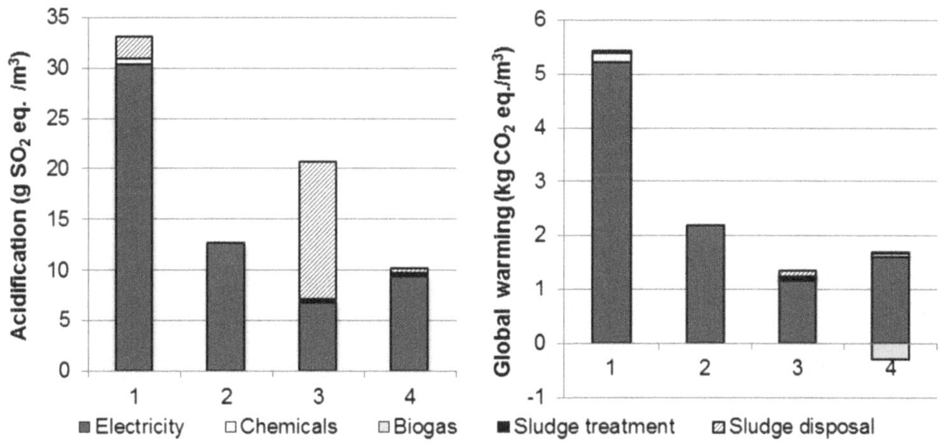

Figure 8.6 (a) Acidification and (b) global warming potential impact the different stages of the treatment process (PI_{TP}) of the MBRs under study per FU (1 m^3 permeate).

8.5.3.4 Global warming

As acidification, global warming results in a net environmental loss (Figure 8.5a), that is completely associated with the treatment process. Also as for acidification, emissions during electricity production, mainly CO_2 from coal and gas combustion, are the main aspects affecting this impact category (Figure 8.6b). The electricity generated during the anaerobic stage of MBR 4 results in an environmental benefit since the emissions associated with it, from the biogas combustion, are lower than the ones expected from the Spanish Electricity Mix, where 16% of the electricity came from coal and 40% from natural gas in the year 2008.

8.5.3.5 Human toxicity

As for the real scale plants assessed in Section 8.4, Figure 8.5 (a) shows human toxicity to be an environmental loss rather than a benefit. The more detailed analysis found in Figure 8.8 (a) shows indirect impact of the treatment process to be the main variable affecting this result.

Focusing on influent and effluent quality, Figure 8.7 (b) highlights the removal efficiency, ranging from 69% for MBR 4 to the 77% of MBR 1 and the principal substances affecting this impact. Zn is the most relevant pollutant in all four reactors, both before and after treatment, even when its concentration is lower than the one reported in other studies such as Carletti *et al.* (2008). The hormone EE2 and the tranquilizer DZP rank second and third in those reactors where PPCPs were analyzed, Ni and Hg holding those positions for MBR 3, where PPCPs were not analyzed. According to the USEtox methodology (Rosenbaum *et al.*, 2008), Zn is the second most toxic heavy metal of those analyzed after Hg (CF 1.3×10^{-3} and 1.4×10^{-2} CTU_H/kg emitted, respectively) but its presence in much higher concentrations than any other metal (Table 8.4) makes it the most relevant substance. On the other hand,

EE2 is relevant because of its toxicity (1.14×10^{-2} CTU_H/kg emitted) although its concentration plays a very relevant role in the case of MBR 4, where much higher concentrations were found in its influent than in the one from MBR 1. It is necessary to indicate that despite the importance of EE2, its relevance, as well as of the other PPCPs, is magnified in Figure 8.7(a) due to the increased concentration in the influent. If concentrations more similar to the ones found in WwTP were used, approximately 100 times lower, EE2 would rank 3rd after Zn and Hg and the impact of other PPCPs would be even lower.

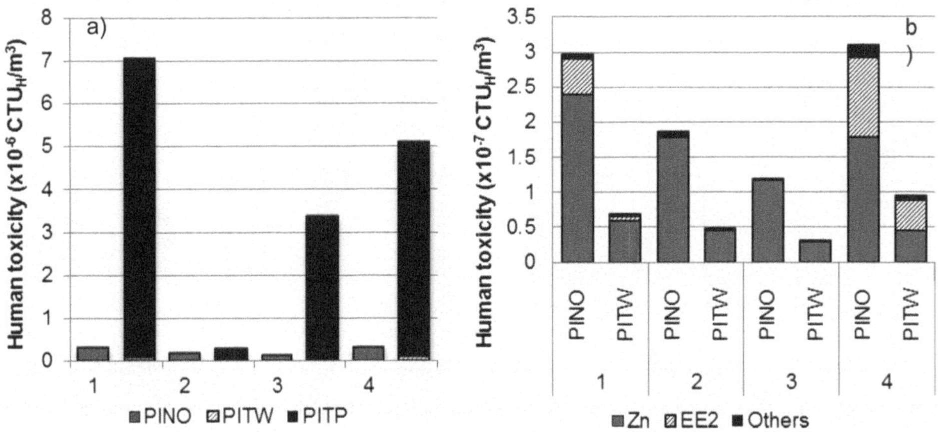

Figure 8.7 (a) Human toxicity potential impact of the influent (PI_{NO}), the effluent (PN_{TW}) and the treatment process (PI_{TP}) of the MBRs under study per FU (1 m^3 permeate), (b) detailed impact for the influent and the effluent per FU (1 m^3 permeate).

The application of sludge is responsible for at least 91% of the impact associated with the treatment process (Figure 8.7a). As in Section 8.4.2.3, this impact is mainly affected by the presence of Zn in the sludge and the high CF given by the USEtox methodology to metals emitted to soil. For those reasons, MBR 1 is the least efficient, since it presents a higher concentration of Zn in the influent (Table 8.4 and Figure 8.7a) and thus a larger fraction is transferred to soil. MBR 4 presented an average concentration of the one found in MBR 1 and MBR 3 and thus it falls between both of them. For MBR 2, sludge application was not modeled due to lack of data regarding sludge production, and thus it presents no increased impact due to treatment aside from the one associated with electricity generation.

8.5.3.6 Freshwater ecotoxicity

The results shown in Figure 8.5(a) indicate the treatment of wastewater using MBRs to be a net environmental benefit, especially for MBR 4, and, to a lesser extent, for MBR 1. As seen in Figure 8.8 (a) this is due to the reduction on the impact generated by the influent.

The impact from the influent and effluent are compared in Figure 8.8(b). It can be seen that the hormone E2 is responsible for the largest share of the impact in both MBR 1 and 4, especially in the latter due to its high concentration, much higher that the theoretically introduced concentration (1 mg/m^3), which is by itself substantially higher than the values found in WwTPs. On the one hand this is due to its toxicity factor (1.84×10^8 compared to 3.9×10^4 CTU_E/kg for Zn, the second most relevant substance) but the high concentration found in the influent MBR 1 (Figure 8.8b) makes possible such a large benefit. As for human toxicity, Zn is the metal with the second highest CF, this time after Cu, but its higher concentration makes its impact more relevant. However, as explained in Section 8.5.3.2.6, the impact of metals does not increase when transferred to soil, reducing the relevance of these substances after treatment.

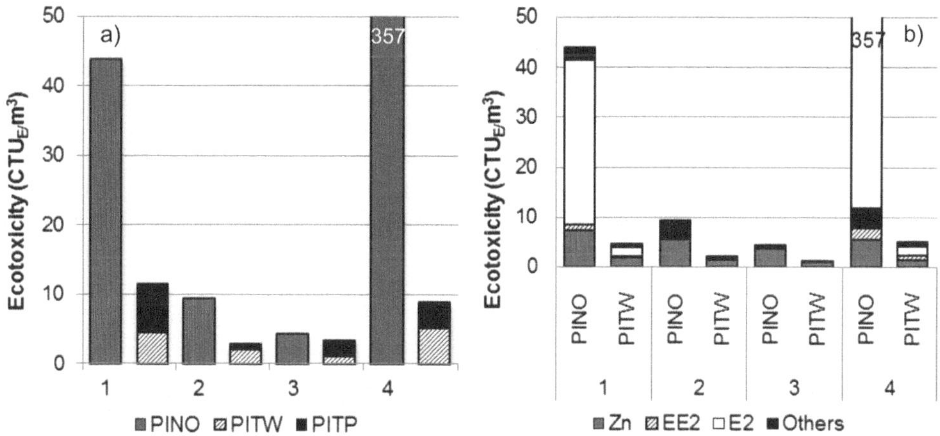

Figure 8.8 (a) Ecotoxicity potential impact of the influent (PI_{NO}), the effluent (PN_{TW}) and the treatment process (PI_{TP}) of the MBRs under study per FU (1 m^3 permeate), (b) detailed impact for the influent and the effluent per FU (1 m^3 permeate).

Cu is the second most relevant source of impact for MBR 2 and 3, where hormones were not introduced. As for human toxicity, the impact of E2 and other PPCPs is influenced by their increased concentration. If values 100 lower are used, E2 becomes the second most relevant substance after Zn for MBR 4 and third more relevant after Zn and Cu for MBR 1.

8.5.4 Result interpretation

Considering the NEB results presented in Figure 8.5(a), MBR 2 presents the best overall profile while MBR 1 presents the worst, the overall classification being MBR 2>MBR 4>MBR 3>MBR 1. Several aspects suggest these results should be considered cautiously. MBR 2 presents a very good profile on the eutrophication category due to a P removal that is not achieved by any other MBR. Similarly, its low impact on the human toxicity; much lower than the other reactors, is due to lack of data regarding sludge production and thus heavy metal emissions to soil. Contrarily, it cannot present a high benefit in the ecotoxicity impact category since hormones were not added to the influent of this reactor and thus, they cannot be removed. Considering the other end of the line, MBR 1 suffers from its lack of N removal, which can be related with the technology itself, but it is also severely affected by exogenous factors such as the higher concentration of Zn in the influent and some that can be attributed to the scale factor such as electricity use.

With regard to toxicity-related categories; environmental concentration has been proved to play a role almost as important as the toxicity potential, as is the case for Zn shown for both human and environmental toxicity. PPCPs in general and hormones in particular can be an issue if not treated correctly, but it would be necessary to conduct similar studies at full-scale plants to ensure that their relevance is not increased by the high concentrations found in the influents under study. According to the results presented here, the proposition of the inclusion of hormones E2 and EE2 as priority pollutants within the EU is justified. The inclusion of Zn would also be desired, if not for its toxicity per se, because of its abundance.

8.6 CONCLUSIONS AND FUTURE OUTLOOK

CBA and LCA-NEB analysis consistently highlight the benefits of the eutrophication reduction achieved by MBR. Influent quality played a major role on both assessments. Important also was the effect of electricity use, which was responsible for one of the largest environmental losses and economic costs of MBR. Even when the two methodologies could benefit from a more standardized comparison (same influent and same scale), they have proven their usefulness when assessing large-scale membrane technologies from economic and environmental points of view.

LCA has been proven to be a valuable tool for the environmental assessment of MBR at both full scale and pilot-plant level. NEB is a suitable approach to systematically assess the benefits and drawbacks of water treatment technologies. The similar tendencies found at both full and pilot-plant scale indicates that the LCA of the latter could be used as a first glimpse for the former. Increased nutrient removal is an aspect that should be considered when choosing a membrane technology. Electricity use has been found to be the main parameter MBR operators can optimize to increase the environmental benefit of the technology. Yet the quality of the influent, both in regard to eutrophication potential and the presence of toxic substances, has been found to be a key parameter to determine the sustainability of an MBR. Neither plant designers nor plant operators control these variables and thus it becomes even more important to reduce pollution sources before these compounds reach the wastewater treatment plant.

REFERENCES

Alfonsin C., Hospido A., Moreira M. T. and Feijoo G. (2012). Assessment of a Hybrid Membrane Technology Focusing on the Environmental Impact Associated to Pharmaceutical and Personal Care Products. ECOSTP 2012: Ecotechnologies for the Wastewater Treatment, IWA Specialist Conference, Santiago de Compostela, Spain, 25–27 June 2012.

Bateman I. J., Cole M. A., Georgiou S. and Hadley D. J. (2006). Comparing contingent valuation and contingent ranking: a case study considering the benefits of urban river quality improvements. *Journal of Environmental Management*, **79**(3), 221–231.

Baumann H. and Tillman A. M. (2004). The Hitch Hiker's Guide to LCA. Studentlitteratur AB, Lund, Sweden.

Becker A. M., Yu K., Stadler L. B. and Smith A. L. (2017). Co-management of domestic wastewater and food waste: a life cycle comparison of alternative food waste diversion strategies. *Bioresource Technology*, **223**, 131–140.

Bengtsson M., Lundin M. and Molander S. (1997). Life Cycle Assessment of Wastewater Systems: Case Studies of Conventional Treatment, Urine Sorting and Liquid Composting in Three Swedish Municipalities. PhD Thesis, Chalmers University of Technology.

Bertanza G., Canato M., Laera G., Vaccari M., Svanström M. and Heimersson S. (2017). A comparison between two full-scale MBR and CAS municipal wastewater treatment plants: techno-economic-environmental assessment. *Environmental Science and Pollution Research*, **24**(21), 17383–17393.

Birol E., Karousakis K. and Koundouri P. (2006). Using economic valuation techniques to inform water resources management: a survey and critical appraisal of available techniques and an application. *Science of the Total Environment*, **365**(1–3), 105–122.

Boyle K. J., Desvousges W. H., Johnson R. H., Dunford S. and Hudson S. H. (1993). An investigation of part-whole biases in contingent valuation studies. *Journal of Environmental Economics and Management*, **27**(1), 64–83.

Brannock M. W. D., Wang Y. and Leslie G. (2010). Evaluation of full-scale membrane bioreactor mixing performance and the effect of membrane configuration. *Journal of Membrane Science*, **350**(1–2), 101–108.

Buntner D., Sánchez Sánchez A. and Garrido J. M. (2011). Three stages MBR (methanogenic, aerobic biofilm and membrane filtration) for the treatment of low-strength wastewaters. *Water Science and Technology*, **64**(2), 397.

Carletti G., Fatone F., Bolzonella D. and Cecchi F. (2008). Occurrence and fate of heavy metals in large wastewater treatment plants treating municipal and industrial wastewaters. *Water Science and Technology*, **57**(9), 1329.

Clauson-Kaas J.C., Dahl F., Dalgaard O., Høibye L., Kjølholt J., Wenzel H. and Larsen H. F. (2006). Videregående renseteknologier for kommunalt spildevand (Advanced treatment technologies for municipal wastewater). DANVA, Lynettefællesskabet og Spildevandscenter. Miljøstyrelsen, Denmark.

Corominas L., Foley J., Guest J. S., Hospido A., Larsen H. F. and Shaw A. (2011). Towards a Standard Method for Life Cycle Assessments (LCA) of Wastewater Treatment Watermatex 2011, 8th IWA Symposium on System Analysis and Integrated Assessment, San Sebastian, Spain, 20–22 June 2011.

Corominas L., Flores-Alsina X., Snip L. and Vanrolleghem P. A. (2012). Comparison of different modeling approaches to better evaluate greenhouse gas emissions from whole wastewater treatment plants. *Biotechnology and Bioengineering*, **109**(11), 2854–2863.

Côté P., Masini M. and Mourato D. (2004). Comparison of membrane options for water reuse and reclamation. *Desalination*, **167**(0), 1–11.

Côté P., Siverns S. and Monti S. (2005). Comparison of membrane-based solutions for water reclamation and desalination. *Desalination*, **182**(1–3), 251–257.

Diamond M., Gandhi N., Adams W., Atherton J., Bhavsar S., Bulle C., Campbell P., Dubreuil A., Fairbrother A., Farley K., Green A., Guinee J., Hauschild M., Huijbregts M., Humbert S., Jensen K., Jolliet O., Margni M., McGeer J., Peijnenburg W., Rosenbaum R., van de Meent D. and Vijver M. (2010). The clearwater consensus: the estimation of metal hazard in fresh water. *The International Journal of Life Cycle Assessment*, **15**(2), 143–147.

Doka G. (2009). Wastewater Treatment. Life Cycle Inventories on Waste Treatment. Swiss Centre for Life Cycle Inventories, Dübendorf, Switzerland.

Dominguez S., Laso J., Margallo M., Aldaco R., Rivero M. J., Irabien Á. and Ortiz I. (2018). LCA of greywater management within a water circular economy restorative thinking framework. *Science of the Total Environment*, **621**, 1047–1056.

Doorn M. R. J., Towprayoon S., Manso-Vieira S. M., Irving W., Palmer C., Pipatti R. and Wang C. (2006). Wastewater treatment and discharge. In: Waste. 2006 IPCC Guidelines for National Greenhouse Gas Inventories, S. Eggleston, L. Buendia, K. Miwa and K. Tanabe (eds), IGES, Japan, 6.1–6.28. [online] http://www.ipcc-nggip.iges.or.jp/public/2006gl/pdf/5_Volume5/V5_6_Ch6_Wastewater.pdf.

EC (1998). Commission Directive 98/15/EC of 27 February 1998 amending Council Directive 91/271/EEC with respect to certain requirements established in Annex I thereof. Official Journal of the European Communities L 067, 07/03/1998, 29–30.

EC (2000). Directive 2000/60/EC of the European Parliament and of the Council of 23 October 2000 establishing a framework for Community action in the field of water policy Official Journal of the European Communities L 327, 22/12/2000 1–73.

EC (2001). Decision No 2455/2001/EC of the European Parliament and of the Council of 20 November 2001 establishing the list of priority substances in the field of water policy and amending Directive 2000/60/EC. Official Journal of the European Communities. L 331, 15/12/2001, 1–5.

EC (2008). Directive 2008/105/EC of the European Parliament and of the Council of 16 December 2008 on environmental quality standards in the field of water policy, amending and subsequently repealing Council Directives 82/176/EEC, 83/513/EEC, 84/156/EEC, 84/491/EEC, 86/280/EEC and amending Directive 2000/60/EC of the European Parliament and of the Council. Official Journal of the European Communities L 348, 24/12/2008, 84–97.

EC (2012). Directive of the European Parliament and of the Council amending Directives 2000/60/EC and 2008/105/EC as regards priority substances in the field of water policy. [online] http://ec.europa.eu/smart-regulation/impact/ia_carried_out/docs/ia_2012/com_2011_0876_en

EEC (1991). Council Directive 91/271/EEC of 21 May 1991 concerning urban waste-water treatment. Official Journal of the European Communities. L 135, 30/05/1991, 40–52.

El Samrani A. G., Lartiges B. S. and Villiéras F. (2008). Chemical coagulation of combined sewer overflow: heavy metal removal and treatment optimization. *Water Research*, **42**(4–5), 951–960.

Emmerson R. H. C., Morse G. K., Lester J. N. and Edge D. R. (1995). The life-cycle analysis of small-scale sewage-treatment processes. *Water and Environment Journal*, **9**(3), 317–325.

European Commission-Joint Research Group-Institute for Environment and Sustainability (2010). International Reference Life Cycle Data System (ILCD) Handbook – General guide for Life Cycle Assessment, Luxemburg. Publications Office of the European Union, Luxemburg.

Expósito A. (2010). Eliminación de Compuestos Cosméticos y Farmacéuticos (CCFs) en un Biorreactor Híbrido de Membrana (HMBR) a Escala Piloto [Removal of pharmaceutical and personal care products (PPCPs) in Hybrid Membrane Bioreactor (HMBR) at pilot-plant scale]. Master Thesis. Department of Chemical Engineering, University of Santiago de Compostela. Santiago de Compostela, Spain

Färe R., Grosskopf S., Lovell C. A. and Yaisawarng S. (1993). Derivation of shadow prices for undesirable outputs: a distance function approach. *Review of Economics and Statistics*, **75**(2), 374–380.

Färe R., Grosskopf S. and Weber W. L. (2001). Shadow prices of Missouri public conservation land. *Public Finance Review*, **29**(6), 444–460.

Färe R., Grosskopf S. and Weber W. L. (2006). Shadow prices and pollution costs in U.S. agriculture. *Ecological Economics*, **56**(1), 89–103.

Flückiger P. and Gubler D. (1994). Grundlagen für die Integration des Downstreamprozesses Abwasserreinigung in Ökobilanzen. Analyse von zwei Fallbeispielen ARA Twann/Ligerz/Lüscherz und ARA Werdhölzli/Zürich [Basis for the integration of the downstream wastewater treatment process in ecobalances. Analysis of two case studies: WwTP Twann/Ligerz/Lüscherz and WwTP Werdhölzli/Zurich]. Diploma work ETH Zürich.

Foley J. (2009). Life cycle assessment of wastewater treatment systems. PhD thesis. Advanced Water Management Center. University of Queensland, Brisbane, Australia

Foley J., de Haas D., Hartley K. and Lant P. (2010). Comprehensive life cycle inventories of alternative wastewater treatment systems. *Water Research*, **44**(5), 1654–1666.

Fu F. and Wang Q. (2011). Removal of heavy metal ions from wastewaters: a review. *Journal of Environmental Management*, **92**(3), 407–418.

Gabarron S., Ferrero G., Moragas L., Robusté J., Comas J. and Rodriguez-Roda I. (2011). Review of Design and Operational Parameters of Full-Scale Membrane Bioreactors for Municipal Wastewater Treatment in Catalonia Water and Industry 2011, IWA Specialist Conference Valladolid, Spain, 1–4 May 2011.

Gallego A., Hospido A., Moreira M. T. and Feijoo G. (2008). Environmental performance of wastewater treatment plants for small populations. *Resources, Conservation and Recycling*, **52**(6), 931–940.

Garrido J. M., Iglesias A. and Fernández R. (2009). Operation of a Membrane Bioreactor Fed with Primary Treated Municipal Wastewater. 8th Aachen Conference Water and Membranes, Aachen, Germany, 27–28 October 2009.

Gil J. A., Túa L., Rueda A., Montaño B., Rodríguez M. and Prats D. (2010). Monitoring and analysis of the energy cost of an MBR. *Desalination*, **250**(3), 997–1001.

Godfrey S., Labhasetwar P. and Wate S. (2009). Greywater reuse in residential schools in Madhya Pradesh, India. A case study of cost-benefit analysis. *Resources, Conservation and Recycling*, **53**(5), 287–293.

Godin D., Bouchard C. and Vanrolleghem P. A. (2012). Net environmental benefit: introducing a new LCA approach on wastewater treatment systems. *Water Science and Technology*, **65**(9), 16–24.

Gonzalez-Serrano E., Rodriguez-Mirasol J., Cordero T., Koussis A. D. and Rodriguez J. J. (2006). Cost of reclaimed municipal wastewater for applications in seasonally stressed semi-arid regions. *Journal of Water Supply: Research and Technology- AQUA*, **54**(6), 355–369.

Guinée J. B., Heijungs R., Huppes G., Keijn R., de Koning A., van Oers L., Sleeswijk A. W., Suh S., Udo de Haes H. A., de Bruij H., van Duin R. and Huijbregts M. A. J. (2002). Handbook on Life Cycle Assessment: Operational Guide to the ISO Standards, Dordrecht (The Netherlands). Kluwer Academic Publishers. Dordrecht, The Netherlands.

Günder B. and Krauth K. (1999). Replacement of secondary clarification by membrane separation – results with tubular, plate and hollow fibre modules. *Water Science and Technology*, **40**(4–5), 311–320.

Hernández-Sancho F., Molinos-Senante M. and Sala-Garrido R. (2010). Economic valuation of environmental benefits from wastewater treatment processes: an empirical approach for Spain. *Science of the Total Environment*, **408**(4), 953–957.

Hernández-Sancho F., Molinos-Senante M. and Sala-Garrido R. (2011). Cost modelling for wastewater treatment processes. *Desalination*, **268**(1–3), 1–5.

Hobson J. (2003). *CH_4 and N_2O emissions from waste water handling*, International Panel on Climate Change. [online] http://www.ipcc-nggip.iges.or.jp/public/gp/english/ (Accessed May 9, 2012).

Høibye L., Clauson-Kaas J., Wenzel H., Larsen H. F., Jacobsen B. N. and Dalgaard O. (2008). Sustainability assessment of advanced wastewater treatment technologies. *Water Science and Technology*, **58**(5), 963.

Holloway R. W., Miller-Robbie L., Patel M., Stokes J. R., Munakata-Marr J., Dadakis J. and Cath T. Y. (2016). Life-cycle assessment of two potable water reuse technologies: MF/RO/UV–AOP treatment and hybrid osmotic membrane bioreactors. *Journal of Membrane Science*, **507**, 165–178.

Hospido A., Moreira M.T., Fernández-Couto M. and Feijoo G. (2004). Environmental performance of a municipal wastewater treatment plant. *The International Journal of Life Cycle Assessment*, **9**(4), 261–271.

Hospido A., Moreira M. T., Martin M., Rigola M. and Feijoo G. (2005). Environmental evaluation of different treatment processes for sludge from urban wastewater treatments: anaerobic digestion versus thermal processes. *The International Journal of Life Cycle Assessment,* **10**(5), 336–345.

Hospido A., Moreira M. T. and Feijoo G. (2008). A comparison of municipal wastewater treatment plants for big centres of population in Galicia (Spain). *The International Journal of Life Cycle Assessment*, **13**(1), 57–64.

Hospido A., Carballa M., Moreira M., Omil F., Lema J. M. and Feijoo G. (2010). Environmental assessment of anaerobically digested sludge reuse in agriculture: potential impacts of emerging micropollutants. *Water Research*, **44**(10), 3225–3233.

Hospido A., Sanchez I., Rodriguez-Garcia G., Iglesias A., Buntner D., Reif R., Moreira M. T. and Feijoo G. (2012). Are all membrane reactors equal from an environmental point of view? *Desalination*, **285**(0), 263–270.

Houillon G. and Jolliet O. (2005). Life cycle assessment of processes for the treatment of wastewater urban sludge: energy and global warming analysis. *Journal of Cleaner Production*, **13**(3), 287–299.

Hutton G. and Haller L. (2004). Evaluation of the Costs and Benefits of Water and Sanitation Improvements at the Global Level. World Health Organization, Geneva.

Ioannou-Ttofa L., Foteinis S., Chatzisymeon E. and Fatta-Kassinos D. (2016). The environmental footprint of a membrane bioreactor treatment process through life cycle analysis. *Science of the Total Environment*, **568**, 306–318.

ISO (2006a). ISO 14040:2006 Environmental management – Life cycle assessment – Principles and framework.

ISO (2006b). ISO 14044:2006 Environmental management – Life cycle assessment – Requirements and guidelines.

Jeong H., Broesicke O. A., Drew B. and Crittenden J. C. (2018). Life cycle assessment of small-scale greywater reclamation systems combined with conventional centralized water systems for the City of Atlanta, Georgia. *Journal of Cleaner Production*, **174**, 333–342.

Jolliet O, Margni M, Charles R, Humbert S, Payet J, Rebitzer G and Rosenbaum R (2003). IMPACT 2002+ a new life cycle impact assessment methodology. *The International Journal of Life Cycle Assessment*, **8**(6), 234–330.

Jolliet O., Rosenbaum R., McKone T. E., Scheringer M., Straalen N. van and Wania F. (2006). Establishing a framework for life cycle toxicity assessment. Findings of the Lausanne review workshop. *The International Journal of Life Cycle Assessment*, **11**(3), 209–212.

Judd S. (2011). The MBR Book, Second Edition: Principles and Applications of Membrane Bioreactors for Water and Wastewater Treatment. Butterworth-Heinemann, https://www.elsevier.com/books/the-mbr-book/judd/978-0-08-096682-3.

Kalbar P. P., Karmakar S. and Asolekar S. R. (2012). Technology assessment for wastewater treatment using multiple-attribute decision-making. *Technology in Society*, **34**(4), 295–302.

Lam C.-M., Leng L., Chen P.-C., Lee P.-H. and Hsu S.-C. (2017). Eco-efficiency analysis of non-potable water systems in domestic buildings. *Applied Energy*, **202**, 293–307.

Larsen H. F., Hauschild M., Wenzel H. and Almemark M. (2007). Homogeneous LCA methodology agreed by Neptune and INNOWATECH Deliverable 4.1. EC Project "NEPTUNE", contract No.: 036845, [online] www.eu-neptune.org.

Larsen H. F., Olsen S. I., Hauschild M. and Laurent A. (2009). Methodology for including speficic biological effects and pathogens aspects into LCA Deliverable 4.2. EC Project "NEPTUNE", contract No.: 036845 [online] www.eu-neptune.org.

Larsen H. F., Hansen P. A. and Boyer-Souchet F. (2010). Decision support guideline based on LCA and cost/efficiency assessment. Deliverable 4.3. EC Project "NEPTUNE", contract No.: 036845. [online] www.eu-neptune.org.

Lassaux S., Renzoni R. and Germain A. (2007). Life cycle assessment of water: from the pumping station to the wastewater treatment plant (9 pp). *The International Journal of Life Cycle Assessment*, **12**(2), 118–126.

Lazarova V., Ruel S. M., Barillon B. and Dauthuille P. (2012). The role of MBR technology for the improvement of environmental footprint of wastewater treatment. *Water Science and Technology*, **66**(10), 2056–2064.

Lesjean B. and Huisjes E. H. (2008). Survey of the European MBR market: trends and perspectives. *Desalination*, **231** (1–3), 71–81.

Lesjean B., Ferre V., Vonghia E. and Moeslang H. (2009). Market and design considerations of the 37 larger MBR plants in Europe. *Desalination and Water Treatment*, **6**(1–3), 227–233.

Life Cycle Initiative (2008). Life Cycle Assessment Training Manual. [online] http://www.estis.net/sites/lcinit/default.asp?site=lcinit&page_id=56666AB6-E732-45F2-A89E-640951EA5F59 (Accessed June 6, 2012).

Lundie S., Peters G. M. and Beavis P. C. (2004). Life cycle assessment for sustainable metropolitan water systems planning. *Environmental Science & Technology*, **38**(13), 3465–3473.

Lundin M., Bengtsson M. and Molander S. (2000). Life cycle assessment of wastewater systems: influence of system boundaries and scale on calculated environmental loads. *Environmental Science & Technology*, **34**(1), 180–186.

Mahgoub M. E.-S. M., van der Steen N. P., Abu-Zeid K. and Vairavamoorthy K. (2010). Towards sustainability in urban water: a life cycle analysis of the urban water system of Alexandria City, Egypt. *Journal of Cleaner Production*, **18** (10–11), 1100–1106.

Memon F., Zheng Z., Butler D., Shirley-Smith C., Lui S., Makropoulos C. and Avery L. (2007). Life cycle impact assessment of greywater recycling technologies for new developments. *Environmental Monitoring and Assessment*, **129**(1), 27–35.

MITYC (2009). La Energia en Espana 2008 (Energy in Spain 2008), Ministerio de Industria. Turismo y Comercio, Madrid, Spain.

MMA (Spanish Ministry of Environment) (2007). Metodologia de analisis de las opciones de oferta para la gestion del agua en la cuenca hidrográfica del Segura (*Methodology for analyzing supply options for the management of water in the Segura River Basin,* in Spanish). Available from: http://www.mma.es/secciones/agua.htm

Molinos-Senante M., Hernández-Sancho F. and Sala-Garrido R. (2010). Economic feasibility study for wastewater treatment: a cost-benefit analysis. *Science of the Total Environment*, **408**(20), 4396–4402.

Molinos-Senante M., Hernández-Sancho F. and Sala-Garrido R. (2011a). Cost-benefit analysis of water reuse projects for environmental purposes: a case study for Spanish wastewater treatment plants. *Journal of Environmental Management*, **92**(12), 3091–3097.

Molinos-Senante M., Hernández-Sancho F., Sala-Garrido R. and Garrido-Baserba M. (2011b). Economic feasibility study for phosphorus recovery processes. *AMBIO*, **40**(4), 408–416.

Molinos-Senante M., Hernández-Sancho F. and Sala-Garrido R. (2012). Economic feasibility study for new technological alternatives in wastewater treatment process: a review. *Water Science and Technology*, **65**(5), 898–906.

Monclús H., Sipma J., Ferrero G., Comas J. and Rodriguez-Roda I. (2010). Optimization of biological nutrient removal in a pilot plant UCT-MBR treating municipal wastewater during start-up. *Desalination*, **250**(2), 592–597.

Muñoz I., Gómez M. J., Molina-Díaz A., Huijbregts M. A. J., Fernández-Alba A. R. and García-Calvo E. (2008). Ranking potential impacts of priority and emerging pollutants in urban wastewater through life cycle impact assessment. *Chemosphere* **74**(1), 37–44.

Muñoz I., Rodríguez A., Rosal R. and Fernández-Alba A. R. (2009). Life cycle assessment of urban wastewater reuse with ozonation as tertiary treatment: a focus on toxicity-related impacts. *Science of the Total Environment*, **407**(4), 1245–1256.

Ni B.-J., Yuan Z., Chandran K., Vanrolleghem P. A. and Murthy S. (2012). Evaluating four mathematical models for nitrous oxide production by autotrophic ammonia-oxidizing bacteria. *Biotechnology and Bioengineering*, **110**(1), 153–163.

Nogueira R., Ferreira I., Janknecht P., Rodríguez J. J., Oliveira P. and Brito A. G. (2007). Energy-saving wastewater treatment systems: formulation of costs functions. *Water Science and Technology*, **56**(3), 85–92.

ONE (2008). Rapport d'activités 2007 (Activities report 2007). Office National de l'Electricité, Casablanca, Morocco.

Ort C., Lawrence M. G., Reungoat J. and Mueller J. F. (2010). Sampling for PPCPs in wastewater systems: comparison of different sampling modes and optimization strategies. *Environmental Science & Technology*, **44**(16), 6289–6296.

Ortiz M., Raluy R. G. and Serra L. (2007). Life cycle assessment of water treatment technologies: wastewater and water-reuse in a small town. *Desalination*, **204**(1–3), 121–131.

Papadopoulos B., Tsagarakis K. P. and Yannopoulos A. (2007). Cost and land functions for wastewater treatment projects: typical simple linear regression versus fuzzy linear regression. *Journal of Environmental Engineering, ASCE*, **133**(6), 581–586.

Pertl A., Mostbauer P. and Obersteiner G. (2010). Climate balance of biogas upgrading systems. *Waste Management*, **30** (1), 92–99.

Pizzol M., Christensen P., Schmidt J. and Thomsen M. (2011). Impacts of "metals" on human health: a comparison between nine different methodologies for life cycle impact assessment (LCIA). *Journal of Cleaner Production*, **19**(6–7), 646–656.

Porro J., Guo L., Nopens I., Sharma K., Benedetti L., Van Hulle S., Vanrolleghem P. A., Amerlinc Y., Yuan Z., Shaw A. and Nopen I. (2011). Towards a Benchmarking Tool for Minimizing Wastewater Utility Greenhouse Gas Footprints. 8th IWA Symposium on System Analysis and Integrated Assessment, San Sebastian, Spain, 20–22 June 2011.

Pretel R., Robles A., Ruano M. V., Seco A. and Ferrer J. (2013). Environmental impact of submerged anaerobic MBR (SAnMBR) technology used to treat urban wastewater at different temperatures. *Bioresource Technology*, **149**, 532–540.

Pretel R., Shoener B. D., Ferrer J. and Guest J. S. (2015). Navigating environmental, economic, and technological trade-offs in the design and operation of submerged anaerobic membrane bioreactors (AnMBRs). *Water Research*, **87**, 531–541.

Pretel R., Robles A., Ruano M. V., Seco A. and Ferrer J. (2016a). Economic and environmental sustainability of submerged anaerobic MBR-based (AnMBR-based) technology as compared to aerobic-based technologies for moderate-/high-loaded urban wastewater treatment. *Journal of Environmental Management*, **166**, 45–54.

Pretel R., Moñino P., Robles A., Ruano M. V., Seco A. and Ferrer J. (2016b). Economic and environmental sustainability of an AnMBR treating urban wastewater and organic fraction of municipal solid waste. *Journal of Environmental Management*, **179**, 83–92.

REE (2009). El Sistema Electrico Espanol 08 (Spanish Electric System 08), Red Electrica de Espana, Alcobendas, Spain.

Reif R., Suárez S., Omil F. and Lema J. M. (2008). Fate of pharmaceuticals and cosmetic ingredients during the operation of a MBR treating sewage. *Desalination*, **221**(1–3), 511–517.

Rodriguez-Garcia G., Molinos-Senante M., Hospido A., Hernandez-Sancho F., Moreira M. T. and Feijoo G. (2011). Environmental and economic profile of six typologies of wastewater treatment plants. *Water Research*, **45**(18), 5997–6010.

Rodriguez-Garcia G., Hospido A., Bagley D. M., Moreira M. T. and Feijoo G. (2012). A methodology to estimate greenhouse gases emissions in life cycle inventories of wastewater treatment plants. *Environmental Impact Assessment Review*, **37**(0), 37–46.

Rodriguez-Roda I., Comas J., Ferrero G., Sancho L., Garrido J. M., Tejero I. and Díaz P. (2009). Estudio económico e identificación de las oportunidades de los BioReactores de Membrana para el tratamiento de aguas residuales (Economic assessment and opportunity identification for membrane bioreactors treating wastewater) In: El Reto de la Eficiencia Económica en la EDAR: Integrando la Economía en la Concepción, Rediseño y Gestión de EDAR, F. Hernández-Sancho and F. Fernández-Polanco Fernández de Moreda (eds), Lapices 4, Santiago de Compostela, Spain.

Rosenbaum R., Bachmann T., Gold L., Huijbregts M., Jolliet O., Juraske R., Koehler A., Larsen H., MacLeod M., Margni M., McKone T., Payet J., Schuhmacher M., van de Meent D. and Hauschild M. (2008). USEtox – the

UNEP-SETAC toxicity model: recommended characterisation factors for human toxicity and freshwater ecotoxicity in life cycle impact assessment. *The International Journal of Life Cycle Assessment*, **13**(7), 532–546.

Sánchez Sánchez A., Garrido J. M. and Méndez R. (2010). A comparative study of tertiary membrane filtration of industrial wastewater treated in a granular and a flocculent sludge SBR. *Desalination*, **250**(2), 810–814.

Santos A. and Judd S. (2010a). The commercial status of membrane bioreactors for municipal wastewater. *Separation Science and Technology*, **45**(7), 850–857.

Santos A. and Judd S. (2010b). The fate of metals in wastewater treated by the activated sludge process and membrane bioreactors: a brief review. *Journal of Environmental Monitoring*, **12**(1), 110.

Seijas M. T., Lago D. and Omil F. (2003). Identificacion de problemas durante la operacion de una planta de tratamiento de aguas municipal y propuesta de medidas correctoras (Identifying problems during the operation of a municipal waste water treatment plant and proposing corrective actions). *Tecnologia del Agua*, (234), 33–40.

Shabman L. and Stephenson K. (2000). Environmental valuation and its economic critics. *Journal of Water Resources Planning and Management-ASCE*, **126**(6), 382–388.

Shoener B. D., Zhong C., Greiner A. D., Khunjar W. O., Hong P.-Y. and Guest J. S. (2016). Design of anaerobic membrane bioreactors for the valorization of dilute organic carbon waste streams. *Energy & Environmental Science*, **9**(3), 1102–1112.

Sipala S., Mancini G. and Vagliasindi F. G. A. (2005). Development of a web-based tool for the calculation of costs of different wastewater treatment and reuse scenarios. *Water Science and Technology: Water Supply*, **3**(4), 89–96.

Smith A. L., Stadler L. B., Cao L., Love N. G., Raskin L. and Skerlos S. J. (2014). Navigating wastewater energy recovery strategies: a life cycle comparison of anaerobic membrane bioreactor and conventional treatment systems with anaerobic digestion. *Environmental Science & Technology*, **48**(10), 5972–5981.

Sonnemann G. (2004). Strengthening capacity building through regional networks. *The International Journal of Life Cycle Assessment*, **9**(5), 334–334.

Suh S. (2002). Missing Inventorv Estimation Tool: An Inventory Estimation Tool for Missing Flows Using Input-Output Technique, Version 2.0. Centre of Environmental Science, Leiden University. Leiden, The Netherlands.

Suh Y. J. and Rousseaux P. (2001). Considerations in life cycle inventory analysis of municipal wastewater treatment systems. Oral presentation at COST, 624.

Swiss Centre for Life Cycle Inventories (2012). Ecoinvent Data 2.2. [online] http://www.ecoinvent.org/ (Accessed May 15, 2012).

Tangsubkul N., Beavis P., Moore S. J., Lundie S. and Waite T. D. (2005). Life cycle assessment of water recycling technology. *Water Resources Management*, **19**(5), 521–537.

Tangsubkul N., Parameshwaran K., Lundie S., Fane A. G. and Waite T. D. (2006). Environmental life cycle assessment of the microfiltration process. *Journal of Membrane Science*, **284**(1–2), 214–226.

Tchobanoglous G., Burton F. L. and Stensel H. D. (2002). Wastewater Engineering: Treatment and Reuse. McGraw-Hill Science/Engineering/Math, New York, NY, US.

Tillman A.M., Svingby M. and Lundström H. (1998). Life cycle assessment of municipal waste water systems. *The International Journal of Life Cycle Assessment*, **3**(3), 145–157.

Verrecht B., Maere T., Nopens I., Brepols C. and Judd S. (2010a). The cost of a large-scale hollow fibre MBR. *Water Research*, **44**(18), 5274–5283.

Verrecht B., Maere T., Benedetti L., Nopens I. and Judd S. (2010b). Model-based energy optimisation of a small-scale decentralised membrane bioreactor for urban reuse. *Water Research*, **44**(14), 4047–4056.

Wenzel H., Hauschild M. and Alting L. (1997). Environmental Assessment of Products, Vol. **1**. 1st ed. Chapman & Hall, London, Great Britain.

Wenzel H., Larsen H. F., Clauson-Kaas J., Høibye L. and Jacobsen B. N. (2008). Weighing environmental advantages and disadvantages of advanced wastewater treatment of micro-pollutants using environmental life cycle assessment. *Water Science and Technology*, **57**(1), 27.

Yilmaz G., Lemaire R., Keller J. and Yuan Z. (2007). Effectiveness of an alternating aerobic, anoxic/anaerobic strategy for maintaining biomass activity of BNR sludge during long-term starvation. *Water Research*, **41**(12), 2590–2598.

Zhang Z. and Wilson F. (2000). Life-cycle assessment of a sewage-treatment plant in South-East Asia. *Journal Chartered Institution Water Environmental Management*, **14**(1), 51–56.

Chapter 9

MBR modeling studies

G. Andreottola[1]*, G. Guglielmi[2] and M. Langone[1]

[1]Dept. Civil and Environmental Engineering, University of Trento, Italy
[2]ETC Engineering srl, Trento, Italy
*Corresponding author: gianni.andreottola@unitn.it

ABSTRACT

In recent years, membrane bioreactors (MBRs) have been increasingly employed for municipal and industrial wastewater treatment. MBR modeling studies to date have focused on the biological processes as well as the engineering aspects (e.g., filtration, membrane fouling and tank hydrodynamics).

In this chapter, the developments and the current state-of-the-art in biological modeling of MBR is presented. This review explores both single mechanisms (e.g., biological, filtration and hydrodynamic models) and multiple mechanisms (hybrid models). The specific features and advantages of each model are presented and discussed. Unmodified and modified applications of Activated Sludge Models (ASMs) to model biological processes in MBR are presented, taking into account features such as influent characteristics, biokinetics, and formation, release and destruction of soluble microbial products (SMPs)/extracellular polymeric substances (EPS). Models which describe filtration and fouling phenomena as well as the ones investigating the hydrodynamic conditions in MBRs are presented. Finally a brief overview of control and operational strategies of MBRs is given.

9.1 INTRODUCTION

MBRs are being increasingly implemented to treat and reuse wastewater, due to their many advantages over the conventional activated sludge (CAS) processes, in terms of high effluent quality, low sludge production and very compact plant size. MBR systems combine conventional biological treatment processes with membrane filtration to provide efficient organics, nitrogen and suspended solids removal.

 Membrane Biological Reactors: Theory, Modeling, Design, Management and Applications to Wastewater Reuse
Editor(s): Faisal I. Hai, Kazuo Yamamoto and Chung-Hak Lee
doi: 10.2166/9781780409177_0269

Recent studies on MBR technology have underlined several crucial specificities of MBR processes compared to CAS systems, such as a different microbial composition (Jiang *et al.* 2005), a higher mixed liquor concentration and a higher viscosity (Germain *et al.* 2007), an elevated sludge retention time, a higher retention of slower growing bacteria, an accumulation of SMP (Gao *et al.* 2004) and a higher aeration rate (Judd, 2006). Coverage of these specificities requires new approaches to modeling (Fenu *et al.* 2010a; Naessens *et al.* 2012; Zuthi *et al.* 2012).

MBRs are characterized by complexities resulting from the interaction of dynamic biological, membrane filtration, fouling and hydrodynamic processes. Due to their intrinsic complexity, integrated models that take into account the different processes involved (biological processes, filtration, fouling and hydrodynamics) would be strongly needed in order to increase the accuracy of MBR design and operation.

Currently, only few attempts have been made to propose integrated models, while, in most cases, models have been developed for simulating specific mechanisms in MBR systems separately: biokinetic, filtration and hydrodynamic models.

This chapter aims to review available studies on MBR modeling with special regard to some specificities of the treatment system. A review of modeling approaches considering both single mechanisms (e.g., biological, filtration and hydrodynamic models) and multiple mechanisms (hybrid models) have been carried out.

The chapter aims at providing a structured overview of the literature on the modeling studies applied to MBRs. Section 9.2 deals with biological models. The development of the Activated Sludge Models (ASMs) made an important contribution toward modeling the biological processes involved in the CAS systems. When the ASMs models are applied to the MBR technology, adaptations are necessary in order to meet the specific behaviour of MBRs. A summary of the past and the current state-of-the-art of the biological modeling of MBR systems is proposed, distinguishing the unmodified/conventional models and the modified/integrated models. Section 9.3 treats filtration and fouling phenomena in MBRs, describing different models developed. Section 9.4 reviews the hydrodynamic models applied to MBR systems, describing the computational fluid dynamics (CFD) modeling, which is a widely applied tool to investigate flow conditions in MBRs. In Section 9.5, a brief overview of control and operational strategies of MBRs is given. Finally, the conclusion section discusses knowledge gaps and perspectives in MBR modeling.

9.2 BIOLOGICAL MODELS

9.2.1 Introduction to ASM models

The development of the family of Activated Sludge Models has been a milestone in biological wastewater treatment modeling. Known today as **ASMs**, (ASM1 to ASM3, see Henze *et al.* (2000) for details of the models, and Gujer (2006), for a description of their development), ASMs were developed by the Task Group on Mathematical Modeling for Design and Operation of Activated Sludge Processes established by the International Association on Water Pollution Research and Control (IAWPCR), as it was then called. In this chapter, standardized notations in wastewater treatment modeling, according to Corominas *et al.* (2010), are adopted (Table 9.1).

The ASMs models have been built on the basic concepts originally introduced in a steady state model by Marais and Ekama (1976) and later proposed as a well-established dynamic model (Dold *et al.* 1980). The ASMs family represents a standard set of mathematical models, describing biological transformations carried out by microorganisms. The biological processes are described as biochemical reactions defined by stoichiometric relations and kinetic equations, based on the accepted Monod-type kinetics, using a matrix format (the Petersen or Gujer table). The matrix format is composed of a stoichiometric and a composition matrix, a rate vector and extra information as units and names. The

Table 9.1 List of components of ASMs.

Components	Standard Notation	Unit	ASM1	ASM2/ ASM2d	ASM3	ASM2d2N
			Henze *et al.* (2000)			Sin and Vanrolleghem, (2006); Hao *et al.* (2002); Henze *et al.* (2000)
State Variables: ***Dissolved Components***						
Soluble biodegradable organics	S_B	g COD/m^3	S_S		S_S	
Fermentable organic matter	S_F	g COD/m^3		S_F		S_F
Fermentation product (considered to be acetate)	S_{Ac}	g COD/m^3		S_A		S_A
Soluble undegradable organics	S_U	g COD/m^3	S_I	S_I	S_I	S_I
Dissolved oxygen	S_{O2}	g COD/m^3	S_O	S_{O2}	S_O	S_{O2}
Ammonium and ammonia nitrogen ($NH_4 + NH_3$)	S_{NHx}	g N/m^3	S_{NH}	S_{NH4}	S_{NH}	S_{NH}
Nitrate and nitrite ($NO_3 + NO_2$) (considered to be NO_3 only for stoichiometry)	S_{NOx}	g N/m^3	S_{NO}	S_{NO3}	S_{NO}	S_{NO3}
Nitrite and acid nitrous ($NO_2 + HNO_2$)	S_{NOx}	g N/m^3	S_{NO}	S_{NO3}	S_{NO}	S_{NO3}
Soluble biodegradable organic N	$S_{B,N}$	g N/m^3	S_{ND}			S_{ND}
Dissolved nitrogen gas	S_{N2}	g N/m^3		S_{N2}	S_{N2}	S_{N2}
Soluble inorganic phosphorus	S_{PO4}	g P/m^3		S_{PO4}		S_{PO4}
Alkalinity (HCO_3^-)	S_{Alk}	mol HCO_3^-/m^3	S_{Alk}	S_{Alk}	S_{Alk}	S_{Alk}
Particulate Components						
Particulate biodegradable organics	XC_B	g COD/m^3	X_S	X_S	X_S	X_S
Particulate undegradable organics	X_U	g COD/m^3		X_I	X_I	X_I
Particulate undegradable organics from the influent	$X_{U,Inf}$	g COD/m^3	X_I			
Particulate undegradable endogenous products	$X_{U,E}$	g COD/m^3	X_P			
Particulate biodegradable organic N	$XC_{B,N}$	g N/m^3	X_{ND}			X_{ND}
Ordinary heterotrophic organisms	X_{OHO}	g COD/m^3	$X_{B,H}$	X_H	X_H	X_H
Autotrophic nitrifying organisms (NH_4 to NO_3^-)	X_{ANO}	g COD/m^3	$X_{B,A}$	X_{AUT}	X_A	
Autotrophic ammonia oxidizer (NH_4 to NO_2^-)						X_{NH}
Autotrophic nitrite oxidizer (NO_2^- to NO_3^-)						X_{NO}
Phosphorus accumulating organisms	X_{PAO}	g COD/m^3		X_{PAO}		
Storage compound in PAOs	$X_{PAO,PHA}$	g COD/m^3		X_{PHA}		
Stored polyphosphates in PAOs	$X_{PAO,PP}$	g P/m^3		X_{PP}		
Metal hydroxide compounds	X_{MeOH}	g TSS/m^3		X_{MeOH}		
Metal phosphate compounds	X_{MeP}	g TSS/m^3		X_{MeP}		
Storage compound in OHOs	$X_{OHO,Stor}$	g COD/m^3			X_{STO}	
Total suspended solids	X_{TSS}	g TSS/m^3		X_{TSS}	X_{SS}	

Source: Adapted from Hauduc *et al.* (2010) and Corominas *et al.* (2010).

matrix notation, first developed by Gujer (1985), allows for communication in a well-organized and elegant format rather than complicated integrated mathematical models.

The Activated Sludge Model No. 1 (**ASM1**) was developed in 1987 (Henze *et al.* 1987) for municipal activated sludge WWTPs to model carbon oxidation, nitrification and denitrification, meaning autotrophic and heterotrophic reactions, with a facultative consumption of oxygen or nitrate as the electron acceptor, without phosphorus removal. In the ASM1, four basic processes are considered: (*i*) *growth of biomass*: aerobic and anoxic growth of heterotrophs (X_{OHO}); aerobic growth of autotrophs (X_{ANO}); (*ii*) *decay* of heterotrophs and autotrophs biomass; (*iii*) *ammonification* of organic nitrogen ($XC_{B,N}$); (*iv*) *hydrolysis* of particulate organic carbon (XC_B) and nitrogen ($XC_{B,N}$). The main concepts incorporated into the formulation of the ASM1 are briefly reported. The total chemical oxygen demand (COD) is assumed as the suitable parameter for defining the organic matter in wastewaters. The readily biodegradable COD (S_B) and the soluble total ammonium-nitrogen concentration (TAN), (S_{NHx}), are considered as the only substrates for the growth of heterotrophs and autotrophs, respectively. While readily biodegradable COD (S_B) can immediately be used by organisms for synthesis, the slowly biodegradable COD (XC_B) must be broken down and converted into readily biodegradable (S_B) material by hydrolysis reaction. Another important concept in the ASM1 model is growth – death – growth (*death-regeneration*) phenomena (Dold *et al.* 1980), defined as the decay of biomass followed by hydrolysis and growth on secondary substrate arising from decay (Van Loosdrecht & Henze, 1999). The biomass decay implies a recycle of organic and nitrogen substrate through the system. As a consequence, the flow of COD in the ASM1 is rather complex as it incorporates interactions between the decay of heterotrophs and of nitrifiers.

The ASMs models have grown over the years, becoming more complex and covering more biochemical processes. The Activated Sludge Model No. 2 (**ASM2**) was introduced as a further development of the Activated Sludge Model No. 1 (Henze *et al.* 1995), allowing the simulation of enhanced biological removal processes. The ASM2 introduces the phosphorus accumulating organisms (PAO), which are able to accumulate phosphorous in the form of poly-phosphate (X_{PAO_PP}), and can only grow on stored cell internal organic material ($X_{PAO,PHA}$) under aerobic conditions. In addition to the phosphorous biological process, the ASM2 includes the fermentation process and two chemical processes to model the chemical precipitation of phosphorus. In contrast to the ASM1, where all particulate organic materials are expressed as COD, ASM2 replaces the readily biodegradable substrate (S_B) with the sum of the end products of fermentation (S_{AC}) and the truly readily biodegradable substrate, which is fermentable (S_F). Further, in the ASM2 the biomass has an internal cell structure and, as a consequence, the new model includes poly-phosphate (X_{PAO_PP}) as an inorganic fraction of activated sludge, which does not exert any COD. In the ASM2, six basic processes are considered: (i) *growth of biomass*: aerobic and anoxic growth of heterotrophs (X_{OHO}) on fermentable substrate (S_F) and fermentable products (S_{Ac}); aerobic growth of phosphorous-accumulating organisms (X_{PAO}) on cell internal organic storage products ($X_{PAO,PHA}$), aerobic growth of autotrophic nitrifiers (X_{ANO}); (ii) *transformation process*: anaerobic fermentation of heterotrophs; (iii) *storage processes*: storage of cell external fermentation products (S_{Ac}) in the form of cell internal organic storage material ($X_{PAO,PHA}$); storage of ortho-phosphate (S_{PO4}) in the form of cell internal poly-phosphate ($X_{PAO,PP}$); (iv) *decay* of heterotrophs, autotrophs and phosphorous-accumulating biomass and decay of PAO storage products; (v) *hydrolysis* of particulate organic carbon (XC_B); (vi) *chemical precipitation of phosphate*. The decay is modeled in analogy to the ASM1. The **ASM2d** is an extension of the ASM2, which includes the denitrifying activity of the phosphorus accumulating organisms (PAOs), in order to improve the simulation of nitrate and phosphate dynamics (Henze *et al.* 2000).

The subsequent **ASM3** model (Gujer *et al.* 1999; Henze *et al.* 2000) was developed in order to replace the ASM1, correcting a number of shortcomings that have emerged from its applications, as thoroughly assessed by Gernaey *et al.* (2004). One of the major features of ASM3 is the inclusion, similarly to the ASM2, of internal cell storage compounds ($X_{OHO,STOR}$) in heterotrophs, shifting the focus from hydrolysis to the storage of organic substrates. Another significant difference is the replacement of the death-regeneration concept by the *growth-endogenous* respiration model, distinguishing the growth and the decay processes, making the flow of COD in ASM3 clearer than in the ASM1. The ASM3 model assumes that all readily biodegradable substrates (S_B) first become stored materials ($X_{OHO,STOR}$) and then are assimilated to biomass for growth. This hypothesis, however, is not true and a direct growth rather than storage followed by growth can be simulated using a low yield coefficient for storage ($Y_{SB,STO}$) and a higher one for subsequent growth ($Y_{STOR,\ OHO}$). ASM3 includes only biological processes such as: (i) *growth of biomass*: aerobic and anoxic growth of heterotrophs on stored organic materials ($X_{OHO,STOR}$), aerobic growth of autotrophic nitrifiers (X_{ANO}); (ii) *storage processes*: aerobic and anoxic storage of readily biodegradable substrates (S_B); (iii) *respiration of biomass*: aerobic and anoxic endogenous respiration; aerobic and anoxic respiration of storage products; (iv) *hydrolysis* of slowly biodegradable substrates (XC_B).

In the ASM1-ASM3 models both nitrification and denitrification are considered to be a single step process, not taking into consideration intermediates of nitrification and denitrification or other biological processes. To date, some authors propose extensions of ASMs models for nitrite build-up in both aerobic and anoxic conditions, as nitrite can play an important role in microbial conversions, and include new biological pathways, such as the anammox process. The ASM1 was extended with a 2 step nitrification model (Wyffels *et al.* 2004) and an evolution of an ASM2d-like model was proposed (**ASM2d2N**) (Sin & Vanrolleghem, 2006).

The ASM1 described by Wyffels *et al.* (2004), has been implemented with submodels for anammox process (Van Hulle *et al.* 2003; Dapena-Mora *et al.* 2004). Additional extensions of ASMs model have been developed, including either three- (Von Schulthess *et al.* 1994) or four-step (Samie *et al.* 2011) denitrification process.

Application of dynamic models requires a calibration of the chosen ASM to the case under study, in order to adjust any model parameters (physical, operational, kinetic, stoichiometric, settling) and improve the fit to measured data. Calibration of ASMs is tedious and requires expert knowledge of the influent wastewater characterization and on-line/off-line measurements and a good knowledge of the operating conditions and the treatment system itself. A critical comparison of systematic calibration protocols for activated sludge models (BIOMATH, HSG and WERF) through a SWOT analysis (Strengths, Weaknesses, Opportunities, and Threats) has been proposed by Sin *et al.* (2005b).

9.2.2 ASMs to MBR modeling

ASMs were originally developed to describe CAS systems with the continuous flow regime and their typical operative conditions. Nowadays, modeling of CAS systems is a well-established part of the design and operation of wastewater treatment plants in their various configurations (carbon, nitrogen and even phosphorous removal).

ASMs have been adopted, with or without modifications, to simulate biomass kinetics of the MBR systems. The simulation of MBR biological processes has been initially done by a simple transfer of dynamic models developed for CAS systems to MBRs, as the CAS and MBR systems are similar from a biochemical engineering viewpoint. However, the application of ASMs models, in their original form, to MBR biokinetics requires specific attention to the experimental identification and calibration of model parameters.

Recent studies on MBR biological processes have underlined several crucial specificities of MBR processes compared to CAS systems, which should be included in the biological modeling of MBR as reviewed by Ng and Kim (2007). With respect to the CAS systems, MBRs present: (i) a different microbial composition leading to different stoichiometric and kinetic parameters (Jiang *et al.* 2005); (ii) a higher mixed liquor concentration and a higher viscosity leading to changes in oxygen transfer and uptake (Germain *et al.* 2007); (iii) an elevated sludge retention time, due to no loss of solids in the effluent and minimal sludge wasting; (iv) a higher retention of slower growing bacteria such as nitrifying and anammox bacteria; (v) an accumulation of soluble microbial products (SMP), due to membrane filtration processes (Gao *et al.* 2004); (vi) a higher aeration rates applied to improve scouring of membranes and nitrification performance (Judd, 2006).

In order to take into account one or more of these MBR specificities several authors have proposed extensions of the original ASMs. For the sake of clarity, we distinguished between the unmodified/conventional and modified/integrated ASMs models applied to MBR systems. According to previous reviews (Fenu *et al.* 2010a; Naessens *et al.* 2012; Zuthi *et al.* 2012), with unmodified/conventional ASMs models we target all studies and applications, where the ASMs models in their original form, as described by Henze *et al.* (2000), are used to simulate MBR biological processes, modifying kinetic parameters in order to match the model predictions with measured experimental data, improving the model simulation. Thus, the unmodified/conventional ASMs have the same model structure of the ASMs models for CAS systems or slight modified and use only the state variables defined in the original ASMs. Unmodified/conventional ASMs can also include some submodels for aeration or hydrolysis in the original ASMs structure.

The modified/integrated ASMs models are extended versions of ASMs models, taking into account the accumulation and degradation of soluble microbial products (SMP/EPS models), introducing new additional biokinetic variables in ASMs for a better description of MBR processes.

9.2.3 Application of unmodified/conventional ASMs to MBR

The application of ASMs in their original form for MBR systems has some advantages as ASMs are widely applied and many simulation programs include ASMs or models based on ASMs (e.g., WEST®, Aquasim, GPS-X® software).

The ASMs models are usually applied for CAS systems, which are completely mixed, operating under a Sludge Retention Time (SRT) in the range 3–15 d (maximum 30 d (Henze *et al.* 2000)), a Hydraulic Retention Time (HRT) in the range of 3–5 h and with a solid content (MLSS) in the range of 1.5–4 g/L (Metcalf & Eddy, 2003). As the operation conditions of MBRs can vary in a large range, the applicability of ASMs, in their original form, for MBR modeling needs to be verified. Itokawa *et al.* (2008), in their study on different municipal MBRs, reported a range of SRT from 15 d to more than 40 d, an operative HRT between 4 h and 6 h and MLSS concentrations from 7 to 13.5 g/L.

In the last twenty years, several authors have applied unmodified/conventional ASMs to the biological modeling of MBR systems. ASM1 (and ASM3) are designed to deal with activated sludge systems with very low load or high SRT (>1 day) where flocculation/adsorption of XC_B and storage are not limiting (Henze *et al.* 2000). The high SRTs, which are typically encountered in the operation of MBRs for municipal wastewater treatment, allow the application of both the ASM1 and the ASM3.

MBR systems were firstly modeled with the ASM1, using default values of parameters (Henze *et al.* 2000) by Chaize and Huyard (1991). The ASM1 parameters relevant for the long-term biological behaviour in MBR systems were calibrated by Jiang *et al.* (2005). Spérandio and Espinosa (2008) applied both the unmodified ASM1 and the ASM3 to a lab scale MBR, using default values of parameters (Henze *et al.* 2000). The authors concluded that both models provided satisfactory prediction for MBR systems, even if the models should be improved for MBRs working at high SRT. Wyffels *et al.* (2004) extended an ASM1 with the two steps of nitrification to model a nitrifying membrane-assisted bioreactor (MBR) for the treatment of sludge reject waters, operating under continuous aeration at low dissolved oxygen (DO) concentrations, with the purpose of generating nitrite accumulation.

The MBR modeling through unmodified/conventional ASMs confirmed mismatches between ASMs prediction and experimental data, especially in cases of high SRTs and high MLSS concentration. Thus, some authors applied ASM1 and ASM3 with slight modifications to the structure in order to better explain MBR behaviours: in ASM1, the death-regeneration concept has been replaced by the endogenous decay concept introduced in ASM3 (Sarioglu *et al.* 2008, 2009a; Lubello *et al.* 2009); in ASM1, the unique kinetic of the particulate biodegradable fraction of COD (X_U) has been substituted with two different hydrolysis processes, distinguishing between the particulate originates from both biomass decay and the particulate of the influent (Lubello *et al.* 2009) (see Section 9.2.3.1).

However, in order to better explain MBR behaviours, some submodels were integrated in the ASMs original structure, such as the dedicated aeration model. Verrecht *et al.* (2010) modelled a community-scale MBR for urban reuse and used ASM2d to take account of phosphate accumulating organism (PAO) activity, although no anaerobic tank was provided. The observed phosphorous removal suggested that part of the anoxic tank may be (intermittently) anaerobic. To obtain better representation of P removal, the ASM2d biomass decay rates modifications proposed by Gernaey and Jørgensen (2004) were adopted. Further, to realistically simulate the aeration demand, a dedicated aeration model was used incorporating the dependency of the oxygen transfer on the mixed liquor concentration and allowing differentiation between coarse and fine bubble aeration, both typically present in MBRs. The calibrated model was able to predict MLSS and effluent nutrient concentrations accurately and was successfully used for optimizing the plant. Similarly, Fenu *et al.* (2010b) adopted the ASM2d to model a full-scale MBR in order to include the inorganic fraction and to account for PAO activity even though an anaerobic reactor was not present. The model was successfully capable of modeling the TN removal but the phosphate removal could not be modelled successfully.

Recently, a significant number of articles have been dedicated to the simultaneous nitrification and denitrification (SNDN) process at low oxygen concentrations. He *et al.* (2009) investigated the impact of DO, food to microorganism ratio (F/M), carbon to nitrogen ratio (C/N) and pH on the performance of the SNDN on a lab scale using a simplified ASM1 without detailed calibration. Sarioglu *et al.* (2009b), in their study, on a small pilot scale – MBR (SRT = 36 d and HRT = 7.2 h), used a model based on the template of the ASM1, modified for endogenous aerobic and anoxic decay to simulate the SNDN process. In addition, unlike the original ASM1, all of the growth and decay processes used in the matrix were coupled with switch functions, defining the diffusion limitation of substrate and oxygen through the flocs. Hocaoglu *et al.* (2011a) modeled the SNDN in a pilot-scale MBR operated as a sequencing batch reactor (SBR) treating black-water at low DO conditions. They used an adapted ASM1 model similar to Sarioglu *et al.* (2009b); further, the soluble and biodegradable organic nitrogen fractions were included into the model as components of COD fractions, the way it was defined in ASM2/2d (Henze *et al.* 2000).

The investigation of the appropriate parameters for MBR modeling set requires careful calibration of the ASM model chosen, taking into account the MBR specificities. The adoption of the default values for stoichiometric and kinetic parameters of the ASMs models have lead to some discrepancies between prediction and experimental data. Jimenez *et al.* (2010) proposed an ASM1 calibration procedure to correctly predict MBR performances at different SRT (SRT = 15 d and 40 d). The authors performed a first model calibration decreasing the inert particulate COD as the SRT increased. Then, regarding nitrification parameters, the main modifications concerned the half-saturation constants (substrate transfers), which depend on floc size distribution and mass transfer limitations. Delrue *et al.* (2010) developed a five step iterative calibration procedure based on the experience accumulated in CAS model calibration. The step 1 was useful in selecting the parameters to be calibrated. The wastewater fractionation was performed through experimental tests in Step 2. From Step 3 to Step 5, the simulated results were compared with the observed data in terms of maximum nitrification rates, oxygen transfer rates and actual nitrification/denitrification rates, in order to calibrate the autotrophic growth and decay rates, the oxygen transfer capacity and the half saturation coefficients. In the MBR, a SNDN process was achieved. A calibrated set of parameters was proposed, but validation was not very successful, mainly due to the fact that the dominant aeration configuration had changed during the validation period. Parameters calibration (such as aerobic heterotrophic and autotrophic growth rate and half-saturation constants) can also be obtained through model calibration with oxygen uptake rate (OUR) profiles of respirometric tests run in parallel (Munz *et al.* 2008b). Moreover, the real MBR behaviours not always could not always be simulated correctly in a steady state, indicating the need for a dynamic calibration taking into account the influent variations (Verrecht *et al.* 2010).

In general, the application of unmodified/conventional ASMs needs a careful calibration, especially for sludge production and nitrification modeling (Fenu *et al.* 2010a). Instead one advantage of the modified/integrated models over the ASMs is their capability to accurately model biomass in MBR without the need for calibration using experimental data (Ng & Kim, 2007).

The main typical discrepancies associated with the application of the unmodified/conventional ASMs to MBR systems are described in the next subsections, such as the overestimation of sludge production and the mismatching of the nitrogen and phosphorous removal rates and oxygen transfer rate. An overview of the unmodified/conventional ASMs applied to MBR and of the parameter set is reported in Table 9.2.

9.2.3.1 Estimation of sludge production

The first attempt to apply unmodified/conventional ASMs was made by Chaize and Huyard (1991), who applied the basic ASM1 to a laboratory scale MBR fed with domestic wastewaters the at a very high SRT (nearly 100 d) and two different HRT (8 h and 2 h). The aim of the research was to model effluent COD, TKN and sludge production, using the default parameter values of the ASM1, with neither calibration nor influent characterization. The predicted effluent COD was slightly lower than that observed, while the predicted TKN was quite close to the observed value. The model overestimated the sludge production, predicting a lower solids concentration in the reactor compared to experimental data. The mismatch was higher at very low HRT and very high SRT in the MBR system.

The overestimation of sludge production in ASMs has also been observed in other MBR studies (e.g., Wagner & Rosenwinkel, 2000), where the SRT was significantly high. For SRTs higher than the default parameter values defined for the ASM1 (maximum 30 d), appropriate parameters should be revisited, especially in terms of influent characteristics, unbiodegradable particulate COD fraction of influent ($X_{U,inf}$), conversion yield of heterotrophs (Y_{OHO}) and biomass decay (b_{OHO} and b_{ANO}). Further, some modeling approaches should be readapted, such as the decay and the hydrolysis modeling.

Influent characterization

One of the main reasons that could explain the difficulty that the ASM1 has in estimating the sludge production of biological processes that operate at high SRTs, is linked to the influent COD

Table 9.2 Parameters set in different studies for MERs using the unmodified/conventional ASMs.

Model	Model Changes	Reactor Type	Process	Operating Conditions					Autotrophs Kinetics		Autotrophs			Heterotrophs Kinetics					Bio-P – Organisms Kinetics			References
				MLSS (g/L)	T (°C)	SRT (d)	HRT (d)	DO (mg/L)	Y_{ANO} (g/g)	$\mu_{ANO,max}$ (1/d)	b_{ANO} (1/d)	$K_{NHx,ANO}$ (mg/L)	$K_{O2,ANO}$ (mg/L)	Y_{OHO} (g/g)	$\mu_{OHO,max}$ (1/d)	b_{OHO} (1/d)	$K_{NOx,OHO}$ (mg/L)	$K_{O2,OHO}$ (mg/L)	Y_{PAO} (g/g)	μ_{PAO} (1/d)	b_{PAO} (1/d)	
ASM1	–	CAS		3–5		<30 d			0.24	0.8	0.15	1	0.4	0.67	6	0.62	0.5	0.2				1
	–	MBR	(OX + NIT)	10–36	20	50–100	2–8	4														2
	–	MBR	(OX + NIT)	8–12	22–28	20	8	>4	**0.25**		**0.08**			**0.72**		**0.25**						3
	–	MBR	(OX + NIT)	4–10	20	10–110	–	2–6		**0.45**	**0.04**	**0.25–0.6**										4
	Aerobic and anoxic Y_{OHO}, as in ASM3	MBR	(OX + NIT) + DEN	7–10	27	30–60	19–48	<1.5		**0.8**	**0.1**	**0.25**	**0.3**	**Aer Anox 0.67 0.54**	**6**	**0.62**	**3**	**0.03**				5
	inorganic solids solubilisation + T influence	MBR	((OX + NIT) + DEN	5	–	15–40	–	aer + anox		**0.85–1**	**0.15**		**0.25–0.5**					**0.1**				6
	endog. decay, as in ASM3 + hydrolysis	MBR	(OX + NIT) + DEN	4–10	20	30–105	22–48	2–3	**0.24**	**0.32–0.38**	**0.06–0.08**	**0.14–0.15**		**0.59–0.61**	**5.1–5.7**	**0.19–0.23**						7
ASM2	–	CAS				<30 d			0.24	1	0.15	1	0.5	0.63	6	0.4	0.5	0.2	0.62	1	0.2	1
	ASM2d + aeration model	MBR	(OX + NIT) + DEN	7–8		47	26	2											0.2	2	0.1	8
	ASM2d + aeration model			9–11								0.2	0.2									
											Aer Ax			Aer Ax								
ASM3	–	CAS				<30 d			0.24	1	0.15 0.05	1	0.5	0.63 0.54	2	0.2 0.1	0.5	0.2				1
	–	MBR	(OX + NIT)	4–10	20	10–110	–	2–6		**0.45**	**0.04**	**0.25–0**										4

1. Henze *et al.* (2000)
2. Chaize and Huyard (1991); ASM1 and default values lead to a sludge overestimation.
3. Jiang *et al.* (2005); Parameters estimated through respirometric tests.
4. Spérandio *et al.* (2005); Parameters estimated through model calibration, ASM1 lead to a sludge overstimation at SRT = 110 d; ASM3 lead to a sludge understimation at SRT <30 d.
5. Delrue *et al.* (2010); Parameters estimated through model calibration.
6. Jimenez *et al.* (2010); Parameters estimated through model calibration.
7. Lubello *et al.* (2009); Parameters estimated through respirometric tests.
8. Verrecht *et al.* (2010); Parameters estimated through model calibration. Focus on energy consumption. However the phosphate removal could not be modeled successfully with ASM2d.

Y_{ANO}, Y_{OHO}, Y_{PAO} = Yield for X_{ANO}, X_{OHO} and X_{PAO} growth.
$\mu_{ANO,max}$, $\mu_{OHO,max}$, $\mu_{PAO,max}$ = Maximum growth rate of X_{ANO}, X_{OHO} and X_{PAO}.
b_{ANO}, b_{OHO}, b_{PAO} = Decay rate for X_{ANO}, X_{OHO} and X_{PAO}.
$K_{NHx,ANO}$ = Half saturation parameter for S_{NHx} for ANO.
$K_{NOx,OHO}$ = Half saturation parameter for S_{NOx} for OHO.
$K_{O2,ANO}$, $K_{O2,OHO}$ = Half saturation parameter for S_{O2} for ANO and OHO.
OX = oxidation of organic carbon; **NIT** = Denitrification; **DEN** = Denitrification.

characterization and to the fate of unbiodegradable compounds (Lubello *et al.* 2009). The separation between biodegradable and unbiodegradable fractions is a crucial aspect in MBR modeling. In MBR systems working at very high SRTs, suspended solids are mainly composed of unbiodegradable matter, originating from both influent and biomass decay. At high SRTs, it can be supposed that some organic compounds, which are considered unbiodegradable at conventional SRTs, become biodegradable in MBRs and a slow but significant hydrolysis of the unbiodegradable particulate fraction may take place. In modeling MBR systems with very long SRTs, the influent wastewater characterization and in particular the unbiodegradable particulate COD fraction of influent ($X_{U,inf}$) plays an important role.

The estimation of the influent unbiodegradable ($X_{U,inf}$) and biodegradable (X_{CB}) particulate fractions are influenced by the experimental procedures. As recently reviewed by Fenu *et al.* (2010a) and Zuthi *et al.* (2012), the choice of the proper technique to determine the particulate $X_{U,inf}$ and X_{CB} may depend on the SRT. Influent COD characterization has conventionally been done using two different approaches: (i) application of "trial- and-error" procedures (fitting methods); and (ii) experimental procedures, integrating chemical, physical and biological and respirometric methods.

Fenu *et al.* (2010a), in their review, suggested for high SRT (50–60 d) MBR systems the "trial- and-error" procedures in order to estimate both the $X_{U,inf}$ and X_{CB}, while the respirometry-based fractionation seems to be valid when the MBRs operate at lower SRTs (<30–40 d).

The influent COD have been estimated combining the trial- and-error procedures and chemical-physical analysis by Delrue *et al.* (2010), using the ASM1 on a full scale – MBR (SRT = 30–60 d) where the simultaneous nitrification and denitrification (SNDN) was achieved. The authors compared three different chemical – physical methods for soluble/particulate fractionation: filtration at 0.1 μm, membrane filtration and coagulation-flocculation followed by filtration at 0.1 μm. Then, the influent unbiodegradable and biodegradable fractions were estimated using both BOD tests and trial- and-error procedures (fittings method). They showed that the total BOD tests were not able to correctly estimate the biodegradable fraction of the wastewater for high SRT MBR systems. The combination of filtration at 0.1 μm and fitting method was chosen as the most reliable protocol for influent COD characterization. Using the fitting method, the wastewater fractionation was identified by adjusting the simulated sludge production rate to the measured value.

Jiang *et al.* (2005) compared two experimental procedure for influent COD characterization in order to calibrate the ASM1 in a side-stream membrane bioreactor (SRT = 20 d and HRT = 8 h). The authors used a 'physical-biological' method, coupling physical filtration (0.45 μm) with respirometric technique (Spanjers & Vanrolleghem, 1995) and a 'physical-chemical' method, combining physical filtration (0.45 μm) and BOD analysis (Hulsbeek *et al.* 2002). Significant differences between the two approaches were observed, especially for the unbiodegradable particulate organics fraction ($X_{U,inf}$), which was found significantly higher when determined with the "physical–chemical" method, causing an overestimation of the MLSS, under steady state conditions.

Sarioglu *et al.* (2009b), in their study, on a small pilot scale – MBR (SRT = 36 d and HRT = 7.2 h), used a model based on the template of the ASM1 and performed a proper influent characterization using a combined chemical, physical biological method. In particular, biodegradable COD components were determined using BOD analysis, according to the method proposed by Roeleveld and Van Loosdrecht (2002). The readily biodegradable COD was determined as the difference between the influent total biodegradable COD and the effluent inert soluble COD. The particulate and soluble unbiodegradable COD of the influent were determined according to the methods proposed by Germirli *et al.* (1991) and Orhon *et al.* (1999). The authors obtained a good fit of MLVSS and MLSS in the bioreactor.

Spérandio and Espinosa (2008) modeled a lab scale MBR using both ASM1 and ASM3, with focus on suspended solids, excess sludge production and wastewater characterization. The authors coupled chemical, physical and respirometric procedures in order to determine the influent organic fractions concentration: unbiodegradable particulate COD ($X_{U,inf}$) was measured at the end of long term BOD measurement (30 days), the soluble unbiodegradable COD (S_U) was supposed to be equal to the outlet concentration (30–50 mg COD/L); the total amount of biodegradable substrate was confirmed by the respirometric technique (Spérandio & Paul, 2000), while the readily (supposed soluble) (S_B) and slowly (supposed particulate) (X_{CB}) biodegradable substrates were separated through 0.45 μm filtration. Organic soluble and particulate nitrogen were deduced from TKN and TAN measurement on total and filtrated samples (0.45 μm). Satisfactory predictions for volatile suspended solids (MLVSS) were obtained using the ASM1 with default kinetic parameters, simulating a proper sludge production at SRT shorter than 50 d, while an overestimation was observed at SRT of 110 d. In comparison, a slightly better prediction was obtained with the ASM3 model at high retention time, but MLVSS and excess sludge production were underestimated in some periods at lower SRT (10–30 days).

Guglielmi *et al.* (2009) applied on a large pilot-scale MBR (SRT = 20–25 d) an extended version of the ASM3 that includes the simultaneous growth of heterotrophs on both storage products ($X_{OHO,STOC}$) and readily biodegradable substrate (S_B), originally proposed by Sin *et al.* (2005a). The authors used a respirometry-based approach for wastewater fractionation, achieving successfully results in terms of sludge production under dynamic conditions. The total biodegradable COD was determined according to Ekama *et al.* (1986), whereas the soluble biodegradable COD (S_B) was estimated according to the single – Oxygen Uptake Rate (OUR) method proposed by Ziglio *et al.* (2001), after the calibration curve with sodium acetate as standard had been determined. Afterwards, S_U was calculated as the difference between the filtered COD in the influent (0.45 mm) and the soluble biodegradable COD, being usually equal to 90–95% of COD concentration in the permeate. Finally, X_U was quantified with a mass balance.

Inert organics accumulation

For high SRTs, the mineral suspended solid concentration (MSS) of the sludge, not included in original ASM models, is also severely overestimated at high SRTs (>50 d), which, consequently, leads to an overestimation of total mixer liquor suspended solids (MLSS) and underestimation of the volatile/total suspended solids ratio (MLVSS:MLSS) ratio (Spérandio & Espinosa 2008). The authors explained that the discrepancy between simulated and experimental data indicated that a significant solubilisation of mineral matter was achieved in the MBR.

Modeling approaches

Focusing on the overestimation of the sludge in MBR modeling using the unmodified/conventional ASMs, some aspects of the original ASMs modeling have been corrected and adapted.

Spérandio and Espinosa (2008) showed some differences in applying ASM1 and ASM3 on their MBR system. The ASM1 properly simulated the sludge production at SRTs shorter than 50 d, while an overestimation was observed at an SRT of 110 d. The ASM3 slightly better predicted the sludge production at higher SRTs, but MLVSS and excess sludge production were underestimated at lower SRTs (10–30 days). The authors explained that this difference can be due to the higher quantity (around 40% more) of unbiodegradable particulate matter (X_U) generated by decay with the ASM1 compared to the ASM3. The ASM1 is based on death-regeneration principle, whereas ASM3 used endogenous respiration theory. Thus, especially at high SRTs, a larger amount of organic matter is generated by decay with ASM1 model. Secondly, a larger quantity of active autotrophic biomass (X_{ANO}) is predicted with the ASM1 compared to the ASM3, due to the fact that the default value of the decay coefficient is lower in the ASM1 than in the ASM3. Further, the authors showed that as the SRT increases, sludge is mainly composed of unbiodegradable particulate material ($X_{U,Inf} + X_{U,E}$) for ASM1 and X_U for ASM3, whereas active biomass and stored substrate become very low (X_{OHO}, X_{ANO}, XC_B $X_{OHO,STOC}$). This is a major limitation of the unbiodegradable particulate material (X_U and $X_{U,inf}$) concept, which is valuable only on low SRT values, while, for higher SRT values, organic compounds which are considered unbiodegradable at conventional SRT become biodegradable (Nowak *et al.* 1999). Therefore, the authors suggested to include in ASM1 and ASM3 a slow hydrolysis mechanism for predicting volatile solids production at high sludge retention in an MBR.

Lubello *et al.* (2009) proposed a revised version of the ASM1 model (ASM1-S) for correctly estimating sludge production at high SRTs. In the ASM-S, the endogenous death – regeneration process of the ASM1 was first substituted with a simple decay process as is used in the ASM3 model. Then, in the ASM1-S, it is supposed that the kinetics of the endogenous decay process and of the particulate hydrolysable COD are of a first-order kinetics and independent, while in the original ASM1 the particulate biodegradable fraction of COD (XC_B) is hydrolysed with a unique process (and kinetics) if it is originated from both biomass decay and COD in the influent. The ASM1 considers that a fixed part of the biomass (indicated by $f_{XU\_Bio,lys}$, fraction of particulate undegradable organics (X_U) generated in biomass decay) cannot be biodegraded any further. This assumption can be considered correct only for limited values of sludge age (Van Loosdrecht & Henze, 1999; Henze *et al.* 2000). Indeed, Lubello *et al.* (2009) suggested that a fraction of particulates produced by the endogenous decay of biomass consists of new organic material. On the basis of the experimental results and their subsequent processing, it appeared that the production of solids may be satisfactorily estimated using the ASM1-S model and performing both a suitable influent COD characterisation and a calibration of the model.

Bio-kinetic parameters calibration

The outcomes of the Chaize and Huyard (1991) study, showed that a non-calibrated ASM1 is able to reasonably simulate the effluent COD and TKN only for MBRs working at low SRT, but leads to an overestimation of the sludge production. Hence, other authors have applied conventional ASM1 models to MBR systems, but taking particular care with both the bio-kinetic parameters calibration and the influent wastewater characterization (Jiang *et al.* 2005; Baek *et al.* 2009; Delrue *et al.* 2010).

In MBR modeling, changes in the decay (b_{OHO}) and yield (Y_{OHO}) of heterotrophic biomass and in the fraction of X_U generated in biomass decay ($f_{XU\_Bio,lys}$) can severely modify the predicted MLSS (Jiang *et al.* 2005; Spérandio & Espinosa 2008) and the excess sludge production, as in MBR technology the MLSS concentration impacts the excess sludge production.

Jiang *et al.* (2005) measured the conversion yield of heterotrophs (Y_{OHO}) in MBR processes, through respirometric methods using acetate. The obtained Y_{OHO} (0.72 gCOD/gCOD at 23°C) was higher than the default value in the ASM1 (0.67 gCOD/gCOD at 20°C), but it was probably due to the storage phenomena with acetate, which is easily converted to PHA (Majone *et al.* 1999). The default values for ASMs are in range 0.63–0.67 gCOD/gCOD. They are valuable for domestic wastewater, where a large variety of carbon source (carbohydrates, proteins, alcohols, carboxylic acids.) are present. The Y_{OHO} obtained in Jiang *et al.* (2005) study can be considered substrate-specific, as it was obtained with a single carbon source.

Delrue *et al.* (2010) even with the use of the ASM1, introduced the reduction of the heterotrophic yield ($Y_{Stor\_OHO,Ax}$) as proposed in the ASM3 in anoxic conditions. In their model simulation the authors adopted $Y_{Stor\_OHO,Ox} = 0.67$ gCOD/gCOD and $Y_{Stor\_OHO,Ax} = 0.54$ gCOD/gCOD, considering that the heterotrophic yield (Y_{OHO}) was dependent on electron acceptor conditions (nitrate or oxygen, as in the ASM3).

Concerning the heterotrophic decay rate (b_{OHO}), it seems to be similar to the ASMs default values. In a preliminary study, Jiang *et al.* (2005) measured b_{OHO} with respirometric tests (Spanjers & Vanrolleghem, 1995) and found a value of 0.25/d, lower than the ASM1 value (0.62/d). The authors explained this lower value with the decrease of predation in the MBR. Nevertheless, Jiang (2007) used the default value for b_{OHO} in the ASM2 (0.4/d), with a good fit of the COD prediction and the experimental data.

Lubello *et al.* (2009), using an ASM1 simplified with a *growth-endogenous* respiration as in the ASM3, reported an optimal value of b_{OHO} of 0.25/d, calculated from the calibration procedure, which is higher than the estimate reached using the respirometric tests (0.19/d). The values were similar to the values in the ASM3 (0.2/d). The authors also reported an optimal value of the fraction of particulate undegradable organics (XU) generated in biomass decay ($f_{XU\_Bio,lys}$) of 0.2, which is similar to the value reported in the literature concerning ASM3 (Henze *et al.* 2000).

Jiang *et al.* (2005) calibrated the ASM1 for modeling a MBR through defining Relative Sensitivity Function (RSF) for several model variables toward all model parameters. The stoichiometric parameter (Y_{OHO}) and kinetic parameters (b_{OHO}) were found to influential the MLSS concentration and effluent quality. Further, Fenu *et al.* (2010a) underlined that modifications of Y_{OHO}, b_{OHO}, and $f_{XU\_Bio,lys}$ parameters play an important role in other processes, such as oxygen consumption or nitrate removal. Consequently, the calibration of those parameters needed to be carefully verified, while it is preferable to vary $X_{U,Inf}$ levels in calibration procedure, because it varies from different wastewaters.

9.2.3.2 Nitrogen and phosphorous removal process performance

Several studies showed the difference between nitrification kinetics and performance in CAS and MBR systems (Aguilera Soriano *et al.* 2003; Frederickson & Cicek, 2005; Holakoo *et al.* 2007; Munz *et al.* 2008b), probably due to the higher maintenance and retention of nitrifiers and the higher bioavailability of substrates in the MBR at higher SRTs.

Discrepancies have been found when modeling MBR processes with ASMs models using the default kinetic and stoichiometric values for nitrification (Henze *et al.* 2000), such as the conversion yield (Y_{ANO}), the decay rate (b_{ANO}) and the autotrophic growth rate (μ_{ANO}). Usually, at high SRTs, a lower maximal growth rate is predicted with respect of the experimental observation, suggesting that in MBRs operated at high SRTs a selection of slow growing bacteria, comparable to an activated sludge process, is taking place, or that there has been a modification of metabolic state. In contrast to the nitrification process, denitrification and phosphorous removal rates are apparently less modified by the membrane configuration.

Regarding the kinetic parameters, the main modifications concern the half-saturation constants (substrate transfers), which depend on both floc size distribution and mass viscosity and may directly influence the biological processes performance predictions.

Nitrification

Although nitrifying activity is known to be bacteria-specific and variable from one process to another even for CAS systems, specific nitrifying biomass differences exist between MBR and CAS systems. Analysis of microbial species (AOB, NOB) by molecular techniques in MBRs would probably help in determining the specific nitrifying bacteria and the proper stoichiometric and kinetic values (Munz *et al.* 2008b).

Spérandio and Espinosa (2008) used ASM1 and ASM3 to calibrate an MBR on a large range of SRT (10–110 days). The default values proposed for the ASM1 ($\mu_{ANO} = 0.8/d$, $b_{ANO} = 0.04/d$) leaded to an overestimation of the nitrification rate. On the other hand, the default values proposed for the ASM3 ($\mu_{ANO} = 1/d$, $b_{ANO} = 0.15/d$) gave better results, but minimized the SRT influence. The best simulation, was achieved considering, especially for high SRTs, the following set parameters: $\mu_{ANO} = 0.45/d$, $b_{ANO} = 0.04/d$. This result may suggest that a lower maximal growth rate ($\mu_{ANO} = 0.45/d$) gives a better description of nitrifiers activity on a large range of SRTs than an increase of the decay rate (b_{ANO}). Recent studies reported lower μ_{ANO} values, in the range 0.16–0.26/d, for MBR systems at high SRTs (Sarioglu *et al.* 2009a), probably due to the inhibitory action of SMPs on the maximum autotrophic growth rate (Ichihashi *et al.* 2006). Nevertheless a recent study showed a good fit of the observed and the predicted effluent ammonia nitrogen concentrations using the ASM1 and the default value at 20°C ($\mu_{ANO} = 0.8/d$, $b_{ANO} = 0.05/d$) (Baek *et al.* 2009).

Concerning the autotrophic decay rate (b_{ANO}), using the ASM1, the determined value was 0.08/d (Jiang *et al.* 2005) and 0.10/d (Delrue *et al.* 2010). However, the values of b_{ANO} reported in literature are in the default range used for ASMs.

Munz *et al.* (2008a) applied an ASM1 modified with a simple decay process derived from the ASM3. Further, the authors, considering nitrification process as a two step process (nitritation and nitratation and, by introducing the state variable for ammonium-oxidizing (X_{AOB}) and nitrite-oxidizing (X_{NOB}) biomass (Iacopozzi *et al.* 2007), characterized the nitrifying biomass: $\mu_{AOB} = 0.25/d$ and $\mu_{NOB} = 0.23/d$; $b_{AOB} = 0.08/d$ and $b_{NOB} = 0.04/d$; $Y_{AOB} = 0.17$ $gCOD_{(XAOB)}/gN$ and $Y_{NOB} = 0.078$ $gCOD_{(XNOB)}/gN$. Manser *et al.* (2006), applying an ASM3, did not detected significant differences between CAS and MBR concerning the ammonia-oxidizing biomass (AOB, $b_{ANO} = 0.13/d$), while a slight difference concerning the nitrite-oxidizing biomass (NOB): 0.28/d and 0.17/d for CAS and MBR, respectively.

Wyffels *et al.* (2004) simulated the partial nitritation MBR using an ASM1 model. A good fit of experimental data with the simulation of the partial nitritation was achieved using the following parameters for the nitrifying biomass: $\mu_{AOB} = 2.025/d$ and $\mu_{NOB} = 1.36/d$; $b_{AOB} = 0.19/d$ and $b_{NOB} = 0.092/d$; $Y_{AOB} = 0.15$ $gCOD_{(XAOB)}/gN$ and $Y_{NOB} = 0.041$ gCOD X_{NOB}/gN. Except for the estimation of the μ_{AOB} through a calibration procedure, the other parameters were adopted according to Wiesmann (1994). The values of the parameters considered in this study are different from other studies where a full nitrification process is achieved. This is due to the different operating conditions of the MBR (higher temperature, lower DO, higher ammonia concentration), which increase the AOB activity over the NOB activity.

The conversion yield (Y_{ANO}) does not seem to be influential. Moreover, this parameter was measured by Jiang *et al.* (2005), yielding the value (0.25 gCOD/gN), close to the ASM1 default value (0.24 gCOD/gN).

Concerning the half-saturation constants for ammonia ($K_{NHx,ANO}$) and oxygen ($K_{O2,ANO}$) many studies showed their influence on the nitrification performance predictions.

The ASMs default value for the half saturation parameter for S_{NHx} ($K_{NHx,ANO}$) is 1 mg N/L. In experiments on both CAS and MBR systems, the literature values reported for MBRs are usually lower than the ASMs default values. Although $K_{NHx,ANO}$ is a biomass-specific parameter, several studies conducted on MBR systems indicated that $K_{NHx,ANO}$ can vary in a wide range, from 0.15 to 2 mg N/L (Jiang *et al.* 2005; Sarioglu *et al.* 2009b; Delrue *et al.* 2010). This could be explained considering the internal and external mass transfer resistance, due to the variable solid concentration in MBR systems and the floc size, which both induce a limitation on the diffusion of soluble organics through the flocs. Further, it should be considered that the flocs size may be influenced by several factors, such as hydrodynamic and aeration regimes, reduction of F/M ratio and SRT. Ammonia and oxygen concentration decrease across the flocs, thus as small as the flocs are, a smaller value of both ammonia and oxygen needed to achieve nitrification. As for the half saturation constant for oxygen, the lower $K_{NHx,ANO}$ in MBR systems may be explained by the fact that the average flocs size (30–40 μm) in

MBRs is lower than that measured in CAS systems (140 μm) (Jiménez *et al.* 2010). However, if the aeration and the shear stress are limited, Spérandio *et al.* (2005) showed that the half saturation constant for nitrification was significantly higher in the MBR process than the value in CAS system, suggesting a higher substrate transfer limitation, due to the large flocs developed.

Regarding the half saturation parameter for oxygen ($K_{O2,ANO}$), the default value is 0.4 mg O_2/L for ASM1 and 0.5 mg O_2/L for ASM2 and ASM3. The $K_{O2,ANO}$ values calculated using ASMs to simulate MBR systems are lower in comparison with the CAS systems, in the range of 0.18–0.4 mg O_2/L (Manser *et al.* 2005; Delrue *et al.* 2010), due to a better oxygen transfer improved by floc morphology. Jimenez *et al.* (2010) studied that MBR and CAS systems have different behaviors. Especially at low oxygen concentration (<2 mg O_2/L), in MBRs, the nitrifying activity is higher than that measured in CAS systems. Consequently the $K_{O2,ANO}$ is lower in the MBR compared to CAS systems. The authors explained this observation considering the lower average flocs size (30–40 μm) in MBR compared to that measured in CAS systems.

Recently, Sarioglu *et al.* (2008) introduced in the ASM1 additional switching functions, including the dependency of biomass decay to the electron acceptor and the incorporation of the endogenous decay concept (ASM3), to account for nitrification–denitrification kinetics under different DO concentrations.

The ASM2d applied by Fenu *et al.* (2010b) was able to correctly simulate the nitrification process only lowering the $K_{O2,ANO}$ from 0.5 to 0.2 mg O_2/L and the half saturation coefficient $K_{NHx,ANO}$ from 1 to a 0.2 mg N/L.

However, the conclusion on half saturation constants ($K_{NHx,ANO}$ and $K_{O2,ANO}$) depends on the hydrodynamic and configuration of the process.

Denitrification

Denitrification kinetics is apparently less modified by the membrane configuration, and the default parameters of ASMs model for CAS systems could be applied directly to the MBRs (Parco *et al.* 2007). In their study, on a MBR (SRT = 20 d), Parco *et al.* (2007) obtained a conventional denitrification rates of 0.25 mg NO_3/mg SS d, similar to CAS systems. Thus, the default values of the reduction factor for anoxic growth ($n_{\mu OHO,Ax}$), and the anoxic heterotrophic yield ($Y_{Stor\_OHO,Ax}$) may be applied using ASMs. However, as specific denitrification rates measured in MBRs are often very low at high SRT (Han *et al.* 2005), design MBRs guidelines recommend a ratio $V_{AER}/V_{AX} = 50/50\%$, whereas it could be at least 75/25% in CAS (Pinnekamp, 2006).

The half saturation constant for nitrate ($K_{NOx,OHO}$) and oxygen ($K_{O2,OHO}$) may be adjusted to apply the unmodified ASMs to the MBR systems, even if it has not been yet specifically determined for MBR using unmodified/conventional ASMs. Recently, Hocaoglu *et al.* (2011b) in an MBR modeling study with the modified ASM1 found that both nitrification and denitrification kinetics vary as a function of the sludge age. Further, they found that the calibrated values of the half saturation constants need to be reduced when the sludge age is lower than 20 days. The authors explained those results by the fact that the high levels of biomass concentrations sustained in the MBR, which are a function of the SRT, affect the mass transfer limitations of nitrogen compounds and DO and define a specific balance between biomass fractions responsible for nitrifications and denitrification.

Sarioglu *et al.* (2008), in their study, showed the impact of diffusion through biomass on the system performance and denitrification process. The authors used an ASM1 approach modified for endogenous decay model and for the anoxic decay of autotrophs and heterotrophs. They used switch functions that regulate nitrification-denitrification with respect to dissolved oxygen concentration in the bulk liquid. In particular, the authors simulated the oxygen diffusion limitation from the bulk liquid into the flocs by assigning specific values to half saturation constants in the corresponding switching functions, much higher than the default ASM values: $K_{NOx,OHO} = 2$ mg N/L (default 0.5 mg N/L) and $K_{O2,OHO} = 1.0$ mg O_2/L (default 0.2 mg O_2/L).

Manser *et al.* (2005) found for MBRs, a $K_{O2,OHO}$ value of 0.05 mg O_2/L, lower than the 0.2 mg O_2/L for a CAS process working at the same conditions. However, this parameter also depends on system hydrodynamics and configuration.

Further, as for the nitrification, Manser *et al.* (2005) showed that the floc size distribution had an influence on oxygen transfer and consequently on oxygen transfer in the denitrification zone. When MBR configurations present sludge recirculation from the aerated membrane tank to the denitrification zone, the level of dissolved oxygen has an important effect on denitrification potential, as dissolved oxygen level, through the parameter $K_{O2,OHO}$, can inhibit denitrification in the anoxic growth reaction modeling.

Delrue *et al.* (2010) found a significantly larger value for the half-saturation coefficient for nitrate ($K_{NOx,OHO} = 3$ mg N/L) caused by the SNDN process and non-ideal mixing. Similar results were achieved by He *et al.* (2009) for the simultaneous nitrification and denitrification (SND) process in a membrane bioreactor.

Phosphorous removal process

Parco *et al.* (2007) concluded that kinetic parameters for biological Phosphorous (P) - removal are comparable in MBR and CAS. Further, no effect of the sludge concentration on anaerobic P-release and acetate consumption rates, and the anoxic and aerobic P-uptake rates were observed. Jiang *et al.* (2008) used the ASM2d to predict phosphorous removal. Using the default ASMs parameters, the model overestimated nitrate concentration while underestimating phosphorous concentration.

In their study, Verrecht *et al.* (2010) concluded that the modeling of biological phosphorus (Bio-P) removal needed more attention, requiring a dynamic calibration. To obtain a better representation of P removal, the authors adopted the ASM2d model with default parameters proposed by Henze *et al.* (2000) and the biomass decay rates modifications proposed by Gernaey and Jørgensen (2004). Nevertheless, the phosphate removal could not be modeled successfully with the ASM2d adopted by Fenu *et al.* (2010b).

Oxygen transfer rate

The DO concentration in the mixed liquor of MBR plays an important role on the biological process. As a consequence, the modeling of oxygen concentration and oxygen transfer should be dealt with carefully. The high MLSS concentration in MBRs influences the aeration efficiency (Germain *et al.* 2007). In the ASMs family the oxygen mass transfer/diffusion phenomena are not considered. Usually switch functions are used in controlling the related processes depending upon oxygen as an electron acceptor. The half saturation coefficient in the switch functions acts as a conversion factor and reflects the lumped effect of diffusional impact on microbial growth. However the aeration modeling should also include oxygen transfer mechanisms. In order to take into account the effect of the elevated MLSS concentrations on oxygen transfer and on the other dependencies of oxygen transfer, aeration submodels have been applied to MBR modeling (Germain *et al.* 2007; Delrue *et al.* 2010; Verrecht *et al.* 2010; Insel *et al.* 2011; Hocaoglu *et al.* 2011a). The aeration models introduced the α-factor, defined as the ratio of the volumetric transfer coefficient under process conditions (i.e., with mixed liquor) to the clean water transfer coefficient. The modeling of the biological aeration demand in MBRs should account for the effect of specific process conditions on the α-factor (Fenu *et al.* 2010a). In an MBR, where the MLSS concentration is 10–20 gMLSS/L, the α-factor is lower (0.2–0.5) with respect to the α-factor in

CAS systems (0.7–1), where the MLSS concentration is <5 gMLSS/L. More details about the aeration model can be found in Maere *et al.* (2011).

9.2.4 Application of modified/integrated ASMs models to MBR

In order to obtain more accurate simulation results, a large effort has been made to understand what aspects should be modified or integrated in unmodified/conventional ASMs models to been represent the MBR technology.

A first aspect not considered in conventional ASMs models is the retention of small organic and inorganic particles larger than the membrane pore size. As a consequence, there is a much higher concentration of dispersed bacteria, organic colloids and biopolymers (such as polysaccharides and proteins) inside the biological tank of a MBR, in comparison to that found in a CAS. The higher concentration of these organic compounds (both EPS – Extracellular Polymeric Substances and SMP – Soluble Microbial Products, as described in Subsection 9.2.4.1) has a direct impact on:

- stimulating the growth of heterotrophic bacteria and, thus, reducing the expected ratio of nitrifiers to total active biomass, provided by conventional ASMs;
- playing a relevant role in the membrane permeability decrease, which can be observed in membrane bioreactors;
- concentrating in the biological tank, especially in case of treatment of industrial wastewaters, inhibitors compounds, associated to fine dispersed particles (e.g., hydrocarbons) that can negatively affects the growth rate of bacteria, especially the most sensitive like AOB (Ammonia-Oxidizing Bacteria) and NOB (Nitrite-Oxidizing Bacteria).

For these reasons, a big effort has been made in the scientific community to develop stand-alone EPS and SMP models on one side and on the other side ASM extensions with EPS/SMP concepts, as better explained in next section.

9.2.4.1 Modeling of SMP/EPS formation and degradation

EPSs are a mixture of organics including proteins, polysaccharides, lipids, humics and DNA residuals. They play an irreplaceable role in bacterial growth (both suspended and attached) by surrounding cells and thus allowing micro-organisms to live in high-density cell communities (Laspidou & Rittmann, 2002a, b). In suspended biomass processes, EPSs are typically distinguished into two separate components, namely bound EPS and free EPS; bound EPS are linked to the floc structure while the free form of EPS freely moves in the liquid bulk between suspended cells and/or biofilm.

SMP (Soluble microbial Products) are soluble components excreted by cells, during cell lysis, cell synthesis and/or other metabolic activities. SMPs are generally divided into two different components:

- utilization-associated products (UAP), that is, SMP deriving from substrate utilization by micro-organisms and therefore produced at a rate proportional to substrate consumption;
- biomass-associated products (BAPs), that is, the SMP associated to the biomass decay, whose production is proportional to the biomass concentration and biomass decay rate.

Although no general agreement has been achieved yet among scientists and researchers about the mutual interaction between SMP and EPS, a widely accepted assumption is represented by the first hypothesis of the unified theory proposed by Laspidou and Rittmann (2002b), according to which free EPS correspond to SMP. In MBRs, EPS and SMPs have been demonstrated to play a relevant role in the slow permeability decrease which can be observed under constant flux operation of membrane bioreactors (Cho & Fane, 2002; Guglielmi *et al.* 2007a; Zhang *et al.* 2010) (see Section 9.3). Moreover, the role of specific components of EPS/SMP in membrane fouling has been demonstrated in a recent work by Sweity *et al.* (2011). The authors investigated the change in EPS adherence and viscoelastic properties of sludge under different sludge ages (3–30 days) by combining QMC-D (Quartz Crystal Microbalance with Dissipation monitoring) and CLSM (Confocal Laser Scanning Microscopy). EPS from an ultrafiltration fouled membrane was extracted and analyzed using QMC-D, and the higher EPS adherence was observed when shorter sludge ages were operated: at shortest SRT (3 days) the highest content of polysaccharides was measured in the biofouling layer on the membrane surface (Sweity *et al.* 2011). The impact of EPS on membrane fouling has been recently studied also by Tian *et al.* (2011), who compared the EPS content and the filtration performance of two identical systems filtering bulking sludge and non-bulking sludge respectively. The membrane fouling due to bulking sludge was much more severe in comparison with the normal sludge, because of different protein composition, higher content of EPS and higher protein/carbohydrate ratio.

Fenu *et al.* (2010a) have recently reviewed the different approaches to modeling EPS and SMP formation and degradation, by differentiating stand-alone models (i.e., explicitly focused on production and elimination of either EPS or SMP) and integrated models where the EPS/SMP concepts are implemented within the well-known ASM structure. An early model for EPS formation was proposed in late '50s by Luedeking and Piret (1959), where the EPS formation rate includes a first-order term associated with bacterial growth. This simplified model has recently been criticized by Laspidou and Rittmann (2002a, b) and Aquino and Stuckey (2008), since it does not include any EPS degradation mechanism. According to the approach proposed by Laspidou and Rittmann (2002a, b), the EPS formation depends on the substrate utilization rate r_s by the active biomass X_a while the EPS loss due to hydrolysis leading to soluble organics is described by a first-order reaction.

Conversely Aquino and Stuckey (2008), on the basis of their experimental observations, suggest the EPS production term not to be dependent on the substrate consumption rate.

The stand-alone SMP models typically split the production/degradation mechanisms of UAP and BAP between each other. However, while there is a general consensus about the UAP formation and degradation, the same does not happen for the BAP formation/degradation, the latter being influence by both EPS hydrolysis and death cells lysis. UAP formation is widely accepted to be a consequence of substrate utilization and to be in a proportion with substrate and active biomass concentration (Laspidou & Rittmann, 2002a, b; Namkung & Rittmann, 1986; Lu *et al.* 2001). Some authors include both heterotrophs and autotrophs in the active biomass involved in UAP formation, assuming that the autotrophic microorganisms are able to produce but not consume SMP. On the other hand, BAP formation/degradation mechanisms are not unequivocally identified. Concerning BAP formation, while some researchers (Laspidou & Rittmann, 2002a, b) assume these compounds to be generated by EPS hydrolysis only, several experimental evidences seem to contradict such hypothesis (Ramesh *et al.* 2006; Jang *et al.* 2006). BAP degradation is typically considered as a direct degradation process, though Jiang *et al.* (2008) add on a intermediate hydrolysis step in their model. Ni *et al.* (2010) showed UAP to be organic molecules usually lower than 290 kDa, whereas BAP consist in macromolecules with a MW (Molecular Weight) ranging between 290 and 5000 kDa, which are further split into a growth-associated component (GBAP) and a endogenous component (EBAP).

The above-reviewed EPS/SMP formation/degradation models are summarized in Table 9.3.

As pointed out by Fenu *et al.* (2010a), introducing the EPS/SMP within the conceptual framework of IWA-ASM for membrane bioreactors mainly aims at linking biological process issues with membrane fouling, predicting the fate of soluble COD and better understanding the MBR process behavior under

Table 9.3 EPS-SMP formation/degradation models.

Process	Equation	References	Notes/Comments
EPS models			
EPS formation	$rEPS = K_1 \mu X + K_2 X$	Leudeking and Piret (1959)	K_1, fraction of substrate electron shunted to EPS formation, in the growth-associated term K_2 coefficient of EPS formation in the non-growth term
EPS formation and degradation	$rEPS = K'_{\text{EPS}} r_s X_a - K_{\text{hyd}} X_{\text{EPS}}$	Laspidou and Rittmann (2002a, b)	X_a, active biomass K'_{EPS}, fraction of electron shunted to EPS formation r_s, substrate utilization rate K_{hyd}, hydrolysis rate of EPS X_{EPS}, EPS concentration
	$rEPS = K_{\text{EPS}} X_a - K_{\text{hyd}} X_{\text{EPS}}$	Aquino and Stuckey (2008)	K_{EPS}, EPS formation rate
SMP models			
UAP formation	$r_{\text{UAP}} = K_1 q_{\text{UAP}} \frac{S}{K_S + S} X_a$	Laspidou and Rittmann (2002a, b)	K_1, kinetic coefficient in EPS model q_{UAP}, maximum specific UAP utilization rate S, substrate concentration K_S, half-saturation constant for substrate S X_a, active biomass concentration
	$r_{\text{UAP}} = \mu_{\text{OHO}} X_{\text{OHO}}$ $r_{\text{UAP}} = \mu_{\text{ANO}} X_{\text{ANO}}$	Lu *et al.* (2001)	μ_{OHO}, heterotrophic growth rate X_{OHO}, heterotrophic active biomass μ_{ANO}, autotrophic growth rate X_{ANO}, autotrophic active biomass Switch-functions depend on the process where UAP is formed
	$r_{\text{UAP}} = k_{\text{UAP}} \frac{\mu_{\text{OHO}}}{Y_{\text{OHO}}} \frac{S_B}{S_B + K_S} \frac{S_{O2}}{S_{O2} + K_{O2\_OHO}} X_{\text{OHO}}$	Ni *et al.* (2010)	k_{UAP}, UAP formation coefficient μ_{OHO}, maximum growth rate of heterotrophic biomass on S_S Y_{OHO}, Yield coefficient for growth on S_S

(*Continued*)

Table 9.3 EPS-SMP formation/degradation models (*Continued*).

Process	Equation	References	Notes/Comments
			S_B, soluble biodegradable substrate K_S, external substrate affinity constant (half-saturation constant) K_{O2_OHO}, dissolved oxygen affinity constant S_{O2}, dissolved oxygen concentration X_{OHO}, active heterotrophic biomass
UAP degradation	$r_{UAP} = -q_{UAP}\frac{S_{UAP}}{K_{UAP}+S_{UAP}}X_{bm}$	Laspidou and Rittmann (2002a, b)	q_{UAP}, maximum specific UAP utilization rate S_{UAP}, UAP concentration K_{UAP}, half-saturation constant for UAP
	$r_{UAP} = -\mu_{SMP}\frac{S_{SMP}}{K_{SMP}+S_{SMP}}X_{OHO}$	Lu *et al.* (2001)	μ_{SMP}, maximum specific growth rate of SMP for heterotrophs S_{SMP}, total soluble microbial products K_{SMP}, SMP half-saturation constant for heterotrophs X_{OHO}, active heterotrophic biomass Switch-function depends on the process where UAP is degraded
	$r_{UAP} = -\frac{\mu_{UAP}}{Y_{UAP}}\frac{S_{UAP}}{S_{UAP}+K_{UAP}}\frac{S_{O2}}{S_{O2}+K_{O2\_OHO}}X_{OHO}$	Ni *et al.* (2010)	μ_{UAP}, maximum rate of UAP degradation Y_{UAP}, Yield coefficient for growth on UAP S_{UAP}, concentration of UAP-related COD K_{UAP}, biomass affinity constant for UAP K_{O2_OHO}, dissolved oxygen affinity constant S_{O2}, dissolved oxygen concentration X_{OHO}, active heterotrophic biomass
BAP formation	$r_{BAP} = K_{hyd}X_{EPS}$ $r_{BAP} = K2^{*} \cdot X_a + K_{hyd}X_{EPS}$	Laspidou and Rittmann (2002a, b)	K_{hyd}, hydrolysis rate X_{EPS}, EPS concentration

		Aquino and Stuckey (2008)	$K2^*$, coefficient of BAP degradation referred to the biomass X_a X_a, heterotrophic active biomass K_{hyd}, hydrolysis rate X_{EPS}, EPS concentration
	$r_{BAP} = k_{hyd1}\frac{S_S}{K_{BAP}+S_S}X_{EPS} + k_{hyd2}\frac{K_{S,BAP}}{K_{BAP}+S_S}X_{EPS}$	Ni *et al.* (2010)	k_{hyd1}, EPS hydrolysis rate for growth-associated BAP S_S, soluble biodegradable substrate K_{BAP}, half-saturation constant for BAP formation X_{EPS}, EPS concentration k_{hyd2}, EPS hydrolysis rate for endogeny-associated BAP
BAP degradation	$r_{BAP} = -q_{BAP}\frac{S_{BAP}}{K_{BAP}+S_{BAP}}X_{OHO}$	Laspidou and Rittmann (2002a, b)	q_{BAP}, maximum specific BAP utilization rate S_{BAP}, BAP concentration K_{BAP}, half-saturation constant for BAP X_{OHO}, heterotrophic biomass concentration
	$r_{BAP} = -K_{hyd,BAP}S_{BAP}X_{OHO}$	Jiang *et al.* (2008)	$K_{hyd\ BAP}$, EPS hydrolysis rate for BAP S_{BAP}, BAP concentration X_{OHO}, heterotrophic active biomass
	$r_{BAP} = -\frac{\mu_{GBAP}}{Y_{GBAP}}\frac{S_{GBAP}}{S_{GBAP}+K_{GBAF}}\frac{S_O}{S_O+K_O}X_H - \frac{\mu_{EBAP}}{Y_{EBAP}}\frac{S}{S_{EBAP}}$	Ni *et al.* (2010)	μ_{GBAP}, maximum rate of GBAP degradation Y_{GBAP}, yield coefficient for growth on GBAP S_{GBAP}, GBAP-associated COD K_{GBAP}, biomass affinity constant for GBAP μ_{EBAP}, maximum rate of EBAP degradation Y_{EBAP}, yield coefficient for growth on EBAP S_{EBAP}, EBAP-associated COD K_{EBAP}, biomass affinity constant for EBAP

Source: Modified after Fenu *et al.* (2010a).

very long SRTs (>30 days). Complex modification of ASM1 and ASM3 have been proposed by Lu and co-workers (Lu *et al.* 2001; Lu *et al.* 2002), which lead to eight additional SMP-related parameters within the standard ASM1 and ASM3 structure (Figure 9.1). However, the main limitation of this approach is related to the SMP-related COD mass balance, which is not closed since the loss of substrate COD is not equal to the sum of formed UAP COD, formed biomass COD and consumed oxygen. The conceptual framework originally proposed by Lu and co-workers has been extensively adopted by other authors in subsequent studies (Zarragoitia-Gonzales *et al.* 2008; Di Bella *et al.* 2008), focusing on the impact of SMP-related COD on the membrane fouling. Further examples of extended ASMs with SMP/EPS concepts have been presented by Jiang *et al.* (2008), Ahn *et al.* (2006) and Oliveira-Esquerre *et al.* (2006). The main difference between the above-mentioned studies is in the way to model the hydrolysis mechanisms of bound EPS, BAP and UAP. Fan and co-workers (2011) proposed two different ASM3-extensions to simulate the SMP production: the former combines the SMP formation and degradation concepts in the conventional ASM3 (ASM3-SMP), the latter introduces the SMP dynamics into a modified version of the ASM3, where growth on soluble substrate and storage occur simultaneously (SSSG-ASM3-SMP). These two approaches were compared at bench scale on a sequencing batch reactor fed with synthetic sewage, by using the modeled DO-profile as a robustness indicator. The SSSG-ASM3-SMP approach was found to better predict the dissolved oxygen consumption during the feast and famine conditions.

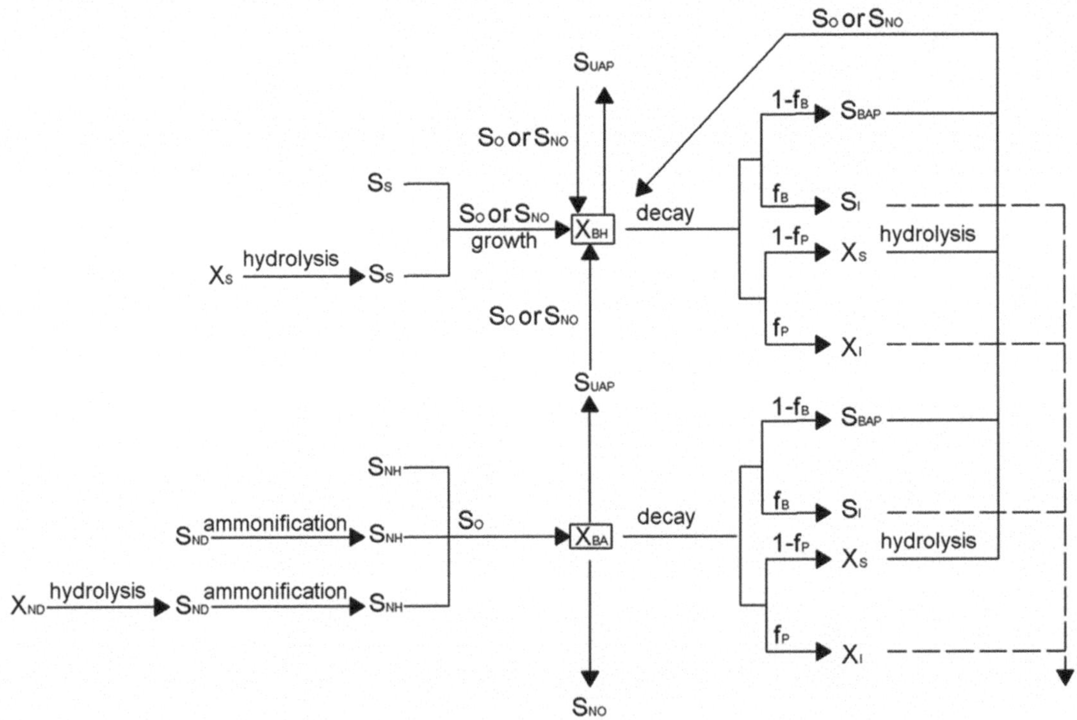

Figure 9.1 Schematic description of the extended ASM1 with SMP proposed by Lu *et al.* (2001).

What is clearly highlighted in reviewing EPS/SMP introduction to ASMs for MBR modeling is that the estimation of the kinetic parameters related to both formation and degradation of such compounds are not easily measured experimentally. Moreover, by trying to better model the experimental observation, a general tendency toward over-parameterization is clearly evident, thus hindering the model applicability and its prediction capability. Some new steps ahead in such sense have been taken by Chen and co-workers (2012), by carrying out the sensitivity and uncertainty analyses of an extended ASM3-SMP model by means of the extended Fourier amplitude sensitivity test. A summary of SMP/EPS parameters within the ASM framework has been reviewed by Fenu *et al.* (2010a and Table 9.4).

Chen *et al.* (2012) developed a model based on the ASM1 that incorporates considerations for the roles of SMP, EPS, and SRT (ASM1-SMP-EPS-SRT). The model agreed qualitatively well with the experiment data, and showed that the concentrations of VSS, SMP, and the endogenous decay coefficient were sensitive to the system SRT.

Recently, a modeling study using modified ASMs model indicated that both nitrification and denitrification kinetics varied as a function of the sludge age (Hocaoglu *et al.* 2011b): the half saturation constants of nitrification and denitrification were not any more constant but were significantly reduced when the sludge age was lowered to 20 days. The model developed by Hocaoglu *et al.* (2011b) essentially defines MBR as an activated sludge process with specific characteristics. It includes the basic backbone of ASM1 modified for endogenous respiration. The model structure was adapted to MBR by providing process description for the following features: (a) mass transfer limitation for DO diffusion and utilization; (b) generation of SMP; (c) accumulation and partial hydrolysis of soluble residual organics. The authors explained the lower half saturation constants for nitrification and denitrification by the fact that high levels of biomass concentrations are sustained in the MBR, which are a function of the SRT. The resulting floc viscosity affects mass transfer limitations of nitrogen compounds and DO and defines a specific balance between biomass fractions responsible for nitrifications and denitrification. This implies that different half saturation parameters are used in the switching functions of DO and nitrogen compounds, defining the balance between nitrification and denitrification rates.

9.3 FILTRATION MODELS

Compared with membrane fouling in conventional membrane applications with single- or quasi-single foulants, fouling in membrane bioreactors is widely accepted to be a more difficult phenomenon to be described due to (i) the matrix complexity and (ii) the dynamic interactions between primary foulants (such as suspended solids) and secondary foulants (e.g., colloids, SMP and EPS) in the bulk and in the resulting cake layer. In understanding fouling mechanisms in membrane bioreactors a key-role is played by the critical flux concept, first introduced by Field *et al.* (1995) and then extensively applied in most of the membrane processes, including MBRs. Although inappropriate to be used in membrane bioreactors in its purest form (the flux value below which no fouling occurs), the critical flux theory has been successfully used to interpret experimental observations at both pilot- and full-scale, especially when measured by means of the flux-stepping method (Le-Clech *et al.* 2003). Considering the most common operation mode of MBRs (i.e., constant flux), several studies have demonstrated that fouling take places in two different steps: a first-step with a slow increase of transmembrane pressure (TMP) followed by a second-step characterized by a sudden increase of TMP, also referred to as "TMP-jump", leading to permeability collapse and process stopping. In all these works (inter alia Liu *et al.* 2005; Ognier *et al.* 2004; Le-Clech *et al.* 2003; Guglielmi *et al.* 2007b; Wintgens, 2003), the fouling rate (dTMP/dt) before and after the TMP-jump and the time to reach the permeability collapse have been used as indicators of the fouling behavior under different sub-critical fluxes conditions. A conceptual model for membrane fouling mechanisms in MBRs has been proposed by Zhang *et al.* (2006), and is schematically represented in Figure 9.2.

Table 9.4 SMP/EPS biokinetics.

EPS/SMP Modeling Affinity	**Affinity Constants**		**Formation**			**Degradation**			**EPS**	
	K_{UAP}	K_{BAP}	k_{SMP}	k_{UAP}	k_{BAP}	K_{UAP}	K_{BAP}	f_{SMP}	K_{EPS}	K_{hyd}
Reference	g_{CODSMP}/L		$mg_{CODcell}$/mg_{CODSMP}	mg_{CODUAP}/mg_{CODSUB}	1/d	$mgCOD_{product}$/$mgCOD_{cell}$/d		–	1/d	1/d
Furumai and Rittmann (1992)			0.5	0.2	0.1			1		
Laspidou and Rittmann (2002a, b)	100	85		0.05	0.17	0.07	1.27		0.18	0.17
Jang *et al.* (2006)	100	85		0.05	0.02	0.07	0.4		0.18	0.02
Aquino and Stuckey (2008)	500	500		0.02	0.0034	0.03	1.2		0.03	0.03
Ni *et al.* (2010)	100	85,106								

ASM Related Work	**Affinity Constants**	**Formation**					**Degradation**		
	K_{SMP}	μ_{SMP}	K_{BAP}	f_{SMP}	f_{UAP}	f_{BAP}	K_{SMP}	$K_{hyd,BAP}$	$K_{hyd,UAP}$
Reference	$gCOD_{SMP}$/L	1/d	1/d	–	–	–	1/d	1/d	1/d
Lu *et al.* (2001)	30	0.7	0.4		0.38				
Lu *et al.* (2002)	60	2.5	0.01		0.3				
Leet *et al.* (2002)	30	0.7		0.4			0.1		
Oliveira-Esquerre *et al.* (2006)	30			0.4					
Zarragoitia-Gonzales *et al.* (2008)	30	0.7	0.22		0.38				
Di Bella *et al.* (2008)	133	8.3	0.33		0.82				
Jiang *et al.* (2008)					0.0963	0.0215		7.4E-7	0.0102

K_{UAP} = Half-saturation constant for UAP.
K_{BAP} = Half-saturation constant for BAP formation.
K_{SMP} = SMP half-saturation constant for heterotrophs.
K_{EPS} = EPS formation rate.
K_{hyd} = Hydrolysis rate of EPS.
f_{SMP} = Fraction of SMP generated in biomass decay.
f_{UAP} = Fraction of UAP generated in biomass decay.
f_{BAP} = Fraction of BAP generated in biomass decayed.
$K_{hdy,BAP}$ = EPS hydrolysis rate for BAP.
$K_{hdy,UAP}$ = EPS hydrolysis rate for UAP.
k_{SMP} = SMP formation coefficient.
k_{UAP} = UAP formation coefficient.
k_{BAP} = BAP formation coefficient.
Source: Adapted from Fenu *et al.* (2010a).

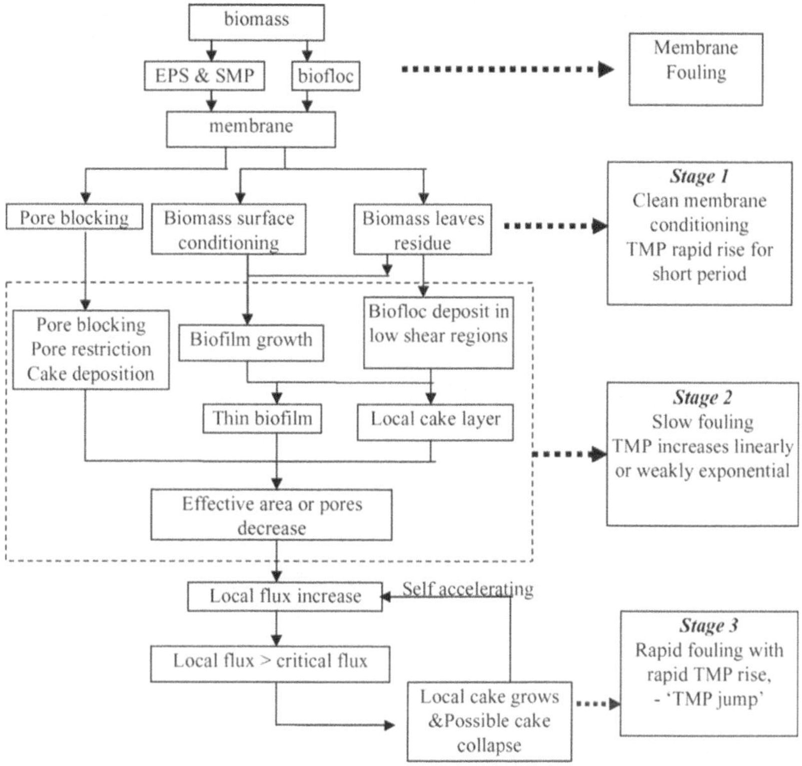

Figure 9.2 Three-steps fouling mechanisms in membrane bioreactors (Zhang *et al.* 2006).

The so-called conditioning fouling (Stage 1) is due to the passive adsorption of colloids and organics and to the initial pore blocking at the very beginning of the filtration; the extent to which this mechanism takes place mainly depends on specific membrane properties such as hydrophobicity and pore size distribution, the Stage 2 (slow fouling) is widely accepted to be caused by the soluble fraction of EPS (i.e., the SMP according to the unified theory proposed by Laspidou and Rittmann); experimental evidence of this assumption is provided by Wang *et al.* (2008), who characterized both membrane foulants and gel layer formed onto the membrane surface under sub-critical conditions by using gel filtration chromatography (GFC) and scanning electron microscopy (SEM). During this low-rate fouling stage, these foulants, whose specific critical flux is lower than the one of the main foulant (suspended solids) do gradually adsorb onto the membrane surface, thus leading to a slow TMP rise.

Once the permeate flux in the actually working membrane regions reaches the critical one, a self-accelerating exponential increase of TMP is observed (Stage 3). Different approaches have been proposed to explain this stage, which have been extensively reviewed by Le-Clech *et al.* (2006):

(1) the inhomogeneous fouling (area loss) model;
(2) the inhomogeneous fouling (pore loss) model;
(3) the critical suction pressure model;
(4) the percolation theory.

The 'area loss' model was proposed by Cho and Fane (2002) to explain the TMP transients in a membrane bioreactor fed with effluent from an up-flow anaerobic sludge blanket (UASB) reactor. The gradual decrease of non-fouled membrane surface due to EPS deposition resulted in a flux redistribution which caused local super-critical regions and, consequently the TMP sudden increase. On the same basis but focusing on the macro- rather than the micro-scale, the 'pore loss' theory was proposed by Ye *et al.* (2005): the gradual reduction of open pores due to the considered model biopolymer (alginate) determined local pore velocities exceeding the critical flux of alginate aggregates, thus leading to the sudden increase of trans-membrane pressure (TMP).

The critical suction pressure model was proposed by Chang *et al.* (2005) to explain the sudden increase of TMP during dead-end filtration experiments of a fine colloid; the fouled membrane autopsy revealed a critical suction pressure to be responsible for the coagulation at the base of the cake-layer. Although conceptually deriving from dead-end filtration experiments, this approach shows some characteristics that make it suitable for explaining the typical TMP-transients observed in real MBR installations. In fact, according to this model fouling continues to take place until the critical suction pressure is reached and this is in fair agreement with the experimental observations at pilot- and full-scale.

The percolation theory (Hermanowicz, 2004) assumes that the porosity of gel layer gradually reduces because of the deposition of the foulants during the filtration process; when a critical condition is reached, the fouling cake resistance increases leading to the TMP jump.

Most probably, these mechanisms are likely to occur all at the same time once during the Stage 3.

The importance of fairly predicting the critical time (i.e., the time at which the sudden TMP increase occurs) is extremely evident in terms of process optimization at full scale, since it can provide support in choosing the suitable flux regime under given conditions of sludge age, MLSS and hydrodynamics. An attempt to model the TMP transient during Stage 2 has been done by Ognier *et al.* (2004), on the basis of the above mentioned 'pore loss' model. The authors propose a simplified single-parameter model according to which:

$$J_P = \frac{J_{P0}}{1 - ((\alpha \cdot J_{P0} \cdot \mu_P \cdot R_P \cdot t^2)/2)}$$

where:

- J_P and J_{P0} is the permeate flow through the single pore at time t and time $t = 0$, respectively;
- α is the model parameter which is function of the solutes concentration and accounts for both convective flow of foulants toward the membrane surface and the variation of the number of open pores;
- μ_P is the permeate dynamic viscosity;
- R_P is the hydraulic resistance of the single pore.

The value of the alpha parameter depends on the specific biological sludge characteristics and membrane properties, and the authors showed a successful application to a purely denitrifying bench-scale membrane bioreactor. Afterwards, Guglielmi *et al.* (2007a) proposed a similar approach emphasizing the mutual role of bound and free EPS, assumed to be in a dynamic equilibrium between themselves. The actual flux variation over time is described by the following equation:

$$J = \frac{1}{1/J_o - k_f \cdot C_f \cdot C_b \cdot Q \cdot t}$$

where

- J_0 (L/m^2 h) is the initial flux obtained from average permeate flow rate Q (m^3/h) and non-fouled membrane area available at the beginning of the filtration;

- k_f ($m^2/h\ g^2$) is the coefficient of fouling propensity which accounting for the effect of free and bound EPS on the reduction of the effective membrane surface area;
- $C_f(g/m^3)$ and C_b (g/m^3) are the concentrations of free and bound EPS, respectively.

Once calibrated the model was successfully validated on experimental data from a large pilot scale membrane bioreactor, showing a good capability in predicting the critical time.

9.4 CFD AND HYDRODYNAMICS – MODELING OF MBR TANKS AND FLUID DYNAMICS

The hydrodynamics within an MBR system have a crucial impact on its performance (Rector *et al.* 2006). Successful modeling of MBR hydrodynamics allows the development of a complete MBR model, that can be used not only for optimising tank design, but also for improving nutrient removal and energy consumption.

Hydrodynamics tools such as the Computational Fluid Dynamics (CFD) provide a very effective tool for investigating and consequently optimizing geometrical and operational aspects within a defined control volume. Therefore, many authors have successfully investigated MBR hydrodynamics at both micro- and macro-scale, the former being related to optimal module design the latter being aimed at optimal process design and operation.

9.4.1 Module design

Many studies have been carried out in the last decade on the hydrodynamics of filtration modules.

Taha and Cui (2002) used CFD to model the movement of a Taylor bubble through a tubular membrane and the related wall shear. Rios *et al.* (2007) conducted a similar study, but also took into account the effect of the sludge rheology. Saalbach and Hunze (2007) used CFD and velocity measurements to study the MBR's filtration stage. The hydrodynamics of a pilot scale MBR with different configurations for air scouring were investigated experimentally and numerically by Drews *et al.* (2008), with the aim to derive a correlation between the aeration flow rate and the circulation velocity in the reactor.

As most energy consumption in MBRs for wastewater treatment comes from scouring aeration for membrane fouling prevention and amelioration, Buetehorn *et al.* (2011) used a novel CFD approach to investigate the impact of irregular fiber arrangement on the aeration efficiency: a commercial hollow fiber module was scanned with X-ray Computer Tomography (CT) to map the instantaneous displacement of fiber in different height of the bundle. The CT revealed an anisotropic resistance to the flow which was implemented in the conceptual framework of a single- and multi-phase CFD model. Single-phase simulations showed the relevant impact of local fiber arrangement, superficial inlet velocity and MLSS concentration on the distribution of cross-flow velocity.

The outcome of these studies highlighted the great potential of CFD in model-aided design of the module geometry, packing density and arrangement and air scouring configurations.

9.4.2 Process design and operation

Most applications of CFD modeling to MBRs have been focused on a single component (e.g., membrane filtration) of the MBR system. Investigations on the hydrodynamics of entire full-scale MBRs have only been carried out in the last few years. Modeling of a full-scale MBR using CFD is very important in order to increase the understanding of the effects of MBR design (e.g., membrane configuration) on energy consumption.

Brannock *et al.* (2009), Brannock *et al.* (2010a) and Wang *et al.* (2008) have examined the entire MBR system using tracer experiments (full-scale and pilot-scale) and CFD analysis (pilot-scale).

In particular, Brannock *et al.* (2010a) developed a CFD model that enables the optimisation of MBR design in terms of achieving preferred mixing regimes (e.g., complete mixing) at the lowest energy input. As a consequence, during the design process, once the energy required by the MBR (e.g., to meet the requirement of aeration due to various viscosity of sludge etc.) has been determined, the mixing regimes of the MBRs can also be predicted rather than assumed. The MBR CFD model was applied to two full-scale MBRs and then validated in terms of the residence time distribution.

A further study by Brannock *et al.* (2010b) investigated the hydrodynamic behavior of two full scale MBR installations, with different configurations in terms of both biological process and membrane geometry, using a CFD model accounting for aeration, sludge rheology and membrane geometry.

The CFD tool was demonstrated to successfully predict the hydrodynamic behavior of even full scale membrane bioreactors, showing a great potential in supporting decision in process design (e.g., energy input through aeration and mixing, size and positions of inlets, baffles and membranes).

The aforementioned conceptual framework was also used by Brannock *et al.* (2009) to compare the internally submerged *versus* externally submerged configurations, both at pilot scale. The externally submerged configuration was found to be closer to the plug flow mixing conditions compared to the internally submerged one, with an expected positive effect on the pollutants removal. The internally submerged configuration showed a higher amount of short circuiting and dead zones in all aeration cases.

9.5 CONTROL AND OPERATIONAL STRATEGIES

While the modified and the unmodified ASMs models are widely used for studying process behavior, system design and process optimization for WWTPs and MBR systems, the COST/IWA Benchmark Simulation Model No. 1 (BSM1) (Copp, 2002; Jeppsson & Pons, 2004), is a standardised simulation procedure for the design and evaluation of control strategies for conventional WWTPs in terms of effluent quality and operational costs: BSM1_LT (Rosen *et al.* 2004) and BSM2 (Jeppsson *et al.* 2006).

Nevertheless, the application of conventional control strategies for aeration, recirculation pumping, carbon addition, and so on to MBRs is yet to be carefully investigated.

Several authors tried to develop simple static spreadsheet models based on the steady-state operation (Yoon *et al.* 2004; Verrecht *et al.* 2008), in order to quantify the operational costs for MBR. Although useful, these models may lead to erroneous conclusions as they do not take into account the dynamic behavior and system configuration.

Fenu *et al.* (2010b) developed a calibrated dynamic biological ASM model of a full scale membrane bioreactor (MBR), which was used to analyze the energy costs of different compartments and devices. The authors used the ASM2d to include the inorganic fraction. The energy consumption data, furnished by suppliers, were also incorporated in the ASM2d model.

Recently, Maere *et al.* (2011) proposed a dynamic benchmark simulation model for MBRs (BSM-MBR) as a platform to evaluate their operational and control strategies. The existing BSM1 for a conventional WWTP was used as starting point and updated in terms of reactor volumes, membrane filtration, aeration capacity and sludge flows. Then, the BSM1 performance criteria were extended for an MBR taking into account MBR characteristics: additional pumping requirements for permeate production; additional aeration requirements for fouling suppression; a dedicated aeration model to incorporate the effects of elevated sludge concentrations on aeration efficiency and costs. Both steady-state and dynamic behavior were considered.

9.6 CONCLUSIONS AND FUTURE OUTLOOK

In this chapter a critical review on MBR modeling studies has been furnished. The main important outcomes can be summarized as follows:

- The application of unmodified ASM based models to MBRs requires a specific attention to the experimental identification and calibration of model parameters, especially for sludge production and nitrification modeling.
- Extensions of ASM based models are necessary in order to take into account the recognized crucial specifities of MBR processes: a different microbial composition, leading to different stoichiometric and kinetic parameters; a higher mixed liquor concentration and a higher viscosity leading to changes in oxygen transfer; higher sludge retention times leading to modified sludge production yields; the accumulation and degradation of SMP due to membrane filtration processes affecting fouling processes. Further research is needed to correctly model SBRs applied to anaerobic conditions, where different biological processes may take place (fermentation, methanogenesis, sulphate reduction, anammox process, etc.) and fouling phenomena may be more severe.
- Fouling in membrane bioreactors is widely accepted to be a difficult phenomenon to be modeled because of the matrix complexity and the dynamic interactions between primary foulants (such as suspended solids) and secondary foulants (e.g., colloids, SMP and EPS) in the bulk and in the cake layer. Many authors demonstrated that the fouling mechanisms take place in three different steps. There have been many attempts to correctly model these three steps, in particular the third one where a sudden TMP increase occurs. Fouling mechanisms comprehension and modeling is extremely important in terms of process optimization at full scale, since it can provide support in choosing the operating conditions in terms of TMP, MLSS, sludge age and hydrodynamics.
- Many studies have been carried out in the last decade on the hydrodynamics of MBR systems, both on single filtration models and on the whole MBR system. The results have shown on one side the great potential of CFD in model-aided design of module geometry, packing density and air scouring configurations; on the other side, the CFD tool has been demonstrated to successfully predict the hydrodynamic behavior of full scale membrane bioreactors, thus showing a great potential in process design (e.g., energy input through aeration and mixing, size and positions of inlets, baffles and membranes).
- There is a dearth of studies on modeling of control strategies for aeration, recirculation pumping, carbon addition, and so on, in MBRs. Some attempts have been made to develop benchmark models for MBRs for evaluating operational and control strategies. Further modeling efforts are necessary in order to develop tools for identifying the best strategies for minimizing energy consumption in MBR applications.

REFERENCES

Aguilera Soriano G., Erb M., Garel C. and Audic J. M. (2003). A comparative pilot-scale study of the performance of conventional activated sludge and membrane bioreactors under limiting operating conditions. *Water Environment Research: A Research Publication of the Water Environment Federation*, **75**, 225–231.

Ahn Y. T., Choi Y. K., Jeong H. S. and Shin S. R. (2006). Modeling of extracellular polymeric substances and soluble microbial products production in a submerged MBR at various SRTs. *Water Science and Technology*, **53**(7), 209–216.

Aquino S. F. and Stuckey D. C. (2008). Integrated model of the production of soluble microbial products (SMP) and extracellular polymeric substances (EPS) in anaerobic chemostats during transient conditions. *Biochemistry Engineering Journal*, **38**, 138–146.

Baek S. H., Jeon S. K. and Pagilla K. (2009). Mathematical modeling of aerobic membrane bioreactor (MBR) using activated sludge model no. *1 (ASM1). Journal of Industrial and Engineering Chemistry*, **15**, 835–840.

Brannock M. W. D., De Wever H., Wang Y. and Leslie G. (2009). Computational fluid dynamics simulations of MBRs: inside submerged versus outside submerged membranes. *Desalination*, **236**, 244–251.

Brannock M., Wang Y. and Leslie G. (2010a). Mixing characterization of full-scale membrane bioreactors: CFD modeling with experimental validation. *Water Research*, **44**, 3181–3191.

Brannock M. W. D., Wang Y. and Leslie G. (2010b). Evaluation of full-scale membrane bioreactor mixing performance and the effect of membrane configuration. *Journal of Membrane Science*, **350**, 101–108.

Buetehorn S., Volmering D., Vossenkaul K., Wintgens T., Wessling M. and Melin T. (2011). CFD simulation of single- and multi-phase flows through submerged membrane units with irregular fiber arrangement. *Journal of Membrane Science*, **384**, 184–197.

Chaize S. and Huyard A. (1991). Membrane bioreactor on domestic wastewater treatment sludge production and modeling approach. *Water Science and Technology*, **23**(7–9), 1591–1600.

Chang S., Fane A. G. and Waite T. D. (2005). Effect of coagulation within the cake-layer on fouling transitions with dead-end hollow fiber membranes. Proceedings of International Congress on Membranes and Membrane Processes (ICOM), Seoul, Korea.

Chen J. C., Ng W. J., Luo R., Mu S., Zhang Z., Andersen M. and Jørgensen P. E. (2012). Membrane bioreactor process modeling and optimization: Ulu Pandan water reclamation plant. *Journal of Environmental Engineering © ASCE*, **138**, 1218–1226.

Cho B. D. and Fane A. G. (2002). Fouling transients in nominally sub-critical flux operation of a membrane bioreactor. *Journal of Membrane Science*, **209**(2), 391–403.

Copp J. B. (2002). The COST Simulation Benchmark: Description and Simulator Manual. Office for Official Publications of the European Communities, Luxembourg ISBN 92-894-1658-0.

Corominas L. L., Rieger L., Takács I., Ekama G., Hauduc H., Vanrolleghem P. A., Oehmen A., Gernaey K. V., Van Loosdrecht M. C. M. and Comeau Y. (2010). New framework for standardized notation in wastewater treatment modeling. *Water Science and Technology: A Journal of the International Association on Water Pollution Research*, **61**(4), 841–857.

Dapena-Mora A., Van Hulle S. W., Luis Campos J., Mendez R., Vanrolleghem P. A. and Jetten M. (2004). Enrichment of Anammox biomass from municipal activated sludge: experimental and modeling results. *Journal of Chemical Technology and Biotechnology*, **79**, 1421–1428.

Delrue F., Choubert J. M., Stricker A. E., Spérandio M., Mietton-Peuchot M. and Racault Y. (2010). Modeling a full scale membrane bioreactor using activated sludge model no.1: challenges and solutions. *Water Science and Technology: A Journal of the International Association on Water Pollution Research*, **62**(10), 2205–2217.

Di Bella G., Mannina G. and Viviani G. (2008). An integrated model for physical-biological wastewater organic removal in a sMBR: model development and parameter estimation. *Journal of Membrane Science*, **322**, 1–12.

Dold P. L., Ekama G. A. and Marais G. V. R. (1980). A general model for the activated sludge process. *Progress in Water Technology*, **12**, 47–77.

Drews A., Prieske H. and Kraume M. (2008). Prediction of the circulation velocity in a membrane bioreactor. *Desalination*, **231**, 219–226.

Ekama G. A., Dold P. L. and Marais G. V. R. (1986). Procedures for determining influent cod fractions and the maximum specific growth rate of heterotrophs in activated sludge systems. *Water Science and Technology*, **18** (6), 91–114.

Fan J., Ding Y., Qiu Z., Li W. and Lu S. (2011). Development of mechanistically based model for simulating soluble microbial products generation in an aerated/non-aerated SBR. *Bioprocess Biosystems Engineering*, **34**, 1151–1161.

Fenu A., Guglielmi G., Jimenez J., Spèrandio M., Saroj D., Lesjean B., Brepols C., Thoeye C. and Nopens I. (2010a). Activated sludge model (ASM) based modeling of membrane bioreactor (MBR) processes: a critical review with special regard to MBR specificities. *Water Research*, **44**, 4272–4294.

Fenu A., Roels J., Wambecq T., De Gussem K., Thoeye C., De Gueldre G. and Van De Steene B. (2010b). Energy audit of a full scale MBR system. *Desalination*, **262**, 121–128.

Field R. W., Wu D., Howell J. A. and Gupta B. B. (1995). Critical flux concept for microfiltration fouling. *Journal of Membrane Science*, **100**, 259–272.

Frederickson K. and Cicek N. (2005). Performance comparison of a pilot-scale membrane bioreactor and a full-scale sequencing batch reactor with sand filtration: treatment of low strength wastewater from a northern Canadian Aboriginal Community. *Water Science and Technology: A Journal of the International Association on Water Pollution Research*, **51**(6–7), 483–490.

Gao M., Yang M., Li H., Yang Q. and Zhang Y. (2004). Comparison between a submerged membrane bioreactor and a conventional activated sludge system on treating ammonia-bearing inorganic wastewater. *Journal of Biotechnology*, **108**, 265–269.

Germain E., Nelles F., Drews A., Pearce P., Krame M., Reid E., Judd S. J. and Stephenson T. (2007). Biomass effects on oxygen transfer in membrane bioreactors. *Water Research*, **41**, 1038–1044.

Germirli F., Orhon D. and Artan N. (1991). Assessment of the initial inert soluble COD in industrial wastewater. *Water Science and Technology*, **23**(Kyoto), 1077–1086.

Gernaey K. V. and Jørgensen S. B. (2004). Benchmarking combined biological phosphorus and nitrogen removal wastewater treatment processes. *Control Engineering Practice*, **12**, 357–373.

Gernaey K. V., Van Loosdrecht M. C. M., Henze M., Lind M. and Jorgensen B. (2004). Activated sludge wastewater treatment plant modeling and simulation: state of the art. Environmental Modeling and Software, **19**, 763–783.

Guglielmi G., Saroj D. P., Chiarani D. and Andreottola G. (2007a). Sub-critical fouling in a membrane bioreactor for municipal wastewater treatment: experimental investigation and mathematical modeling. *Water Research*, **41** (17), 3903–3914.

Guglielmi G., Chiarani D., Judd S. J. and Andreottola G. (2007b). Flux criticality and sustainability in a hollow fibre submerged membrane bioreactor for municipal wastewater treatment. *Journal of Membrane Science*, **289**(1–2), 241–248.

Guglielmi G., Avesani D., Brepols C., Foxon K., Brouckaert C. and Buckley C. (2009). Sludge Production in Aerobic and Anaerobic Membrane Bioreactors. Proceedings of WISA Membrane technology Conference. 13e15 May, Stellenbosch, South Africa.

Gujer W. (1985). Ein dynamisches Modell fur die Simulation von komplexen Belebtschlammverfahren (A dynamic model for the simulation of complex processes of activated sludge). EAWAG, Dubendorf, Switzerland.

Gujer W. (2006). Activated sludge modeling: past, present and future. *Water Science and Technology*, **53**(3), 111.

Gujer W., Henze M., Mino T. and Van Loosdrecht M. C. M. (1999). Activated sludge model no. 3. *Water Science and Technology*, **39**(1), 183.

Han S.-S., Bae T.-H., Jang G.-G. and Tak T.-M. (2005). Influence of sludge retention time on membrane fouling and bioactivities in membrane bioreactor system. *Process Biochemistry*, **40**, 2393–2400.

Hao X., Heijnena J. J. and Van Loosdrecht M. C. M. (2002). Model-based evaluation of temperature and inflow variations on a partial nitrification–ANAMMOX biofilm process. *Water Research*, **36**, 4839–4849.

Hauduc H., Rieger L., Takàcs I., Héduit A., Vanrolleghem P. A. and Gillot S. (2010). A systematic approach for model verification: application on seven published activated sludge models. *Water Science and Technology*, **61**(4), 825–839.

He S., Xue G. and Wang B. (2009). Factors affecting simultaneous nitrification and de-nitrification (SND) and its kinetics model in membrane bioreactor. *Journal of Hazardous Materials*, **168**, 704–710.

Henze M., Grady C. P. L., Jr, Gujer W., Marais G. V. R. and Matsuo T. (1987). Activated Sludge Model No.1. IAWPRC Scientific and Technical Report No. 1. IAWPRC, London.

Henze H., Gujer W., Mino T., Matsuo T., Wentzel M. C. and Marais G. V. R. (1995). Activated Sludge Model No.2, IAWQ Scientific and Technical Report, No.3, IAWQ, London.

Henze M., Gujer W., Mino T. and Van Loosdrecht M. (2000). Activated Sludge Models ASM1, ASM2, ASM2d and ASM3. IWA Publishing, London.

Hermanowicz S. W. (2004). Membrane Filtration of Biological of Biological Solids: A Unified Framework and Its Application to MBR. Proc. Water Environment-Membrane Technology Conference, Seoul, South Korea.

Hocaoglu S. M., Insel G., Cokgor E. U. and Orhon D. (2011a). Effect of low dissolved oxygen on simultaneous nitrification and denitrification in a membrane bioreactor treating black water. *Bioresource Technology*, **102**, 4333–4340.

Hocaoglu S. M., Insel G., Cokgor E. U. and Orhon D. (2011b). Effect of sludge age on simultaneous nitrification and denitrification in membrane bioreactor. *Bioresource Technology*, **102**, 6665–6672.

Holakoo L., Nakhla G., Bassi A. S. and Yanful E. K. (2007). Long term performance of MBR for biological nitrogen removal from synthetic municipal wastewater. *Chemosphere*, **66**, 849–857.

Hulsbeek J. J. W., Kruit J., Roeleveld P. J. and Van Loosdrech M. C. M. (2002). A practical protocol for dynamic modeling of activated sludge systems. *Water Science and Technology: A Journal of the International Association on Water Pollution Research*, **45**(6), 127–136.

Iacopozzi I., Innocenti V., Marsili-Libelli S. and Giusti E. (2007). A modified activated sludge model no. 3 (ASM3) with two-step nitrification–denitrification. Environmental Modeling and Software, **22**, 847–861.

Ichihashi O., Satoh H. and Mino T. (2006). Sludge–sludge interaction in the enhanced biological phosphorus removal process. *Water Science and Technology*, **53**(6), 1.

Insel G., Hocaoğlu S. M., Cokgor E. U. and Orhon D. (2011). Modeling the effect of biomass induced oxygen transfer limitations on the nitrogen removal performance of membrane bioreactor. *Journal of Membrane Science*, **368**, 54–63.

Itokawa H., Thiemig C. and Pinnekamp J. (2008). Design and operating experiences of municipal MBRs in Europe. *Water Science and Technology: A Journal of the International Association on Water Pollution Research*, **58**(12), 2319–2327.

Jang N., Ren X., Cho J. and Kim I. S. (2006). Steady-state modeling of bio-fouling potential with respect to the biological kinetics in the sMBR. *Journal of Membrane Science*, **284**, 352–360.

Jeppsson U. and Pons M.-N. (2004). The COST benchmark simulation model–current state and future perspective. *Control Engineering Practice*, **12**, 299–304.

Jeppsson U., Rosen C., Alex J., Copp J., Gernaey K. V., Pons M.-N. and Vanrolleghem P. A. (2006). Towards a benchmark simulation model for plant-wide control strategy performance evaluation of WWTPs. *Water Science and Technology*, **53**(1), 287.

Jiang T. (2007). Characterization and Modeling of Soluble Microbial Products in Membrane Bioreactors. PhD thesis, Ghent University, Belgium, pp. 241.

Jiang T., Liu X., Kennedy M. D., Schippers J. C. and Vanrolleghem P. A. (2005). Calibrating a side-stream membrane bioreactor using activated sludge model no. *1. Water Science and Technology*, **52**(10–11), 359–367.

Jiang T., Myngheer S., De Pauw D. J. W., Spanjers H., Nopens I., Kennedy M. D., Amy G. and Vanrolleghem P. A. (2008). Modeling the production and degradation of soluble microbial products (SMP) in membrane bioreactors (MBR). *Water Research*, **42**, 4955–4964.

Jimenez J., Grelier P., Meinhold J. and Tazi-Pain A. (2010). Biological modeling of MBR and impact of primary sedimentation. *Desalination*, **250**, 562–567.

Judd S. (2006). *The MBR Book: Principles and Applications of Membrane Bioreactors in Water and Wastewater Treatment*, 1st edn, Elsevier, Amsterdam, Boston, London. ISBN 978-1-85617-481-7.

Laspidou C. S. and Rittmann B. E. (2002a). Non-steady state modeling of extracellular polymeric substances, soluble microbial products and active and inert biomass. *Water Research*, **36**(8), 1983–1992.

Laspidou C. S. and Rittmann B. E. (2002b). A unified theory for extracellular polymeric substances, soluble microbial products, and active and inert biomass. *Water Research*, **36**(11), 2711–2720.

Le-Clech P., Jefferson B., Chang I.-S. and Judd S. J. (2003). Critical flux determination by the flux-step method in a submerged membrane bioreactor. *Journal of Membrane Science*, **227**(1–2), 81–93.

Le-Clech P., Chen V. and Fane A. G. (2006). Fouling in membrane bioreactors used in wastewater treatment. *Journal of Membrane Science*, **284**, 17–53.

Liu R., Huang X., Chen L., Wen X. and Qian Y. (2005). Operational performance of a submerged membrane bioreactor for reclamation of bath wastewater. *Process Biochemistry*, **40**, 125–130.

Lu S. G., Imai T., Ukita M., Sekine M., Higuchi T. and Fukagawa M. (2001). A model for membrane bioreactor process based on the concept of formation and degradation of soluble microbial products. *Water Research*, **35**(8), 2038–2048.

Lu S. G., Imai T., Ukita M., Sekine M. and Higuchi T. (2002). Modeling prediction of membrane bioreactor process with the concept of soluble microbial product. *Water Science and Technology*, **46**(11–12), 63–70.

Lubello C., Caffaz S., Gori R. and Munz G. (2009). A modified activated sludge model to estimate solids production at low and high solid retention time. *Water Research*, **43**, 4539–4548.

Luedeking R. and Piret E. C. (1959). A kinetic study of lactic acid fermentation batch process at controlled pH. *Journal of Biochemical and Microbiological Technology and Engineering*, **1**(4), 393–412.

Maere T., Verrecht B., Moerenhout S., Judd S. and Nopens I. (2011). BSM-MBR: a benchmark simulation model to compare control and operational strategies for membrane bioreactors. *Water Research*, **45**, 2181–2190.

Majone M., Dircks K. and Beun J. J. (1999). Aerobic storage under dynamic conditions in activated sludge processes. The state of the art. *Water Science and Technology*, **39**(1), 61–73.

Manser R., Gujer W. and Siegrist H. (2005). Consequences of mass transfer effects on the kinetics of nitrifiers. *Water Research*, **39**, 4633–4642.

Manser R., Gujer W. and Siegrist H. (2006). Decay processes of nitrifying bacteria in biological wastewater treatment systems. *Water Research*, **40**, 2416–2426.

Marais G. V. R. and Ekama G. A. (1976). The activated sludge process. Part 1 – Steady state behaviour. *Water SA*, **2**, 163–199.

Metcalf and Eddy. (2003). Wastewater Engineering: Treatment and Reuse, 4th edn, McGraw-Hill. ISBN: 9780070418783.

Munz G., Gori R., Cammilli L. and Lubello C. (2008a). Characterization of tannery wastewater and biomass in a membrane bioreactor using respirometric analysis. *Bioresource Technology*, **99**, 8612–8618.

Munz G., Gualtiero M., Salvadori L., Claudia B. and Claudio L. (2008b). Process efficiency and microbial monitoring in MBR (membrane bioreactor) and CASP (conventional activated sludge process) treatment of tannery wastewater. *Bioresource Technology*, **99**, 8559–8564.

Naessens W., Maere T. and Nopens I. (2012). Critical review of membrane bioreactor models–part 1: biokinetic and filtration models. *Bioresource Technology*, **122**, 95–106.

Namkung E. and Rittmann B. E. (1986). Soluble microbial products (SMP) formation kinetics by biofilm. *Water Research*, **20**(6), 795–806.

Ng A. N. L. and Kim A. S. (2007). A mini-review of modeling studies on membrane bioreactor (MBR) treatment for municipal wastewaters. *Desalination*, **212**, 261–281.

Ni B.-J., Zeng R. J., Fang F., Xie W.-M., Sheng G.-P. and Yu H.-Q. (2010). Fractionating soluble microbial products in the activated sludge process. *Water Research*, **44**, 2292–2302.

Nowak O., Svardal K., Franz A. and Kuhn V. (1999). Degradation of particulate organic matter–a comparison of different model concepts. *Water Science and Technology*, **39**(1), 119–127.

Ognier S., Wisniewski C. and Grasmick A. (2004). Membrane bioreactor fouling in sub-critical filtration conditions: a local critical flux concept. *Journal of Membrane Science*, **229**, 171–177.

Oliveira-Esquerre K. P., Narita H., Yamato N., Funamizu N. and Watanabe Y. (2006). Incorporation of the concept of microbial products formation into ASM3 and the modeling of a membrane bioreactor for a wastewater treatment. *Brazilian Journal of Chemical Engineering*, **23**(4), 461–471.

Orhon D., Genceli E. A. and Cogkor E. U. (1999). Characterization and modeling of activated sludge for tannery wastewater. *Water Environment Research*, **71**, 50–63.

Parco V., Du Toit G., Wentzel M. and Ekama G. (2007). Biological nutrient removal in membrane bioreactors: denitrification and phosphorus removal kinetics. *Water Science and Technology: A Journal of the International Association on Water Pollution Research*, **56**(6), 125–134.

Pinnekamp J. (2006). Municipal water and waste management: membrane technology for waste water treatment. In: H. Friedrich (ed.). FIW Verlag, Aachen, ISBN 3-939377-01-5.

Ramesh A., Duu-Jong L. and Hong S. G. (2006). Soluble microbial products (SMP) and soluble extracellular polymeric substances (EPS) fromwastewater sludge. *Environmental Biotechnology*, **73**, 219–225.

Rector T. J., Garland J. L. and Starr S. O. (2006). Dispersion characteristics of a rotating hollow fiber membrane bioreactor: effects of module packing density and rotational frequency. *Journal of Membrane Science*, **278**(1–2), 144–150.

Rios N., Nopens I., de Schepper V., Jiang T., Verstraete W. and Vanrolleghem P. (2007). A Rheological Model for Activated Sludge in a Side-stream MBR. IWA, Harrogate, UK.

Roeleveld P. J. and Van Loosdrecht M. C. M. (2002). Experience with guidelines for wastewater characterisation in The Netherlands. *Water Science and Technology: A Journal of the International Association on Water Pollution Research*, **45**(6), 77–87.

Rosen C., Jeppsson U. and Vanrolleghem P. A. (2004). Towards a common benchmark for long-term process control and monitoring performance evaluation. *Water Science and Technology*, **50**(11), 41–49.

Saalbach J. and Hunze M. (2007). Flow Structures in MBR-Tanks. IWA, Berlin, Germany.

Samie G., Bernier J., Rocher V. and Lessard P. (2011). Modeling nitrogen removal for a denitrification biofilter. *Bioprocess Biosystem Engineering*, **34**, 747–755.

Sarioglu M., Insel G., Artan N. and Orhon D. (2008). Modeling of long-term simultaneous nitrification and denitrification (SNDN) performance of a pilot scale membrane bioreactor. *Water Science and Technology: A Journal of the International Association on Water Pollution Research*, **57**(11), 1825–1833.

Sarioglu M., Insel G., Artan N. and Orhon D. (2009a). Effect of biomass concentration on the performance and modeling of nitrogen removal for membrane bioreactors. *Journal of Environmental Science and Health, Part A*, **44**, 733–743.

Sarioglu M., Insel G., Artan N. and Orhon D. (2009b). Model evaluation of simultaneous nitrification and denitrification in a membrane bioreactor operated without an anoxic reactor. *Journal of Membrane Science*, **337**, 17–27.

Sin G. and Vanrolleghem P. A. (2006). Evolution of an ASM2d-like model structure due to operational changes of an SBR process. *Water Science and Technology*, **53**(12), 237.

Sin G., Guisasola A., De Pauw D. J. W., Baeza J. A., Carrera J. and Vanrolleghem P. A. (2005a). A new approach for modeling simultaneous storage and growth processes for activated sludge systems under aerobic conditions. *Biotechnology and Bioengineering*, **92**, 600–613.

Sin G., Van Hulle S., De Pauw D., Van Griensven A. and Vanrolleghem P. (2005b). A critical comparison of systematic calibration protocols for activated sludge models: a SWOT analysis. *Water Research*, **39**, 2459–2474.

Spanjers H. and Vanrolleghem P. A. (1995). Respirometry as a tool for rapid characterization of wastewater and activated sludge. *Water Science and Technology*, **31**(2), 105–114.

Spérandio M. and Paul E. (2000). Estimation of wastewater biodegradable COD fractions by combining respirometric experiments in various So/Xo ratios. *Water Research*, **34**, 1233–1246.

Spérandio M. and Espinosa M. C. (2008). Modeling an aerobic submerged membrane bioreactor with ASM models on a large range of sludge retention time. *Desalination*, **231**, 82–90.

Spérandio M., Masse A., Espinosa-Bouchot M. C. and Cabassud C. (2005). Characterization of sludge structure and activity in submerged membrane bioreactor. *Water Science and Technology: A Journal of the International Association on Water Pollution Research*, **52**(10–11), 401–408.

Sweity A., Ying W., Ali-Shtayeh M. S., Yang F., Bick A., Oron G. and Herzberg M. (2011). Relation between EPS adherence, viscoelastic properties, and MBR operation: biofouling study with QMC-D. *Water Research*, **45**, 6430–6440.

Taha T. and Cui Z. F. (2002). CFD modelling of gas-sparged ultrafiltration in tubular membranes. *Journal of Membrane Science*, **210**(1), 13–27.

Tian Y., Chen L., Zhang S., Cao C. and Zhang S. (2011). Correlating membrane fouling with sludge characteristics in membrane bioreactors: an especial interest in EPS and sludge morphology analysis. *Bioresource Technology*, **102**, 8820–8827.

Van Hulle S. W. H., Dapena-Mora A. N. A., Campos J. L., Mendez R., Jetten M. S. M. and Vanrolleghem P. A. (2003). Modeling start-up and operation of an anammox sequencing batch reactor. *Medical Faculty Landbouww, University of Gent*, **68**, 295–298.

Van Loosdrecht M. C. M. and Henze M. (1999). Maintenance endogeneous respiration, lysis, decay and predation. *Water Science and Technology*, **39**(1), 107–117.

Verrecht B., Judd S., Guglielmi G., Brepols C. and Mulder J. W. (2008). An aeration energy model for an immersed membrane bioreactor. *Water Research*, **42**, 4761–4770.

Verrecht B., Maere T., Benedetti L., Nopens I. and Judd S. (2010). Model-based energy optimisation of a small-scale decentralised membrane bioreactor for urban reuse. *Water Research*, **44**, 4047–4056.

Von Schulthess R., Wild D. and Gujer W. (1994). Nitric and nitrous oxides from denitrifying activated sludge at low oxygen concentration. *Water Science and Technology*, **30**(6), 123–132.

Wagner J. and Rosenwinkel K. (2000). Sludge production in membrane bioreactors under different conditions. *Water Science and Technology*, **41**(10), 251–258.

Wang Z., Wu Z., Yin X. and Tian L. (2008a). Membrane fouling in a submerged membrane bioreactor (MBR) under sub-critical flux operation: membrane foulant and gel layer characterization. *Journal of Membrane Science, 325* (1), 238–244.

Wang Y., Ong K. W., Brannock M. W. D. and Leslie G. L. (2008b). Evaluation of membrane bioreactor performance via residence time distribution: effects of membrane configuration and mixing. *Water Science and Technology*, **57**(3), 353–359.

Wintgens T., Rosen J., Melin T., Brepols C., Drensla K. and Engelhardt N. (2003). Modeling of a membrane bioreactor system for municipal wastewater treatment. *Journal of Membrane Science*, **216**, 55–65.

Wyffels S., Van Hulle S. W. H., Boeckx P., Volcke E. I. P., Van Cleemput O., Vanrolleghem P. A. and Verstraete W. (2004). Modeling and simulation of oxygen-limited partial nitritation in a membrane-assisted bioreactor (MBR). *Biotechnology and Bioengineering*, **86**, 531–542.

Ye Y., Le-Clech P., Chen V. and Fane A. G. (2005). Evolution of fouling during cross-flow filtration of model EPS solutions. *Journal of Membrane Science*, **264**, 190–199.

Yoon S.-H., Kim H.-S. and Yeom I.-T. (2004). The optimum operational condition of membrane bioreactor (MBR): cost estimation of aeration and sludge treatment. *Water Research*, **38**, 37–46.

Zarragoitia-Gonzales A., Schetrite S., Alliet M., Jaregui-Haza U. and Albasi C. (2008). Modeling of SMBRs: conceptual study about link between activated sludge biokinetics, aeration and fouling process. *Journal of Membrane Science*, **325**, 612–625.

Zhang J., Chua H. C., Zhou J. and Fane A. G. (2006). Factors affecting the membrane performance in submerged membrane bioreactors. *Journal of Membrane Science*, **284**, 54–66.

Zhang J., Zhou J., Liu Y. and Fane A. G. (2010). A comparison of membrane fouling under constant and variable organic loadings in submerged membrane bioreactors. *Water Research*, **44**(18), 5407–5413.

Ziglio G., Andreottola G., Foladori P. and Ragazzi M. (2001). Experimental validation of a single-OUR method for wastewater RBCOD characterisation. *Water Science and Technology: A Journal of the International Association on Water Pollution Research*, **43**(11), 119–126.

Zuthi M. F. R., Ngo H. H. and Guo W. S. (2012). Modeling bioprocesses and membrane fouling in membrane bioreactor (MBR): a review towards finding an integrated model framework. *Bioresource Technology*, **122**, 119–129.

Chapter 10

Gas-diffusion, extractive, biocatalytic, and electrochemical membrane biological reactors

Oskar Modin[1], Faisal I. Hai[2], Long D. Nghiem[2], Angelo Basile[3] and Kensuke Fukushi[4]*

[1]*Division of Water Environment Technology, Department of Civil and Environmental Engineering, Chalmers University of Technology, Gothenburg, Sweden*
[2]*Strategic Water Infrastructure Laboratory, School of Civil Mining and Environmental Engineering, the University of Wollongong, NSW 2522, Australia*
[3]*ITM-CNR, University of Calabria, Italy*
[4]*Integrated Research System for Sustainability Science, the University of Tokyo, Japan*
**Corresponding author: oskar.modin@chalmers.se*

ABSTRACT

Several emerging technologies for treatment or resource recovery from wastewater are based on the combination of membranes and biochemical conversions. In this chapter, we review gas-diffusion, extractive, biocatalytic, and electrochemical membrane bioreactors (MBRs). Gas-diffusion MBRs facilitate efficient delivery of a gaseous oxidant or reductant such as oxygen, hydrogen, or methane to microbial biofilms treating wastewater. Extractive MBRs completely separate the wastewater stream from the microbes and make toxic wastewaters amenable to biological treatment. Biocatalytic MBRs utilize enzymes or immobilized microbial cells for degradation of persistent xenobiotics or for synthesis of fine chemicals. Electrochemical MBRs make it possible to recovery energy, chemicals, and nutrients from wastewater. Some technologies have already been commercialized. Many are currently in the research and development stage and could potentially contribute to more efficient removal of pollutants and recovery of resources from wastewater in the future.

 Membrane Biological Reactors: Theory, Modeling, Design, Management and Applications to Wastewater Reuse
Editor(s): Faisal I. Hai, Kazuo Yamamoto and Chung-Hak Lee
doi: 10.2166/9781780409177_0305

10.1 INTRODUCTION

In addition to the biosolids separation membrane bioreactor (MBR), the combination of membranes and biochemical conversions has led to a range of innovative environmental biotechnology applications. This chapter reviews gas-diffusion, extractive, biocatalytic, and electrochemical membrane biological reactors focusing on applications for wastewater treatment (Figure 10.1). In gas-diffusion MBRs, the membrane delivers a gaseous substrate, such as oxygen, hydrogen, or methane, to a biofilm growing on the membrane surface. The gaseous substrate supports oxidation or reduction of compounds in the liquid. Gas-diffusion MBRs are often called membrane biofilm reactors (MBfRs) as the biofilm growing on the membrane surface plays an integral role for reactor performance (Celmer-Repin *et al.*, 2010; Hwang *et al.*, 2009; Martin & Nerenberg, 2012). MBfRs typically offer high substrate utilization efficiency and have been investigated, for example, for simultaneous nitrification and denitrification and for high-rate removal of organic matter. In extractive MBRs (EMBRs), the membrane serves as a barrier that allows selective extraction of compounds from a wastewater stream into a bioreactor compartment where degradation takes place. The EMBR can be used to treat wastewater streams with characteristics such as high toxicity or fluctuating pH that would otherwise make biological treatment unsuitable

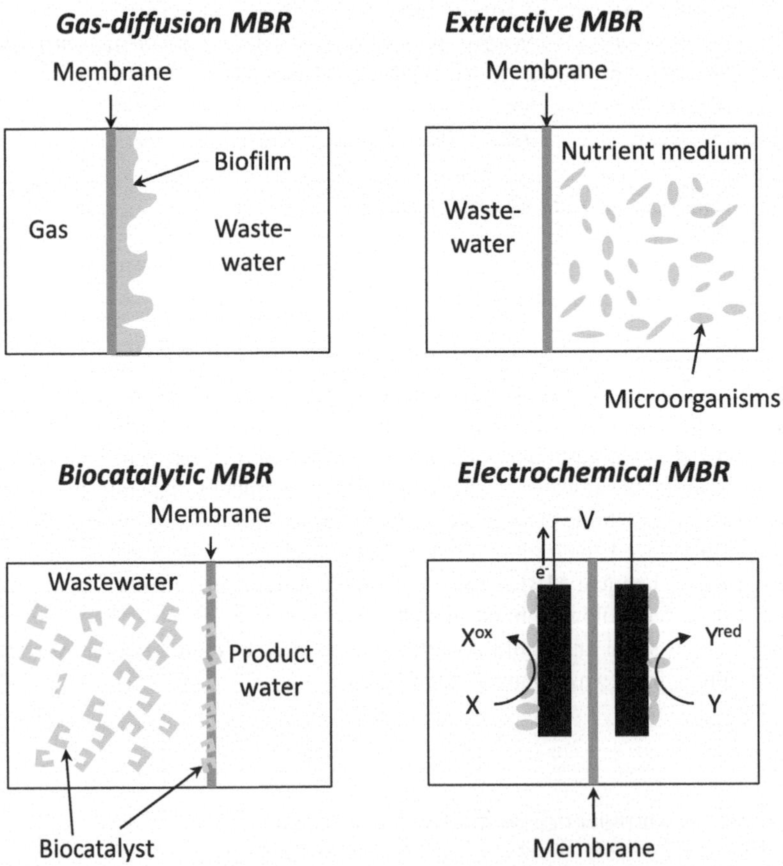

Figure 10.1 Four types of membrane biological reactors reviewed in Chapter 10.

(Livingston, 1994). Biocatalytic MBRs refer to either enzymatic membrane reactors (EMRs) or immobilized-cell membrane reactors (ICMRs), which in the context of wastewater treatment have been investigated primarily for removal of persistent chemicals (Hai *et al.*, 2013). In EMRs, enzymes achieve specific biochemical conversions and membranes are used to separate the enzymes from the treated water. The enzymes may be immobilized on the membrane surface or present in solution on the feed side of the membrane. In ICMRs, whole cells of specific microorganisms immobilized in a membrane carry out the biochemical conversions. Electrochemical MBRs refer to microbial bioelectrochemical systems (BESs), which have recently emerged as potentially useful technologies for energy recovery or chemical product formation from organic compounds in wastewater (Logan & Rabaey, 2012). Microbial oxidation of organics generates an electrical current flowing between an anode and a cathode. By placing one or several ion exchange membranes between the two electrodes, BESs have been used, for example, to desalinate water and produce alkaline solutions. The combination of electrochemistry and membranes has also been used to extract specific components, such as ammonium and fermentation products, from wastewater.

This chapter focuses on the applications and performance of gas diffusion, extractive, biocatalytic, and electrochemical MBRs. In Section 10.2, we review MBfRs fed with air, hydrogen, or methane gas for removal of contaminants in the liquid. Sections 10.3–10.5 cover extractive, biocatalytic, and electrochemical MBRs, respectively. In Section 10.6, we summarize the entire chapter and discuss challenges that should be addressed for the different technologies.

10.2 MEMBRANE BIOFILM REACTORS

10.2.1 Overview

Membrane aeration of bioreactors for wastewater treatment, as opposed to air bubbling, has several advantages. The membrane allows a high partial pressure of oxygen to be kept within the membrane, which results in a high oxygen transfer driving force. The membrane also allows 100% efficient, bubble-free transfer of oxygen into the liquid, which prevents stripping of volatile compounds as well as foaming (Ahmed & Semmens, 1992a; Côté *et al.*, 1988). Researchers who explored membranes for aeration of wastewater reactors discovered biofilm formation on the membrane surfaces (Schaffer *et al.*, 1960; Wilderer *et al.*, 1985). Although the development of biofilms on membranes limits the transfer of oxygen into the bulk liquid, the MBfRs have several advantages. Biofilm systems accomplish high biomass retention time, which allows slow-growing microorganisms such as nitrifiers to be retained in the reactor. Depending on the available surface area, biofilm reactors can also have high amount of biomass per unit volume compared to the suspended growth reactors. In conventional biofilm reactors, the activity of the biomass is typically limited by diffusion of oxygen from the bulk liquid. In aeration MBfRs, a high oxygen partial pressure within the membrane provides a large driving force for oxygen transfer into the biofilm, which results in better utilization of the available biomass. The first aeration MBfR was developed by Yeh and Jenkins (1978) who used Teflon tubing to deliver pure oxygen to biofilm growing on the outer surface of the tubing. They called the reactor an aerobic media trickling filter and investigated its performance for organics removal and nitrification. Onishi *et al.* (1980) patented an aeration MBfR with microporous hollow fiber membranes. Biofilms growing on oxygen-permeating membranes can also develop different microenvironments, which, for example, can promote simultaneous nitrification and denitrification. The biofilm layer closest to the membrane will be aerobic whereas the layer near the bulk liquid will be anoxic. Timberlake *et al.* (1988) was the first to investigate this concept for simultaneous removal of organics and nitrogen.

10.2.2 Membrane materials and configurations

Membranes of various materials and configurations have been used in MBfRs. The modules can be of either flat sheet or hollow fiber configuration. For practical applications, hollow fiber membrane modules are preferable because they provide a large surface area per unit volume. Flat sheet membranes may be easier to use in specific cases such as studies regarding concentration profiles of various substrates within a biofilm using microsensors (e.g. Modin *et al.*, 2010).

Membrane materials can be classified as porous, dense, or composite. Porous membranes contain pores through which the gas can permeate. Since porous membranes have low resistance to gas transfer, only a very low intramembrane pressure can be kept without the formation of bubbles (Ahmed & Semmens, 1992a). This limits the rate of oxygen transfer using porous membranes. Dense membranes do not contain pores. Instead, the gas dissolves into the dense membrane polymer and diffuses through the membrane into the liquid, and thus higher intramembrane gas pressures can be kept without the formation of bubbles (Syron & Casey, 2008b). Silicone is a commonly used dense membrane that is highly permeable to oxygen. However, a drawback is that silicone tubing typically has quite thick walls (several hundred micrometres), which provide high resistance to gas transfer. Composite membranes consist of a very thin dense layer (~1 μm) supported by porous membrane layers. The porous layers provide structural support to the thin dense layer that could otherwise easily be damaged. The dense layer allows high intramembrane gas pressures without the formation of bubbles. Composite membranes have been used in many studies on gas-diffusion MBRs fed with hydrogen (e.g. Lee & Rittmann, 2000), but also in aeration MBfR studies (e.g. Brindle *et al.*, 1999).

The attachment and retention of biofilm on the membrane surface is crucial in MBfRs. Several researchers have tried to improve the retention of biofilm by dressing the membrane in a porous support layer (e.g. Hirasa *et al.*, 1991) or by chemical modification of the membrane surface (e.g. Hibiya *et al.*, 2003; Terada *et al.*, 2003).

Membranes can be operated in either dead-end or open-end configuration. The dead-end configuration means that one end of the membrane module is sealed so that all the supplied gas has to permeate through the membrane surface into the attached biofilms. This type of configuration makes it possible to reach 100% oxygen utilization efficiency; however, the gas composition along the length of hollow fiber membranes may change due to back diffusion of gases such as nitrogen, carbon dioxide, and water vapor from the liquid into the membrane, which leads to reduced oxygen transfer rate (Ahmed & Semmens, 1992a, b). A way to solve this problem is to use periodic venting of the gas in the hollow fibers (Perez-Calleja *et al.*, 2017). Open-end configuration means that both ends of a membrane module are open and the gas flows into one end and out from the other. Only a portion of the supplied gas permeates through the membrane surface; however, the partial pressure of oxygen can be kept nearly constant along the length of membrane fibers (Syron & Casey, 2008b). Dead-end configuration is preferable when expensive gases such as pure oxygen are used. Open-end configuration has been used when the feed gas is air.

10.2.3 Aeration MBfRs

Aeration MBfRs have previously been reviewed by Syron and Casey (2008b) and Casey *et al.* (1999). Below we provide an overview of the performance of aeration MBfRs for removal of organic matter, nutrients, and xenobiotic compounds from wastewater.

10.2.3.1 Removal of organic matter

Removal of organic matter is one of the most important objectives in wastewater treatment. The possibility of having high oxygen transfer rate and efficiency in aeration MBfRs make them suitable for high-rate removal of organic matter. The results from several studies are summarized in Table 10.1. These studies show that, at least when fed with pure oxygen, aeration MBfRs can reach very high volumetric removal rates. Brindle *et al.* (1999) obtained a removal rate of 27 kg COD/m^3 d in a pilot-scale reactor fed with wastewater from an alcohol-cider production facility. Pankhania *et al.* (1999) obtained 21.8 kg COD/m^3 d

Table 10.1 Removal of organic matter (as COD or TOC) and total nitrogen (TN) in aeration MBfRs.

Membrane Type **Specific Surface Area and Feed Gas** **Wastewater**	**Removal Rate (kg/m^3 d)** **Removal Efficiencies** **HRT**	**References**
Microporous dead-end hollow fiber, 511 m^2/m^3, pure oxygen Synthetic containing peptone and meat extract	COD: 7.69 86% 36 min	Pankhania *et al.* (1994)
Microporous dead-end hollow fiber, 511 m^2/m^3, pure oxygen Synthetic containing peptone and meat extract	COD: 21.8 89% 34 min	Pankhania *et al.* (1999)
Composite dead-end hollow fiber, 447 m^2/m^3, pure oxygen Alcoholic-cider production wastewater	COD: 27 81% 1.8 h	Brindle *et al.* (1999)
Porous, dead-end flat sheet 18.6 m^2/m^3, air and pure oxygen Primary effluent supplemented with nutrients	TOC: 0.08, TN: 0.02 TOC: 33–56%, TN: 17–24% 0.4–1.1 days	Timberlake *et al.* (1988)
Porous, open-end tube 60 m^2/m^3, air synthetic (peptone, meat-, yeast extract)	TOC: 0.24, TN: 0.12 TOC: 83%, TN: 55% 3.5 h	Suzuki *et al.* (1993)
Microporous, dead-end hollow fiber, 422 m^2/m^3, air Synthetic (ammonium acetate)	COD: 4.5, TN: 1.0 COD: 95%, TN: $>$90% 6 h	Semmens *et al.* (2003)
Microporous, dead-end hollow fiber 50 m^2/m^3, air Synthetic (peptone, dextrin, meat extract)	TOC: 0.29, TN: 0.22 TOC: 96%, TN: 85% 15 days	Terada *et al.* (2003)
Porous, open-end carbon tubes 37.3 m^2/m^3, air Synthetic (glucose and ammonium)	COD: 0.60, TN: 0.13 COD: 86%, TN: 84% 12 h	Hu *et al.* (2008)
Porous, open-end carbon tubes 72 m^2/m^3, air at 25 kPa Synthetic (ammonium acetate)	COD: 0.7, TN: 0.14 COD: 85%, TN: 86% 8 h	Liu *et al.* (2010)
Dense, open-end hollow fiber, 810 m^2/m^3, air at 41 kPa Synthetic industrial wastewater containing glycerol and *N*-methyl pyrrolidone	COD: 2.9, TN: 0.08 COD: 98–99%, TN: 89% 24.6 h	Stricker *et al.* (2011)
Dense, open-end hollow fiber 67.5–90 m^2/m^3, air Synthetic (glucose and ammonium)	COD: 0.5, TN: 0.11 COD: 71%, TN: 80.7% 12 h	Wei *et al.* (2012)

with synthetic wastewater containing peptone and meat extract as carbon source. Both these studies used hollow fiber membrane modules with specific surface areas of approximately 500 m^2/m^3 fed with pure oxygen gas. The removal rates achieved by MBfRs are more than an order of magnitude larger than typical loading rates for the aerobic (completely mixed) activated sludge process, which ranges from 0.3 to 1.6 kg BOD/m^3 d (Metcalf & Eddy *et al.*, 2004).

A drawback of the high removal rates is that they are accompanied by excessive biofilm growth, which leads to problems with channeling and clogging of the reactors. Various cleaning protocols using either air or water to scour the membrane have been developed (Syron & Casey, 2008b). Thick biofilms require very high intramembrane oxygen pressures to keep the entire biofilm aerobic. The activity of thick biofilm may also be limited by the rate of diffusion of organic matter from the bulk liquid. In a modelling study, Syron and Casey (2008a) showed that the highest volumetric removal rates can be obtained with thin membrane fibers and thin biofilms providing a large surface area per unit volume. Stricker *et al.* (2011) used a dense hollow fiber membrane with extremely small fiber diameter (50–70 μm) and high specific surface area (810 m^2/m^3). They applied a relatively low organic surface loading (3–4 g COD/m^2 d) to prevent the development of excessively thick biofilms, which led to biofilm thicknesses of about 250–350 μm. The biofilm did not, however, reach a steady state but kept growing at a rate of about 0.020 g total solids (TS)/m^2 d. Although a relatively modest overall volumetric removal rate of 2.9 kg COD/m^3 d was achieved, it should be noted that the reactor was divided into several stages in which the first stage in fact received most of the organic load and yielded a removal rate of 5.3 kg COD/m^3 d. The reactor also had the advantage of being operated with air instead of pure oxygen. Liao and Liss (2007) compared thermophilic and mesophilic aerated MBfRs for organics removal and found that the biofilm thickness was thinner in the thermophilic reactor operated at 55°C, suggesting that thermophilic operation could be an option to manage biofilm growth when the process is applied to treat warm wastewater.

10.2.3.2 Removal of nutrients

Aeration MBfRs are especially well-suited for simultaneous removal of nitrogen and organic matter through nitrification and denitrification. The biofilm–membrane interface has high oxygen concentration and low concentration of organic carbon, which favors the development of nitrifiers near the membrane surface. Being located at the base of the biofilm, the slow-growing nitrifiers are also protected from erosion. The biofilm near the bulk liquid interface is anoxic and has high concentration of organic carbon, which allows denitrification to proceed (Semmens & Essila, 2001). An intermediate aerobic heterotrophic biofilm layer may contribute to the removal of organic matter. Timberlake *et al.* (1988) demonstrated this concept and showed that simultaneous removal of organic matter and nitrogen was possible, although quite low removal rates were obtained (Table 10.1) (Timberlake *et al.*, 1988). The development of stratified biofilms has been shown in several studies using microbial community analysis techniques (e.g. Cole *et al.*, 2004; LaPara *et al.*, 2006; Terada *et al.*, 2003; Wang *et al.*, 2011; Yamagiwa *et al.*, 1998). Satoh *et al.* (2004) used microelectrodes for O_2, NO_2^-, NO_3^-, NH_4^+, and pH to investigate activities in biofilms attached onto polyurethane membranes fed with oxygen at a flux of 0.28 m^3/m^2 h. They found that nitrification occurred in a zone between the membrane surface and 60 μm into the biofilm. As the biofilm thickness increased, an anoxic outer zone developed, which coincided with increased denitrification rate. Denitrification took place just outside the nitrification zone. Suzuki *et al.* (1993) showed the importance of restricting the thickness of the denitrifying biofilm zone. A too thick denitrifying zone will limit diffusion of ammonium from the bulk liquid into the nitrifying biofilm zone near the membrane surface. This will reduce the nitrification rate, and consequently, the total nitrogen removal rate in the system (Suzuki *et al.*, 1993). A hybrid reactor consisting of a membrane-attached nitrifying biofilm and denitrifying activated sludge was developed by Downing *et al.* (2010). Landes *et al.* (2011) investigated an aerated MBfR containing membranes to support nitrifiers and inert biofilm carriers to support denitrifiers. Sequencing batch MBfRs have also been developed for combined nitrogen and biological phosphorous removal (Castillo *et al.*, 1999; Terada *et al.*, 2006). An advantage is MBfRs compared to conventional biofilms processes is that they produce less N_2O, which is a potent greenhouse gas. N_2O is generated both during nitrification and denitrification. A membrane-attached counter-diffusion biofilm leads to spatial separation of aerobic nitrification and anoxic denitrification, which was shown to reduce N_2O generation in comparison to conventional co-diffusion biofilms when fed with a synthetic nutrient medium mimicking food-processing wastewater (Kinh *et al.*, 2017).

Aeration MBfRs could be suitable for high-rate treatment of nitrogenous wastewater poor in organic content (Table 10.2). Most studies have focused on nitrification of secondary effluent with ammonium concentration of <50 mg $NH_4{}^+ - N/L$ (Brindle & Stephenson, 1996; Brindle *et al.*, 1998; Motlagh *et al.*, 2008). Walter *et al.* (2005) investigated nitrification under loading rates ranging from 1 to 15 kg N/m^3 d. At the highest loading, the nitrification rate was 7 kg N/m^3 d, showing that high rates of ammonium oxidation are possible in MBfRs. If the goal is nitritation, i.e., oxidation of ammonium to nitrite without further oxidation to nitrate, studies by Wang *et al.* (2009) and Lackner *et al.* (2010) suggest that conventional co-diffusion biofilms perform better than the MBfR. A low dissolved oxygen (DO) concentration is needed to prevent oxidation of nitrite by nitrite-oxidizing bacteria. This was difficult to maintain in the MBfR, which has a high DO concentration in the membrane–biofilm interface (Lackner *et al.*, 2010). However, intermittent aeration can be a way to suppress nitrite oxidation in MBfRs (Ma *et al.*, 2017). Completely autotrophic nitrogen removal through partial nitritation and anammox has been obtained in aeration MBfRs (Gong *et al.*, 2008; Pellicer-Nacher *et al.*, 2010). Partial ammonium oxidation to nitrite takes place in the aerobic biofilm layer near the membrane surface whereas anammox activity, i.e., the conversion of ammonium and nitrite to nitrogen gas, takes place in the anoxic biofilm layer near the bulk liquid. In a modelling study, Terada *et al.* (2007) compared counter- and co-diffusion biofilms from completely autotrophic nitrogen removal. For biofilm thicknesses between 450 and 1400 μm, the counter-diffusion biofilm (i.e., aeration MBfR) had higher removal efficiency than the co-diffusion biofilm (i.e., conventional biofilm reactor). Thicker biofilms led to deteriorating performance due to $NH_4{}^+$ diffusion limitation. Experimental studies on completely autotrophic nitrogen removal have achieved rates of 0.73–0.78 kg N/m^3 d (Gong *et al.*, 2008; Pellicer-Nacher *et al.*, 2010), which are high compared to total nitrogen (TN) removal rates for aeration MBfR with heterotrophic denitrification.

Research on the use of aeration MBfR for high-rate nitrification or complete autotrophic nitrogen removal from wastewater streams such as sludge liquor and landfill leachate, which are rich in nitrogen and poor in organics, deserves further attention. In a pilot-scale experiment using a 60 L reactor and real landfill leachate as feed, a nitrification efficiency of 80–99% was achieved (Syron *et al.*, 2015).

10.2.3.3 Removal of xenobiotics

Aeration MBfRs have been investigated for degradation of several groups of xenobiotic compounds (Hai *et al.*, 2013). The bubble-free transfer of oxygen makes MBfRs especially suited for degradation of volatile organic compounds (VOCs) that may escape to the atmosphere if conventional air sparging is used as oxygen supply. Debus and Wanner (1992) investigated xylene degradation in biofilms attached on a silicone membrane tube. The biofilm that developed on the membrane degraded xylene and prevented it from entering the gas phase. However, a too thick biofilm retarded degradation because of

Table 10.2 Nitrification and completely autotrophic nitrogen removal in aeration MBfRs.

Membrane Type **Specific Surface Area and Feed Gas** **Wastewater**	**Removal Rate (kg/m^3 d)** **Removal Efficiencies** **HRT**	**References**
Microporous dead-end hollow fiber, 185 m^2/m^3, pure oxygen Synthetic (NH_4HCO_3)	NH_4^+–N: 0.20 99% 3.6–6 h	Brindle and Stephenson (1996)
Microporous dead-end hollow fiber, 185 m^2/m^3, pure oxygen Synthetic (NH_4HCO_3)	NH_4^+–N: 1.0 83% 1 h	Brindle *et al.* (1998)
Dense, open-end silicone tube 15.7 m^2/m^3, O_2+N_2 mixture Synthetic ($(NH_4)_2SO_4$)	NH_4^+–N: 0.03 95% 8 h	Hsieh *et al.* (2002)
Porous, dead-end hollow fiber 318 m^2/m^3, air Synthetic inorganic (NH_4^+)	NH_4^+–N: 0.15 >90%	Terada *et al.* (2004)
Dense, dead-end tube 220 m^2/m^3, oxygen at 0.63 bar Synthetic (ammonium and acetate)	NH_4^+–N: 7 47% 2 h	Walter *et al.* (2005)
Dense, open-end silicone tubes Air at 29.4 kPa Synthetic (glucose and NH4$^+$)	NH_4^+–N: 0.14 90% 8 h	Wu *et al.* (2006)
Composite open-end hollow fiber fabric 23 m^2/m^3, air or gas mix with 0.6 atm O_2 Synthetic (ammonium and bicarbonate)	NH_4^+–N: 0.23–0.55 ~40–90% 85–260 min	Motlagh *et al.* (2008)
Dense, hollow fiber membranes, 69 m^2/m^3, air or pure oxygen Landfill leachate	NH_4^+–N: 0.10–0.62[a] 70% to >90% 5–20 days	Syron *et al.* (2015)
Porous, open-end carbon tubes 70 m^2/m^3, air at 0.015 MPa Synthetic (ammonium and bicarbonate)	TN: 0.73 84% 5.5 h	Gong *et al.* (2008)
Composite, open-end hollow fiber 141 m^2/m^3, air Synthetic (ammonium and bicarbonate)	TN: 0.78 72% 7.5 h	Pellicer-Nacher *et al.* (2010)

[a]Calculated nitrogen loading rates based on data given in the article.

slower diffusion of xylene from the bulk liquid to the inner aerobic biofilm layer. In a follow-up study, Debus *et al.* (1994) used microelectrodes and showed that the aerobic biofilm layer extended 410 μm from the membrane surface. The degradation of acetonitrile, another VOC, was investigated by Li *et al.* (2008b) using a microporous hollow fiber membrane reactor with a specific surface area of 84.5 m^2/m^3. They achieved TOC and TN removal efficiencies of 98.6 and 83.3%, respectively, at a loading rate of 0.95 kg acetonitrile/m^3 d. Aeration MBfRs have also been investigated for the removal of fluoroacetate and pharmaceutical compounds (Heffernan *et al.*, 2009; Kim *et al.*, 2010). The development of stratified biofilms with aerobic and anaerobic layers make aeration MBfRs suitable for degradation of certain xenobiotic compounds such as perchloroethylene, which require anaerobic conditions for dechlorination and aerobic conditions for degradation of intermediates (Ohandja & Stuckey, 2007). An overview of xenobiotic compounds whose removal has been investigated in various types of MBfRs is provided in Table 10.3.

Table 10.3 An overview of xenobiotic compounds removed in various types of MBfRs (modified after Hai *et al.*, 2013).

Reactor	Target Compound	References
Aeration MBfR	2,4-Dichloro-phenoxy-acetic acid	Rothemund *et al.* (1994)
	Acetonitrile	Li *et al.* (2008a), Li *et al.* (2008b, 2009)
	Benzene, 2-chloro-phenol, 2,4-dichloro-toluene	Kolb and Wilderer (1995)
	Chlorophenol	Wobus *et al.* (1995), Wobus and Röske (2000)
	Fluoroacetate	Heffernan *et al.* (2009)
	4-Fluorobenzoate	Misiak *et al.* (2011)
	Perchloroethylene	Ohandja and Stuckey (2007)
	Phenol	Woolard and Irvine (1994), Woolard and Irvine (1995)
	Sulfamethazine, sulfathiazole	Kim *et al.* (2010)
	Xylenes, ethylbenzene	Debus and Wanner (1992), Debus *et al.* (1994), Debus (1995)
Hydrogen MBfR	Arsenate	Chung *et al.* (2006a)
	Bromate	Downing and Nerenberg (2007), Martin *et al.* (2009)
	Chromate	Chung *et al.* (2006b)
	Perchlorate	Nerenberg and Rittmann (2002), Nerenberg *et al.* (2002), Chung *et al.* (2007), Nerenberg *et al.* (2008), Van Ginkel *et al.* (2008), Ziv-El and Rittmann (2009b), Ziv-El and Rittmann (2009a), Sahu *et al.* (2009)
	Perchlorate, chlorate, chlorite, bromate, chromate, selenate, selenite, arsenate, dichloromethane	Nerenberg and Rittmann (2004)
	Selenite	Chung *et al.* (2006c)
	Selenite, chromate	Chung *et al.* (2009)
	Trichlorethene	Edstrom *et al.* (2005)
Methane MBfR	Alachlor, malathion, aldicarb	Modin *et al.* (2008c)
	Trichloroethylene	Clapp *et al.* (1999)

10.2.4 Hydrogen MBfRs

Hydrogen can be used as an electron donor for biological reduction of a range of water contaminants. Compared to air, hydrogen gas is expensive, so efficient transfer of hydrogen gas into the liquid phase is important for practical applications. The MBfR operated in dead-end mode is thus an excellent reactor configuration for hydrogen-based degradation processes. The hydrogen MBfR was developed by Lee and Rittmann (2000), who used it for denitrification of nitrate-contaminated water. Using a composite hollow fiber membrane module with a specific surface area of 167 m^2/m^3 and a HRT of 42 min, 92% nitrate removal was achieved from an influent containing 12.5 mg NO_3-N/L. Several other oxidized contaminants in drinking water have also been removed in hydrogen MBfRs (Table 10.3). Nerenberg and Rittmann (2004) screened several compounds and showed that perchlorate, chlorate, chlorite, bromate, chromate, selenate, selenite, arsenate, and dichloromethane could be reduced to varying extent by hydrogen-fed biofilms acclimated to oxygen or nitrate. The use of hydrogen MBfRs for removal of oxidized contaminants in water has been reviewed by Rittmann *et al.* (2004), Rittmann (2006), and Rittmann, (2018). For the purpose of wastewater treatment, hydrogen MBfRs have primarily been investigated for denitrification following a nitrification reactor. The combination of an aeration MBfR for nitrification and a hydrogen MBfR for denitrification for complete removal of nitrogen from wastewater was first reported in 2005 (Cowman *et al.*, 2005; Shin *et al.*, 2005). Using two sequential membrane reactors, the first for nitrification and the second for denitrification, a total nitrogen removal rate and efficiency of 1.20 kg N/m^3 d and 99%, respectively, were obtained (Shin *et al.*, 2005). Smith *et al.* (2008) constructed a membrane module consisting of juxtaposed hydrogen- and oxygen-permeating membrane fibers. This had the benefit of achieving nitrification and denitrification in the same reactor volume. A hydrogen MBfR followed by a biosolids separation MBR was investigated for denitrification of aquaculture wastewater, which typically has high nitrate concentration. A removal rate of 0.19 kg N/m^3 d was obtained at a salinity of 30 ppm (Visvanathan *et al.*, 2008).

10.2.5 Methane MBfRs

Aerobic methanotrophic bacteria are well-known for their co-metabolic degradation of trichloroethylene. Clapp *et al.* (1999) fed a silicone tube membrane with a mixture of methane and oxygen to degrade trichloroethylene in the liquid. Removal efficiencies of 80–90% were obtained at loading rates of 100–320 $\mu mol/m^2$ d. Methane MBfRs have also been shown to remove the pesticides aldicarb, alachlor, and malathion (Table 10.3); however, atrazine was not degraded (Modin *et al.*, 2008c). Several other oxidized contaminants such as chromate and bromate have also been reduced in methane MBfRs (Luo *et al.*, 2017; Lv *et al.*, 2018). Denitrification has been investigated in a series of studies. Aerobic methanotrophs can support denitrification by producing intermediate organic compounds that can be used by coexisting denitrifiers. MBfRs fed with either a mixture of methane and oxygen or only methane from the membrane and oxygen from the bulk liquid have been investigated for denitrification (Modin *et al.*, 2008a, 2010). In those studies, nitrate removal rates of up to 6.59 g/m^2 d were obtained; however, the volumetric removal rates were quite low due to very low specific membrane surface area of the reactor (0.002 m^2/m^3). The removal efficiency in terms of mol nitrate per mol methane was shown to be much higher in a MBfR compared to a suspended culture (Modin *et al.*, 2008b). This was explained by the development of aerobic and anoxic microenvironments in the membrane-attached biofilm, which favoured coexistence of aerobic methanotrophs and denitrifiers (Modin *et al.*, 2010). Methane-driven denitrification has also been investigated under anaerobic conditions. This process is accomplished by very slow-growing microorganisms (Ettwig *et al.*, 2008; Haroon *et al.*, 2013; Raghoebarsing *et al.*, 2006). However, using MBfRs, high nitrogen removal rates have been obtained (Luo *et al.*, 2018;

Shi *et al.*, 2013; Xie *et al.*, 2017). A study of an MBfR integrating both anaerobic methane-driven denitrification and anammox reported a nitrogen removal rate of 1 $kgN/m^3\ d^{-1}$ (Xie *et al.*, 2017).

10.3 EXTRACTIVE MBRS FOR CORROSIVE/TOXIC WASTEWATER TREATMENT

An EMBR system consists of an elastomeric membrane separating a wastewater stream from a bioreactor (Figure 10.2). The elastomeric membrane selectively transports the target organic compounds while preventing the passage of ionic species such as salts, acids, and alkalis from the wastewater phase into a separate bioactive medium of the bioreactor. The target organic compounds are then degraded in the bioreactor by a consortium of bacteria, and the bioreactor can be manipulated independent of the original wastewater to maximize the biological degradation process.

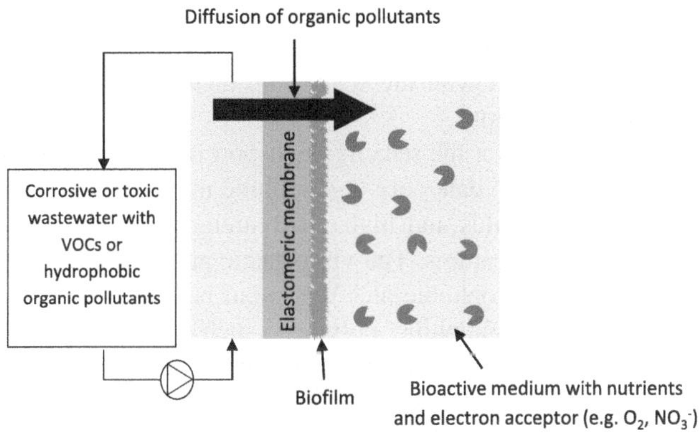

Figure 10.2 A simplified schematic of EMBRs (modified after Hai *et al.*, 2013).

EMBRs have several notable advantages over the conventional MBR approach. First of all, because biodegradation of the target pollutants occurs in the bioactive medium rather than the original wastewater, the EMBR process allows for the treatment of organic pollutants even in hostile wastewater matrices (e.g. corrosive or toxic) that are not compatible for biological treatment. Second, since EMBR does not require aeration, the transfer of VOCs into the atmosphere during the treatment process can be prevented (Freitas Dos Santos & Livingston, 1995b; Livingston, 1994). In the EMBR process, favorable biological conditions in the bioactive medium lead to the establishment of a biofilm on the bioactive side of the membrane. The oxygen enters the biofilm from the bioactive medium side while the VOCs diffuse through the membrane and enter the biofilm in the opposite direction. The biofilm becomes a reactive zone and the thickness of the biofilm ensures that there is no direct contact between the VOCs and the aerating gas, and thus air-stripping does not occur. In fact, extensive laboratory studies over a wide range of operating conditions have confirmed the absence of any VOCs in the biomedium and the aerated air stream (Livingston, 1994). Last but not least, since biodegradation occurs outside the original wastewater matrix and in a separate reactor, various bioaugmentation techniques can be applied to enhance biological degradation, endowing an excellent opportunity for the treatment of persistent organic pollutants (Almeida *et al.*, 1999).

The original EMBR process was introduced by Livingston and co-workers at Imperial College, UK, in the early 1990s. Membrane Extraction Technology Ltd (a spinoff company from Imperial College) was established soon after Livingston filed the first patent for this technology. Since then about 50 scientific articles have been published on this topic as recorded by Scopus. A major breakthrough was noted in 2000, when Membrane Extraction Technology, with assistance from the UK government, commissioned a pilot EMBR plant at the Elf Atochem UK Ltd's site at Widnes, UK, to treat industrial waste containing benzene and other organic contaminants in an aqueous matrix of 30% w/v aluminum trichloride (Mathers, 2000). The pilot plant was run for six months, treating 3 tonnes of aluminum trichloride solution per day. The removal efficiencies of benzene and benzophonone were reported to be more than 99.9 and 90.0%, respectively (Mathers, 2000). Although no further commercial development of the EMBR has been recorded since this very successful pilot demonstration, it is noteworthy that in March 2010, Membrane Extraction Technology Ltd was sold to Evonik Industries AG which is a multinational specialty chemical manufacturer for €4.5 million.

EMBRs present an attractive option for the treatment of corrosive and toxic industrial wastewater. However, full-scale applications of the EMBR technology are not without challenges. The first hurdle is the high cost and the scarcity of suitable elastomeric membranes for EMBR applications. In addition, there is a range of difficulties associated with the scaling up of the process to transfer this technology from the laboratory to full-scale applications.

In theory, any membranes that allow for the selective transport of the target organic compounds can be used in EMBR applications. However, to date only organophilic materials which are typically elastomers (i.e., polymers with low Young's modulus and high yield strength compared to other materials) have been successfully used for EMBR applications. The viscoelastic nature of organophilic materials is ideal for the diffusive transport of small, hydrophobic and VOCs but not water, inorganic salts and any other ionic species. Amongst the many organophilic polymers such as nitrile, butadiene rubber, styrene butadiene rubber, and siloxane that are commercially available, siloxane-based polymers (which are also known as silicone rubber) have been most frequently used in EMBR studies reported in the literature. Very few non-siloxane polymeric membranes including polypropylene (Chung *et al.*, 2004) and polyvinylidene fluoride (Juang *et al.*, 2009) have been tested for EMBR applications. Silicone rubber is a mixture of inorganic–organic polymers with the chemical formula $[R_2SiO]_n$, where R is an aliphatic group such as methyl and ethyl. It has excellent physical and chemical stability and is generally non-reactive and resistant to extreme environments and temperatures while still maintaining the desirable engineering properties. Silicone rubber membrane can be manufactured in either flat sheet or tubular/hollow fiber configurations. While silicone rubber has been used in many consumer products, it is still considered quite expensive for water and wastewater application when compared to other polymeric materials.

Recent research has also progressively addressed the various fundamental aspects that are essential for a commercial realization of EMBRs. This includes both mathematical modeling as well as efforts to utilize the advantages of EMBRs over the conventional processes, while mitigating the limitations of EMBRs. Sanaeepur *et al.* (2012) applied computational fluid dynamics to model key transport phenomena throughout the EMBR process. Since the bioreactor can be manipulated independent of the incoming wastewater in an EMBR, defined cultures of specific bacterial strains can be maintained to selectively degrade persistent organic pollutants. A notable example of this approach is the use of the pure bacterial, *Alcaligenes faecalis*, culture in the bioreactor of an EMBR to treat toluene containing wastewater (Mehdizadeh *et al.*, 2011). The establishment of a biofilm on the membrane prevents the release of VOCs into the atmosphere (Freitas Dos Santos & Livingston, 1995a). However, the biofilm layer can also be a major drawback of the EMBRs since it limits mass transfer of both the target organic pollutants and oxygen. Thus, it is necessary to limit the thickness of the biofilm to 200–400 μm (Freitas Dos Santos & Livingston, 1995a). Emanuelsson and co-workers addressed this challenge by developing an anaerobic EMBR using nitrate as an alternative electron acceptor (Emanuelsson & Livingston, 2002; Emanuelsson *et al.*, 2003). They developed a technique using nitrate concentration in the bioactive medium to control the biofilm thickness (Emanuelsson & Livingston, 2002).

In summary, the EMBR is still an emerging technology. Potential applications of this technology are in the treatment of hazardous wastewater and toxic waste clean-up for which the conventional activated sludge treatment processes are not suitable. With further research and development, it is likely that full-scale applications of EMBRs will be realized in the near future.

10.4 BIOCATALYTIC MBRs

10.4.1 Types and applications of biocatalytic MBRs

Membranes have long been an integral part of the biotechnology processes (Charcosset, 2006). Apart from the well-established methods such as sterile filtration of fermentation media and recovery and purification of fermentation products (Nabais & Cardoso, 1999; Rathore & Shirke, 2011), other applications of membrane processes have been introduced more recently. Examples of such emerging processes include biocatalytic MBRs for food processing and fine chemical synthesis, and pollution remediation. A few recent articles (Jornitz *et al.*, 2002; Rathore & Shirke, 2011; Saxena *et al.*, 2009; van Reis & Zydney, 2007) have comprehensively reviewed the application of membrane technology in the upstream and the downstream of the biotechnology processes (i.e., clarification and sterilization; recovery, purification and concentration of fermentation products, etc.). This section focuses on two types of biocatalytic MBRs, namely, EMRs and ICMRs.

The EMR is a type of MBR which combines selective mass transport with enzyme-catalyzed biochemical reactions (Hai & Yamamoto, 2011; Hai *et al.*, 2013). The commercialized form of MBR, i.e., the biosolids separation MBR differs from the EMR in that the former involves whole-cell microbes in suspended form and the membrane there acts mainly as a solid–liquid separator. By contrast, as the name suggests, an EMR utilizes enzymes harvested from a culture, not the whole cell. An EMR accomplishes separation of enzymes from products and/or substrates by a semi-permeable membrane (Jochems *et al.*, 2011; López *et al.*, 2002). Permeable substrates and products can be selectively separated from the reaction mixture, and the selective removal of products from the reaction site increases the conversion of product-inhibited or thermodynamically unfavorable reactions (Giorno & Drioli, 2000). EMRs are particularly attractive because compared with the chemical catalysts, the catalytic action of enzymes is extremely efficient and selective, and reduces the formation of by-products (Giorno & Drioli, 2000). An EMR can be simply a reactor containing enzymes in suspension and combined with a membrane unit. However, EMRs can be designed in a way that the membrane acts both as a catalytic and separation unit (Figure 10.3a). Enzymes can be flushed along a membrane module, segregated within a membrane module, or immobilized in or on the membrane by various methods such as entrapment, gelification, physical adsorption, ionic binding, covalent binding or cross-linking. The choice of the reactor configuration depends on the properties of the reaction system (Giorno & Drioli, 2000). On the other hand, in an ICMR, the membrane is a separator between the wastewater and the nutrient-enriched cell medium, and simultaneously acts as a support for biofilm growth (Chung *et al.*, 2004; Hai *et al.*, 2013; Juang *et al.*, 2008; Li & Loh, 2007a) (Figure 10.3b). The wastewater flows through the lumen side of the hollow fiber membrane module and the nutrient-enriched cell medium

(a)

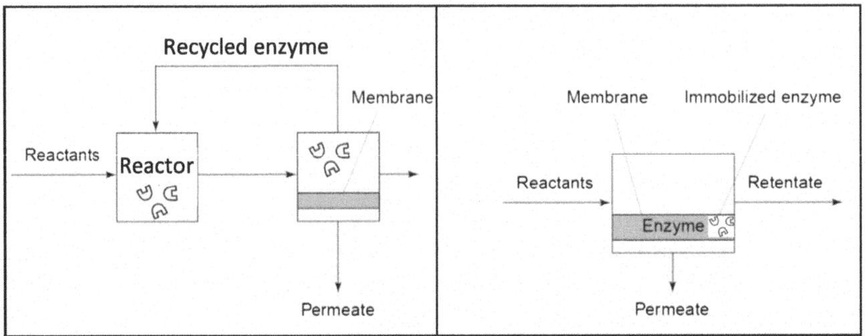

(b)

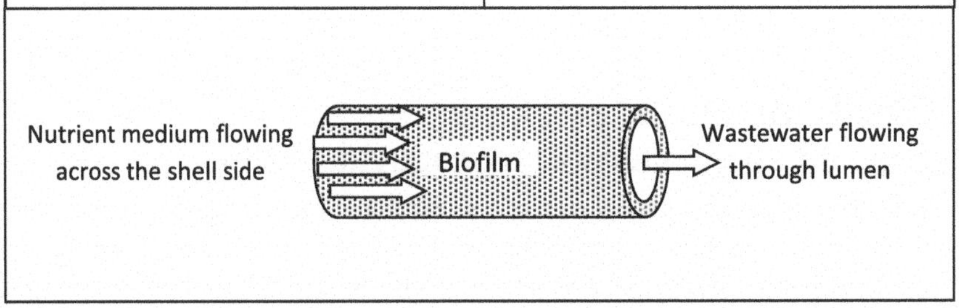

Figure 10.3 Simplified schematic of biocatalytic MBRs. (a) Configurations of EMR, i.e., a reactor with suspended enzyme combined with a membrane unit vs a reactor with the membrane active as a catalytic and separation unit, (b) flow of different streams and location of biofilm in an ICMR (modified after Giorno & Drioli, 2000 and Hai *et al.*, 2013).

flows across the shell side. In this way, the microbes can treat the effluent on one side and can be kept active via the nutrient stream at the other side without the nutrient medium contaminating the effluent.

Research on ICMRs has focused exclusively on wastewater treatment whereas research on EMRs has focused on two major streams of applications: synthesis of fine chemicals (Charcosset, 2006; Giorno & Drioli, 2000; Hai & Modin, 2013) and wastewater treatment (Hai *et al.*, 2013). Several applications of EMRs in the agro-food sector have been documented in the literature (Giorno & Drioli, 2000; Giorno *et al.*, 2010; Mazzei *et al.*, 2010). One of the first cases of the use of EMRs in the food sector was the production of milk with low lactose content (Illanes *et al.*, 1990). Nowadays, the hydrolysis of lactose (present in whole milk or cheese whey) by the use of EMRs is an effective technique running on a large scale (Mazzei *et al.*, 2010). EMRs have also been used in reduction of viscosity of fruit juice by hydrolyzing pectins (Alvarez *et al.*, 1996), treatments of musts and wine by the conversion of polyphenolic compounds and anthocyanes (Cantarelli *et al.*, 1989), hydrolysis of oil (Pronk *et al.*, 1988), production of sugars (Silman *et al.*, 1980), and the removal of peroxides from dairy products. Furthermore, several studies have demonstrated the technical feasibility of EMRs for the production of fine chemicals such as amino acids (Shi *et al.*, 1997), amines (Hoehn & Bornscheuer, 2009), carboxylic acids (Ceynowa & Rauchfleisz, 2003), antibiotics (Schroen *et al.*, 2009), anti-inflammatory drugs (Giorno *et al.*, 2007), personal care products (Zhang *et al.*, 2008) and alternative food or food ingredients that confer health benefits (Giorno & Drioli, 2000; Mazzei *et al.*, 2010). Fewer studies have investigated the application of biocatalytic MBRs for wastewater treatment. Hai *et al.* (2013) have comprehensively reviewed the studies, which have tested various forms of biocatalytic MBRs for problematic pollutants which are not effectively removed from wastewater by the conventional reactors. A succinct overview of such applications will be presented in this section.

10.4.2 Membranes for biocatalytic MBRs

Membranes in several configurations, namely, hollow fiber (Li & Wang, 2008), flat sheet (Akay *et al.*, 2002), or spiral wound (Lante *et al.*, 2000) have been used in studies related to biocatalytic MBRs; however, predominantly hollow fiber modules have been used (Hai *et al.*, 2013). Micro- or ultrafiltration membranes fabricated with a wide variety of organic polymers, namely, cellulose acetate (Chung *et al.*, 1998), polyacrylonitrile (Kowalska *et al.*, 1998), polyethersulfone (Li & Wang, 2008), polysulfone (Li & Loh, 2007b), polypropylene (Chung *et al.*, 2005), polyamide (Godjevargova *et al.*, 2006), and polyvinyledene fluoride (Juang & Kao, 2009) have been used in ICMRs. The studies that have explored EMRs show a similar trend of utilization of varieties of polymeric (predominantly) ultrafiltration membranes (Akay *et al.*, 2002; Bohdziewicz, 1998; Lante *et al.*, 2000). Membrane material may have an important impact on the efficiency of biocatalytic MBRs (Edwards *et al.*, 1999a). Available commercial membranes often require modifications to make them suitable for enzyme immobilization (Vasileva *et al.*, 2009). Jochems *et al.*(2011) have presented an in-depth review on the membrane modification strategies for enzyme immobilization. They reported that researchers have used different immobilization techniques, but to date no general set of rules have been proposed for making it easier to choose between the techniques.

10.4.3 EMRs for xenobiotics removal

Table 10.4 summarizes the salient features of the studies related to EMRs. To date most of the studies on EMRs have been conducted on the removal of phenols and phenolic derivatives. Various configurations of EMRs utilizing different types of enzymes have been tested in relation to phenolic compound removal. A few studies using direct contact EMRs (enzymes in suspension), however, applying different types of enzymes for phenol removal, could be identified (Calabrò *et al.*, 2009; Flock *et al.*, 1999; Gómez *et al.*, 2012). Flock *et al.* (1999) observed greater than 96% removal of phenol by four sequential batch reactors containing raw soybean seed-hulls (source of soybean peroxidase) with a retention time of 20 min in each reactor. Tyrosinase-catalyzed degradation of polyphenols in an EMR was reported by Calabrò *et al.* (2009). Comparing three different plant peroxidases, namely, horseradish, artichoke and soybean peroxidases, Gomez et al. (2012) reported better degradation of 4-chlorophenol by an EMR using soybean peroxidases.

A greater number of studies have investigated phenol removal performance of EMRs with enzymes immobilized on the membrane. Bohdziewicz (1998) reported efficient biodegradation of phenol by an EMR system wherein a crude mixture of enzymes obtained from a *Pseudomonas* strain was immobilized on an ultrafiltration membrane. Burton *et al.* (1998) used an EMR wherein polyphenol oxidase obtained from *Agaricus bisporus* was immobilized onto capillary membranes, and reported hydroxylation of a range of phenolic substrates to yield catechols, which were then oxidized by the enzyme to give *o*-quinone products. Moeder *et al.* (2004) demonstrated degradation of selected hydroxylated aromatic compounds by using a microporous polypropylene hollow fiber membrane impregnated with horseradish peroxidase and laccase. Akay *et al.* (2002) developed an EMR by immobilizing a crude enzyme extract obtained from a strain of *Pseudomonas syringae* on membrane and reported that phenol degradation rate was dependent on the initial concentration of phenol and the membrane flux. Lante *et al.* (2000) studied the biodegradation of 18 phenolic derivatives including chlorophenols, cresols and methoxyphenols by a laccase-immobilized EMR, and reported that the type and/or the position of substituent group affected the level of substrate oxidation. Edwards *et al.* (1999a) proposed post-treatment by a chitosan-packed adsorption column in order to remove the colored, quinone-type by-products, which were produced during removal of polyphenols by an EMR and was not retained by the membrane. In another study,

Table 10.4 Removal of xenobiotics by EMRs (modified after Hai *et al.* (2013)).

Target Pollutant	Membrane	Enzyme	Type	Reference
Phenols and phenolic derivatives				
Phenol	–	Manganese peroxidase	Enzyme in suspension	López *et al.* (2011)
Phenol	Polyacrylonitrile ultrafiltration membrane	Catechol 2, 3-dioxygenase	Enzyme immobilized	Bohdziewicz (1998)
Phenol	Ultrafiltration membrane prepared from acrylonitrile copolymer	Horseradish peroxidase	Enzyme immobilized	Vasileva *et al.* (2009)
Phenol and 2-chlorophenol	Ultrafiltration (molecular weight cutoff of 10 kDa)	Soybean peroxidase	Enzyme in suspension	Flock *et al.* (1999)
4-chlorophenol	Ultrafiltration	Soybean, horseradish and artichoke-peroxidases	Enzyme in suspension	Gómez *et al.* (2012)
Polyphenols	Flat polyethersulfone membrane	Tyrosinase	Enzyme in suspension	Calabrò *et al.* (2009)
18 phenolic substrates (chlorophenols, cresols and methoxyphenols)	Spiral-wound asymmetric polyethersulfone membrane (molecular weight cutoff of 3 kDa)	Laccase	Enzyme immobilized	Lante *et al.* (2000)
Phenol and catechol	Flat polyamide membrane with nominal pore size of 0.2 μm	Crude mixed enzyme extract (from *Pseudomonas syringae*)	Enzyme immobilized	Akay *et al.* (2002)
Phenol and cyanide	Flat polyacrylonitrile membrane	Crude mixed enzyme extract (from a *Pseudomonas* strain)	Enzyme immobilized	Bodzek *et al.* (1996)
Phenol and catechol	Capillary membranes (nylon and polysulfone), 0.45 μm	Polyphenol oxidase	Enzyme immobilized	Burton *et al.* (1998)
Phenols (phenol, *p*-cresol, 4-methoxyphenol, and 4-chlorophenol)	Polyethersulfone and polysulfone membranes	Polyphenol oxidase	Enzyme immobilized	Edwards *et al.* (1999a)
Phenols	Chitosan gel-coated polysulfone membranes	Polyphenol oxidase	Enzyme immobilized	Edwards *et al.* (1999b)

3,4-dimethylphenols, 4-ethylphenol, 2-hydroxy-1,2,3,4-tetrahydronaphthalene, 2-hydroxy-decahydronaphthalene and 4-hydroxy-biphenyl	Microporous polypropylene hollow fiber membranes	Horseradish peroxidase and laccase	Enzyme immobilized	Moeder *et al.* (2004)
Other xenobiotics				
Dye	Ultrafiltration polyethersulfone (molecular weight cutoff of 10 kDa)	Manganese peroxidase	Enzyme in suspension	López *et al.* (2004)
Dye	Ultrafiltration polyacrylonitrile (molecular weight cutoff of 20 kDa).	Laccase	Enzyme in suspension	Chhabra *et al.* (2009)
Dye	Chitosan membrane on nylon support	Laccase	Enzyme immobilized	Katuri *et al.* (2009)
Cyanide	Flat membrane	Cyanidase	Enzyme immobilized	Basheer *et al.* (1993)
Quinol	Nylon membrane	Laccase	Enzyme immobilized	Durante *et al.* (2004)
Xylan	Polysulfone membrane	Xylanase	Enzyme immobilized	Edward *et al.* (2003)
Aromatic hydrocarbon from petrochemical-based effluent	Polysulfone ultrafiltration membranes	Laccase and manganese-dependant peroxidase	Enzyme immobilized	Edwards *et al.* (2002)
Phenylurea pesticide	Microfiltration PVDF	Laccase	Enzyme immobilized	Jolivalt *et al.* (2000)
Polygalacturonic acid	YM30 ultrafiltration membranes (molecular weight cutoff of 30 kDa)	Pectolyase Y23	Enzyme in suspension	Gallifuoco *et al.* (2001)
Estrogen (e.g. estrone and estradiol)	Ultrafiltration polyethersulfone (molecular weight cutoff of 10 kDa)	Laccase	Enzyme in suspension	Lloret *et al.* (2012a)

Edwards *et al.* (1999b) utilized chitosan gel as an immobilization matrix for polyphenol oxidase on polysulfone capillary membranes. During the treatment of industrial phenolic effluents, adsorption onto chitosan coating facilitated the removal of the by-products following phenol conversion by the immobilized enzyme.

As noted earlier, relatively fewer studies reporting on the removal of other groups of compounds, apart from the phenol derivatives, by EMRs can be found. Bodzek *et al.* (1996) reported the importance of membrane operating parameters as well as of the concentration of the target compounds on the decomposition of phenol and cyanides by an EMR equipped with a polyacrylonitrile membrane on which an enzymatic fraction isolated from a bacterial strain of *Pseudomonas* was immobilized. Basheer *et al.* (1993) applied a cyanidase-immobilized EMR for the treatment of cyanide-containing wastewaters from the food industry. An EMR with xylanase from *Thermomyces lanuginosus* SSBP immobilized on polysulfone membranes was reported to degrade synthetic xylan effluent (Edward *et al.*, 2003). By using an EMR with laccase immobilized onto a PVDF membrane, Jolivalt *et al.* (2000) achieved degradation of a herbicide derivative, *N*′,*N*′-(dimethyl)-*N*-(2-hydroxyphenyl)urea, and simultaneous separation of the insoluble product by membrane filtration. Edwards *et al.* (2002) explored immobilization of laccase and manganese-dependent peroxidase (MnP) from *Trametes versicolor* onto ultrafiltration membranes to degrade aromatic hydrocarbons from a petrochemical industrial effluent. López *et al.* (2004) developed a direct contact EMR (enzyme in suspension) for the oxidation of azo dyes by MnP. Chhabra *et al.* (2009) used laccase and a mediator, namely, 2,2-azino-di-(-ethylbenzothiazoline-6-sulfonic acid) (ABTS), for the treatment of a textile dye solution in a direct contact EMR. In another study, Mendoza *et al.* (2011) demonstrated azo dye degradation by an EMR containing laccase and a mediator, namely, 2,2,6,6-tetramethylpiperidine 1-oxyl coupled to polyethylene glycol. Lloret *et al.* (2012a) reported efficient degradation of estrone and estradiol by an EMR containing laccase in suspension.

Only a few studies have systematically compared the performance of EMRs with free and immobilized enzymes. Edwards *et al.* (2002) reported that immobilization of enzymes in an EMR facilitated enzyme–substrate contact. Quinol degradation by an EMR containing a nylon membrane as the immobilization support for enzymes (laccase from *Rhus vernicifera)* was reported by Durante *et al.* (2004). They reported that compared with the affinity of the free enzyme, the affinity of the immobilized enzyme for its substrate (target compound) was higher. On the other hand, López *et al.* (2011) compared continuous degradation of phenol in EMRs with free and immobilized enzymes, respectively. They contended that although an immobilized enzyme system may involve lower enzyme consumption, enzyme immobilization is a costly process, and, therefore, operating with free enzymes may be a cheaper option. The performance of various EMR configurations such as direct enzyme contact, enzyme impregnated, immobilized enzyme and a reactor system involving laccase immobilization in chitosan membranes for decolorization of an azo dye were examined using laccase purified from white-rot fungi, *Pleurotus ostreatus* 1804 (Katuri *et al.*, 2009). Among the different reactor configurations employed, laccase encapsulated in chitosan membrane showed advantages such as requirement of shorter HRT and reusability of the enzyme for a number of repeated cycles.

10.4.4 Membrane fouling in EMRs for xenobiotics removal

When an enzyme solution is flushed through an ultrafiltration membrane that rejects the enzyme molecules, the enzymes can form a porous, thin gel layer ('secondary membrane') on the membrane. Wastewaters also contain colloidal suspended particles as well as water-soluble macromolecules, which may co-deposit with the enzymes, thus reducing the enzymatic activity and increasing the filtration pressure as a result of the concentration–polarization/fouling phenomena (Giorno & Drioli, 2000; Hai *et al.*, 2013; Jochems *et al.*,

2011). Concentration–polarization phenomena and fouling strongly affect the performance of EMRs, and appropriate fluid dynamic conditions (e.g. cross flow filtration mode) and reactor designs are necessary to control performance at a steady state (Akay *et al.*, 2002; Giorno & Drioli, 2000). Examples of new membrane module design and operation mode include the rotary disk membranes and the use of baffles to create a Dean vortex (Giorno & Drioli, 2000).

Membrane modification for enzyme immobilization can provide an additional advantage in terms of reducing membrane fouling by reducing undesired interactions (Jochems *et al.*, 2011). Edwards *et al.* (2002) reported that the phenomenon of fouling typically associated with the operation of ultrafiltration systems was reduced by the action of the immobilized enzyme layer. However, in immobilized enzyme systems, diffusion limitations may arise due to the steric hindrance exerted by the activated solid matrix to the free diffusion of substrate and reaction products towards or away from the catalytic site (Durante *et al.*, 2004). Furthermore, the formation of insoluble reaction products due to non-enzymatic reactions (oxidative coupling) is another important technical constraint impairing the development of immobilized enzyme EMRs. In fact, prolonged and repeated use of an immobilized enzyme system, either in a packed-bed, a fluidized bed, or in a membrane reactor format, can result in the accumulation of precipitate on the outlet filter of the reactor (fouling) leading to significant reductions in the flow rates or to complete plugging (Duran *et al.*, 2002). For example, Jolivalt *et al.* (2000) reported that the transformation of a herbicide by an immobilized laccase was accompanied by the formation of an insoluble purple-colored product, which accumulated on the membrane surface. At longer HRTs, more aggressive membrane fouling was observed by Chhabra *et al.* (2009), and this was attributed to the formation of an insoluble polymer layer over the membrane surface due to slower mass transport under the longer HRT.

10.4.5 Inhibition of enzymatic activity in EMRs for xenobiotics removal

One main drawback of EMRs is that they require a large amount of enzymes to achieve high removal efficiency due to enzyme deactivation (Erhan *et al.*, 2002; Rios *et al.*, 2004). Gradual loss of enzymatic activity due to various physical, chemical and biological inhibitors can occur. Apart from a few exceptions (Basheer *et al.*, 1993, Calabrò *et al.*, 2009; Chhabra *et al.*, 2009; Lloret *et al.*, 2012b), available studies have reported rapid loss of enzymatic activity during EMR operation. For example, repeated pulses of MnP were required to be added every 10 h to maintain stable dye decoloration in a continuous flow EMR (López *et al.*, 2004). In a study by Katuri *et al.* (2009) the residual laccase activity in the EMR gradually reduced at the end of each batch run and finally 75% of the activity was lost by the end of the 13th run (30 h). Hata *et al.* (2010) reported a loss of 90% of the laccase activity within 8 h in a batch test. Enzymatic activity can be very sensitive to the presence of several compounds including intermediates and final reaction products that can potentially cause inhibition (Gallifuoco *et al.*, 2001). Significant diffusional limitations can be observed when the polymers formed during enzymatic reactions cover the active center of the enzymes (Bodalo *et al.*, 2006). Deformation of the enzyme molecules during filtration, depending on the shear rate and flow geometry, can also cause deactivation (Akay *et al.*, 2002; Katuri *et al.*, 2009). Peroxidases require the presence of hydrogen peroxide for enhanced performance; however, deactivation due to high concentrations of hydrogen peroxide has also been noted (Flock *et al.*, 1999; López *et al.*, 2011). An additive such as polyethylene glycol has been reported to act as enzyme protectors (Bodalo *et al.*, 2006); however, such addition increases cost and raises the concern of secondary pollution.

Enzyme immobilization has been reported to increase thermostability and resistance to inhibitors and chemical reagents, and allow continuous use of the enzyme (Bohdziewicz, 1998). For example, in cases where the reaction products can inactivate the enzyme, immobilization facilitates separation of the products from the enzyme and reduces inhibition. Compared with the non-immobilized enzyme reaction, the inactivation of polyphenol oxidase was shown to be significantly reduced by immobilizing the enzyme onto the membrane in an EMR (Edwards *et al.*, 1999a, b). In a study by Lante *et al.* (2000), free laccase retained only 18% of its initial activity after 30 h of repeated runs. By contrast, immobilized laccase showed a promising half-life of more than 150 h. Erhan *et al.* (2002) reported that the stability of the immobilized enzyme was better than that of the free enzyme in solution and the immobilized enzyme maintained 70% of its initial activity for approximately 100 h. However, immobilization is an expensive process (Gómez *et al.*, 2012). Furthermore, although immobilization has been proved to increase stability of the enzymes, this may cause changes in their catalytic activity and enantioselectivity (Jochems *et al.*, 2011; Majeau *et al.*, 2010). Enzymatic activity can be significantly decreased due to different orientations of the enzyme with respect to the membrane or due to multiple point attachments (Butterfield *et al.*, 2001). Additional impediments arise during application of EMR for wastewater treatment. The stabilities and catalytic properties of free or immobilized enzymes within an EMR are dramatically affected by the harsh high-strength wastewater conditions (Katuri *et al.*, 2009). Investigations to date regarding application of EMRs for wastewater treatment have been mainly carried out on low-strength wastewater with limited total organic carbon and total suspended solid loadings (Hai *et al.*, 2013; Majeau *et al.*, 2010).

10.4.6 ICMRs for xenobiotics removal

Table 10.5 summarizes the salient features of the studies related to ICMRs. Li and Loh (2007b) studied the effect of feed concentration (1000–2000 mg/L) and feeding rate on the removal of phenol by an ICMR with *Pseudomonas putida* ATCC49451 immobilized in asymmetric, polysulfone hollow fiber membranes. The results confirmed that continuous operation of the bioreactor could be conducted through fine tuning of a number of operating conditions, depending on whether complete phenol degradation or high degradation capacity is desirable. Similarly, Juang *et al.* (2009) reported that phenol degradation efficiency was dependent on both initial phenol concentration and initial cell concentration. In the case of cometabolic transformation of 4-chlorophenol in the presence of phenol, the removal rates of the substrates were dependent not only on absolute concentration of the substrates, but also on the concentration ratios of 4-chlorophenol to phenol. During continuous operation at high substrate concentrations, DO was found to be insufficient, and the substrate degradation was enhanced by aerating the feed with pure oxygen (Li & Loh, 2005, 2006a).

A few studies have focused on the relative advantages of suspended and immobilized cultures (Hai *et al.*, 2013). Chung *et al.* reported that compared with immobilized biomass, suspended cells demonstrated relatively faster phenol degradation rates, however, suffered substrate inhibition at relatively lower phenol levels (<800 mg/L) (Chung *et al.*, 2004, 2005). Accordingly, the degradation of high-strength phenol solutions (>1000 mg/L) by *Pseudomonas putida* immobilized on hollow fibers was higher than by freely suspended cells. Loh *et al.* (1999) reported that in immobilized systems, cells often detach from the membranes when phenol concentrations reach non-inhibitory levels; in such cases, the time taken for completion of degradation may be shorter, suggesting a beneficial effect of having both suspended and immobilized cultures. Chung *et al.* (1998) reported similar observations. In a system containing both suspended and immobilized biomass, the contribution of biofilm towards net degradation may vary (Juang & Kao, 2009; Juang & Tsai, 2006); nevertheless, for maintaining a stable degradation at high phenol concentration, the importance of an immobilized system cannot be overlooked. In addition to the target toxic compounds, other physico-chemical factors, such as media

Table 10.5 Removal of xenobiotics by ICMR (modified after Hai *et al.*, 2013).

Target Pollutant	Membrane	Microbes	References
Phenol	Cellulose acetate membranes	*Pseudomonas putida* (ATCC49451)	Chung *et al.* (1998)
Phenol	Activated carbon impregnated polyethersulfone hollow fiber	*Pseudomonas putida*	Zhang *et al.* (2008)
Phenol	Polysulfone hollow fiber membranes	*Pseudomonas putida* ATCC49451	Li and Loh (2007b)
Phenol	Microporous polypropylene hollow fiber	*Pseudomonas putida* CCRC14365	Chung *et al.* (2005)
Phenol	Activated carbon-filled cellulose acetate hollow-fiber membranes	*Pseudomonas putida* ATCC17484	Zhu *et al.* (2000)
Phenol	Polyamide and polyacrylonitrile membrane	*Trichosporon cutaneum* R57	Godjevargova *et al.* (2006)
Phenol	Polyvinylidene fluoride	*Pseudomonas putida* BCRC14365	Juang *et al.* (2009)
Phenol and 4-chlorophenol	Activated carbon impregnated polysulfone hollow fiber	*Pseudomonas putida* ATCC49451	Li and Loh (2006b)
Phenol and 4-chlorophenol	Polysulfone hollow fiber	*Pseudomonas putida* ATCC49451	Li and Loh (2006a)
Phenols and cyanides	Polyacrylonitrile ultrafiltration membrane	*Agrobacterium radiobacter*, *Staphylococcus seiuri* and *Pseudomonas diminuta*	Kowalska *et al.* (1998)
Phenol and sodium salicylate	Microporous polypropylene hollow fiber module	*Pseudomonas putida* CCRC14365	Juang and Tsai (2006)
Heavy metals, 3-chlorobenzoate and 2,4-dichlorophenoxy acetic acid	Flat and tubular polysulfone membrane	*Alcaligenes eutrophus*	Diels *et al.* (1995)
Chromate	Cellulose acetate ultrafiltration and microfiltration membranes with hydraulic pore size of 30 and 400 nm, respectively	Six strains of *Pseudomonas*	Konovalova *et al.* (2003)
Preservative substances, cephalosporins and ions	Cellulose acetate microfiltration membrane	*Escherichia coli*	Madaeni and Zarbakhsh (2009)

salinity can have detrimental effect on suspended cultures. For instance, in a study conducted by Juang and Wu (2007), the suspended cells grew well only at NaCl concentrations below 0.44 M, while the cells immobilized on microporous hollow fibers could sustain a NaCl concentration of up to 1.52 M.

Membrane properties such as material and surface area can pose important impact on phenol degradation by ICMRs. Godjevargova *et al.* (2006) studied the effect of membrane materials on cell immobilization and on subsequent phenol degradation performance, and observed better performance of a polyacrylonitrile-immobilized system over a polyamide-immobilized system. Juang and Huang (2008) reported that a polyvinylidene fluoride hollow fiber was more suitable than polypropylene because polyvinylidene fluoride fiber had lower extent of phenol sorption, thus facilitating faster mass transfer to the biomass medium. While studying the cometabolic transformation of 4-chlorophenol in the presence of phenol, Li and Loh reported that, to maintain an optimum balance between the membrane surface area available for cell immobilization and the extra-capillary space available for suspended cell growth, the volume covered by the membrane should be within approximately 30% of the reactor volume (Li & Loh, 2005; 2006a). The importance of membrane surface area was also reported by Juang *et al.* (2009).

Incorporation of granular activated carbon (GAC) into hollow fiber membranes has been reported to result in enhanced degradation of phenol by ICMRs. Better biodegradation of phenol by *Ps. putida* in an ICMR with GAC-incorporated hollow fibers over a GAC-free bioreactor during start-up and under elevated loading phases was demonstrated (Li & Wang, 2008; Wang & Li, 2007). It was reported that sorption, biodegradation, and bioregeneration were the key steps involved in phenol removal in such an ICMR. Li *et al.* highlighted the importance of the relative ratio of polysulfone (membrane polymer) and GAC during fabrication of the hybrid hollow fibers (Li & Loh, 2006a, 2007a). In their study, the hollow fibers containing polysulfone and GAC in 2:1 ratio demonstrated the best performance for the removal of phenol and 4-chlorophenol, both under batch and continuous operations. Zhu *et al.* (2000) compared phenol degradation performance of suspended biomass with that of the biomass immobilized on cellulose acetate hollow fibers with or without incorporation of GAC. No discernible cell growth or phenol degradation occurred in the case of free suspension up to 35 h. Although the biomass immobilized on both pure cellulose acetate fiber and activated carbon-filled fiber completely degraded phenol, the biodegradation rate of the activated carbon-filled fiber system was higher than that of the activated carbon-free fiber system.

Only a few studies have investigated the removal of other groups of compounds apart from the phenol derivatives by ICMRs. Madaeni and Zarbakhsh (2009) designed a prototype of an ICMR wherein colonies of bacteria (*Escherichia coli*) were subjected to sedimentation on the surface of a microfiltration membrane. The hybrid system effectively removed preservative substances, cephalosporins and ions, which was not possible by microfiltration alone. The suitability of different bacterial strains for the removal of different toxic target compounds, namely, heavy metals and organic xenobiotics, was focused in a study by Diels *et al.* (1995). They used *Alcaligenes eutrophus* CH34 bacteria for precipitation of heavy metals, while a strain *Alcaligenes eutrophus* AE1308 was found suitable for organic xenobiotics (such as 3-chlorobenzoate and 2,4-dichlorophenoxyacetic acid) degradation. Konovalova *et al.* (2003) demonstrated remediation of Cr(VI)-contaminated water in semi-batch operation of an ICMR with bacteria from the *Pseudomonas* group. Kowalska *et al.* (1998) isolated three types of microorganisms from a mixed population of activated sludge adapted to phenol and cyanide. The membrane with a mixture of microorganisms from the strains *Agrobacterium radiobacter*, *Staphylococcus seiuri* and *Pseudomonas diminuta* immobilized on its surface appeared to be the most effective in biodegradation of phenol and cyanide. To maintain optimal biofilm thickness and protect against biofouling of the hollow fibers, a sequestering agent was used by Chung *et al.* (2005).

10.5 ELECTROCHEMICAL MBRS

Electrochemical MBRs refer to reactors in which electrochemical-, biological-, and membrane processes are combined. The purpose of the electrochemical processes can be to convert chemical energy into electrical energy or vice versa, or to force migration of ions towards the anode or the cathode. The purpose of the biological processes is to remove pollutants from wastewater, convert them into compounds that could be extracted and valorized, or to catalyze reactions at the electrode surfaces. The purpose of the membrane is either to function as a separator between the anode and the cathode or to facilitate extraction of specific compounds from the wastewater. Electrochemical MBRs for a wide variety of applications are described in the scientific literature.

One category of electrochemical MBRs are called microbial BES and have recently received a lot of attention for their potential role in resource recovery from wastewater (Logan & Rabaey, 2012; Rabaey *et al.*, 2007). A BES typically consists of an anaerobic anode compartment in which microorganisms oxidize organic compounds and use the anode as the electron acceptor. The electrons flow through an external circuit to the cathode, where a reduction reaction takes place. The reduction reaction at the cathode may also be catalyzed by microorganisms. The most investigated type of BES is the microbial fuel cell (MFC), which directly converts wastewater organics into electrical energy (Liu *et al.*, 2004). In an MFC, oxygen is reduced to water at the cathode. Electrical energy can be recovered from the system because the overall reaction, i.e., oxidation of organic compounds and reduction of oxygen, is thermodynamically favorable (Logan *et al.*, 2006). Another type of BES is the microbial electrolysis cell (MEC), which requires an input of electrical energy to drive the reactions at the anode and the cathode. An MEC can be used, for example, to produce hydrogen from organic compounds in wastewater. Microorganisms in the anode compartment oxidize organic compounds and use the anode as the electron acceptor. The electrons flow to the cathode where hydrogen ions are reduced to hydrogen gas. Input of electrical energy is required to drive the system because the overall reaction, i.e., oxidation of organics and reduction of hydrogen ions, is thermodynamically unfavorable (Logan *et al.*, 2008). In addition to electrical energy and hydrogen, BESs can also utilize the energy content of organic compounds in wastewater to generate, for example, hydrogen peroxide (Rozendal *et al.*, 2009), methane (Clauwaert *et al.*, 2008), and alkaline solutions (Rabaey *et al.*, 2010). They could also be used as BOD sensors (Modin & Wilen, 2012; Kim *et al.*, 2003) and to recover metals (Modin *et al.*, 2012; Ter Heijne *et al.*, 2010).

Membranes play an important role in BESs. As electrons flow from the anode to the cathode in the external circuit, ions must migrate between the electrodes in the liquid to maintain charge balance. Membranes make it possible to have separate anode and cathode compartments with different conditions but still allow ion migration between the electrodes. For example, in an MFC, the anode compartment should be anaerobic and the cathode compartment aerobic. Typically, an ion exchange membrane is used to separate the compartments. Ion exchange membranes consist of a polymer backbone with charged functional groups. Anion exchange membranes (AEMs) have positively charged groups, which allow migration of anions through the membrane. Cation exchange membranes (CEMs) have negatively charged groups allowing migration of cations. There are also charge mosaic membranes that have functional groups of both positive and negative charge, and bipolar membranes which consist of an AEM and a CEM adjacent to each other (Rozendal *et al.*, 2008).

The use of ion exchange membranes in BESs has led to reduced pH in the anode compartment and increased pH in the cathode compartment (Gil *et al.*, 2003). The reason for this is that the oxidation reaction at the anode releases hydrogen ions (Equation 1) and the reduction reaction at the cathode consumes hydrogen ions (Equation 2). Unless the flow of electrons through the external circuit is balanced by an equal flow of hydrogen or hydroxide ions through the ion exchange membrane, there will be a pH shift between the two compartments. In wastewater, other ions such as Na^+, K^+, and Cl^- are present in much higher concentrations than H^+ or OH^-. Thus, these other ions will be responsible for most of the ionic charge transfer. In an MFC study using a Nafion CEM, Rozendal *et al.* (2006) found that Na^+, K^+, Mg^{2+}, Ca^{2+}, and $NH_4{}^+$ accounted for practically the entire transfer of ionic charge.

$$CH_2O + H_2O \rightarrow CO_2 + 4e^- + 4H^+ \tag{10.1}$$

$$O_2 + 4e^- + 4H^+ \rightarrow 2H_2O \tag{10.2}$$

Although pH shifts are problematic in both MFCs and MECs, the combination of ion exchange membranes and BESs has led to the development of several new applications. In a microbial desalination cell, an AEM is placed facing the anode and a CEM facing the cathode. The compartment that is formed between the two membranes is used to desalinate water. In a proof-of-concept study, Cao *et al.* (2009) achieved 93% desalination of a salt solution containing 35 g/L NaCl. They fed a nutrient medium containing acetate to the anode and a ferricyanide solution to the cathode and produced a maximum of 31 W/m^3 of power while simultaneously desalinating the salt solution. By stacking AEMs and CEMs, Chen *et al.* (2011) were able to increase the rate of desalination. A reactor containing two compartments for desalination had 1.4 times larger desalination rate than a reactor containing one compartment. A schematic of a stacked microbial desalination cell is shown in Figure 10.4. The flow of one electron from the anode to the cathode results in the migration of one Na^+ and one Cl^- from each of the stacked desalination compartments. Although stacking compartments results in a higher desalination rate, it can also increase the internal resistance of the reactor, which reduces the energy recovery (or increase energy consumption if the reactor is operated as a MEC). If the cathode compartment is kept anaerobic, hydrogen gas can be produced by the reduction of hydrogen ions while water is being desalinated. With a voltage input of 0.55 V, Mehanna *et al.* (2011) produced 0.16 m^3 H_2/m^3 d, while simultaneously desalinating a NaCl solution. Microbial desalination cells have been suggested as pretreatment for reverse osmosis (Mehanna *et al.*, 2010). A challenge is to minimize the volume of wastewater needed per volume of desalinated water.

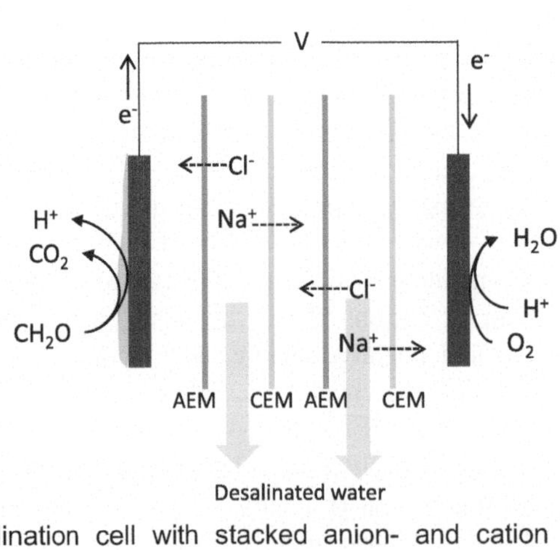

Figure 10.4 Microbial desalination cell with stacked anion- and cation exchange membranes (AEM and CEM).

Stacks of alternating AEMs and CEMs can also be used to recover electrical energy from salinity gradients in a process called reverse electrodialysis (RED). The spaces between the membranes are fed with alternating sea water and fresh water that cause a salinity gradient across each membrane. This salinity gradient generates a voltage, which means that electric power can be harvested. By combining RED stacks with BESs, electrical energy production from salinity gradients can be combined with electrical energy recovery from organic compounds in wastewater. This is a suitable combination considering many wastewater treatment plants are located near the coast line. RED stacks have been used to enhance the power output from MFCs (Kim & Logan, 2011a) and to produce hydrogen gas in MECs with little or zero energy input (Kim & Logan, 2011b).

BESs have also been designed to generate alkaline solutions at the cathode. If a CEM is used to separate the anode and cathode in a BES fed with wastewater to the anode chamber, the transfer of one electron will likely be accompanied by the migration of one cation ion through the membrane. Since Na^+ ion are present in the highest concentration in wastewater, they will make up the majority of the ionic charge transfer. On the cathode, oxygen could be reduced to water (MFC operation) or hydrogen ions could be reduced to hydrogen gas (MEC operation). Both reactions consume hydrogen ions, which mean that a surplus of hydroxide ions will accompany the Na^+ ions that accumulate in the cathode compartment. In the anode compartment, there will be a generation of acidity as the oxidation of organic matter on the biological anode releases hydrogen ions. If the flow of wastewater through the anode compartment is much higher than the flow of liquid through the cathode compartment, the increase in alkalinity concentration at the cathode will be much higher than the increase in acidity concentration at the anode. This concept was utilized by Rabaey *et al.* (2010) to produce an alkaline solution in the cathode compartment. Using a synthetic wastewater containing acetate as anode feed, they were able to produce a 3.4% wt/wt caustic solution at the cathode. In an extension of this concept, Modin *et al.* (2011) investigated using a BES to redistribute alkalinity between process streams in a wastewater treatment plant. Reject water from sludge dewatering process is rich in ammonium and is commonly treated using biological nitrification before being fed back into the main treatment plant. The nitrification process consumes alkalinity so, for example, NaOH has to be added to the reject water. A BES could potentially be used to redistribute alkalinity from the influent wastewater to the reject water. Since the reject water flow is much smaller than the flow of influent wastewater, a small decrease in alkalinity concentration in the wastewater would correspond to a large increase in alkalinity concentration in the reject water (Modin *et al.*, 2011). This concept could also be used to recover nitrogen from reject water by converting ammonium to ammonia, which can be stripped off as a gas (Wu & Modin, 2013).

Electrochemical MBRs can be used to extract and concentrate valuable components from wastewater. One example is nitrogen, which is typically present as ammonium ions. Electrochemical systems in which the anode and cathode are separated by a CEM can be used to drive ammonium to the cathode compartment where a concentrated solution can be obtained (Kuntke *et al.*, 2018). Several types of systems have been investigated. For example, MFCs or MECs could be used to recover ammonium from urine. Organic compounds in the urine are oxidized at the biological anodes. At the cathode, either oxygen reduction (MFC) or hydrogen evolution (MEC) takes places. The current flow drives ammonium from the anode compartment to the cathode compartment (Kuntke *et al.*, 2012; Zamora *et al.*, 2017). The system does not necessarily need to rely on a biocatalyzed anode oxidizing organics. Sludge liquor from anaerobic digesters contains high ammonium concentrations but low levels of biodegradable organics. Desloover *et al.* (2012) used an abiotic electrochemical reactor to concentrate ammonium at an energy consumption rate of 5 kWh/kg N.

Electrochemical systems can also be used to extract organic acids from fermentation reactors. Fermentation of biomass leads to the production of volatile fatty acids (VFAs) such as acetic-, propionic-, and butyric acid. At a pH higher than about 5, these are mainly present as anions in the water. Jones *et al.* (2015) used electrodialysis to extract VFAs from sucrose and grass fermentation broths. The reactor consisted of a stack of alternating AEMs and CEMs placed between two electrodes. An applied electrical current forced ions to move through the membranes. Anions such as acetate, propionate, and butyrate moved through AEMs towards the anode and become concentrated. Andersen *et al.* (2015) used an electrochemical system to extract VFAs from a fermentation reactor fed with thin stillage. In this study, the electrochemical system served several purposes. Hydrogen generated at the cathode drove the fermentation processes towards generation of VFAs with longer carbon chains (C4–C6), the cathodic consumption of H^+ allowed pH control without the addition of chemicals, and the current drove anions through an AEM to the anode, which allowed to VFAs to be concentrated in the anode compartment. Tao *et al.* (2016) developed a system of reactors combining biological-, chemical-, electrochemical-, and membrane processes to produce polyhydroxyalkanoates and recover nutrients from waste sludge. The sludge was first fed to an anaerobic fermenter. A microfiltration unit was used to separate the biomass from the VFA-containing water. The VFAs and ammonium was concentrated using electrodialysis. Then, ammonium was chemically precipitated as struvite and the VFA-containing water was fed to a biological reactor where it was used as substrate for production of polyhydroxyalkanoates (Tao *et al.*, 2016).

10.6 SUMMARY AND FUTURE OUTLOOK

The combination of membranes and biochemical conversions has led to the development of a range of innovative wastewater treatment processes. Most of the technologies described in this chapter are still in the research and development stage; however, they offer the potential for scale up. Although gas-diffusion MBfRs, EMBRs, biocatalytic MBRs, and electrochemical MBRs are four different technologies, they all rely on the combination of a biocatalyst with a membrane and share some similar challenges.

It is important to maintain a biocatalyst that carries out the desired biochemical transformations over time. Particularly in MBfRs and EMBRs, biofilm management is a crucial parameter. A too thin biofilm will not provide the desired conversion rates, while a too thick biofilm will lead to diffusional resistance, limiting performance. Aerobic MBfRs have been demonstrated to achieve very high rate removal of organic matter; however, maintaining an appropriate biofilm thickness can be challenging. Interestingly, the reactors fed with only ammonium seem to be free from the problems with excess biomass (Hwang *et al.*, 2009). Autotrophs grow with lower yield than heterotrophs, which means that problems with excess biomass and thick biofilms should be easier to control in reactors for nitrification or completely autotrophic nitrogen removal. Further studies on aeration MBfRs for reject water (sludge liquor) or other nitrogen-rich and organics-poor wastewater streams are therefore warranted. Hydrogen MBfRs also operate based on the activities of autotrophic microorganisms. These reactors are very versatile and have been proven for several oxidized water contaminants such as nitrate and perchlorate (Nerenberg & Rittmann, 2002). Both aeration- and hydrogen MBfRs are being commercialized (Martin & Nerenberg, 2012; Oxymem, 2018; Peeters *et al.*, 2016). Methane MBfRs have been investigated for denitrification and removal of oxidized contaminants and xenobiotic compounds. Aerobic methanotrophs grow fast which means biofilm management will be a problem in methane MBfRs (Rishell *et al.*, 2004), just like in aeration MBfRs for organics removal. However, the MBfR appears to be a suitable reactor configuration for the slow-growing denitrifying methanotrophs (Shi *et al.*, 2013). In EMRs, membrane fouling and enzyme activity decay are two major limitations (Hai *et al.*, 2013; Rios *et al.*, 2004). The high cost of enzyme production remains another important issue precluding the practical application of

EMRs for wastewater treatment. Unlike nanotechnology or biosensor use, large amounts of enzymes are required for bioremediation applications (Majeau *et al.*, 2010). Thus, in comparison to other applications, the applications of enzyme-based wastewater treatment systems remain a less studied stream. Compared to EMRs, a limited number of studies on ICMR have been conducted. Studies only from certain research groups who targeted a limited number of recalcitrant compounds (predominantly phenol) are available. Currently only a few number of publications on these topics are published each year. The development and implementation of these systems at an industrial scale requires a broader as well as an in-depth understanding of the core processes.

The choice of membrane material and configuration is an important issue for all the MBRs described in this chapter. In electrochemical MBRs, prudent use of ion exchange membranes and biocatalyzed electrodes has expanded possible applications from wastewater treatment and energy recovery to desalination and production of chemicals. Another future avenue of research could be onsite electrochemical generation of membrane cleaning chemicals at MBR wastewater treatment plants (e.g. Modin & Fukushi, 2012). In EMBRs, the choice of membrane determines which compounds will diffuse from the wastewater stream into the bioreactor compartment. Inexpensive membranes that allow efficient transfer of target pollutants are needed. In MBfRs, it is desirable that the membrane material promotes attachment and retention of biofilm. The configuration should provide a large surface area for biofilm attachment, allow effective biofilm management, and avoid hydraulic short-circuiting. Most of the studies reviewed in this chapter have utilized commercially available membranes which were originally not developed for the technologies described here. Membranes materials and configurations developed specifically for application in, for example, MBfRs or electrochemical MBRs are rare. Closer collaborations between researchers in environmental biotechnology and membrane developers could potentially lead to improved performance of the MBR technologies described in this chapter.

REFERENCES

Ahmed T. and Semmens M. J. (1992a). Use of sealed end hollow fibers for bubbleless membrane aeration: experimental studies. *Journal of Membrane Science*, **69**, 1–10.

Ahmed T. and Semmens M. J. (1992b). The use of independently sealed microporous hollow fiber membranes for oxygenation of water: model development. *Journal of Membrane Science*, **69**, 11–20.

Akay G., Erhan E., Keskinler B. and Algur O. F. (2002). Removal of phenol from wastewater using membrane-immobilized enzymes: Part II. Cross-flow filtration. *Journal of Membrane Science*, **206**(1–2), 61–68.

Almeida J. S., Reis M. M. and Crespo J. G. (1999). Development of extractive membrane bioreactors for environmental applications. *Environment Protection Engineering*, **25**(1), 111–121.

Alvarez V., Andres L. J., Riera F. A. and Alvarez R. (1996). Microfiltration of apple juice using inorganic membranes: process optimization and juice stability. *Canadian Journal of Chemical Engineering*, **74**(1), 156–162.

Andersen S. J., Candry P., Basadre T., Khor W. C., Roume H., Hernandez-Sanabria E., Coma M. and Rabaey K. (2015). Electrolytic extraction drives volatile fatty acid chain elongation through lactic acid and replaces chemical pH control in thin stillage fermentation. *Biotechnology for Biofuels*, **8**, 221.

Basheer S., Kut Ö. M., Prenosil J. E. and Bourne J. R. (1993). Development of an enzyme membrane reactor for treatment of cyanide-containing wastewaters from the food industry. *Biotechnology and Bioengineering*, **41**(4), 465–473.

Bodalo A., Gomez J. L., Gomez E., Hidalgo A. M., Gomez M. and Yelo A. M. (2006). Removal of 4-chlorophenol by soybean peroxidase and hydrogen peroxide in a discontinuous tank reactor. *Desalination,* **195**(1–3), 51–59.

Bodzek M., Bohdziewicz J. and Kowalska M. (1996). Immobilized enzyme membranes for phenol and cyanide decomposition. *Journal of Membrane Science*, **113**(2), 373–384.

Bohdziewicz J. (1998). Biodegradation of phenol by enzymes from *Pseudomonas* sp. immobilized onto ultrafiltration membranes. *Process Biochemistry*, **33**(8), 811–818.

Brindle K. and Stephenson T. (1996). Nitrification in a bubbleless oxygen mass transfer membrane bioreactor. *Water Science and Technology*, **34**(9), 261–267.

Brindle K., Stephenson T. and Semmens M. (1998). Nitrification and oxygen utilization in a membrane aeration bioreactor. *Journal of Membrane Science*, **144**, 197–209.

Brindle K., Stephenson T. and Semmens M. (1999). *Pilot-plant treatment of a high-strength brewery wastewater using a membrane-aeration bioreactor. Water Environment Research*, **71**(6), 1197–1204.

Burton S. G., Boshoff A., Edwards W. and Rose P. D. (1998). Biotransformation of phenols using immobilised polyphenol oxidase. *Journal of Molecular Catalysis B: Enzymatic*, **5**(1–4), 411–416.

Butterfield D. A., Bhattacharyya D., Daunert S. and Bachas L. (2001). Catalytic biofunctional membranes containing site-specifically immobilized enzyme arrays: a review. *Journal of Membrane Science*, **181**(1), 29–37.

Calabrò V., Curcio S., De Paola M. G. and Iorio G. (2009). Optimization of membrane bioreactor performances during enzymatic oxidation of waste bio-polyphenols. *Desalination*, **236**(1–3), 30–38.

Cantarelli C., Brenna O., Giovanelli G. and Rossi M. (1989). Beverage stabilization through enzymatic removal of phenolics. *Food Biotechnology*, **3**(2), 203–213.

Cao X., Huang X., Liang P., Xiao K., Zhou Y., Zhang X. and Logan B. E. (2009). A new method for water desalination using microbial desalination cells. *Environmental Science & Technology*, **43**, 7148–7152.

Casey E., Glennon B. and Hamer G. (1999). Review of membrane aerated biofilm reactors. *Resources, Conservation and Recycling*, **27**, 203–215.

Castillo P. A., Gonzalez-Martinez S. and Tejero I. (1999). Biological phosphorous removal using a biofilm membrane reactor: operation at high organic loading rates. *Water Science and Technology*, **40**(4–5), 321–329.

Celmer-Repin D., Hwang J. H., Cicek N. and Oleszkiewicz J. A. (2010). Autotrophic nitrogen-removing biofilms on porous and non-porous membranes. *Environmental Technology*, **31**(12), 1391–1401.

Ceynowa J. and Rauchfleisz M. (2003). High enantioselective resolution of racemic 2-arylpropionic acids in an enzyme membrane reactor. *Journal of Molecular Catalysis B: Enzymatic*, **23**(1), 43–51.

Charcosset C. (2006). Membrane processes in biotechnology: an overview. *Biotechnology Advances*, **24**(5), 482–492.

Chen X., Xia X., Liang P., Cao X., Sun H. and Huang X. (2011). Stacked microbial desalination cells to enhance water desalination efficiency. *Environmental Science & Technology*, **45**(6), 2465–2470.

Chhabra M., Mishra S. and Sreekrishnan T. R. (2009). Laccase/mediator assisted degradation of triarylmethane dyes in a continuous membrane reactor. *Journal of Biotechnology*, **143**(1), 69–78.

Chung J., Li X. and Rittmann B. (2006a). Bio-reduction of arsenate using a hydrogen-based membrane biofilm reactor. *Chemosphere*, **65**(1), 24–34.

Chung J., Nerenberg R. and Rittmann B. (2006b). Bio-reduction of soluble chromate using a hydrogen-based membrane biofilm reactor. *Water Research*, **40**(8), 1634–1642.

Chung J., Nerenberg R. and Rittmann B. E. (2006c). Bioreduction of selenate using a hydrogen-based membrane biofilm reactor. *Environmental Science & Technology*, **40**, 1664–1671.

Chung J., Nerenberg R. and Rittmann B. E. (2007). Evaluation for biological reduction of nitrate and perchlorate in brine water using the hydrogen-based membrane biofilm reactor. *Journal of Environmental Engineering*, **133**(2), 157–164.

Chung J., Rittmann B. E., Her N., Lee S.-H. and Yoon Y. (2009). Integration of H2-based membrane biofilm reactor with RO and NF membranes for removal of chromate and selenate. *Water, Air, and Soil Pollution*, **207**(1–4), 29–37.

Chung T. P., Wu P. C. and Juang R. S. (2004). Process development for degradation of phenol by *Pseudomonas putida* in hollow-fiber membrane bioreactors. *Biotechnology and Bioengineering*, **87**(2), 219–227.

Chung T. P., Wu P. C. and Juang R. S. (2005). Use of microporous hollow fibers for improved biodegradation of high-strength phenol solutions. *Journal of Membrane Science*, **258**(1–2), 55–63.

Chung T. S., Loh K. C. and Goh S. K. (1998). Development of cellulose acetate membranes for bacteria immobilization to remove phenol. *Journal of Applied Polymer Science*, **68**(10), 1677–1688.

Clapp L. W., Regan J. M., Ali F., Newman J. D., Park J. K. and Noguera D. L. (1999). Activity, structure, and stratification of membrane-attached methanotrophic biofilms cometabolically degrading trichloroethylene. *Water Science and Technology*, **39**(7), 153–161.

Clauwaert P., Toledo R., Van Der Ha D., Crab R., Verstraete W., Hu H., Udert K. M. and Rabaey K. (2008). Combining biocatalyzed electrolysis with anaerobic digestion. *Water Science and Technology*, **57**(4), 575–579.

Cole A. C., Semmens M. J. and Lapara T. M. (2004). Stratification of activity and bacterial community structure in biofilms grown on membranes transferring oxygen. *Applied and Environmental Microbiology*, **70**(4), 1982–1989.

Côté P., Bersillon J.-L., Huyard A. and Faup G. (1988). Bubble-free aeration using membranes: process analysis. *Journal WPCF*, **60**(11), 1986–1992.

Cowman J., Torres C. I. and Rittmann B. E. (2005). Total nitrogen removal in an aerobic/anoxic membrane biofilm reactor system. *Water Science and Technology*, **52**(7), 115–120.

Debus O. (1995). Transport and reaction of aromatics, O2 and CO2 within a membrane bound biofilm in competition with suspended biomass. *Water Science and Technology*, **31**(1), 129–141.

Debus O. and Wanner O. (1992). Degradation of xylene by a biofilm growing on a gas-permeable membrane. *Water Science and Technology*, **26**(3–4), 607–616.

Debus O., Baumgärtl H. and Sekoulov I. (1994). Influence of fluid velocities on the degradation of volatile aromatic compounds in membrane bound biofilms. *Water Science and Technology*, **29**(10–11), 252–262.

Desloover J., Woldeyohannis A. A., Verstraete W., Boon N. and Rabaey K. (2012). Electrochemical resource recovery from digestate to prevent ammonia toxicity during anaerobic digestion. *Environmental Science & Technology*, **46** (21), 12209–12216.

Diels L., Van Roy S., Somers K., Willems I., Doyen W., Mergeay M., Springael D. and Leysen R. (1995). The use of bacteria immobilized in tubular membrane reactors for heavy metal recovery and degradation of chlorinated aromatics. *Journal of Membrane Science*, **100**(3), 249–258.

Downing L. S. and Nerenberg R. (2007). Kinetics of microbial bromate reduction in a hydrogen-oxidizing, denitrifying biofilm reactor. *Biotechnology and Bioengineering*, **98**(3), 543–550.

Downing L. S., Bibby K. J., Esposito K., Fascianella T., Tsuchihashi R. and Nerenberg R. (2010). Nitrogen removal from wastewater using a hybrid membrane-biofilm process: pilot-scale studies. *Water Environment Research*, **82**(3), 195–201.

Duran N., Rosa M. A., D'annibale A. and Gianfreda L. (2002). Applications of laccases and tyrosinases (phenoloxidases) immobilized on different supports: a review. *Enzyme and Microbial Technology*, **31**(7), 907–931.

Durante D., Casadio R., Martelli L., Tasco G., Portaccio M., De Luca P., Bencivenga U., Rossi S., Di Martino S., Grano V., Diano N. and Mita D. G. (2004). Isothermal and non-isothermal bioreactors in the detoxification of waste waters polluted by aromatic compounds by means of immobilised laccase from *Rhus vernicifera*. *Journal of Molecular Catalysis B: Enzymatic*, **27**(4–6), 191–206.

Edstrom J. A., Semmens M., Hozalski R. M., Clapp L. W. and Novak P. J. (2005). Stimulation of dechlorination by membrane-delivered hydrogen: small field demonstration. *Environmental Engineering Science*, **22**(3), 281–293.

Edward V. A., Pillay V. L., Swart P., Jacobs E. and Singh S. (2003). Degradation of synthetic xylan effluent using a membrane bioreactor. *South African Journal of Science*, **99**(7–8), 315–317.

Edwards W., Bownes R., Leukes W. D., Jacobs E. P., Sanderson R., Rose P. D. and Burton S. G. (1999a). A capillary membrane bioreactor using immobilized polyphenol oxidase for the removal of phenols from industrial effluents. *Enzyme and Microbial Technology*, **24**(3–4), 209–217.

Edwards W., Leukes W. D., Rose P. D. and Burton S. G. (1999b). Immobilization of polyphenol oxidase on chitosan-coated polysulphone capillary membranes for improved phenolic effluent bioremediation. *Enzyme and Microbial Technology*, **25**(8–9), 769–773.

Edwards W., Leukes W. D. and Bezuidenhout J. J. (2002). Ultrafiltration of petrochemical industrial wastewater using immobilised manganese peroxidase and laccase: application in the defouling of polysulphone membranes. *Desalination*, **149**(1–3), 275–278.

Emanuelsson E. a. C. and Livingston A. G. (2002). Study of membrane attached biofilm performance with nitrate as electron acceptor. *Desalination*, **149**(1–3), 211–215.

Emanuelsson E. a. C., Arcangeli J. P. and Livingston A. G. (2003). The anoxic extractive membrane bioreactor. *Water Research*, **37**(6), 1231–1238.

Erhan E., Keskinler B., Akay G. and Algur O. F. (2002). Removal of phenol from water by membrane-immobilized enzymes: Part I. *Dead-end filtration. Journal of Membrane Science*, **206**(1–2), 361–373.

Ettwig K. F., Shima S., Van De Pas-Schoonen K. T., Kahnt J., Medema M. H., Op Den Camp H. J., Jetten M. S. and Strous M. (2008). Denitrifying bacteria anaerobically oxidize methane in the absence of Archaea. *Environmental Microbiology*, **10**(11), 3164–3173.

Flock C., Bassi A. and Gijzen M. (1999). Removal of aqueous phenol and 2-chlorophenol with purified soybean peroxidase and raw soybean hulls. *Journal of Chemical Technology & Biotechnology*, **74**(4), 303–309.

Freitas Dos Santos L. M. and Livingston A. G. (1995a). Membrane-attached biofilms for VOC wastewater treatment. II: Effect of biofilm thickness on performance. *Biotechnology and Bioengineering*, **47**(1), 90–95.

Freitas Dos Santos L. M. and Livingston A. G. (1995b). Novel membrane bioreactor for detoxification of VOC wastewaters: Biodegradation of 1,2-dichloroethane. *Water Research*, **29**(1), 179–194.

Gallifuoco A., Alfani F., Cantarella M. and Viparelli P. (2001). Studying enzyme-catalyzed depolymerizations in continuous reactors. *Industrial & Engineering Chemistry Research*, **40**(23), 5184–5190.

Gil G.-C., Chang I.-S., Kim B. H., Kim M., Jang J.-K., Park H. S. and Kim H. J. (2003). Operational parameters affecting the performannce of a mediator-less microbial fuel cell. *Biosensors and Bioelectronics*, **18**(4), 327–334.

Giorno L., D'amore E., Mazzei R., Piacentini E., Zhang J., Drioli E., Cassano R. and Picci N. (2007). An innovative approach to improve the performance of a two separate phase enzyme membrane reactor by immobilizing lipase in presence of emulsion. *Journal of Membrane Science*, **295**(1–2), 95–101.

Giorno L., De Bartolo L. and Drioli E. (2010). Membrane bioreactors. In: Comprehensive Biotechnology, E. Drioli and L. Giorno (eds), 2nd edn, Academic Press, Burlington, Country, pp. 263–288.

Giorno L. and Drioli E. (2000). Biocatalytic membrane reactors: applications and perspectives. *Trends in Biotechnology*, **18**(8), 339–349.

Godjevargova T., Ivanova D., Aleksieva Z. and Burdelova G. (2006). Biodegradation of phenol by immobilized *Trichosporon cutaneum* R57 on modified polymer membranes. *Process Biochemistry*, **41**(11), 2342–2346.

Gómez M., Murcia M. D., Ortega S., Barbosa D. S., Vayá G. and Hidalgo A. M. (2012). Removal of 4-chlorophenol in a continuous membrane bioreactor using different commercial peroxidases. *Desalination and Water Treatment*, **37** (1–3), 97–107.

Gong Z., Liu S., Yang F., Bao H. and Furukawa K. (2008). Characterization of functional microbial community in a membrane-aerated biofilm reactor operated for completely autotrophic nitrogen removal. *Bioresource Technology*, **99**(8), 2749–2756.

Hai F. I. and Modin O. (2013). Application of membrane technology for the development of sustainable biotechnology processes. In: Membrane Processes for Sustainable Growth, A. Basile and A. Cassano (eds), Nova Science Publishers Inc., NY, Country.

Hai F. I. and Yamamoto K. (2011). Membrane biological reactors. In: Treatise on Water Science, P. Wilderer (ed.), Elsevier, Oxford, Country, pp. 571–613.

Hai F. I., Nghiem L. D. and Modin O. (2013). Biocatalytic membrane reactors: removal of recalcitrant and emerging pollutants In: Handbook of Membrane Reactors, A. Basile (ed.), Woodhead Publishing, Cambridge, Country, Vol. **2**.

Haroon M. F., Hu S., Shi Y., Imelfort M., Keller J., Hugenholtz P., Yuan Z. and Tyson G. W. (2013). Anaerobic oxidation of methane coupled to nitrate reduction in a novel archaeal lineage. *Nature*, **500**(7464), 567–570.

Hata T., Shintate H., Kawai S., Okamura H. and Nishida T. (2010). Elimination of carbamazepine by repeated treatment with laccase in the presence of 1-hydroxybenzotriazole. *Journal of Hazardous Materials*, **181**(1–3), 1175–1178.

Heffernan B., Murphy C. D., Syron E. and Casey E. (2009). Treatment of fluoroacetate by a *Pseudomonas fluorescens* biofilm grown in membrane aerated biofilm reactor. *Environmental Science & Technology*, **43**, 6776–6785.

Hibiya K., Terada A., Tsuneda S. and Hirata A. (2003). Simultaneous nitrification and denitrification by controlling vertical and horizontal microenvironment in a membrane-aerated biofilm reactor. *Journal of Biotechnology*, **100**, 23–32.

Hirasa O., Ichijo H. and Yamauchi A. (1991). Preparation of new support for immobilization of activated sludge. *Journal of Fermentation and Bioengineering*, **71**(5), 376–378.

Hoehn M. and Bornscheuer U. T. (2009). Biocatalytic routes to optically active amines. *Chemcatchem*, **1**(1), 42–51.

Hsieh Y. L., Tseng S. K. and Chang Y. J. (2002). Nitrification using polyvinyl alcohol-immobilized nitrifying biofilm on an O2-enriching membrane. *Biotechnology Letters*, **24**, 315–319.

Hu S., Yang F., Sun C., Zhang J. and Wang T. (2008). Simultaneous removal of COD and nitrogen using a novel carbon-membrane aerated biofilm reactor. *Journal of Environmental Sciences*, **20**(2), 142–148.

Hwang J. H., Cicek N. and Oleszkiewicz J. A. (2009). Membrane biofilm reactors for nitrogen removal: state-of-the-art and research needs. *Water Science and Technology*, **60**(11), 2739–2747.

Illanes A., Ruiz A., Zuniga M. E., Aguirre C., O'reilly S. and Curotto E. (1990). Immobilization of lactase for the continuous hydrolysis of whey permaete. *Bioprocess Engineering*, **5**(6), 257–262.

Jochems P., Satyawali Y., Diels L. and Dejonghe W. (2011). Enzyme immobilization on/in polymeric membranes: Status, challenges and perspectives in biocatalytic membrane reactors (BMRs). *Green Chemistry*, **13**(7), 1609–1623.

Jolivalt C., Brenon S., Caminade E., Mougin C. and Pontié M. (2000). Immobilization of laccase from *Trametes versicolor* on a modified PVDF microfiltration membrane: characterization of the grafted support and application in removing a phenylurea pesticide in wastewater. *Journal of Membrane Science*, **180**(1), 103–113.

Jones R. J., Massanet-Nicolau J., Guwy A., Premier G. C., Dinsdale R. M. and Reilly M. (2015). Removal and recovery of inhibitory volatile fatty acids from mixed acid fermentations by conventional electrodialysis. *Bioresource Technology*, **189**, 279–284.

Jornitz M., Soelkner P. and Meltzer T. (2002). Sterile filtration – a review of the past and present technologies. *PDA Journal of Pharmaceutical Science and Technology*, **56**(4), 192–195.

Juang R. S. and Huang W. C. (2008). Use of membrane contactors as two-phase bioreactors for the removal of phenol in saline and acidic solutions. *Journal of Membrane Science*, **313**(1–2), 207–216.

Juang R. S. and Kao H. C. (2009). Estimation of the contribution of immobilized biofilm and suspended biomass to the biodegradation of phenol in membrane contactors. *Biochemical Engineering Journal*, **43**(2), 122–128.

Juang R. S. and Tsai S. Y. (2006). Role of membrane-attached biofilm in the biodegradation of phenol and sodium salicylate in microporous membrane bioreactors. *Journal of Membrane Science*, **282**(1–2), 484–492.

Juang R. S. and Wu C. Y. (2007). Microbial degradation of phenol in high-salinity solutions in suspensions and hollow fiber membrane contactors. *Chemosphere*, **66**(1), 191–198.

Juang R. S., Chung T. P., Wang M. L. and Lee D. J. (2008). Experimental observations on the effect of added dispersing agent on phenol biodegradation in a microporous membrane bioreactor. *Journal of Hazardous Materials*, **151**(2–3), 746–752.

Juang R. S., Huang W. C. and Hsu Y. H. (2009). Treatment of phenol in synthetic saline wastewater by solvent extraction and two-phase membrane biodegradation. *Journal of Hazardous Materials*, **164**(1), 46–52.

Katuri K. P., Venkata Mohan S., Sridhar S., Pati B. R. and Sarma P. N. (2009). Laccase-membrane reactors for decolorization of an acid azo dye in aqueous phase: process optimization. *Water Research*, **43**(15), 3647–3658.

Kim B. H., Chang I. S., Gil G. C., Park H. S. and Kim H. J. (2003). Novel BOD (biological oxygen demand) sensor using mediator-less microbial fuel cell. *Biotechnology Letters*, **25**, 541–545.

Kim J., Song I., Lee S., Kim P., Oh H., Park J. and Choung Y. (2010). Decomposition of pharmaceuticals (sulfamethazine and sulfathiazole) using oxygen-based membrane biofilm reactor. *Desalination*, **250**(2), 751–756.

Kim Y. and Logan B. E. (2011a). Microbial reverse electrodialysis cells for synergistically enhanced power production. *Environmental Science & Technology*, **45**(13), 5834–5839.

Kim Y. and Logan B. E. (2011b). Hydrogen production from inexhaustible supplies of fresh and salt water using microbial reverse-electrodialysis electrolysis cells. *Proceedings of the National Academy of Sciences*, **108**(39), 16176–16181.

Kinh C. T., Riya S., Hosomi M. and Terada A. (2017). Identification of hotspots for NO and N_2O production and consumption in counter- and co-diffusion biofilms for simultaneous nitrification and denitrification. *Bioresource Technology*, **245**, 318–324.

Kolb F. R. and Wilderer P. A. (1995). Activated carbon membrane biofilm reactor for the degradation of volatile organic pollutants. *Water Science and Technology*, **31**(1), 205–213.

Konovalova V. V., Dmytrenko G. M., Nigmatullin R. R., Bryk M. T. and Gvozdyak P. I. (2003). Chromium(VI) reduction in a membrane bioreactor with immobilized *Pseudomonas* cells. *Enzyme and Microbial Technology*, **33**(7), 899–907.

Kowalska M., Bodzek M. and Bohdziewicz J. (1998). Biodegradation of phenols and cyanides using membranes with immobilized microorganisms. *Process Biochemistry*, **33**(2), 189–197.

Kuntke P., Smiech K. M., Bruning H., Zeeman G., Saakes M., Sleutels T. H., Hamelers H. V. and Buisman C. J. (2012). Ammonium recovery and energy production from urine by a microbial fuel cell. *Water Research*, **46**(8), 2627–36.

Kuntke P., Sleutels T., Rodriguez Arredondo M., Georg S., Barbosa S. G., Ter Heijne A., Hamelers H. V. M. and Buisman C. J. N. (2018). (Bio)electrochemical ammonia recovery: progress and perspectives. *Applied Microbiology and Biotechnology*, **102**(9), 3865–3878.

Lackner S., Terada A., Horn H., Henze M. and Smets B. F. (2010). Nitritation performance in membrane-aerated biofilm reactors differs from conventional biofilm systems. *Water Research*, **44**(20), 6073–6084.

Landes N. C., Jackson W. A. and Morse A. N. (2011). Limitations encountered for the treatment of a low C:N waste using a modified membrane-aerated biofilm reactor. *Water Environment Research*, **83**(2), 128–139.

Lante A., Crapisi A., Krastanov A. and Spettoli P. (2000). Biodegradation of phenols by laccase immobilised in a membrane reactor. *Process Biochemistry*, **36**(1–2), 51–58.

Lapara T. M., Cole A. C., Shanahan J. W. and Semmens M. J. (2006). The effects of organic carbon, ammoniacal-nitrogen, and oxygen partial pressure on the stratification of membrane-aerated biofilms. *Journal of Industrial Microbiology & Biotechnology*, **33**(4), 315–323.

Lee K.-C. and Rittmann B. E. (2000). A novel hollow-fibre membrane biofilm reactor for autohydrogenotrophic denitrification of drinking water. *Water Science and Technology*, **41**(4–5), 219–226.

Li Y. and Loh K. C. (2005). Cometabolic transformation of high concentrations of 4-chlorophenol in an immobilized cell hollow fiber membrane bioreactor. *Journal of Environmental Engineering*, **131**(9), 1285–1292.

Li Y. and Loh K. C. (2006a). Continuous cometabolic transformation of 4-chlorophenol in the presence of phenol in a hollow fiber membrane bioreactor. *Journal of Environmental Engineering*, **132**(3), 309–314.

Li Y. and Loh K. C. (2006b). Activated carbon impregnated polysulfone hollow fiber membrane for cell immobilization and cometabolic biotransformation of 4-chlorophenol in the presence of phenol. *Journal of Membrane Science*, **276** (1–2), 81–90.

Li Y. and Loh K.-C. (2007a). Hybrid-hollow-fiber membrane bioreactor for cometabolic transformation of 4-chlorophenol in the presence of phenol. *Journal of Environmental Engineering*, **133**(4), 404–410.

Li Y. and Loh K. C. (2007b). Continuous phenol biodegradation at high concentrations in an immobilized-cell hollow fiber membrane bioreactor. *Journal of Applied Polymer Science*, **105**(4), 1732–1739.

Li Y. and Wang C. (2008). Phenol biodegradation in hybrid hollow-fiber membrane bioreactors. *World Journal of Microbiology and Biotechnology*, **24**(9), 1843–1849.

Li T., Bai R. and Liu J. (2008a). Distribution and composition of extracellular polymeric substances in membrane-aerated biofilm. *Journal of Biotechnology*, **135**(1), 52–57.

Li T., Liu J., Bai R. and Wong F. S. (2008b). Membrane-aerated biofilm reactor for the treatment of acetonitrile wastewater. *Environmental Science & Technology*, **42**, 2099–2104.

Li T., Bai R., Ohandja D.-G. and Liu J. (2009). Biodegradation of acetonitrile by adapted biofilm in a membrane-aerated biofilm reactor. *Biodegradation*, **20**(4), 569–580.

Liao B. Q. and Liss S. N. (2007). A comparative study between thermophilic and mesophilic membrane aerated biofilm reactors. *Journal of Environmental Engineering and Science*, **6**(2), 247–252.

Liu H., Ramnarayanan R. and Logan B. E. (2004). Production of electricity during wastewater treatment using a single chamber microbial fuel cell. *Environmental Science & Technology*, **38**, 2281–2285.

Liu H., Yang F., Shi S. and Liu X. (2010). Effect of substrate COD/N ratio on performance and microbial community structure of a membrane aerated biofilm reactor. *Journal of Environmental Sciences*, **22**(4), 540–546.

Livingston A. G. (1994). Extractive membrane bioreactors: a new process technology for detoxifying chemical industry wastewaters. *Journal of Chemical Technology and Biotechnology*, **60**(2), 117–124.

Lloret L., Eibes G., Feijoo G., Moreira M. T. and Lema J. M. (2012a). Continuous biotransformation of estrogens by laccase in an enzymatic membrane reactor. *Chemical Engineering Transactions*, **27**, 31–36.

Lloret L., Eibes G., Feijoo G., Moreira M. T. and Lema J. M. (2012b). Degradation of estrogens by laccase from *Myceliophthora thermophila* in fed-batch and enzymatic membrane reactors. *Journal of Hazardous Materials*, 213–214, 175–183.

Logan B. E. and Rabaey K. (2012). Conversion of wastes into bioelectricity and chemicals by using microbial electrochemical technologies. *Science*, **337**(6095), 686–690.

Logan B. E., Hamelers H. V. M., Rozendal R. A., Schröder U., Keller J., Freguia S., Aelterman P., Verstraete W. and Rabaey K. (2006). Microbial fuel cell: methodology and technology. *Environmental Science & Technology*, **40** (17), 5181–5192.

Logan B. E., Call D. F., Cheng S., Hamelers H. V. M., Sleutels T. H. J. A., Jeremiasse A. W. and Rozendal R. A. (2008). Microbial electrolysis cells for high yield hydrogen gas production from organic matter. *Environmental Science & Technology*, **42**(23), 8630–8640.

Loh K. C., Chung T. S. and Ang W. F. (1999). Immobilized-cell membrane bioreactor for high-strength phenol wastewater. *Journal of Environmental Engineering*, **126**(1), 75–79.

López C., Mielgo I., Moreira M. T., Feijoo G. and Lema J. M. (2002). Enzymatic membrane reactors for biodegradation of recalcitrant compounds. Application to dye decolourisation. *Journal of Biotechnology*, **99**(3), 249–257.

López C., Moreira M. T., Feijoo G. and Lema J. M. (2004). Dye decolorization by manganese peroxidase in an enzymatic membrane bioreactor. *Biotechnology Progress*, **20**(1), 74–81.

López C., Moreira M. T., Feijoo G. and Lema J. M. (2011). Economic comparison of enzymatic reactors and advanced oxidation processes applied to the degradation of phenol as a model compound. *Biocatalysis and Biotransformation*, **29**(6), 344–353.

Luo J. H., Wu M., Yuan Z. and Guo J. (2017). Biological bromate reduction driven by methane in a membrane biofilm reactor. *Environmental Science and Technology Letters*, **4**(12), 562–566.

Luo J. H., Chen H., Yuan Z. and Guo J. (2018). Methane-supported nitrate removal from groundwater in a membrane biofilm reactor. *Water Research*, **132**, 71–78.

Lv P. L., Zhong L., Dong Q. Y., Yang S. L., Shen W. W., Zhu Q. S., Lai C. Y., Luo A. C., Tang Y. and Zhao H. P. (2018). The effect of electron competition on chromate reduction using methane as electron donor. *Environmental Science and Pollution Research*, **25**(7), 6609–6618.

Ma Y., Domingo-Félez C., Plósz B. G. and Smets B. F. (2017). Intermittent aeration suppresses nitrite-oxidizing bacteria in membrane-aerated biofilms: a model-based explanation. *Environmental Science and Technology*, **51**(11), 6146–6155.

Madaeni S. S. and Zarbakhsh M. (2009). Removal of organics and ions from wastewater using a hybrid system of microfiltration membrane and microorganisms. *Engineering in Life Sciences*, **9**(4), 324–330.

Majeau J. A., Brar S. K. and Tyagi R. D. (2010). Laccases for removal of recalcitrant and emerging pollutants. *Bioresource Technology*, **101**(7), 2331–2350.

Martin K. J. and Nerenberg R. (2012). The membrane biofilm reactor (MBfR) for water and wastewater treatment: principles, applications, and recent developments. *Bioresource Technology*, **122**, 83–94.

Martin K. J., Downing L. S. and Nerenberg R. (2009). Evidence of specialized bromate-reducing bacteria in a hollow fiber membrane biofilm reactor. *Water Science and Technology*, **59**(10), 1969–1974.

Mathers M. (2000). Novel membrane bioreactor converts waste to commodity. *Filtration and Separation*, **37**(7), 30–32.

Mazzei R., Drioli E. and Giorno L. (2010). Biocatalytic membranes and membrane bioreactors. In: Comprehensive Membrane Science and Engineering E. Drioli and L. Giorno (ed.), Elsevier Oxford, Country, 195–212.

Mehanna M., Saito T., Yan J., Hickner M., Cao X., Huang X. and Logan B. E. (2010). Using microbial desalination cells to reduce water salinity prior to reverse osmosis. *Energy & Environmental Science*, **3**(8), 1114.

Mehanna M., Kiely P. D., Call D. F. and Logan B. E. (2011). Microbial electrodialysis cell for simultaneous water desalination and hydrogen gas production. *Environmental Science & Technology*, **44**, 9578–9583.

Mehdizadeh S. N., Mehrnia M. R., Abdi K. and Sarrafzadeh M. H. (2011). Biological treatment of toluene contaminated wastewater by *Alcaligenes faecalis* in an extractive membrane bioreactor; experiments and modeling. *Water Science and Technology*, **64**(6), 1239–1246.

Mendoza L., Jonstrup M., Hatti-Kaul R. and Mattiasson B. (2011). Azo dye decolorization by a laccase/mediator system in a membrane reactor: enzyme and mediator reusability. *Enzyme and Microbial Technology*, **49**(5), 478–484.

Metcalf & Eddy, Tchobanoglous G., Burton F. L. and Stensel H. D. 2004. Wastewater Engineering, Treatment and Reuse, 4th edn. McGraw-Hill, New York.

Misiak K., Casey E. and Murphy C. D. (2011). Factors influencing 4-fluorobenzoate degradation in biofilm cultures of *Pseudomonas knackmussii* B13. *Water Research*, **45**(11), 3512–3520.

Modin O. and Fukushi K. (2012). Development and testing of bioelectrochemical reactors converting wastewater organics into hydrogen peroxide. *Water Science and Technology*, **66**(4), 831–836.

Modin O. and Wilen B. M. (2012). A novel bioelectrochemical BOD sensor operating with voltage input. *Water Research*, **46**(18), 6113–6120.

Modin O., Fukushi K., Nakajima F. and Yamamoto K. (2008a). Performance of a membrane biofilm reactor for denitrification with methane. *Bioresource Technology*, **99**(17), 8054–8060.

Modin O., Fukushi K., Nakajima F. and Yamamoto K. (2008b). A membrane biofilm reactor achieves aerobic methane oxidation coupled to denitrification (AME-D) with high efficiency. *Water Science and Technology*, **58**(1), 83–87.

Modin O., Fukushi K. and Yamamoto K. (2008c). Simultaneous removal of nitrate and pesticides from groundwater using a methane-fed membrane biofilm reactor. *Water Science and Technology*, **58**(6), 1273–1279.

Modin O., Fukushi K., Nakajima F. and Yamamoto K. (2010). Nitrate removal and biofilm characteristics in methanotrophic membrane biofilm reactors with various gas supply regimes. *Water Research*, **44**(1), 85–96.

Modin O., Fukushi K., Rabaey K., Rozendal R. A. and Yamamoto K. (2011). Redistribution of wastewater alkalinity with a microbial fuel cell to support nitrification of reject water. *Water Research*, **45**(8), 2691–2699.

Modin O., Wang X., Wu X., Rauch S. and Fedje K. K. (2012). Bioelectrochemical recovery of Cu, Pb, Cd, and Zn from dilute solutions. *Journal of Hazardous Materials*, **235–236**, 291–297.

Moeder M., Martin C. and Koeller G. (2004). Degradation of hydroxylated compounds using laccase and horseradish peroxidase immobilized on microporous polypropylene hollow fiber membranes. *Journal of Membrane Science*, **245**(1–2), 183–190.

Motlagh A. R. A., Lapara T. and Semmens M. (2008). Ammonium removal in advective-flow membrane-aerated biofilm reactors (AF-MABRs). *Journal of Membrane Science*, **319**(1–2), 76–81.

Nabais A. M. A. and Cardoso J. P. (1999). Purification of benzylpenicillin filtered broths by ultrafiltration and effect on solvent extraction. *Bioprocess Engineering*, **21**(2), 157–163.

Nerenberg R. and Rittmann B. E. (2002). Perchlorate as a secondary substrate in a denitrifying, hollow-fiber membrane biofilm reactor. *Water Science and Technology: Water Supply*, **2**(2), 259–265.

Nerenberg R. and Rittmann B. E. (2004). Hydrogen-based, hollow-fiber membrane biofilm reactor for reduction of perchlorate and other oxidized contaminants. *Water Science and Technology*, **49**(11–12), 223–230.

Nerenberg R., Rittmann B. E. and Najm I. (2002). Perchlorate reduction in a hydrogen-based membrane-biofilm reactor. *AWWA Journal*, **94**(11), 103–114.

Nerenberg R., Kawagoshi Y. and Rittmann B. E. (2008). Microbial ecology of a perchlorate-reducing, hydrogen-based membrane biofilm reactor. *Water Research*, **42**(4–5), 1151–1159.

Ohandja D.-G. and Stuckey D. C. (2007). Biodegradation of PCE in a hybrid membrane aerated biofilm reactor. *Journal of Environmental Engineering*, **133**(1), 20–27.

Onishi H., Numazawa R. and Takeda H. 1980. Process and apparatus for wastewater treatment. U.S. patent application.

Oxymem. (2018) Oxymem Ltd [Online]. Ireland. https://oxymem.com/ (accessed 5 September 2018).

Pankhania M., Stephenson T. and Semmens M. (1994). Hollow fiber bioreactor for wastewater treatment using bubbleless membrane aeration. *Water Research*, **28**(10), 2233–2236.

Pankhania M., Brindle K. and Stephenson T. (1999). Membrane aeration bioreactors for wastewater treatment: completely mixed and plug-flow operation. *Chemical Engineering Journal*, **73**, 131–136.

Peeters J., Vicevic G., Koops G. H. and Côté P. (2016). The role of innovative technologies in achieving energy-neutral wastewater treatment. *Water Practice and Technology*, **11**(4), 691–701.

Pellicer-Nacher C., Sun S.-P., Lackner S., Terada A., Schreiber F., Zhou Q. and Smets B. F. (2010). Sequential aeration of membrane-aerated biofilm reactors for high rate autotrophic nitrogen removal: experimental demonstration. *Environmental Science & Technology*, **44**, 7628–7634.

Perez-Calleja P., Aybar M., Picioreanu C., Esteban-Garcia A. L., Martin K. J. and Nerenberg R. (2017). Periodic venting of MABR lumen allows high removal rates and high gas-transfer efficiencies. *Water Research*, **121**, 349–360.

Pronk W., Kerkhof P. J. a. M., Van Helden C. and Van't Riet K. (1988). Hydrolysis of triglycerides by immobilized lipase in a hydrophilic membrane reactor. *Biotechnology and Bioengineering*, **32**(4), 512–518.

Rabaey K., Rodríguez J., Blackall L. L., Keller J., Gross P., Batstone D., Verstraete W. and Nealson K. H. (2007). Microbial ecology meets electrochemistry: electricity-driven and driving communities. *The ISME Journal*, **1**(1), 9–18.

Rabaey K., Bützer S., Brown S., Keller J. and Rozendal R. A. (2010). High current generation coupled to caustic production using a lamellar bioelectrochemical system. *Environmental Science & Technology*, **44**, 4315–4321.

Raghoebarsing A. A., Pol A., Van De Pas-Schoonen K. T., Smolders A. J., Ettwig K. F., Rijpstra W. I., Schouten S., Damste J. S., Op Den Camp H. J., Jetten M. S. and Strous M. (2006). A microbial consortium couples anaerobic methane oxidation to denitrification. *Nature*, **440**(7086), 918–921.

Rathore A. S. and Shirke A. (2011). Recent developments in membrane-based separations in biotechnology processes: Review. *Preparative Biochemistry and Biotechnology*, **41**(4), 398–421.

Rios G. M., Belleville M. P., Paolucci D. and Sanchez J. (2004). Progress in enzymatic membrane reactors – a review. *Journal of Membrane Science*, **242**(1–2), 189–196.

Rishell S., Casey E., Glennon B. and Hamer G. (2004). Characteristics of a methanotrophic culture in a membrane-aerated biofilm reactor. *Biotechnology Progress*, **20**(4), 1082–1090.

Rittmann B. E. (2006). The membrane biofilm reactor: the natural partnership of membranes and biofilm. *Water Science and Technology*, **53**(3), 219–225.

Rittmann B. E. (2018). Biofilms, active substrata, and me. *Water Research*, **132**, 135–145.

Rittmann B. E., Nerenberg R., Lee K. C., Najm I., Gillogly T. E., Lehman G. E. and Adham S. S. (2004). Hydrogen-based hollow-fiber membrane biofilm reactor (MBfR) for removing oxidized contaminants. *Water Science and Technology*, **4**(1), 127–133.

Rothemund C., Camper A. and Wilderer P. A. (1994). Biofilms growing on gas permeable membranes. *Water Science and Technology*, **29**(10–11), 447–454.

Rozendal R. A., Hamelers H. V. M. and Buisman C. J. N. (2006). Effects of membrane cation transport on pH and microbial fuel cell performance. *Environmental Science & Technology*, **40**, 5206–5211.

Rozendal R. A., Sleutels T. H., Hamelers H. V. and Buisman C. J. (2008). Effect of the type of ion exchange membrane on performance, ion transport, and pH in biocatalyzed electrolysis of wastewater. *Water Science and Technology*, **57**(11), 1757–1762.

Rozendal R. A., Leone E., Keller J. and Rabaey K. (2009). Efficient hydrogen peroxide generation from organic matter in a bioelectrochemical system. *Electrochemistry Communications*, **11**(9), 1752–1755.

Sahu A. K., Conneely T., Nüsslein K. and Ergas S. J. (2009). Hydrogenotrophic denitrification and perchlorate reduction in ion exchange brines using membrane biofilm reactors. *Biotechnology and Bioengineering*, **104**(3), 483–491.

Sanaeepur H., Hosseinkhani O., Kargari A., Ebadi Amooghin A. and Raisi A. (2012). Mathematical modeling of a time-dependent extractive membrane bioreactor for denitrification of drinking water. *Desalination*, **289**(0), 58–65.

Satoh H., Ono H., Rulin B., Kamo J., Okabe S. and Fukushi K.-I. (2004). Macroscale and microscale analyses of nitrification and denitrification in biofilms attached on membrane aerated biofilm reactors. *Water Research*, **38** (6), 1633–1641.

Saxena A., Tripathi B. P., Kumar M. and Shahi V. K. (2009). Membrane-based techniques for the separation and purification of proteins: an overview. *Advances in Colloid and Interface Science,* **145**(1–2), 1–22.

Schaffer R. B., Ludzack F. J. and Ettinger M. B. (1960). Sewage treatment by oxygenation through permeable plastic films. *Journal WPCF*, **32**(9), 939–941.

Schroen C. G. P. H., Van Roon J. L., Beefink H. H., Tramper J. and Boom R. M. (2009). Membrane applications for antibiotics production. *Desalination*, **236**(1–3), 78–84.

Semmens M. J. and Essila N. J. (2001). Modeling biofilms on gas-permeable supports: flux limitations. *Journal of Environmental Engineering*, **127**(2), 126–133.

Semmens M., Dahm K., Shanahan J. and Christianson A. (2003). COD and nitrogen removal by biofilms growing on gas permeable membranes. *Water Research*, **37**(18), 4343–4350.

Shi Y., Hu S., Lou J., Lu P., Keller J. and Yuan Z. (2013). Nitrogen removal from wastewater by coupling anammox and methane-dependent denitrification in a membrane biofilm reactor. *Environmental Science & Technology*, **47**(20), 11577–11583.

Shi Y. F., Jin F. X. and Wu Y. Y. (1997). Microfiltration membrane bioreactor in stirred backflush operation for biotransformation using intact cells. *Process Biochemistry*, **32**(5), 387–390.

Shin J., Sang B., Chung Y. and Choung Y. (2005). The removal of nitrogen using an autotrophic hybrid hollow-fiber membrane biofilm reactor. *Desalination*, **183**(1–3), 447–454.

Silman R. W., Black L. T., Mcghee J. E. and Bangley E. B. (1980). Hydrolysis of raffinose in a hollow-fiber reactor using an unrefined mixture of α-galactosidase and invertase. *Biotechnology and Bioengineering*, **22**(3), 533–541.

Smith D. P., Rector T., Reid-Black K., Hummerick M., Strayer R., Birmele M., Roberts M. S. and Garland J. L. (2008). Redox control bioreactor: a unique biological water processor. *Biotechnology and Bioengineering*, **99**(4), 830–845.

Stricker A.-E., Lossing H., Gibson J. H., Hong Y. and Urbanic J. C. (2011). Pilot scale testing of a new configuration of the membrane aerated biofilm reactor (MABR) to treat high-strength industrial sewage. *Water Environment Research*, **83**(1), 3–14.

Suzuki Y., Miyahara S. and Takeishi K. (1993). Oxygen supply method using gas-permeable film for wastewater treatment. *Water Science and Technology*, **28**(7), 243–250.

Syron E. and Casey E. (2008a). Model-based comparative performance analysis of membrane aerated biofilm reactor configurations. *Biotechnology and Bioengineering*, **99**(6), 1361–1373.

Syron E. and Casey E. (2008b). Membrane-aerated biofilms for high rate biotreatment: performance appraisal, engineering principles, scale-up, and development requirements. *Environmental Science & Technology*, **42**(6), 1833–1844.

Syron E., Semmens M. J. and Casey E. (2015). Performance analysis of a pilot-scale membrane aerated biofilm reactor for the treatment of landfill leachate. *Chemical Engineering Journal*, **273**, 120–129.

Tao B., Passanha P., Kumi P., Wilson V., Jones D. and Esteves S. (2016). Recovery and concentration of thermally hydrolysed waste activated sludge derived volatile fatty acids and nutrients by microfiltration, electrodialysis and struvite precipitation for polyhydroxyalkanoates production. *Chemical Engineering Journal*, **295**, 11–19.

Ter Heijne A., Liu F., Van Der Weijden R., Weijma J., Buisman C. J. N. and Hamelers H. V.M. (2010). Copper reduction combined with electricity production in a microbial fuel cell. *Environmental Science & Technology*, **44**(11), 4376–4381.

Terada A., Hibiya K., Nagai J., Tsuneda S. and Hirata A. (2003). Nitrogen removal characteristics and biofilm analysis of a membrane-aerated biofilm reactor applicable to high strength nitrogenous wastewater treatment. *Journal of Bioscience and Bioengineering*, **95**(2), 170–178.

Terada A., Yamamoto T., Tsuneda S. and Hirata A. (2004). Enhancement of biofilm formation onto surface-modified hollow-fiber membranes and its application to a membrane-aerated biofilm reactor. *Water Science and Technology*, **49**(11–12), 263–268.

Terada A., Yamamoto T., Tsuneda S. and Hirata A. (2006). Sequencing batch membrane biofilm reactor for simultaneous nitrogen and phosphorus removal: novel application of membrane-aerated biofilm. *Biotechnology and Bioengineering*, **94**(4), 730–739.

Terada A., Lackner S., Tsuneda S. and Smets B. F. (2007). Redox-stratification controlled biofilm (ReSCoBi) for completely autotrophic nitrogen removal: the effect of co- versus counter-diffusion on reactor performance. *Biotechnology and Bioengineering*, **97**(1), 40–51.

Timberlake D. L., Strand S. E. and Williamson K. J. (1988). Combined aerobic heterotrophic oxidation, nitrification and denitrification in a permeable-support biofilm. *Water Research*, **22**(12), 1513–1517.

Walter B., Haase C. and Räbiger N. (2005). Combined nitrification/denitrification in a membrane reactor. *Water Research*, **39**(13), 2781–2788.

Van Ginkel S. W., Ahn C. H., Badruzzaman M., Roberts D. J., Lehman S. G., Adham S. S. and Rittmann B. E. (2008). Kinetics of nitrate and perchlorate reduction in ion-exchange brine using the membrane biofilm reactor (MBfR). *Water Research*, **42**(15), 4197–4205.

Van Reis R. and Zydney A. (2007). Bioprocess membrane technology. *Journal of Membrane Science*, **297**(1–2), 16–50.

Wang C. and Li Y. (2007). Incorporation of granular activated carbon in an immobilized membrane bioreactor for the biodegradation of phenol by *Pseudomonas putida*. *Biotechnology Letters*, **29**(9), 1353–1356.

Wang R., Terada A., Lackner S., Smets B. F., Henze M., Xia S. and Zhao J. (2009). Nitritation performance and biofilm development of co- and counter-diffusion biofilm reactors: modeling and experimental comparison. *Water Research*, **43**(10), 2699–2709.

Wang R., Zhan X., Zhang Y. and Zhao J. (2011). Nitrifying population dynamics in a redox stratified membrane biofilm reactor (RSMBR) for treating ammonium-rich wastewater. *Frontiers of Environmental Science & Engineering in China*, **5**(1), 48–56.

Vasileva N., Godjevargova T., Ivanova D. and Gabrovska K. (2009). Application of immobilized horseradish peroxidase onto modified acrylonitrile copolymer membrane in removing of phenol from water. *International Journal of Biological Macromolecules*, **44**(2), 190–194.

Visvanathan C., Phong D. D. and Jegatheesan V. (2008). Hydrogenotrophic denitrification of highly saline aquaculture wastewater using hollow fiber membrane bioreactor. *Environmental Technology*, **29**(6), 701–707.

Wei X., Li B., Zhao S., Qiang C., Zhang H. and Wang S. (2012). COD and nitrogen removal in facilitated transfer membrane-aerated biofilm reactor (FT-MABR). *Journal of Membrane Science*, **389**, 257–264.

Wilderer P. A., Bräutigam J. and Sekoulov I. (1985). Application of gas permeable membranes for auxiliary oxygenation of sequencing batch reactors. *Conservation & Recycling*, **8**(1/2), 181–192.

Wobus A. and Röske I. (2000). Reactors with membrane-grown biofilms: Their capacity to cope with fluctuating inflow conditions and with shock loads of xenobiotics. *Water Research*, **34**(1), 279–287.

Wobus A., Ulrich S. and Röske I. (1995). Degradation of chlorophenols by biofilm on semi-permeable membranes in two types of fixed bed reactors. *Water Science and Technology*, **32**(8), 205–212.

Woolard C. R. and Irvine R. L. (1994). Biological treatment of hypersaline wastewater by a biofilm of halophilic bacteria. *Water Environment Research*, **66**(3), 230–235.

Woolard C. R. and Irvine R. L. (1995). Response of periodically operated halophilic biofilm reactor to changes in salt concentration. *Water Science and Technology*, **31**(1), 41–50.

Wu C. Y., Ushiwaka S., Horii H. and Yamagiwa K. (2006). Boosting nitrification by membrane-attached biofilm. *Water Science and Technology*, **54**(9), 121–128.

Wu X. and Modin O. (2013). Ammonium recovery from reject water combined with hydrogen production in a bioelectrochemical reactor. *Bioresource Technology*, **146**, 530–536.

Xie G. J., Cai C., Hu S. and Yuan Z. (2017). Complete nitrogen removal from synthetic anaerobic sludge digestion liquor through integrating anammox and denitrifying anaerobic methane oxidation in a membrane biofilm reactor. *Environmental Science and Technology*, **51**(2), 819–827.

Yamagiwa K., Yoshida M., Ito A. and Ohkawa A. (1998). A new oxygen supply method for simultaneous organic carbon removal and nitrification by a one-stage biofilm process. *Water Science and Technology*, **37**(4–5), 117–124.

Yeh S.-J. and Jenkins C. R. (1978). Pure oxygen fixed film reactor. *Journal of the Environmental Engineering Division*, **4**, 611–623.

Zamora P., Georgieva T., Ter Heijne A., Sleutels T. H. J. A., Jeremiasse A. W., Saakes M., Buisman C. J. N. and Kuntke P. (2017). Ammonia recovery from urine in a scaled-up microbial electrolysis cell. *Journal of Power Sources*, **356**, 491–499.

Zhang L., Liang S. S., Hellgren L. I., Jonsson G. E. and Xu X. B. (2008). Phospholipase C-catalyzed sphingomyelin hydrolysis in a membrane reactor for ceramide production. *Journal of Membrane Science*, **325**(2), 895–902.

Zhu G., Chung T. S. and Loh K. C. (2000). Activated carbon-filled cellulose acetate hollow-fiber membrane for cell immobilization and phenol degradation. *Journal of Applied Polymer Science*, **76**(5), 695–707.

Ziv-El M. C. and Rittmann B. E. (2009a). Water quality assessment of groundwater treated with a membrane biofilm reactor. *AWWA Journal*, **101**(12), 77–83.

Ziv-El M. C. and Rittmann B. E. (2009b). Systematic evaluation of nitrate and perchlorate bioreduction kinetics in groundwater using a hydrogen-based membrane biofilm reactor. *Water Research*, **43**(1), 173–181.

Chapter 11

Anaerobic MBRs

Weijue Gao[1], Bao-Qiang Liao[1] and Faisal I. Hai[2]*

[1]Department of Chemical Engineering, Lakehead University, 955 Oliver Road, Thunder Bay, Ontario, Canada P7B 5E1

[2]Strategic Water Infrastructure Laboratory, The University of Wollongong, Northfields Ave, NSW 2522, Australia

**Corresponding author: bliao@lakeheadu.ca*

ABSTRACT

Anaerobic membrane bioreactor (AnMBR) has been shown to be one of the most promising technologies for sustainable wastewater treatment. It offers several advantages over aerobic MBRs, namely, requirement of no energy for aeration, low sludge production and energy resource production, and low nutrient demand. Compared to conventional anaerobic processes, AnMBRs can maintain higher biomass concentrations, have higher treatment capacity, excellent effluent quality, and smaller footprint. Compared with conventional anaerobic digestion, they are relatively robust to cope with variations in organic loading and inhibitory conditions due to complete retention of biomass by membranes. The AnMBR technology is now gaining acceptance by a wide range of industries. The achievements of AnMBRs in pilot-scale studies and full-scale applications are surveyed in this chapter. Lab- and pilot-scale studies provide important information to scale up AnMBRs and upgrade the existing anaerobic process into full-scale AnMBR systems. However, membrane fouling is still the major issue limiting its popularity and development in commercialization. The factors affecting the treatment performance and membrane fouling are reviewed with the discussion of fouling mechanisms, characteristics and control strategies. Moreover, an overview of its commercial potential in water reuse, energy production, and costs of AnMBRs in wastewater treatments is covered. The combination of AnMBRs and other effluent polishing treatments is desirable for the purpose of water reuse. Meanwhile, it is appealing to take advantage of the

 Membrane Biological Reactors: Theory, Modeling, Design, Management and Applications to Wastewater Reuse
Editor(s): Faisal I. Hai, Kazuo Yamamoto and Chung-Hak Lee
doi: 10.2166/9781780409177_0343

biogas produced as an energy resource. The lower total cost of AnMBR than that of aerobic MBRs shows its economic feasibility. Finally, the research needs for future developments are summarized based on the state of the art of AnMBR technology.

11.1 INTRODUCTION

Anaerobic digestion is a unique biological treatment alternative particularly suitable for the high-strength organic effluents that are biodegradable and non-toxic to methanogenic bacteria. It has been applied in waste and wastewater treatment for over a hundred years. It is of great interest that anaerobic treatment can achieve pollution control and resource recovery (i.e., methane production, reductive dechlorination and sulphur recovery) at the same time (Rintala & Puhakka, 1994). It is well known that aerobic treatment can be easily used and is generally effective in breaking down waste products. However, total costs are high due to factors such as energy consumption by aeration (Gander *et al.*, 2000), high sludge yield (Liu & Tay, 2001), and high cost for excessive sludge disposal (Wei *et al.*, 2003). By contrast, the anaerobic treatment process transforms the waste products into carbon dioxide, methane and smaller amount of biomass. The operating costs of anaerobic process are much lower because of its competitive advantages, such as absence of oxygen, low biomass growth, and renewable energy source production.

Notwithstanding the potential advantages over the aerobic processes, the conventional anaerobic reactors have certain inherent limitations. Anaerobic digestion is a complex biochemical reaction that occurs in four steps (i.e., hydrolysis, fermentation/acidogenesis, acetogenesis, and methanogenesis) carried out by several types of microorganisms. The stages of hydrolysis and acidogenesis are more robust than acetogenesis and methanogenesis. Because of the complicated metabolic pathway of anaerobic processes, the symbiotic relationship among all these microorganisms is vital to achieve process stability. Any disturbances in the balance between acid-forming and methane-forming microorganisms will cause reactor instability. The two groups have different sensitivity to environmental conditions and diverse physiology, nutritional needs and growth kinetics. For example, the optimal pH range for methanogens is 6.8–7.2 while a more acidic pH is desirable for acid-forming microorganisms (Rajeshwari *et al.*, 2000). Adjustment of pH is usually necessary to maintain a proper acidic/basic environment to protect biological metabolism. Hence, anaerobic digestion is more sensitive to changes in operating and environmental conditions such as temperature and pH and presence of inhibitory compounds in wastewater: biological activities can be altered and the recovery time after upsets may be long. Besides, anaerobic wastewater treatment suffers from biomass separation problems. A post-treatment is usually required to meet the effluent-quality criteria (Stahl *et al.*, 2004).

In recent years, considerable attention has been paid to the use of membranes technologies in conjunction with anaerobic reactors, namely anaerobic membrane bioreactors (AnMBRs). The membranes replace the traditional gravity sedimentation unit (clarifier), thereby simplifying the multi-unit operations to a single unit operation and allowing a smaller footprint. Membranes are used as filtration units so that biomass can be fully retained within the reactor. Complete biomass retention eliminates the biomass separation problems, allows proliferation of slow-growing microorganisms in the reactor, improves the effluent quality (solids free), allows higher biomass concentrations and desired organic loading rates (OLRs), thus greatly increasing the system stability and productivity.

The applications and opportunities of AnMBRs were reviewed comprehensively by Liao *et al.* (2006). Berube *et al.* (2006) focused on application of AnMBRs to low-strength municipal wastewater and discussed the parameters that govern permeate flux of membrane during such applications.

More recently, a general overview regarding the role of the membrane in AnMBRs was presented by Szentgyorgyi and Belafi-Bako (2010). This chapter looks at the recent developments in topics such as membrane fouling mechanisms, factors affecting system performance and membrane fouling, the potential of water reuse and renewable energy production, and economical evaluation. Special attention is paid to their successful full-scale and pilot-scale applications. Finally, the future research needs and opportunities are discussed.

11.2 HISTORY

The separation of the final effluent from the activated sludge by an ultrafiltration membrane was first reported in 1969 (Smith *et al.*, 1969). Membrane filtration was not introduced to anaerobic wastewater treatment until 1978 when Grethlein (1978) tested an external cross-flow membrane module to process septic tank effluent. Not long after that, in the early 1980s, Dorr-Oliver developed the first commercially available AnMBR to treat high-strength whey processing wastewater (Li *et al.*, 1985; Sutton *et al.*, 1983). However, the early AnMBR systems were not upgraded to full-scale due to high membrane cost. In 1987, an anaerobic treatment system with the incorporation of ultrafiltration membrane (anaerobic digestion ultrafiltration, ADUF) was developed in South Africa (Ross *et al.*, 1990). Some of these pilot- and full-scale ADUF systems are still playing a role in industrial wastewater treatment. A 6-year research and development program ('Aqua Renaissance 90 project') was initiated by the Japanese Ministry of International Trade and Industry, aiming at the development of AnMBRs with different configurations for water reuse and energy recovery by treating industrial and domestic wastewater (Kimura, 1991). The results indicated that the COD removals for the treatment of concentrated wastewater by membrane separation units were generally over 90%, and the power requirement for the test plants was only less than a third of that for the conventional wastewater treatment.

The last 20 years have seen a rapid development in the field of AnMBRs. Considerable efforts have been focused on reducing the energy consumption as well as promoting renewable resource recovery and water reuse. A variety of wastewaters has been treated by AnMBRs (Liao *et al.*, 2006, 2012). Other than the external cross-flow configuration, submerged membrane bioreactor has been developed and tested with different membrane materials and membrane module configurations. Research activities have been conducted for elucidation of membrane fouling mechanisms, characterization of membrane foulants, and membrane fouling control in AnMBRs.

11.3 SYSTEM CONFIGURATIONS

AnMBR systems are based on two main configurations: anaerobic sidestream/external membrane bioreactor (AnsMBR) (Figure 11.1a) and anaerobic immersed/submerged membrane bioreactor (AniMBR) (Figure 11.1b and c). For AnsMBR, the membrane modules are placed outside the bioreactor and a pump is required to push bioreactor effluent through the membrane unit. To date, most of the reported AnMBR studies have focused on AnsMBRs. However, this configuration needs high cross-flow velocity to alleviate membrane fouling, which may result in breakage of sludge flocs, reduction in sludge activity, high-energy cost, and consequently reduced economic feasibility of this technology (Hu & Stuckey, 2006). Besides, frequent physicochemical cleaning is required to maintain the performance of such an AnsMBR. Modified external MBRs have been developed by applying gas scouring of membrane placed in a separate tank to control the membrane fouling (Torres *et al.*, 2011). Still, it was difficult to reduce the energy consumption significantly for the AnsMBR operation.

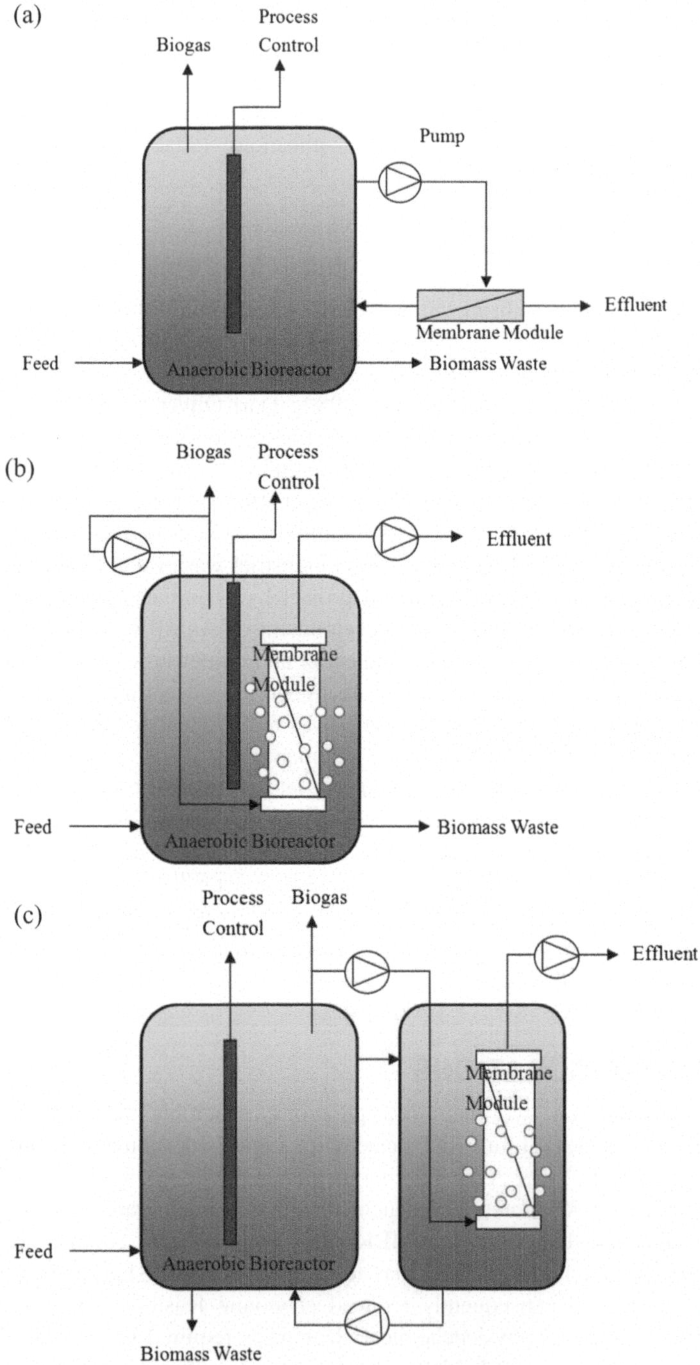

Figure 11.1 Schematic of (a) Pressure-driven anaerobic sidestream membrane bioreactor (ansMBR) (b) Vacuum-driven anaerobic immersed membrane bioreactor (aniMBR) (c) Vacuum-driven aniMBR with the membrane immersed in an external chamber.

One interesting alternative to the conventional side-stream process is the AniMBR, which employs membranes directly submerged into the bioreactor (Figure 11.1b) or a separate tank (Figure 11.1c). The basic concept of (aerobic) submerged/immersed MBRs was first developed by Professor Yamamoto in Japan (Yamamoto *et al.*, 1989). In the submerged/immersed MBRs, the membrane is operated under a vacuum and the operating conditions are much milder than in external MBR systems because of the lower tangential velocities. Indeed, the energy demand of this configuration can be up to two orders of magnitude lower than that of external systems (Chang *et al.*, 2002), and the shear stress on biomass associated with recirculation can be reduced. In addition, AniMBRs can allow for self-cleaning of the membrane surface by recirculating the biogas produced (Liao *et al.*, 2006).

11.4 APPLICATIONS OF AnMBRS

Both AnsMBR and AniMBR have been increasingly employed for treating synthetic, municipal and industrial wastewaters (Liao *et al.*, 2006, 2012), demonstrating the advantages of AnMBRs over other treatment technologies in the treatment of a wide range of wastewaters.

11.4.1 Municipal wastewater treatment

Municipal wastewater is characterized as low-strength wastewater (chemical oxygen demand (COD), 250–800 mg/L; suspended solids (SS), 120–400 mg/L) that is mainly treated by aerobic activated sludge processes (Liao *et al.*, 2006). Large amount of municipal wastewater requires high-rate treatment at low (ambient or psychrophiloic) temperatures with short HRTs (Smith *et al.*, 2012). Aerobic MBRs have been widely used for municipal wastewater treatment. Despite the common concern that AnMBRs may not be suitable for treatment of low-strength municipal wastewater treatment, in recent years, there has been a growing interest in the adoption of AnMBRs for this purpose because they possess the potential to simultaneously recover energy (biogas) (Berube *et al.*, 2006; Kocadagistan & Topcu, 2007; Smith *et al.*, 2012). In addition, the existence of membrane eliminates sludge washout and ensures a long SRT irrespective of the HRT to overcome the limitation of low biomass production for anaerobic processes. By coupling membrane technology, sufficient biomass concentration can be maintained for effective treatment of low-strength wastewaters, especially at cold climates when biomass growth is greatly reduced. The residual levels of soluble and colloidal organic contaminants could also be retained in the AnMBR system, and thus be hydrolysed and biodegraded (Berube *et al.*, 2006). The typical lab- and pilot-scale studies on application of AnMBR for municipal wastewater treatment are summarized in Table 11.1. In order to reduce the operation cost for heating, AnMBRs were operated at the normal temperatures rather than at mesophilic or thermophilic temperatures; however, COD removal efficiencies of above 80% were achieved (An *et al.*, 2009; Martinez-Sosa *et al.*, 2012). With an OLR of 2.36 kg COD/m^3d, the effluent COD concentration of the AnMBR process was about 77.5 mg/L (An *et al.*, 2009). Moreover, an AnMBR and an aerobic MBR demonstrated similar COD removal under the same operational conditions for the treatment of domestic wastewater, although AnMBRs required longer acclimation period compared to aerobic MBRs (Achilli *et al.*, 2011; Baek & Pagilla, 2006). On the other hand, anaerobic biomass may fail to adapt to the wastewater under specific conditions (e.g., extremely low wastewater strength, toxicity, etc.) (Saddoud *et al.*, 2006, 2009). AnMBRs offer the possibility for energy recovery, however, the challenge still remains for low-strength wastewater treatment as a great amount of methane generated is dissolved and lost in the liquid phase. About 70 and 50% of the total methane exited with the gas phase (Kim *et al.*, 2011; Smith *et al.*, 2011). Although several methods (e.g., air stripping of the effluent, degassing membrane, and down-flow hanging sponge reactor) were

Table 11.1 AnMBRs (lab and pilot scale) for municipal wastewater treatment.

Type of Reactor[a]	Configuration, Scale[b] and Volume	Characteristics of Membrane[c]	Flux (L/m^2 h)	Temp (°C)	HRT (d)	SRT (d)	OLR (kg COD/m^3 d)	MLSS (g/L)	Feed COD (mg/L)	Effluent COD (mg/L)	COD Removal
CSTR (Baek & Pagilla, 2006)	External, L (10 L)	PVDF, pore size: 0.1 μm, MWCO: 200 kDa	–[d]	32	0.5–2	19–233	0.02–2.11	0.72–7.45	84[e]	24–38[e]	44–76%
CSTR (Baek *et al.*, 2010)	External, L (10 L)	PVDF, pore size: 0.1 μm	–	–	0.5–2	19–217	0.03–0.16	3–7	38–131[e]	18–37[e]	55–72%
CSTR (Kocadagistan & Topcu, 2007)	External, L (15 L)	Cellulose acetate, MF, pore size: 0.2 μm	450	35	0.7	–	–	1–1.5	350–500	<30	98.1%
CSTR (An *et al.*, 2009)	Submerged, L (12.9 L)	Tubular non-woven filter, pore size: 0.64 μm	5	15–20	0.11	–	2.36	–	259.5 ± 343.8	77.5 ± 29.5	–
CSTR/AniMBR (Giménez *et al.*, 2011)	Submerged, P (1.3 m^3/0.8 m^3)	hollow-fibre UF membrane, pore size: 0.05 μm	20	33	0.25–0.83	70	–	6–22	445 ± 95	77	87%
CSTR (Martinez-Sosa *et al.*, 2011a)	Submerged, P (350 L)	flat sheet polyether sulphone UF membrane, pore size: 38 nm	7	35/28/20	0.8	680	0.5–1.1	15–21	630 ± 82	<80	<90%
CSTR (Dagnew *et al.*, 2011)	Submerged, P (630 L)	Hollow-fibre, ZeeWeed™	17	22	0.35	80–100			224	47	79%
UASB (Herrera-Robledo *et al.*, 2011)	External, P (849 L)	PVDF, tubular UF membranes, MWCO: 100 kDa	45–50	22	0.25	180	–	–	445 ± 138	33	93%
UASB (Gouveia *et al.*, 2015)	Submerged, P (284 L)	Hollow-fibre (ZW-10 Zenon, GE), pore size: 0.045 μm	10–14	18	0.53–0.59	–	1.6–2.0	23–26	978 ± 210	93.7–147	90%
Staged fluidized AnMBR (Shin *et al.*, 2014)	Submerged, P (1 m^3 and 0.77 m^3)	PVDF, hollow-fibre, pore size: 0.03 μm	7.5	8–30	4.6–6.8	–	–	47	198–362	4–25	>90%

[a]CSTR = completely stirred tank reactor, UASB = upflow anaerobic sludge blanket bioreactor.
[b]L = laboratory/bench scale, P = pilot scale.
[c]MF = microfiltration, UF = ultrafiltration, PVDF = polyvinylidene fluoride.
[d]–indicates value not reported.
[e]Units are SCOD (soluble chemical oxygen demand).
[f]∞ indicates no biomass wastes

attempted for dissolved methane recovery, none of them is both practically and economically feasible (Bandara *et al.*, 2011; Hatamoto *et al.*, 2010; McCarty *et al*, 2011).

Pilot-scale AnMBR studies treating municipal wastewater with high COD removal efficiencies and low biosolids productions achieved at HRTs comparable to conventional aerobic processes have been recently reviewed by Shin and Bae (2018). The AniMBRs treating low-strength municipal wastewater achieved close to 90% COD removal (Giménez *et al.*, 2011; Herrera-Robledo *et al.*, 2011; Martinez-Sosa *et al.*, 2011b). An AniMBR demonstration plant was operated at sub-critical fluxes and high mixed liquor total solids (MLTS) levels (>22 g/L) for almost 2 years (Robles *et al.*, 2012). By placing the membrane module on top of the UASB reactor, continuous operation under psychrophilic conditions for three years had been achieved without physical or chemical cleaning (Gouveia *et al.*, 2015). A pilot-scale staged anaerobic fluidized membrane bioreactor was able to obtain COD removals of 94 and 90% and BOD5 removals of 98 and 90% for summer and winter seasons, respectively, when fully acclimated (Shin *et al.*, 2014). All of the above pilot-scale studies did not encounter any irreversible fouling problems. These pilot-scale results indicate the potential of AnMBRs for municipal wastewater treatment.

11.4.2 Industrial wastewater treatment

Anaerobic biological processes are mainly applied to high-strength wastewaters which are difficult to treat with aerobic processes. Unlike municipal wastewater, industrial wastewater is generally characterized by high organic strength (1–200 g COD/L), non-neutral pH, different temperatures, salinity, SS, turbidity, heavy metals, colour, etc. (Lin *et al.*, 2012). Additionally, some constituents in the industrial wastewaters may be inhibitory or toxic to anaerobic degradation. The harsh conditions induced by these streams can largely limit the feasibility of conventional anaerobic processes, and AnMBRs stand out as another promising option for treating industrial wastewaters (Dereli *et al.*, 2012a). Therefore, before making the decision to apply an AnMBR, lab-scale studies are helpful to identify the anaerobic treatability of wastewater (i.e., physical, chemical and biological characteristics) as well as the potential problems and their solutions.

To date AnMBRs have been tested with food processing, pulp and paper, textile, landfill leachate, pharmaceutical, oily and petrochemical wastewaters (Cornelissen *et al.*, 2001; Li, 2010; Maleki *et al.*, 2018; Saravanane & Sundararaman, 2009; Stamatelatou *et al.*, 2009). Table 11.2 presents treatment performance and operational parameters of AnMBR systems treating different industrial wastewater. Food processing wastewaters are readily biodegradable and contain high organic strengths (1,000–85,000 mg COD/L) with a wide range of SS concentrations. AnMBRs are extremely suitable for food processing wastewaters allowing complete solid removal and high COD removal (>90%) under high OLRs (>12 kg COD/m^3 d) (Fuchs *et al.*, 2003; Galib *et al.*, 2016; Maleki *et al.*, 2018; Saddoud *et al.*, 2007; Torres *et al.*, 2011). For the petrochemical effluent consisting mostly of C2 to C6 short-chain fatty acids, a high OLR of up to 25 kg COD/m^3 d was achieved with 98% of the COD converted into methane and biomass (Van Zyl *et al.*, 2008). Due to the presence of high fat oil and suspended matters in the slaughterhouse wastewater, the addition of a fixed-bed reactor (acidogenesis step) prior to an AnMBR (methanogenesis step) was needed at high OLRs (Saddoud & Sayadi, 2007). Pulp and paper wastewaters are anaerobically treatable. In recent years, Liao and colleagues have conducted in-depth studies on AniMBR systems for treatments of different types of pulp and paper wastewaters (Gao *et al.*, 2010b, 2011a, b, 2016; Lin *et al.*, 2009, 2010, 2011b; Xie *et al.*, 2010). In those studies, the AniMBR process was technically feasible for the treatment of kraft evaporator condensate, thermomechanical pulping whitewater and pressate with COD removal of 93–99%, 87%, and 76–83%, respectively (Gao *et al.*, 2011a, b; Xie *et al.*, 2010). The biogas recovered was of excellent fuel quality with 80–90%

Table 11.2 AnMBRs (lab and pilot scale) for industrial wastewater treatment.

Type of Wastewater	Type of Reactor[a]	Configuration, Scale[b] and Volume	Characteristics of Membrane[c]	Temp (°C)	HRT (d)	SRT (d)	OLR (kg COD/m^3 d)	MLSS (g/L)	Feed COD (mg/L)	Effluent COD (mg/L)	COD Removal
Food processing wastewaters											
Cheese whey (Saddoud *et al.*, 2007)	Two-phase (acidogenic/ methanogenic) CSTR + M	External, L (5 L/15 L)	Ceramic membrane, pore size: 0.2 µm	37	1/4	–/ 29.7–78.6	–/3–19.78	~8.5	68,600	–[d]	98.5%
Brewery wastewater (Torres *et al.*, 2011)	CSTR	External, L (4.5 L)	Tubular membrane module, ceramic (pore size: 0.2 µm) and polymeric (pore size: 30 nm) membrane.	30	–	–	12	–	21,000	190	>97%
Malting wastewater (Maleki *et al.*, 2018)	CSTR	Submerged, L (4.5 L)	Flat sheet, PVDF, pore size: 0.1 µm	36	1.6–9.4	200	1.36–3.18	12	4,630–5,150	283–506	90–94%
Meat processing wastewater (Galib *et al.*, 2016)	CSTR	Submerged, L (5 L)	Hollow-fibre, ZeeWeed 500 D, pore size: 0.04 µm	24	5/2/1	50	0.4–3.2	11.5	2,050–6,880	96–170	88–95%
Food processing and washing wastewater (He *et al.*, 2005)	CSTR	External, P (0.4 m^3)	PES UF membranes, MWCO: 20000–70000 Da	37	0.67–4.17	50	0.88–4.83	6–8	2,000–15,000	141–3,600	73–94.2%
Snacks factory wastewater (Diez *et al.*, 2012)	UASB	Submerged, P (0.76)	Hollow fibre, PVDF, pore size: 0.4 µm	35	–	–	5.1	7.9–10.4	8,200–22,000	110–250	97%
Food processing wastewater (Lu *et al.*, 2011)	CSTR	Submerged, P (20 m^3)	Rotating membrane mocule.	–	1.37–2	–	20 m^3/d (flow rate)	10–30	6,000	50	>90%
Pulp and paper wastewaters											
Thermo-mechanical pulping pressate (Gao *et al.*, 2011a)	CSTR	Submerged, L (10 L)	PVDF, UF membrane, MWCO: 70000 Da	37/45/55	–	–	2.59 ± 0.53	10.9 ± 0.5	2,120–3,600	380–530/500 –720/600–810	76–83%

Thermo-mechanical pulping whitewater (Gao *et al.*, 2010b)	CSTR	Submerged, L (0.01 m^3)	PVDF, UF membrane, MWCO: 70000 Da	37	~2.5	280	2.4 ± 0.4	5.7–10.0	2,600,5500, 10,000	112 ± 35	>95%
Kraft evaporator condensate (Lin *et al.*, 2009)	CSTR	Submerged, L (0.01 m^3)	PVDF UF membrane, MWCO: 70000 Da	37, 55	–	~230	–	8	10,000	–	97–99%
Landfill leachate											
Landfill leachate (Zayen *et al.*, 2010)	CSTR	External, L (50 L)	UF, MWCO = 100000 Da	37	7	–	1–6.27	0.44–3	15,000–41,000	–	90%
Diluted landfill leachate (Bohdziewicz *et al.*, 2008)	CSTR	Submerged, L (29 L)	UF membrane, pore size: 0.1 μm	35	1–7	–	0.7–4.9	10	5,000	417	<90%
Other industrial wastewaters											
Petrochemical effluent (Van Zyl *et al.*, 2008)	CSTR	Submerged, L (50 L)	Flat sheet, pore size = 0.45 μm	37	1.31	175	>25	>30	19,101	<500	>97%
Slaughterhouse wastewater (Saddoud & Sayadi, 2007)	CSTR	External, L (50 L)	UF membrane, MWCO: 100 kDa	37	1.25–3.33	–	4.37–16.32	~2–10	10,174–12,430	338–4,556	62–96.4%
Slaughterhouse wastewater (Saddoud & Sayadi, 2007)	Two-phase (acidogenic/methanogenic) CSTR + M	External, L (50 L/50 L)	UF membrane, MWCO: 100 kDa	37	1.25	–	12.7 ± 1.71	~8–10	10,580 ± 990	196 ± 4	98.75 ± 0.44%
Thin stillage wastewater (Dereli *et al.*, 2012b)	CSTR + M	Submerged, P (12 m^3)	Flat-sheet UF membrane, PVDF, pore size: 0.08 μm	37	17 ± 4	300	4.5–7	26	98,000	1700	>98%
Animal waste (Wong *et al.*, 2009)	CMAD + AnMBR	External, P (100 L/100 L)	Tubular UF membrane, PVDF, pore size: 0.03 μm	mesophilic conditions	18	28	3.3/2.4	–	44,900	3440	92%
Pharmaceutical wastewater (Huang *et al.*, 2018)	UASB	External, P (180 L)	Hollow fibre, PVDF, pore size: 0.02 μm	10–38	1/3/4	250	2.37–4.46	16.7	2,700–6,760	215–814	87–94%

[a]CSTR = completely stirred tank reactor, CMAD = complete mix anaerobic digester, UASB = upflow anaerobic sludge blanket bioreactor; M designates the location of the membrane.
[b]L = laboratory/bench scale, P = pilot scale.
[c]UF = ultrafiltration, PVDF = polyvinylidene fluoride, PES = polyethersulphone.
[d]–indicates value not reported.

methane for the kraft evaporator condensate and 62–80% for thermomechanical pulping whitewater and pressate (Gao *et al.*, 2011a; Liao *et al.*, 2010, 2011b; Xie *et al.*, 2010). However, feed toxic shock, pH shock and temperature shock caused deterioration of biological activity and biogas production, which will be further discussed in Section 11.6.2. Pulp and paper wastewaters are normally discharged at high temperatures (50–70°C); therefore, thermophilic AnMBRs are feasible and favourable to avoid energy consumptions during the process of pre-cooling and post-heating prior to and after biological treatment. However, compared to the mesophilic AnMBRs, severe membrane fouling problems have been reported in case of thermophilic AnMBRs (Lin *et al.*, 2009). Landfill leachate is a high-strength, complex mixture containing dissolved organic compounds (alcohols, acids, aldehydes, short-chain sugars, etc.), ammonia, halogenated hydrocarbons, heavy metals, and inorganic salts (Bohdziewicz *et al.*, 2008). It originates from percolation of rainwater and moisture through waste within the landfill which undergoes chemical and/or biological transformations (Xu *et al.*, 2008). Landfill leachate was efficiently treated by an AnMBR without any physical or chemical pretreatment. The average COD removal and biogas yield were as high as 90% and 0.46 L/g COD removed, respectively (Zayen *et al.*, 2010). Nevertheless, the applicable permeate flux was only 2.5–3.5 L/m^2 h (Zayen *et al.*, 2010). In another study, Bohdziewicz *et al.* (2008) obtained 90% COD removal by an AniMBR at a hydraulic retention time (HRT) of 2 days and an OLR of 2.5 kg COD/m^3 d. Because of the high effluent COD concentration (417 mg/L), they employed reverse osmosis as a post-treatment.

Pilot testing is required to formulate optimal design from removal efficiency and economical considerations. Studies regarding industrial wastewater treatment by pilot-scale AnMBRs have focused on both high-strength (Chen *et al.*, 2008b; He *et al.*, 2005; Minami, 1994) as well as low-strength wastewaters (COD < 1000 mg/L) (Fan *et al.*, 2000; Kato *et al.* 1994; Lin & Peng, 1994). An anaerobic/oxic membrane bioreactor (A/O MBR) achieved efficient COD (82%), BOD_5 (96%), colour (71%), and turbidity (99%) removal while treating a dyeing wastewater from a woolen mill (Fan *et al.*, 2000). Chen *et al.* (2008b) conducted a pilot-scale test with a two-phase anaerobic digestion (TPAD) system and a subsequent aerobic MBR treating real pharmaceutical wastewater (COD: 5,789 and 58,792 mg/L) from the processes of product manufacturing and equipment cleaning. The quality of the effluent (COD: 40 mg/L; pH: 6.8–7.6) complied with the standard of direct discharge to natural waters. A pilot-scale AnMBR showed significant treatment of real pharmaceutical wastewater containing amoxicillin, ceftriaxone, cefoperazone and ampicillin with removal efficiencies of 73.2 $\pm$ 4.3, 47.7 $\pm$ 2.2, 79.4 $\pm$ 4.1 and 34.6 $\pm$ 3.3%, respectively (Huang *et al.*, 2018). A successful pilot-scale AnMBR study treating thin stillage (influent COD: 70–150 g/L) achieved superior effluent quality and increased biogas production compared to conventional anaerobic processes (Dereli *et al.*, 2012b). Lu *et al.* (2011) reported a pilot-scale (20 m^3/d) AniMBR treating high-strength food processing wastewater (influent COD: 6,000 mg/L). The COD removal reached 95% with a methane production rate of 0.18 L CH_4/g COD (Lu *et al.*, 2011). Wong *et al.* (2009) reported successful treatment of animal waste by a pilot AnMBR system with the mean removals of total Kjeldahl nitrogen, total phosphate, COD, and total solids of 31, 96, 92, and 82%, respectively. These valuable baseline data gained from lab- and pilot-scale studies lay the foundation for the scale up of AnMBRs.

AnMBRs have been developing from an advanced laboratory technology into a technologically acceptable and commercially feasible process for full-scale application, as shown in Table 11.3. To date, however, full-scale operations have been mainly limited to wastewater from food industry (Butcher, 1989; Christian *et al.*, 2010; Ross *et al.*, 1992). From 2006 to 2008, five full-scale AnMBR systems were constructed by Kubota Corporation in Japan for stillage treatment for Shouchu (Japanese spirits made from sweet potato, rice or other grains), potato processing sites, garbage bio-gasification facilities, and food factory treatment plants (Grant *et al.*, 2008; Kanai *et al.*, 2010). The COD, BOD, and total

Table 11.3 Full-scale applications of AnMBR.

Type of Wastewater	Type of Reactor[a]	Configuration and Volume	Characteristics of Membrane[b]	Temp (°C)	HRT (d)	SRT (d)	OLR (kg COD/m^3 d)	MLSS (g/L)	Feed COD (g/L)	Effluent COD (mg/L)	COD Removal
Wheat starch waste (Butcher, 1989)	CSTR	External (2000 m^3)	UF membrane	40	–[c]	–	2.1	10	–	–	78%
Maize-processing effluent (Ross *et al.*, 1992)	CSTR	External (2610 m^3)	Tubular UF membrane, pore size <0.1 μm	35	5.2	–	2.9	21	15	400	97%
Wastewater from the production of salad dressings and barbeque sauces (Christian *et al.*, 2010)	Anaerobic reactor + AnMBR, CSTR	Submerged (8330 m^3 + 4 × 103 m^3)	Flat-sheet membrane, pore size = 0.2 μm	35	–	–	310 m^3/d (flow rate)	11.4–48	34	225	99.4%
Alcohol stillage (Grant *et al.*, 2008)	AnMBR + aerobic MBR, CSTR	Submerged	Pore size = 0.4 μm	55	–	–	10–15	30–50	77–110	<10[d]	>98%

[a]CSTR = completely stirred tank reactor.
[b]UF = ultrafiltration.
[c]–Indicates value not reported.
[d]Units are BOD.

suspended solids (TSS) removals were typically over 98% in the AnMBR process despite high OLRs (10–15 kg COD/m^3 d). The ADI Systems Inc. has employed membranes (Kubota Corporation, Japan) in full-scale AnMBR systems since 2000 and in full-scale trials since the mid-1990s with over a dozen installations in Japan and the United States (McMahon, 2010). In July 2008, ADI Systems Inc. (Fredericton, Canada) built the first AnMBR installation in North America (Ken's Foods, Massachusetts, USA) to treat raw wastewater from the production of salad dressings and barbeque sauces (Christian *et al.*, 2010). It has been a success to upgrade the previous anaerobic process into an AnMBR system. This plant is currently the largest AnMBR installation in the world with a design influent flow rate of 475 m^3/d. A high COD removal efficiency of 99.4% is achieved with an influent COD, BOD, and TSS of 39,000 mg/L, 18,000 mg/L, and 12,000 mg/L, respectively. The full-scale applications illustrated the merits of the process for the treatment of high-strength wastewaters and the production of a solids-free effluent.

11.5 MEMBRANE FOULING

The AnMBR technology is gradually shaping up into a reliable and efficient technology and is expected to become an option for many domestic and industrial wastewater treatments. However, membrane fouling is regarded as one of the most challenging issues that restrict their widespread applications. It results in a reduction of permeate flux or an increase in transmembrane pressure (TMP), reduced productivity and increased operating costs (e.g., added energy consumption, increased membrane cleaning and replacement cost).

Compared to the aerobic MBRs, a limited number of studies have been conducted on the fouling of AnMBRs; however, it is worthy of our reference through the findings regarding membrane fouling on aerobic MBRs. Skouteris *et al.* (2012) concluded in a recent review that membrane fouling in AnMBRs is more severe than in aerobic MBRs because of lower sludge filterability. Contrary to these findings, similar or less fouling propensity for AnMBRs than aerobic MBRs were also reported by Achili *et al.* (2011) and Beak and Pagilla (2006), respectively. Different modes and levels of fouling in aerobic and anaerobic MBRs may be attributed to the fact that the sludge characteristics, including physicochemical structures and biological activities, in anaerobic and aerobic systems are different. Under aerobic conditions, sludge is produced at high specific growth rates (Judd, 2011). The anaerobic sludge mainly originates from its low biomass yield and influent particulates (Judd, 2011; Lant & Hartley, 2007), and it is more susceptible to any possible change in operational and environmental conditions as mentioned in Section 11.1. For instance, Sheng *et al.* (2006) found that the stability of aerobic flocs were far above the anaerobic flocs under an elevated shear stress. The particle-size distribution of anaerobic sludge was found to have a smaller modal size with a larger distribution range (Lant & Hartley, 2007). According to Van Voorthuizen *et al.* (2008) and Martin-Garcia *et al.* (2011), a substantially greater amount of colloidal particles were accumulated in an AniMBR than in a submerged aerobic MBR due to the slower biodegradation rates for the anaerobic process. The three-dimensional excitation-emission matrix (EEM) fluorescence spectroscopy applied to extracellular polymeric substances (EPS) extracted from anaerobic and aerobic sludge revealed the difference in their chemical structures (Sheng & Yu, 2006). Novak *et al.* (2003) reported that the concentration of soluble protein in the supernatant was substantially greater than that of polysaccharide under anaerobic conditions, while polysaccharides were the major biopolymer in the supernatant under aerobic conditions.

Summarizing the effect of different wastewater characteristics and operational conditions on membrane performance, Lin *et al.* (2012) concluded that more attention should be paid to membrane fouling control in MBRs treating industrial wastewaters. However, unified and well-structured theories on membrane fouling are not currently available because of the inherent complexity of the system. As a result, considerable efforts have been made to explore the mechanisms of membrane fouling in AnMBRs and the control strategies. The state of the art of membrane fouling monitoring, characterization and control in MBRs have been presented in Chapter 3. This section briefly mentions the basics and highlights the information more relevant to the AnMBRs for the fundamental understanding of the development of membrane fouling in AnMBRs.

11.5.1 Membrane fouling mechanisms

According to removability of the foulants, membrane fouling can be classified into removable, irremovable and irreversible fouling (Meng *et al.*, 2009). For removable fouling, foulants can be readily removed from the membrane surface by appropriate physical cleaning (e.g., backwashing). Irremovable fouling caused by stronger interactions between foulants and membrane materials needs chemical cleaning to remove. The irreversible fouling (or permanent fouling) represents fouling remaining on the membrane surface or within the membrane pores even after intensive chemical cleaning. Of the two fouling modes, namely, internal fouling (pore clogging/narrowing) and external fouling (cake/gel layer formation), generally, irremovable fouling and irreversible fouling are attributed to pore blocking, and cake layer formation makes main contribution to removable fouling (Hai & Yamamoto, 2011). Cake layer formation was identified as the dominant mechanism limiting the membrane performance of AnMBRs (Choo & Lee, 1996a; Jeison & Van Lier, 2007a; Torres *et al.*, 2011). Membrane fouling can also be classified into biofouling, organic fouling and inorganic fouling according to the components of foulants (Liao *et al.*, 2004). Biofouling refers to the deposition of cell debris and colloidal particles, growth and metabolism of bacteria cells or flocs on the membranes, as well as adsorption of extracellular polymeric substances (EPS) and soluble microbial products (SMP) (Liao *et al.*, 2006; Meng *et al.*, 2009). Organic fouling is defined as the accumulation and adsorption of organic constituents on membranes. Inorganic fouling is due to the chemical or biological precipitation of a great number of metal ions due to the presence of a great number of cations (i.e., Ca^{2+}, Mg^{2+}, Al^{3+}, and Fe^{3+}), anions and ionizable groups of biopolymers (i.e., $CO_3{}^{2-}$, $SO_4{}^{2-}$, $PO_4{}^{3-}$, OH^-, COO^-) (Kim *et al.*, 2007; Meng *et al.*, 2009). These three types of fouling are usually observed simultaneously, although the relative contribution of each one depends on membrane characteristics, sludge characteristics, environmental conditions, reactor design, and the operating strategy (Liao *et al.*, 2006).

Notably, short- and long-term fouling undergo different fouling mechanisms (Charfi *et al.*, 2012; Geng & Hall, 2007). Short-term fouling is mostly reversible due to sludge deposition and superficial pore blocking, while long-term fouling is comprised of reversible and irreversible fouling caused by the long contact time of deposited compounds with the membrane surface. Challenges still remain to predict long-term fouling using short-term experimental data (Le-Clech *et al.*, 2006). A schematic of membrane fouling mechanisms is presented in Figure 11.2. As noted above, at the initial stage of filtration, membrane fouling is caused by the deposition of solutes or colloids in pore space and accumulation of sludge flocs onto membrane surface. If the shear force at the membrane surface is not adequate to remove the foulants, cake formation occurs. Spatial and temporal changes of the foulants composition during long-term operation are anticipated (Meng *et al.*, 2009).

11.5.2 Membrane fouling characterization

11.5.2.1 Physical characterization

A number of techniques such as scanning electron microscopy (SEM), confocal laser scanning microscopy (CLSM) and atomic force microscopy (AFM) have been used for physical characterization of membrane fouling (Meng *et al.*, 2010). SEM is one of the most common instruments providing

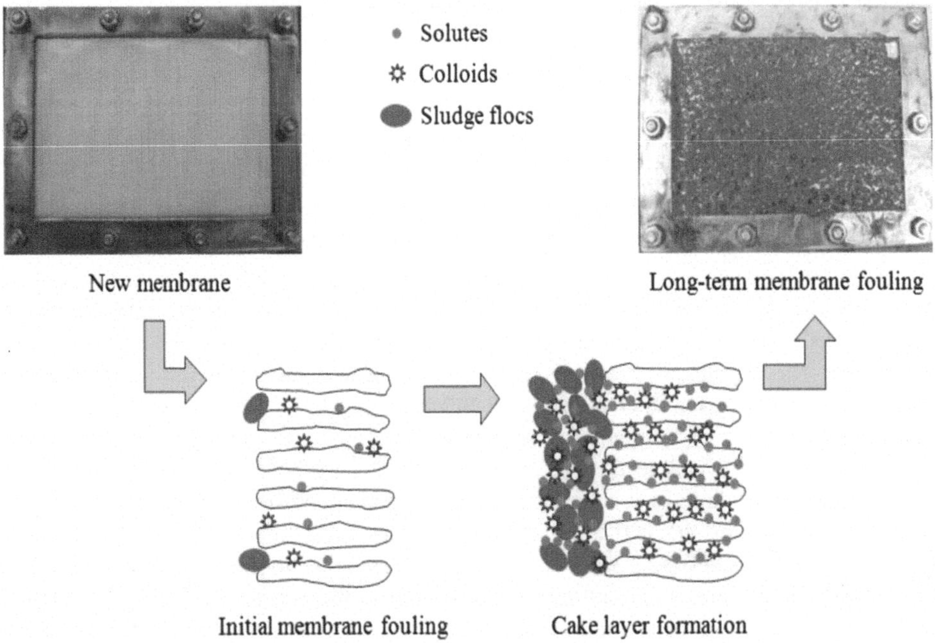

Figure 11.2 Membrane fouling mechanisms.

high-resolution images at nano/micro-metre scale. SEM was utilized to observe the fouling layer directly and evaluate the effectiveness of fouling control methods (Vyrides & Stuckey, 2009a; Wen *et al.*, 2008). However, SEM fails to keep the natural state of the fouling layer due to the pre-treatment protocols of SEM samples including sample dehydration and gold coating. There has been improvement by environmental scanning electron microscopy (ESEM) which allows the observation of wet samples in their natural state with a relatively low lateral resolution (10–20 nm) (Zhang *et al.*, 2007). Special specimen pre-treatment is not needed for AFM, which can provide information of cake surface morphology with three-dimensional (3D) images. By observations using SEM and AFM, the cake layers formed in the thermophilic AniMBR were observed to be more compact and less porous than that in the mesophilic AniMBR (Lin *et al.*, 2009). CLSM has recently become a powerful approach for characterization of membrane fouling. The key feature of CLSM, optical sectioning, allows 3D reconstructions of complex objects. The combination of CLSM and image analysis can not only visualize the 3D structure of the fouling layer, but also quantify its architecture. CLSM images revealed a heterogeneous structure of the cake layer formed on the flat-sheet membrane in an AniMBR (Gao *et al.*, 2011b).

11.5.2.2 Chemical characterization

The chemical components of membrane foulants have been extensively characterized by energy-dispersive X-ray spectroscopy (EDX), Fourier transform infrared (FTIR) spectroscopy (Gao *et al.*, 2011b), X-ray photoelectron spectroscopy (XPS) (Choo & Lee, 1996a), X-ray fluorescence technology (XRF) (Trzcinski & Stuckey, 2016), three-dimensional EEM fluorescence spectroscopy (Sheng & Yu, 2006; Zhang *et al.*, 2011), solid-state 13C-nuclear magnetic resonance (NMR) spectroscopy (Vyrides *et al.*, 2010), size-exclusion chromatography (SEC) (Trzcinski *et al.*, 2011; Zhang *et al.*, 2010), and inductively coupled plasma (ICP) (Kang *et al.*, 2002). EEM fluorescence spectroscopy is a sensitive technique that has been successfully utilized to obtain the structural information of organic substances by capture of their specific fluorescence features. Protein-like substances were identified as an important role in organic fouling EEM fluorescence analyses (An *et al.*, 2009; Wang *et al.*, 2010). SEC can separate organic foulants on the basis of size exclusion to understand the molecular weight distribution or size of EPS/SMP compounds. Based on SEC analysis, Sheng (2011) reported that the SMP in the supernatant of the mixed liquor of a full-scale MBR mainly consisted of macromolecules (9,900–277,000 Da), humic substance (1,000–4,000 Da), and the organic acids (<1,000 Da). High molecular weight (MW) (over 761.3 kDa) compounds were completely degraded or retained in the AniMBR, while low MW (between 1.9 and 0.180 Da) and medium MW (down to 58.4 kDa) materials were present in the permeate (Trzcinski & Stuckey, 2009a). The membrane foulants in AnMBR can comprise not only organic substances but also inorganic elements. Based on SEM-EDX analysis, Berube *et al.* (2006) reported that inorganic precipitates in an AnMBR consisted of struvite (e.g., $MgNH_4PO_4 \cdot 6H_2O$), calcite, and clay, which can be attributed to the production of ammonium and phosphate ions during anaerobic decomposition of organics. ICP and EDX are usually used for inorganic fouling characterization with the limitation of deliquescent samples and surface detection, respectively. Lew *et al.* (2009) found an accumulation of sulphate and aliphatic material on the membrane by ICP and FTIR spectroscopy of the influent, permeate, and retentate samples. By using XRF and EDX, calcium precipitate, namely, monohydrocalcite was found to cause severe fouling during leachate treatment (Trzcinski & Stuckey, 2016). An *et al.* (2009) demonstrated that Mg, Ca, Na, Fe, Al, K, and Si were the major inorganic elements in cake layer by EDX analysis. It is also suggested that bridging between deposited biopolymers and these inorganic foulants could enhance the compactness of fouling layer (An *et al.*, 2009).

11.5.2.3 Microbiological characterization

The microbial community structures have been investigated to explore the biodiversity and population dynamics of microorganisms in MBRs using molecular techniques such as polymerase chain reaction-denaturing gradient gel electrophoresis (PCR–DGGE) (Gao *et al.*, 2011b) and fluorescence in situ hybridization (FISH) (Jeison *et al.*, 2009a). PCR-DGGE is a useful deoxyribonucleic acid (DNA) fingerprinting procedure tool in studying complex microbial communities. Specialized 16s PCR primers amplify a portion of the gene (~400 base pairs) and include a high 'G + C clamp' that anchors one end of the double-stranded DNA to itself. These PCR-amplified 16s ribosomal ribonucleic acid (rRNA) gene fragments are all the same size, but they move to different locations in a denaturing gradient gel because different sequences are immobilized (via denaturation) at different locations in the gel. The separated individual DNA fragments (bands) can be isolated and sequenced (Madsen, 2008). In a full-scale anaerobic bioreactor treating paper mill wastewater, the majority of the prominent bands from the archaeal DGGE profile as well as a major proportion of the archaeal clone library matched with sequences from *Methanosaeta* spp (Roest *et al.*, 2005). Some bacterial members belonging to the family Cellulomonadaceae (*Cellulomonas*, *Oerskovia*) capable of hydrolysing cellulose and carbohydrates, the genus *Propionibacterium* which produces propionate, the sulphate-reducing bacteria, and the syntrophic fatty acid-oxidizing microorganisms can be affiliated with all steps along the anaerobic degradation pathway. FISH applies microscopic detection of cells whose biomakers (e.g., DNA, rRNA) hybridize to fluorescently tagged probe molecules of known binding specificity, showing the composition and spatial structure of the microbial community in the reactor (Yuzir & Ali, 2008).

Some researchers have characterized the microbial communities in AnMBRs those are involved in cake layer formation. The selective enrichments of certain species on membrane surface were observed, while some species played a major role in the metabolism of substrates (Calderón *et al.*, 2011; Gao *et al.*, 2010a; Lin *et al.*, 2011). The structures of microbial communities in the SS and on the membrane surface were different, and it was revealed that there were significant changes in population densities along the cake layer depth (Gao *et al.*, 2011b). Identifying the triggers of selective cell deposition and the dominant species in cake layers are of interest to develop strategies for membrane fouling control.

11.6 FACTORS AFFECTING THE TREATMENT PERFORMANCE AND MEMBRANE FOULING

The main categories of factors strongly affecting treatment performance and membrane fouling are membrane properties and reactor design, sludge characteristics and microbial activity, hydrodynamic conditions, operating conditions, and environmental conditions (Table 11.4). The complex interactions of membrane material, sludge properties and operational conditions are still not fully understood.

In an AnMBR, the fouling behaviour is considered to be directly influenced by sludge characteristics, membrane properties and their interactions (Figure 11.3). Hydrodynamic conditions (i.e. permeate flux, TMP and biogas sparging intensity) exert direct influence on the fouling layer and sludge itself. The biological system design and operating parameters, e.g., sludge retention time (SRT), HRT or OLR, food to microorganism (F/M) ratio, nutrient conditions, etc., play significant roles in the membrane filtration performance (Figure 11.3). The operating and environmental conditions can modify physicochemical properties of the sludge and consequently influence mode and intensity of membrane fouling (Figure 11.3). The treatment performance of AnMBRs is significantly governed by the operating conditions, microbial activity, membrane properties and reactor design, and influent properties. Also, the environmental conditions are other important factors affecting the microbial activity, and thus influence the treatment performance.

11.6.1 Membrane properties

Membrane characteristics such as materials, pore size, porosity, surface charge, roughness, and hydrophobicity have a direct impact on membrane performance and fouling. In the order of descending pore size (or molecular weight cut-off (MWCO)), membranes are generally classified as microfiltration (MF) (pore size: 0.1–5 µm), ultrafiltration (UF) (pore size: 0.01–0.1 µm), nanofiltration (NF) (pore size: 0.001–0.01 µm), and reverse osmosis (RO) (pore size: 0.0001–0.001 µm) membranes (Judd, 2006). In order to achieve good water purification by a bioreactor coupled with MF or UF membranes, a longer HRT or lower OLR needs to be applied (i.e., requires larger tanks), while NF or RO membranes necessitate higher capital and operating costs for such membrane operation (Arros-Alileche *et al.*, 2008). MF and UF membranes in the form of flat sheet or hollow fibre are the most common membranes for AnMBRs. MF membranes have pores large enough to retain materials such as solids and bacteria, while UF membrane will further retain more macromolecules and colloids. Meanwhile, larger pore size usually leads to increased membrane flux in AnMBRs (Hernandez *et al.*, 2002). Comparing the performance of UF membranes having a MWCO ranging from 20,000 to 70,000 Da in an AnsMBR, He *et al.* (2005) reported steeper initial flux decline in case of the membrane with lower MWCO as a result of concentration polarization. The membrane with the largest MWCO showed the largest initial water flux, however for long-term operation, it exhibited the most severe flux decline due to membrane pore clogging. With comparison of various MF and UF membranes in anaerobic digestion through stirred-cell filtration, Choo & Lee (1996b) concluded that an optimal pore size exists depending on the sizes of membrane pore and sludge suspension constituents. It was suggested that highly porous membranes with

Table 11.4 Summary of factors affect treatment performance and membrane fouling in AnMBRs.

Reactor Types[a]	Influent Types	Membrane[b] Types	Operating and Environmental Conditions[c]	Hydrodynamic [d] Conditions	Sludge Properties[f]	Membrane and Treatment Performance[g]	Major Factors
AniMBR							
UASB (Wu *et al.*, 2009)	Synthetic municipal wastewater	FS, PVDF Pore size: 0.22 μm Area: 0.05 m^2	Temp: 35°C HRT: 6 h OLR: 0.3 kg TOC/m^3 d	Gas sparging rate: 1.5 L/min	–[e]	Flux: 25 LMH Effluent COD: 5–10 mg/L Fouling rate: 0.33–2.52 kPa/h	Gas sparging; floc formation
CSTR (Robles *et al.*, 2013)	Municipal wastewater	HF Pore size: 0.05 μm Area: 30 m^2	Temp: 20, 25, and 33°C SRT: 70 d HRT: 5.5–24.5 h Filtration/relaxation: 250/50 s; 30 s of back-flush every 10 cycles; 40 s of ventilation every 10 cycles; 30 s of degasification every 50 cycles.	SGDm: 0.23 Nm^3/m^2 h	MLSS: 8–30 g/L EPS_c: 34 (33°C) and 24 (20°C) mg/gMLVSS EPS_p: 121 (33°C) and 74 (20°C) mg/gMLVSS SMP_c: 5 (33°C) and 2 (20°C) mg/gMLVSS SMP_p: 82 (33°C) and 14 (20°C) mg/gMLVSS	Flux: 10, 12, and 13.3 LMH Membrane permeability: 25–250 LMH/bar COD removal: 85%	Temperature; biomass concentration; EPS; SMP
CSTR (Chen *et al.*, 2017)	Synthetic low-strength wastewater	FS, chlorinated polyethylene Pore size: 0.2 μm Area: 0.116 m^2	Temp: 25°C HRT: 0.4–3.8 d OLR: 0.352.10 kg COD/m^3d	Specific gas flow rate: 2.6 m^3/m^2 h	MLSS: 6.38–9.26 g/L SMP_c: 5–8 mg/gMLVSS SMP_p: 15–30 mg/gMLVSS EPS_c: 6–17 mg/gMLVSS EPS_p: 35–90 mg/g MLVSS	Flux: 1.08–6.46 LMH Fouling rate: 1.2–3.4 × 10^{11} per m/d COD removal: 95%	EPS; SMP
CSTR and polishing aerobic MBR (Trzcinski & Stuckey, 2016)	Leachate	FS, polyethylene Pore size: 0.4 μm Area: 0.1 m^2	Temp: 35°C HRT: 48, 24, 16, 12 and 8 h OLR: 1–19.8 kg COD/m^3 d	Gas sparging rate: 5 L/min	MLSS: 3 g/L	COD removal: 96%	Inorganic fouling of calcium precipitates
CSTR + membrane tank (Huang *et al.*, 2008)	Glucose	FS Pore size: 0.45 μm Area: 0.118 × 2 m^2	Temp: 25–30°C SRT: 30 and 60 d HRT: 12 h OLR: 1.1 kg COD/m^3d Filtration/relaxation: 8/2 min	Gas sparging rate: 1.5 L/min SGDm: 30 Nm^3/m^2 h	MLSS: 4.3–5.6 and 5.7 g/L	Flux: 5.3 LMH COD removal: 99% Biogas production: 0.023 and 0.028 L CH_4/g MLVSS d	SRT; fine particles

(*Continued*)

Table 11.4 Summary of factors affect treatment performance and membrane fouling in AnMBRs (*Continued*).

Reactor Types[a]	Influent Types	Membrane[b] Types	Operating and Environmental Conditions[c]	Hydrodynamic [d] Conditions	Sludge Properties[f]	Membrane and Treatment Performance[g]	Major Factors
CSTR (Ding *et al.*, 2015)		HF,PVDF Pore size: 0.2 μm Area: 0.2 m^2	Temp: 35°C SRT: 300 d HRT: 12 h OLR: 1.1 kg COD/m^3 d	–	EPS_c: 1.06–1.27 mg/gMLVSS EPS_p: 6.76–7.96 mg/gMLVSS	–	EPS
CSTR + membrane tank	Domestic wastewater	HF, PVDF Pore size: 0.08 μm Area: 12.5 m^2	SRT: 100 d HRT: 16 d	UG: 0.02–0.078 m/s	MLSS: 7.7 g/L SMP: 598 mg COD/L SMP_c: 47 mg/L SMP_p: 108 mg/L	Flux: 6 LMH COD removal: 84%	Membrane configuration; seed sludge (flocculated and granulated); SMP; colloidal particles
EGSB + membrane tank (Martin-Garcia *et al.*, 2011)		HF, PVDF Pore size: 0.04 μm Area: 0.93 m^2	SRT: ∞ HRT: 16 d	UG: 0–0.057 m/s	MLSS: 0.1–0.6 g/L SMP: 198 mg COD/L SMP_c: 18 mg/L SMP_p: 50 mg/L	Flux: 6 LMH COD removal: 86%	
CSTR (Gao *et al.*, 2013)	Thermo-mechanical pulping pressate	FS, PVDF MWCO: 70 kDa	Temp: 37, 45, and 55°C OLR: 2.6 kg COD/m^3d	Gas sparging rate: 1.5 L/min	MLSS: 10.9 g/L EPS_c: 2–5 mg/gMLSS EPS_p: 25–35 mg/gMLSS SMP_c: 35–60 mg/L SMP_p: 400–600 mg/L	Flux: 5.6–6.9 LMH COD removal: 76–83%	Operating temperature; temperature shocks; microbial community; floc size; EPS; SMP; cake layer structure
EGSB (Chu *et al.*, 2005)	Synthetic domestic wastewater	HF, PE Pore size: 0.1 μm Area: 0.1 m^2	Temp: 11–25°C SRT: 145 d HRT: 3.5–5.7 h OLR: 1.6–4.5 kg COD/ m^3d Filtration/relaxation: 3/1.5 min	UG: 2.5–8 m/h	TSS: 14–21 g/L EPS_c: 4.4–6.1 mg/gVSS EPS_p: 0.6–1.6 mg/gVSS	Flux: 10.4 LMH TMP < 100 kPa COD removal: 85–96%	Operating temperature; gas sparging; suction mode; EPS
UASB (Wen *et al.*, 1999)	Domestic wastewater	HF, PE Pore size: 0.03 μm Area: 0.3 m^2	HRT: 6 and 4 h SRT: 150 d OLR: 0.5–12.5 kg COD/m^3d Filtration/relaxation: 4/1, 3/2, 2/0.5, 8/2 min	–	MLSS: 16–22.5 g/L	Flux: 5 and 10 LMH TMP < 70 kPa COD removal: 97% Fouling Rate: 0.2 and 0.5 kPa/h	Intermittent suction mode; membrane flux
CSTR (Aquino *et al.*, 2006)	Synthetic	PE Pore size: 0.4 μm Area: 0.1 m^2	Temp: 35°C SRT: 150 d HRT: 6 h	Gas sparging rate: 5 L/min	VSS: 2.6 (no PAC addition) and 3.7 (with PAC addition) g/L SMP: 116 (no PAC addition) and 180 (with PAC addition) mg COD/L SMP_c: 18 mg/L SMP_p: 52 (no PAC addition) and 58 (with PAC addition) mg/L	Flux: 10 and 20 LMH TMP: ~25 (no PAC addition) and ~10 (with PAC addition) kPa COD removal: ~96%	High-MW protein and carbohydrates from cell lysis and EPS

UASB (Lin *et al.*, 2009)	Kraft evaporator condensate	FS, PVDF MWCO: 70 kDa	Temp: 37 and 55°C HRT: 19.7 and 77 h OLR: 12.2 and 3.1 kg COD/m^3 d	Gas sparging rate: 0.75 L/min	MLSS: 8 g/L Total EPS: 52 and 57 mg/gVSS SMP: 250 and 850 mg COD/L	Flux: 7.2 and 2.4 LMH TMP: 10 and 25 kPa	Floc size; SMP; BPC; bound EPS; cake layer structure
UASB + membrane tank (Van Voorthuizen *et al.*, 2008)	Black water	HF, PVDF, MWCO: 250 kDa Area: 0.04 m^2	Temp: 37°C Filtration/relaxation: 8/1, plus 1 min backwash HRT: 12 h	UG: 8–16 m/h	SMP: 327 mg COD/L SMP_c: 81 mg/L SMP_p: 70 mg/L	Flux: 10 LMH COD removal: 91% CH_4 production: 0.27 g CH_4–COD/g COD	Feed composition; gas velocity; soluble and colloidal compounds
CSTR (Van Voorthuizen *et al.*, 2008)	Black water	HF, PVDF, MWCO: 250 kDa Area: 0.04 m^2	Temp: 37°C Filtration/relaxation: 8/1, plus 1 min backwash HRT: 12 h	UG: 40 m/h	SMP: 269 mg COD/L SMP_c: 45 mg/L SMP_p: 69 mg/L	Flux: 8 LMH COD removal: 86% CH_4 production: 0.35 g CH_4–COD/g COD	Soluble and colloidal compounds
CSTR (Lee *et al.*, 2008)	Glucose	FS, PE Pore size: 0.45 μm Area: 0.1 m^2	Temp: 35°C SRT: 450 d HRT: 9 h OLR: 8.8 kg TOC/m^3 d Filtration/relaxation: 7/3 min	Gas sparging rate: 5 L/min	MLSS: 9.5 g/L Total EPS: 2.3–235 mg TOC/L EPS_c: 5.6–58 mg TOC/L EPS_p: 3–179 mg TOC/L	Flux: 5 LMH TMP: 2–10 kPa Glucose removal: 99.6% Fouling rate: 0.0083 kPa/h	EPS
AnsMBR							
UASB (Cho & Fane, 2002)	Biomass	FS, PVDF Pore size: 0.22 μm	OLR: 4–12 kg COD/m^3d No relaxation	CFV: 0.93 m/s	EPS_c on membrane: 180–1500 mg/m^2	Flux: 30, 40 and 50 LMH TMP: 2.5–1.5, 2.5–40 and 2.5–10 kPa COD removal: 98% Fouling rate: <0.05, 0.38 and 0.75 kPa/h	EPS; cake layer formation
CSTR (Dereli *et al.*, 2018)	Whey permeate obtained from a cheese production plant	Tubular UF membranes Pore size: 0.03 μm Area: 0.014 m^2	Volume loading rate: 3.0 ± 1.3 5.0 ± 0.3 kg COD m^{-3} d^{-1} SRT: 300 and 50 d HRT: 10 6 and 6.0 d	CFV: 0.5 m/s	TSS : 460 and 730 mg/L VSS: 340 and 700 mg/L	Flux: 9–20 LMH COD removal: around 96 and 98%	Substrate composition
UASB (Jeison *et al.*, 2009b)	VFA mixture	Tubular, ceramic Pore size: 0.2 μm Area: 0.013 m^2	Temp: 55°C OLR: 10–55 kg COD/m^3d Filtration cycle: 5 min of filtration plus 20s of backflush	CFV: 1.5 m/s Gas superficial velocity: 0.1 m/s	VSS: 6–24 g/L	Flux: 20–40 LMH Fouling rate: <3 kPa/h	Cake formation; CFV; gas superficial velocity

(Continued)

Table 11.4 Summary of factors affect treatment performance and membrane fouling in AnMBRs (*Continued*).

Reactor Types[a]	Influent Types	Membrane[b] Types	Operating and Environmental Conditions[c]	Hydrodynamic [d] Conditions	Sludge Properties[f]	Membrane and Treatment Performance[g]	Major Factors
UASB (Salazar-Peláez *et al.*, 2011)	Synthetic municipal wastewater	Tubular, PVDF, MWCO: 100 kDa	HRT: 4, 8, and 12 h Filtration cycle: 6 h of filtration and 20 min of cleaning with NaOCl (300 mg/L).	CFV: 2 m/s	SMP and EPS concentrations increased with HRT reduction	TMP: 103.35 kPa COD removal: around 80%	HRT; EPS; SMP; floc size
CSTR (Sui *et al.*, 2008)	Synthetic	HF, PE Pore size: 0.4 µm Area: 0.05 × 2 m^2	HRT: 10.4 h OLR: 1.84–5.76 kg COD/m^3d Ultrasonic power intensity: 0.122 W/cm^2	CFV: 0.75 m/s	MLSS: 8–10 g/L	Flux: 24 LMH COD removal >80%	Ultrasounic irradiation; biomass concentration
CSTR (Zhang *et al.*, 2007)	Swine manure	Tubular, PES MWCO: 20 kDa	HRT: 6 d SRT: 118 and 211 d Loading rate: 1 and 2 kg VS/m^3d	CFV: 1.5–1.9 m/s	SS: 27 and 49 g/L	Flux: 5–10 LMH TMP: 20–70 kPa COD removal: 95.4 and 93.6%	CFV; cleaning frequency; gel layer formation; inorganic fouling
CSTR (Beaubien *et al.*, 1996)	Acetate	Ceramic, Pore size: 0.2 µm Area: 0.2 m^2	HRT: 1 d SRT: 30 d	CFV: 2 m/s	MLSS: 1.6, 7.2 and 22 g/L	Flux: 65, 47 and 27 LMH TMP: 35 kPa COD removal: 65–95%	TMP; CFV; biomass concentration
CSTR (Harada *et al.*, 1994)	Synthetic (high portion of particulate COD)	PS, 3000 Da Area: 0.02 m^2	HRT: 120, 80, and 48 h SRT: ∞ OLR: 1, 1.5 and 2.5 kg COD/m^3d Temp: 35°C No relaxation	CFV: 0.8 m/s	MLSS: 15 g/L SMP: 1,200 mg COD/L SMP_c: 80 mg/L SMP_p: 400 mg/L	Flux: 21–42 LMH TMP: 49 kPa Permeate COD <80 mg/L COD removal >98%	Sludge rheological properties: soluble organics and biomass concentrations
CSTR (Ho & Sung, 2010)	Synthetic municipal wastewater	Tubular, PTFE Pore size: 1 µm Area: 0.09 m^2	Temp: 15 and 25°C HRT: 12 h OLR: 1 kg COD/m^3d SRT: ∞	–	MLSS: 6–10 g/L SMP: 150–420 and 50–120 mg COD/L	Flux: 5 LMH TMP: 6.0–55.2 kPa COD removal: >95 and 85%	Operating temperature; cake layer formation

[a]CSTR = completely stirred tank reactor, UASB = upflow anaerobic sludge blanket bioreactor, EGSB = expended granular sludge bed reactor.
[b]FS = flat sheet, HF = hollow fibre, PVDF = polyvinylidene fluoride, PE = polyethylene, PES = polyethersulphone, PTFE = polytetrafluoroethylene.
[c]∞ = infinite.
[d]SGD_m = specific gas demand, UG = superficial gas velocity, CFV = cross-flow velocity.
[e]– indicates value not reported.
[f]EPS_c = carbohydrate content in EPS, EPS_p = protein content in EPS, SMP_c = carbohydrate content in SMP, SMP_p = protein content in SMP.
[g]LMH = L/m^2 h, PAC = powdered activated carbon.

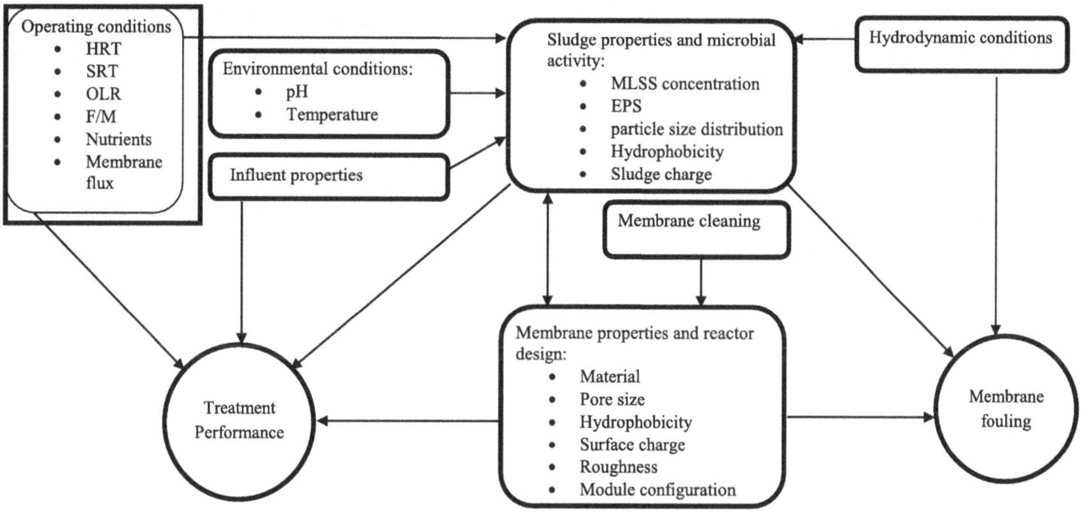

Figure 11.3 Factors affecting membrane fouling and treatment performance.

even pore-size distribution that have as little overlap as possible with the size distribution of the particles to be filtered can limit fouling due to pore blocking and enhance filtration efficiency (Stephenson *et al.*, 2000).

According to the nature of the membrane material, membranes can be categorized as polymeric (e.g. polyamide, polysulphone, polypropylene, polyethylene (PE), polyethersulphone (PES), polyvinylidene fluoride (PVDF), and polyvinyl chloride (PVC)), inorganic (e.g. ceramic, metallic, and porous glass), and modified materials (e.g. membrane modified by grafting of hydrophilic monomers on surface, organic/inorganic composite membranes) (Kang *et al.*, 2002). There is however limited use of inorganic membranes in AnMBRs because of their high cost and severe inorganic fouling (Choo & Lee, 1998; Choo *et al.*, 2000), in spite of the high initial flux (200–250 L/m^2 h) achievable (Ghyoot & Verstraete, 1997; Kim *et al.*, 2007). Polymeric membranes have been widely used in AnMBRs for wastewater treatment due to their low costs. Woven and non-woven filters have been used to reduce manufacturing and operating cost of the AnMBR. In spite of more than 99% of solids retention, the flux could only reach 3 L/m^2 h, accompanying unstable hydraulic operation, severe cake layer formation, and sudden increases in filtration resistance (Jeison *et al.*, 2008a). In general, hydrophobic membranes tend to favour the strong attachment of hydrophobic molecules (i.e., lipids, protein), thus reducing the applicable fluxes. Several studies have focused on the surface modification of hydrophobic polymeric membranes by grafting hydrophilic groups. The polypropylene (PP) membrane surface turned from strongly hydrophobic to hydrophilic by ozone treatment followed by graft polymerization with 2-hydroxy-ethyl methacrylate, and the modified membrane with 70% degree of grafting showed 13.5% of flux enhancement compared with virgin PP membrane (Sainbayar *et al.*, 2001). Other hydrophilic monomers, such as acrylic acid, 2,4-phenylenediamine, and ethylene diamine were grafted on the PVDF membrane surface by UV photo-grafting to improve the antifouling properties and flux recovery ability (Rahimpour *et al.*, 2009). On the other hand, membranes with negative or neutral charge have lower fouling potential than positively charged membranes, as most bacterial flocs are negatively charged with a zeta potential from −10 to −30 mV. The membrane with a hydrophilic, more negatively charged, and smooth surface would be more effective in reducing the initial adhesion of microorganisms (Kang *et al.*, 2006).

However, the membranes will not be in direct contact with the suspension anymore when cake or gel layers form to act as a secondary membrane, reducing pore blocking and influencing the filtration performance. Actually, a cake layer could help to improve the effluent quality by reducing the amount of organics passing through membrane (Akram & Stuckey, 2008b). Therefore, dynamic membrane technology is under development by taking advantage of this secondary membrane formed on the support material (meshes or fabrics) instead of using actual membrane (Zhang *et al.*, 2010).

11.6.2 Effects of operating and environmental conditions

Anaerobic digestion is a complex process and is greatly influenced by many factors. These factors include wastewater characteristics, HRT, influent organic concentration, OLR, pH, temperature, and nutritional requirements. It is well known that sludge properties are affected by operating and environmental conditions. Therefore, a change in operating and/or environmental conditions in AnMBRs may lead to changes in sludge properties, membrane fouling behaviour, and finally membrane filtration performance.

11.6.2.1 Effects of SRT and HRT

SRT variations directly alter sludge composition and MLSS concentration (Huang *et al.*, 2011). Longer SRT may achieve better treatment performance in terms of COD removal and biogas production (Huang *et al.*, 2011; Trzcinski & Stuckey, 2010). At the same HRT of 12 h, the methane production rate in an AniMBR with an SRT of infinite days (1.29 LCH_4/d) was almost double the value with a SRT of 30 days (0.67 LCH_4/d) (Huang *et al.*, 2011). An AniMBR operated at an SRT of 300 days could maintain 95% soluble chemical oxygen demand (SCOD) removal at both 35 and 20°C, whereas AniMBR operated at an SRT of 30 days could only achieve this performance at 35°C (Trzcinski & Stuckey, 2010). Additionally, at an SRT of 30 days the AniMBR resulted in extensive membrane fouling due to lack of hydrolysis of organic matter (Charfi *et al.*, 2012; Trzcinski & Stuckey, 2010). However, SRTs (>50 days) were found to increase the MLSS, SMP concentrations, and fine particles, which deteriorated effluent quality and increased membrane fouling (Chu *et al.*, 2005; He *et al.*, 2005; Hu & Stuckey, 2006; Huang *et al.*, 2008, 2011). In another study, no clear difference in membrane fouling was detected under the SRT of 60 and 100 days (Herrera-Robledo *et al.*, 2010). Noting the difference in observations made in independent studies regarding the effect of SRT, Meng *et al.* (2009) argued that the optimum SRT is governed by the specific operating conditions at which an MBR is operated.

OLR is determined by the influent organic concentration and HRT. Long-term starvation will induce total loss of cell activity and alter sludge properties (Wu & Lee, 2011). On the other hand, a shorter HRT (<15 h) or higher OLR may cause a drastic decrease in the removal performance or increase in membrane fouling rates of AnMBR systems (Saddoud & Sayadi, 2007; Salazar-Peláez *et al.*, 2011). An HRT of 2 days led to a higher biogas yield (0.45 L/g COD removed) compared with an HRT of 5–7 days (0.3 L/g COD removed) (Bohdziewicz *et al.*, 2008). The reason was that the proper range of redox potential for biogas production during fermentation process was determined as −350 to −450 mV. The redox potential (−370 mV) in the AnMBR at an HRT of 2 days was in this range, while either a higher (−320 mV at an HRT of 1 day) or lower (−466 to −417 mV at an HRT of 7 days) redox value can inhibit methanogenesis. Total VFA removal efficiency of a two-stage thermophilic AnMBR decreased from 96.3 to 82% as the OLR increased from 5 to 12 kg COD/m^3 d (Wijekoon *et al.*, 2011). In addition, a shorter HRT or higher OLR provides more nutrients to the biomass, leading to a change in biological activity and a higher MLSS (Huang *et al.*, 2011; Wijekoon *et al.*, 2011). Therefore, the decrease of HRT or increase of OLR will induce the production of more EPS because bound EPS is growth related and is produced in direct proportion to substrate utilization (Baek & Pagilla, 2006; Laspidou & Rittmann,

2002), and this may lead to aggravated membrane fouling (Salazar-Peláez *et al.*, 2011). At the start-up period of AnMBRs, long acclimatization period under long SRTs and low OLRs are required for the successful development of slow-growing populations (Akram & Stuckey, 2008a).

11.6.2.2 Effects of temperature and pH

Anaerobic digestion can be conducted in psychrophilic (<25°C), mesophilic (25–40°C), and thermophilic (>45 °C) temperature ranges (El-Mashad *et al.*, 2004). Temperature is commonly believed to play a significant role in the biological wastewater treatment performance and stability, especially for the sensitive anaerobic processes where sudden transient changes can lead to deterioration of the reactor's performance (Ahring *et al.*, 2001; Bouskova *et al.*, 2005; Chae *et al.*, 2008; Ndegwa *et al.*, 2008; Yilmaz *et al.*, 2008). The increase in temperature results in much faster chemical and biological reactions, but not necessarily a better treatment performance of AnMBRs. The COD removal efficiency of an AnMBR was higher at 25°C (>95%) than that at 15°C (>85%) (Ho & Sung, 2010). However, the average COD removal efficiencies of an AniMBR treating thermo-mechanical pulping pressate at 55°C (76%) was slightly lower than that at 37°C (83%) (Gao *et al.*, 2011a). Meanwhile, the extent of membrane fouling increased with an increase in the operating temperature. The fouling resistance had a close relationship with the combined effect of all the factors, such as floc size and the amount of cake sludge (Gao *et al.*, 2012). Temperature variations can occur due to seasonal fluctuations, frequent temperature transits of wastewater streams, or the accordant junction of them or failure to control temperature (unexpected cooling and heating problems). It is important for a system to deal with these situations in industrial applications where wastewaters are discharged at a high strength and a wide range of temperatures. In conventional anaerobic reactors, the effects of temperature variations have been linked with poor sludge settling that brings high effluent SS, effluent turbidity and biomass washout (Ahn & Forster, 2002). In addition, temperature has significant impact on microbial activity and community composition (El-Mashad *et al.*, 2004). The effect of temperature on acetogenesis and methanogenesis is more significant than hydrolysis and acidogenesis, as those bacteria are more specialized (Rajeshwari *et al.*, 2000). By the incorporation of the membrane filtration technology with complete solid–liquid separation, AnMBRs are ideal candidates to alleviate the problems associated with conventional anaerobic reactors. A few studies have assessed the impact of temperature on the performance of AnMBRs. A higher viscosity and soluble COD concentrations in the AniMBRs associated with a reduction in flux was observed at 20°C than at 35°C (Martinez-Sosa *et al.*, 2011a; Trzcinski & Stuckey, 2010). Under psychrophilic conditions, some compounds (e.g., long-chain acids and alkanes) became recalcitrant, however, satisfactory treatment efficiencies were still achieved if high SRTs could be maintained to compensate for the low activity at lower temperatures (Trzcinski & Stuckey, 2010). Gao *et al.* (2011a) reported a COD removal efficiency of 76–83% to treat thermomechanical pulping pressate by using an AniMBR at the operating temperatures of 37, 45 and 55°C. In that study, the AniMBR could tolerate 5/10°C temperatures shocks at 37°C and the temperature variations from 37 to 45°C. The 5°C shocks at 45°C had minor impact, while only larger magnitudes of temperature shock (10°C) at 45°C followed by the temperature variation from 45°C to 55°C caused severe deterioration in COD removal and biogas production of the AniMBR (Gao *et al.*, 2011a), showing significant resistance of the AniMBR process against temperature variation.

The optimum pH for anaerobic digestion appears to be 6–8 (Ward *et al.*, 2008). The variation in pH can greatly influence the cell morphology, structure, and microbial metabolism including utilization of carbon and energy sources, efficiency of substrate degradation, synthesis of proteins and various types of storage materials, and release of metabolic products from cells, and thus sludge properties (e.g., flocculation and adhesion phenomena) (Baily & Ollis, 1986; Gottschalk, 1986). The pH condition may also affect membrane functions directly (membrane morphology, surface charge and the pore structure of membranes) (Dong *et al.*, 2006; Nanda *et al.*, 2010; Tadkaew *et al.*, 2010). In practice, the combination of various wastewater streams with different pH values in an integrated plant will result in changes in pH with time and cause pH shocks in downstream biological treatment systems (Sipma *et al.*, 2010). In addition, the failure of pH control systems may also occur, causing incidental pH shocks. Some studies were carried out on the effect of acidic pH on the anaerobic digestion system (Yu & Fang, 2003). The production of propionate and hydrogen was favoured at pH 4.0–5.0 compared with the predominant production of acetate, butyrate, and i-butyrate at pH 6.0–7.0, demonstrating the significant impact of pH on performance. The optimum pH for the overall acidogenic activity was found to be 6.0. Gao *et al.* (2010b) found that a pH 8.0 shock had limited impact on the performance while significantly alkaline shocks (pH 9.0 and 10.0) had significant impact on the performance of an AniMBR for pulp and paper effluent treatment.

11.6.2.3 Wastewater composition

The wastewater composition, i.e., types and characteristics of contaminant and nutrients can significantly influence treatment performance and membrane fouling. The presence of micronutrients and trace elements is crucial for microbial growth and performance, although the required concentrations are very low. For example, the concentrations of iron, nickel, and cobalt are relatively high in methanogenic bacteria (Rajeshwari *et al.*, 2000). The substrate type and the protein/polysaccharide ratio (PN/PC) can affect EPS and SMP production as well as the formation of colloids (Arabi, 2009). Toxic/inhibitory substances (e.g., heavy metals, ammonia, and sulphide) present in wastewater of several industries (e.g., metal finishing, hydrometallurgical, pulp and paper, textile, refining, petrochemical, tanneries, and battery manufacturing companies) may strongly inhibit the activity of microorganisms and hence deteriorate the performance of anaerobic digestion (Chen *et al.*, 2008a). For instance, the wood processing effluents containing resin compounds, chlorinated phenolics, and tannins have high toxicity on methanogens (Vidal & Diez, 2005). An AnMBR was unable to reach its steady-state because of the toxicity in the domestic wastewater composition (Saddoud *et al.*, 2009). Liao *et al.* (2010) suggested pre-treatment for the removal of toxic sulphur compounds in kraft evaporator condensate before treatment by an AniMBR. Furthermore, wastewater type results in distinct treatment efficiency and membrane fouling behaviour. For the same AniMBR system, the sustainable membrane flux (4.6 L/m^2 h) for thermomechanical pulping whitewater treatment was much lower than that (6.9 L/m^2 h) for thermomechanical pulping pressate treatment (Gao *et al.*, 2011a, b). Significant differences in particle-size distribution, colloidal particle content, the proteins/polysaccharides ratio, and soluble compounds molecular weight distribution were observed between the two types of wastewaters. The higher PN/PC ratio in the supernatants as well as the presence of a fraction of colloids (0.1–10 μm) in feed waters and mixed liquors strongly corresponded to the lower sustainable membrane flux and higher membrane filtration resistance in the AniMBR (Gao *et al.*, 2013). Nitrogen deprivation was found to result in a reduced sludge particle size and decreased sludge filterability, and thus adversely affected the biological treatment performance and reactor stability (Dereli *et al.*, 2018).

11.6.3 Hydrodynamic conditions

Hydrodynamic conditions such as shear intensity have a complex influence on membrane performance. Applications of air sparging in AnMBRs can only be found in special system designs (Judd, 2011), for instance, two-phase (acidogenic/methanogenic) anaerobic reactors with aeration and filtration only in the acidogenic phase (Lee *et al.*, 2001), nitrification/denitrification MBR systems with intermittent aeration to achieve an aerobic–anaerobic cycle (Nagaoka & Nemoto, 2005). In AniMBRs, biogas can be recirculated to carry out the functions of maintaining the solids in suspension and promoting scouring of the membrane surface (Vyrides & Stuckey, 2009a; Xie *et al.*, 2010). The size and motion of bubbles have effects on generation of secondary liquid flows that enhance the mass transfer between the liquid and the membrane surface. In most cases, the intensity of biogas scouring was simply described as gas flow rate (Akram & Stuckey, 2008b), although superficial gas velocity (UG) values (Jeison & Van Lier, 2006a) and specific gas demand (SGD_m) (Huang *et al.*, 2008) have been also used for AnMBRs. The membrane fluxes in AniMBRs are influenced by sparging intensities, TMPs, and membrane filtration/relaxation modes (Berube *et al.*, 2006). Increasing the sparging intensity may be able to improve membrane permeability by only to some extent. For example, doubling the applied gas sparging rate resulted in an increase in critical flux of only 1–3 L/m^2 h (Jeison *et al.*, 2008b). The contribution of increasing UGs (3–8 m/h) and relaxation time (1–2 min) to membrane permeability was only limited for short-term (<10 days) operation (Chu *et al.*, 2005). A practical limit of sparging intensity appears to exist above which no more flux improvement and membrane fouling control is observed (Jeison *et al.*, 2008b; Yang *et al.*, 2011). Recently, Robles *et al.* (2012) conducted a systematic study on the effect of biogas sparging intensity on membrane flux and fouling in a pilot-scale AniMBR treating municipal wastewater. They found that there is a linear dependency between critical flux and the specific gas demand per unit of membrane area (Robles *et al.*, 2012); no chemical cleaning was required during almost 2 years of operation under sub-critical flux conditions, indicating absence of irreversible fouling.

In AnsMBRs, membrane surface shear is performed by cross-flow velocity (CFV) which is usually employed to provide membrane driving force and maintain high membrane permeability. The membrane fluxes in AnsMRRs are mainly governed by CFVs and TMPs (Berube *et al.*, 2006). Beaubien *et al.* (1996) concluded that at low TMP, the flux depended on applied TMP; at high TMP, the flux is determined by hydrodynamic conditions. Increasing shear intensity could increase energy cost and disrupt sludge flocs, producing small size particles and releasing more EPS which deteriorate membrane performance (Jeison *et al.*, 2009b). Furthermore, some researchers observed negative effects of high shear conditions over microbial activity (Choo & Lee, 1996a; Padmasiri *et al.*, 2007). Brockmann and Seyfried (1997) observed that microbial activity reduced to half after 20 times of sludge circulation, and 90% of microbial activity was lost within 100 cycles.

11.6.4 Sludge properties

A complex system like AnMBR is basically an enclosed system that concentrates organic foulants in the sludge suspension. Membrane fouling is influenced by the components in the mixed liquor (i.e., cells, cell debris, microbial metabolites, and substrate components). Not only SS, but also many other fractions like colloids, soluble organic matter, inorganic precipitates and EPS can contribute to the overall performance (Charfi *et al.*, 2012; Trzcinski & Stuckey, 2016). The resistance-in-series (RIS) model has been used frequently to analyse the membrane fouling phenomenon and to characterize the relative significance of each resistance component in heterogeneous MBRs (Khalili-Garakani *et al.*, 2011). However, it should be noted that a simple addition of individual resistances, as is done in RIS model, is questionable (Chang *et al.*, 2009).

11.6.4.1 Mixed liquor suspended solids

The concentration of mixed liquor suspended solids (MLSS) can significantly influence membrane filtration performance. MLSS concentration is considered to directly cause the increase of cake layer resistance (Rc) which is often expressed by the following equation (Chang *et al.*, 2001):

$$Rc = \alpha \cdot V \cdot C_b \tag{11.1}$$

where α is specific cake resistance (m/kg); V is the permeate volume per unit area (m^3/m^2); and C_b is the bulk MLSS concentration (kg/m^3). A number of studies have reported drop in permeate flux as biomass concentration increased. For example, Beaubien *et al.* (1996) reported the membrane flux, which decreased from 4.68 L/m^2 h at 2.5 g/L SS concentration to 2.52 L/m^2 h at 22 g/L, depended on SS concentration in a AnsMBR. The rapid flux decline turned to be moderate when the SS increased to above 2.5 g/L (Beaubien *et al.*, 1996). In a AniMBR, Jeison and van Lier (2006a) reported a first-order relation between critical flux and biomass concentration under mesophilic conditions, while under thermophilic conditions, biomass concentration presented a value below which no further effect on critical flux was observed. Robles *et al.* (2013) found membrane permeability proportionally decreased with any increase in the MLSS level when the AniMBR was operated at both psychrophilic and mesophilic conditions. Ghyoot and Verstraete (1997) also observed decreased permeate fluxes from around 450 to 250 L/m^2 h with increased TSS concentration from 6 to 25 g/L at a TMP of 180–200 kPa. When the biomass concentration was increased, the convective flow of solids towards the membrane surface increased correspondingly (Jeison & van Lier, 2006a). The flux decline was mainly caused by a notable increase in colloidal particle concentration at higher TSS concentrations (Ghyoot & Verstraete, 1997). The present colloidal particles tends to cause pore clogging because of their size ranges, to have slower back transport velocities than other coarser particles due to the lower diffusion rates, and thus to increase membrane fouling tendency (Choo & Lee, 1998). In a study by Baek *et al.* (2010), the deposition rate of the SS on the membrane surface decreased as the MLSS concentrations decreased from about 6 to 3 g/L (Baek *et al.*, 2010). By contrast, Van Zyl *et al.* (2008) and Zhang *et al.* (2007) reported no reduction of membrane flux up to an MLSS concentration of 30 and 106.1 g/L, respectively. More recently, Robles *et al.* (2012) successfully operated a pilot-scale AniMBR system for municipal wastewater treatment for almost 2 years at a high MLSS level (>22 g/L). MLSS concentration is intimately related to viscosity of biomass suspension, and an increase in sludge viscosity will weaken the shear stress close to the membrane, resulting in a substantial TMP increase (Berube *et al.*, 2006). Akram and Stuckey (2008b) found the increase in TSS (from 5 to 23 g/L) led to an increase in viscosity (from 7 to 14 cP), resulting in a significant flux drop from 5 to 2 L/m^2 h. For each MBR system, a critical MLSS concentration exists under which the viscosity remains low and rises only slowly with the MLSS concentration and above which viscosity tends to increase exponentially with the solids concentration (Le-Clech *et al.*, 2006).

11.6.4.2 Particle-size distribution

The particle-size distributions were compared between AnsMBRs and AniMBRs by Judd (2011) through a number of publications. It seems that the mean particle sizes (3–16 μm) in AnsMBRs was much smaller than that (65–90 μm) in the AniMBRs due to the floc breakage during pumping for sludge recirculation in side-stream configurations. The Carmen–Kozeny equation describes the impact of particle-size distribution on the cake layer resistance: the smaller the floc sizes are, the greater the cake resistance is. The magnitude of back-transport mechanisms is strongly related to particle size as the shear-induced hydrodynamic diffusivity is proportional to the square of the particle diameter multiplied by the shear rate (Tardieu *et al.*, 1998). Consequently, a particle with a larger size on the membrane surface can be detached much more easily at the same shear conditions. During the study of anaerobic biosolids deposition on membrane surface in an AnsMBR, an exponential flux decline was caused by the continuous floc size reduction from a mean diameter of about 16 to 7 μm (Choo & Lee, 1998). The characteristics of different fractions of microbial flocs (small flocs vs. large flocs) from an AniMBR were determined to identify their role in membrane fouling (Lin *et al.*, 2011). A higher level of bound EPS was associated with smaller flocs experiencing a higher tendency to deposit over the membrane surface (Lin *et al.*, 2011). In an AniMBR treating low-strength wastewater, the fine flocs was found promoting denser cake layer formation due to their smaller back transport velocity and denser structure when compared with large flocs (Liu *et al.*, 2012). Recent studies have shown that smaller flocs in bulk sludge are accumulated on membrane surfaces to form cake layers in AniMBRs (Gao *et al.*, 2011b; Lin *et al.*, 2009, 2011).

11.6.4.3 Extracellular polymeric substances and soluble microbial products

Due to the presence of the membrane in the AnMBR process, the large macromolecules (polysaccharides, proteins (including enzymes), DNA, lipids, nucleic acids, and humic substances) are retained in the sludge suspension by the membrane. EPS are high MW secretions biosynthesized and released by microbial cells due to cell metabolism and autolysis. EPS enhances the integrity and stability of flocs or biofilm as well as provides a suitable and protective environment for the survival and robustness of microorganisms. EPS and SMPs have received considerable attention in recent years with many reports indicating that they are one of the most significant factors affecting fouling in AnMBRs. Harada *et al.* (1994) concluded that soluble organics, rather than MLSS, primarily determined the formation of fouling layer. The dramatic increase in bound EPS contents (polysaccharides: from below 5.6 to 58 mg/L; protein: from below 3 to 179 mg/L) was deemed responsible for decreased flux and fast bio-fouling during hydrogen fermentation from an AniMBR (Lee *et al.*, 2008). Lin *et al.* (20011b) revealed that small flocs, abundant bound EPS and inorganic materials play important role in the formation process and the high specific filtration resistance of cake layer in an AniMBR. Aquino *et al.* (2006) determined the high-MW fraction of the backflush water as protein and polysaccharide. These SMP originating from cell lysis seemed to contribute to the internal fouling (Aquino *et al.*, 2006). Stuckey and colleagues have conducted a series of studies about SMP production in anaerobic digesters. Stuckey (2012) concluded that in anaerobic systems, any type of stress (e.g., high SRTs (Vyrides & Stuckey, 2009b), nutrient deficiencies with the presence of toxic compounds (Aquino & Stuckey, 2004a), organic and hydraulic shock loads (Aquino & Stuckey, 2004b)), change in the type of feed (Barker & Stuckey, 2001)) can lead to dramatic increase in the production of SMPs causing increase in fouling. Recently, Chen *et al.* (2017) noted that high OLR could lead to SMP-induced pore blockages as well as EPS-induced fast-growth and compact cake layer on the membrane surface, causing severe fouling. However, it is worth pointing out that the information regarding the role of EPS and SMPs in membrane fouling in AnMBRs is still limited. In aerobic MBRs, the effect of EPS/SMPs is rather debatable as discussed in previous chapters (Drew *et al.*, 2006). Therefore, further research is needed with respect to the correlation between EPS/SMPs and fouling in AnMBRs with the factors influencing their production and composition.

The components of EPS are either tightly bound to cells (TB-EPS) or loosely associated with cells (LB-EPS), and recently proposed modified EPS extraction methods (Liu *et al.*, 2012) allow for selective extraction of LB-EPS and TB-EPS. Methods of bound EPS extraction include chemical methods (ethylenediaminetetraacetic acid (EDTA), formaldehyde + NaOH, glutaraldehyde), physical methods (heating, sonication, cation exchange resin, sonication + cation exchange resin, and centrifugation alone) (Comte *et al.*, 2006). Chemical extraction usually results in most effective extraction of bound EPS, whereas the cation exchange resin and heating are preferred methods because of their simplicity (Le-Clech *et al.*, 2006). The dominant components in extracted EPS, polysaccharides and proteins, are commonly determined according to the phenol-sulphuric acid method with glucose as standard (Dubois *et al.*, 1956) and Folin method with bovine serum albumin as standard (Lowry *et al.*, 1951), respectively. In an AniMBR, cake layer had higher EPS content (both proteins and polysaccharides) than that of bulk sludge, indicating significant EPS accumulation in the cake layer (Lin *et al.*, 2011b). The EPS extracted from the membrane foulants exhibited broader distributions of MW compared with the EPS in the influent and the effluent (An *et al.*, 2009). Moreover, the methanogenic activity of cake sludge was much lower than that of bulk sludge (Ho & Sung, 2010). Two possible reasons were proposed by the authors: because of the intense shear force, the active microorganisms tend to grow in suspension. The other reason could be the prolonged SRT in anaerobic process, so that inert or endogenous phase microorganisms increased in the system (Ho & Sung, 2010). Ding *et al.* (2015) found that the LB-EPS extracted from bulk sludge and soluble EPS extracted from cake sludge induced the largest flux decline in a mesophilic AniMBR. Besides, it has been reported that the biofilm in a low dissolved oxygen (DO) MBR (DO < 0.1 mg/L) had higher amount of EPS and lower porosity than the biofilm in a high DO MBR (DO $= 6.0$ mg/L), resulting in faster decline of membrane permeability and membrane fouling rate under anoxic or anaerobic conditions (Kim *et al.*, 2006). However, Yun *et al.*, (2006) reported that for two MBRs operated at aerobic (DO $= 6.0$ mg/L) and anoxic (DO $= 0.3$ mg/L) conditions, the EPS were spread out more uniformly in the anoxic biofilm in spite of lower amount (22 mg-polysaccharide/g-MLVSS) compared to that in the aerobic biofilm (55 mg-polysaccharide/g-MLVSS). The higher membrane fouling rate in the anoxic MBR was caused by the anoxic biofilm with a uniform and dense structure. Thus, not only the quantity and composition of EPS, but also its spatial distribution may affect membrane filterability and requires more attention in future research.

SMP, often termed as 'soluble EPS', are defined as the pool of organic compounds that are released into solution from substrate metabolism (substrate-utilization-associated products (UAP)) and biomass decay (biomass-associated products (BAP)) (Barker & Stuckey, 1999; Laspidou & Rittmann, 2002). They can also originate from the feed. Although SMPs were anaerobically biodegradable, a relatively long residence time (11 days in some cases) was required (Schiener *et al.*, 1998). In MBR systems, part of them can be retained and accumulated by membrane or even cake layer. There are mainly three methods of separating the liquid phase from the biomass: filtration through a filter paper (0.45 μm), centrifugation and sedimentation. After extraction, the measurement of SMP (protein and polysaccharide components) is similar as EPS. Dissolved organic carbon (DOC) was suggested as an alternative to complex and costly measurements of SMP components (Lyko *et al.*, 2008). Accumulated SMP in the mixed liquor generally causes inhibition to the metabolic activity of the activated sludge and has a negative influence to the membrane permeability (Tian *et al.*, 2011). Particularly in AnMBRs, SMP can block membrane pores, adsorb on the membrane surface, form a gel layer, and/or build up on cake layers through physical and chemical adsorption, leading to smaller filtration areas, greater hydraulic resistance and finally a decrease in membrane permeability (Aquino *et al.*, 2006; Liao *et al.*, 2004; Herrera-Robledo *et al.*, 2011).

11.6.5 Strategies for performance stability and membrane fouling control

11.6.5.1 Reducing the fouling rate

Several common strategies to reduce membrane fouling rate and maintain sustainable operation have been reported: membrane modification (Bae *et al.*, 2006), sustainable flux operation (sub-critical flux operation) (Robles *et al.*, 2012), flux stoppage (relaxation) (Wen *et al.*, 1999), gas sparging (Huang *et al.*, 2011), control of operating conditions such as optimized SRT, HRT and biomass concentration (Liao *et al.*,

2006), and addition of mixed liquor modifying additives such as activated carbon/coagulant/flocculent (Baêta *et al.*, 2012).

Membrane fouling rate can be reduced by operating a membrane below the critical flux. The critical flux was introduced over 15 years ago (Field *et al.*, 1995) and was originally defined as the flux below which no fouling occurs. However, in complex systems fouling is inevitable, even at no flux conditions, because fouling occurs as soon as the membranes touch the bulk sludge. Consequently, the definition of the critical flux has been modified as the flux above which the relation between flux and TMP becomes non-linear (Jeison, 2007). Among the several proposed critical flux determination methods (Tiranuntakul *et al.*, 2011), a flux-step method can be easily applied in any type of membrane process, both at lab and full scale (Le Clech *et al.*, 2003). In order to refer to a certain flux for long-term operation, the concept of sub-critical flux or sustainable flux, the flux below critical flux with acceptable value dTMP/dt, is considered for gaining tolerable levels of fouling to operate MBRs for extended periods of time (Jeison & Van Lier, 2006b). It is feasible to control membrane fouling by selecting suitable operating parameters since they govern biomass growth and decay as discussed in Section 11.6.2. Prolonged stable operation periods can be maintained by other methods such as membrane relaxation (flux stoppage) that is typically applied to encourage diffusive back transport of foulants away from the membrane surface under a concentration gradient. The effect of back transport can be enhanced by shear force created by gas scouring.

The modification of suspension properties has been tested in order to control membrane fouling and enhance membrane filterability. The addition of powdered activated carbon (PAC), granular activated carbon (GAC), membrane flux enhancer (MFE, the modified cationic polyelectrolyte and $FeCl_3$), or zeolite will lead to the flocculation of activated sludge, entrapment of colloids and soluble compounds, and formation of a more porous cake layer (Akram & Stuckey, 2008b; Wu *et al.*, 2009). PAC can also provide solid support for biomass growth to form strong and dense biologically activated carbon which enhances flocculation ability and prevent accumulation on membranes (Hu & Stuckey, 2007). Akram and Stuckey (2008b) observed that when applied at a lower dosage (1.67 g/L), PAC addition can improve membrane performance in terms of cost, flux and soluble COD removal; however, addition of 3.4 g/L PAC significantly increased the sludge viscosity and consequently led to reduced flux. The monomeric and polymeric coagulants (e.g., alum, $Al_2(SO_4)_3$, $FeCl_3$, cationic polyelectrolytes, polymeric aluminium chloride, etc.) that are extensively applied in water treatment, can be added directly into MBRs for removing colloidal and soluble organic fractions, and thereby reduce fouling (Meng *et al.*, 2009).

Ultrasound has been proved to be effective to enhance membrane permeability in an AnMBR due to cavitation effect and acoustic streaming (Xu *et al.*, 2011). However, ultrasonic irradiation may have slightly negative effect on the activity of anaerobic bacteria (Sui *et al.*, 2008). A number of other new methods, such as ozonation, UV irradiation, and electric field have been applied for aerobic MBRs with positive results. They can be potentially feasible to mitigate membrane fouling for AnMBRs. For instance, sludge was effectively moved away from the membrane by an electric repulsive force, because the surface of the activated sludge is negatively charged (Akamatsu *et al.*, 2010).

11.6.5.2 Membrane cleaning

Reducing the rate of fouling can prolong the length of time between cleanings. However, at some point, when further filtration is no longer sustainable due to elevated TMP or low flux, the membrane must be cleaned. Physical cleaning is normally achieved by reversing permeate through the membrane whilst continuing to scour the membrane with bubbling. Backwashing has been found to successfully remove most of the removable fouling (Choo *et al.*, 2000; Lee *et al.*, 2001; Lew *et al.*, 2009). Chemical cleaning has been widely applied for cleaning membranes in MBRs either in situ or ex situ. The chemical agents include NaOCl, citric acid, NaOH, H_2O_2, EDTA, Cl_2, etc. (Lee *et al.*, 2001; Zhang *et al.*, 2007). For example, alkaline cleaning (NaOH) has been used to remove biological fouling. NaOH and ethanol were able to eliminate inner pore fouling, and ethanol could also restore the hydrophilicity of the PVC membrane (Tian *et al.*, 2010). The prevalent cleaning agents are sodium hypochlorite (for organic foulants) and citric acid (for inorganic foulants). Alternatively, a low concentration of chemical agents can be added to the backflush water to produce 'chemically enhanced backfulsh'. Citric acid has been found to be an effective cleaning agent for inorganic fouling removal in AnMBRs (Jeison & Van Lier, 2007a; Vallero *et al.* 2005). Compared to backwash-driven chemical cleaning, dynamic submerged chemical cleaning in external tanks was found more effective cleaning non-uniform fouling layer on the membrane (Ramos *et al.*, 2014).

11.7 COMMERCIAL POTENTIAL OF AnMBRs

11.7.1 Water reuse and energy production

Due to the uncontrolled increase of water consumption, the importance of wastewater reuse has been recognized as one of the most efficient alternatives for mitigating the worldwide pressure on water crisis. For example, wastewater is reused for irrigation, groundwater recharge, fountains in parks, sanitary flushing, different industrial processes as well as indirect and even direct potable water supply. On the other hand, because a certain amount of energy is stored in wastewater, recently co-generation of energy with wastewater treatment has appeared as an attractive option. Anaerobic digestion provides the opportunity to recover energy from wastewater through biogas production in the form of converting complex organic matters into carbon dioxide and methane. Meanwhile, membrane technology plays a key role in a number of prominent water reuse schemes worldwide, for the purpose of complete biomass retention allowing higher SRT and harbouring microbes of high degradation capacity. Therefore, AnMBRs present potential for producing energy and reusable water in combination with other treatment options.

Wastewater contains various contaminants (e.g., pathogens, hazardous chemicals). To meet the water quality requirements for wastewater reuse, the removal of nitrogen, phosphorus, trace organic contaminants (e.g., endocrine disrupting chemicals (EDCs) and pharmaceuticals and personal care products (PPCPs)), and bacteria are essential. In AnMBRs, membrane can readily retain bacteria. However, anaerobic process alone cannot achieve effective removal of nutrients and colour. Further treatment, for example by aerobic processes, is needed for water use (Lu *et al.*, 2011; Rezania *et al.*, 2006; Trzcinski & Stuckey, 2009b). Effluent polishing steps can only be skipped for special water reuse needs when nutrients are desirable (e.g., agricultural irrigation (Martinez-Sosa *et al.*, 2011a)). The conventional removal of nitrogen is nitrification followed by denitrification in activated sludge process. The major issues during this process reside in the energy consumption of aeration, slow growth of nitrifying bacteria, and the needs to add external carbon source for denitrification. Also, trace organic contaminants are not effectively removed by most conventional wastewater treatment plants (WWTPs). Thus, many combined techniques have been proposed to produce reusable water (Chan *et al.*, 2009). For instance, the elimination of selected trace organic contaminants by a full-scale anaerobic/anoxic/aerobic-membrane bioreactor was investigated by Xue *et al.* (2010). This process removed more than 90% of most selected EDCs (bisphenol A, estrone, 17ß-estradiol, 17α-estradiol, estriol, 17α-ethinylestradiol) and PPCPs (caffeine, *N*, *N*-diethyl-*m*-toluamide, erythromycin, trimethoprim, and ketoprofen). However, the removal efficiencies of two other EDCs (4-octylphenol and 4-nonylphenol), two musk fragrances (galaxolide and tonalide) and three other PPCPs (carbamazepine, diclofenac, and sulpiride) were only less than 80, 60 and 20%, respectively. A novel approach of partial aeration of an AnMBR using oxygen-transferring membranes was applied to improve nitrogen removal efficiency (Kappell *et al.*, 2005). An AOBF/MF system (anaerobic–oxic–anoxic biofilm filtration followed by membrane filtration) was evaluated to treat a blended wastewater comprising domestic wastewater, black water, and landfill leachate (Hyun & Lee, 2009). The treated effluent met the reuse standard for industrial and agricultural water. Denitrification was accomplished by the parts under anaerobic and anoxic conditions, and nitrification and phosphorus intake were achieved by the parts under aerobic conditions.

Biogas production from anaerobic digestion has received the greatest attention, especially for the industries generating high-strength wastewaters. The higher strength wastewater stores greater energy because of the higher organic contents. The methane potential of anaerobic sludge from six Swedish pulp and paper mills varied between 100 and 200 mL CH_4/g VS, and the addition of cobalt (Co) improved the transformation of acetate (Karlsson *et al.*, 2011). But it does not mean that AnMBRs cannot be employed to treat low-strength wastewaters for energy recovery and water reuse (Sutton *et al.*, 2010). Approximately 70% of COD was available for methane recovery in an AnMBR treating synthetic municipal wastewater with a permeate COD of lower than 40 mg/L (Ho & Sung, 2009). The recent research activities for biogas has been on recovering maximum energy, and on improving the quality of the biogas. Jeong *et al.* (2010) optimized the acidogenesis process in an AnsMBR by using response surface methodology. Applying the optimum conditions for maximum volatile fatty acids (VFA) production (an HRT of 2.01 days and a substrate concentration of 29.30 g COD/L) they achieved high effluent quality (COD removal of over 99%) and enhanced biogas production (average CH_4 production rate of 0.28 $m^3/m^3\ d^{-1}$). It is worth mentioning that methane leakages should be avoided since the global warming potential (GWP) of CH_4 is 23 times the GWP of CO_2.

11.7.2 Reduced energy consumption

Although AnMBRs do not incur aeration cost and offer the capability of recovering biogas energy, energy is consumed through biogas sparging to mitigate membrane fouling. A modelling study was developed to compare the energy requirements of aerobic and anaerobic MBRs (Martin *et al.*, 2011). The estimated energy consumptions of AnMBRs were lower (0.80 kWh/m^3 for influent of 1.14 g COD/L) than aerobic MBRs (2 kWh/m^3 for influent up to 1 g COD/L). The produced CH_4 is sufficient to offset a significant proportion of the total energy demand (Lin *et al.*, 2011a; Martin *et al.*, 2011). Van Zyl *et al.* (2008) estimated that the electricity produced (2.02 kWh/kg COD removed) could be ~7 times more than required for system operation based on the biogas production from a lab-scale AniMBR for treating synthetic Sasol's Fischer–Tropsch acid water stream, a high strength dissolved petrochemical effluent containing mostly C_2–C_6 short-chain fatty acids and some methanol and ethanol. However, the fluxes of AnMBRs are lower due to much lower sludge filterability (Spagni *et al.*, 2010) than aerobic MBR implying higher capital cost in the form of cost of membranes. A wide variation in energy demand was found in both AniMBRs (0.03–5.7 kWh/m^3) and AnsMBRs (0.23 and 16.52 kWh/m^3) due to the different gas sparging intensities and crossflow velocities, respectively (Martin *et al.*, 2011). Kim *et al.* (2011) used an anaerobic reactor treating dilute wastewater followed by a GAC-amended anaerobic fluidized-bed bioreactor with submerged membranes directly in contact with the GAC. They indicated that the energy consumption (total energy required, 0.058 kWh/m^3; energy required for the AniMBR alone, 0.028 kWh/m^3) is significantly less than other AniMBRs with gas sparging for fouling control (Kim *et al.*, 2011). The energy demands for pilot-scale AnMBRs (0.04–1.35 kWh/m^3) was reported to be lower than those of lab-scale AnMBRs and aerobic MBRs (Shin & Bae, 2018). Although five out of nine pilot-scale AnMBRs was energy positive, further improvements are required for future implementation of AnMBRs (Shin & Bae, 2018).

11.7.3 Economic analysis

Cost is one of the most decisive considerations for the implementation of a new technology. The total costs of an AnMBR system include the capital (e.g., reactor design and construction, membrane, and equipments installation), operating costs (e.g., energy consumption, sludge disposal, and chemical consumption), and maintenance costs (e.g., membrane cleaning and replacement, electrical equipment replacement). The most important fractions of total life cycle costs were found to be the membrane costs and the cost of the energy required for biogas sparging (Jeison & Van Lier, 2007b; Lin *et al.*, 2011a). However, a recent cost analysis of a full-scale AniMBR based on the data obtained from a lab-scale study for municipal secondary wastewater treatment showed that the operational costs can be totally offset by benefits from biogas recovery (Lin *et al.*, 2011a). Maleki *et al.* (2018) also reported that for AnMBR treatment of malting wastewater in warm climates, the revenue generated by produced methane would offset the majority of the wastewater treatment costs. It is impressive that an anaerobic digester with ultrafiltration plant installed was reported running for over 10 years for the treatment of soluble wheat starch at Tenstar (Butcher, 1989). The project led to an annual savings in effluent charges of over £100 k and a biogas benefit of £30 k (Butcher, 1989).

The total cost is significantly affected by the reactor configuration. The costs of the AniMBR have been consistently reported to be several times lower than the costs of the AnsMBR (Jeison & Van Lier, 2008; Lu *et al.*, 2011), because AnsMBR needs high cross-flow velocity to alleviate membrane fouling which reduces its economic feasibility. Moreover, a high cross-flow velocity may also decrease treatment performance due to a high biomass stress associated with recirculation. Christian *et al.* (2010) reported a 50% reduction of operating expenses following the upgrade of an existing anaerobic process to an AnMBR. The cost reduction was attributed to increased system capacity, ability to treat wastewater with higher biomass, and elimination of dewatering and solids disposal.

Compared to the aerobic MBRs, limited information is available on the cost assessment of AnMBRs. There has not been a direct comparison of energy demand between aerobic and anaerobic MBRs; however due to the absence of aeration and the potential of biogas recovery AnMBRs are likely to be a more cost-competitive option. AnMBRs bond the advantages of both membrane technology and anaerobic treatment in terms of high effluent quality, low operation and maintenance costs, and high loading potentials. It is interesting to note that the total costs for treating the evaporator condensate discharged from a kraft pulp production process were reduced by using an AnsMBR followed by the conventional aerobic process compared with the aerobic process only, which was attributed to less electric power consumption for aeration, less production of excess sludge, and energy recovery by evolved methane gas (Minami *et al.*, 1991; Minami, 1994).

11.8 CONCLUSION AND FUTURE OUTLOOK

The AnMBR technology is an option for sustainable wastewater treatment owing to several appealing benefits compared to the conventional anaerobic technologies, such as excellent effluent quality, low excess sludge production, high-treatment efficiency, small footprint, and high net energy production. However, research is still required with the intention of enhancing its commercial feasibility. There are three ways to achieve this purpose: (1) To extend the applicable range (i.e., type of wastewater, operating conditions, and reactor design); (2) To improve the treatment efficiency and stability; (3) To reduce the operating and maintenance costs.

Currently, AnMBRs have been tested with synthetic, food processing, industrial, high solids content, and municipal wastewaters at laboratory-, pilot-, and full-scales. Each wastewater type results in distinct treatment efficiency and membrane fouling behaviour. Further research is needed to access other potential applications, to reveal the impact of influent characteristics on membrane fouling and performance of AnMBRs, and to sufficiently understand the AnMBR technology for wastewater treatment.

Anaerobic treatment is such a susceptible process that its stability and efficiency can be greatly influenced by many factors. In addition to factors studied previously (e.g., pH and operating temperature (Gao *et al.*, 2010b, 2011a)), the occurrence and effects of other transient conditions encountered in industrial wastewater treatment, such as organic loading rate, salinity, toxic compound load need to be investigated for AnMBRs.

Although many MBR publications deal with fouling, it still is an unresolved problem requiring special attention. More full-scale studies are needed to fully understand the long-term membrane fouling and enhance filtration performance. First, membrane fouling is significantly affected by sludge constituents; thus, further research needs to be conducted on the modification of the mixed liquor by the use of flocculants or adsorbents that may also be effective to reduce the concentration of SMP and mitigate membrane fouling. Secondly, as the dominant mechanism of membrane fouling, the process of cake formation on membrane surfaces and its structure are of importance for regulating the development of membrane fouling. The TMP evolution was correlated to some extent with cake layer porosity. However, the initial fouling stage and TMP jump remain poorly explained. Thirdly, existing mathematical models allow prediction of membrane fouling based on a number of fouling factors (e.g., operating parameters, membrane properties, sludge properties, and reactor design). Membrane fouling can be better simulated with the consideration of the inhomogeneous structure of cake layer. Modified fouling models should be developed to examine and predict membrane fouling based on this new insight. Fourthly, additional research is required with regard to the identification of pioneering bacteria that are prone to attach onto the membrane surface. It is necessary that future research studies evaluate these aspects to achieve a full understanding of membrane fouling mechanisms.

Higher flux without deteriorating wastewater treatment efficiency implies higher productivity accompanied by low unit cost. Thus, pursuance of flux enhancement is always crucial for the broad application of AnMBRs in the future. Better and more detailed understanding of membrane fouling processes and mechanisms will help to develop better fouling control and cleaning strategies.

There is much less information on the energy and cost issues of AnMBR systems for wastewater treatment. The methane produced can be used to offset part of the energy demand, while investigations are still needed to maximize biogas generation rates and to improve the biogas clean-up technology. Better-controlled pilot-scale AnMBR studies and even full-scale applications are needed to address these issues.

REFERENCES

Achilli A., Marchand E. A. and Childress A. E. (2011). A performance evaluation of three membrane bioreactor systems: aerobic, anaerobic, and attached-growth. *Water Science and Technology*, **63**(12), 2999–3005.

Ahn J. H. and Forster C. F. (2002). The effect of temperature variations on the performance of mesophilic and thermophilic anaerobic filters treating a simulated paper mill wastewater. *Process Biochemistry*, **37**(6), 589–594.

Ahring B. K., Ibrahim A. A. and Mladenovska Z. (2001). Effect of temperature increase from 55 to 65°C on performance and microbial population dynamics of an anaerobic reactor treating cattle manure. *Water Research*, **35**(10), 2446–2452.

Akamatsu K., Lu W., Sugawara T. and Nakao S.-I. (2010). Development of a novel fouling suppression system in membrane bioreactors using an intermittent electric field. *Water Research*, **44**(3), 825–830.

Akram A. and Stuckey D. C. (2008a). Biomass acclimatisation and adaptation during start-up of a submerged anaerobic membrane bioreactor (SAMBR). *Environmental Technology*, **29**(10), 1053–1065.

Akram A. and Stuckey D. C. (2008b). Flux and performance improvement in a submerged anaerobic membrane bioreactor (SAMBR) using powdered activated carbon (PAC). *Process Biochemistry*, **43**(1), 93–102.

An Y., Wang Z., Wu Z., Yang D. and Zhou Q. (2009). Characterization of membrane foulants in an anaerobic non-woven fabric membrane bioreactor for municipal wastewater treatment. *Chemical Engineering Journal*, **155**(3), 709–715.

Aquino S. F. and Stuckey D. C. (2004a). Soluble microbial products formation in anaerobic chemostats in the presence of toxic compounds. *Water Research*, **38**(2), 255–266.

Aquino S. F. and Stuckey D. C. (2004b). The effect of organic and hydraulic shock loads on the production of soluble microbial products in anaerobic digesters. *Water Environment Research*, **76**(7), 2628–2636.

Aquino S. F., Hu A. Y., Akram A. and Stuckey D. C. (2006). Characterization of dissolved compounds in submerged anaerobic membrane bioreactors (SAMBRs). *Journal of Chemical Technology & Biotechnology*, **81**(12), 1894–1904.

Arabi S. (2009). Impact of Influent Feed and Mixed Liquor Characteristics on Fouling in Submerged Membrane Bioreactors for Wastewater Treatment. PhD thesis, Department of Chemical and Biochemical Engineering, The University of Western Ontario. London, Canada.

Arros-Alileche S., Merin U., Daufin G. and Gesan-Guiziou G. (2008). The membrane role in an anaerobic membrane bioreactor for purification of dairy wastewaters: a numerical simulation. *Bioresour Technol*, **99**(17), 8237–8244.

Bae T.-H., Kim I.-C. and Tak T.-M. (2006). Preparation and characterization of fouling-resistant TiO_2 self-assembled nanocomposite membranes. *Journal of Membrane Science*, **275**(1–2), 1–5.

Baek S. H. and Pagilla K. R. (2006). Aerobic and anaerobic membrane bioreactors for municipal wastewater treatment. *Water Environment Research*, **78**(2), 133–140.

Baek S.H., Pagilla K. and Kim H.-J. (2010). Lab-scale study of an anaerobic membrane bioreactor (AnMBR) for dilute municipal wastewater treatment. *Biotechnology and Bioprocess Engineering*, **15**(4), 704–708.

Baêta B., Ramos R., Lima D. and Aquino S. (2012). Use of submerged anaerobic membrane bioreactor (SAMBR) containing powdered activated carbon (PAC) for the treatment of textile effluents. *Water Sci Technol.* , **65**(9), 1540–1547.

Baily J. E. and Ollis D. F. (1986). Biochemical Engineering Fundamentals. McGraw-Hill, New York.

Bandara W. M. K. R. T. W., Satoh H., Sasakawa M., Nakahara Y., Takahashi M. and Okabe S. (2011). Removal of residual dissolved methane gas in an upflow anaerobic sludge blanket reactor treating low-strength wastewater at low temperature with degassing membrane. *Water Research*, **45**(11), 3533–3540.

Barker D. J. and Stuckey D. C. (1999). A review of soluble microbial products (SMP) in wastewater treatment systems. *Water Research*, **33**(14), 3063–3082.

Barker D. J. and Stuckey D. C. (2001). Modeling of soluble microbial products in anaerobic digestion: the effect of feed strength and composition. *Water Environment Research*, **73**(2), 173–184.

Beaubien A., Baty M., Jeannot F., Francoeur E. and Manem J. (1996). Design and operation of anaerobic membrane bioreactors: development of a filtration testing strategy. *Journal of Membrane Science*, **109**(2), 173–184.

Berube P. R., Hall E. R. and Sutton P. M. (2006). Parameters governing permeate flux in an anaerobic membrane bioreactor treating low-strength municipal wastewaters: a literature review. *Water Environment Research*, **78**(8), 887–896.

Bohdziewicz J., Neczaj E. and Kwarciak A. (2008). Landfill leachate treatment by means of anaerobic membrane bioreactor. *Desalination*, **221**(1–3), 559–565.

Bouskova A., Dohanyos M., Schmidt J. E. and Angelidaki I. (2005). Strategies for changing temperature from mesophilic to thermophilic conditions in anaerobic CSTR reactors treating sewage sludge. *Water Research*, **39** (8), 1481–1488.

Brockmann M. and Seyfried C. F. (1997). Sludge activity under the conditions of crossflow microfiltration. *Water Science and Technology*, **35**(10), 173–181.

Butcher G. J. (1989). Experiences with anaerobic digestion of wheat starch processing waste. *International Biodeterioration*, **25**(1–3), 71–77.

Calderón K., Rodelas B., Cabirol N., González-López J. and Noyola A. (2011). Analysis of microbial communities developed on the fouling layers of a membrane-coupled anaerobic bioreactor applied to wastewater treatment. *Bioresource Technology*, **102**(7), 4618–4627.

Chae K. J., Jang A., Yim S. K. and Kim I. S. (2008). The effects of digestion temperature and temperature shock on the biogas yields from the mesophilic anaerobic digestion of swine manure. *Bioresource Technology*, **99**(1), 1–6.

Chan Y. J., Chong M. F., Law C. L. and Hassell D. G. (2009). A review on anaerobic-aerobic treatment of industrial and municipal wastewater. *Chemical Engineering Journal*, **155**(1–2), 1–18.

Chang I.-S., Bag S.-O. and Lee C.-H. (2001). Effects of membrane fouling on solute rejection during membrane filtration of activated sludge. *Process Biochemistry*, **36**(8–9), 855–860.

Chang I.-S., Le Clech P., Jefferson B. and Judd S. (2002). Membrane fouling in membrane bioreactors for wastewater treatment. *Journal of Environmental Engineering*, **128**, 1018–1029.

Chang I.-S., Fieldb R. and Cuib Z. (2009). Limitations of resistance-in-series model for fouling analysis in membrane bioreactors: a cautionary note. *Desalination and Water Treatment,* **8** 31–36.

Charfi A., Ben Amar N. and Harmand J. (2012). Analysis of fouling mechanisms in anaerobic membrane bioreactors. *Water Research*, **46**(8), 2637–2650.

Chen R., Nie Y., Hu Y., Miao R., Utashiro T., Li Q., Xu M. and Li Y.-Y. (2017). Fouling behaviour of soluble microbial products and extracellular polymeric substances in a submerged anaerobic membrane bioreactor treating low-strength wastewater at room temperature. *Journal of Membrane Science*, **531**, 1–9.

Chen Y., Cheng J. J. and Creamer K. S. (2008a). Inhibition of anaerobic digestion process: a review. *Bioresource Technology*, **99**(10), 4044–4064.

Chen Z., Ren N., Wang A., Zhang Z.-P. and Shi Y. (2008b). A novel application of TPAD-MBR system to the pilot treatment of chemical synthesis-based pharmaceutical wastewater. *Water Research*, **42**(13), 3385–3392.

Cho B. D. and Fane A. G. (2002). Fouling transients in nominally sub-critical flux operation of a membrane bioreactor. *Journal of Membrane Science*, **209**(2), 391–403.

Choo K.-H. and Lee C.-H. (1996a). Membrane fouling mechanisms in the membrane-coupled anaerobic bioreactor. *Water Research*, **30**(8), 1771–1780.

Choo K. H. and Lee C. H. (1996b). Effect of anaerobic digestion broth composition on membrane permeability. *Water Science and Technology*, **34**(9), 173–179.

Choo K.-H. and Lee C.-H. (1998). Hydrodynamic behavior of anaerobic biosolids during crossflow filtration in the membrane anaerobic bioreactor. *Water Research*, **32**(11), 3387–3397.

Choo K.-H., Kang I.-J., Yoon S.-H., Park H., Kim J.-H., Adiya S. and Lee C.-H. (2000). Approaches to membrane fouling control in anaerobic membrane bioreactors. *Water Science and Technology*, **41**(10–11), 363–371.

Christian S., Grant S., Wilson D., Mccarthy P., Mills D. and Kolakowski M. (2010). The first two years of full-scale Anaerobic Membrane Bioreactor (AnMBR) operation treating a high strength industrial wastewater at Kens Foods Inc. *Proceedings of the Water Environment Federation*, **2010**(13), 4019–4033.

Cho L.-B., Yang F.-L. and Zhang X.-W. (2005). Anaerobic treatment of domestic wastewater in a membrane-coupled expended granular sludge bed (EGSB) reactor under moderate to low temperature. *Process Biochemistry*, **40**(3–4), 1063–1070.

Comte S., Guibaud G. and Baudu M. (2006). Relations between extraction protocols for activated sludge extracellular polymeric substances (EPS) and EPS complexation properties: Part I. Comparison of the efficiency of eight EPS extraction methods. *Enzyme and Microbial Technology*, **38**(1–2), 237–245.

Cornelissen E. R., Van Buggenhout S., Van Ermen S., De Smedt M., Van Impe J. and Koning J. (2001). Anaerobic treatment of brewery wastewater with an internal membrane bioreactor. *Meded Rijksuniv Gent Fak Landbouwkd Toegep Biol Wet*, **66**(3a), 135–138.

Dagnew M., Parker W., Seto P., Waldner K., Hong Y., Bayly R. and Cumin J. (2011). Pilot testing of an AnMBR for municipal wastewater treatment. *Proceedings of the Water Environment Federation*, **2011**(11), 4931–4941.

Dereli R. K., Ersahin M. E., Ozgun H., Ozturk I., Jeison D., Van Der Zee F. and Van Lier J. B. (2012a). Potentials of anaerobic membrane bioreactors to overcome treatment limitations induced by industrial wastewaters. *Bioresource Technology*, **122**, 160–170.

Dereli R. K., Urban D. R., Heffernan B., Jordan J. A., Ewing J., Rosenberger G. T. and Dunaev T. I. (2012b). Performance evaluation of a pilot-scale Anaerobic Membrane Bioreactor (AnMBR) treating ethanol thin stillage. *Environmental Technology*, **33**, 1–6.

Dereli R.K., Wang X., van der Zee F.P. and van Lier J.B. (2018). Biological performance and sludge filterability of anaerobic membrane bioreactors under nitrogen limited and supplied conditions. *Water Research*, **137**, 164–172.

Diez V., Ramos C. and Cabezas J. (2012). Treating wastewater with high oil and grease content using an Anaerobic Membrane Bioreactor (AnMBR). Filtration and cleaning assays. *Water Science and Technology*, **65**(10), 1847–1853.

Ding Y., Tian Y., Li Z., Zuo W. and Zhang J. (2015). A comprehensive study into fouling properties of extracellular polymeric substance (EPS) extracted from bulk sludge and cake sludge in a mesophilic anaerobic membrane bioreactor. *Bioresource Technology*, **192**, 105–114.

Dong B. Z., Chen Y., Gao N. Y. and Fan J. C. (2006). Effect of pH on UF membrane fouling. *Desalination*, **195**(1–3), 201–208.

Drews A., Lee C.-H. and Kraume M. (2006). Membrane fouling – a review on the role of EPS. *Desalination*, **200**(1–3), 186–188.

Dubois M., Gilles K. A., Hamilton J. K., Rebers P. A. and Smith F. (1956). Colorimetric method for determination of sugars and related substances. *Analytical Chemistry*, **28**(3), 350–356.

El-Mashad H. M., Zeeman G., Van Loon W. K., Bot G. P. and Lettinga G. (2004). Effect of temperature and temperature fluctuation on thermophilic anaerobic digestion of cattle manure. *Bioresource Technology*, **95**(2), 191–201.

Fan Y.-B., Wang J.-S., Jiang Z.-C., Chen M.-X., Xu K. and Jia Z.-P. (2000). Treatment of a dyeing wastewater from a woolen mill using an A/O membrane bio-reactor. *Journal of Environmental Sciences (IOS Press)*, **12**(3), 344.

Field R. W., Wu D., Howell J. A. and Gupta B. B. (1995). Critical flux concept for microfiltration fouling. *Journal of Membrane Science*, **100**(3), 259–272.

Fuchs W., Binder H., Mavrias G. and Braun R. (2003). Anaerobic treatment of wastewater with high organic content using a stirred tank reactor coupled with a membrane filtration unit. *Water Research*, **37**(4), 902–908.

Galib M., Elbeshbishy E., Reid R., Hussain A. and Lee H.-S. (2016). Energy-positive food wastewater treatment using an anaerobic membrane bioreactor (AnMBR). *Journal of Environmental Management*, **182**, 477–485.

Gander M., Jefferson B. and Judd S. (2000). Aerobic MBRs for domestic wastewater treatment: a review with cost considerations. *Separation and Purification Technology*, **18**(2), 119–130.

Gao D.-W., Zhang T., Tang C.-Y. Y., Wu W.-M., Wong C.-Y., Lee Y. H., Yeh D. H. and Criddle C. S. (2010a). Membrane fouling in an anaerobic membrane bioreactor: differences in relative abundance of bacterial species in the membrane foulant layer and in suspension. *Journal of Membrane Science*, **364**(1–2), 331–338.

Gao W. J. J., Lin H. J., Leung K. T. and Liao B. Q. (2010b). Influence of elevated pH shocks on the performance of a submerged anaerobic membrane bioreactor. *Process Biochemistry*, **45**(8), 1279–1287.

Gao W. J., Leung K. T., Qin W. S. and Liao B. Q. (2011a). Effects of temperature and temperature shock on the performance and microbial community structure of a submerged anaerobic membrane bioreactor. *Bioresource Technology*, **102**(19), 8733–8740.

Gao W. J., Lin H. J., Leung K. T., Schraft H. and Liao B. Q. (2011b). Structure of cake layer in a submerged anaerobic membrane bioreactor. *Journal of Membrane Science*, **374**(1–2), 110–120.

Gao W. J., Qu X., Leung K. T. and Liao B. Q. (2012). Influence of temperature and temperature shock on sludge properties, cake layer structure, and membrane fouling in a submerged anaerobic membrane bioreactor. *Journal of Membrane Science*, **421–422**, 131–144.

Gao W. J., Han M. N., Qu X., Xu C. and Liao B. Q. (2013). Characteristics of wastewater and mixed liquor and their role in membrane fouling. *Bioresource Technology*, **128**, 207–214.

Gao W., Han M., Xu C. C., Liao B., Hong Y., Cumin J. and Dagnew M. (2016). Performance of submerged anaerobic membrane bioreactor for thermomechanical pulping wastewater treatment. *Journal of Water Process Engineering*, **13**, 70–78.

Geng Z. and Hall E. R. (2007). A comparative study of fouling-related properties of sludge from conventional and membrane enhanced biological phosphorus removal processes. *Water Research*, **41**(19), 4329–4338.

Ghyoot W. R. and Verstraete W. H. (1997). Coupling membrane filtration to anaerobic primary sludge digestion. *Environmental Technology*, **18**(6), 569–580.

Giménez J. B., Robles A., Carretero L., Durán F., Ruano M. V., Gatti M. N., Ribes J., Ferrer J. and Seco A. (2011). Experimental study of the anaerobic urban wastewater treatment in a submerged hollow-fibre membrane bioreactor at pilot scale. *Bioresource Technology*, **102**(19), 8799–8806.

Gottschalk G. (1986). Bacterial Metabolism. Springer-Verlag, New York.

Gouveia J., Plaza F., Garralon G., Fdz-Polanco F. and Peña M. (2015). A novel configuration for an anaerobic submerged membrane bioreactor (AnSMBR). Long-term treatment of municipal wastewater under psychrophilic conditions. *Bioresource Technology*, **198**, 510–519.

Grant S., Page I., Moro M. and Yamamoto T. (2008). Full-scale applications of the anaerobic membrane bioreactor process for treatment of Stillage from alcohol production in Japan. *Proceedings of the Water Environment Federation*, **2008**(7), 7556–7570.

Grethlein H. E. (1978). Anaerobic digestion and membrane separation of domestic wastewater. *Water Pollution Control Federation*, **50**(4), 754–763.

Hai F. I. and Yamamoto K. (2011). Membrane biological reactors. In: Treatise on Water Science, P. Wilderer (ed.), Elsevier, UK, Vol. **4**, pp. 571–613.

Harada H., Momonoi K., Yamazaki S. and Takizawa S. (1994). Application of anaerobic UF membrane reactor for treatment of a wastewater containing high strength particulate organics. *Water Science and Technology*, **30**(12), 307–319.

Hatamoto M., Yamamoto H., Kindaichi T., Ozaki N. and Ohashi A. (2010). Biological oxidation of dissolved methane in effluents from anaerobic reactors using a down-flow hanging sponge reactor. *Water Research*, **44**(5), 1409–1418.

He Y., Xu P., Li C. and Zhang B. (2005). High-concentration food wastewater treatment by an anaerobic membrane bioreactor. *Water Research*, **39**(17), 4110–4118.

Hernandez A. E., Belalcazar L. C., Rodriguez M. S. and Giraldo E. (2002). Retention of granular sludge at high hydraulic loading rates in an anaerobic membrane bioreactor with immersed filtration. *Water Science and Technology*, **45**(10), 169–174.

Herrera-Robledo M., Morgan-Sagastume J. M. and Noyola A. (2010). Biofouling and pollutant removal during long-term operation of an anaerobic membrane bioreactor treating municipal wastewater. *Biofouling*, **26** (1), 23–30.

Herrera-Robledo M., Cid-León D. M., Morgan-Sagastume J. M. and Noyola A. (2011). Biofouling in an anaerobic membrane bioreactor treating municipal sewage. *Separation and Purification Technology*, **81**(1), 49–55.

Ho J. and Sung S. (2009). Anaerobic membrane bioreactor treatment of synthetic municipal wastewater at ambient temperature. *Water Environment Research*, **81**(9), 922–928.

Ho J. and Sung S. (2010). Methanogenic activities in anaerobic membrane bioreactors (AnMBR) treating synthetic municipal wastewater. *Bioresource Technology*, **101**(7), 2191–2196.

Hu A. Y. and Stuckey D. C. (2006). Treatment of dilute wastewaters using a novel submerged anaerobic membrane bioreactor. *Journal of Environmental Engineering*, **132**(2), 190–198.

Hu A. Y. and Stuckey D. C. (2007). Activated carbon addition to a submerged anaerobic membrane bioreactor: effect on performance, transmembrane pressure, and flux. *Journal of Environmental Engineering*, **133**(1), 73–80.

Huang B., Wang H.-C., Cui D., Zhang B., Chen Z.-B. and Wang A.-J. (2018). Treatment of pharmaceutical wastewater containing β-lactams antibiotics by a pilot-scale anaerobic membrane bioreactor (AnMBR). *Chemical Engineering Journal*, **341**, 238–247.

Huang Z., Ong S. L. and Ng H. Y. (2008). Feasibility of submerged anaerobic membrane bioreactor (SAMBR) for treatment of low-strength wastewater. *Water Science and Technology*, **58**(10), 1925–1931.

Huang Z., Ong S. L. and Ng H. Y. (2011). Submerged anaerobic membrane bioreactor for low-strength wastewater treatment: Effect of HRT and SRT on treatment performance and membrane fouling. *Water Research*, **45**(2), 705–713.

Hyun K. S. and Lee S. J. (2009). Biofilm/membrane filtration for reclamation and reuse of rural wastewaters. *Water Science and Technology*, **59**(11), 2145–2152.

Jeison D. (2007). Anaerobic Membrane Bioreactors for Wastewater Treatment. PhD thesis, Wageningen University. Wageningen, The Netherlands.

Jeison D. and Van Lier J. B. (2006a). Cake layer formation in anaerobic submerged membrane bioreactors (AnSMBR) for wastewater treatment. *Journal of Membrane Science*, **284**(1–2), 227–236.

Jeison D. and Van Lier J. B. (2006b). On-line cake-layer management by trans-membrane pressure steady state assessment in anaerobic membrane bioreactors for wastewater treatment. *Biochemical Engineering Journal*, **29** (3), 204–209.

Jeison D. and Van Lier J. B. (2007a). Cake formation and consolidation: main factors governing the applicable flux in anaerobic submerged membrane bioreactors (AnSMBR) treating acidified wastewaters. *Separation and Purification Technology*, **56**(1), 71–78.

Jeison D. and Van Lier J. B. (2007b). Thermophilic treatment of acidified and partially acidified wastewater using an anaerobic submerged MBR: factors affecting long-term operational flux. *Water Research*, **41**(17), 3868–3879.

Jeison D. and Van Lier J. B. (2008). Feasibility of thermophilic anaerobic submerged membrane bioreactors (AnSMBR) for wastewater treatment. *Desalination*, **231**(1–3), 227–235.

Jeison D., Días I. and Lier J. B. V. (2008a). Anaerobic membrane bioreactors: are membranes really necessary?. *Electronic Journal of Biotechnology*, **11**(4), 1–9.

Jeison D., Van Betuw W. and Van Lier J. B. (2008b). Feasibility of anaerobic membrane bioreactors for the treatment of wastewaters with particulate organic matter. *Separation Science and Technology*, **43**(13), 3417–3431.

Jeison D., Plugge C. M., Pereira A. and Lier J. B. V. (2009a). Effects of the acidogenic biomass on the performance of an anaerobic membrane bioreactor for wastewater treatment. *Bioresource Technology*, **100**(6), 1951–1956.

Jeison D., Telkamp P. and Van Lier J. B. (2009b). Thermophilic sidestream anaerobic membrane bioreactors: the shear rate dilemma. *Water Environment Research*, **81**(11), 2372–2380.

Jeong E., Kim H.-W., Nam J.-Y. and Shin H.-S. (2010). Enhancement of bioenergy production and effluent quality by integrating optimized acidification with submerged anaerobic membrane bioreactor. *Bioresource Technology*, **101** (1, Supplement 1), S7–S12.

Judd S. (2006). The MBR Book: Principles and Applications of Membrane Bioreactors in Water and Wastewater Treatment. Elsevier Ltd., London.

Judd S. (2011). Chapter 2 – Fundamentals, The MBR Book, 2nd edn. Oxford, Butterworth-Heinemann, Oxford, pp. 55–207.

Kanai M., Ferre V., Wakahara S., Yamamoto T. and Moro M. (2010). A novel combination of methane fermentation and MBR – Kubota Submerged Anaerobic Membrane Bioreactor process. *Desalination*, **250**(3), 964–967.

Kang I.-J., Yoon S.-H. and Lee C.-H. (2002). Comparison of the filtration characteristics of organic and inorganic membranes in a membrane-coupled anaerobic bioreactor. *Water Research*, **36**(7), 1803–1813.

Kang S., Hoek E. M. V., Choi H. and Shin H. (2006). Effect of membrane surface properties during the fast evaluation of cell attachment. *Separation Science and Technology*, **41**(7), 1475–1487.

Kappell A. S., Semmens M. J., Novak P. J. and Lapara T. M. (2005). Novel application of oxygen-transferring membranes to improve anaerobic wastewater treatment. *Biotechnology and Bioengineering*, **89**(4), 373–380.

Karlsson A., Truong X. B., Gustavsson J., Svensson B. H., Nilsson F. and Ejlertsson J. (2011). Anaerobic treatment of activated sludge from Swedish pulp and paper mills – biogas production potential and limitations. *Environmental Technology*, **32**(14), 1559–1571.

Kato M. T., Field J. A., Versteeg P. and Lettinga G. (1994). Feasibility of expanded granular sludge bed reactors for the anaerobic treatment of low-strength soluble wastewaters. *Biotechnology and Bioengineering*, **44**(4), 469–479.

Khalili-Garakani A., Mehrnia M. R., Mostoufi N. and Sarrafzadeh M. H. (2011). Analyze and control fouling in an airlift membrane bioreactor: CFD simulation and experimental studies. *Process Biochemistry*, **46**(5), 1138–1145.

Kim J., Lee C. H. and Choo K. H. (2007). Control of struvite precipitation by selective removal of NH(4)+ with dialyzer/zeolite in an anaerobic membrane bioreactor. *Applied Microbiology and Biotechnology*, **75**(1), 187–193.

Kim J., Kim K., Ye H., Lee E., Shin C., Mccarty P. L. and Bae J. (2011). Anaerobic fluidized bed membrane bioreactor for wastewater treatment. *Environmental Science & Technology*, **45**(2), 576–581.

Kim H. Y., Yeon K. M., Lee C. H., Lee S. and Swaminathan T. (2006). Biofilm structure and extracellular polymeric substances in low and high dissolved oxygen membrane bioreactors. *Separation Science and Technology*, **41**(7), 1213–1230.

Kimura S. (1991). Japan's Aqua Renaissance 90 project. *Water Science and Technology*, **23**(7–9), 1573–1582.

Kocadagistan E. and Topcu N. (2007). Treatment investigation of the Erzurum city municipal wastewaters with anaerobic membrane bioreactors. *Desalination*, **216**(13), 367–376.

Lant P. and Hartley K. (2007). Solids characterisation in an anaerobic migrating bed reactor (AMBR) sewage treatment system. *Water Research*, **41**(11), 2437–2448.

Laspidou C. S. and Rittmann B. E. (2002). A unified theory for extracellular polymeric substances, soluble microbial products, and active and inert biomass. *Water Research*, **36**(11), 2711–2720.

Le-Clech P., Chen V. and Fane T. A. G. (2006). Fouling in membrane bioreactors used in wastewater treatment. *Journal of Membrane Science*, **284**(1–2), 17–53.

Le Clech P., Jefferson B., Chang I. S. and Judd S. J. (2003). Critical flux determination by the flux-step method in a submerged membrane bioreactor. *Journal of Membrane Science*, **227**(1–2), 81–93.

Lee D.-Y., Li Y.-Y., Noike T. and Cha G.-C. (2008). Behavior of extracellular polymers and bio-fouling during hydrogen fermentation with a membrane bioreactor. *Journal of Membrane Science*, **322**(1), 13–18.

Lee S. M., Jung J. Y. and Chung Y. C. (2001). Novel method for enhancing permeate flux of submerged membrane system in two-phase anaerobic reactor. *Water Research*, **35**(2), 471–477.

Lew B., Tarre S., Beliavski M., Dosoretz C. and Green M. (2009). Anaerobic membrane bioreactor (AnMBR) for domestic wastewater treatment. *Desalination*, **243**(1–3), 251–257.

Li A., Kothari D. and Corrado J. J. (1985). Application of Membrane Anaerobic Reactor System for the Treatment of Industrial Wastewaters. Proceedings of the 39th Industrial Waste Conference, Purdue University, Ann Arbor Science, Ann Arbor, MI.

Li Z. (2010). Integrated submerged Membrane Bioreactor Anaerobic/Aerobic (ISMBR-A/O) for nitrogen and phosphorus removal during oil refinery wastewater treatment. *Petroleum Science and Technology*, **28**(3), 286–293.

Liao B. Q., Bagley D. M., Kraemer H. E., Leppard G. G. and Liss S. N. (2004). A review of biofouling and its control in membrane separation bioreactors. *Water Environment Research*, **76**(5), 425–436.

Liao B.-Q., Kraemer J. T. and Bagley D. M. (2006). Anaerobic membrane bioreactors: applications and research directions. *Critical Reviews in Environmental Science and Technology*, **36**(6), 489–530.

Liao B. Q., Xie K., Lin H. J. and Bertoldo D. (2010). Treatment of kraft evaporator condensate using a thermophilic submerged anaerobic membrane bioreactor. *Water Science and Technology*, **61**(9), 2177–2183.

Lin H., Chen J., Wang F., Ding L. and Hong H. (2011a). Feasibility evaluation of submerged anaerobic membrane bioreactor for municipal secondary wastewater treatment. *Desalination*, **280**(1–3), 120–126.

Lin H., Liao B.-Q., Chen J., Gao W., Wang L., Wang F. and Lu X. (2011b). New insights into membrane fouling in a submerged anaerobic membrane bioreactor based on characterization of cake sludge and bulk sludge. *Bioresource Technology*, **102**(3), 2373–2379.

Lin H. J., Xie K., Mahendran B., Bagley D. M., Leung K. T., Liss S. N. and Liao B. Q. (2009). Sludge properties and their effects on membrane fouling in submerged anaerobic membrane bioreactors (SAnMBRs). *Water Research*, **43** (15), 3827–3837.

Lin H. J., Xie K., Mahendran B., Bagley D. M., Leung K. T., Liss S. N. and Liao B. Q. (2010). Factors affecting sludge cake formation in a submerged anaerobic membrane bioreactor. *Journal of Membrane Science*, **361**(1–2), 126–134.

Lin H. J., Gao W. J., Leung K. T. and Liao B. Q. (2011). Characteristics of different fractions of microbial flocs and their role in membrane fouling. *Water Science and Technology*, **63**(2), 262–269.

Lin H.J., Gao W., Meng F., Liao B.-Q., Leung K.-T., Zhao L., Chen J. and Hong H. (2012). Membrane bioreactors for industrial wastewater treatment: a critical review. *Critical Reviews in Environmental Science and Technology*, **42** (7), 677–740.

Lin S. H. and Peng C. F. (1994). Treatment of textile wastewater by electrochemical method. *Water Research*, **28**(2), 277–282.

Liu Y. and Tay J.-H. (2001). Strategy for minimization of excess sludge production from the activated sludge process. *Biotechnology Advances*, **19**(2), 97–107.

Liu Y., Liu H., Cui L. and Zhang K. (2012). The ratio of food-to-microorganism (F/M) on membrane fouling of anaerobic membrane bioreactors treating low-strength wastewater. *Desalination*, **297**(0), 97–103.

Lowery O. H., Rosebrough N., Farr A. L. and Randall R. J. (1951). Protein measurement with the folin phenol reagent. *Journal of Biological Chemistry*, **193**, 265–275.

Lu X., Zhou B., Shen F., Li J., Bian X., Liu Z., Shi L., Hou Z. and Deng B. (2011). 厌氧膜生物反应器在食品废水处理中的应用研究 (Treatment of food processing wastewater with anaerobic membrane bioreactor). *Membrane Science and Technology*, **31**(3), 234–238.

Lyko S., Wintgens T., Al-Halbouni D., Baumgarten S., Tacke D., Drensla K., Janot A., Dott W., Pinnekamp J. and Melin T. (2008). Long-term monitoring of a full-scale municipal membrane bioreactor – characterisation of foulants and operational performance. *Journal of Membrane Science*, **317**(1–2), 78–87.

Madsen E. L. (2008). Environmental Microbiology: From Genomes to Biogeochemistry. Blackwell publishing Ltd., Malden, USA.

Maleki E., Catalan L. J. and Liao B. (2018). Effect of organic loading rate on the performance of a submerged anaerobic membrane bioreactor (SAnMBR) for malting wastewater treatment and biogas production. *Journal of Chemical Technology & Biotechnology*, **93**(6), 1636–1647.

Martin I., Pidou M., Soares A., Judd S. and Jefferson B. (2011). Modelling the energy demands of aerobic and anaerobic membrane bioreactors for wastewater treatment. *Environmental Technology*, **32**(9), 921–932.

Martin-Garcia I., Monsalvo V., Pidou M., Le-Clech P., Judd S. J., Mcadam E. J. and Jefferson B. (2011). Impact of membrane configuration on fouling in anaerobic membrane bioreactors. *Journal of Membrane Science*, **382** (12), 41–49.

Martinez-Sosa D., Helmreich B., Netter T., Paris S., Bischof F. and Horn H. (2011a). Anaerobic submerged membrane bioreactor (AnSMBR) for municipal wastewater treatment under mesophilic and psychrophilic temperature conditions. *Bioresource Technology*, **102**(22), 10377–10385.

Martinez-Sosa D., Helmreich B., Netter T., Paris S., Bischof F. and Horn H. (2011b). Pilot-scale anaerobic submerged membrane bioreactor (AnSMBR) treating municipal wastewater: the fouling phenomenon and long-term operation. *Water Science and Technology*, **64**(9), 8.

Martinez-Sosa D., Helmreich B. and Horn H. (2012). Anaerobic submerged membrane bioreactor (AnSMBR) treating low-strength wastewater under psychrophilic temperature conditions. *Process Biochemistry*, **47**(5), 792–798.

Mccarty P. L., Bae J. and Kim J. (2011). Domestic wastewater treatment as a net energy producer–can this be achieved? *Environmental Science & Technology*, **45**(17), 7100–7106.

McMahon J. (2010). Anaerobic membrane bioreactor system treats high strength wastewater. https://www.waterworld.com/articles/mem/articles/print/volume-2/issue-3/feature/anaerobic-membrane-bioreactor-system-treats-high-strength-wastewater.html

Meng F., Chae S.-R., Drews A., Kraume M., Shin H.-S. and Yang F. (2009). Recent advances in membrane bioreactors (MBRs): membrane fouling and membrane material. *Water Research*, **43**(6), 1489–1512.

Meng F., Liao B., Liang S., Yang F., Zhang H. and Song L. (2010). Morphological visualization, componential characterization and microbiological identification of membrane fouling in membrane bioreactors (MBRs). *Journal of Membrane Science*, **361**(1–2), 1–14.

Minami K. (1994). A trial of high performance anaerobic treatment on wastewater from a kraft pulp mill. *Desalination*, **98**(1–3), 273–283.

Minami K., Okamura K., Ogawa S. and Naritomi T. (1991). Continuous anaerobic treatment of wastewater from a kraft pulp mill. *Journal of Fermentation and Bioengineering*, **71**(4), 270–274.

Nagaoka H. and Nemoto H. (2005). Influence of extracellular polymeric substances on nitrogen removal in an intermittently-aerated membrane bioreactor. *Water Science and Technology*, **51**(11), 151–158.

Nanda D., Tung K.-L., Li Y.-L., Lin N.-J. and Chuang C.-J. (2010). Effect of pH on membrane morphology, fouling potential, and filtration performance of nanofiltration membrane for water softening. *Journal of Membrane Science*, **349**(1–2), 411–420.

Ndegwa P. M., Hamilton D. W., Lalman J. A. and Cumba H. J. (2008). Effects of cycle-frequency and temperature on the performance of anaerobic sequencing batch reactors (ASBRs) treating swine waste. *Bioresource Technology*, **99**(6), 1972–1980.

Novak J. T., Sadler M. E. and Murthy S. N. (2003). Mechanisms of floc destruction during anaerobic and aerobic digestion and the effect on conditioning and dewatering of biosolids. *Water Research*, **37**(13), 3136–3144.

Padmasiri S. I., Zhang J., Fitch M., Norddahl B., Morgenroth E. and Raskin L. (2007). Methanogenic population dynamics and performance of an anaerobic membrane bioreactor (AnMBR) treating swine manure under high shear conditions. *Water Research*, **41**(1), 134–144.

Rahimpour A., Madaeni S. S., Zereshki S. and Mansourpanah Y. (2009). Preparation and characterization of modified nano-porous PVDF membrane with high antifouling property using UV photo-grafting. *Applied Surface Science*, **255**(16), 7455–7461.

Rajeshwari K. V., Balakrishnan M., Kansal A., Lata K. and Kishore V. V. N. (2000). State-of-the-art of anaerobic digestion technology for industrial wastewater treatment. *Renewable and Sustainable Energy Reviews*, **4**(2), 135–156.

Ramos C., Zecchino F., Ezquerra D. and Diez V. (2014). Chemical cleaning of membranes from an anaerobic membrane bioreactor treating food industry wastewater. *Journal of Membrane Science*, **458**, 179–188.

Rezania B., Oleszkiewicz J. A. and Cicek N. (2006). Hydrogen-driven denitrification of wastewater in an anaerobic submerged membrane bioreactor: potential for water reuse. *Water Science and Technology*, **54**(11–12), 207–214.

Rintala J. A. and Puhakka J. A. (1994). Anaerobic treatment in pulp- and paper-mill waste management: a review. *Bioresource Technology*, **47**(1), 1–18.

Robles A., Ruano M. V., García-Usach F. and Ferrer J. (2012). Sub-critical filtration conditions of commercial hollow-fibre membranes in a submerged anaerobic MBR (HF-SAnMBR) system: the effect of gas sparging intensity. *Bioresource Technology*, **114**(0), 247–254.

Robles A., Ruano M. V., Ribes J. and Ferrer J. (2013). Performance of industrial scale hollow-fibre membranes in a submerged anaerobic MBR (HF-SAnMBR) system at mesophilic and psychrophilic conditions. *Separation and Purification Technology*, **104**(0), 290–296.

Roest K., Heilig H. G. H. J., Smidt H., De Vos W. M., Stams A. J. M. and Akkermans A. D. L. (2005). Community analysis of a full-scale anaerobic bioreactor treating paper mill wastewater. *Systematic and Applied Microbiology*, **28**(2), 175–185.

Ross W. R., Barnard J. P., Le Roux J. and De Villiers H. A. (1990). Application of ultrafiltration membranes for solids-liquid separation in anaerobic digestion systems: the ADUF process. *Water S.A.*, **16**, 85.

Ross W. R., Barnard J. P., Strohwald N. K., Grobler C. J. and Sanetra J. (1992). Practical application of the ADUF process to the full-scale treatment of maize-processing effluent. *Water Science and Technology*, **25**(10), 27.

Saddoud A. and Sayadi S. (2007). Application of acidogenic fixed-bed reactor prior to anaerobic membrane bioreactor for sustainable slaughterhouse wastewater treatment. *Journal of Hazardous Materials*, **149**(3), 700–706.

Saddoud A., Ellouze M., Dhouib A. and Sayadi S. (2006). A comparative study on the anaerobic membrane bioreactor performance during the treatment of domestic wastewaters of various origins. *Environmental Technology*, **27**(9), 991–999.

Saddoud A., Hassairi I. and Sayadi S. (2007). Anaerobic membrane reactor with phase separation for the treatment of cheese whey. *Bioresource Technology*, **98**(11), 2102–2108.

Saddoud A., Abdelkafi S. and Sayadi S. (2009). Effects of domestic wastewater toxicity on anaerobic membrane-bioreactor (MBR) performances. *Environmental Technology*, **30**(13), 1361–1369.

Sainbayar A., Kim J. S., Jung W. J., Lee Y. S. and Lee C. H. (2001). Application of surface modified polypropylene membranes to an anaerobic membrane bioreactor. *Environmental Technology, 22(9), 1035* - 1042.

Salazar-Peláez M. L., Morgan-Sagastume J. M. and Noyola A. (2011). Influence of hydraulic retention time on fouling in a UASB coupled with an external ultrafiltration membrane treating synthetic municipal wastewater. *Desalination*, **277**(1–3), 164–170.

Saravanane R. and Sundararaman S. (2009). Effect of loading rate and HRT on the removal of cephalosporin and their intermediates during the operation of a membrane bioreactor treating pharmaceutical wastewater. *Environmental Technology*, **30**(10), 1016–1022.

Schiener P., Nachaiyasit S. and Stuckey D. C. (1998). Production of soluble microbial products (SMP) in an anaerobic baffled reactor: composition, biodegradability, and the effect of process parameters. *Environmental Technology*, **19** (4), 391–399.

Sheng C. (2011). Application of submerged hollow fiber membrane in membrane bioreactors: filtration principles, operation, and membrane fouling. *Desalination*, **283**(0), 31–39.

Sheng G.-P. and Yu H.-Q. (2006). Characterization of extracellular polymeric substances of aerobic and anaerobic sludge using three-dimensional excitation and emission matrix fluorescence spectroscopy. *Water Research*, **40** (6), 1233–1239.

Sheng G. P., Yu H. Q. and Li X. Y. (2006). Stability of sludge flocs under shear conditions: roles of extracellular polymeric substances (EPS). *Biotechnology and Bioengineering*, **93**(6), 1095–1102.

Shin C. and Bae J. (2018). Current status of the pilot-scale anaerobic membrane bioreactor treatments of domestic wastewaters: a critical review. *Bioresource Technology*, **247**, 1038–1046.

Shin C., McCarty P. L., Kim J. and Bae J. (2014). Pilot-scale temperate-climate treatment of domestic wastewater with a staged anaerobic fluidized membrane bioreactor (SAF-MBR). *Bioresource Technology*, **159**, 95–103.

Sipma J., Osuna M. B., Emanuelsson M. A. E. and Castro P. M. L. (2010). Biotreatment of industrial wastewaters under transient-state conditions: process stability with fluctuations of organic load, substrates, toxicants, and environmental parameters. *Critical Reviews in Environmental Science and Technology*, **40**(2), 147–197.

Skouteris G., Hermosilla D., López P., Negro C. and Blanco Á. (2012). Anaerobic membrane bioreactors for wastewater treatment: a review. *Chemical Engineering Journal*, **198–199**(0), 138–148.

Smith A. L., Stadler L. B., Love N. G., Skerlos S. J. and Raskin L. (2012). Perspectives on anaerobic membrane bioreactor treatment of domestic wastewater: a critical review. *Bioresource Technology*, **122**, 149–159.

Smith A. L., Love N. G., Skerlos S. J. and Raskin L. (2011). Role of Membrane Biofilm in Psychrophilic Anaerobic Membrane Bioreactor for Domestic Wastewater Treatment. Proceedings of 84th Annual Water Environment Federation Technical Exhibition and Conference (WEFTEC), Los Angeles, California, USA.

Smith C. W., Gregorio D. and Taleott R. M. (1969). The Use of Ultrafiltration Membrane for Activated Sludge Separation. Proceedings of the 24th Annual Purdue Industrial Waste Conference, Purdue University, West Lafayette, Indiana, USA.

Spagni A., Casu S., Crispino N. A., Farina R. and Mattioli D. (2010). Filterability in a submerged anaerobic membrane bioreactor. *Desalination*, **250**(2), 787–792.

Stahl N., Tenenbaum A. and Galil N. I. (2004). Advanced treatment by anaerobic process followed by aerobic membrane bioreactor for effluent reuse in paper mill industry. *Water Science and Technology*, **50**(3), 245–252.

Stamatelatou K., Kopsahelis A., Blika P. S., Paraskeva C. A. and Lyberatos G. (2009). Anaerobic digestion of olive mill wastewater in a periodic anaerobic baffled reactor (PABR) followed by further effluent purification via membrane separation technologies. *Journal of Chemical Technology & Biotechnology*, **84**(6), 909–917.

Stephenson T., Judd S., Jefferson B. and Brindle K. (2000). Membrane Bioreactors for Wastewater Treatment. IWA Publishing, London, UK.

Stuckey D. C. (2012). Recent developments in anaerobic membrane reactors. *Bioresource Technology*, **122**(0), 137–148.

Sui P., Wen X. and Huang X. (2008). Feasibility of employing ultrasound for on-line membrane fouling control in an anaerobic membrane bioreactor. *Desalination*, **219**(1–3), 203–213.

Sutton P. M., Li R. R. and Korchin S. R. (1983). Dorr-Oliver's Fixed Film Suspended Growth Anaerobic Systems for Industrial Wastewater Treatment and Energy Recovery. Proceedings of the 37th Industrial Waste Conference, Purdue University, Lafayette, IN.

Sutton P. M. S., Melcer H., Schraa O. J. and Togna A. P. (2010). Treating municipal wastewater with the goal of resource recovery. *Water Science and Technology*, **63**(1), 25–31.

Szentgyorgyi E. and Belafi-Bako K. (2010). Anaerobic membrane bioreactors. *Hungarian Journal of Industrial Chemistry*, **38**(2), 181–185.

Tadkaew N., Sivakumar M., Khan S. J., Mcdonald J. A. and Nghiem L. D. (2010). Effect of mixed liquor pH on the removal of trace organic contaminants in a membrane bioreactor. *Bioresour Technol*, **101**(5), 1494–1500.

Tardieu E., Grasmick A., Geaugey V. and Manem J. (1998). Hydrodynamic control of bioparticle deposition in a MBR applied to wastewater treatment. *Journal of Membrane Science*, **147**(1), 1–12.

Tian J.-Y., Chen Z.-L., Yang Y.-L., Liang H., Nan J. and Li G.-B. (2010). Consecutive chemical cleaning of fouled PVC membrane using NaOH and ethanol during ultrafiltration of river water. *Water Research*, **44**(1), 59–68.

Tian Y., Chen L., Zhang S. and Zhang S. (2011). A systematic study of soluble microbial products and their fouling impacts in membrane bioreactors. *Chemical Engineering Journal*, **168**(3), 1093–1102.

Tiranuntakul M., Schneider P. A. and Jegatheesan V. (2011). Assessments of critical flux in a pilot-scale membrane bioreactor. *Bioresource Technology*, **102**(9), 5370–5374.

Torres A., Hemmelmann A., Vergara C. and Jeison D. (2011). Application of two-phase slug-flow regime to control flux reduction on anaerobic membrane bioreactors treating wastewaters with high suspended solids concentration. *Separation and Purification Technology*, **79**(1), 20–25.

Trzcinski A. P. and Stuckey D. C. (2009a). Anaerobic digestion of the organic fraction of municipal solid waste in a two-stage membrane process. *Water Science and Technology*, **60**(8), 1965–1978.

Trzcinski A. P. and Stuckey D. C. (2009b). Continuous treatment of the organic fraction of municipal solid waste in an anaerobic two-stage membrane process with liquid recycle. *Water Research*, **43**(9), 2449–2462.

Trzcinski A. P. and Stuckey D. C. (2010). Treatment of municipal solid waste leachate using a submerged anaerobic membrane bioreactor at mesophilic and psychrophilic temperatures: analysis of recalcitrants in the permeate using GC-MS. *Water Research*, **44**(3), 671–680.

Trzcinski A. P. and Stuckey D. C. (2016). Inorganic fouling of an anaerobic membrane bioreactor treating leachate from the organic fraction of municipal solid waste (OFMSW) and a polishing aerobic membrane bioreactor. *Bioresource Technology*, **204**, 17–25.

Trzcinski A. P., Ofoegbu N. and Stuckey D. C. (2011). Post-treatment of the permeate of a submerged anaerobic membrane bioreactor (SAMBR) treating landfill leachate. *Journal of Environmental Science and Health, Part A*, **46**(13), 1539–1548.

Vallero M. V. G., Lettinga G. and Lens P. N. L. (2005). High rate sulfate reduction in a submerged anaerobic membrane bioreactor (SAMBaR) at high salinity. *Journal of Membrane Science*, **253**(1–2), 217–232.

Van Voorthuizen E., Zwijnenburg A., Van Der Meer W. and Temmink H. (2008). Biological black water treatment combined with membrane separation. *Water Research*, **42**(16), 4334–4340.

Van Zyl P. J., Wentzel M. C., Ekama G. A. and Riedel K. J. (2008). Design and start-up of a high rate anaerobic membrane bioreactor for the treatment of a low pH, high strength, dissolved organic waste water. *Water Science and Technology*, **57**(2), 291–295.

Vidal G. and Diez M. C. (2005). Methanogenic toxicity and continuous anaerobic treatment of wood processing effluents. *Journal of Environmental Management*, **74**(4), 317–325.

Vyrides I. and Stuckey D. C. (2009a). Saline sewage treatment using a submerged anaerobic membrane reactor (SAMBR): effects of activated carbon addition and biogas-sparging time. *Water Research*, **43**(4), 933–942.

Vyrides I. and Stuckey D. C. (2009b). Effect of fluctuations in salinity on anaerobic biomass and production of soluble microbial products (SMPs). *Biodegradation*, **20**(2), 165–175.

Vyrides I., Santos H., Mingote A., Ray M. J. and Stuckey D. C. (2010). Are compatible solutes compatible with biological treatment of saline wastewater? batch and continuous studies using Submerged Anaerobic Membrane Bioreactors (SAMBRs). *Environmental Science & Technology*, **44**(19), 7437–7442.

Wang Z., Wu Ye S. and Tang Z. S. (2010). Application of Anaerobic Membrane Bioreactor to the Treatment of Low-Strength Municipal Wastewater, Bioinformatics and Biomedical Engineering (iCBBE). 2010 4th International Conference, June 18–20, 2010, Chengdu, China.

Ward A. J., Hobbs P. J., Holliman P. J. and Jones D. L. (2008). Optimisation of the anaerobic digestion of agricultural resources. *Bioresource Technology*, **99**(17), 7928–7940.

Wei Y., Van Houten R. T., Borger A. R., Eikelboom D. H. and Fan Y. (2003). Minimization of excess sludge production for biological wastewater treatment. *Water Research*, **37**(18), 4453–4467.

Wen C., Huang X. and Qian Y. (1999). Domestic wastewater treatment using an anaerobic bioreactor coupled with membrane filtration. *Process Biochemistry*, **35**(3–4), 335–340.

Wen X., Sui P. and Huang X. (2008). Exerting ultrasound to control the membrane fouling in filtration of anaerobic activated sludge–mechanism and membrane damage. *Water Science and Technology*, **57**(5), 773–779.

Wijekoon K. C., Visvanathan C. and Abeynayaka A. (2011). Effect of organic loading rate on VFA production, organic matter removal and microbial activity of a two-stage thermophilic anaerobic membrane bioreactor. *Bioresource Technology*, **102**(9), 5353–5360.

Wong K., Xagoraraki I., Wallace J., Bickert W., Srinivasan S. and Rose J. B. (2009). Removal of viruses and indicators by anaerobic membrane bioreactor treating animal waste. *Journal of Environmental Quality*, **38**(4), 1694–1699.

Wu B., An Y., Li Y. and Wong F. S. (2009). Effect of adsorption/coagulation on membrane fouling in microfiltration process post-treating anaerobic digestion effluent. *Desalination*, **242**(1–3), 183–192.

Wu S. C. and Lee C. M. (2011). Correlation between fouling propensity of soluble extracellular polymeric substances and sludge metabolic activity altered by different starvation conditions. *Bioresource Technology*, **102**(9), 5375–5380.

Xie K., Lin H. J., Mahendran B., Bagley D. M., Leung K. T., Liss S. N. and Liao B. Q. (2010). Performance and fouling characteristics of a submerged anaerobic membrane bioreactor for kraft evaporator condensate treatment. *Environmental Technology*, **31**(5), 511–521.

Xu M., Wen X., Yu Z., Li Y. and Huang X. (2011). A hybrid anaerobic membrane bioreactor coupled with online ultrasonic equipment for digestion of waste activated sludge. *Bioresource Technology*, **102**(10), 5617–5625.

Xu Y., Zhou Y., Wang D., Chen S., Liu J. and Wang Z. (2008). Occurrence and removal of organic micropollutants in the treatment of landfill leachate by combined anaerobic-membrane bioreactor technology. *Journal of Environmental Science (China)*, **20**(11), 1281–1287.

Xue W., Wu C., Xiao K., Huang X., Zhou H., Tsuno H. and Tanaka H. (2010). Elimination and fate of selected micro-organic pollutants in a full-scale anaerobic/anoxic/aerobic process combined with membrane bioreactor for municipal wastewater reclamation. *Water Research*, **44**(20), 5999–6010.

Yamamoto K., Hiasa H., Talat M. and Matsuo T. (1989). Direct solid liquid separation using hollow fiber membranes in activated sludge aeration tank. *Water Science and Technology*, **21**, 43–54.

Yang J., Spanjers H. and Van Lier J. B. (2011). Pulse shear stress for anaerobic membrane bioreactor fouling control. *Water Science and Technology*, **64**(2), 355–360.

Yilmaz T., Yuceer A. and Basibuyuk M. (2008). A comparison of the performance of mesophilic and thermophilic anaerobic filters treating papermill wastewater. *Bioresource Technology*, **99**(1), 156–163.

Yu H. Q. and Fang H. H. P. (2003). Acidogenesis of gelatin-rich wastewater in an upflow anaerobic reactor: influence of pH and temperature. *Water Research*, **37**(1), 55–66.

Yun M.-A., Yeon K.-M., Park J.-S., Lee C.-H., Chun J. and Lim D. J. (2006). Characterization of biofilm structure and its effect on membrane permeability in MBR for dye wastewater treatment. *Water Research*, **40**(1), 45–52.

Yuzir M. and Ali M. (2008). Effect of Redox Conditions in the Degradation of (RS)-MCPP during Biological Treatment of Synthetic Wastewater in an Anaerobic Membrane Bioreactor (AMBr). PhD thesis, Newcastle University.

Zayen A., Mnif S., Aloui F., Fki F., Loukil S., Bouaziz M. and Sayadi S. (2010). Anaerobic membrane bioreactor for the treatment of leachates from Jebel Chakir discharge in Tunisia. *Journal of Hazardous Materials*, **177**(1–3), 918–923.

Zhang J., Padmasiri S. I., Fitch M., Norddahl B., Raskin L. and Morgenroth E. (2007). Influence of cleaning frequency and membrane history on fouling in an anaerobic membrane bioreactor. *Desalination*, **207**(1–3), 153–166.

Zhang X., Wang Z., Wu Z., Lu F., Tong J. and Zang L. (2010). Formation of dynamic membrane in an anaerobic membrane bioreactor for municipal wastewater treatment. *Chemical Engineering Journal*, **165**(1), 175–183.

Zhang X., Wang Z., Wu Z., Wei T., Lu F., Tong J. and Mai S. (2011). Membrane fouling in an anaerobic dynamic membrane bioreactor (AnDMBR) for municipal wastewater treatment: characteristics of membrane foulants and bulk sludge. *Process Biochemistry*, **46**(8), 1538–1544.

Chapter 12

Hybrid processes, new generation membranes and novel MBR designs

*Olivier Lefebvre[1], Kok Kwang Ng[2], Kai Yin Tang[3], Tze Chiang Albert Ng[4] and How Yong Ng[5]**

[1]Assistant Professor, Civil and Environmental Engineering, National University of Singapore; ceelop@nus.edu.sg
[2]Research Fellow, Civil and Environmental Engineering, National University of Singapore; ceengkk@nus.edu.sg
[3]Post-graduate Student, Civil and Environmental Engineering, National University of Singapore
[4]Senior Research Fellow, Civil and Environmental Engineering, National University of Singapore; albertng@nus.edu.sg
[5]Professor, Civil and Environmental Engineering, National University of Singapore; howyongng@nus.edu.sg
**Corresponding author: howyongng@nus.edu.sg*

ABSTRACT

This chapter focuses on innovative applications of membrane bioreactors (MBR). With the reduction in membrane cost and the introduction of more stringent regulations for wastewater discharge to natural bodies, MBR is becoming a viable solution for water reclamation in many situations. It may become a commonly adopted technology in the near future, as water scarcity becomes a serious issue in many parts of the world and concerns related to health and safety of treated wastewater for water reuse rises. This chapter focuses on four areas of development for MBRs: (i) hybrid MBR processes integrated with conventional biological and physico-chemical water treatment options; (ii) novel membrane designs for MBRs; (iii) innovative MBR designs; and (iv) emerging concepts for energy recovery based on the MBR process.

Editor(s): Faisal I. Hai, Kazuo Yamamoto and Chung-Hak Lee
doi: 10.2166/9781780409177_0387

12.1 INTRODUCTION

Sanitation is one of the challenges of the 21st century which must be met in order to combat poverty. In addition, water reuse will likely be favoured in many densely populated parts of the world where the availability of fresh water will become a concern. The membrane bioreactor (MBR) may become the norm for water reclamation in the near future because of its ability to increase both the amount of water available (water security) and the quality of the reclaimed water (water safety).

The basis of modern water reclamation has been laid out in the past century and therefore, the first step shall be the integration of the MBR technology with those well-understood technologies to form hybrid MBR systems. For example, the combination of MBR and biofilm reactors might help develop specific functions in a bioreactor, while physico-chemical processes may further improve the quality of reclaimed water, in order to broaden the possibilities for reuse. A second step will be the development of improved MBR systems that can stand on their own and produce reclaimed water suitable for reuse in a more compact design. The development of innovative membranes, making use of carbon nanotubes, titanium dioxide, as well as grafted polymers and electrospun nanofibers will help reach that target when they become affordable. Meanwhile, nanofiltration, forward osmosis (FO) and membrane distillation membranes are already commercially available and being used to produce water of excellent quality. They are also starting to be combined with bioreactors to form innovative MBRs at the research and development stage.

Finally, looking further into the future, the escalating cost of energy will compel the development of an energy efficient water reclamation system. At the moment, high-energy consumption is one of the main drawbacks of the MBR technology, but innovative concepts are being explored in order to reduce energy consumption or even net a positive gain of energy using the MBR technology. These include hydrodynamic arrangements within the MBR, pressure retarded osmosis (PRO) and microbial fuel cells.

12.2 INTEGRATED MBR SYSTEMS FOR WATER RECLAMATION

The trend in the bioreactor design is moving towards hybrid bioreactors with many functions in a single vessel, with the MBR often being at the core of the system because of its unique advantages such as good effluent quality and compact structure. This chapter reviews some of the most recent advances in the field including biofilm MBRs, aerobic granular sludge MBRs and MBRs integrated with physico-chemical processes. Although not detailed in this chapter, there is an even wider range of hybrid-MBR processes, that have been successfully combined with an upflow anaerobic sludge blanket (UASB) reactor (Hua *et al.*, 2009) or with a constructed wetland (Xiao *et al.*, 2010) to achieve excellent COD, suspended solids and nutrient removal efficiencies at low cost and competitive hydraulic retention times (HRT).

12.2.1 Biofilm MBR

A hybrid biofilm MBR makes use of a fluidized or immobile support for microbial growth. This allows a high biomass concentration to be maintained in the MBR, favouring a more compact module and a smaller footprint. The efficacy and fouling propensity of biofilm MBRs relies upon the combination of a suitable carrier, MBR configuration, balance between the fixed and suspended biomass, carrier packing ratio, nature of the wastewater, etc. A range of biofilm MBR designs is displayed in Figure 12.1, while typical operating conditions and performance achieved are provided in Table 12.1.

The selection of moving-bed vs fixed-bed MBR has significant implications on the fouling propensity of the MBR. The use of a fluidized medium in a membrane-coupled moving-bed reactor was indeed shown to provide physical cleaning of the membrane by scrubbing (Lee *et al.*, 2006), but this effect can be countered by the higher fouling potential of attached biomass as compared to suspended sludge (Lee *et al.*, 2001). Similarly, a combination of anaerobic fixed biofilm and aerobic submerged MBR with or without mobile carriers was evaluated by Yang *et al.* (2009). They found that the addition of mobile carriers efficiently reduced membrane fouling in spite of an alteration of the sludge biological activity. They attributed this effect to the shear stress imposed by the carrier on the suspended biomass.

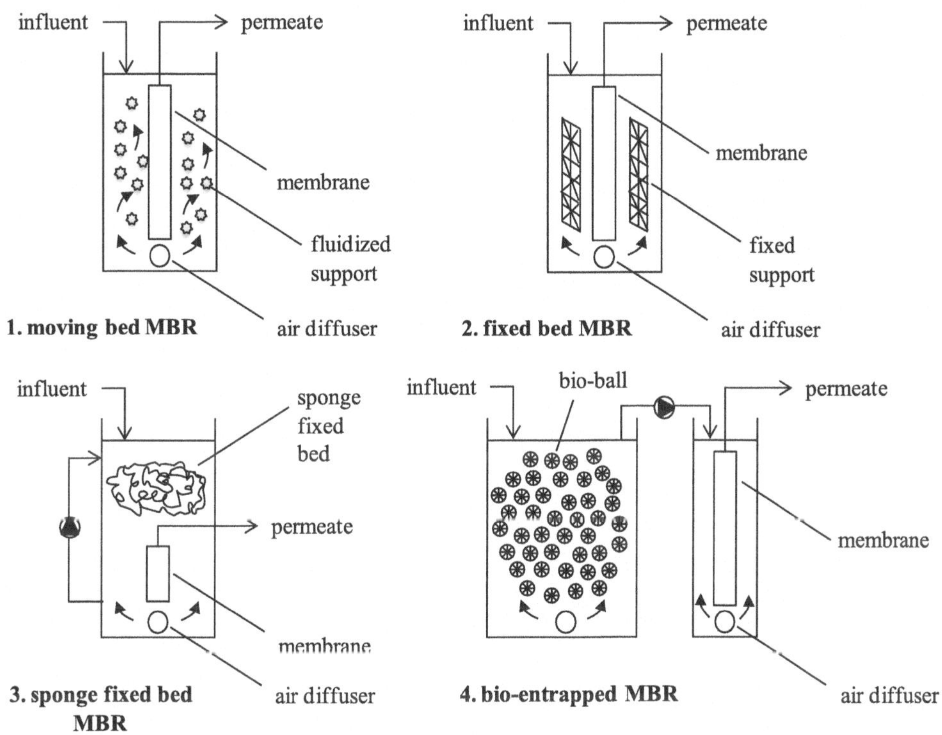

Figure 12.1 Examples of biofilm-MBR designs.

Among, the innovative designs for biofilm MBRs, Table 12.1 highlights the efficacy of the sponge fixed-bed MBR, which produces a treated effluent near reuse standard (except for nitrogen) (Rodriguez-Hernandez *et al.*, 2012). Similarly, the bio-entrapped MBR (BEMBR) is another recent development in hybrid MBRs. While similar to the MBMBR, which allows biofilms to develop naturally (albeit slowly) on moving carriers, the principle of BEMBR is based on physically entrapping microbial biomass inside a carrier that is fluidized inside the bioreactor (Ng *et al.*, 2011). The main advantage of these bio-balls is faster start-up, improved mass-transfer distribution and reduced clogging. The BEMBR also allows for the augmentation of slow-growing organisms, which would otherwise be outcompeted under normal planktonic operations. Cellulose triacetate was found to be the most suitable chemical to trap microorganism on the carrier because it provides the necessary mechanical strength and durability under both aerobic and anaerobic conditions. The new BEMR offers effective organics removal while reducing membrane fouling.

Table 12.1 Operating conditions and performance of biofilm MBRs for wastewater treatment.

Reactor	Membrane Configuration	Packing		Type of Wastewater	Volume (L)	HRT (h)	SRT (d)	OLR (kg m^{-3} d^{-1})	Removal (%)			Ref
		Material	Ratio (%)						COD	NH_4^+	TN	
BEMBR	External	CTA	55	Food	50	6–12	500	1–5	93–98	NA*	NA	1
FBMBR	Submerged	Sponge	~50	Domestic	70	288	60	2.7	91	98	80	2
FBMBR	Submerged	HDPE	20	Mixed	190	4.5–5.4	27	~2.8	84	42	21	3
MBMBR	Submerged	HDPE	20	Mixed	190	4.5–5.4	27	~2.8	89	40	14	3
MBMBR	External	HDPE	20	Tannery	6.7	NA	1–10	4.5	95	76	NA	4
MBMBR	External	HDPE	20	Food	6.7	2.6	NA	6.8	99	>98	NA	4
MBMBR	Submerged	HDPE	5–20	Glucose	6	10	10	2.4	>98	NA	NA	5
MBMBR	External	HDPE	60–67	Domestic	116	7.2	NA	0.92	82	NA	46	6
MBMBR	External	HDPE	50	Domestic	2000	10	10	~1.2	94	97	51	7

*NA = not available.

Abbreviations: BEMBR = bio-entrapped membrane bioreactor; CTA = cellulose triacetate; HDPE = high-density polyethylene; FBMBR = fixed-bed membrane bioreactor; MBMBR = moving-bed membrane bioreactor.

Ref: 1. Ng *et al.* (2011); 2. Rodriguez-Hernandez *et al.* (2012); 3. Yang *et al.* (2012); 4. Artiga *et al.* (2005); 5. Lee *et al.* (2006); 6. Phattaranawik and Leiknes (2009); 7. Liu *et al.* (2010).

The use of a carrier in an MBR also renders possible several original applications such as combined nitrification/denitrification in the same reactor, whereby nitrification occurs on the outer layer of the biofilm while denitrification happens in the core of the biofilm (Li *et al.*, 2008). This is made possible because of the gradient of oxygen that occurs inside a biofilm as evidenced by several models and verified experimentally on numerous occasions (Pochana *et al.*, 1999). In Artiga *et al.* (2005), the use of small rough plastic carriers made it further feasible to decouple the sludge retention time (SRT) for nitrifiers and for heterotrophs, which resulted in the presence of a higher fraction of nitrifiers in the reactor.

12.2.2 Granular sludge MBR

Granulation is a phenomenon that occurs naturally in sequencing batch reactors (SBR) under specific conditions with short sedimentation and draw phases in order to wash out slow settling biomass (Morgenroth *et al.*, 1997). Granules (up to 0.8 mm of diameter) display unique physical and chemical characteristics, such as compact structure, good flocculation and settling ability that allows production of good-quality effluent. Interestingly, aerobic granules formed under stress conditions were found to have a different extracellular polymeric substances (EPS) composition and distribution as compared to conventional flocculating sludge, with a higher protein content mostly located near the core of the granules (McSwain *et al.*, 2005). Considering the contribution of EPS to membrane fouling, granulation can increase membrane permeability when a membrane module is added to the system, forming the granular sludge MBR (Li *et al.*, 2005). Owing to their low compressibility, aerobic granules were also found to limit biofilm build up in the interior surface of the membrane, a phenomenon that contributes to irreversible fouling resistance (Juang *et al.*, 2011). There are specific advantages in incorporating membranes in SBRs: enrichment of fastidious or slow-growing organisms such as phosphate accumulating organisms or anaerobic ammonium oxidizing (anammox) bacteria will benefit from the hybrid membrane sequencing batch reactor (MSBR). The SBR operating conditions allow biomass to naturally granulate with a membrane to retain and concentrate the biomass. MSBR was found to be suitable for rapid startup of an anammox community (Trigo *et al.*, 2006).

However, because of limited biofilm growth on the membrane surface, direct deposition of colloids and solutes can cause pore blocking in the absence of a cake layer (Jun *et al.*, 2007). Granular sludge disaggregation was also observed following the change of operating mode from SBR to MBR and this could be explained by filamentous bacteria overgrowth following the abrupt change of environmental conditions (SRT, F:M ratio, etc.) (Wang *et al.*, 2008) and by high shear forces resulting from higher aeration rate needed for membrane scouring in MBR mode. Such disaggregation releases EPS into the soluble phase, which is detrimental to membrane permeability.

12.2.3 MBR integrated with physico-chemical processes

Physico-chemical processes are increasingly the focus of research in conjunction with membrane treatment in order to (i) improve the quality of the final product for drinking water or water reuse applications; and (ii) address membrane fouling issues. Some examples of hybrid MBR/physico-chemical processes for various types of wastewater are presented in Table 12.2 and detailed in this section.

12.2.3.1 MBR with Fenton oxidation process

The Fenton oxidation process is based on the generation of highly reactive hydroxyl radicals from H_2O_2 using an iron catalyst as an advanced wastewater treatment. Fenton oxidation can help make the MBR technology more economical by reducing sludge production, while maintaining excellent permeate

Table 12.2 Operating conditions and performance of MBRs combined with physico-chemical processes for wastewater treatment.

Physico-Chemical Process	Type of Wastewater	Volume (L)	HRT (h)	SRT (d)	OLR (kg m^{-3} d^{-1})	Removal (%)			Ref
						COD	TOC	Color	
Fenton	Dyeing	6	18	30	1.4–1.8	NA*	88	91	1
Photo-Fenton	Pesticides	20	8.3	NA	NA	NA	98	NA	2
Photocatalysis/ozonation	Dyeing	10	17	NA	0.2–0.6	NA	90	>99	3
Ozonation	Pharmaceutical	10	150	60	1.7	99	NA	NA	4
Ozonation	Pharmaceutical	6	90	30	0.5	97	95	NA	5
UV/H_2O_2	Pharmaceutical	6	90	30	0.5	97	93	NA	5
PAC	Distillery	8	168	NA	5.7–6.9	31–41	NA	NA	6
PAC	Tannery	260	100	95	1.0	80	NA	NA	7
Coagulation (PACl)	Dairy	NA	10	NA	3.4–5.1	98	NA	NA	8
Coagulation ($FeCl_3$)	Piggery	12	48	∞	2.5	>99	NA	NA	9

*NA = not available.

Abbreviation: PACl = polyaluminum chloride.

Ref: 1. Feng *et al.* (2010); 2. Ballesteros Martin *et al.* (2011); 3. Qu *et al.* (2009a); 4. Mascolo *et al.* (2010); 5. Laera *et al.* (2012); 6. Satyawali and Balakrishnan (2009); 7. Munz *et al.* (2007); 8. Chen and Liu (2012); 9. Kornboonraksa *et al.* (2009).

quality (He & Wei, 2010). MBRs combined with Fenton oxidation have further proved successful for advanced TOC and colour removal from dyeing wastewater (Feng *et al.*, 2010). Finally, the photo-Fenton oxidation process – which uses solar irradiation for catalyst regeneration – combined with MBR was shown to be a viable option to treat toxic wastewater containing pesticides, yielding permeate quality of reuse standards (Ballesteros Martin *et al.*, 2011). In that study, little change was detected in the activated sludge population as monitored by denaturing gradient gel electrophoresis (DGGE), indicating that it was not affected by photo-treated intermediates present in the water.

12.2.3.2 MBR with ozonation

Ozonation is recognized as an excellent means to reduce the biorefractory portion of industrial (e.g., pharmaceutical) wastewater. The efficacy of a combined ozonation and MBR process has been assessed in different operational modes, where ozonation was applied as a pre- or post-treatment, and results have been found to vary accordingly. For example, pre-ozonation of simulated surface water polluted by sewage discharge was found to slightly improve organic removal and reduce membrane fouling in the subsequent MBR treatment (Liu *et al.*, 2011). The reduction of membrane fouling was also confirmed by the injection of ozone directly into the MBR (Wu & Huang, 2010). In another study, integrating ozonation in the recirculation stream of the MBR effluent improved the removal efficiency of the antibacterial nalidixic acid as compared to ozone polishing after MBR (Laera *et al.*, 2012). A similar effect, which could be explained by synergistic effects of chemical oxidation and bioprocesses, was observed on a variety of pharmaceutical degradation by-products. Ozonation was finally found an efficient mean to achieve discoloration of the permeate of small-scale MBRs treating gray water (Abegglen *et al.*, 2009).

12.2.3.3 MBR with activated carbon

The use of activated carbon in various forms (biological, granular or powdered) has been extensively employed in drinking water treatment all over the world. Because of more stringent legislation on health and environmental protection, their combination with MBR is increasingly becoming the focus of research, not only for drinking water but also for wastewater treatment, including tannery (Munz *et al.*, 2007) and distillery (Satyawali & Balakrishnan, 2009) effluents. When applied at very low doses and long SRTs, powdered activated carbon (PAC) was found to reduce membrane fouling in MBRs (Remy *et al.*, 2009). The benefits of PAC can be explained by its incorporation into bioflocs – altering their physical and mechanical properties – along with a decrease of protein and carbohydrate concentrations in the MBR supernatant (Lesage *et al.*, 2008). Ng *et al.* (2010) further demonstrated that biological activated carbon (BAC) reduces membrane fouling via three synergistic mechanisms including adsorption of fine foulants, membrane scouring and biodegradation phenomena. PAC was also found to reduce foaming in an MBR treating distillery effluents (Satyawali & Balakrishnan, 2009) and to achieve discoloration of MBR permeates (Abegglen *et al.*, 2009). Finally, MBR–granular activated carbon (GAC) and MBR–PAC systems proved particularly efficient at removing trace organic contaminants, especially hydrophilic and persistent compounds (e.g. carbamazepine, diclofenac, etc.) that the MBR technology alone is unable to treat well (Serrano *et al.*, 2011).

12.2.3.4 MBR with coagulation

Similarly as PAC – with which it is frequently combined – coagulation has been proven useful for drinking water applications (Tian *et al.*, 2010). Increasingly, chemical coagulation/flocculation (e.g., using $FeCl_3$, polyaluminum chloride, etc.) is combined with MBR for industrial wastewater treatment, such as dairy

(Chen & Liu, 2012) or piggery (Kornboonraksa *et al.*, 2009) effluents. Coagulants including alum, ferric chloride ($FeCl_3$), polymeric ferric sulfate were found to reduce membrane fouling in an MBR due to a modification of the particle-size distribution (Wu & Huang, 2008; Zhang *et al.*, 2008). Specifically, coagulants act by enlarging the size of small colloidal particles, which are known to be a major contributor to membrane fouling (Fan *et al.*, 2007).

12.3 INNOVATIVE MEMBRANE DESIGN FOR MBR

The emergence of nanotechnologies can benefit water treatment via the functionalization of microfiltration and ultrafiltration (UF) membranes, by nanoengineered particles deposited on their surface or embedded inside their matrices. When coupled with the MBR technology, these innovative membranes can provide new functionalities to MBRs to bind specific contaminants or act as nanocatalysts. Innovative membrane designs surveyed in this section include carbon nanotubes (CNTs) and titanium dioxide (TiO_2) modified membranes, as well as grafted polymer membranes and electrospun nanofiber membranes.

12.3.1 CNT-doped membranes

CNTs are defined as allotropes of carbon with a multi-cylindrical nanostructure that brings significant benefits to many areas of nanotechnology. To date, CNTs have found applications for heavy metal separation from aqueous environments because of their exceptional properties, which, on top of their high affinity for metal ions, include a large specific surface area, high tensile strength, high flexibility, high chemical, electrochemical and thermal stability, corrosion resistance and biocompatibility.

Despite the multiple advantages of membranes as a separation technology, polymeric membrane materials are still limited by their low selectivity, high- temperature instability and decomposition in organic solvent. Therefore, significant efforts are being conducted towards the development of CNT-doped membranes. For example, Kim and Van der Bruggen (2010) incorporated multi-walled CNTs into polymeric membranes with the following advantages to the membrane: (i) enhanced sorption capacity, (ii) improved antifouling properties and (iii) increased mechanical strength of the membrane.

12.3.2 TiO_2-doped membranes

In recent years, TiO_2 photocatalysis has received attention as a novel advanced oxidation process (AOP) for detoxification of water, wastewater and air. On top of their antibacterial and photostability properties, nanoparticles of TiO_2 are particularly efficient at destructuring contaminants and pollutants in aqueous environments (Pekakis *et al.*, 2006). Furthermore, in an MBR, suspended TiO_2 can significantly reduce membrane fouling (Xi & Geissen, 2001). In order to avoid the need to separate the nanoparticles of TiO_2 from the treated water, a novel approach consists of incorporating TiO_2 onto the membrane, which confers interesting anti-microbial properties to the membrane, while allowing the degradation of retained compounds.

Figure 12.2 shows the mechanism of detoxification by photocatalytic TiO_2-doped membranes. Briefly, valence band holes (h_{vb}^+) and conduction band electrons (e_{cb}^-) are generated during the illumination of TiO_2 surface with ultraviolet (UV) irradiation greater than the band gap energy of TiO_2. Hydroxyl radicals (HO·) are generated by the reaction of valence band holes with water (H_2O/OH^-) and peroxide radicals are formed by the reaction of electrons with adsorbed molecular oxygen (O_2). Thereafter, organic pollutants are oxidized by hydroxyl radical and turn into non-toxic material (Damodar *et al.*, 2009). Self-cleaning/antifouling mechanisms of TiO_2-doped membrane surface are also presented in Figure 12.2. This also involves the generation of electrons and holes but the process is different. Electrons (ecb-) reduce Ti^{4+} cations to Ti^{3+} and holes ($h_{vb}{}^{+}$) oxidize $O_2{}^{-}$ anions to O_2. As a result of oxygen vacancy, water occupies the free sites, which improves surface hydrophilicity (Damodar *et al.*, 2009).

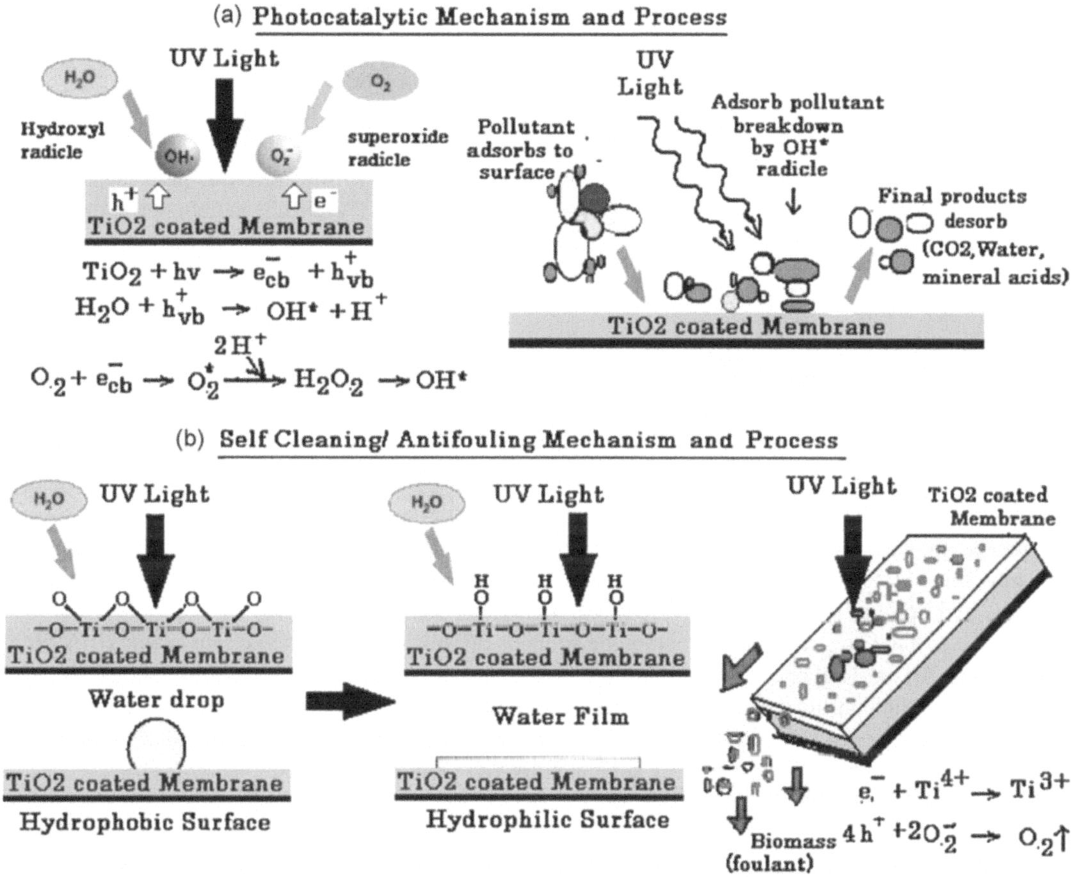

Figure 12.2 (a) Photocatalysis mechanism and process. (b) Self- cleaning/antifouling mechanism and process of PVDF/TiO_2 membrane. Reprinted from *Journal of Hazardous Materials*, 172/2–3, Damodar, R. A., You, S.-J. & Chou, H.-H. Study the self-cleaning, antibacterial and photocatalytic properties of TiO_2 entrapped PVDF membranes, 1321–1328, Copyright (2009), with permission from Elsevier.

The advantages of TiO_2-doped membranes further include (i) excellent self-cleaning/antifouling properties, (ii) prevention of membrane fouling and (iii) integrated separation and photocatalytic degradation of target compounds in a single membrane device.

12.3.3 Grafted polymer membranes

Graft co-polymerization has been identified as a promising method for attaching a variety of functional groups onto a polymer membrane (Bhattacharya and Misra, 2004). Grafting hydrophilic groups onto UF membranes has been used to improve their fouling resistance by increasing the wettability of the membrane surfaces in order to reduce their ability to interact with species present in the solution. Grafting presents the advantage over coating that graft monomers are chemically bonded to the membrane surface, increasing the stability and lifetime of the membrane. However, Asatekin *et al.* (2006) reported that the high-energy cost associated with grafting can considerably increase membrane fabrication costs and the polymerization of ungrafted chains can block surface pores and reduce the intrinsic membrane permeability, while leaving the membrane susceptible to internal pore fouling. To eliminate internal pore fouling, they developed a novel type of nanofiltration (NF) membrane by grafting poly (vinylidene fluoride)-*graft*-poly(oxyethylene) methacrylate (PVDF-*g*-POEM) onto the surface of a commercial polyvinylidene fluoride (PVDF) UF membrane. This generated hydrophilic and charge-neutral nanochannels (~2 nm), which allowed water and small molecules to pass through. The grafted NF membrane exhibited lower fouling than the commercial PVDF UF membrane with no flux decline throughout the experimental period.

12.3.4 Graphene oxide membranes

Graphene oxide (GO) and GO-coated membranes have been receiving much attention in recent years due to the many benefits it has to offer to the filtration process in biological tanks. GO is a two-dimensional nanomaterial that is easy to assemble on or as membranes due to various oxygen-containing functional groups and large surface area (Faria *et al.*, 2017; Hu & Mi, 2013). Fabrication methods include (1) vacuum-assisted filtration (Han *et al.*, 2013), (2) layer-by-layer deposition (Hu & Mi, 2013) and (3) spin-coating (Kim *et al.*, 2013). The chemical stability and high mechanical strength make it physically suitable to be used in many applications. Moreover, GO membranes have shown to be extremely hydrophilic, which not only resulted in higher flux recovery (Ko *et al.*, 2018), but also anti-adhesive surface properties since membrane foulants are mostly hydrophobic (Hu *et al.*, 2016; Rana & Matsuura, 2010). Higher biological fouling resistance was also observed from GO membranes because of a reduction in surface roughness, which minimizes biofilm formation, and the innate antibacterial properties of GO. Bacterial cells attached onto GO membranes are inactivated by oxidative stresses caused by reactive oxygen species generated by GO (Zou *et al.*, 2016).

12.3.5 Electrospun nanofiber membranes

Electrospining is a novel fabrication method that produces continuous nanofibers from 80 nm to several hundred nanometers in a non-woven structure (Bilad *et al.*, 2011). Electrospun nanofiber membranes display better performance than polymeric and ceramic membranes because of their unique architecture, high surface porosity and permeability (Bilad *et al.*, 2011; Wang *et al.*, 2012a). They have been successfully adopted in MF, UF and NF. Wang *et al.* (2012a) showcased the exceptional properties of electrospun nanofiber membranes for high-flux MF application, being able to achieve substantially (2–3 times) higher flux than conventional MF membranes. They further indicated that reducing the diameter of the fibers favored the formation of smaller pore sizes and narrower pore-size distribution, resulting in increased micro-particle rejection rate. Bilad *et al.* (2011) also reported the importance of fiber diameter and sheet thickness as key issues that affect the pore size and material cost. Finally, Singh *et al.* (2010) carbonized nanofiber membranes at a temperature of 400–500°C under a nitrogen atmosphere to increase their adsorption capacity, making them suitable for the removal of disinfection by-products from water bodies.

12.3.6 Metal-organic framework membranes

Over the last decade, membranes modified with a layer of metal-organic framework (MOF) have been actively researched. MOF layer is prepared by the self-assembly of metal ions or metal clusters with organic ligands, forming a fine networked structure of various defined shapes such as nanowires or nanosheets. There are many sub-classes of MOF based on the type of metal used, with titanium and zinc being the most common (Hou *et al.*, 2018; Zhu *et al.*, 2018). Because of the high surface area (Farha *et al.*, 2012), low density and highly selective surface (Furukawa *et al.*, 2011), MOF membranes have been shown to be successful for gas separation (Rui *et al.*, 2018). These innate characteristics coupled with increased hydrophilicity, high pore volume, an abundance of adsorption sites and the ability to form MOFs with different structures are capable of enhancing the membrane's filtration performance (Farha *et al.*, 2012; James, 2003; Li *et al.*, 2015). MOF membranes have been demonstrated as MF, NF and UF membranes with applications such as desalination and oil–water separation (Kadhom & Deng, 2018). However, much more work is needed before MOF membranes reach maturity for full-scale commercialization. Focus on economical large-scale fabrication methods and the required thickness of the active layer are currently lacking. In addition, adhesion stability of the MOF coating on the substrate and operational stability under various physico-chemical conditions are required before such membrane may be tested for complex wastewater treatment systems.

12.4 INNOVATIVE MBR DESIGNS

Innovative MBR designs are driven by the multiple applications of MBRs engendered by the democratization and novel developments of membrane technology. This has resulted in a multitude of novel MBR configurations designed to produce excellent water quality for water reuse (e.g., nanofiltration or FO MBR, membrane distillation bioreactor, etc.). Another sector of interest is the development of MBRs that excel at fouling control such as air-sparging or air-lift MBR. Finally, MBRs designed for special applications, such as anammox are also in development.

12.4.1 NF-MBR

Most MBRs rely on MF and UF membranes for biomass retention in the reactor vessel. Yet, NF membranes can reject organic matters with much lower MW of approximately 200 Da and therefore, address the removal of emerging contaminants such as micropollutants and pharmaceuticals. NF membranes mainly separate the contaminants by size exclusion (for uncharged molecules), charge effects and electrostatic interactions (for ionic species). Membrane properties, feed composition and operating conditions also affect the rejection rate of solutes by NF membranes (Bellona & Drewes, 2005).

The advantages of NF-MBR include elimination of phase changes or chemical additives, simplicity of concept and operation, modularity and easy scale-up, rejection of lower molecular weight contaminants (e.g., pharmaceutical and endocrine hormones) and reduction of equipment size. Drawbacks are related to insufficient separation depending on the compound molecular size, post-treatment of the concentrate, membrane lifetime and chemical tolerance to alkaline and acid cleaning. As of date, the application of NF membranes has been limited to drinking water and water reuse. Yet, Choi *et al.* (2002) reported that the NF-MBR could provide extra-clean permeate and reasonable water productivity for the treatment of synthetic wastewater. Cellulose acetate membranes further allowed higher permeate productivity than polyamide NF membrane (Choi *et al.*, 2005). On the long term, NF-MBR could produce consistently very good-quality permeate with 40–60% and 70–90% rejection rates for monovalent and divalent ions, respectively.

12.4.2 FO-MBR

Permeate production through MF or UF membranes is obtained via vacuum suction and while this provides effluent of drinking water quality, the energy consumption is intensive. Yet, the advances in membrane engineering during the past few decades have brought about a resurgence of interest in processes based on forward – or direct – osmosis (FO). In the FO process, water molecules move from a region of higher water concentration to a region of lower concentration through a tight membrane, thus providing an alternative water production pathway. In addition, water movement due to osmotic pressure differences does not require any input of external energy and as such offers an alternative manner through which water can be extracted from wastewater or seawater with minimal energy requirement.

Henceforth, a recent novel integration of the FO process with MBR configuration yielded the innovative FO membrane bioreactor (FO–MBR). In an FO–MBR system, permeate production is driven by differences in osmotic pressure instead of hydraulic pressure, resulting in lower fouling propensities (Achilli *et al.*, 2009; Cornelissen *et al.*, 2008). By recirculating a highly concentrated salt solution (known as draw solution) within the submerged membrane module, pure water is drawn from the wastewater feed stream across the FO membrane and into the draw solution, leaving the contaminants within the bioreactor mixed liquor. Permeate flux causes dilution of the draw solution, which consists of a highly concentrated salt solution. Subsequently, drinking water is recovered via pressurized filtration, such as reverse osmosis (RO) or NF, and reject stream is recaptured for reuse (Figure 12.3).

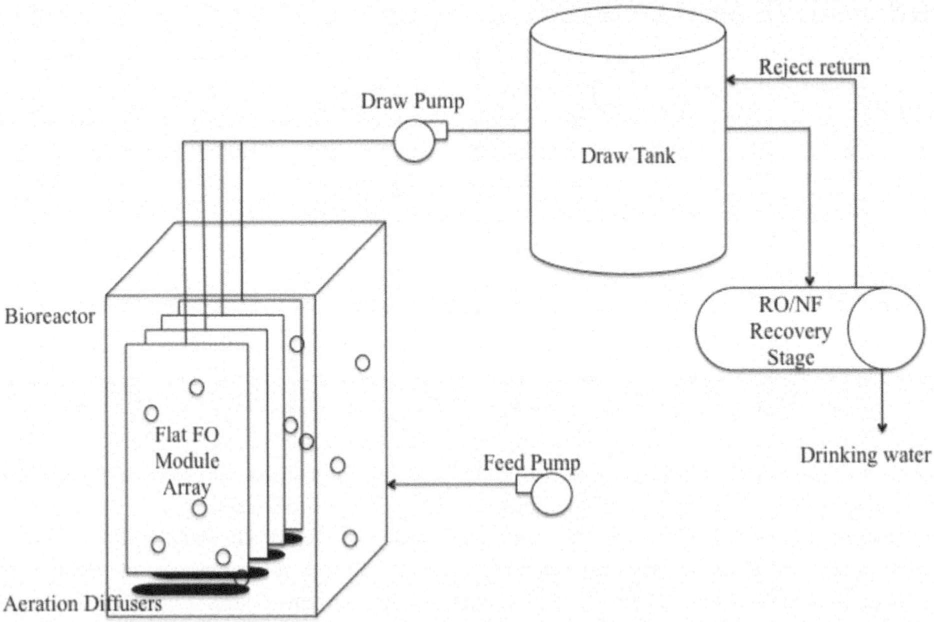

Figure 12.3 Schematics of an FO-MBR based on flat-sheet FO membrane module configuration.

The main drawback of FO-MBR is related to the increase of the total dissolved solids (TDS) concentration over time within the bioreactor, which can adversely impact sludge viability. Hence, while conventional MBRs are usually operated with a SRT ranging from 5 to 50 days, high SRT in FO-MBRs can become detrimental due to excessive TDS accumulation. Another consequence of high membrane rejection is that the HRT is not decoupled from the operational SRT. This implies that operational flux is in fact controlled by the SRT because flux magnitude is dependent on the osmotic pressure difference, i.e., higher bioreactor TDS concentrations resulting from prolonged SRT reduce the osmotic pressure difference between the draw solution and the wastewater feed stream, and therefore reduce the flux. Finally, it must be noted that while the activity of extracting clean water from the feed stream requires almost no energy, the succeeding recovery of drinking water from the diluted draw solution requires considerable amounts of energy, by means of either NF or RO. Yet, there is hope that with continual improvements to the FO membrane structure and process design, the energy consumption associated with the recovery step can be counterbalanced, therefore making FO-MBR an exciting potential candidate that is both energy- and economically efficient.

Extraction of freshwater from the diluted draw solution while reconcentrating the diluted raw solution for reuse in the FO-MBR process can be avoided if the diluted draw solution can be directly used for downstream application. An example is the application of fertigation, where the crop nutrients/fertilizers in dissolved form are supplied to the crops through an irrigation network system. In this application, concentrated fertilizer is used as the draw solution of the FO process because the diluted draw solution can be directly applied for fertigation (Moody & Kessler, 1976).

12.4.3 MD-MBR

Membrane distillation (MD) involves a combination of water evaporation and vapor condensation within an integrated process, encompassing water vapor transportation through the pores of a hydrophobic membrane via a transmembrane temperature gradient. In detail, permeate production starts from feed evaporation at the pore entrance and through the pores, finally condensing on the other side of the membrane (collected as the cooler permeate). MD can further be combined with an MBR to yield the MD-MBR, which benefits from the following advantages: (i) 100% (theoretical) ionic rejection and other non-volatiles, (ii) lower operating pressures than conventional pressure-driven processes, (iii) lower operating temperatures than conventional distillation processes, and (iv) performance insensitive to feed concentration (unlike RO). Likewise, the MD-MBR also inherits the challenges associated with MD in considerations with the current state of technological advancement, which include membrane wetting and scaling, temperature polarization and salinity accumulation within the bioreactor. However, looking beyond the drawbacks of the MD configuration, the MDMBR provides a unique manner through which high permeate quality can be obtained independently of the biological treatment performance of the MBR.

In the MD-MBR configuration, the MD unit is usually submerged within the bioreactor with the membrane unit in direct contact with the mixed liquor. Permeate production within such systems are driven by the vapor pressure difference across an unwetted membrane pore. The operational temperature is maintained between 30 and 80°C. Because of the higher operating temperatures, MD-MBR requires the implementation of a thermophilic microflora within the mixed liquor. Unlike FO-MBR, the permeate is obtained directly without the need for downstream pressurized filtration recovery. Not only is this trait similar to conventional MBR configurations, but reported fluxes for MD-MBRs are also higher (ranging from 2 to 15 L mh) than pioneering batches of high retention MBRs (i.e. NF-MBRs at 0.042 L mh) (Phattaranawik *et al.*, 2008). A schematic of the MD-MBR is provided in Figure 12.4.

To enhance the feasibility of the MD-MBR concept, the issue of temperature polarization has to be tackled by continual advancements in MD membrane science and technology. In a MD process, mass transfers take place simultaneously with heat conduction across the membrane matrix, and consequently, the temperature of the boundary layer at the feed side is lowered, while that of the boundary layer on the permeate side is increased. This phenomenon, known as temperature polarization, causes the temperatures at the membrane surface to differ from that of the bulk solutions, resulting in reduced driving force available for mass transfers (Martínez-Díez & Vázquez-González, 1999). Hence, it is imperative that the process be made more efficient by making future MD membranes with materials that are insulating and thermally stable. One method for accomplishing this is to create higher membrane porosity (during membrane fabrication) to take advantage of air's lower heat conductivity to lower the membrane matrix's overall heat-transfer ability.

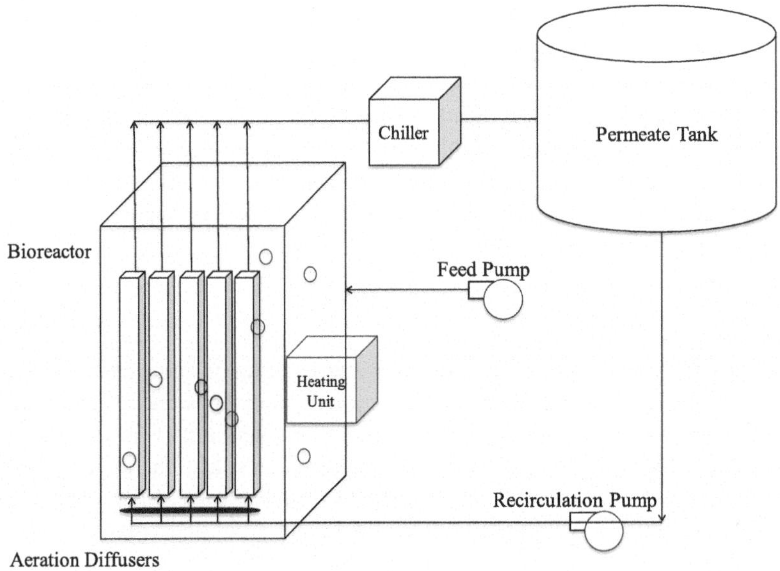

Figure 12.4 Schematics of an MD-MBR.

Finally, the viability of the MD process depends on an efficient use of energy and the availability of cheap and waste energy as well. If waste heat is not readily available, the applicability of MD processes such as MD-MBR is moot. Issues such as heat-transfer inefficiencies (from waste heat sources to feed stream) will play a critical role in determining the potential of scaling up the MD-MBR process.

12.4.4 Anammox-MBR

The anaerobic ammonium oxidation (anammox) process has recently received much attention and been put forward as a new and promising way to treat wastewater containing high concentrations of ammonium without additional sources of organic carbon. Anammox can achieve high nitrogen removal rates at lower operational costs and smaller footprint requirements than conventional nutrient removal plants. Anammox consists of the denitrification of nitrite using ammonium as the electron donor to yield nitrogen gas. A partial nitrification step must be performed beforehand to convert only one half of the ammonium to nitrite. The main product of anammox is N_2, however; approximately 10% of the fed nitrogen (ammonium and nitrite) is converted to nitrate. The stoichiometry of the anammox reaction is accepted to be as follows:

$$NH_4^+ + 1.32NO_2^- + 0.066HCO_3^- + 0.13H^+ \rightarrow 1.02N_2 + 0.26NO_3^- + 2.03H_2O + 0.066CH_2O_{0.5}N_{0.15}$$

The performance of anammox is highly dependent on environmental factors such as HRT, dissolved oxygen concentration, temperature, wastewater composition and nitrogen concentration (Tsushima *et al.*, 2007). Although the anammox process has been successfully applied in laboratory-scale, pilot-scale and full-scale, the application of anammox is still limited by its long start-up period due to the very low growth rate of anammox bacteria. For example, Kuenen (2008) indicated that the first full-scale anammox reactor took approximately 2 years for the bacteria to grow. In order to solve this issue, various types of reactors (fluidized-bed reactor, SBR, UASB, etc.) have been assessed to enrich anammox bacteria. Among them, the MBR could prove to be the best alternative for the installation of a quick and stable anammox ecosystem, due to its capacity to completely retain the biomass, associated with high loading rates to promote growth of anammox bacteria (van der Star *et al.*, 2008). In their study, Van der Star *et al.* (2008) showed that the doubling time of anammox bacteria in an MBR could be reduced below 10 days. Wang *et al.* (2009) also reached the conclusion that MBR was a suitable system for the start-up of anammox process. In their study, anammox bacteria grew and reached a steady state within 16 days of operation, achieving an average of more than 90% of ammonia and nitrite removal efficiency. Tao *et al.* (2012) demonstrated the advantages of MBR over SBR for the cultivation of slow-growing anammox organisms. In their MBR, anammox activity was increased by 19 times as compared to that of the SBR.

12.4.5 Bioaugmented MBR

Bioaugmented reactors are based on using bacteria, yeast or fungi of special degrading capacity to either improve the treatment performance and stability of the process or allow faster start-up. One of the main issues is the washout of the selected strain and its replacement by endogenous species. For example, a strain of *Citrobacter braakii* was found to degrade a wide range of anionic surfactants but could not be maintained in a completely stirred tank reactor. Upon the addition of a membrane, *Citrobacter braakii* became the dominant bacterial strain again and help generate a permeate of reuse standard (Dhouib *et al.*, 2005). (Qu *et al.*, 2009b) compared the efficiency of bromoamine acid (BAA) removal by MBR bioaugmented with a strain of *Sphingomonas xenophaga* as compared to a non-augmented system. They found that in both cases, BAA-utilizing populations gradually became dominant but the bioaugmented BAA displayed more stable performance. The same strain of *Sphingomonas xenophaga* was later found capable to decolorize 1-amino-4-bromoanthraquinone-2-sulfonic acid in another bioaugmented MBR (Wang *et al.*, 2009). Other applications include the biodegradation of 4-aminobenzenesulfonate, an aromatic amine and recalcitrant toxic pollutant, by *Pannonibacter* sp. (Zhang *et al.*, 2011a) or the enhanced removal of dyes from a textile effluent by the white-rot fungus *Coriolus versicolor* (Hai *et al.*, 2011).

Finally, bioaugmentation using genetically engineered microorganisms also constitutes a promising way to remove recalcitrant chemicals from wastewater. For example, the incorporation of an atrazine chlorohydrolase gene of *Pseudomonas* sp. into an *Escherichia coli* strain allowed complete removal of atrazine in an MBR after a start-up period of 2 days, during which the *E. coli* cell density first dropped but then remained stable for the remaining experimental period (Liu *et al.*, 2008).

12.4.6 Bioaugmented MBR for fouling control

Bacteria share information to coordinate their behaviour such as biofilm formation through intercellular signalling molecules in a process known as quorum sensing (QS). These signal molecules are known as autoinducers, with acylated homoserine lactone (AHL) being the most common type found in Gram-negative bacteria populations. In general, there are three methods to inhibit the QS system: (1)

prevention of signal synthesis, (2) inactivation of signals, and (3) interference of signal receptor. Of the three methods, inactivation of signal molecules is considered to be the most effective. This is typically done with quorum quenching (QQ) bacteria capable to metabolizing the QS signals such as AHLs.

Bioaugmentation of BH4 (*Rhodococcus* sp.) has mostly been used in MBR studies. Capable of degrading a broad spectrum of AHLs, BH4 was isolated from a full-scale MBR plant (Oh *et al.*, 2013). However, the relative abundance of *Rhodococcus* sp. in MBRs was found to be very low (<0.02%) in 10 real MBR plants, which served to indicate that BH4 did not compete well with other microorganisms in treating municipal wastewaters (Oh *et al.*, 2012). As such, bioaugmentation of QQ bacteria was required to artificially increase the population in order to realise a positive impact on membrane fouling reduction. Bioaugmentation of BH4 in MBR was found to lead to significantly reduced attachment of biomass and organic foulants onto membrane surfaces (Weerasekara *et al.*, 2016). It is not expected that QQ alone will be sufficient for fouling control, but has synergistic impacts when used with other control methods such as air sparging, backwashing and/or chlorination.

12.4.7 Aeration imposed shear for fouling control

Membrane fouling is one of the major issues associated with MBR systems, resulting in increased operating costs and potentially shorter membrane life span because of frequent chemical cleaning. As described in Chapter 4, various mitigation methods such as membrane relaxation, forward flushing, backwashing and chemical cleaning can be implemented. Even though a suitable combination of mitigation methods can be efficient at increasing the membrane permeability and prolonging the life of the membrane, some of these options are expensive, time-consuming and result in ceased activity during the cleaning period.

In the recent years, air sparging has been developed as a novel in-situ cleaning technology to control and reduce membrane fouling. During air sparging, air bubbles are injected into the lumen or outside the membrane to enhance the permeate flux. Membrane air sparging provides effective turbulence around the membrane and membrane movement that scours the particles and removes other deposited materials on the surface of the membrane (Park *et al.*, 2010).

Membrane air sparging has proved to be an effective and easy technique for enhancing physical separation in MF and UF processes, but little information is available to quantitatively and qualitatively assess flux improvements for submerged MBRs. It is believed that the membrane surface contacts intimately with the air bubbles, which might induce a shear stress when the bubbles are suppressing the cake on the MBR (Chang & Judd, 2002). During the air sparging process, the upward aerator is generally located under or near the bottom in order to generate air bubbles to sparge and scour the foulants deposition on the membrane surface. Other studies have demonstrated a variety of flow patterns to increase the flux improvement including injecting large air bubbles into tubular or hollow fiber membranes or combining the liquid- and air- flows together in a pipe (Psoch & Schiewer, 2008). However, the efficacy of such configurations highly depends on the ratio of air-to-liquid flow rate in the MBR system, and Cabassud *et al.* (1997) have indicated that slug flow is the most suitable regime to limit particle deposition, due to higher shear stress.

Several studies have demonstrated the efficacy of air sparging technique for MBR membrane cleaning in short- and long-term experiments. In particular, Psoch and Schiewer (2005) found that the permeability of an MBR combining air sparging and backflushing over a 56-day period was approximately 4 times higher than that achieved with conventional filtration. They attributed this effect to increased shear stress and reduced membrane fouling caused by air sparging. Finally, Park *et al.* (2010) showed that simultaneous upward and downward air sparging was more efficient than upward air sparging alone at enhancing the membrane permeability and reducing membrane fouling.

Despite being widely applied, continuous air sparging requires high-energy consumption. As such, several aeration strategies have been investigated and applied to reduce power consumption while maintaining the effectiveness of fouling control. Intermittent aeration (IA) has been investigated in both lab- and pilot-scale MBRs (Capodici *et al.*, 2015; Shao *et al.*, 2015). Importantly, results have shown that such an IA strategy did not reduce biological performance, particularly in removal efficiencies of organic matter and ammonia. IA had the added advantage of creating cyclical patterns of shear stresses, which resulted in smaller floc sizes (Campo *et al.*, 2017), reducing more irreversible fouling (Yang *et al.*, 2017). Modification of the IA strategy included using slug bubbling, which at regular intervals, bursts of gas/air introduced into an MBR would cause the air bubbles to collide and form large bubbles known as gas slugs (Ghosh & Cui, 1999). Periodic slug bubbling was also found to be an effective method for fouling control while reducing energy requirements by 40–50% compared to current industry norms (Wang *et al.*, 2018; Zhang *et al.*, 2009, 2011a, b).

12.4.8 Mechanically imposed shear for fouling control

Imparting shear forces along membrane surfaces is the predominant method in controlling membrane fouling. Shear can be imposed either by aeration with air or other gasses, or by mechanically moving the membrane module. Mechanically imposed shear has been applied in numerous variations, but are all based on the same concept of creating shear forces along the membrane surfaces by imposing some forms of movements on the membrane modules, as opposed to the liquid in aeration-imposed shear. These variations include (1) disc filters that rotate or vibrate (Zsirai *et al.*, 2016), (2) hollow fiber modules that vibrate (Zsirai *et al.*, 2016) or swing (Qin *et al.*, 2015), (3) flat sheet modules that mechanically induce oscillation (Buzatu *et al.*, 2017) or magnetically induced vibration (Bilad *et al.*, 2012).

A submerged hollow fiber MBR with pendulum-type oscillation was found to improve fouling control and enhance membrane permeability by 11 times compared to the conventional MBR system in treating oily wastewater (Qin *et al.*, 2015). And magnetically induced membrane vibration had markedly higher fluxes and lower fouling rates than conventional MBR systems (Bilad *et al.*, 2012), thereby demonstrating the effectiveness of such mechanically imposed shearing systems for fouling control. The main advantage in these systems lies in the energy costs required to achieve the same result as air scouring. The cost associated with shear imposed mechanically was found to be 20–70% less than that of conventional aeration (Buzatu *et al.*, 2017).

12.5 INNOVATIVE CONCEPTS

This section deals with the possibilities to simultaneously produce clean water and recover energy in an MBR. One of the main drawbacks of the MBR technology is its elevated energy consumption mostly driven by the very high aeration rate used for mixing, scouring and to promote high nitrification rates. Utilization of the excess aeration energy, for example, to drive a propeller would thus be beneficial for the overall energy optimization of MBRs. Another way to recover energy via MBR lies in PRO, which uses the concentration gradient across a membrane to produce osmotic power. The pressurized water can run a turbine and generate electricity.

12.5.1 Mechanical recovery of energy in MBR

Phattaranawik and Leiknes (2011) recently disclosed the hydrodynamic arrangements of liquid, air and solids in an MBR that can make use of the excess aeration energy to activate a propeller. Their set-up is presented in Figure 12.5. Briefly, the concept is based on a succession of chambers, baffles and perforated plates. In the lower (aerobic) chamber, aeration is provided in the form of fine air bubbles which later coalesce into larger bubbles upon approaching the upper (anoxic) chamber. There, the large bubbles with higher momentum are used to drive the main propeller (represented with 4 blades in the inset of Figure 12.5), which then activates two mixing propellers (1 blade each). These propellers are used to mix the anoxic chamber. With proper engineering control of the mixing conditions, the authors were able to increase the speed of the mixing propellers from 14 to 62 rpm. The minimum air flowrate to drive the propeller was 8 L/min but better mixing in the upper chamber was achieved at 18 L/min. Phattaranawik and Leiknes (2011) concluded the possibility of using the excess aeration energy to drive a dynamo for electricity generation.

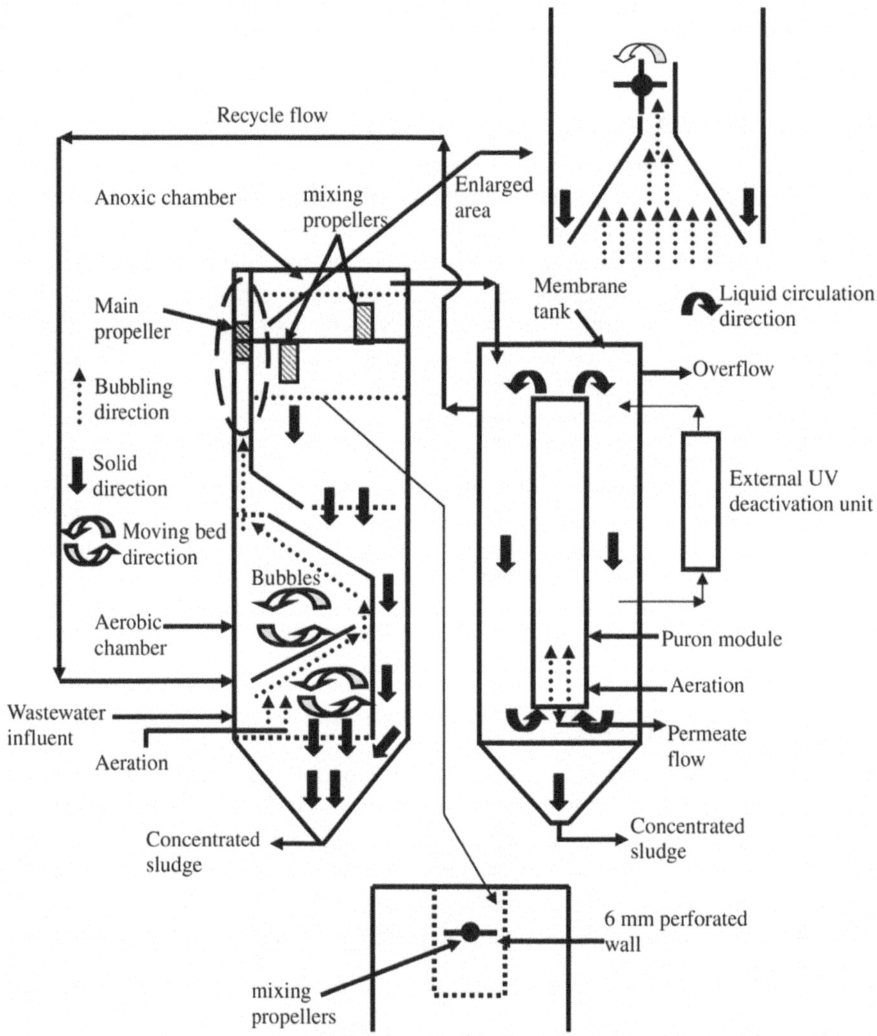

Figure 12.5 Experimental set-up for mechanical recovery of excess aeration energy in an MBR. Reprinted from *Bioresource Technology*, 102/3, Phattaranawik, J. & Leiknes, T. Extractive biofilm membrane bioreactor with energy recovery from excess aeration and new membrane fouling control, 2301–2307, Copyright (2011), with permission from Elsevier.

12.5.2 PRO-MBR

In PRO, a semi-permeable membrane delivers a pressurized permeate water which is capable of generating electric energy by depressurization using a hydroturbine and a generator. Although the main idea of producing electrical energy from salinity-gradient was recognized more than half a century ago (Achilli and Childress, 2010), PRO is still considered a rather novel technology that has been boosted by the progress in the development of semi-permeable membranes. Using a FO membrane, it was estimated that it is possible to recover up to 0.3 W/m^2 of membrane from seawater by PRO (Kim *et al.*, 2010). Energy that can be extracted from paring wastewater and RO brine as feed and draw solutions, respectively, was found to be up to 0.4 kWh/m^3 (Lin *et al.*, 2014). Even though power generation is hindered at this time by high concentration polarization, there is hope that the development of more suitable membranes shall result in higher energy recovery by PRO. Furthermore, the concepts of MBR and PRO could be combined in order to achieve wastewater treatment and renewable energy production using osmotic membranes with low pressure and energy requirements (Achilli *et al.*, 2008). The design of the prospective PRO-MBR is given in Figure 12.6.

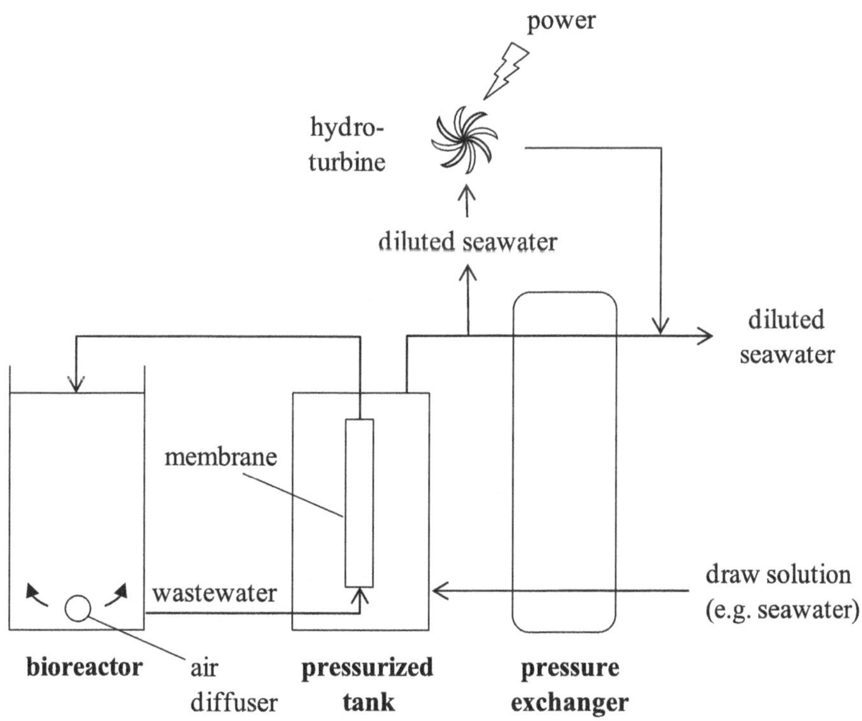

Figure 12.6 Schematic representation of the prospective pressure retarded osmotic MBR.

12.5.3 MFC-MBR

A microbial fuel cell (MFC) is a bioelectrochemical device that allows the recovery of bioelectricity from a substrate such as a wastewater. It typically comprises an anode – covered by a biofilm of electrophilic bacteria that oxidize the substrate and split it into electrons and protons – linked by an electrical circuit to a cathode – where the protons and electrons produced at the anode combine with oxygen to form water (Lefebvre *et al.*, 2011). The anode chamber and the cathode chamber are typically separated by a cation – or in some cases an anion – exchange membrane but unlike in MBRs, the membrane is merely used as a separation device and filtration does not generally occur within an MFC. However, the anode and the cathode, which are typically made of a porous carbon or stainless-steel material have occasionally been used as filtration material, making the MFC work as an MBR (Figure 12.7). In the study by Cheng *et al.* (2006), advective flow of the wastewater through the anode led to reduced internal resistance and consequently increased power density (up to 51 W/m^3), as proton transfer was forced towards the cathode. In the alternative configuration developed by Wang *et al.* (2011), the anode-treated effluent was filtered through the cathode, processing a final effluent mostly devoid of suspended solids, while generating moderate power density of 4 W/m^3. It should be noted here that direct comparison of these power densities is impossible as the operational conditions and designs were extremely different. Another combination of MFC and MBR is finally showcased in Figure 12.7, in which the aerobic MBR is used as the cathode chamber for the MFC. This final configuration allowed the production of a maximum power density of 6 W m^{-3} (Wang *et al.*, 2012b).

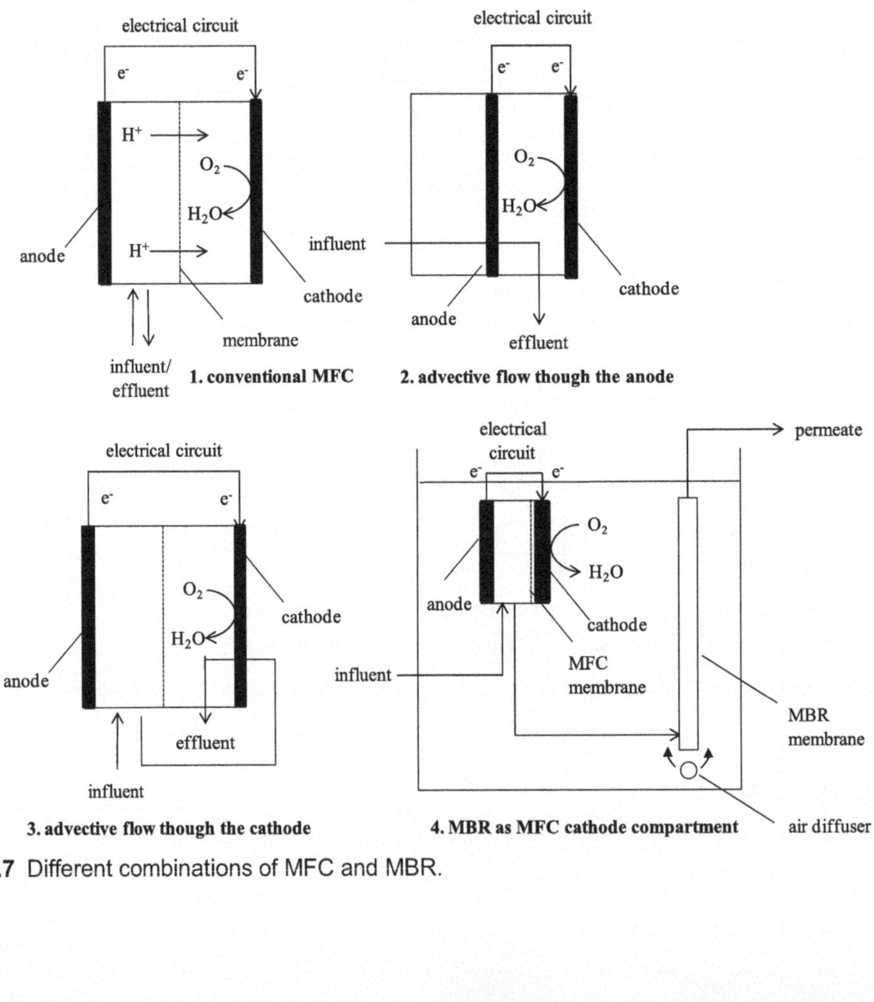

Figure 12.7 Different combinations of MFC and MBR.

12.6 CONCLUSION AND FUTURE OUTLOOK

Taking into account the water and energy challenges that await humanity in the 21st century, the MBR technology is likely to become the option of choice; however, its main drawbacks (membrane fouling and high-energy consumption) will have to be tackled via research and development. This chapter introduces the near future of the MBR technology – from its integration with conventional water reclamation technologies to the development of novel membranes, novel practices and novel concepts. The trend in the bioreactor design is moving towards hybrid bioreactors with many functions in a single vessel, with the MBR often being at the core of the system because of its unique advantages such as good effluent quality and compact structure. This chapter reviews some of the most recent advances in the field including biofilm MBRs, aerobic granular sludge MBRs and MBRs integrated with physico-chemical processes. The emergence of nanotechnologies can benefit water treatment via the functionalization of microfiltration and UF membranes, by nanoengineered particles deposited on their surface or embedded inside their matrices. When coupled with the MBR technology, these innovative membranes can provide new functionalities to MBRs to bind specific contaminants or act as nanocatalysts. Recent developments have seen the introduction of a multitude of novel MBR configurations designed to produce excellent water quality for water reuse (e.g., nanofiltration or FO MBR, membrane distillation bioreactor, etc.). Another sector of interest is the development of MBRs that excel at fouling control such as air-sparging or air-lift MBR. This chapter also highlights the possibilities to simultaneously produce clean water and recover energy in an MBR: utilization of the excess aeration energy, for example, to drive a propeller; energy recovery via application of PRO; combining a microbial fuel cell and MBR.

REFERENCES

Abegglen C., Joss A., Boehler M., Buetzer S. and Siegrist H. (2009). Reducing the natural color of membrane bioreactor permeate with activated carbon or ozone. *Water Science and Technology*, **60**(1), 155–165.

Achilli A. and Childress A. E. (2010). Pressure retarded osmosis: From the vision of Sidney Loeb to the first prototype installation – Review. *Desalination*, **261**(3), 205–211.

Achilli A., Cath T. Y., Marchand E. A. and Childress A. E. (2008). Osmotic Membrane Bioreactor and Pressure Retarded Osmotic Membrane Bioreactor for Wastewater Treatment and Water Desalination. ICOM, Honolulu, Hawaii.

Achilli A., Cath T. Y., Marchand E. A. and Childress A. E. (2009). The forward osmosis membrane bioreactor: A low fouling alternative to MBR processes. *Desalination*, **239**(1–3), 10–21.

Artiga P., Oyanedel V., Garrido J. M. and Méndez R. (2005). An innovative biofilm-suspended biomass hybrid membrane bioreactor for wastewater treatment. *Desalination*, **179**(1–3), 171–179.

Asatekin A., Menniti A., Kang S., Elimelech M., Morgenroth E. and Mayes A. M. (2006). Antifouling nanofiltration membranes for membrane bioreactors from self-assembling graft copolymers. *Journal of Membrane Science*, **285**(1–2), 81–89.

Ballesteros Martin M. M., Garrido L., Casas Lopez J. L., Sanchez O., Mas J., Maldonado M. I. and Sanchez Perez J. A. (2011). An analysis of the bacterial community in a membrane bioreactor fed with photo-Fenton pre-treated toxic water. *Journal of Industrial Microbiology & Biotechnology*, **38**(9), 1171–1178.

Bellona C. and Drewes J. E. (2005). The role of membrane surface charge and solute physico-chemical properties in the rejection of organic acids by NF membranes. *Journal of Membrane Science*, **249**(1–2), 227–234.

Bhattacharya A. and Misra B. N. (2004). Grafting: A versatile means to modify polymers: Techniques, factors and applications. *Progress in Polymer Science*, **29**(8), 767–814.

Bilad M. R., Westbroek P. and Vankelecom I. F. J. (2011). Assessment and optimization of electrospun nanofiber-membranes in a membrane bioreactor (MBR). *Journal of Membrane Science*, **380**(1–2), 181–191.

Bilad M. R., Mezohegyi G., Declerck P. and Vankelecom I. F. J. (2012). Novel magnetically induced membrane vibration (MMV) for fouling control in membrane bioreactors. *Water Research*, **46**, 63–72.

Buzatu P., Qiblawey H., Nasser M. S. and Judd S. (2017). Comparative power demand of mechanical and aeration imposed shear in an immersed membrane bioreactor. *Water Resarch*, **126**, 208–215.

Cabassud C., Laborie S. and Lainé J. M. (1997). How slug flow can improve ultrafiltration flux in organic hollow fibres. *Journal of Membrane Science*, **128**(1), 93–101.

Campo R., Capodici M., Di Bella G. and Torregrossa M. (2017). The role of EPS in the foaming and fouling for a MBR operated in intermittent aeration conditions. *Biochemical Engineering Journal*, **118**, 41–52.

Capodici M., Di Bella G., Di Trapani D. and Torregrossa M. (2015). Pilot scale experiment with MBR operated in intermittent aeration condition: Analysis of biological performance. *Bioresource Technology*, **177**, 398–405.

Chang I. S. and Judd S. J. (2002). Air sparging of a submerged MBR for municipal wastewater treatment. *Process Biochemistry*, **37**(8), 915–920.

Chen W. W. and Liu J. R. (2012). The possibility and applicability of coagulation-MBR hybrid system in reclamation of dairy wastewater. *Desalination*, **285**, 226–231.

Cheng S., Liu H. and Logan B. E. (2006). Increased power generation in a continuous flow MFC with advective flow through the porous anode and reduced electrode spacing. *Environmental Science & Technology*, **40**(7), 2426–2432.

Choi J. H., Dockko S., Fukushi K. and Yamamoto K. (2002). A novel application of a submerged nanofiltration membrane bioreactor (NF MBR) for wastewater treatment. *Desalination*, **146**(1–3), 413–420.

Choi J. H., Fukushi K. and Yamamoto K. (2005). Comparison of treatment efficiency of submerged nanofiltration membrane bioreactors using cellulose triacetate and polyamide membrane. *Water Science and Technology*, **51** (6–7), 305–312.

Cornelissen E. R., Harmsen D., de Korte K. F., Ruiken C. J., Qin J.-J., Oo H. and Wessels L. P. (2008). Membrane fouling and process performance of forward osmosis membranes on activated sludge. *Journal of Membrane Science*, **319**(1–2), 158–168.

Damodar R. A., You S.-J. and Chou H.-H. (2009). Study the self cleaning, antibacterial and photocatalytic properties of TiO_2 entrapped PVDF membranes. *Journal of Hazardous Materials*, **172**(2–3), 1321–1328.

Dhouib A., Hdiji N., Hassairi I. and Sayadi S. (2005). Large scale application of membrane bioreactor technology for the treatment and reuse of an anionic surfactant wastewater. *Process Biochemistry*, **40**(8), 2715–2720.

Fan F., Zhou H. and Husain H. (2007). Use of chemical coagulants to control fouling potential for wastewater membrane bioreactor processes. *Water Environment Research*, **79**(9), 952–957.

Farha O. K., Eryazici I., Jeong N. C., Hauser B. G., Wilmer C. E., Sarjeant A. A., Snurr R. Q., Nguyen S. T., Yazaydin A. Ö. and Hupp J. T. (2012). Metal-organic framework materials with ultrahigh surface areas: Is the sky the limit? *Journal of the American Chemical Society*, **134**, 15016–15021.

Faria A. F., Liu C., Xie M., Perreault F., Nghiem L. D., Ma J. and Elimelech M. (2017). Thin-film composite forward osmosis membranes functionalized with graphene oxide–silver nanocomposites for biofouling control. *Journal of Membrane Science*, **525**, 146–156.

Feng F., Xu Z., Li X., You W. and Zhen Y. (2010). Advanced treatment of dyeing wastewater towards reuse by the combined Fenton oxidation and membrane bioreactor process. *Journal of Environmental Sciences-China*, **22**(11), 1657–1665.

Furukawa H., Go Y. B., Ko N., Park Y. K., Uribe-Romo F. J., Kim J., O'Keeffe M. and Yaghi O. M. (2011). Isoreticular expansion of metal-organic frameworks with triangular and square building units and the lowest calculated density for porous crystals. *Inorganic Chemistry*, **50**, 9147–9152.

Ghosh R. and Cui Z. F. (1999). Mass transfer in gas-sparged ultrafiltration: Upward slug flow in tubular membranes. *Journal of Membrane Science*, **162**, 91–102.

Hai F. I., Yamamoto K., Nakajima F. and Fukushi K. (2011). Bioaugmented membrane bioreactor (MBR) with a GAC-packed zone for high rate textile wastewater treatment. *Water Research*, **45**(6), 2199–2206.

Han Y., Xu Z. and Gao C. (2013). Ultrathin graphene nanofiltration membrane for water purification. *Advanced Functional Materials*, **23**, 3693–3700.

He M. H. and Wei C. H. (2010). Performance of membrane bioreactor (MBR) system with sludge Fenton oxidation process for minimization of excess sludge production. *Journal of Hazardous Materials*, **176**(1–3), 597–601.

Hou J., Wei Y., Zhou S., Wang Y. and Wang H. (2018). Highly efficient H_2/CO_2 separation via an ultrathin metal-organic framework membrane. *Chemical Engineering Science*, **182**, 180–188.

Hu M. and Mi B. (2013). Enabling graphene oxide nanosheets as water separation membranes. *Environmental Science & Technology*, **47**, 3715–3723.

Hu M., Zheng S. and Mi B. (2016). Organic fouling of graphene oxide membranes and its implications for membrane fouling control in engineered osmosis. *Environmental Science & Technology*, **50**, 685–693.

Hua J., Zhang L. and Li Y. (2009). Application of UASB-MBR system for landfill leachate treatment. 2009 International Conference on Energy and Environment Technology, Vol. **3**, Proceedings.

James S. L. (2003). Metal-organic frameworks. *Chemical Society Reviews*, **32**, 276.

Juang Y. C., Su A., Fang L. H., Lee D. J. and Lai J. Y. (2011). Fouling with aerobic granule membrane bioreactor. *Water Science and Technology*, **64**(9), 1870–1875.

Jun Z., Yang F. L., Meng F. G., Peng A. and Di W. (2007). Comparison of membrane fouling during short-term filtration of aerobic granular sludge and activated sludge. *Journal of Environmental Sciences-China*, **19**(11), 1281–1286.

Kadhom M. and Deng B. (2018). Metal-organic frameworks (MOFs) in water filtration membranes for desalination and other applications. *Applied Materials Today*, **11**, 219–230.

Kim J. and Van der Bruggen B. (2010). The use of nanoparticles in polymeric and ceramic membrane structures: Review of manufacturing procedures and performance improvement for water treatment. *Environmental Pollution*, **158**(7), 2335–2349.

Kim H., Choi J. S. and Lee S. (2010). Pressure-Retarded Osmosis for Energy Production: Membrane Materials and Operating Conditions. IWA Regional Conference and Exhibition on Membrane Technology & Water Reuse, Istanbul, Turkey, IWA publishing.

Kim H. W., Yoon H. W., Yoon S., Yoo B. M., Ahn B. K., Cho Y. H., Shin H. J., Yang H., Paik U. and Kwon S. (2013). Selective gas transport through few-layered graphene and graphene oxide membranes. *Science*, **342**, 91–96.

Ko K., Yu Y. J., Kim M. J., Kweon J. and Chung H. (2018). Improvement in fouling resistance of silver-graphene oxide coated polyvinylidene fluoride membrane prepared by pressurized filtration. *Separation and Purification Technology*, **194**, 161–169.

Kornboonraksa T., Lee H. S., Lee S. H. and Chiemchaisri C. (2009). Application of chemical precipitation and membrane bioreactor hybrid process for piggery wastewater treatment. *Bioresource Technology*, **100**(6), 1963–1968.

Kuenen J. G. (2008). Anammox bacteria: From discovery to application. *Nature Reviews Microbiology*, **6**(4), 320–326.

Laera G., Cassano D., Lopez A., Pinto A., Pollice A., Ricco G. and Mascolo G. (2012). Removal of organics and degradation products from industrial wastewater by a membrane bioreactor integrated with ozone or UV/H_2O_2 treatment. *Environmental Science & Technology*, **46**(2), 1010–1018.

Lee J., Ahn W. Y. and Lee C. H. (2001). Comparison of the filtration characteristics between attached and suspended growth microorganisms in submerged membrane bioreactor. *Water Research*, **35**(10), 2435–2445.

Lee W. N., Kang I. J. and Lee C. H. (2006). Factors affecting filtration characteristics in membrane-coupled moving bed biofilm reactor. *Water Research*, **40**(9), 1827–1835.

Lefebvre O., Uzabiaga A., Chang I. S., Kim B. H. and Ng H. Y. (2011). Microbial fuel cells for energy self-sufficient domestic wastewater treatment – a review and discussion from energetic consideration. *Applied Microbiology and Biotechnology*, **89**(2), 259–270.

Lesage N., Sperandio M. and Cabassud C. (2008). Study of a hybrid process: Adsorption on activated carbon/membrane bioreactor for the treatment of an industrial wastewater. *Chemical Engineering and Processing*, **47**(3), 303–307.

Li X. F., Gao F. S., Hua Z. Z., Du G. C. and Chen J. (2005). Treatment of synthetic wastewater by a novel MBR with granular sludge developed for controlling membrane fouling. *Separation and Purification Technology*, **46**(1-2), 19–25.

Li Y. Z., He Y. L., Ohandja D. G., Ji J., Li J. F. and Zhou T. (2008). Simultaneous nitrification-denitrification achieved by an innovative internal-loop airlift MBR: Comparative study. *Bioresource Technology*, **99**(13), 5867–5872.

Li W., Zhang Y., Li Q. and Zhang G. (2015). Metal-organic framework composite membranes: Synthesis and separation applications. *Chemical Engineering Science*, **135**, 232–257.

Lin S., Straub A. P. and Elimelech M. (2014). Thermodynamic limits of extractable energy by pressure retarded osmosis. *Energy & Environmental Science*, **7**, 2706–2714.

Liu C., Huang X. and Wang H. (2008). Start-up of a membrane bioreactor bioaugmented with genetically engineered microorganism for enhanced treatment of atrazine containing wastewater. *Desalination*, **231**(1-3), 12–19.

Liu Q., Wang X. C., Liu Y., Yuan H. and Du Y. (2010). Performance of a hybrid membrane bioreactor in municipal wastewater treatment. *Desalination*, **258**(1–3), 143–147.

Liu T., Chen Z. L., Yu W. Z. and You S. J. (2011). Characterization of organic membrane foulants in a submerged membrane bioreactor with pre-ozonation using three-dimensional excitation-emission matrix fluorescence spectroscopy. *Water Research*, **45**(5), 2111–2121.

Martínez-Díez L. and Vázquez-González M. I. (1999). Temperature and concentration polarization in membrane distillation of aqueous salt solutions. *Journal of Membrane Science*, **156**(2), 265–273.

Mascolo G., Laera G., Pollice A., Cassano D., Pinto A., Salerno C. and Lopez A. (2010). Effective organics degradation from pharmaceutical wastewater by an integrated process including membrane bioreactor and ozonation. *Chemosphere*, **78**(9), 1100–1109.

McSwain B. S., Irvine R. L., Hausner M. and Wilderer P. A. (2005). Composition and distribution of extracellular polymeric substances in aerobic flocs and granular sludge. *Applied and Environmental Microbiology*, **71**(2), 1051–1057.

Moody C. D. and Kessler J. O. (1976). Forward osmosis extractors. *Desalination*, **18**(3), 283–295.

Morgenroth E., Sherden T., van Loosdrecht M. C. M., Heijnen J. J. and Wilderer P. A. (1997). Aerobic granular sludge in a sequencing batch reactor. *Water Research*, **31**(12), 3191–3194.

Munz G., Gori R., Mori G. and Lubello C. (2007). Powdered activated carbon and membrane bioreactors (MBR-PAC) for tannery wastewater treatment: Long term effect on biological and filtration process performances. *Desalination*, **207**(1–3), 349–360.

Ng C. A., Sun D., Zhang J. S., Wu B. and Fane A. G. (2010). Mechanisms of fouling control in membrane bioreactors by the addition of powdered activated carbon. *Separation Science and Technology*, **45**(7), 873–889.

Ng K.-K., Lin C.-F., Panchangam S. C., Hong P.-K. A. and Yang P.-Y. (2011). Reduced membrane fouling in a novel bio-entrapped membrane reactor for treatment of food and beverage processing wastewater. *Water Research*, **45**(14), 4269–4278.

Oh H. S., Yeon K. M., Yang C. S., Kim S. R., Lee C. H., Park S. Y., Han J. Y. and Lee J. K. (2012). Control of membrane biofouling in MBR for wastewater treatment by quorum quenching bacteria encapsulated in microporous membrane. *Environmental Science & Technology*, **46**, 4877–4884.

Oh H. S., Kim S. R., Cheong W. S., Lee C. H. and Lee J. K. (2013). Biofouling inhibition in MBR by *Rhodococcus* sp. BH4 isolated from real MBR plant. *Applied Microbiology and Biotechnology*, **97**, 10223–10231.

Park H.-D., Lee Y. H., Kim H.-B., Moon J., Ahn C.-H., Kim K.-T. and Kang M.-S. (2010). Reduction of membrane fouling by simultaneous upward and downward air sparging in a pilot-scale submerged membrane bioreactor treating municipal wastewater. *Desalination*, **251**(1-3), 75–82.

Pekakis P. A., Xekoukoulotakis N. P. and Mantzavinos D. (2006). Treatment of textile dyehouse wastewater by TiO_2 photocatalysis. *Water Research*, **40**(6), 1276–1286.

Phattaranawik J. and Leiknes T. (2009). Double-deck aerated biofilm membrane bioreactor with sludge control for municipal wastewater treatment. *Aiche Journal*, **55**(5), 1291–1297.

Phattaranawik J. and Leiknes T. (2011). Extractive biofilm membrane bioreactor with energy recovery from excess aeration and new membrane fouling control. *Bioresource Technology*, **102**(3), 2301–2307.

Phattaranawik J., Fane A. G., Pasquier A. C. S. and Bing W. (2008). A novel membrane bioreactor based on membrane distillation. *Desalination*, **223**(1–3), 386–395.

Pochana K., Keller J. and Lant P. (1999). Model development for simultaneous nitrification and denitrification. *Water Science and Technology*, **39**(1), 235–243.

Psoch C. and Schiewer S. (2005). Long-term study of an intermittent air sparged MBR for synthetic wastewater treatment. *Journal of Membrane Science*, **260**(1–2), 56–65.

Psoch C. and Schiewer S. (2008). Long-term flux improvement by air sparging and backflushing for a membrane bioreactor, and modeling permeability decline. *Desalination*, **230**(1–3), 193–204.

Qin L., Fan Z., Xu L., Zhang G., Wang G., Wu D., Long X. and Meng Q. (2015). A submerged membrane bioreactor with pendulum type oscillation (PTO) for oily wastewater treatment: Membrane permeability and fouling control. *Bioresource Technology*, **183**, 33–41.

Qu Y.-y., Yang Q., Zhou J.-t., Gou M., Xing L.-l. and Ma F. (2009a). Combined MBR with photocatalysis/ozonation for bromoamine acid removal. *Applied Biochemistry and Biotechnology*, **159**(3), 664–672.

Qu Y. Y., Zhou J. T., Wang J., Xing L. L., Jiang N., Gou M. and Uddin M. S. (2009b). Population dynamics in bioaugmented membrane bioreactor for treatment of bromoamine acid wastewater. *Bioresource Technology*, **100**(1), 244–248.

Rana D. and Matsuura T. (2010). Surface modifications for antifouling membranes. *Chemical Reviews*, **110**, 2448–2471.

Remy M., van der Marel P., Zwijnenburg A., Rulkens W. and Temmink H. (2009). Low dose powdered activated carbon addition at high sludge retention times to reduce fouling in membrane bioreactors. *Water Research*, **43**(2), 345–350.

Rodriguez-Hernandez L., Esteban-Garcia A. L., Lobo A., Temprano J., Alvaro C., Mariel A. and Tejero I. (2012). Evaluation of a hybrid vertical membrane bioreactor (HVMBR) for wastewater treatment. *Water Science and Technology*, **65**(6), 1109–1115.

Rui Z., James J. B. and Lin Y. S. (2018). Highly CO_2 perm-selective metal-organic framework membranes through CO_2 annealing post-treatment. *Journal of Membrane Science*, **555**, 97–104.

Satyawali Y. and Balakrishnan M. (2009). Performance enhancement with powdered activated carbon (PAC) addition in a membrane bioreactor (MBR) treating distillery effluent. *Journal of Hazardous Materials*, **170**(1), 457–465.

Serrano D., Suarez S., Lema J. M. and Omil F. (2011). Removal of persistent pharmaceutical micropollutants from sewage by addition of PAC in a sequential membrane bioreactor. *Water Research*, **45**(16), 5323–5333.

Shao S., Qu F., Liang H., Li K., Yu H., Chang H. and Li G. (2015). Powdered activated carbon-membrane bioreactor operated under intermittent aeration and short sludge retention times for micro-polluted surface water treatment. *International Biodeterioration & Biodegradation*, **102**, 81–88.

Singh G., Rana D., Matsuura T., Ramakrishna S., Narbaitz R. M. and Tabe S. (2010). Removal of disinfection byproducts from water by carbonized electrospun nanofibrous membranes. *Separation and Purification Technology*, **74**(2), 202–212.

Tao Y., Gao D.-W., Fu Y., Wu W.-M. and Ren N.-Q. (2012). Impact of reactor configuration on anammox process start-up: MBR versus SBR. *Bioresource Technology*, **104**(0), 73–80.

Tian J.-y., Chen Z.-l., Nan J., Liang H. and Li G.-b. (2010). Integrative membrane coagulation adsorption bioreactor (MCABR) for enhanced organic matter removal in drinking water treatment. *Journal of Membrane Science*, **352**(1–2), 205–212.

Trigo C., Campos J. L., Garrido J. M. and Méndez R. (2006). Start-up of the Anammox process in a membrane bioreactor. *Journal of Biotechnology*, **126**, 475–487.

Tsushima I., Ogasawara Y., Kindaichi T., Satoh H. and Okabe S. (2007). Development of high-rate anaerobic ammonium-oxidizing (anammox) biofilm reactors. *Water Research*, **41**(8), 1623–1634.

van der Star W. R. L., Miclea A. I., van Dongen U., Muyzer G., Picioreanu C. and van Loosdrecht M. C. M. (2008). The membrane bioreactor: A novel tool to grow anammox bacteria as free cells. *Biotechnology and Bioengineering*, **101**(2), 286–294.

Wang J. F., Wang X., Zhao Z. G. and Li J. W. (2008). Organics and nitrogen removal and sludge stability in aerobic granular sludge membrane bioreactor. *Applied Microbiology and Biotechnology*, **79**(4), 679–685.

Wang J., Lu H., Jin R. F., Zhou J. T., Liu G. F. and Xing L. L. (2009). Decolorization of 1-amino-4-bromoanthraquinone-2-sulfonic acid in bioaugmented membrane bioreactor. *Process Biochemistry*, **44**(8), 812–816.

Wang Y.-K., Sheng G.-P., Li W.-W., Huang Y.-X., Yu Y.-Y., Zeng R. J. and Yu H.-Q. (2011). Development of a novel bioelectrochemical membrane reactor for wastewater treatment. *Environmental Science & Technology*, **45**(21), 9256–9261.

Wang R., Liu Y., Li B., Hsiao B. S. and Chu B. (2012a). Electrospun nanofibrous membranes for high flux microfiltration. *Journal of Membrane Science*, **392–393**(0), 167–174.

Wang Y.-P., Liu X.-W., Li W.-W., Li F., Wang Y.-K., Sheng G.-P., Zeng R. J. and Yu H.-Q. (2012b). A microbial fuel cell–membrane bioreactor integrated system for cost-effective wastewater treatment. *Applied Energy*, **98**(0), 230–235.

Wang B., Zhang K. and Field R. W. (2018). Optimization of aeration variables in a commercial large-scale flat-sheet MBR operated with slug bubbling. *Journal of Membrane Science*, **567**, 181–190.

Weerasekara N. A., Choo K. H. and Lee C. H. (2016). Biofouling control: Bacterial quorum quenching versus chlorination in membrane bioreactors. *Water Research*, **103**, 293–301.

Wu J. L. and Huang X. (2008). Effect of dosing polymeric ferric sulfate on fouling characteristics, mixed liquor properties and performance in a long-term running membrane bioreactor. *Separation and Purification Technology*, **63**(1), 45–52.

Wu J. L. and Huang X. (2010). Use of ozonation to mitigate fouling in a long-term membrane bioreactor. *Bioresource Technology*, **101**(15), 6019–6027.

Xi W. and Geissen S.-u. (2001). Separation of titanium dioxide from photocatalytically treated water by cross-flow microfiltration. *Water Research*, **35**(5), 1256–1262.

Xiao E.-R., Liang W., He F., Cheng S.-P. and Zhen-Bin W. (2010). Performance of the combined SMBR–IVCW system for wastewater treatment. *Desalination*, **250**(2), 781–786.

Yang Q.-Y., Yang T., Wang H.-J. and Liu K.-Q. (2009). Filtration characteristics of activated sludge in hybrid membrane bioreactor with porous suspended carriers (HMBR). *Desalination*, **249**(2), 507–514.

Yang F., Wang Y., Bick A., Gilron J., Brenner A., Gillerman L., Herzberg M. and Oron G. (2012). Performance of different configurations of hybrid growth membrane bioreactor (HG-MBR) for treatment of mixed wastewater. *Desalination*, **284**, 261–268.

Yang M., Liu M., Yu D., Zheng J., Wu Z., Zhao S., Chang J. and Wei Y. (2017). Numerical simulation of scaling-up for AEC-MBRs regarding membrane module configurations and cyclic aeration modes. *Bioresource Technology*, **245**, 933–943.

Zhang H. F., Sun B. S., Zhao X. H. and Gao Z. H. (2008). Effect of ferric chloride on fouling in membrane bioreactor. *Separation and Purification Technology*, **63**(2), 341–347.

Zhang K., Cui Z. and Field R. W. (2009). Effect of bubble size and frequency on mass transfer in flat sheet MBR. *Journal of Membrane Science*, **332**, 30–37.

Zhang J. S., Wang Y. Q., Zhou J. T., Wang R., Goh S. W. and Fane A. G. (2011a). Enhanced biodegradation of 4-aminobenzenesulfonate in membrane bioreactor by *Pannonibacter* sp. W1. *Water Science and Technology*, **63**(11), 2752–2758.

Zhang K., Wei P., Yao M., Field R. W. and Cui Z. (2011b). Effect of the bubbling regimes on the performance and energy cost of flat sheet MBRs. *Desalination*, **283**, 221–226.

Zhu J., Li P. Z., Guo W., Zhao Y. and Zou R. (2018). Titanium-based metal–organic frameworks for photocatalytic applications. *Coordination Chemistry Reviews*, **359**, 80–101.

Zou X., Zhang L., Wang Z. and Luo Y. (2016). Mechanisms of the antimicrobial activities of graphene materials. *Journal of the American Chemistry Society*, **138**, 2064–2077.

Zsirai T., Qiblawey H., A-Marri M. J. and Judd S. (2016). The impact of mechanical shear on membrane flux and energy demand. *Journal of Membrane Science*, **516**, 56–63.

Chapter 13

Removal of emerging trace organic contaminants (TrOC) by MBR

Faisal I. Hai[1], Long D. Nghiem[1], Stuart J. Khan[2], Muhammad B. Asif[1], William E. Price[3] and Kazuo Yamamoto[4,5,6]*

[1]*Strategic Water Infrastructure Laboratory, School of Civil, Mining and Environmental Engineering, The University of Wollongong, Wollongong, Australia*
[2]*Water Research Centre, The University of New South Wales, Sydney, Australia*
[3]*Strategic Water Infrastructure Laboratory, School of Chemistry, The University of Wollongong, Wollongong, Australia*
[4]*Environmental Science Center, University of Tokyo, Japan*
[5]*Research and Development Initiative, Chuo University, Japan*
[6]*Kindai University, Japan*
**Corresponding author: faisal@uow.edu.au*

ABSTRACT

This chapter reviews current and new technologies for the removal of emerging trace organic contaminants (TrOC) from wastewater for safe reuse. In particular, it compares the relative performance of membrane bioreactors (MBR) with other conventional biological processes. It is shown that MBR has many intrinsic advantages over other methods leading to superior performance. Fundamental to this is the presence of a membrane barrier which provides a range of other adsorption mechanisms for TrOC. However, in practice there are many factors affecting TrOC removal by MBR, leading to sometimes conflicting reports in the literature from both laboratory-scale studies as well as full-scale plants. These factors include both structural and physicochemical characteristics of the TrOC as well as operational parameters of the MBR process. It is concluded that further work is required, particularly in areas such as improved understanding of the overall fate and degradation pathways of TrOC undergoing MBR processes, investigation of specific microbial–TrOC interactions and development of hybrid systems having MBR at the core to treat the wide varieties of TrOC from wastewater. In this context, the working principle and lab-based performance of the emerging 'high retention' MBRs have been discussed.

 Membrane Biological Reactors: Theory, Modeling, Design, Management and Applications to Wastewater Reuse
Editor(s): Faisal I. Hai, Kazuo Yamamoto and Chung-Hak Lee
doi: 10.2166/9781780409177_0413

13.1 INTRODUCTION

Trace organic contaminants (TrOC) in municipal wastewater consist of a wide range of naturally occurring and synthetic chemicals (Deblonde *et al.*, 2011). They include industrial chemicals, chemicals used in households, chemicals excreted by people, and chemicals formed during wastewater and drinking-water treatment processes. Some key classes of TrOC include pharmaceuticals and personal care products (PPCPs), estrogenic and androgenic hormones, perfluorochemicals and nanoparticles, pesticides or their metabolites and synthetic industrial organic chemicals. TrOC are biologically active and can thus present a threat to the aquatic environment with effects such as acute and chronic toxicity to aquatic organisms, accumulation in the ecosystem and loss of habitats and biodiversity, as well as a range of possible adverse effects on human health. Modern sewage treatment plants (STPs) can effectively remove total organic carbon (TOC) and total nitrogen (TN) as well as achieve some degree of disinfection. However, these plants have not been specifically designed to remove TrOCs (Reif *et al.*, 2008). This chapter provides an in-depth overview of the factors, mechanisms and extent of TrOC removal by classical MBR and its emerging formats.

13.2 TrOC IN WATER AND THEIR POTENTIAL IMPACT ON REUSE

PPCPs (and their active metabolites) are excreted to sewage by people as well as direct disposal of unused drugs by households (Shon *et al.*, 2006). Since pharmaceuticals are designed to evoke biological responses, their inherent biological activity and the diverse range of compounds identified in sewage (and the environment) have been the cause of considerable concern during the past decade (Boxall *et al.*, 2012; Huerta *et al.*, 2012). Specific problems have not been encountered for most classes of drugs, but issues regarding potent endocrine disrupting compounds (Balabanic *et al.*, 2011; Liu *et al.*, 2011), their aquatic toxicity (Connon *et al.*, 2012; Fatta-Kassinos *et al.*, 2011) and the spread of antibacterial resistance (Jury *et al.*, 2011) may have significant ecological implications. A broad range of pharmaceutically active compounds have been reported in US drinking water as a consequence of unplanned indirect potable reuse (Benotti *et al.*, 2009). Antiseptics such as triclosan and triclocarban are commonly used in face washes and anti-gum-disease toothpaste. They are increasingly being used in a wider range of household products, including deodorants, antiperspirants, detergents, dishwashing liquids, cosmetics and anti-microbial creams, lotions, and hand soaps. The presence of triclosan and triclocarban in reclaimed sewage effluent has led to concerns regarding their potential to accumulate in irrigated soils and runoff (Heidler & Halden, 2008; Xu *et al.*, 2009).

Natural steroidal hormones such as estradiol (E2), estrone (E1) and testosterone are also excreted to sewage by people. During the last two decades, natural steroidal hormones have been widely implicated in a range of endocrinological abnormalities in aquatic species that are affected by sewage effluent (Leet *et al.*, 2011). Impacts have been identified by a number of bio-indicators, most commonly elevated production of the protein vitellogenin, which is an essential precursor for egg production in fish (Matozzo *et al.*, 2008; Rotchell & Ostrander, 2003).

Perfluorochemicals such as perfluorobutanoic acid (PFBA), perfluoropentanoic acid (PFPeA), perfluorohexanoic acid (PFHxA), perfluorooctanoic acid (PFOA), perfluorohexane sulfonate (PFHxS), and perfluorooctane sulfonate (PFOS) are persistent and toxic chemicals that have recently emerged as drinking water contaminants of concern. These perfluorochemicals are highly water soluble and are used in the production of water- and stain-resistant products, including cookware and clothing, as well as in fire-fighting foams. They also arise from the breakdown of fluorotelomer alcohols, which are widely used in consumer products such as greaseproof food wrappers and stain-resistant carpet treatments.

A range of perfluoronated chemicals have recently been reported in drinking water that was not known to be affected by discharge from a facility that manufactures them (Mak *et al.*, 2009; Post *et al.*, 2009). Perfluorinated residues (predominantly PFOA and PFOS) have been reported in the reclaimed effluents from tertiary-treatment wastewater treatment plants at total concentrations of 90–470 ng/L (Plumlee *et al.*, 2008).

Pesticides may enter municipal wastewater systems by a variety of means, including storm water influx and illegal direct disposal to sewage systems. Some leftover household pesticides are known to be disposed of into municipal sewers (Slack *et al.*, 2005). Additional routes, of unknown significance, include washing fruit and vegetables prior to household consumption; insect repellents washed from human skin; flea-rinse shampoos for pets; washing clothes and equipment used for applying pesticides. Pesticides have been designed and used for their detrimental effects on a wide range of biological species (Buchanan *et al.*, 2012)

Depending on the catchment area, and the extent of the trade waste program to control chemicals at the wastewater source, a very wide range of synthetic industrial chemicals are often measurable in urban municipal wastewater. Examples include plasticizers and heat stabilizers, biocides, epoxy resins, bleaching chemicals and by-products, solvents, degreasers, dyes, chelating agents, polymers, polyaromatic hydrocarbons, polychlorinated biphenyls and phthalates. Many of these chemicals are known to be toxic to a diverse range of organisms including humans.

13.3 RELATIVE PERFORMANCE OF MBR AND OTHER BIOLOGICAL PROCESSES

13.3.1 Conceptual expectations

MBRs can potentially achieve better removal of TrOC than conventional activated sludge (CAS) processes due to their ability to operate under longer solids retention times (SRTs), higher biomass concentrations and complete retention of suspended solids. Spring *et al.* (2007) argued that compared to CAS processes, MBR may provide three additional removal mechanisms for TrOC: firstly, the membrane serves as an effective barrier to solids, onto which many TrOC will adsorb (Boonyaroj *et al.*, 2012; Johnson & Sumpter, 2001; Sahar *et al.*, 2011); secondly, the membrane surface may also adsorb TrOC (Chang *et al.*, 2003); and finally, the longer SRT in MBR may facilitate additional biological transformation of TrOC.

Usually, SRTs longer than 25 dd are applied in MBR, while typical SRT values for CAS processes can vary from 8 up to 25 d (Çiçek *et al.*, 1999; Hai & Yamamoto, 2011; Visvanathan *et al.*, 2000). Longer SRT values can favor the development of slow-growing bacteria, the presence of more diverse microbial communities with broader physiological capabilities and the adaptation of these communities to degrade specific organic compounds (Aubenneau *et al.*, 2011). The possibility of genetic mutation and adaptation of microorganisms to assimilate persistent organic compounds also increases at longer SRT (Cicek *et al.*, 2001; Rosenberger *et al.*, 2002). Higher biomass concentrations in the bioreactor may lead to an intensification of the biological processes and may increase the interaction between microorganisms and the chance of genetic information exchange (De Wever *et al.*, 2007). Furthermore, biomass in an MBR has been shown to have a higher viable fraction compared to CAS processes (Çiçek *et al.*, 1999). This can be attributed to improved mass-transfer conditions in the MBR favored by: (i) smaller flocs (between 10 and 100 μm, as compared to 100 and 500 μm in the CAS process), (ii) larger specific floc surface per unit reactor volume (10 times higher in MBR than in CAS systems) (Zhang *et al.*, 1997), and (iii) the presence of many free-living bacteria. In addition to improved mass transfer and higher sludge activity that enhance biodegradation, the cake layer on the membrane can potentially prevent the escape of some extracellular enzymes and soluble oxidants creating a more active biological mixture capable of degrading a wider range of carbon sources (Çiçek *et al.*, 1999). Furthermore, because of the higher biomass concentrations, the food to microorganism (F/M) ratio in MBR is lower, which could result in more complete mineralization of the TrOC (Figure 13.1).

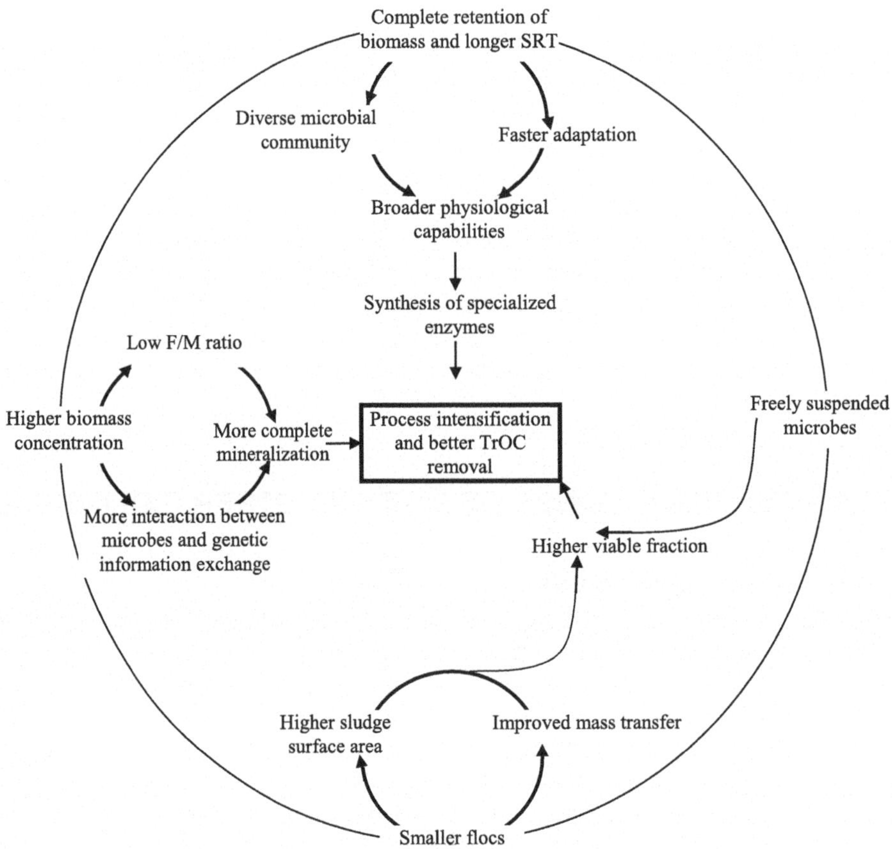

Figure 13.1 Conceptual schematic of enhanced TrOC removal by MBR.

Despite the conceptual expectation of better removal of resistant compounds such as some TrOC by MBR, this improved performance is not always observed. Although better and/or more stable removal of TrOC possessing moderate-to-high biodegradability by MBR treatment in comparison to that of CAS processes has been reported, significant variation in MBR removal performance has also been noted in several recent studies, particularly for biologically persistent hydrophilic compounds (Clara *et al.*, 2005c; Joss *et al.*, 2005b; Tadkaew *et al.*, 2011). Section 13.2.2 provides a detailed account of the literature reports on comparative TrOC removal performance of CAS and MBR systems.

13.3.2 Reported comparative performance of CAS and MBR

A number of investigations have been conducted comparing the removal efficiency of TrOC from wastewater by the CAS processes and advanced MBR treatment. These two treatment options demonstrate compound-specific comparative performance. Occasionally, TrOC removal efficiency has been observed to be very similar, and high for both treatments (e.g., for ibuprofen, naproxen, acetaminophen and paroxetine) (Cirja *et al.*, 2008a; Oulton *et al.*, 2010), while some compounds such as the anti-epileptic drug carbamazepine and diuretic hydrochlorothiazide can pass through both the systems untransformed (Radjenovic *et al.*, 2007, 2009b). However, some studies point to improved removal efficiency by MBR compared to CAS treatment. Therefore, TrOC can be grouped into three categories based on the degree of their removal (Kimura *et al.*, 2005): (i) easily removed by both CAS and MBR (e.g., ibuprofen); (ii) not efficiently removed by either of them (e.g., carbamazepine, clofibric acid, dichloprop, and diclofenac); (iii) better removed by MBR (e.g., ketoprofen, mefenamic acid, and naproxen).

As noted earlier, some authors report comparable elimination rates in MBR and CAS for some TrOC belonging to different groups such as endocrine disrupting chemicals (EDC), pharmaceuticals and PPCPs, and industrial precursors (Clara *et al.*, 2005c; De Wever *et al.*, 2007; Joss *et al.*, 2005a). For instance, musk fragrances tonalide and galaxolide were reported to be removed to the same extent (85%) in MBR and CAS treating the same effluent, with slightly lower effluent concentrations from the MBR unit (Clara *et al.*, 2004, 2005c). For 4-t-octylphenol and bisphenol A, the removal efficiency was very similar when CAS followed by tertiary filtration or MBR treatments were applied (Cases *et al.*, 2011). In batch experiments with MBR and CAS sludge, Joss *et al.* (2006) observed no difference between their kinetic biodegradation constants (k_{biol}) for diclofenac, which were low in both cases (≤ 0.1 L/(g_{SS} d)).

Contrary to the reports of similar TrOC removal by MBR and CAS systems, several studies have confirmed better removal of some TrOC by MBR over the CAS processes (Gobel *et al.*, 2007; Kimura *et al.*, 2005; Lesjean *et al.*, 2005; Quintana *et al.*, 2005; Radjenovic *et al.*, 2007). For example, Bernhard *et al.* (2006) reported significantly better removal of linear alkylbenzene sulfonates (LAS) by an MBR compared with a CAS process. De Wever *et al.* (2004) showed that the removal of LAS by MBR was slightly better; additionally MBR effluent contained lower amounts of the more recalcitrant sulfophenylcarboxylate metabolites. Gonzalez *et al.* (2007) reported similar elimination (96–98%) of LAS and coconut diethanol amides by CAS and MBR processes; nevertheless, for all studied compounds the MBR effluent concentrations were consistently lower and independent of the influent concentrations. Additionally, MBR effluent quality in terms of chemical oxygen demand (COD), NH_4^+ concentration and total suspended solids (TSS) was always superior to that of the CAS system and also independent of the influent quality.

De Wever *et al.* (2007) reported that although a very similar removal of most of the studied isomers of naphthalene disulfonates (NDSA) was observed by MBR and CAS treatments, 1,7- and 2,7-NDSA were more completely degraded by the MBR system, probably due to the lower F/M ratio and a longer SRT. In a study by Weiss and Reemtsma (2008), for half of the studied compounds, such as benzotriazole, 5-tolyltriazole (5-TTri), benzothiazole-2-sulfonate and 1,6-NDSA, removal by MBR was significantly better than by CAS, while no improvement was recorded for 1,5-NDSA, 1,3-NDSA and 4-TTri (low removal) and naphthalene-1-sulfonate (>90% removal) (Weiss & Reemtsma, 2008). Significant improvement in biodegradation of insect repellents and metabolites, namely, icaridin (trade name Bayrepel), icaridin acid and *N*, *N*-diethyl-meta-toluamide (DEET) was observed when MBR, instead of CAS treatment, was applied (Bernhard *et al.*, 2006). Some studies have also presented MBR as a process that leads to higher estrogen removal when compared with CAS processes (Clara *et al.*, 2005c; Kreuzinger *et al.*, 2004; Lesjean *et al.*, 2005; Radjenovic *et al.*, 2009a; Weiss & Reemtsma, 2008). Kloepfer *et al.* (2005) reported significant difference in efficiency between MBR and CAS for benzothiazole removal.

In a study by Radjenovic *et al.* (2007), for most of the pharmaceutical compounds investigated, the performance of MBR treatment was better (removal efficiency >80%) and effluent concentrations of, for example, diclofenac, ketoprofen, ranitidine, gemfibrozil, bezafibrate, pravastatin, and ofloxacin were more uniform than for the CAS system (Radjenovic *et al.*, 2007). Better or more uniform removal of TrOC by MBR could be a result of the smaller floc size of the MBR sludge, which enhances mass transfer by diffusion and therefore increases elimination (Radjenovic *et al.*, 2007). Sulfamethoxazole, including the fraction present as N4-acetyl-sulfamethoxazole, was eliminated by approximately 80% in MBR as compared to about 60% in CAS processes, independent of the SRT (Gobel *et al.*, 2007). The removal of trimethoprim, clarithromycin and dehydro-erythromycin was highly variable in CAS systems, while in MBR a higher elimination was observed at longer SRTs (up to 90% at the SRT of 60–80 d in comparison to up to 50% at the SRT 16–33 d) (Gobel *et al.*, 2007). Two pilot-scale MBRs exhibited enhanced elimination of several pharmaceutical residues poorly removed by the CAS treatment (e.g., mefenamic acid, indomethacin, diclofenac, propyphenazone, pravastatin, gemfibrozil) (Radjenovic *et al.*, 2009b). Radjenovic *et al.* (2007) observed significantly improved removal of lipid regulators and cholesterol-lowering statin drugs (gemfibrozil, bezafibrate, clofibric acid and pravastatin), β-blockers (atenolol and metoprolol), antibiotics (ofloxacin and erythromycin), anti-ulcer agent (ranitidine) and some analgesics and anti-inflammatory drugs (propyphenazone, mefenamic acid, and diclofenac).

MBR was more efficient in the removal of diethyl phthalate, di-*n*-butyl phthalate and butyl benzyl phthalate as compared to CAS followed by conventional tertiary filtration (Cases *et al.*, 2011). In another study, the addition of an ultrafiltration post-treatment unit to the CAS system significantly increased (by up to 28%) the removal of all tested antibiotics (Sahar *et al.*, 2011). Sorption (to both suspended and membrane-attached biomass) was shown to be a significant removal mechanism (Sahar *et al.*, 2011). In a related study, Boonyaroj *et al.* (2012) confirmed the role of different fractions of membrane-attached biomass in effective retention of the target TrOC on the membrane. Urase *et al.* (2005) also reported that in the filtration of mixed liquor of activated sludge, additional removal of hydrophobic TrOC by membranes may be expected due to the additional barrier provided by the particle deposition on the membrane surface. However, Reemtsma *et al.* (2002) suggested that such additional removal is uncertain for hydrophilic compounds in conventional MBR. Li *et al.* (2011a) provided additional insight into this matter. They confirmed that direct addition of powdered activated carbon into MBR can result in significant additional removal of TrOC, namely, sulfamethoxazole and carbamazepine.

A generalization for the contrasting findings of "enhanced removal in MBR" vs "no additional benefit of MBR relative to CAS" was offered by Weiss and Reemtsma (2008): MBR is only superior for compounds with an intermediate biodegradation potential, whereas no benefits are apparent for easily biodegradable or recalcitrant compounds. However, it is worth noting that removal of TrOCs by MBR is more stable and consistent as compared to that by the CAS process (Asif *et al.*, 2018a) (Figure 13.2). Indeed, from an analysis conducted by Oulton *et al.* (2010), half of all available data for TrOC treated by MBR corresponded to treatment efficiencies between 41–98%. The range was broader for CAS, with half of all reports for TrOC falling between removal efficiencies of 23 and 91%, suggesting only a modest improvement in the extent of TrOC removal by MBR relative to that by CAS process. Clara *et al.* (2005a) summarized that, as in CAS, the TrOC removal performance of MBR is dependent on the SRT, and that no significant differences in the treatment efficiency of CAS and MBR exists when operated at comparable SRT. De Wever *et al.* (2007), on the other hand, argued that while MBR may not necessarily yield higher removal efficiencies, it was nevertheless more advantageous because it exhibits a more consistent performance and shorter lag times, indicating a superior response to fluctuating influent concentrations. Besides, MBR offers the advantage of increased plant flexibility through more compact installation and allows modification of design to fine-tune biological performance. Thus, it is important to know the mode of influence of various factors governing TrOC removal by MBR. Section 13.4 provides a detailed discussion on this topic.

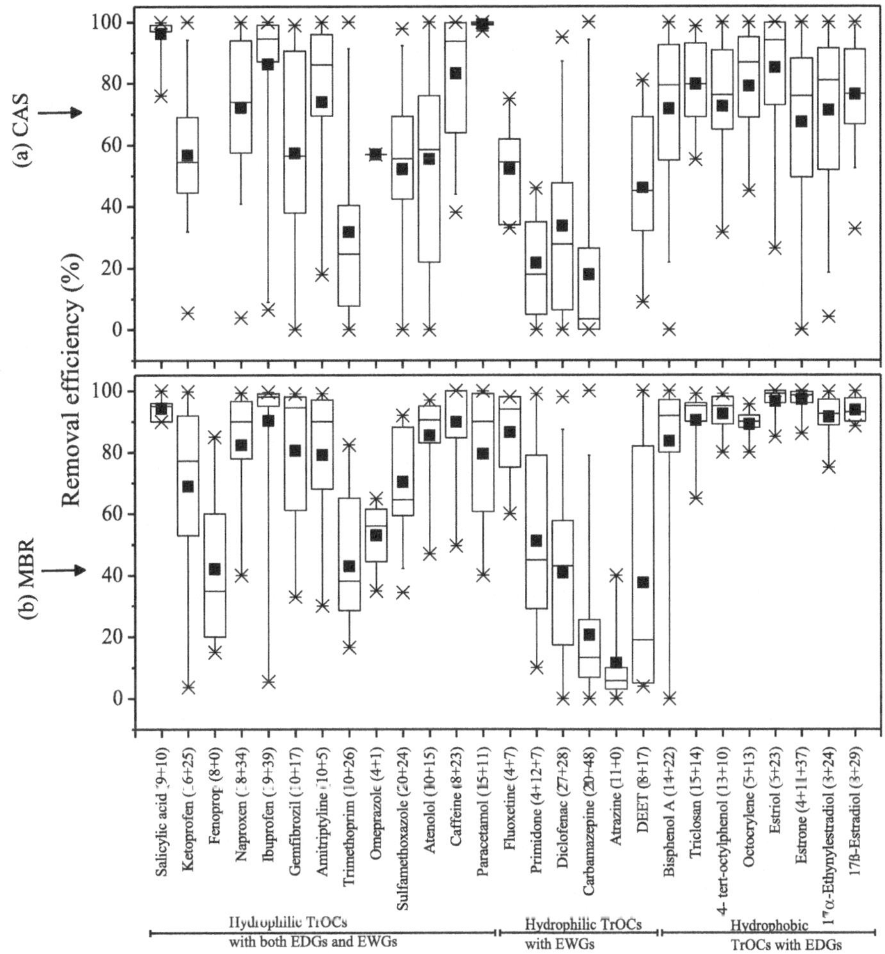

Figure 13.2 Variations in the removal of TrOCs by (a) CAS, and (b) MBR. Box-and-whisker plot shows information about the interquartile range; median (horizontal line in the box); min and max (whiskers); and average (block square in the box). Numbers within parentheses on the x-axis represent the number of data points (MBR+CAS). Adapted with permission from Asif *et al.* (2018a).

13.4 EFFECT OF TROC PRESENCE IN WASTEWATER ON BASIC PERFORMANCE OF MBR

In a biological STP, chemical perturbations may cause toxicity, leading to deflocculation, disrupted nitrification and incomplete treatment (Bott & Love, 2001). For example, a toxic shock of the xenobiotics pentachlorphenol (PCP) (0.16 mMol), ortho-nitrophenol (oNP) (0.58 mMol) and a combination of PCP (0.08 mMol) and oNP (0.29 mMol) led to a remarkable reduction of NH_3 oxidizing (nitrifying) bacteria and those from the family of Enterobacteriaceae (Topalova *et al.*, 1999). However, TrOC seem not to cause any significant perturbations in basic wastewater treatment performance by CAS or MBR systems. For instance, the influence of pharmaceuticals on bacterial community structure in CAS was assessed by Kraigher *et al.* (2008) in small-scale bioreactors containing different concentrations (5, 50, 200 and 500 μg/L) of several commonly used pharmaceuticals (e.g., ibuprofen, naproxen, ketoprofen, diclofenac and clofibric acid). They showed the disappearance of certain groups of bacteria (e.g., Phylum *Nitrospira*) due to the presence of the pharmaceuticals; however, there was no discernible effect on the level of basic water quality parameters. Reif *et al.* (2008) confirmed the absence of any inhibitory effects on biomass growth, COD removal and nitrification due to the addition of 12 TrOC including pharmaceuticals belonging to different therapeutic groups and three polycyclic musk fragrances into the synthetic wastewater fed to an MBR at environmentally relevant concentrations ranging from 10 to 20 μg/L. Similar negligible influence of TrOC on the basic treatment performance of MBR has been reported in other studies where TrOC were added into feed at concentrations ranging from few μg/L (Tadkaew *et al.*, 2011) to several hundred μg/L (Li *et al.*, 2011a).

Navaratna *et al.* (2012a), however, reported an increased concentration of extracellular polymeric substances (both protein and carbohydrates) in the bioreactor and a reduction in biomass production after the addition of ametryn (1 mg/L), a herbicide, in an MBR. Also, Delgado *et al.* (2010) observed that microorganisms in an MBR exposed to the anti-cancer drug cyclophosphamide and its principal metabolites showed increase in the endogenous respiration rate as compared to a control MBR, suggesting that the stress caused by these chemicals led to a diversion of carbon and/or energy from growth to adaptive and protective responses. Additionally, relatively smaller floc size and the overproduction of polysaccharides led to greater membrane fouling in that study. However, despite the higher COD concentration in the supernatant, permeate COD concentrations were almost the same for both MBR, highlighting the beneficial role of the membrane (Delgado *et al.*, 2010). In a study by Aubenneau *et al.* (2011), in spite of momentary changes in endogenous and exogenous oxygen needs no inhibition was observed in the presence of carbamazepine in a batch reactor. This result was confirmed by an MBR experiment where no significant difference in COD removal, sludge production and oxygen requirement was observed due to the presence of carbamazepine in the feed (Aubenneau *et al.*, 2011).

13.5 FACTORS AFFECTING TROC REMOVAL BY MBR

The mechanisms of TrOC removal do not follow a general rule since their relative contribution depends on the physicochemical properties of the TrOC, the origin and composition of the wastewater, and the operating parameters of the wastewater treatment facility (Cirja *et al.*, 2008a).

13.5.1 Characteristics of the TrOC

13.5.1.1 Categorization based on usage

Generalizations regarding TrOC removal based on the intended applications or origins can be difficult due to the disparity of reported removal efficiencies for the TrOC within the class (Onesios *et al.*, 2009). For example, Tadkaew *et al.* (2011) reported that no generalization could be made for any of the six therapeutic classes of pharmaceuticals investigated in that study: their removal efficiencies by MBR treatment varied widely (10–98%) even within the same class of compounds. These widely varying removals within a category suggest that accurately predicting biodegradability based on the intended therapeutic function may not be possible. This is not surprising as compounds within the same TrOC class can have vastly different chemical structures. For example, anticonvulsant drugs have chemical structures as diverse as branched chain carboxylic acids and complex azepines (Onesios *et al.*, 2009). Again, although all antibiotics are designed to destroy bacteria, different categories of antibiotics such as sulfonamides and macrolides possess different structural characteristics and mechanisms of action. While the macrolides each have one ring with side chains or sugars, every sulfonamide has two relatively small rings connected by a sulfur atom and C–N bonds. Their structural differences are reflected in the more polar and hydrophilic nature of sulfonamides relative to the macrolides (log $K_{ow} < 1$ and log $K_{ow} > 3$, respectively), which have important implications for their removal by MBRs (Sahar *et al.*, 2011). Since biodegradation involves enzymatic reactions specific to chemical structures, the biodegradability of TrOC with different structures grouped in the same class is expected to vary, thwarting efforts to observe general trends. Contrary to these difficulties, Onesios *et al.* (2009) noted the case of antiseptic compounds. They observed antiseptic removals of greater than 70% in majority of the studies reviewed in their article. Similarly, Tadkaew *et al.* (2011) reported very low removal of three pesticides (atrazine, linuron, DEET) and near complete removal or removal to below the analytical limit of all eight steroid hormones and three other EDCs selected for investigation (bisphenol A, nonylphenol, and t-octylphenol).

13.5.1.2 Physicochemical properties

TrOC can be removed from aqueous solution by four broad mechanisms, namely volatization, sorption, photolysis, and biodegradation. These mechanisms are governed by the physicochemical properties of the TrOC and the operating condition of the MBR system. In general, the contribution of photolysis to the overall removal of TrOC during MBR treatment is negligible. This is because of the very high suspended solids content in the reactor and the limited exposure to sunlight (ultraviolet radiation) during MBR treatment. TrOC can be transferred from the aqueous phase to the gas phase, if they are sufficiently volatile, or to the solid phase, if they are sufficiently hydrophobic. The volatility of an organic compound can be described by the Henry's law constant, $k_H = (C_{gas}/C_{aq})$, where C_{gas} and C_{aq} are equilibrium concentrations or molar fractions of the compound in the gas and aqueous phases of a closed system. The unit of the Henry's law constant depends on the units of C_{gas} and C_{aq}. According to Sternstrom *et al.* (1989), volatization becomes significant when the Henry's law constant of the compound is more than 10^{-3} (atm.m^3/mol), particularly under aeration conditions. If the Henry's law constant is not available, the vapour pressure of the compound, which can be obtained from databases such as SciFinder (https://www.cas.org/products/scifinder), can be a good alternative means to estimate the potential of volatization. However, it is noteworthy that the extent of TrOC removal via volatilization depends also on the compound hydrophobicity, which is discussed below.

The hydrophobicity of an organic compound can be estimated by the octanol–water partitioning coefficient (K_{ow}) or the solid water partitioning coefficient (K_d). The former uses *n*-octanol as the reference organic phase. Thus, K_{ow} is essentially the ratio of the compound's concentration in a known volume of *n*-octanol to its concentration in a known volume of water after partitioning of the chemical between the octanol and water has reached equilibrium. The K_{ow} value of any given compound can be readily obtained from SciFinder or estimated using commercially available molecular simulation software with reasonable accuracy. On the other hand, the K_d value is specific to a particular solid phase and thus it can only be experimentally determined. Due to typical values ranging across numerous orders of magnitude, K_{ow} is usually expressed in a log scale as log K_{ow} (or sometimes denoted log *P*). It is noteworthy that, instead of determining experimentally, researchers often estimate log K_{ow} from quantitative structure–property relationships models, which do not always take ionization into account. However, some TrOC contain ionizable functional groups and thus they can ionize to become either negatively or positively charged depending on the solution pH. To account for this possibility, log *D* or the effective octanol–water partitioning coefficient can be used to reflect the dependence of hydrophilicity on pH. In this case, log *D* can be calculated as $\log D_{(pH)} = \log K_{ow} - \log (1+10^{(pH - pKa)})$. In general, sorption to biosolids has been observed as a major removal mechanism when log *D* value of the TrOC is about 3 or higher.

The sorption of a specific TrOC onto the mixed liquor suspended solids (MLSS) can occur via several different routes: adsorption onto bacterial lipid structure, sorption onto polysaccharide structures outside bacterial cells, and chemical binding to bacterial proteins and nucleic acids (Lester, 1999). In addition to hydrophobic interactions, a number of other reactions such as complex formations with metal ions, ion exchange, and hydrogen bindings can also govern the partition of the organic compound between the solid and the liquid phase (Boxall *et al.*, 2002). These non-hydrophobic interaction mechanisms can occur simultaneously with the hydrophobic adsorption process. For example, fluoroquinolone antibiotics are very hydrophilic, but are mainly eliminated from the aqueous phase by sorption to sludge presumably via electrostatic interactions (Göbel *et al.*, 2007). Thus, in a specific system, where the solid phase is well defined, K_d (instead of K_{ow}), as mentioned above, is commonly used to experimentally describe the sorption of the TrOC onto solids. Stevens-Garmon *et al.* (2011) noted that the positively charged pharmaceutically active compounds they studied, namely, amitriptyline, clozapine, verapamil, risperidone, and hydroxyzine, had the highest sorption potential ($\log K_d = 2.8$–3.8), as compared to the neutral and negatively charged chemicals. K_d can also be normalized by the organic fraction of the solid phase to obtain K_{oc}, which is the organic carbon water partitioning coefficient. Hyland *et al.* (2012) reported that $\log K_{oc}$ correlated well with $\log K_{ow}$ for the non-ionized TrOC, and with $\log D$ for the anionic TrOC, where $\log D$ is greater than 2. No other obvious trend in sorption was observed for the remaining anionic TrOC or for the cationic TrOC. Stevens-Garmon *et al.* (2011) observed that although the K_d value correlated with a compound's hydrophobicity, higher K_d values for the positively charged compounds for a given $\log D$ indicated the importance of additional sorption mechanisms such as electrostatic interactions. Following a thorough literature review, Fernandez-Fontaina *et al.* (2012a) noted that K_d values determined in batch assays are significantly lower than those determined from continuous reactors, highlighting the importance of measuring sorption coefficients under the real operating conditions.

Although a coupling effect between biosorption and biodegradation is possible because higher biosorption means longer time for biodegradation to occur, it has been suggested that for compounds showing a $\log K_d$ below 2.5, sorption onto biosolids is not relevant and their removal will consequently be governed by biodegradation (Joss *et al.*, 2005a). Biodegradation is mediated by microbial activity, and the degradation rates are strongly dependent on the environmental conditions such as the redox potential of the systems and the microbial populations present as well as the TrOC chemical structure (Cirja *et al.*, 2008b). Suarez *et al.* (2010) proposed that compounds could be classified according to their k_{biol} into very highly ($k_{biol} > 5$ L/(g_{SS} d)), highly ($1 < k_{biol} < 5$ L/(g_{SS} d)), moderately ($0.5 < k_{biol} < 1$ L/(g_{SS} d)) and hardly ($k_{biol} < 0.5$ L/(g_{SS} d)) biodegradable. Similarly, Joss *et al.* (2006) identified three TrOC groups according to their biodegradation constant: compounds with $k_{biol} < 0.1$ L/(g_{SS} d) are not removed to a significant extent (<20%), compounds with $k_{biol} > 10$ L/(g_{SS} d) are transformed by >90% and a moderate removal is expected to be in between. It is noted that the poor removal by MBR treatment of the TrOC listed in Table 13.1 appears consistent with their low hydrophobicity ($\log D$, $\log K_d$) and/or low biodegradability (k_{biol}).

To some extent, the removal of TrOC by MBR treatment can be related to their molecular weight. Tadkaew *et al.* (2011) examined the removal of 40 TrOC by an aerobic MBR and reported a weak, but discernible correlation between the removal efficiency of TrOC and their molecular weights. Compounds with molecular weight of more than 300 g/mol were relatively well removed (>60%), while the removal efficiencies of those with molecular weight of less than 300 g/mol varied from almost zero to more than 98%. They attributed this observation to the relatively higher hydrophobicity of the tested compounds having molecular weight of more than 300 g/mol. Another plausible explanation is that compounds with higher molecular weight tend to have more branches and functional groups, which would offer more opportunities for the microbes to selectively cleave a certain target site and initiate degradation.

Table 13.1 Physicochemical properties of some TrOC showing low and/or highly variable removal by MBR treatment.

Class	Compound	Log D_{ow} (pH 7)[a]	Charge at pH 7	Dissociation Constant (pK_a)[a]	Water Solubility (mg/L)[b]	Log K_d	k_{biol} (L/(g $_{SS}$· d))	Chemical Structure	Removal, % (Reference)
Analgesic	Hydrocodone ($C_{18}H_{21}NO_3$)	1.05	Neutral	8.52 ± 0.20	6,870			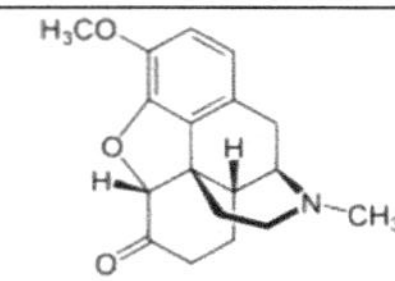	0 Kim *et al.* (2007)
Non-steroidal anti-inflammatory	Diclofenac ($C_{14}H_{11}Cl_2NO_2$)	1.77	Negative	4.18 ± 0.10 −2.26 ± 0.50	2.4	1.2–2.7 Ternes *et al.* (2004a)	<0.1 Ternes *et al.* (2004a)	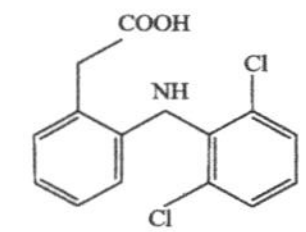	0–58 Bernhard *et al.* (2006), Kim *et al.* (2007), Kimura *et al.* (2005), Tadkaew *et al.* (2011)
	Indomethacin ($C_{19}H_{16}ClNO_4$)	1.29	Negative	3.96 ± 0.30	0.937	1.45 Joss *et al.* (2006)	<0.21 Joss *et al.* (2006)	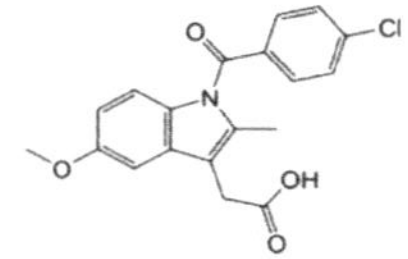	47 Radjenovic *et al.* (2007), Radjenovic *et al.* (2009a)
Anti-depressants and mood stabilizers	Carbamazepine ($C_{15}H_{12}N_2O$)	1.89	Neutral	13.94 ± 0.20 −0.49 ± 0.20	18	0.09 Ternes *et al.* (2004a); 1.31—1.83 Carballa *et al.* (2008)	<0.01 Suarez *et al.* (2008)	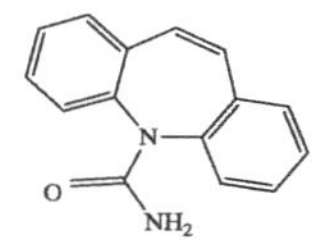	0–13 Bernhard *et al.* (2006), Kim *et al.* (2007), Tadkaew *et al.* (2011)
	Dilantin ($C_{15}H_{12}N_2O_2$)	2.5	Positive	8.33 ± 0.50	32	1.65–<2.01 Stevens-Garmon *et al.* (2011)		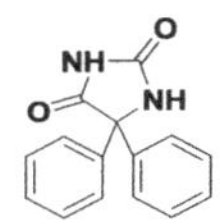	0–12 Snyder *et al.* (2007), Tadkaew *et al.* (2011)
Antibiotic	Metronidazole ($C_6H_9N_3O_3$)	-0.14	Neutral	14.44 ± 0.10 2.58 ± 0.34	9500	Undetectable Loke *et al.* (2002)		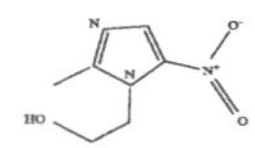	10–50 Hai *et al.* (2011c), Nguyen *et al.* (2012)
	Trimethoprim ($C_{14}H_{18}N_4O_3$)	0.38	Positive	7.20 ± 0.12	400	2.3 Suarez *et al.* (2008)	0.65 Suarez *et al.* (2012)	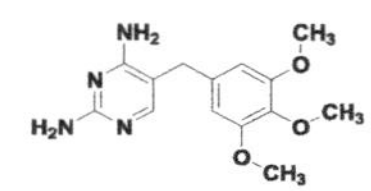	0–90 Gobel *et al.* (2007), Kim *et al.*

(*Continued*)

Table 13.1 Physicochemical properties of some TrOC showing low and/or highly variable removal by MBR treatment. *(Continued)*.

Class	Compound	Log D_{ow} (pH 7)[a]	Charge at pH 7	Dissociation Constant (pK_a)[a]	Water Solubility (mg/L)[b]	Log K_d	k_{biol} (L/(g $_{SS}$· d))	Chemical Structure	Removal, % (Reference)
									(2007), Tadkaew *et al.* (2011)
	Erythromycin ($C_{37}H_{67}NO_{13}$)	0.81	Neutral	13.09 ± 0.70 8.16 ± 0.70	1.440	2.2 Suarez *et al.* (2008)	2.4 Suarez *et al.* (2012) 0.5—1 Suarez *et al.* (2008)	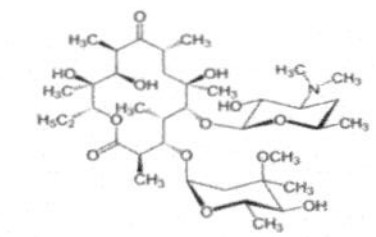	4–67 Kim *et al.* (2007), Radjenovic *et al.* (2007)
Antihistamine	Loratidine ($C_{22}H_{23}ClN_2O_2$)	3.89	Neutral	4.27 ± 0.20	0.011	3.35–3.5 Radjenovic *et al.* (2009a)		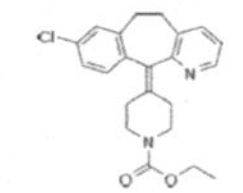	0–33 Radjenovic *et al.* (2009a)
Diuretics	Hydrochlorothiazide ($C_7H_8ClN_3O_4S_2$)	−0.03	Negative	8.95 ± 0.20 −4.08 ± 0.20	722	1.38 Radjenovic *et al.* (2009a)		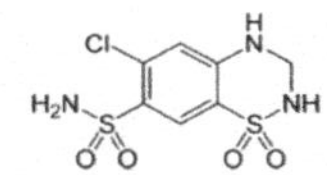	0–66 Radjenovic *et al.* (2007), Radjenovic *et al.* (2009a)
	Triamterene ($C_{12}H_{11}N_7$)	1.26	Negative	6.30 ± 0.50	48.2			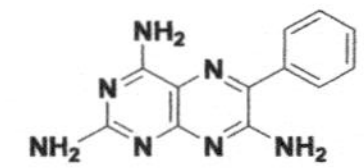	27 Tadkaew *et al.* (2011)
Anxiolytic	Meprobamate ($C_9H_{18}N_2O_4$)	0.7	Neutral	13.09 ± 0.50	4700	<1.48–<2.2 Stevens-Garmon *et al.* (2011)		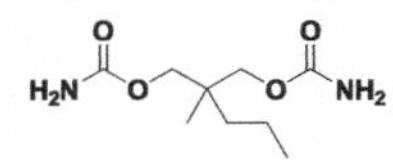	0–18 Snyder *et al.* (2007), Tadkaew *et al.* (2011)
Pesticide/herbicide	Fenoprop ($C_9H_7Cl_3O_3$)	−0.13	Negative	2.93	71			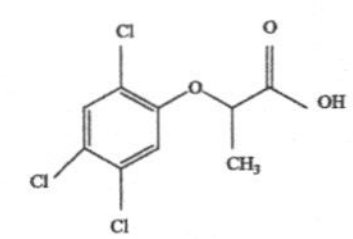	5–50 Hai *et al.* (2011a), Nguyen *et al.* (2012)
	Atrazine ($C_8H_{14}ClN_5$)	2.63	Neutral	2.35 ± 0.50	34.7	0.7 Wick *et al.* (2011), <1.48—2.1 Stevens-Garmon *et al.* (2011)		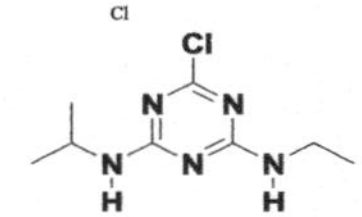	5–40 Bernhard *et al.* (2006), Bouju *et al.* (2008), Tadkaew *et al.* (2011)
	Linuron ($C_9H_{10}Cl_2N_2O_2$)	3.20	Neutral	12.13 ± 0.7	75			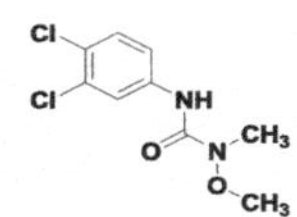	20 Tadkaew *et al.* (2011)

	Diethyltoluamide (DEET) ($C_{12}H_{17}NO$)	1.96	Neutral	−1.37 ± 0.7	912	<1.48–2 Stevens-Garmon *et al.* (2011)	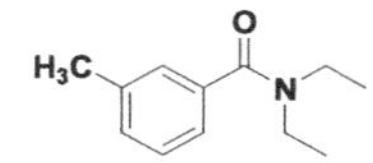	0–78 Bernhard *et al.* (2006), Kim *et al.* (2007), Snyder *et al.* (2007)
	Bentazone ($C_{10}H_{12}N_2O_3S$)	0.81	Negative	3.28 ± 0.70	570		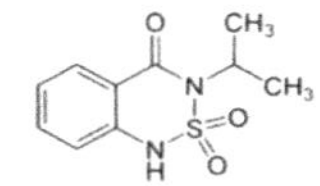	16 Bernhard *et al.* (2006)
	Isoproturon ($C_{12}H_{18}N_2O$)	2.89	Positive	15.06 ± 0.70 0.85 ± 0.50	65	0.95 Wick *et al.* (2011)	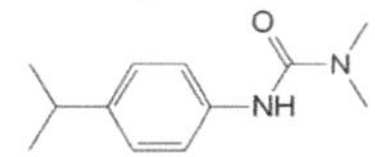	25 Bernhard *et al.* (2006)
Industrial precursor/ additive	Ethylenediaminetetraacetic acid (EDTA) ($C_{10}H_{16}N_2O_8$)	−6.29	Negative	1.782 ± 0.10 11.24 ± 0.60	1,000		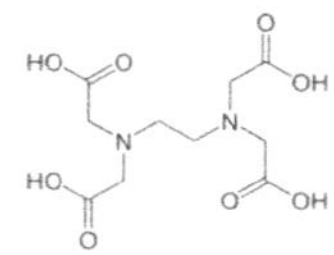	0 Bernhard *et al.* (2006)
Flame retardant	Tris (1-chloro-2-propyl) phosphate (TCPP) ($C_9H_{18}Cl_3O_4P$)	2.53	Neutral	na	1,600	<1.48–2.3 Stevens-Garmon *et al.* (2011)	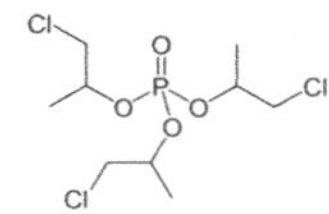	12 Bernhard *et al.* (2006)

[a]Source: SciFinder database https://scifinder.cas.org/scifinder/view/scifinder/scifinderExplore.jsf.
[b]Source: http://chem.sis.nlm.nih.gov/chemidplus/.
na: data not available.
K_d: sludge-water distribution coefficient (L/kg_{SS}); k_{biol}: pseudo-first-order degradation constant ($L/(g_{SS}\cdot d)$).

No single physicochemical parameter can be used to predict the biodegradation of TrOC. Instead, recent research has successfully demonstrated the connection between the molecular structure and/or functional moieties of TrOC and their biodegradability during MBR treatment (Kümmerer & Al-Ahmad, 1997; Reemtsma *et al.*, 2002; Tadkaew *et al.*, 2011). Compounds such as esters, nitriles and aromatic alcohols have functional groups that may increase biodegradability, whereas compounds with complex structure (e.g. alkyl chain branch) (Corvini *et al.*, 2006) and toxic groups (e.g. halogen groups (Hai *et al.*, 2011b)) show higher resistance to complete biodegradation. Tadkaew *et al.* (2011) systematically studied the influence of electron-withdrawing groups (EWGs) and electron-donating groups (EDGs) on TrOC removal. Examples of these EWGs and EDGs are shown in Figure 13.3. Tadkaew *et al.* (2011) subsequently proposed a qualitative framework for prediction of TrOC removal by MBR (Figure 13.4). Tadkaew *et al.* (2011) showed that all hydrophilic and moderately hydrophobic (log $D < 3.2$) compounds possessing strong electron-withdrawing functional groups had removal efficiency of <20%. By contrast, high removal efficiencies were observed with most compounds bearing electron-donating functional groups such as hydroxyl and primary amine groups. Notably, 'removal' in that study indicated the disappearance of the parent molecules, and not necessarily complete mineralization. Interestingly, Hai *et al.* (2011b) reported that even the compounds with high halogen content (halogens are strong electron withdrawing and toxic groups) were removed from the aqueous phase when they were significantly hydrophobic (Hai *et al.*, 2011b). Therefore, the simultaneous influence of hydrophobicity and chemical structure on the removal of TrOC should be taken into account.

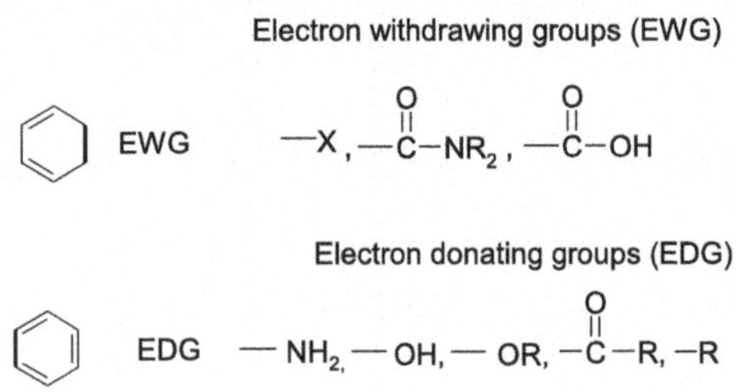

Figure 13.3 Examples of electron-withdrawing and -donating functional groups.

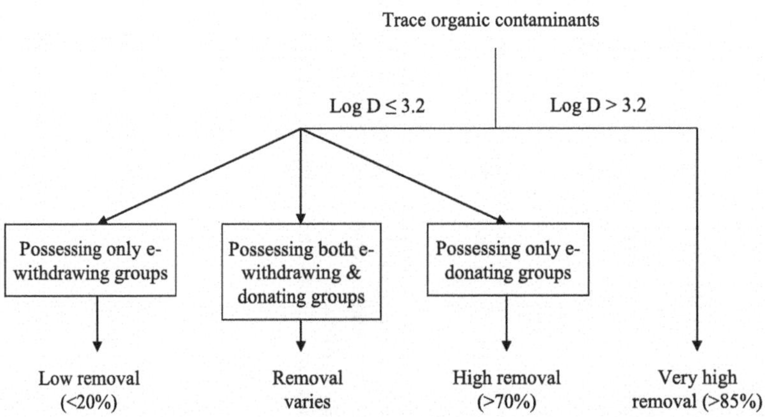

Figure 13.4 Qualitative framework based on hydrophobicity and chemical structure (functional groups) for prediction of TrOC removal by MBRs. Reproduced with permission from Tadkaew *et al.* (2011).

Andreozzi *et al.* (2006) studied the effects of the presence of a single functional group (e.g. –OH, $–CH_3$, –Cl, $–NO_2$, etc.) within a compound on its aerobic biodegradation. The results showed that increasing the number of the nitro groups or chlorine atoms, causes a decrease in the value of k_{biol}. In contrast, for an already activated structure (i.e. containing phenol, benzoic acid, etc.), the addition of hydroxyl and methyl groups did not improve biodegradability significantly, probably because the whole process then became transport controlled (i.e. inhibited diffusion of the larger molecule through the microbial cell membrane). Thus, the effects of each group also depend on the presence of other groups in the molecular structure (Andreozzi *et al.*, 2006). Kümmerer and Al-Ahmad (1997) reported that the presence of sugar moieties and fluorine atoms on a compound could increase and decrease the relative biodegradability, respectively. The removal efficacy of polar compounds such as naphthalene sulfonates (anionic surfactants) during MBR treatment depends strongly on their respective molecular structure (Reemtsma *et al.*, 2002): the removal of the naphthalene monosulfonates was almost complete, while the removal of naphthalene disulfonates was limited to about 40%. The precise chemical structures of the alkyl chains of nonylphenol (NP) and LAS can have a significant impact on biodegradation pathways. For instance, branched isomers of NP are very recalcitrant and resulting metabolites possess incompletely degraded alkyl chains while ultimate degradation of linear NP is faster (Cirja *et al.*, 2008a). The removal rate of steroids was reported to be influenced by their chemical structure. 17β-Estradiol (E2) and 17α-ethinylestradiol (EE2) have basically the same chemical structure, except the ethinyl group present in EE2, which leads to drastic differences in biodegradability. In wastewater treatment systems, microorganisms are able to degrade E2 quite easily, while EE2 is very resistant, although similar aqueous phase removal of both compounds may be observed due to their significant hydrophobicity (Ternes *et al.*, 1999).

13.5.2 Operating parameters

13.5.2.1 Concentration, characteristics and acclimatization of biomass

While reviewing the contemporary literature, Auriol *et al.* (2006) observed that MLSS concentration is an important factor governing the removal of estrogens in biological wastewater treatment plants. A higher MLSS concentration resulted in better removal of estrogens, while an increase in influent estrogen concentration caused a decrease in removal, probably due to the fact that EDC sorption on the suspended solids was a major mechanism of removal (Auriol *et al.*, 2006). Yi and Harper (2007a) reported that owing to smaller particle size, MBR MLSS generally has higher EE2 partitioning coefficients and smaller sorption hysteresis index (deviation of adsorption and desorption values from one another) than CAS MLSS. Results of batch elimination tests conducted by Kimura *et al.* (2007) revealed that the MLSS in an MBR had larger specific sorption capacities mainly due to its larger specific surface area.

MLSS concentration has been reported to be important in not only biosorption but also for improving biodegradation. Li *et al.* (2005) found that the biodegradation of E2 depended markedly upon the levels of mixed liquor volatile suspended solids (MLVSS) concentration in aerobic reactors at 1.75, 0.85 and 0.435 g/L. In a study by Shariati *et al.* (2010) the removal of acetaminophen (paracetamol) was observed to improve from 20 to 40% when the MLSS changed from 2 to 15 g/L. Because acetaminophen is a hydrophilic compound, the removal performance improvement could be attributed to the improvement in biodegradation. A sequencing batch biofilter reactor characterized by high MLSS concentration (up to 40 g/L) and a long SRT (up to 6 months) achieved 6% (estrone, E1) to 32% (E2) better removal over a CAS system of selected EDC, namely, E1, E2, bisphenol A and 4-tert-octylphenol (Balest *et al.*, 2008). Li *et al.* (2011b) studied the effect of MLSS concentration on the removal of carbamazepine by an MBR at a carbamazepine loading of 750 µg/L d. The % removal rate of carbamazepine did not increase much beyond a MLSS concentration of 5 g/L. This indicated that due to the insignificant adsorption of carbamazepine onto MLSS, biodegradation, in contrast to biosorption, played the main role in carbamazepine removal by the MBR. Interestingly, under an MLSS concentration of approximately 1 g/L the removal rate of carbamazepine was the lowest. This underscored the importance of maintenance of an adequate amount of biomass in the reactor to achieve satisfactory degree of recalcitrant pollutant degradation. Sahar *et al.* (2011) also observed that at MLSS concentrations over 4 g/L TrOC removal is quite effective and does not change much with further increase of MLSS concentration.

Biodegradation rates are strongly dependent upon environmental conditions such as the redox potential of the systems and the microbial populations present. Acclimatization of microorganisms to the substrate requires time, and the affinity of the bacterial enzymes for the TrOC influences the pollutant transformation or decomposition (Spain *et al.*, 1980). Gonzalez *et al.* (2006) attributed the faster diminution of diclofenac concentration in their study to better acclimation of the microorganisms to the influent. Although not directly proven, Reif *et al.* (2008) explained lower biodegradation of musk fragrances in their study as compared to that in a study by Clara *et al.* (2005c) by incomplete acclimation of sludge during their short-term study. Sahar *et al.* (2011) reported improvement in removal of macrolide antibiotics during stable operating conditions as compared to that during the period of biomass buildup in the MBR. The naproxen removal performance of a pilot plant increased from 27% up to 99% during the first 300 d, indicating a possible acclimation of bacteria to this pharmaceutical (Suarez *et al.*, 2012). The importance of adapted microbial populations in the removal of estrogen was shown by the dramatic differences in mineralization of ^{14}C-17β-estradiol by biosolids from a municipal plant compared to that from an industrial plant, 84% vs 4%, respectively (Layton *et al.*, 2000). While operating an anoxic bioreactor, Suarez *et al.* (2010) observed gradual increase in k_{biol} (0.1 L/g_{SS}.d between day 0 and 200; 0.3 L/g_{SS} *d* from day 200 to 350 and 1.5 L/g_{SS} *d* afterwards), and attributed this to the gradual adaptation of bacteria.

13.5.2.2 Solids retention time and hydraulic retention time

The affinity of bacterial enzymes in the mixed liquor for a TrOC influences its transformation or decomposition. The improvement of biodegradation of various TrOC with the increase in SRT can be explained by two possible mechanisms (Cirja *et al.*, 2008a; Leclercq *et al.*, 2009; Ternes *et al.*, 2004b): (i) the bacterial population may become more diversified with increasing SRT, possibly because slow-growing bacteria eventually reach effective population numbers; (ii) the microorganisms may diversify their metabolic activity in response to the lower sludge loading with bulk organics (i.e., lower substrate availability); in this case, increased TrOC removal might be due only to the broadened enzyme spectrum and not necessarily to the microbial community (Ternes *et al.*, 2004b).

In a survey conducted in full-scale CAS plants for both gemfibrozil and diclofenac, removal did not appear to be strongly influenced by SRT of up to 15 d, while SRT of over 30 d were associated with more frequent non-quantifiable effluent levels of diclofenac (Lishman *et al.*, 2006). A similar observation on the removal of nonylphenol and bisphenol A was described by Ivashechkin *et al.* (2004) during the operation of MBR and CAS processes at SRT of 12 and 25 d. The high removal efficiency (95%) of nonylphenol ethoxylates associated with the application of long SRT in MBR was confirmed as well by Terzic *et al.* (2005).

In a study by Kimura *et al.* (2007), the MBR operated with an SRT of 65 d clearly showed better performance in removing pharmaceuticals than the MBR with an SRT of 15 d. The difference between the two MBRs was particularly significant in terms of elimination of ketoprofen and diclofenac (Kimura *et al.*, 2007). Better removal of E1, E2 and EE2 with the increase in SRT has been reported in several studies (Baronti *et al.*, 2000; Estrada-Arriaga & Mijaylova, 2011; Johnson *et al.*, 2000; Spengler *et al.*, 2001). Over the range of SRT (35–95 d) and HRT (7–12 h) tested, Estrada-Arriaga & Mijaylova (2011) observed the best removal (>98%) of estrogens (*E*1, *E*2, and *EE*2) under an HRT of 12 h and SRT of 60 d. Joss *et al.* (2004) studied the degradation of *E*1, *E*2, and *EE*2 in batch experiments using sludge from CAS plants with an SRT of 11 d and MBR plants with 30 d. For the natural estrogens *E*1 and *E*2, degradation activity seemed to be higher in MBR than in CAS by a factor of 2–3 with respect to the applied SRT. By comparing the effluent from two laboratory-scale MBR systems with SRT of 17 and 33 d, Hu *et al.* (2007b) reported that higher SRT was better for the removal of overall estrogenicity in MBR systems. In a study by Tambosi *et al.* (2010) an antibiotic (roxithromycin) registered lower removal efficiencies compared to the nonsteroidal anti-inflammatory drugs (NSAID), which were eliminated or metabolized with efficiencies of around 100%, and the removal of the former was better for the case of a SRT of 30 d (81%) than in 15 d (57%).

The minimum SRT required for high removal tends to vary depending on the TrOC. Caffeine and oxybenzone, two compounds classified as consistently exhibiting excellent removal, had critical SRT less than 5 d (Oppenheimer *et al.*, 2007). On the other hand, TrOC classified as having poor removal, such as galaxolide and tris(2- chloroethyl)phosphate, had critical SRTs greater than 15 d. Clara *et al.* (2005a) also reported 10–15 d as the critical SRT value for the removal of TrOC with the exception of a few essentially non-biodegradable compounds. Based on a critical review of the available studies, Cirja *et al.* (2008a) showed that a diversified microbial community, including nitrifying bacteria, may develop at SRTs higher than 8 d. Ternes *et al.* (2004b) reported that the lipid regulator bezafibrate, the antibiotic sulfamethoxazole, and the antiphlogistics ibuprofen and acetylsalicylic acid require a SRT of 2–5 d for significant degradation, while EE2, the anti-inflammatory drug diclofenac, the contrast medium iopromide, and the antibiotic roxithromycin need 5–15 d. In laboratory-scale continuous-flow CAS systems, at an HRT of 10 h and SRT ranging between 3 and 20 d, the majority of EDCs (>90%) could be removed mainly via biodegradation (Stasinakis *et al.*, 2010).

Studies where an SRT longer than the above-mentioned critical values were applied, often showed no effect of SRT on TrOC removal. For instance, Gobel *et al.* (2007) found no difference in the removal efficiencies of various antibiotics in two STP with SRT of 21–25 and 10–12 d, respectively. In another study, no significant effects on the removal of a number of TrOC were found for a variation of HRT (7 – 14 h) and SRT (26–102 d), suggesting that the lowest values selected have already been high enough for good removal (Weiss & Reemtsma, 2008). However, longer critical SRT values have been reported for some compounds (up to 60–80 d) (Gobel *et al.*, 2007; Lishman *et al.*, 2006). For compounds like trimethoprim and macrolide antibiotics azithromycin, erythromycin, and clarithromycin, a clear increase in transformation was found at SRT of 60–80 d. In the same study, higher reduction of roxithromycin was observed already at a SRT of 33 d (Gobel *et al.*, 2007). On the other hand, highly persistent compounds tend to show very low removal over a broad range of SRT. For example, carbamazepine was removed at less than 20% under different SRT ranging from 1 to 20 d (Cirja *et al.*, 2008a). Indeed an SRT of more than 500 d also resulted in no carbamazepine removal (Clara *et al.*, 2005a). Furthermore, contradictory effect of SRT on compounds with low inherent biodegradability can be observed in the literature. For instance, Clara *et al.* (2004) reported higher removal of diclofenac with increasing SRT (from 10 to 56 d) in an MBR. However, Joss *et al.* (2005a) found no improvement in degradation of diclofenac with increased SRT (from 12 to 75 d). By contrast, the biodegradation of this compound was always below 15% (i.e. within analytical error) except when the SRT was longer than 150 d, when it reached 70% (Fernandez-Fontaina *et al.*, 2012a). Although there is no apparent relationship between SRT and TrOC removal by MBR, Figure 13.5 indicates that the removal of a few pharmaceuticals including sulfamethoxazole, diclofenac, mefenamic acid and carbamazepine could improve significantly with the increase in SRT. On the other hand, SRT was observed not to affect the removal of readily biodegradable and resistant hydrophilic TrOCs (Asif *et al.*, 2018a).

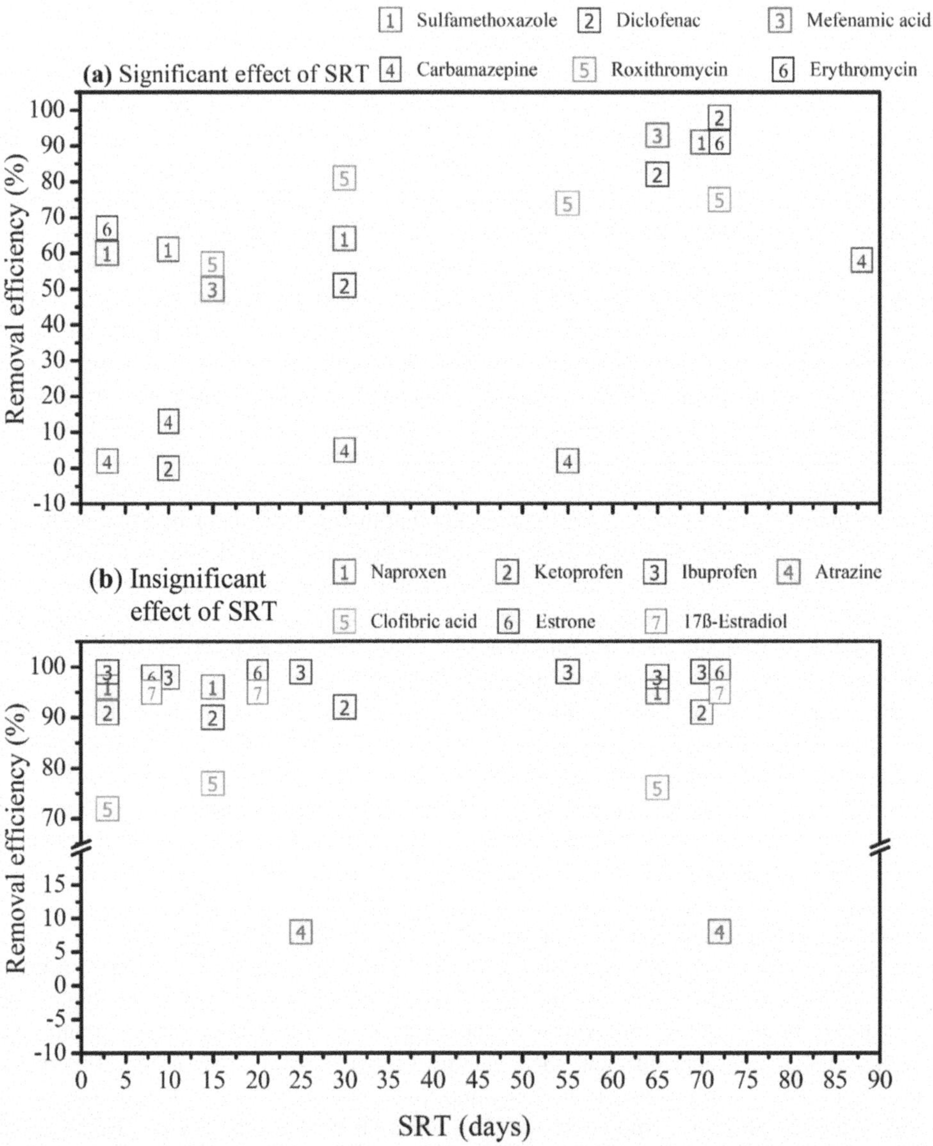

Figure 13.5 Effect of SRT on the aqueous phase removal of the selected TrOCs by MBRs. Reproduced with permission from Asif *et al.* (2018a).

In addition to biodegradation, SRT may influence adsorption of TrOC on sludge. An increase in SRT causes a reduction in the F/M ratio, thereby reducing the excess amount of sludge and the number of renewable binding sites. Therefore, the removal of compounds which are mainly removed by adsorption on sludge could decrease when the SRT increases (Bouju *et al.*, 2008; Sahar *et al.*, 2011). By contrast, some researchers have observed increased biomass hydrophobicity at longer SRT (Lee *et al.*, 2003; Liao *et al.*, 2001), meaning that better adsorption may be observed at a longer SRT, offsetting the limitation imposed by less number of new binding sites. For instance, Suarez *et al.* (2012) attributed the increase in removal of EE2 at SRTs above 20 d to its significant increase in adsorption of MLSS in addition to biodegradation. The adsorption kinetics for tetracyclines was determined at various biomass concentrations in sequencing batch reactors at different SRT and HRT (Kim *et al.*, 2005). Between 75 and 95% of the applied tetracyclines was adsorbed onto the sludge after 1 h. At the longer SRT (10 d) the removal of tetracyclines was 85–86%, while the decrease of SRT to 3 d gave a lower removal (78%). The lower degradation rates were assigned to the reduced biomass concentration once the SRT was shortened (Cirja *et al.*, 2008a). Notably, Hyland *et al.* (2012) observed no correlation of a set of neutral and charged TrOC sorption data with respect to SRT within a SRT range of 3–17 d; probably the opposite impacts of reduced number of renewal binding sites vs increased hydrophobicity mutually cancelled out in this case.

As noted earlier, biological transformation of TrOC is not only a function of their k_{biol} values, but also of their K_d values, since biodegradable compounds with large K_d values ($\log K_d > 2.5$) will be removed when the SRT in the aeration tank is long enough to accomplish their biological degradation. However, as reviewed above, once a certain limit value of SRT is exceeded, the removal efficiencies are not enhanced anymore (Suarez *et al.*, 2008) (Table 13.2). At sufficiently long SRT in the treatment system, the contact time, the diffusion into the flocs, and the adaptation of microorganisms to the substrate are improved. From the studies reviewed here, it can be concluded that SRT ranging between 10 and 30 d may allow for sufficient removal rates concerning most of the investigated biodegradable TrOC.

Table 13.2 Simplified representation of the interrelation between the biodegradation (k_{biol}) and adsorption (K_d) constants and SRT governing aqueous phase removal of TrOC.

K_d	k_{biol}	Dependence on SRT	Removal
High	High	No	High
High	Low	Yes	Moderate to high
Medium	High	Yes/No	Moderate
Low	Low	No	Low

Besides SRT, a relationship between HRT and biodegradation of TrOC may be expected. Compared to the case of MBR, relatively clearer impact of HRT on TrOC removal by CAS plants can be noticed in the literature. For instance, Maurer *et al.* (2007) noted that HRT and reactor design were responsible for the variation they observed in removal efficiencies of β-blockers from one CAS plant to another. In another study, a decrease in removal efficiency of β-blockers was found when the HRT was decreased to accommodate the increased amount of water that needed to be treated following a rainfall event (Vieno *et al.*, 2007). The effect of HRT on the removal of an antibiotic ampicillin was studied in a pilot-scale anaerobic baffled reactor at steady state (Zhou *et al.*, 2006). Ampicillin removal increased from 16% to 42% when the HRT was increased from 1.25 to 2.50 d. With the same HRT increase, the antibiotic aureomycin had a removal-increase from 26 to 31% (Zhou *et al.*, 2006). In a series of different waste stabilization ponds with a total HRT of 78 d, a 73% decrease in carbamazepine concentration was observed (Leclercq *et al.*, 2009). The authors attributed this removal to more efficient adsorption of this hardly biodegradable compound onto sludge under the prolonged HRT. HRT was shown to affect elimination of ibuprofen and ketoprofen in CAS plants (Tauxe-Wuersch *et al.*, 2005) in a way that lower removal was observed for the shorter HRT. Similar effect was also observed during heavy rain periods, when rainwater caused a decrease in HRT (Ternes, 1998). It has been reported that high values of HRT allow adsorption, and biodegradation processes are carried out more easily increasing the estrogen removal by CAS plants (Andersen *et al.*, 2003; Johnson & Sumpter, 2001; Layton *et al.*, 2000; Williams *et al.*, 2009).

While studying the removal of 19 TrOC including EDC and PPCP, Xue *et al.* (2010) observed that the removal rates of most target compounds reached a quasi-plateau in 5 h under aerobic conditions, indicating that an HRT of *ca.* 5 h in aerobic tanks may be sufficient for the elimination of most target TrOC. In a CAS reactor, the decrease in the biodegradation efficiencies was clearly noticed for fluoxetine, erythromycin, roxithromycin and trimethoprim when the HRT was reduced from 4.3 to 1 d (Fernandez-Fontaina *et al.*, 2012a). This observation suggests that these substances suffered significant kinetic (shorter HRT) and/or stoichiometric (higher TrOC loading rates) limitations in their biodegradation. Longer HRT and/or higher nitrogen removal rates (more ammonium monooxygenase activity) are needed to cope with the increased loading rates of these TrOC (Fernandez-Fontaina *et al.*, 2012a). Kim *et al.* (2005), however, noted that shorter HRT can result in an increase in the biomass concentration due to the higher substrate loading, and the higher biomass concentrations in the reduced HRT phase may compensate for the shorter reaction times.

Due to the decoupling of HRT and SRT in MBR, most MBR studies report no obvious effect of HRT under the tested ranges. For instance, no obvious influence of HRT (3.9–8 h) on BPA removal was observed in an MBR (Chen *et al.*, 2008). Weiss and Reemtsma (2008) reported a statistically insignificant effect of HRT (7–14 h) on the removal of a range of polar TrOC including those which are easily degradable and those which are highly persistent. Bernhard *et al.* (2006) reported that the reduction of the HRT from 10 to 7 h did not influence the removal of selected non-adsorbing, persistent TrOC in a laboratory-scale MBR. In fact, the MBR in that study performed better than a CAS plant operated using a HRT of 22 h (Bernhard *et al.*, 2006). However, over the tested ranges of HRT (7–12 h), Estrada-Arriaga and Mijaylova (2011) observed the best removal (>98%) of estrogens (*E*1, *E*2, and *EE*2) under an HRT of 12 h.

13.5.2.3 Cometabolism and TrOC loading

In wastewater, TrOC occur primarily at concentrations of $<10^{-4}$ g/L. At these levels, biological transformation or degradation of TrOC occur only if a primary substrate is available for the corresponding bacteria to grow on. Hence, cometabolic transformation may be the major removal mechanism of TrOC in activated sludge treatment of municipal wastewater, i.e., the bacteria break down or partially convert the TrOC and do not use it as a carbon source. For example, in batch tests, bezafibrate, naproxen, and ibuprofen were degraded only cometabolically (Quintana *et al.*, 2005). On the other hand, De Wever *et al.* (2007) reported that the TrOC investigated in their study were biodegraded rather than being removed through cometabolism. They argued that in the latter case one would expect that the lower F/M ratio in the MBR leads to substrate limitations; however, they did not observe that in their study (De Wever *et al.*, 2007). In another likely scenario, mixed-substrate growth may take place where the bacteria use the TrOC as a carbon and energy source and may mineralize it completely. For instance, ketoprofen was partly mineralized as a sole source of carbon and energy and the metabolites determined by LC/MS suggested that ketoprofen degradation proceeded along the pathway known for biphenyls and related compounds (Ternes *et al.*, 2004b). In a laboratory-scale sequencing batch reactor, cometabolism of sulfamethoxazole with acetate was observed and it was also found that microorganisms may use the degraded antibiotic as a nitrogen source (Drillia *et al.*, 2005).

The characteristic very low concentrations of TrOC limit their biodegradative fate, since for most compounds in an STP this fate is governed by non-growth-limiting (enzyme-saturating) substrate concentrations (copiotrophic metabolism); however, TrOC are present at enzyme-subsaturating levels, which necessitate an oligotrophic metabolism (Daughton & Ternes, 1999). At low concentration, the kinetics of decomposition of TrOC occurs mostly according to a first-order reaction (Cirja *et al.*, 2008a). In a series of batch tests conducted by Chen and Hu (2009), over an observation range of 10–200 μg/L, degradation rates of E2 and its conjugates were higher at the higher initial concentration. Probably, substrate limitation under a low initial concentration was the main reason for such an observation. Similarly, in a study by Li *et al.* (2011b), although the % removal rate of carbamazepine was larger under lower loading rate conditions (2–6 μg/L d) as compared to a higher loading rate (750 μg/L d), the removal rate (in μg/L d) was in fact lower in case of the lower loading rate (2–6 μg/L d). The results in that study indicated that carbamazepine is extraordinarily persistent to biodegradation at low concentrations. De Wever *et al.* (2007) also demonstrated the existence of a minimum concentration level for the degradation of 1,3-naphthalene disulfonate. This again points to a minimal concentration level below which microorganisms do not gain sufficient energy for growth, and cometabolism forms the main mechanism of biodegradation. However, in an experiment conducted with undiluted mixed liquor spiked with varying E2 concentrations (10, 30, and 50 μg/L) a removal trend irrelevant to its spiking concentrations was observed (Li *et al.*, 2005).

13.5.2.4 *Mixed liquor pH*

Wastewater pH can influence TrOC removal by influencing both the physiology of the microorganisms (optimum pH for enzymatic activity) and the TrOC solubility in wastewater (Cirja *et al.*, 2008a). In a study by Reddy *et al.* (2005) acidic pH prevented the dissociation of the steroid conjugates to free steroids. Urase *et al.* (2005) reported improved removal of several pharmaceuticals, such as ibuprofen, fenprofen, naproxen, and diclofenac at an acidic pH, as compared to that in usual neutral pH, whilst the removal of neutral compounds *E*2, *EE*2, and bisphenol A did not improve significantly when the pH was acidic. It was explained that TrOC can exist in different protonation states at different pH, which can lead to different extents of hydrophobicity (log *D*) at different pH values (Cirja *et al.*, 2008b; Tadkaew *et al.*, 2010).

During sludge treatments, at stages such as sludge dewatering and conditioning with lime, the pH is raised to over 9 and thus TrOC can desorb from sludge. For instance, desorption of BPA in aqueous phase took place at $\text{pH} > 12$ due to the increased solubility of the deprotonated form of BPA (Clara *et al.*, 2004). The consequence of such high release was a high backloading of the STP via the recycling of the process water (Cirja *et al.*, 2008a). In another study, the sludge–water partition coefficients (K_p) of the investigated EDC in activated sludge from a STP increased with the decrease of pH value for almost all the investigated compounds (BPA, *E*2, *EE*2) (Cirja *et al.*, 2008a). In the case of compounds containing one carboxyl group, the K_p values at $\text{pH} = 5.6$ were 2.5–30 times higher than those at $\text{pH} = 6.7$, while for compounds having phenol groups such as E1, EE2, BPA the increase of partition coefficient varied to a lower extent within this range of pH values (Cirja *et al.*, 2008a).

Despite the possibility of enhanced adsorption of certain ionizable TrOC on sludge, there are some practical constraints against operation under acidic pH. For instance, acidic pH may have adverse impact on certain microbial groups, which may in turn lead to reduction of TOC, TN, and/or total phosphorous (TP) removal (Baldwin & Campbell, 2001; Zhang *et al.*, 2005). Bo *et al.* (2009) reported that biodegradation of clofibric acid was impaired at low pH operation, and improved removal occurred only after a lag phase following the return of the mixed liquor pH to neutral. It was also demonstrated that the hydrophobicity of norfloxacin varies with pH, with the hydrophobicity being very low at pH < 4 and pH > 10 and the maximal hydrophobicity was reached at a pH of 7.5 (Cirja *et al.*, 2008a).

13.5.2.5 Mixed liquor temperature

Because microbial growth and activity, as well as solubility and other physicochemical properties of organics (Cirja *et al.*, 2008a), are significantly affected by temperature conditions, temperature variability have been related to deterioration in bulk water quality parameters and system instability (Cirja *et al.*, 2008a; Morgan-Sagastume & Allen, 2003). The effects have been dependent on the temperature stability and the magnitude of any fluctuations, and have been linked to sludge deflocculation and decreased sludge metabolic activity. Nevertheless, systematic studies on the effects of temperature variation on TrOC removal in either CAS process or MBR remain very scarce (Hai *et al.*, 2011c). Most of the observations of variation of TrOC removal with ambient temperature have been anecdotal and based on measurement of limited number of samples at full-scale plants, and have been reported as relatively high effluent concentrations of certain TrOC during low winter temperature or vice versa (Sui *et al.*, 2011; Vieno *et al.*, 2005). In addition to temperature, other factors such as overall pollutant loading, precipitation and sunlight availability (important for photodegradation) can also influence the observed seasonal variations in effluent concentration. Therefore in the absence of a controlled experimental design the effect of temperature cannot be accurately ascertained. Suarez *et al.* (2010) observed that E2 and EE2 removal by a laboratory-scale activated sludge reactor was not significantly different at 16 and 26°C. By contrast, Tanghe *et al.* (1998) reported significant deterioration in the removal capacity of nonylphenol by a laboratory CAS reactor due to a temperature shift from 28 to 10°C. Through experiments conducted in a laboratory-scale MBR, Hai *et al.* (2011c) provided unique insight into the effect of dynamic short-term (diurnal) temperature variation on TrOC removal by MBR treatment. With a few exceptions, operation at 45°C clearly exerted detrimental effects on the removal efficiency of the TrOC selected in that study. The removal of most hydrophobic compounds (log $D > 3.2$) was stable during operation under a temperature range of 10–35°C. On the other hand, for the less hydrophobic compounds (log $D < 3.2$) a comparatively more pronounced variation between removals in the lower temperature regimes (10–35°C) was observed. Lower and more variable removal efficiency was seen for certain hydrophilic compounds at 10°C, which have been reported to be moderately recalcitrant to MBR treatment. Cirja *et al.* (2008a) argued that CAS processes may show better stability than MBR during seasonal temperature variations: the larger surface of a CAS plant than MBR would attenuate the variations of temperature, protecting bacterial activity against temperature shock produced in the system. However, no specific study to compare this aspect of CAS and MBR performance could be identified in the literature.

13.5.2.6 Mixed liquor dissolved oxygen concentration

Mixed liquor dissolved oxygen (DO) concentration (redox conditions) affects biodegradation. Degradation can occur under aerobic (molecular oxygen available), denitrifying (no molecular oxygen available, nitrate available), or anaerobic (neither molecular oxygen nor nitrate available) conditions. Different removal efficiencies have been observed for anaerobic, anoxic, and aerobic conditions (Joss *et al.*, 2004). For example, the natural estrogens *E*1 and *E*2 undergo degradation in anoxic tanks of a CAS system, whereas the synthetic contraceptive *EE*2 decomposes only under aerobic conditions (Ternes *et al.*, 2004b). Joss *et al.* (2004) reported that: (i) degradation of *E*1 and *E*2 takes place in anaerobic, anoxic, and aerobic environments, but at significant different rates; (ii) oxidation of *E*2 is faster than of *E*1, and (iii) *EE*2 was only significantly removed under aerobic conditions and at slower rates than natural estrogens. Lee and Liu (2002) showed in batch experiments that $E2$ was more persistent under anaerobic conditions than under aerobic conditions, but was still biodegradable by the culture. In a study by Thompson *et al.* (2005), biodegradation was observed to be responsible for the majority of triclosan removal when the DO level was high, while sorption became the dominant removal mechanism when the DO level was low. Suarez *et al.* (2010) conducted experiments with a set of TrOC and classified them according to their aerobic and anoxic biodegradability as follows: (i) highly biodegradable under aerobic and anoxic conditions, e.g., ibuprofen, fluoxetine, natural estrogens, and musk fragrances (galaxolide, tonalide, and celestolide), (ii) highly biodegradable under aerobic conditions but persistent in the anoxic reactor, e.g., diclofenac, naproxen, $EE2$, roxithromycin, and erythromycin, (iii) moderately biodegradable under aerobic and anoxic conditions, e.g., citalopram, and (iv) resistant to biological transformation, for example, sulfamethoxazole, trimethoprim, carbamazepine, and diazepam.

While investigating the elimination of 19 TrOC including EDC and PPCP in a full-scale anaerobic/anoxic/aerobic-membrane bioreactor process, Xue *et al.* (2010) observed a fast adsorption of all TrOC on sludge in the first compartment of the treatment process (i.e. in the anaerobic tank), leading to rapid decline of the aqueous concentration of the target TrOC; however, aerobic environment was found crucial in promoting the overall degradation of the compounds. In another study, some compounds such as diclofenac, naproxen, EE2, and erythromycin were found to be highly biodegradable under aerobic conditions but persistent in the anoxic reactor. Ying *et al.* (2002) reported that aerobic conditions facilitate biotransformation of the metabolites of alkylphenol carboxylates than anaerobic conditions. The degradation of $E1$, $\beta E2$, and $EE2$ was studied under aerobic and anaerobic conditions in a simulated activated sludge system (Auriol *et al.*, 2006). It was concluded that under anaerobic conditions, the degradation rates for $E1$ and $EE2$ were considerably (10–20 times) lower than under aerobic conditions while the degradation of $\beta E2$ was not significantly different.

A few studies focusing on the effect of DO concentration have found beneficial effects of incorporating anoxic/anaerobic sequence(s) in the treatment train. This section highlights information of the TrOC showing better and/or significant biodegradability under anoxic/anaerobic regimes. In a study conducted by Zwiener and Frimmel (2003) diclofenac was degraded more in an anoxic biofilm reactor. Enhanced removal of alkylphenolic compounds observed in a very low-loaded CAS was attributed by Clara *et al.* (2005b) to the formation of anaerobic zones during aeration limited conditions. On the other hand, Park *et al.* (2009) observed as high as 66% carbamazepine removal by a constructed wetland under anoxic conditions. Maeng *et al.* (2011) attributed better removal of carbamazepine during river bank filtration than during artificial recharge to the anoxic environment prevailing. Xu *et al.* (2008) reported that pretreatment by an anerobic filter led to removal of 17 organochlorine pesticides (94%), 16 polycyclic aromatic hydrocarbons (59%) and 4-nonylphenol (4-NP) by an aerobic MBR from landfill leachate, while only 4-NP was removed without anaerobic pretreatment. In a study by Suarez *et al.* (2012) several TrOC such as tonalide, galaxolide, celestolide, fluoxetine, trimethoprim and erythromycin showed significant transformation rates during the anoxic process, while the remaining TrOC were only removed in the aerobic compartment (Suarez *et al.*, 2012). Chelliapan *et al.* (2006) achieved 90% removal of macrolide antibiotics in an up-flow anaerobic stage reactor. Arikan (2008) observed 75% removal of tetracyclines during anaerobic digestion of cattle manure. Drewes *et al.* (2001) studied the removal of adsorbable organo-iodine (AOI) in laboratory soil columns under different redox conditions. They found that aerobic conditions did not lead to significant biotransformation of AOI, while anoxic conditions yielded only about 20% removal, and anaerobic conditions achieved removals of about 57% (Drewes *et al.*, 2001). Musson *et al.* (2010) examined the anaerobic degradation of a few selected compounds and reported their degradability in the following order: acetylsalicylic acid > metoprolol tartrate > acetaminophen > ibuprofen.

Following a comprehensive literature review, Hai *et al.* (2011a) noted negligible difference in carbamazepine removal by sequential anoxic/anaerobic – aerobic MBR vs conventional aerobic MBR, although evidence of positive effect of anoxic regime on the removal of persistent TrOC including carbamazepine in different forms of treatment processes are available. Interestingly, Hai *et al.* (2011a) reported that near-anoxic conditions ($DO = 0.5$ mg/L) can be a favorable operating regime for the removal of the persistent TrOC carbamazepine by MBR treatment. They explained that the discrepancy of the observations in their study with the outcomes of the previous ones may stem from the difference between the "sequential anoxic-aerobic" and "continuous near-anoxic ($DO = 0.5$ mg/L)" operation modes. Unlike in the 'continuous near-anoxic' mode, in an anoxic–aerobic MBR, oxygen transfer from the aerated compartments to the anoxic zone due to the sludge recirculation may influence the removal efficiency (Cinar *et al.*, 2003).

13.6 CORRELATION OF TrOC REMOVAL WITH NITRIFICATION AND DENITRIFICATION

In addition to the studies on the potential effect of anoxic/aerobic operation regimes on TrOC removal by MBR, a number of recent studies have specifically highlighted the relationship between achievement of nitrification and/or denitrification and TrOC removal.

There is circumstantial evidence linking nitrifying sludge to a unique capability to biologically degrade EE2, i.e., EE2 biotransformation can be cometabolically mediated under operating conditions that allow for enrichment of nitrifiers (Vader *et al.*, 2000; Yi & Harper Jr, 2007b). Ammonium monooxygenase enzyme extracted from a nitrifying mixed culture removed EE2 in batch experiments (Yi *et al.*, 2006). Working at a constant nitrogen loading rate (0.11 g NH_4^+–N/L d) and temperature (25°C), a linear relationship ($r^2 >$ 0.90) between specific TrOC biodegradation rate and nitrification rate was found by Fernandez-Fontaina *et al.* (2012a). They reported a higher affinity of nitrifying activated sludge for antiphlogistics and musk fragrances (>75 mg biodegraded/g N removed) than for antibiotics and fluoxetine (<55 mg biodegraded/g N removed). In a full-scale municipal plant including a nitrification step, degradation rates of estrogens ranged between 79 and 95% and the extended biodegradation was mainly attributed to the nitrifying activity (Vader *et al.*, 2000). In that study, when the sludge was adapted to the nitrification process, the degradation of *EE*2 reached a rate leading to a half-life of approximately 28 h. On the other hand, the degradation of *EE*2 stopped when the sludge lost the nitrification capacity due to the low temperature (Vader *et al.*, 2000). Estrada-Arriaga and Mijaylova (2011) observed a linear relationship between nitrification and removal of *E*1, *E*2, and *EE*2 in an MBR. In a study by Clouzot *et al.* (2010), sorption was confirmed to be the predominant mechanism for *EE*2 removal (88%); however, biodegradation by a sludge acclimated to a media specific to autotrophic biomass resulted in improved *EE*2 removal (99%) by an MBR. Removal of diclofenac in a laboratory-scale activated sludge reactor was positively affected by the development of nitrifying biomass and increased from 0% up to 74% (Suarez *et al.*, 2010). The biodegradability of trimethoprim was studied during different sewage treatment steps using batch systems (Perez *et al.*, 2005). The main outcome of that study was that the activated sludge treatment comprising nitrification process was the only treatment capable of eliminating trimethoprim.

Ammonia oxidation by ammonia monooxygenase to hydroxylamine is only the first step in the biological nitrification process. Enzymes that catalyze the subsequent reactions of hydroxylamine to nitrite (hydroxylamine oxidoreductase) or nitrite to nitrate (nitrite oxidoreductase) could also significantly contribute to the observed oxidative biotransformation reactions (Helbling *et al.*, 2012). Experiments conducted with specific inhibition of the ammonia oxidizing bacteria (AOB) and axenic culture confirmed that the AOB are most likely responsible for the first step of *EE*2 degradation; however, the heterotrophic bacteria present in the nitrifier enrichment culture might play a role in the subsequent removal of the metabolites (De Gusseme *et al.*, 2009). Batt *et al.* (2006) found that the antibiotic trimethoprim was biotransformed by about 70% by a CAS inoculum, but only about 25% by a nitrifying sludge inoculum, indicating the importance of a mixed culture containing the heterotrophs.

Interestingly, De Gusseme *et al.* (2009) observed faster removal of EE2 under a $NH_4^+{-}N$ concentration of 1.5 mg/L, as compared to 52.5 mg/L, suggesting competition between ammonium and EE2 for (enzymatic) removal by the nitrifier enrichment culture. Ammonia monooxygenase may have a significantly stronger affinity for ammonia relative to the TrOC and may preferentially oxidize ammonia when the ammonia concentration remains high. Likewise, substrate limiting conditions may lead to a higher cometabolic activity of ammonia monooxygenase (Helbling *et al.*, 2012). On the other hand, Yi *et al.* (2006) reported that ammonia affected the relative amounts of EE2 removed via sorption and biodegradation in nitrifying sludge. They found that sorption was most important when the initial ammonia concentration was 48 mg/L or less, but at higher ammonia concentrations biodegradation became more influential.

Only a few specific studies regarding the correlation between denitrification and TrOC removal could be identified. Faster degradation kinetics was measured in the nitrifying reactor compared to the denitrifying system for the majority of PPCP studied by Suarez *et al.* (2010). Reungoat *et al.* (2010) attributed insignificant TrOC removal in the denitrification zone of a multi-barrier water reclamation plant to the preferential use of the added methanol as the primary electron donor by the bacteria. On the other hand, gemfibrozil was significantly removed in denitrified effluents, whereas a trickling filter-treated effluent exhibited a concentration of 1,235 ng/L (Drewes *et al.*, 2002). Suarez *et al.* (2010) reported better ibuprofen removal due to changes in the characteristics of the bacterial population (e.g. its enzymatic spectrum) which developed from a standard heterotrophic to a strict denitrifying biomass. In another study, most of *E*1 and *E*2 was reported to be eliminated already in the denitrifying step of a STP, whereas *EE*2 depletion was only observed during the aerobic process (Andersen *et al.*, 2003). EE2 was biodegraded by activated sludge under anaerobic conditions in the presence of nitrate (Zeng *et al.*, 2009); however in the absence of nitrate, the removal of *EE*2 was a result of sorption onto activated sludge. Paetkau (2011) compared *EE*2 and nitrogen removal in a conventional and a simultaneous nitrification–denitrification membrane bioreactor (SND-MBR). The SND-MBR removed greater amounts of TN with no observable impact on *EE*2 reduction and membrane operations.

13.7 EFFECT OF MBR-EFFLUENT DISINFECTION ON TrOC REMOVAL

Although of high quality, the effluent from MBRs is not suitable for direct non-potable reuse under, for example, California Title 22 guidelines, unless a disinfectant is added. A disinfectant will often be added in any case, even for discharge purposes, but few studies have examined the impact of disinfection on EDC removal by either CAS or MBR systems.

Chlorination is the most common mode of disinfection. The fate of well-known estrogenic chemicals during chlorination has been studied recently. For example, the addition of 1 mg/L of chlorine, as sodium hypochlorite, to aqueous solutions of bisphenol A can form several species of chlorinated phenol compounds after an exposure time of 60 min (Yamamoto & Yasuhara, 2002). Additionally, while chlorine can completely remove *EE*2 in Milli-Q water samples at doses of 1 mM over a 60 min contact time, the formation of chlorinated *EE*2 products capable of producing estrogenic responses can still occur (Moriyama *et al.*, 2004).

Compared to drinking water, biologically treated wastewater contains many more types of dissolved organic matter (DOM), namely, soluble microbial products and unknown estrogenic/antiestrogenic chemicals. Therefore, the chance of conversion of the parent EDC to chlorinated by-products that still retain endocrine disrupting capability following addition of chemical disinfectants to the MBR effluent is higher. Indeed, the results of recent studies confirm this (Deborde *et al.*, 2004; Hu *et al.*, 2003; Kuruto-Niwa *et al.*, 2005; Wu *et al.*, 2009). For instance, Wu *et al.* (2009) reported that chlorination can increase the antiestrogenic activity of wastewater. Notably, chlorination may not be similarly effective for all TrOC. In a study by Spring *et al.* (2007) disinfection of MBR effluent with chlorine, chloramines, and chlorine dioxide provided no significant additional removal of *E*1, *E*2, and *EE*2 but could remove 95% of added bisphenol A. Ultraviolet light at a dose appropriate for non-potable reuse did not affect removal of the target EDC (Spring *et al.*, 2007). The formation of potential chlorinated phenol compounds was, however, not investigated in that study.

Most reactions of ozone, another disinfecting agent, with TrOC take place in competition with fast reactions of ozone with the hydrophobic fraction of the DOM (Domenjoud *et al.*, 2011). Ozonation of aerobically treated gray water at an applied ozone dose of 15 mg/L reduced the concentrations of octocrylene, nonylphenol, triclosan, galaxolide, tonalide, and 4-methylbenzylidene camphor to below the limits of quantification, with removal efficiencies of at least 79% (Hernandez-Leal *et al.*, 2011). Ozonation of secondary treated effluent formed oxidation by-products of the TrOC but decreased the level of non-specific toxicity and other specific toxic modes of action, demonstrating that the mixture of oxidation by-products was less potent than the mixture of the parent compounds for the considered effects (Reungoat *et al.*, 2010). In a study investigating post treatment of secondary wastewater, Lee *et al.* (2012) recommended ozonation and biofiltration, wherein the presence of unknown TrOC oxidation products in ozone contactor effluent can be minimized by biofiltration. Based on the high degree of microbial inactivation and TrOC destruction, ozone/H_2O_2 and biological activated carbon (BAC) was proposed by Gerrity *et al.* (2011) as a viable alternative to the standard indirect potable reuse configuration (i.e., membrane filtration, reverse osmosis, UV/H_2O_2, and aquifer injection), particularly for inland applications where brine disposal is an issue. However, several issues, including regrowth of coliform bacteria in the BAC process, must be addressed prior to full-scale implementation (Gerrity *et al.*, 2011).

13.8 OVERALL FATE AND METABOLIC PATHWAYS

Compared with the number of studies conducted on TrOC removal by biological wastewater processes, much fewer studies have focused on their overall fate. In the majority of studies 'removal' has meant the disappearance of the parent molecules from the aqueous phase with no further investigation regarding the ultimate fate of the TrOC. For example, Auriol *et al.* (2006) noted the large ambiguity in the literature on the mechanism (biodegradation/biosorption) of EDC removal. Furthermore, the pathways and metabolites of TrOC degradation have been elucidated only in a selected number of studies. This section provides a brief overview of the related studies.

Joss *et al.* (2005a) presented a comprehensive method for setting up mass balances for TrOC over entire wastewater treatment plants, including an estimation of the accuracy of the quantified fate (i.e. removal by sorption and biological transformation). Abegglen *et al.* (2009) reported biological transformation as the main elimination process of 25 pharmaceuticals (antibiotics, antiphlogistics, lipid regulators, iodinated contrast media, and hormones) spiked into a household MBR. Atorvastatin and simvastatin drugs were removed by a CAS process with high efficiency (85–90%), with sorption accounting for less than 10% of overall removal (Ottmar *et al.*, 2012). In another study, *N*-nitrosamine removal efficiencies by a laboratory-scale MBR ranged from 24 to 94%, depending on their molecular properties, and biotransformation was mainly responsible for the observed removal (Wijekoon *et al.*, 2013). On the other hand, the fate of two differently labeled radioactive forms of EE2 during MBR treatment revealed that real mineralization was only 1%, while radioactivity mainly remained sorbed in the reactor, resulting in a removal of approximately 80% (Cirja *et al.*, 2007). The high biosorption of some TrOC is an indication of the risk of leaching of these compounds during sludge disposal. TrOC fate during sludge digestion has been covered in a number of studies (see, e.g., Citulski & Farahbakhsh, 2010; Dialynas & Diamadopoulos, 2012; Hamid & Eskicioglu, 2012; Radjenovic *et al.*, 2009b; Stasinakis, 2012; Suarez *et al.*, 2008; Xia *et al.*, 2005).

Some TrOC undergo microbially mediated reactions resulting in the formation of transformation products. There is a dearth of information regarding the reaction pathways that govern these transformations. Nevertheless, recent studies have reported the biodegradation products of selected TrOC in batch and reactor (CAS/MBR) biodegradation studies (see, e.g., Bouju *et al.*, 2011; Helbling *et al.*, 2010; Kosjek *et al.*, 2007; Leclercq *et al.*, 2010; Miao *et al.*, 2005; Prasse *et al.*, 2011; Quintana *et al.*, 2005; Schroder *et al.*, 2012). Hashim *et al.* (2010) proposed that due to the variable susceptibility of chiral pharmaceuticals to biodegradation, investigations of the patterns of enantiomeric fractionation of pharmaceuticals in wastewater and environmental samples may provide an increasingly solid understanding of the relationship between biotransformation processes. In a follow-up study, Hashim *et al.* (2011) provided evidence of enantioselective biodegradation of three pharmaceuticals in an MBR. (S)-Ibuprofen was shown to be preferentially degraded compared to (R)-ibuprofen with an average decrease in enantiomeric fraction from 0.52 to 0.39. In contrast, (R)-ketoprofen was preferentially degraded compared to (S)-ketoprofen with a relatively minor increase in enantiomeric fraction from 0.52 to 0.63. The use of a relatively pure enantiomeric solution of (S)-naproxen resulted in a significant change in enantiomeric fraction from 0.99 to 0.65 (Hashim *et al.*, 2011). In another study, slightly higher biodegradation rates for the S-enantiomeric forms of two beta-blockers were reported by Ribeiro *et al.* (2013).

13.9 POST TREATMENTS AND MBR-BASED HYBRID SYSTEMS

MBRs can offer better treatment efficiency of some TrOC than the CAS processes. However, as discussed in this chapter, MBR treatment alone is not effective for all TrOC that are known to occur in domestic and industrial wastewater. The low removal efficiency of biologically persistent and hydrophilic TrOC necessitates the integration of MBR with several other post-treatment processes to ensure adequate removal of TrOC. Chapter 12 provides a greater coverage of the MBR-based hybrid processes; this section highlights the case of TrOC removal.

13.9.1 Combination with physicochemical processes

The concept of combining MBRs with other advanced treatment processes has been already implemented in, for example, indirect potable water recycling schemes. In such a scheme domestic wastewater is first treated to a secondary effluent standard using the CAS process. The effluent is further purified using a combination of two or several advanced treatment processes including reverse osmosis (RO) or nanofiltration (NF), activated carbon adsorption, ozonation, and UV disinfection or advanced oxidation. Because MBR can produce effluent with much lower bulk organic content when compared to CAS, significant synergy can be realised when it is integrated with other advanced treatment processes. In addition, given the small physical footprint of the MBR process, it is also possible to deploy these integrated systems for decentralised water recycling applications. In this section, we review the current literature on the integration of advanced treatment processes including activated carbon adsorption, NF/RO, UV oxidation, and ozonation with MBR treatment for TrOC removal. Schematic diagrams of these MBR-based hybrid systems are shown in Figure 13.6.

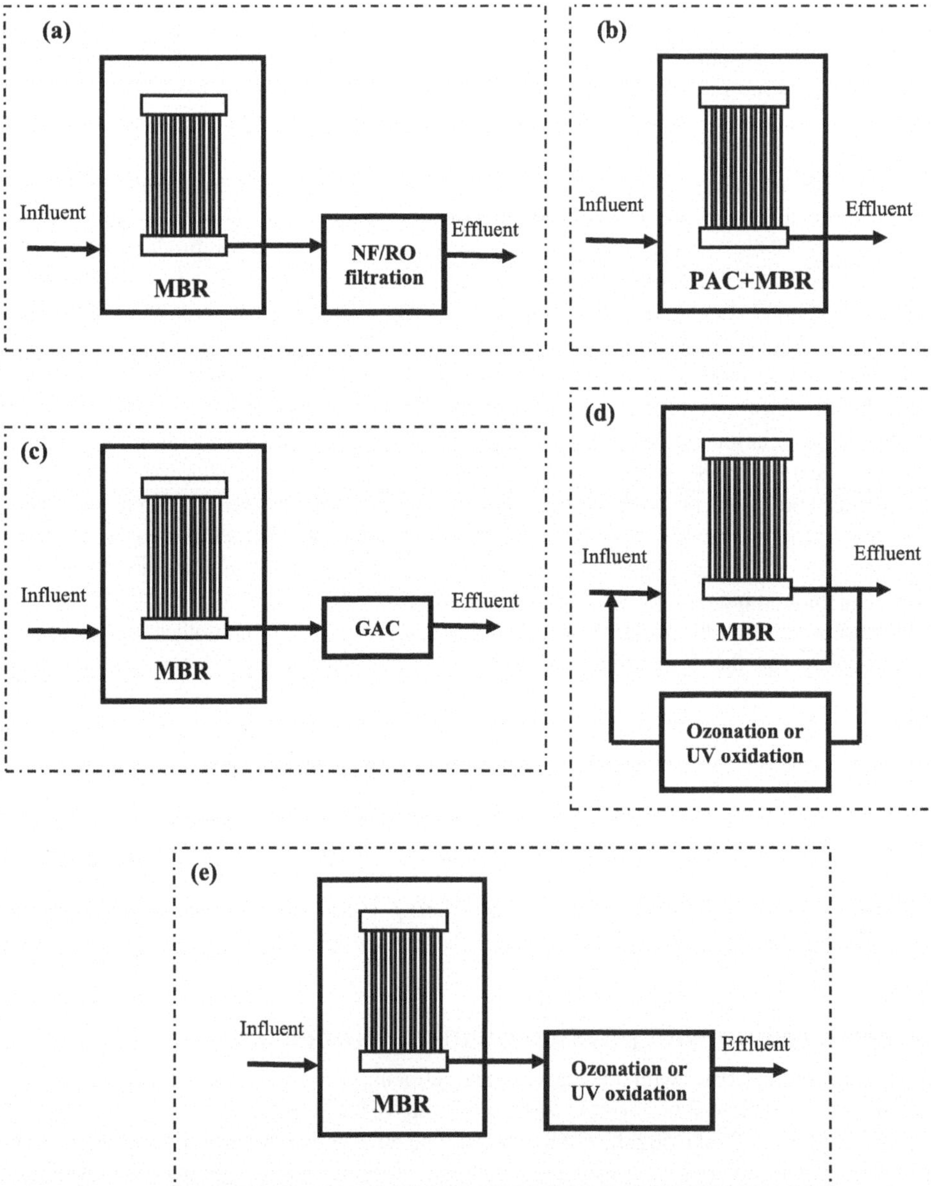

Figure 13.6 Schematic diagrams of MBR-based hybrid treatment systems: (a) MBR-NF/RO; (b) PAC-MBR; (c) MBR-GAC; (d) side stream MBR – oxidation; and (e) post-treatment MBR – oxidation.

The combination of MBR and NF/RO membrane filtration (Figure 13.6a) has been reported to achieve: (i) stable and highly treated water quality with respect to several basic water quality parameters such as TOC,

NH_3^-, and NH_4^+, and (ii) stable permeate flux of NF/RO membranes over extended operating periods (Comerton *et al.*, 2005; Dialynas & Diamadopoulos, 2009; Jacob *et al.*, 2010; Qin *et al.*, 2006; Tam *et al.*, 2007). The advantages of combining MBR and NF/RO treatment with respect to the removal of TrOC have also been demonstrated by a few recent studies (Alturki *et al.*, 2010b; Dolar *et al.*, 2012). Using a set of 40 TrOC, Alturki *et al.* (2010b) reported that MBR could effectively remove hydrophobic and biodegradable TrOC. Additionally, hydrophilic TrOC were effectively removed by the four NF/RO membranes selected in that study. The combination of MBR and the low-pressure RO membrane ESPA2 resulted in more than 95% removal or removal to below the analytical detection limit of all 40 TrOC investigated in that study. They also reported negligible membrane fouling of the NF270 and ESPA2 membranes. These results were confirmed in a subsequent study by Dolar *et al.* (2012), who conducted a pilot study to investigate the complementarities of MBR treatment and the RO process. All 20 TrOC investigated in their study were removed by 99% or more by the pilot MBR-RO system. Apart from post treatment of MBR effluent, recent developments integrating NF (Zaviska *et al.*, 2013) or forward osmosis (Alturki *et al.*, 2012) membranes with bioreactors are also worth-mentioning.

Activated carbon adsorption can be used in conjunction with an MBR system in two different configurations: (i) addition of powdered activated carbon (PAC) directly into the reactor of the MBR (Li *et al.*, 2011a; Serrano *et al.*, 2011; W. Yang *et al.*, 2010; Zhang *et al.*, 2008) as shown in Figure 13.6b and (ii) post-treatment of the MBR permeate using either a granular activated carbon (GAC) column (Navaratna *et al.*, 2012b; Nguyen *et al.*, 2012) or a continuously mixed reactor containing a slurry of PAC (Lipp *et al.*, 2012a) as shown in Figure 13.6c. Because of the complete retention of sludge by the membrane, the TrOC adsorbed onto PAC can be efficiently removed by the PAC–MBR system (Li *et al.*, 2011a, b). On the other hand, GAC can specifically target the residual TrOC in MBR permeate without any significant competition or interference from the bulk organics (Nguyen *et al.*, 2012; Xiao *et al.*, 2017). Indeed, PAC addition into MBR (PAC-MBR) or GAC post-treatment (MBR-GAC) has been reported to significantly complement MBR treatment to obtain high overall removal of less hydrophobic and biologically resistant TrOC. However, over an extended operating period, in both configurations, the breakthrough of TrOC is inevitable and periodic regeneration/replenishment of the activated carbon is necessary. According to Nguyen *et al.* (2013a), charged and hydrophilic TrOC are more likely to breakthrough first. Nguyen *et al.* (2013a) also conducted a preliminary comparison between the PAC-MBR and MBR-GAC configurations and reported that the former can perform better in terms of activated carbon usage.

Similar to activated carbon adsorption, due to low SS and background TOC concentration in MBR permeate, an advanced oxidation process (AOP) can be efficiently coupled with MBR treatment. Two reactor configurations can be perceived for such combinations: (i) side stream treatment with internal recirculation between MBR and AOP (Figure 13.6d), and (ii) post-treatment of the MBR permeate (Figure 13.6e). Although any oxidation treatment process can be integrated with MBR treatment in this fashion, most studies available to date have been concerned with ozonation and UV (or UV in combination with TiO_2 or H_2O_2). Such a combination can minimize some inherent limitations of AOPs such as UV radiation scattering by SS and competition with background TOC. In addition, the degradation of a specific TrOC by biological and oxidation processes can be triggered by different mechanisms. In other words, biological degradation and oxidation can complement each other for the removal of a specific TrOC. For example, Nguyen *et al.* (2013b) investigated the removal of TrOC by an MBR–UV system in a configuration similar to that shown in Figure 13.6e and reported that MBR treatment and direct UV oxidation by themselves achieved a low removal of carbamazepine, whilst their combination resulted in more than 96% removal. Similar observation was also reported by Laera *et al.* (2011) as they studied the integration of MBR and UV+TiO_2 treatment in a side stream recirculation

Table 13.3 Performance of the combined/hybrid MBR treatment process for TrOC removal.

Hybrid Process	TrOCs	Initial Concentration (ng/L)	Operating Conditions	Removal	References
Integrated MBR-UV/TiO_2	Carbamazepine	1×10^7	MLSS = 3.1 g/L SRT = 60 d HRT = 50 h UV wavelength = 360 nm TiO_2 dose = 0.5 g/L	95%	Laera *et al.* (2011)
Integrated MBR-PAC	22 TrOCs including pharmaceuticals, personal care products, pesticides steroid hormones and industrial chemicals	5,000	MLSS = 4.5–5 g/L SRT = infinite HRT = 24 h PAC dose = 0.1 g/L	90–99% (15 TrOCs) and 10–80% (7 TrOCs)	Nguyen *et al.* (2013a)
Integrated MBR-PAC	22 TrOCs including pharmaceuticals, personal care products, pesticides steroid hormones and industrial chemicals	5,000	MLSS = 4.5–5 g/L SRT = infinite HRT = 24 h PAC dose = 0.5 g/L	80–99% (19 TrOCs) and 10–70% (3 TrOCs)	Nguyen *et al.* (2013a)
Integrated MBR-PAC	Sulfamethoxazole (SMX) and Carbamazepine (CBZ)	7.5×10^5	MLSS = 11 g/L SRT = infinite HRT = 24 h PAC dose = 1 g/L	82% (SMX) and 92% (CBZ)	Li *et al.* (2011a)

Integrated MBR-PAC	Fluoxetine, Ibuprofen, Naproxen, Diclofenac, Carbamazepine, Trimethoprim, Roxithromycin, Erythromycin, and Diazepam	2×10^4	MLSS = 3.5 6 g/L SRT = infinite HRT = 24 h PAC dose = 1 g/L	85–99%	Serrano *et al.* (2011)
Integrated MBR-PAC	Carbamazepine (CBZ) and Diclofenac (DCF)	390–1,800	MLSS = 3–4 g/L SRT = 20 d HRT = 24 h PAC dose = 5–10 mg/L	80–99% (CBZ) and 90–99% (DCF)	Lipp *et al.* (2012b)
Integrated AnMBR-PAC	Trimethoprim (TMP), Sulfamethoxazole (SMX), Carbamazepine (CBZ), Diclofenac (DCF) and Triclosan (TCS)	2,000	MLSS = 8–10 g/L SRT = 213 d HRT = 6 h PAC dose = 1 g/L	99% (TMP), 95% (SMX), 92% (CBZ), 83% (DCF) and 93%(TCS)	Xiao *et al.* (2017)
MBR followed by GAC	22 TrOCs including pharmaceuticals, personal care products, pesticides steroid hormones and industrial chemicals	5,000	SRT of MBR = infinite HRT of MBR = 24 h GAC contact time = 7 min	Above 98% for all TrOCs	Nguyen *et al.* (2012)

configuration (Figure 13.6d). Pollice *et al.* (2012) compared the side stream recirculation and the post-treatment configuration for the removal of the antibacterial product nalidixic acid, its degradation products and other organics from a wastewater of a pharmaceutical production facility. They reported that although the post-treatment configuration outperformed the side stream recirculation configuration with respect to basic water quality parameters, the latter resulted in better removal of most TrOC.

AOPs cannot completely mineralize TrOCs, and the metabolites formed following AOPs may be more toxic than the parent compounds. For instance, Alharbi *et al.* (2017) studied the formation of metabolites following the UV-photolysis and UV+H_2O_2 treatment of four pharmaceuticals, namely diclofenac, sulfamethoxazole, carbamazepine and trimethoprim. Based on the toxicity assessment of the treated effluent, they reported that the metabolites of diclofenac, sulfamethoxazole and carbamazepine are more toxic than the parent compound. However, it is worth-mentioning that metabolites formed during the AOPs have been reported to be effectively mineralized by the activated sludge (Hu *et al.*, 2008; Kosjek *et al.*, 2009; Prado *et al.*, 2017). For example, Hübner *et al.* (2014) observed that the metabolites of carbamazepine formed following ozonation process were mineralized by bacteria in biological sand column operated at an HRT of 5–6 d. When an integrated gamma radiation and activated sludge process was investigated by Wang and Wang (2017), up to 79% carbamazepine mineralization by the activated sludge was achieved. Further studies are required to confirm degradation and mineralization pathways of a broad spectrum of TrOCs during the integrated MBR-AOP processes.

In summary, the combination of MBR and an AOP to form an MBR-based hybrid system can significantly enhance the removal of TrOC. However, this approach would lead to an increase in both the investment and operational costs. It is also noteworthy that most of the data available to date are from laboratory-scale studies. Further research is still needed for full-scale validation, scaling up, system optimization, and to improve our overall understanding of the complementarities between MBR and AOPs. A succinct overview of TrOC removal by combined/hybrid MBR treatment process is presented in Table 13.3.

13.9.2 Bioaugmented and enzymatic MBR for TrOC removal

Well-acclimated microbial strains capable of either using TrOC as the sole source of carbon or degrading them by cometabolism can be cultivated and later introduced into large-scale biological reactors, hence implementing bioaugmentation for TrOC removal (Basile *et al.*, 2011). Despite the great potential of such an approach, only a few such studies could be identified in the literature.

Liu and Huang (2008) reported efficient (>90%) and stable atrazine removal for around two months in an MBR bioaugmented with genetically engineered *Escherichia coli* containing an atrazine chlorohydrolase gene. The performance of the proposed system compared favorably with the report of low and/or unstable removal of atrazine by the wild-type atrazine-degrading strain or mixed microbial consortium in the conventional biological treatment processes (Ghosh & Philip, 2004; Kontchou & Gschwind, 1999; Protzman *et al.*, 1999).

Ghyoot *et al.* (2000) examined the behavior of a 3-chlorobenzoate (3CBA) degrading *Pseudomonas putida* BN210 carrying the self-transferable clc-element containing the *clc*-genes for 3CBA metabolism in a CAS and an MBR system. Although molecular techniques indicated that strain BN210 disappeared or survived only in low cell numbers in both the reactors, the MBR showed higher resistance towards shock loading of 3CBA in terms of improved COD removal. Molecular characterization of the isolates strongly suggested that in the MBR the *clc*-element had been *in situ* disseminated from the initial inoculum to contaminant bacteria, which had invaded the reactor and which finally became the dominant strains to continue 3CBA degradation. Ghyoot *et al.* (2000) explained that possibly high biomass concentrations in the MBR stimulated transfer of the *clc*-element from strain BN210 to autochthonous bacteria, and that autochthonous 3CBA degrading bacteria might benefit from specific conditions in the MBR.

Recent studies confirm significant removal of various TrOC by pure culture of white-rot fungi (WRF) under sterile batch test conditions (Yang *et al.*, 2013b). WRF species secrete ligninolytic enzymes, namely, laccase, lignin peroxidase (LiP), manganese peroxidase (MnP) and versatile peroxidase (VP). TrOC removal by whole-cell fungal treatment systems may vary depending on the type of WRF species or strains. Among the tested WRF species, *Trametes versicolor* has been mostly studied for the removal of different TrOC groups such as pharmaceuticals, personal care products, pesticide and steroid hormones, possibly because it can secrete three ligninolytic enzymes i.e., laccase, LiP and MnP (Kang *et al.*, 2008; Marco-Urrea *et al.*, 2006, 2010; Nguyen *et al.*, 2014c; Tran *et al.*, 2010). Other WRF species such as *Trichoderma viride* (laccase, tyrosinase), *Bjerkandera adusta* (laccase, LiP and MnP), *Phanerochaete chrysosporium* (LiP and MnP), *Phanerochaete sordida* (laccase and LiP) *Pleurotus ostreatus* (laccase, MnP) have also been studied, achieving significant removal (>70%) of TrOCs including some pharmaceuticals, e.g. naproxen, diclofenac and ketoprofen, and endocrine disrupting compounds, e.g. steroid hormones, bisphenol A and triclosan (Golan-Rozen *et al.*, 2011, 2015; Hata *et al.*, 2010; Rodarte-Morales *et al.*, 2011).

Yang *et al.* (2013a) tested an MBR containing a mixed microbial community including the white-rot fungus *T. versicolor* for the treatment of a synthetic wastewater containing two TrOC, namely, bisphenol A and diclofenac. Notably, while diclofenac has been widely reported to be highly recalcitrant to bacterial cultures, bisphenol A has been documented to be comparatively easily removed by bacterial cultures. The fungus demonstrated excellent degradation of the parent compounds in pure culture batch tests; however, a reduced removal performance of diclofenac by the MBR was observed. Yang *et al.* (2013a) attributed this to perturbations due to continuous washout of the extracellular enzymes, and bacterial contamination. Relatively stable removal of bisphenol A (80–90%) and diclofenac (~55%) was achieved by applying an HRT of 2 d, at the bisphenol A and diclofenac loadings of 475 ± 25 and $345 \pm 112\ \mu g/L$ d, respectively. In another study by Nguyen *et al.* (2013c), performance of an MBR containing a mixed microbial community including the white-rot fungus *T. versicolor* was assessed for the removal of 30 TrOCs from synthetic wastewater. As compared to conventional MBR, the fungus-augmented MBR achieved 30–55% better removal for six TrOCs, namely, fenoprop, clofibric acid, pentachlorophenol, naproxen, diclofenac, and ketoprofen. It is important to note that bacterial contamination significantly deteriorates the performance of fungal bioreactor during long-term operations (Badia-Fabregat *et al.*, 2017; Cruz-Morató *et al.*, 2013, 2014; Ferrando-Climent *et al.*, 2015; Jelic *et al.*, 2012; Zhang & Geißen, 2012). Bacterial contamination can affect fungal MBR by: (i) reducing the enzyme secretion capacity of WRF due to disruption in its growth, and (ii) consuming the enzymes produced by WRF (Espinosa-Ortiz *et al.*, 2016; Libra *et al.*, 2003; Yang *et al.*, 2013a). Because bacteria are fast-growing prokaryotes, they can easily outgrow WRF and prevail in fungal MBR (Hai *et al.*, 2009; Libra *et al.*, 2003). For the mitigation of bacterial contamination, a number of strategies have been assessed as reviewed by Asif *et al.* (2017c). These strategies include immobilized fungal growth, biomass replacement, influent pretreatment as well as the use of micro-screen to retain fungus, while allowing bacterial washout. However, these strategies could only extend the operation of fungal MBR for a few weeks.

An alternative to whole-cell WRF treatment is the use of crude or purified enzymes. This separates fungal growth and pollutant degradation steps. Enzyme washout (i.e., loss of enzyme along with treated effluent) is a critical limitation of enzyme applications in continuous-flow bioreactors. These limitations can be addressed by immobilizing enzyme onto a support material (Ansari & Husain, 2012; Nguyen *et al.*,

2016a) or by integrating an enzymatic bioreactor with a membrane of appropriate pore size (Lloret *et al.*, 2013; Nguyen *et al.*, 2014a, 2015, 2016a). The enzymatic membrane bioreactors (EMBR) offer several benefits over immobilized enzyme including ease of operation and maintenance as well as reduced mass transfer limitations.

TrOC removal by EMBR has been observed to depend on factors such as oxidation–reduction potential of the enzyme and the presence of EDGs or EWGs in the chemical structure of TrOCs (Asif *et al.*, 2017a; Lloret *et al.*, 2012; Yang *et al.*, 2013b). Lloret *et al.* (2012) investigated the removal of two estrogens (i.e., estrone and 17*β*-estradiol) in an EMBR and achieved 64 and 100% removal of estrone and 17*β*-estradiol, respectively. In addition, EMBR effectively reduced the estrogenic activity of the treated effluent by 97%. Similarly, Nguyen *et al.* (2015) investigated the removal of 30 TrOC in EMBR and observed that the removal of phenolic TrOCs with EDGs ranged between 70 and 99%. On the other hand, removal of phenolic TrOCs with EWGs ranged from 25 to 50%. Non-phenolic TrOCs were poorly (5–20%) removed by EMBR except for a few non-phenolic TrOCs including benzophenone (75%), amitriptyline (95%), octocrylene (99%) and diclofenac (45%) (Nguyen *et al.*, 2015). Performance of EMBR for a broad spectrum of TrOCs is presented in Figure 13.7.

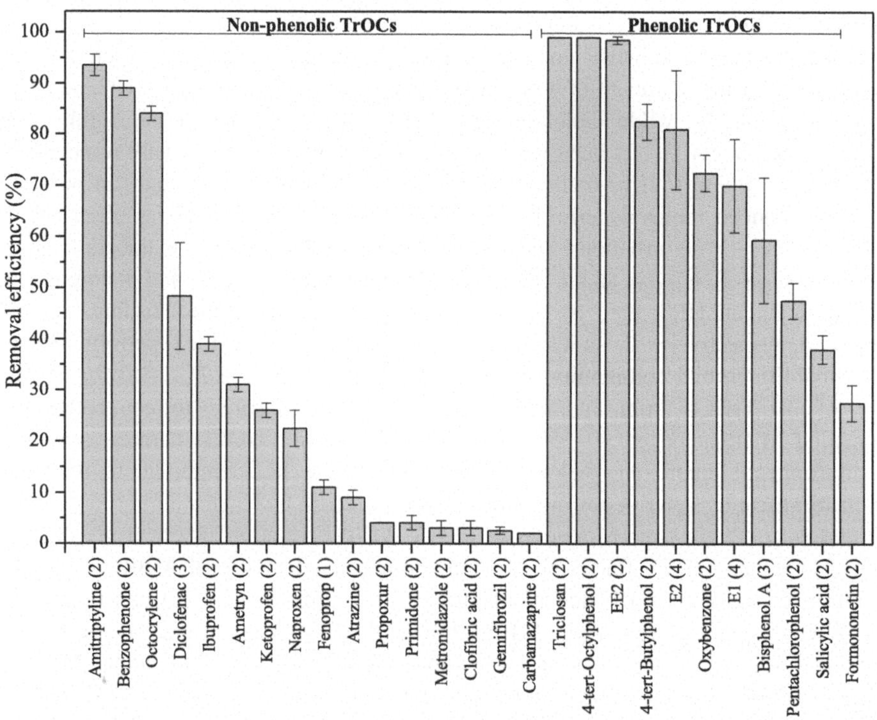

Figure 13.7 Performance of enzymatic membrane bioreactors (EMBR) for the removal of TrOCs. Numbers within parentheses indicate number of data points. *E*1: estrone; *E*2: 17*β* – estradiol; *EE*2: 17*α* – ethinylestradiol. Data source: Lloret *et al.* (2012, 2013; Nguyen *et al.* (2014a, 2015, 2016b).

13.9.3 High retention (HR)-MBR for TrOC removal

As discussed in Section 13.9.1, NF or RO membrane separation process can be combined with MBR for effective removal of TrOCs that are not well removed by MBR alone (Alturki *et al.*, 2010c; Cartagena *et al.*, 2013). In a combined MBR-NF/RO process, MBR permeate is treated by NF/RO membrane, and the organic impurities (e.g., TrOCs) can be recycled back to the MBR for further degradation. Instead of an additional step, high retention membrane processes can be directly integrated with the activated sludge-based bioreactor (Asif *et al.*, 2018a; Luo *et al.*, 2014). High retention membranes decouple organic retention time from HRT due to effective retention of organic impurities within the bioreactor of HR-MBR systems. Thus it was envisaged that the prolonged contact time between the activated sludge and TrOCs may improve their biodegradation (Luo *et al.*, 2014; Yeo *et al.*, 2015). To date, three configurations of HR-MBR, namely, membrane distillation bioreactor (MDBR; Figure 13.8a), osmotic membrane bioreactor (OMBR; Figure 13.8b) and NF/RO-MBR (Figure 13.8c) have been studied. The working principles of HR-MBR are discussed in Chapter 12 of this book.

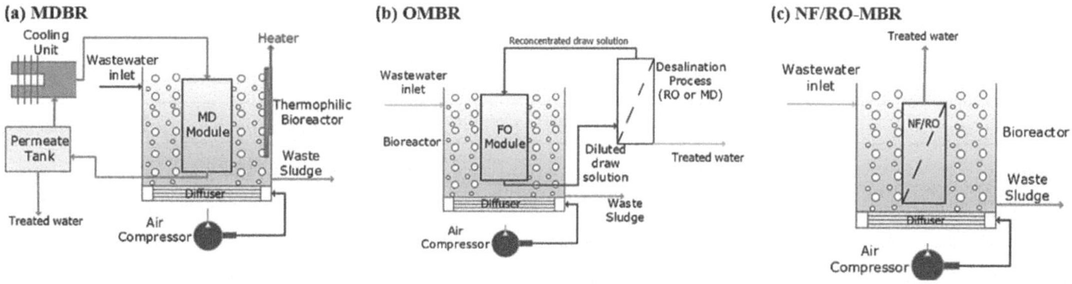

Figure 13.8 Schematics of HR-MBR configurations. (a) membrane distillation bioreactor (MDBR); (b) Osmotic/forward osmosis membrane bioreactor (OMBR); and (c) NF/RO-MBR. Reproduced with permission from Asif *et al.* (2018a).

13.9.3.1 TrOC retention mechanisms of high retention membranes

In addition to the degradation by the activated sludge, membrane retention appears to be a dominant mechanism for removal of TrOCs that are not easily biodegradable (Asif *et al.*, 2018a). The membrane retention mechanisms are briefly discussed here to understand the factors governing the removal of TrOC by MD, FO and NF membranes. Following a thorough literature survey, Asif *et al.* (2018a) suggested that the performance of high retention membranes for TrOC removal could be influenced by the membrane properties (e.g. surface charge and hydrophobicity); wastewater characteristics (e.g. pH, organic and inorganic impurities); and operating parameters (e.g. bioreactor volume and operation duration). Compared to NF and FO membrane separation processes, TrOC retention by MD process is simpler. This is because TrOC retention by MD membranes mainly depends on the volatility and hydrophobicity of pollutants (Luo *et al.*, 2014; Wijekoon *et al.*, 2014a), while their retention by NF and FO membranes has been observed to be governed by a number of factors (Table 13.4) such as TrOC properties, operating conditions and membrane properties (Bellona *et al.*, 2004; Hau *et al.*, 2014; Hu *et al.*, 2007; Nghiem & Coleman, 2008; Nghiem *et al.*, 2005). In general, removal of TrOCs can be classified into two categories based on the $pK_H/\log D$ ratio in a stand-alone MD process: (i) $pK_H/\log D$ ratio < 2.5 = ineffective retention (50–70%); and (ii) $pK_H/\log D$ ratio >2.5 effective retention (90–99%) (Wijekoon *et al.*, 2014a).

Despite the difference in the operating principles, remarkable similarities have been reported in the mechanisms of TrOC retention by NF and FO membranes as discussed below. NF and FO membranes can retain TrOCs by following mechanisms: (i) size exclusion; (ii) sorption on membrane surface; and/or (iii) electrostatic interactions (Agenson *et al.*, 2003; Coday *et al.*, 2014; Luo *et al.*, 2014; Nghiem *et al.*, 2006). Size exclusion is the major retention mechanism for non-ionic hydrophilic (log $D < 3.3$) TrOCs, which depends on the molecular weight cut-off (MWCO) of the NF and FO membrane. In a study by Comerton *et al.* (2008), retention of a non-ionic antibiotic (i.e. carbamazepine, log $D = 1.89$) by a tight NF membrane (MWCO < 00 g/mol) was 97% from a filtered lake water spiked with a mixture of 22 TrOCs. Notably, carbamazepine retention dropped to 50% in that study when the filtered lake water was treated with a loose NF membrane with a MWCO of greater than 300 g/mol (Comerton *et al.*, 2008). Despite the difference in membrane material, retention of carbamazepine by cellulose triacetate (Xie *et al.*, 2012b) and thin film composite polyamide (Jin *et al.*, 2012) FO membranes ranged between 80–90% and 90–95%, respectively. For other hydrophilic TrOCs such as clofibric acid, metronidazole and DEET, both NF and FO membranes reportedly achieved 80–99% retention (Alturki *et al.*, 2010a; Cath *et al.*, 2010; Linares *et al.*, 2011; Valladares *et al.*, 2011; Verliefde *et al.*, 2009).

Table 13.4 Factors influencing the retention of TrOCs by high retention MD, FO and NF membranes. Reproduced with permission from Asif *et al.* (2018a).

Factors	MD Membrane	FO Membrane	NF Membrane
Fouling	*	*	*
Diffusion of solute	–	*	*
Hydrophobicity	*	*	*
Membrane MWCO	–	*	*
Charge on TrOCs	–	*	*
Membrane surface charge	–	*	*
Polarity	–	*	*
Molecular width	–	*	*
Volatility of TrOCs	*	*	*
Temperature and pH	*	*	*

"–": no effect according to available reports.

Adsorption of hydrophobic TrOCs (e.g., bisphenol A and steroid hormones) on the surface of NF and FO membranes has been observed, which results in their effective retention (above 80%) (Alturki *et al.*, 2013; Nghiem & Coleman, 2008; Verliefde *et al.*, 2009; Xie *et al.*, 2012a). However, during long-term operation, hydrophobic TrOCs adsorbed on membrane surface can diffuse into permeate, thereby reducing their extent of retention by NF and FO membranes (Nghiem & Coleman, 2008; Verliefde *et al.*, 2009). Notably, the diffusion of hydrophobic TrOC into membrane permeate continues until an equilibrium between the concentration of TrOC on/near membrane surface and membrane permeate is established. The role of adsorption mechanism for TrOC retention diminishes after the establishment of equilibrium, and TrOC retention by FO and NF membrane is governed mainly by size exclusion and charge repulsion mechanisms (Coday *et al.*, 2014; Taheran *et al.*, 2016; Yoon *et al.*, 2006).

Due to the protonation of the functional groups, surface of NF and FO membranes exhibit negative charge at pH ≤ 7 (Coday *et al.*, 2014; Comerton *et al.*, 2008). Therefore, the interaction of charged TrOCs (e.g. naproxen and diclofenac) with membrane surface can affect their retention by NF/FO membrane. The attraction between the positively charged TrOCs such as steroid hormones and membrane surfaces may lead to their diffusion into the permeate, thereby achieving ineffective retention by NF and FO membranes. On the other hand, the repulsion between the negatively charged pharmaceuticals such as naproxen, diclofenac and ketoprofen and membrane surface resulted in their effective retention by NF and FO membranes (Kimura *et al.*, 2004; Radjenović *et al.*, 2008; Verliefde *et al.*, 2007). Based on the effect of TrOC properties and membrane type as well as operating parameters, a qualitative framework for predicting the TrOC retention by NF and FO membranes is presented in Figure 13.9.

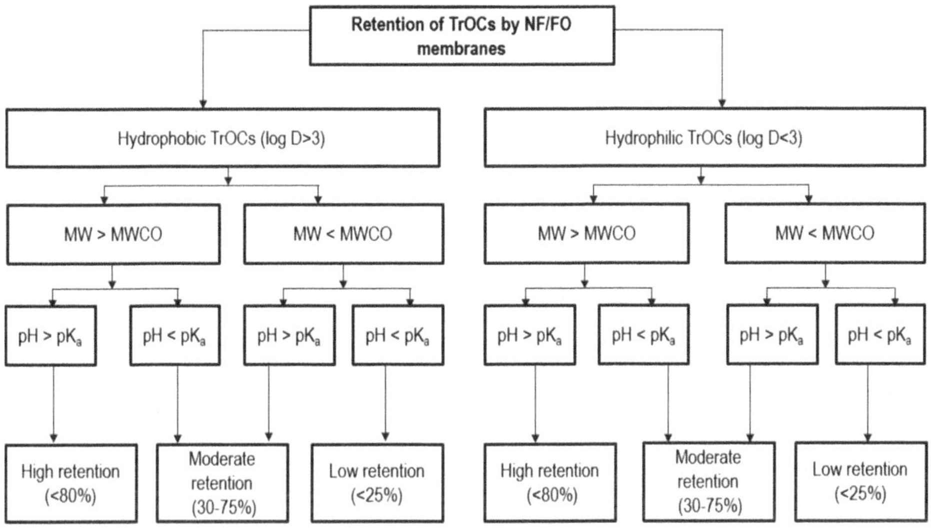

Figure 13.9 Qualitative framework for the prediction of TrOC retention by NF and FO membranes. The case of an acidic compound is presented here. A basic compound will become positively charged at pH $< pK_b$. The transformation of neutral TrOCs to negatively charged at pH $> pK_a$ or positively charged at pH $< pK_b$ can affect their retention by NF and the FO membranes. Reproduced with permission from Asif *et al.* (2018a).

13.9.3.2 Aqueous phase removal of TrOCs

Like classical MBR, the performance of HR-MBR too depends on the HR-MBR configuration and TrOC properties (Fernandez-Fontaina *et al.*, 2012b; Luo *et al.*, 2017; Phan *et al.*, 2016; Wijekoon *et al.*, 2014b). Removal of TrOCs based on the concentration in membrane-permeate of HR-MBR is presented in Table 13.5, indicating that removal ranged between 90 and 99% can be achieved for a range of TrOCs.

Based on a comprehensive literature survey, the aqueous phase removal of TrOCs by CAS, MBR and HR-MBR was compared by Asif *et al.* (2018a). They concluded that HR-MBR is effective for TrOC removal, and it can achieve median removal of almost 90%. On the other hand, median TrOC removal is about 60 and 65% by CAS and conventional MBR, respectively (Figure 13.10).

By reducing TrOC concentration via biodegradation, HR-MBR can achieve better TrOC removal as compared to standalone high retention membrane processes. For example, Wijekoon *et al.* (2014a) studied the rejection of a mixture of 30 TrOCs by a standalone direct contact MD process, and achieved partial retention (50–70%) of volatile TrOCs ($pK_H < 9$) such as 4-tert-octylphenol ($pK_H = 5.06$), benzophenone ($pK_H = 5.88$) and amitriptyline ($pK_H = 8.18$). By contrast, when the MD process was combined with biodegradation, effective removal (95–99%) was achieved for all the selected TrOCs including those partially retained by the standalone MD process (Wijekoon *et al.*, 2014a, b). Similarly, a standalone FO process has been reported to achieve 40–80, 70–75, 80 and 70–90% retention of bisphenol A, oxybenzone, estrone and 17α-ethynylestradiol, respectively (Coday *et al.*, 2014), while

Table 13.5 Performance of different HR-MBR configurations for the removal of TrOCs. Reproduced with permission from Asif *et al.* (2018a).

TrOCs	Chemical Formula[a]	Molecular Weight[a] g/mole	Dissociation Coefficient (pK_a)[a]	Henry Constant (H)[b]	pK_H^b	Log *D* at pH = 7[a]	Removal Efficiency (%)		
							OMBR	MDBR	NF-MBR
Primidone	$C_{12}H_{14}N_2O$	218.25	12.26 ± 0.40	1.164×10^{-14}	13.93	0.83	>99	>99	–
Ketoprofen	$C_{16}H_{14}O_3$	254.28	4.23 ± 0.10	2.005×10^{-14}	13.70	0.19	>99	>99	94
Naproxen	$C_{14}H_{14}O_3$	230.26	4.84 ± 0.30	2.096×10^{-13}	12.68	0.73	>99	>99	98
Gemfibrozil	$C_{15}H_{22}O_3$	250.33	4.75	7.677×10^{-13}	12.11	2.07	>95	>99	99
Metronidazole	$C_6H_9N_3O_3$	171.15	14.44 ± 0.10	2.073×10^{-12}	11.68	−0.14	>95	>99	99
Diclofenac	$C_{14}H_{11}Cl_2NO_2$	296.15	4.18 ± 0.10	3.098×10^{-12}	11.51	1.77	>95	90	45–95
Fenoprop	$C_9H_7Cl_3O_3$	269.51	2.93	3.284×10^{-12}	11.48	−0.13	83–99	95	–
Ibuprofen	$C_{13}H_{18}O_2$	206.28	4.41 ± 0.10	4.066×10^{-11}	10.39	0.94	>99	>99	100
Ametryn	$C_9H_{17}N_5S$	27.33	3.71 ± 0.41	4.418×10^{-10}	9.35	2.97	>99	>99	–
Clofibric acid	$C_{10}H_{11}ClO_3$	214.65	3.18 ± 0.10	2.909×10^{-10}	9.54	−1.06	>99	>99	75
Carbamazepine	$C_{15}H_{12}N_2O$	236.27	13.94 ± 0.20	8.168×10^{-10}	9.09	1.89	50–99	95	18–75
Octocrylene	$C_{24}H_{27}N$	361.48	–	3.382×10^{-9}	8.47	6.89	80–90	90	95
Amitriptyline	$C_{20}H_{23}N$	277.40	9.18 ± 0.28	6.596×10^{-9}	8.18	2.28	>99	>99	83–100
Atrazine	$C_8H_{14}ClN_5$	215.68	2.27 ± 0.10	5.223×10^{-8}	7.28	2.64	75–90	>99	16–80
Propoxur	$C_{11}H_{15}NO_3$	209.24	1.49 ± 0.70	5.265×10^{-7}	6.28	1.54	>99	>99	–
Benzophenone	$C_{13}H_{10}O$	182.22	–	1.316×10^{-6}	5.88	3.21	>99	95	>99
N, N-Diethyl-meta-toluamide (DEET)	$C_{12}H_{17}NO$	191.3	–	1.410×10^{-6}	5.85	2.42	40–90	–	60
Estriol	$C_{18}H_{24}O_3$	298.33	10.25 ± 0.70	1.644×10^{-11}	10.78	1.89	>99	>99	–
17α – Ethynylestradiol	$C_{20}H_{24}O_2$	269.40	10.24 ± 0.60	3.399×10^{-10}	9.47	4.11	>99	>99	–
Oxybenzone	$C_{14}H_{12}O_3$	228.24	7.56 ± 0.35	5.851×10^{-10}	9.23	3.89	>99	>99	–
Estrone	$C_{18}H_{22}O_2$	270.37	10.25 ± 0.40	9.286×10^{-10}	9.03	3.62	>99	>99	95
17β – Estradiol	$C_{18}H_{24}O_2$	272.38	10.27	1.173×10^{-9}	8.93	4.15	>99	>99	–
17β – Estradiol-17-acetate	$C_{20}H_{26}O_3$	314.42	10.26 ± 0.60	2.151×10^{-9}	8.67	5.11	>99	>99	–
Bisphenol A	$C_{15}H_{16}O_2$	228.29	10.29 ± 0.10	2.197×10^{-9}	8.66	3.64	>99	>99	95–97
Salicylic acid	$C_7H_6O_3$	138.12	3.01 ± 0.10	6.653×10^{-9}	8.18	−1.13	>99	95	70
Triclosan	$C_{12}H_7Cl_3O_2$	289.54	7.80 ± 0.35	6.537×10^{-7}	6.18	5.28	>99	>99	82
4-tert-Butylphenol	$C_{10}H_{14}O$	150.22	10.13 ± 0.13	7.136×10^{-6}	5.15	3.40	>99	>99	88
4-tert-Octylphenol	$C_{14}H_{22}O$	206.32	10.15 ± 0.15	8.670×10^{-6}	5.06	5.18	>99	>99	–

[a]Data extracted from SciFinder Scholar.
[b]Henry's law constant (H) = Vapor pressure × molecular weight/water solubility; and $pK_H = -\log_{10} H$.
"–": not available.

their retention by OMBR has been observed to be consistently above 99% (Luo *et al.*, 2015, 2017). Performance of MDBR and OMBR is better than the standalone MD and FO processes, because volatile and hydrophobic TrOCs such as bisphenol A and oxybenzone are readily degraded by activated sludge (Holloway *et al.*, 2014; Luo *et al.*, 2015; Wijekoon *et al.*, 2014b).

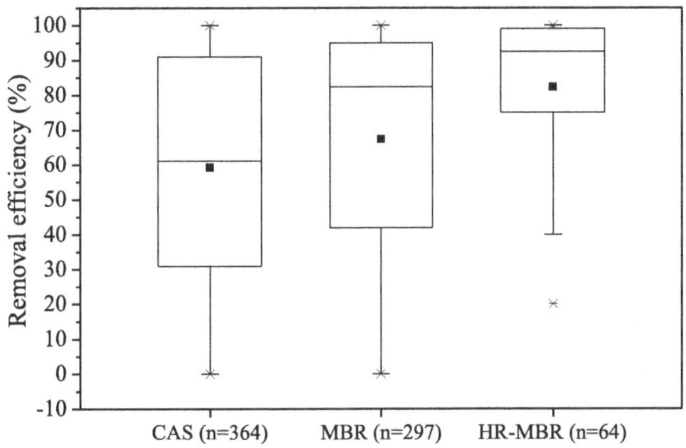

Figure 13.10 Box-and-whisker plot for the variations in the aqueous phase removal of TrOCs by CAS, MBR and HR-MBR. Reproduced with permission from Asif *et al.* (2018a).

13.9.3.3 Fate of TrOCs in HR-MBR

One of the main rationale for the development of HR-MBRs is to improve degradation of TrOCs that are not readily biodegradable. Based on the comparison of removal data from different studies, degradation appears to be more stable in HR-MBR as compared to CAS and MBR (Figure 13.11). TrOC degradation by the activated sludge for some TrOCs improved marginally in HR-MBR. Similar to the performance of conventional MBR, effective degradation (70–99%) has been reported for TrOCs that: (i) contain EDGs in their molecule; and/or (ii) are hydrophobic (Luo *et al.*, 2017; Phan *et al.*, 2016; Wijekoon *et al.*, 2014b). On the other hand, poor degradation of TrOCs containing strong EWGs such as carbamazepine (amide EWG) and atrazine (chloride EWG) (10–30%) was observed (Luo *et al.*, 2017; Phan *et al.*, 2016; Wijekoon *et al.*, 2014b). Irrespective of the level of biodegradation, due to the high retention membrane, the TrOCs will not appear in the permeate.

According to a literature survey carried out by Asif *et al.* (2018a), sorption of TrOCs on to activated sludge varies for hydrophilic and hydrophobic TrOCs, and may range between 1–10% and 2–30%, respectively. This implies that membrane retention and biodegradation are the main removal mechanism, and contribute significantly to the removal of TrOCs by HR-MBR (Figure 13.12). Notably, accumulation of resistant TrOCs in the bioreactor of HR-MBR can make the processes for sludge treatment and disposal complex. Therefore, there is a need to improve the biodegradation of TrOC in HR-MBR. This could be achieved by introducing white-rot fungi and their ligninolytic enzymes (Hai *et al.*, 2006) as discussed in Section 13.9.2. In a recent report, integration of an NF or MD membrane to an enzymatic bioreactor achieved better TrOC degradation than a conventional ultrafiltration-EMBR (Asif *et al.*, 2017b, 2018b, c; Nguyen *et al.*, 2014b, 2015), indicating the benefit of combining white-rot fungal enzyme system with high retention membranes.

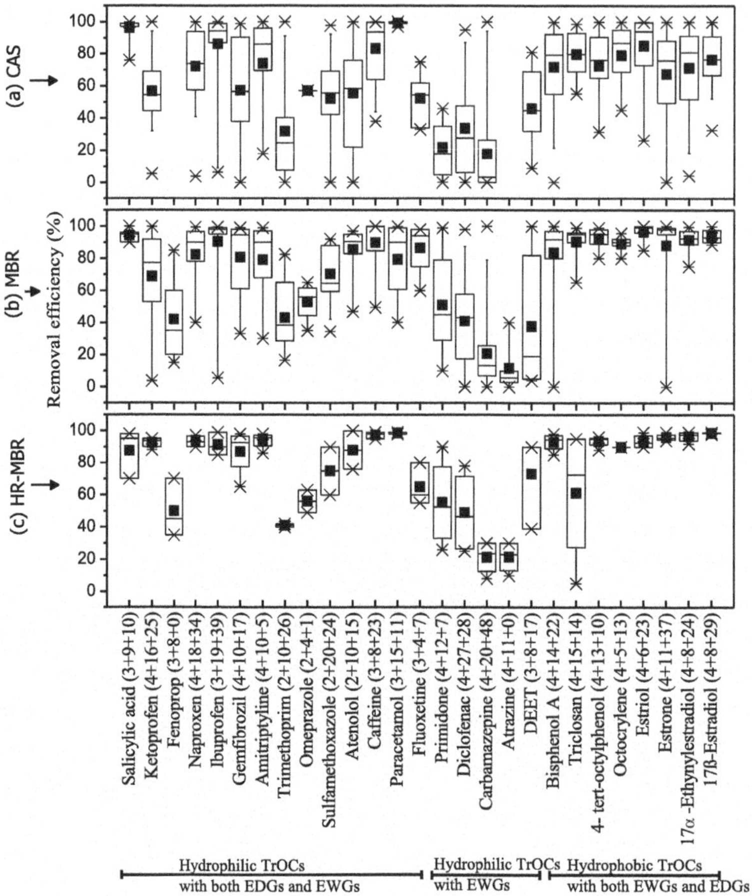

Figure 13.11 Box-and-whisker plot showing the biodegradation of TrOCs in (a) CAS, (b) MBR; and (C) HR-MBR. Reproduced with permission from Asif *et al.* (2018a).

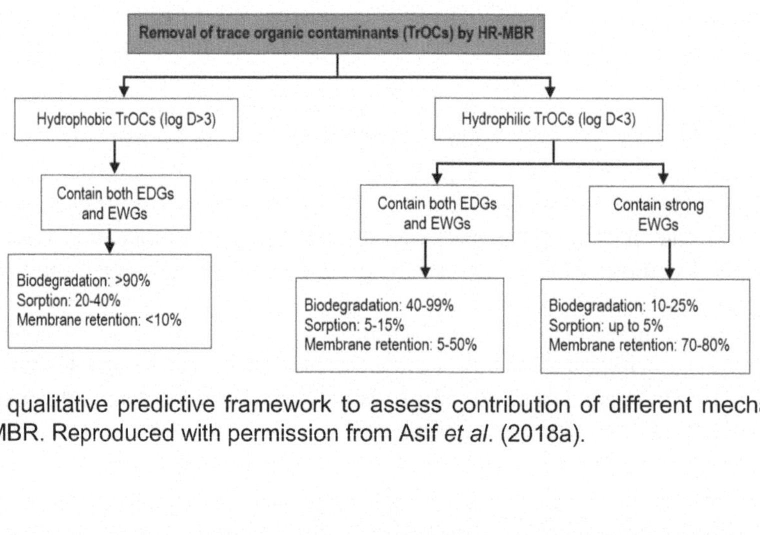

Figure 13.12 A qualitative predictive framework to assess contribution of different mechanisms of TrOC removal in HR-MBR. Reproduced with permission from Asif *et al.* (2018a).

13.10 CONCLUSION AND FUTURE OUTLOOK

This chapter has examined the potential advantages of MBR processes over conventional CAS ones for the biodegradation and treatment of TrOC. It is clear that MBR does offer some advantages, for a range of significant TrOC as well as having operational benefits such as providing a membrane barrier. This enables MBR to operate under longer SRT, higher biomass concentrations and complete retention of suspended solids. Superior performance may also be provided by additional removal mechanisms for TrOC available to MBR. These include adsorption process whereby TrOC are adsorbed either by the membrane directly or onto the retained solids; the longer SRT in MBR may facilitate additional biological transformation of TrOC. However, in practice, performance of MBR over CAS depends greatly on the conditions and the particular TrOC being investigated, and reports in the literature vary.

The situation is often complicated not only by operational parameters but also by the vastly different reported removals for a single TrOC. Describing biotransformation trends even just at the compound level is complex and often currently not fully understood. One of the potential explanations for the diversity of reported values is that removal can denote different degrees of biodegradation in two separate studies. Reasons for this can be many, including differences in initial TrOC concentrations, primary substrate concentrations, incubation times, and microbial inoculum sources.

In order to understand the extent of biodegradation and its pathways, more monitoring and data for metabolites or end products of mineralization is required. This will need to include determination of the actual biodegradability of TrOC, detailed biodegradation studies, such as mineralization experiments and biodegradation pathway studies. This will also allow clearer understanding of the potential of metabolites for ecological and human health impacts compared to the parent TrOC.

In addition, there are a number of possible reasons why laboratory-scale studies do not exactly match removals reported from their full-scale counterparts. These may include the use of arbitrarily chosen TrOC concentrations that can be much higher than those encountered in treatment systems and in the environment. These higher concentrations facilitate analytical detection and quantification, but they also have the potential to induce toxic effects in the microbes present.

Another area that merits further research is on the specific microbial communities degrading TrOC. Whether biodegradation of a compound is carried out by one specific microbial species or whether the process depends on the interactions of an entire microbial community could prove useful in optimizing the removal of particular TrOC. In addition, it would be of interest to further investigate the need for a community to adapt to the presence of a TrOC before it produces the specific enzymes needed to degrade it. The area of bioaugmentation of microbes of higher degradation capacity with the indigenous microbes remains a mostly unexplored topic. In addition to attempts to fine-tune biodegradation of TrOC, development of hybrid systems having MBR at the core will be indispensable to treat the wide varieties of TrOC from wastewater.

REFERENCES

Abegglen C., Joss A., McArdell C. S., Fink G., Schlüsener M. P., Ternes T. A. and Siegrist H. (2009). The fate of selected micropollutants in a single-house MBR. *Water Research*, **43**(7), 2036–2046.

Agenson K. O., Oh J.-I. and Urase T. (2003). Retention of a wide variety of organic pollutants by different nanofiltration/reverse osmosis membranes: controlling parameters of process. *Journal of Membrane Science*, **225**(1), 91–103.

Alharbi S. K., Kang J., Nghiem L. D., Van De Merwe J. P., Leusch F. D. and Price W. E. (2017). Photolysis and UV/H_2O_2 of diclofenac, sulfamethoxazole, carbamazepine, and trimethoprim: identification of their major degradation products by ESI–LC–MS and assessment of the toxicity of reaction mixtures. *Process Safety and Environmental Protection*, **112**, 222–234.

Alturki A. A., Tadkaew N., McDonald J. A., Khan S. J., Price W. E. and Nghiem L. D. (2010a). Combining MBR and NF/RO membrane filtration for the removal of trace organics in indirect potable water reuse applications. *Journal of Membrane Science*, **365**(1–2), 206–215.

Alturki A. A., Tadkaew N., McDonald J. A., Khan S. J., Price W. E. and Nghiem L. D. (2010b). Combining MBR and NF/RO membrane filtration for the removal of trace organics in indirect potable water reuse applications. *Journal of Membrane Science*, **365**(1–2), 206–215.

Alturki A. A., Tadkaew N., McDonald J. A., Khan S. J., Price W. E. and Nghiem L. D. (2010c). Combining MBR and NF/RO membrane filtration for the removal of trace organics in indirect potable water reuse applications. *Journal of Membrane Science*, **365**(1), 206–215.

Alturki A. A., McDonald J., Khan S. J., Hai F. I., Price W. E. and Nghiem L. D. (2012). Performance of a novel osmotic membrane bioreactor (OMBR) system: flux stability and removal of trace organics. *Bioresource Technology*, **113**(0), 201–206.

Alturki A. A., McDonald J. A., Khan S. J., Price W. E., Nghiem L. D. and Elimelech M. (2013). Removal of trace organic contaminants by the forward osmosis process. *Separation and Purification Technology*, **103**, 258–266.

Andersen H., Siegrist H., Halling-Sorensen B. and Ternes T. A. (2003). Fate of estrogens in a municipal sewage treatment plant. *Environmental Science and Technology*, **37**(18), 4021–4026.

Andreozzi R., Cesaro R., Marotta R. and Pirozzi F. (2006). Evaluation of biodegradation kinetic constants for aromatic compounds by means of aerobic batch experiments. *Chemosphere*, **62**(9), 1431–1436.

Ansari S. A. and Husain Q. (2012). Potential applications of enzymes immobilized on/in nano materials: a review. *Biotechnology Advances*, **30**(3), 512–523.

Arikan O. A. (2008). Degradation and metabolization of chlortetracycline during the anaerobic digestion of manure from medicated calves. *Journal of Hazardous Materials*, **158**(2–3), 485–490.

Asif M. B., Hai F. I., Hou J., Price W. E. and Nghiem L. D. (2017a). Impact of wastewater derived dissolved interfering compounds on growth, enzymatic activity and trace organic contaminant removal of white rot fungi – a critical review. *Journal of Environmental Management*, **201**, 89–109.

Asif M. B., Hai F. I., Kang J., Van De Merwe J. P., Leusch F. D., Yamamoto K., Price W. E. and Nghiem L. D. (2017b). Degradation of trace organic contaminants by a membrane distillation – enzymatic bioreactor. *Appl. Sci*, **7**(9), 879.

Asif M. B., Hai F. I., Singh L., Price W. E. and Nghiem L. D. (2017c). Degradation of pharmaceuticals and personal care products by white-rot fungi – a critical review. *Current Pollution Reports*, **3**(2), 88–103.

Asif M. B., Ansari A. J., Chen S.-S., Nghiem L. D., Price W. E. and Hai F. I. (2018a). Understanding the mechanisms of trace organic contaminant removal by high retention membrane bioreactors: a critical review. *Environmental Science and Pollution Research*, DOI: 10.1007/s11356-018-3256-8.

Asif M. B., Hai F. I., Dhar B. R., Ngo H. H., Guo W., Jegatheesan V., Price W. E., Nghiem L. D. and Yamamoto K. (2018b). Impact of simultaneous retention of micropollutants and laccase on micropollutant degradation in enzymatic membrane bioreactor. *Bioresource Technology*, **267**, 473–480.

Asif M. B., Hai F. I., Kang J., Van De Merwe J. P., Leusch F. D., Price W. E. and Nghiem L. D. (2018c). Biocatalytic degradation of pharmaceuticals, personal care products, industrial chemicals, steroid hormones and pesticides in a membrane distillation-enzymatic bioreactor. *Bioresource Technology*, **247**, 528–536.

Aubenneau M., Tahar A., Casellas C. and Wisniewski C. (2011). Membrane bioreactor for pharmaceutically active compounds removal: effects of carbamazepine on mixed microbial communities implied in the treatment. *Process Biochemistry*, **45**(11), 1826–1831.

Auriol M., Filali-Meknassi Y., Tyagi R. D., Adams C. D. and Surampalli R. Y. (2006). Endocrine disrupting compounds removal from wastewater, a new challenge. *Process Biochemistry*, **41**(3), 525–539.

Badia-Fabregat M., Lucas D., Tuomivirta T., Fritze H., Pennanen T., Rodríguez-Mozaz S., Barceló D., Caminal G. and Vicent T. (2017). Study of the effect of the bacterial and fungal communities present in real wastewater effluents on the performance of fungal treatments. *Science of the Total Environment*, **579**, 366–377.

Balabanic D., Rupnik M. and Klemencic A. K. (2011). Negative impact of endocrine-disrupting compounds on human reproductive health. *Reproduction Fertility and Development*, **23**(3), 403–416.

Baldwin D. D. and Campbell C. E. (2001). Short-term effects of low pH on the microfauna of an activated sludge wastewater treatment system. *Water Quality Research Journal of Canada*, **36**(3), 519–535.

Balest L., Mascolo G., Di Iaconi C. and Lopez A. (2008). Removal of endocrine disrupter compounds from municipal wastewater by an innovative biological technology. *Water Science and Technology*, **58**(4), 953–956.

Baronti C., Curini R., D'Ascenzo G., Di Corcia A., Gentili A. and Samperi R. (2000). Monitoring natural and synthetic estrogens at activated sludge sewage treatment plants and in a receiving river water. *Environmental Science and Technology*, **34**(24), 5059–5066.

Basile T., Petrella A., Petrella M., Boghetich G., Petruzzelli V., Colasuonno S. and Petruzzelli D. (2011). Review of endocrine-disrupting-compound removal technologies in water and wastewater treatment plants: an EU perspective. *Industrial & Engineering Chemistry Research*, **50**(14), 8389–8401.

Batt A. L., Kim S. and Aga D. S. (2006). Enhanced biodegradation of lopromide and trimethoprim in nitrifying activated sludge. *Environmental Science and Technology*, **40**(23), 7367–7373.

Bellona C., Drewes J. E., Xu P. and Amy G. (2004). Factors affecting the rejection of organic solutes during NF/RO treatment – a literature review. *Water Research*, **38**(12), 2795–2809.

Benotti M. J., Trenholm R. A., Vanderford B. J., Holady J. C., Stanford B. D. and Snyder S. A. (2009). Pharmaceuticals and endocrine disrupting compounds in US drinking water. *Environmental Science & Technology*, **43**(3), 597–603.

Bernhard M., Maller J. and Knepper T. P. (2006). Biodegradation of persistent polar pollutants in wastewater: comparison of an optimised lab-scale membrane bioreactor and activated sludge treatment. *Water Research*, **40**(18), 3419–3428.

Bo L., Urase T. and Wang X. (2009). Biodegradation of trace pharmaceutical substances in wastewater by a membrane bioreactor. *Frontiers of Environmental Science & Engineering in China*, **3**(2), 236–240.

Boonyaroj V., Chiemchaisri C., Chiemchaisri W., Theepharaksapan S. and Yamamoto K. (2012). Toxic organic micro-pollutants removal mechanisms in long-term operated membrane bioreactor treating municipal solid waste leachate. *Bioresource Technology*, **113**(0), 174–180.

Bott C. B. and Love N. G. (2001). The immunochemical detection of stress proteins in activated sludge exposed to toxic chemicals. *Water Research*, **35**(1), 91–100.

Bouju H., Buttiglieri G. and Malpei F. (2008). Perspectives of persistent organic pollutants (POPS) removal in an MBR pilot plant. *Desalination*, **224**(1–3), 1–6.

Bouju H., Hommes G., Wintgens T., Malpei F. and Corvini P. F. X. (2011). The fate of 14C-radiolabelled diclofenac and 4- hydroxydiclofenac in membrane bioreactor treatment of wastewater. *Water Science and Technology*, **63**(12), 2878–2885.

Boxall A. B.A., Blackwell P., Cavallo R., Kay P. and Tolls J. (2002). The sorption and transport of a sulphonamide antibiotic in soil systems. *Toxicology Letters*, **131**(1–2), 19–28.

Boxall A. B.A., Rudd M. A., Brooks B. W., Caldwell D. J., Choi K., Hickmann S., Innes E., Ostapyk K., Staveley J. P., Verslycke T., Ankley G. T., Beazley K. F., Belanger S. E., Berninger J. P., Carriquiriborde P., Coors A., DeLeo P. C., Dyer S. D., Ericson J. F., Gagne F., Giesy J. P., Gouin T., Hallstrom L., Karlsson M. V., Larsson D. G.J., Lazorchak J. M., Mastrocco F., McLaughlin A., McMaster M. E., Meyerhoff R. D., Moore R., Parrott J. L., Snape J. R., Murray-Smith R., Servos M. R., Sibley P. K., Straub J. O., Szabo N. D., Topp E., Tetreault G. R., Trudeau V. L. and Van Der Kraak G. (2012). Pharmaceuticals and personal care products in the environment: what are the big questions? *Environmental Health Perspectives*, **120**(9), 1221–1229.

Buchanan I., Liang H. C. and Razaviarani V. (2012). Pesticides and herbicides. *Water Environment Research*, **84**(10), 1533–1567.

Carballa M., Fink G., Omil F., Lema J. M. and Ternes T. (2008). Determination of the solid–water distribution coefficient (Kd) for pharmaceuticals, estrogens and musk fragrances in digested sludge. *Water Research*, **42**(1–2), 287–295.

Cartagena P., El Kaddouri M., Cases V., Trapote A. and Prats D. (2013). Reduction of emerging micropollutants, organic matter, nutrients and salinity from real wastewater by combined MBR–NF/RO treatment. *Separation and Purification Technology*, **110**, 132–143.

Cases V., Alonso V., Argandona V., Rodriguez M. and Prats D. (2011). Endocrine disrupting compounds: a comparison of removal between conventional activated sludge and membrane bioreactors. *Desalination*, **272**(1–3), 240–245.

Cath T. Y., Hancock N. T., Lundin C. D., Hoppe-Jones C. and Drewes J. E. (2010). A multi-barrier osmotic dilution process for simultaneous desalination and purification of impaired water. *Journal of Membrane Science*, **362** (1–2), 417–426.

Chang S., Waite T. D., Schäfer A. I. and Fane A. G. (2003). Adsorption of the endocrine-active compound estrone on microfiltration hollow fiber membranes. *Environmental Science & Technology*, **37**(14), 3158–3163.

Chelliapan S., Wilby T. and Sallis P. J. (2006). Performance of an up-flow anaerobic stage reactor (UASR) in the treatment of pharmaceutical wastewater containing macrolide antibiotics. *Water Research*, **40**(3), 507–516.

Chen J., Huang X. and Lee D. (2008). Bisphenol A removal by a membrane bioreactor. *Process Biochemistry*, **43**(4), 451–456.

Chen X. and Hu J. (2009). Degradation of 17-β-estradiol and its conjugates: effects of initial concentration and MLSS concentration. *Process Biochemistry*, **44**(12), 1330–1334.

Çiçek N., Franco J. P., Suidan M. T., Urbain V. and Manem J. (1999). Characterization and comparison of a membrane bioreactor and a conventional activated-sludge system in the treatment of wastewater containing high-molecular-weight compounds. *Water Environment Research*, **71**(1), 64–70.

Cicek N., Macomber J., Davel J., Suidan M. T., Audic J. and Genestet P. (2001). Effect of solids retention time on the performance and biological characteristics of a membrane bioreactor. *Water Science and Technology*, **43**(11), 43–50.

Cinar O., Deniz T. and Grady C. P.L. (2003). Effects of oxygen on anoxic biodegradation of benzoate during continuous culture. *Water Environment Research*, **75**, 434–443.

Cirja M., Zuehlke S., Ivashechkin P., Hollender J., Schaffer A. and Corvini P. F.X. (2007). Behavior of two differently radiolabelled 17α-ethinylestradiols continuously applied to a laboratory-scale membrane bioreactor with adapted industrial activated sludge. *Water Research*, **41**(19), 4403–4412.

Cirja M., Ivashechkin P., Schäffer A. and Corvini P. (2008a). Factors affecting the removal of organic micropollutants from wastewater in conventional treatment plants (CTP) and membrane bioreactors (MBR). *Reviews in Environmental Science and Biotechnology*, **7**(1), 61–78.

Cirja M., Ivashechkin P., Schäffer A. and Corvini P. F.X. (2008b). Factors affecting the removal of organic micropollutants from wastewater in conventional treatment plants (CTP) and membrane bioreactors (MBR). *Reviews in Environmental Science & Biotechnology*, **7**(1), 61–78.

Citulski J. A. and Farahbakhsh K. (2010). Fate of endocrine-active compounds during municipal biosolids treatment: a review. *Environmental Science and Technology*, **44**(22), 8367–8376.

Clara M., Strenn B., Ausserleitner M. and Kreuzinger N. (2004). Comparison of the behaviour of selected micropollutants in a membrane bioreactor and a conventional wastewater treatment plant. *Water Science and Technology*, **50**(5), 29–36.

Clara M., Kreuzinger N., Strenn B., Gans O. and Kroiss H. (2005a). The solids retention time – a suitable design parameter to evaluate the capacity of wastewater treatment plants to remove micropollutants. *Water Research*, **39**(1), 97–106.

Clara M., Kreuzinger N., Strenn B., Gans O. and Kroiss H. (2005b). The solids retention time – A suitable design parameter to evaluate the capacity of wastewater treatment plants to remove micropollutants. *Water Research*, **39**(1), 97–106.

Clara M., Strenn B., Gans O., Martinez E., Kreuzinger N. and Kroiss H. (2005c). Removal of selected pharmaceuticals, fragrances and endocrine disrupting compounds in a membrane bioreactor and conventional wastewater treatment plants. *Water Research*, **39**(19), 4797–4807.

Clouzot L., Doumenq P., Roche N. and Marrot B. (2010). Kinetic parameters for 17α-ethinylestradiol removal by nitrifying activated sludge developed in a membrane bioreactor. *Bioresource Technology*, **101**(16), 6425–6431.

Coday B. D., Yaffe B. G.M., Xu P. and Cath T. Y. (2014). Rejection of trace organic compounds by forward osmosis membranes: a literature review. *Environmental Science & Technology*, **48**(7), 3612–3624.

Comerton A. M., Andrews R. C. and Bagley D. M. (2005). Evaluation of an MBR-RO system to produce high quality reuse water: microbial control, DBP formation and nitrate. *Water Research*, **39**(16), 3982–3990.

Comerton A. M., Andrews R. C., Bagley D. M. and Hao C. (2008). The rejection of endocrine disrupting and pharmaceutically active compounds by NF and RO membranes as a function of compound and water matrix properties. *Journal of Membrane Science*, **313**(1), 323–335.

Connon R. E., Geist J. and Werner I. (2012). Effect-based tools for monitoring and predicting the ecotoxicological effects of chemicals in the aquatic environment. *Sensors*, **12**(9), 12741–12771.

Corvini P., Schäffer A. and Schlosser D. (2006). Microbial degradation of nonylphenol and other alkylphenols – our evolving view. *Applied Microbiology and Biotechnology*, **72**(2), 223–243.

Cruz-Morató C., Ferrando-Climent L., Rodriguez-Mozaz S., Barceló D., Marco-Urrea E., Vicent T. and Sarrà M. (2013). Degradation of pharmaceuticals in non-sterile urban wastewater by *Trametes versicolor* in a fluidized bed bioreactor. *Water Research*, **47**(14), 5200–5210.

Cruz-Morató C., Lucas D., Llorca M., Rodriguez-Mozaz S., Gorga M., Petrovic M., Barceló D., Vicent T., Sarrà M. and Marco-Urrea E. (2014). Hospital wastewater treatment by fungal bioreactor: removal efficiency for pharmaceuticals and endocrine disruptor compounds. *Science of the Total Environment*, **493**, 365–376.

Daughton C. G. and Ternes T. A. (1999). Pharmaceuticals and personal care products in the environment: agents of subtle change. *Environmental Health Perspectives*, **107**(Suppl. 6), 907–938.

Deblonde T., Cossu-Leguille C. and Hartemann P. (2011). Emerging pollutants in wastewater: a review of the literature. *International Journal of Hygiene and Environmental Health*, **214**(6), 442–448.

Deborde M., Rabouan S., Gallard H. and Legube B. (2004). Aqueous chlorination kinetics of some endocrine disruptors. *Environmental Science and Technology*, **38**(21), 5577–5583.

De Gusseme B., Pycke B., Hennebel T., Marcoen A., Vlaeminck S. E., Noppe H., Boon N. and Verstraete W. (2009). Biological removal of 17α-ethinylestradiol by a nitrifier enrichment culture in a membrane bioreactor. *Water Research*, **43**(9), 2493–2503.

Delgado L. F., Faucet-Marquis V., Schetrite S., Pfohl-Leszkowicz A., Paranthoen S. and Albasi C. (2010). Effect of cytostatic drugs on the sludge and on the mixed liquor characteristics of a cross-flow membrane bioreactor: consequence on the process. *Journal of Membrane Science*, **347**(1–2), 165–173.

De Wever H., Van Roy S., Dotremont C., Muller J. and Knepper T. (2004). Comparison of linear alkylbenzene sulfonates removal in conventional activated sludge systems and membrane bioreactors. *Water Science and Technology*, **50**(5), 219–225.

De Wever H., Weiss S., Reemtsma T., Vereecken J., Müller J., Knepper T., Rörden O., Gonzalez S., Barcelo D. and Dolores Hernando M. (2007). Comparison of sulfonated and other micropollutants removal in membrane bioreactor and conventional wastewater treatment. *Water Research*, **41**(4), 935–945.

Dialynas E. and Diamadopoulos E. (2009). Integration of a membrane bioreactor coupled with reverse osmosis for advanced treatment of municipal wastewater. *Desalination*, **238**(1–3), 302–311.

Dialynas E. and Diamadopoulos E. (2012). The effect of biomass adsorption on the removal of selected pharmaceutical compounds in an immersed membrane bioreactor system. *Journal of Chemical Technology & Biotechnology*, **87**(2), 232–237.

Dolar D., Gros M., Rodriguez-Mozaz S., Moreno J., Comas J., Rodriguez-Roda I. and Barcelo D. (2012). Removal of emerging contaminants from municipal wastewater with an integrated membrane system, MBR-RO. *Journal of Hazardous Materials*, **15**, 239–240 (accepted 2012).

Domenjoud B., Cortes-Francisco N., Guastalli A. R., Caixach J., Esplugas S. and Baig S. (2011). Ozonation of municipal secondary effluent; removal of hazardous micropollutants and related changes of organic matter composition. *Journal of Advanced Oxidation Technologies*, **14**(1), 138–146.

Drewes J. E., Fox P. and Jekel M. (2001). Occurrence of iodinated X-ray contrast media in domestic effluents and their fate during indirect potable reuse. *Journal of Environmental Science and Health - Part A Toxic/Hazardous Substances and Environmental Engineering*, **36**(9), 1633–1645.

Drewes J. E., Heberer T. and Reddersen K. (2002). Fate of pharmaceuticals during indirect potable reuse. *Water Science and Technology*, **46**(3), 73–80.

Drillia P., Dokianakis S. N., Fountoulakis M. S., Kornaros M., Stamatelatou K. and Lyberatos G. (2005). On the occasional biodegradation of pharmaceuticals in the activated sludge process: the example of the antibiotic sulfamethoxazole. *Journal of Hazardous Materials*, **122**(3), 259–265.

Espinosa-Ortiz E. J., Rene E. R., Pakshirajan K., van Hullebusch E. D. and Lens P. N. (2016). Fungal pelleted reactors in wastewater treatment: applications and perspectives. *Chemical Engineering Journal*, **283**, 553–571.

Estrada-Arriaga E. and Mijaylova P. (2011). Influence of operational parameters (sludge retention time and hydraulic residence time) on the removal of estrogens by membrane bioreactor. *Environmental Science and Pollution Research*, **18**(7), 1121–1128.

Fatta-Kassinos D., Meric S. and Nikolaou A. (2011). Pharmaceutical residues in environmental waters and wastewater: current state of knowledge and future research. *Analytical and Bioanalytical Chemistry*, **399**(1), 251–275.

Fernandez-Fontaina E., Omil F., Lema J. M. and Carballa M. (2012a). Influence of nitrifying conditions on the biodegradation and sorption of emerging micropollutants. *Water Research*, **46**(16), 5434–5444.

Fernandez-Fontaina E., Omil F., Lema J. M. and Carballa M. (2012b). Influence of nitrifying conditions on the biodegradation and sorption of emerging micropollutants. *Water Research*, **46**(16), 5434–5444.

Ferrando-Climent L., Cruz-Morató C., Marco-Urrea E., Vicent T., Sarrà M., Rodriguez-Mozaz S. and Barceló D. (2015). Nonconventional biological treatment based on *Trametes versicolor* for the elimination of recalcitrant anticancer drugs in hospital wastewater. *Chemosphere*, **136**, 9–19.

Gerrity D., Gamage S., Holady J. C., Mawhinney D. B., Quinones O., Trenholm R. A. and Snyder S. A. (2011). Pilot-scale evaluation of ozone and biological activated carbon for trace organic contaminant mitigation and disinfection. *Water Research*, **45**(5), 2155–2165.

Ghosh P. K. and Philip L. (2004). Atrazine degradation in anaerobic environment by a mixed microbial consortium. *Water Research*, **38**(9), 2277–2284.

Ghyoot W., Springael D., Dong Q., Van Roy S., Nuyts G. and Diels L. (2000). Bioaugmentation with the clc-element carrying *Pseudomonas putida* BN210 in a membrane separation bioreactor. *Water Science and Technology*, **41**(10–11), 279–286.

Gobel A., McArdell C. S., Joss A., Siegrist H. and Giger W. (2007). Fate of sulfonamides, macrolides, and trimethoprim in different wastewater treatment technologies. *Science of the Total Environment*, **372**(2–3), 361–371.

Göbel A., McArdell C. S., Joss A., Siegrist H. and Giger W. (2007). Fate of sulfonamides, macrolides, and trimethoprim in different wastewater treatment technologies. *The Science of the Total Environment*, **372**(23), 361–371.

Golan-Rozen N., Chefetz B., Ben-Ari J., Geva J. and Hadar Y. (2011). Transformation of the recalcitrant pharmaceutical compound carbamazepine by *Pleurotus ostreatus*: role of cytochrome P450 monooxygenase and manganese peroxidase. *Environmental Science & Technology*, **45**(16), 6800–6805.

Golan-Rozen N., Seiwert B., Riemenschneider C., Reemtsma T., Chefetz B. and Hadar Y. (2015). Transformation pathways of the recalcitrant pharmaceutical compound carbamazepine by the white-rot fungus *Pleurotus ostreatus*: effects of growth conditions. *Environmental Science & Technology*, **49**(20), 12351–12362.

Gonzalez S., MÃuller J., Petrovic M., Barcelo D. and Knepper T. P. (2006). Biodegradation studies of selected priority acidic pesticides and diclofenac in different bioreactors. *Environmental Pollution*, **144**(3), 926–932.

Gonzalez S., Petrovic M. and Barcelo D. (2007). Removal of a broad range of surfactants from municipal wastewater – Comparison between membrane bioreactor and conventional activated sludge treatment. *Chemosphere*, **67**(2), 335–343.

Hai F. I. and Yamamoto K. (2011). Membrane biological reactors. In: Treatise on Water Science W. Peter, (ed.), Elsevier, Oxford, pp. 571–613.

Hai F. I., Yamamoto K. and Fukushi K. (2006). Development of a submerged membrane fungi reactor for textile wastewater treatment. *Desalination*, **192**(1), 315–322.

Hai F. I., Yamamoto K., Nakajima F. and Fukushi K. (2009). Factors governing performance of continuous fungal reactor during non-sterile operation–ty2y8he case of a membrane bioreactor treating textile wastewater. *Chemosphere*, **74**(6), 810–817.

Hai F. I., Li X., Price W. E. and Nghiem L. D. (2011a). Removal of carbamazepine and sulfamethoxazole by MBR under anoxic and aerobic conditions. *Bioresource Technology*, **102**(22), 10386–10390.

Hai F. I., Tadkaew N., McDonald J. A., Khan S. J. and Nghiem L. D. (2011b). Is halogen content the most important factor in the removal of halogenated trace organics by MBR treatment? *Bioresource Technology*, **102**(10), 6299–6303.

Hai F. I., Tessmer K., Nguyen L. N., Kang J., Price W. E. and Nghiem L. D. (2011c). Removal of micropollutants by membrane bioreactor under temperature variation. *Journal of Membrane Science*, **383**(1–2), 144–151.

Hamid H. and Eskicioglu C. (2012). Fate of estrogenic hormones in wastewater and sludge treatment: a review of properties and analytical detection techniques in sludge matrix. *Water Research*, **46**(18), 5813–5833.

Hashim N. H., Nghiem L. D., Stuetz R. M. and Khan S. J. (2010). Enantiospecific fate of ibuprofen, ketoprofen and naproxen in a laboratory-scale membrane bioreactor. *Water Research*, **45**(18), 6249–6258.

Hashim N. H., Nghiem L. D., Stuetz R. M. and Khan S. J. (2011). Enantiospecific fate of ibuprofen, ketoprofen and naproxen in a laboratory-scale membrane bioreactor. *Water Research*, **45**(18), 6249–6258.

Hata T., Kawai S., Okamura H. and Nishida T. (2010). Removal of diclofenac and mefenamic acid by the white rot fungus *Phanerochaete sordida* YK-624 and identification of their metabolites after fungal transformation. *Biodegradation*, **21**(5), 681–689.

Hau N. T., Chen S.-S., Nguyen N. C., Huang K. Z., Ngo H. H. and Guo W. (2014). Exploration of EDTA sodium salt as novel draw solution in forward osmosis process for dewatering of high nutrient sludge. *Journal of Membrane Science*, **455**, 305–311.

Heidler J. and Halden R. U. (2008). Meta-analysis of mass balances examining chemical fate during wastewater treatment. *Environmental Science & Technology*, **42**(17), 6324–6332.

Helbling D. E., Hollender J., Kohler H. P. E. and Fenner K. (2010). Structure-based interpretation of biotransformation pathways of amide-containing compounds in sludge-seeded bioreactors. *Environmental Science and Technology*, **44**(17), 6628–6635.

Helbling D. E., Johnson D. R., Honti M. and Fenner K. (2012). Micropollutant biotransformation kinetics associate with WWTP process parameters and microbial community characteristics. *Environmental Science and Technology*, **46**(19), 10579–10588.

Hernandez-Leal L., Temmink H., Zeeman G. and Buisman C. J. N. (2011). Removal of micropollutants from aerobically treated grey water via ozone and activated carbon. *Water Research*, **45**(9), 2887–2896.

Holloway R. W., Regnery J., Nghiem L. D. and Cath T. Y. (2014). Removal of trace organic chemicals and performance of a novel hybrid ultrafiltration-osmotic membrane bioreactor. *Environmental Science & Technology*, **48**(18), 10859–10868.

Hu J., Cheng S., Aizawa T., Terao Y. and Kunikane S. (2003). Products of aqueous chlorination of 17Î^{2}-estradiol and their estrogenic activities. *Environmental Science and Technology*, **37**(24), 5665–5670.

Hu J., Jin X. and Ong S. (2007a). Rejection of estrone by nanofiltration: influence of solution chemistry. *Journal of Membrane Science*, **302**(1), 188–196.

Hu J. Y., Chen X., Tao G. and Kekred K. (2007b). Fate of endocrine disrupting compounds in membrane bioreactor systems. *Environmental Science & Technology*, **41**(11), 4097–4102.

Hu L., Martin H. M., Arce-Bulted O., Sugihara M. N., Keating K. A. and Strathmann T. J. (2008). Oxidation of carbamazepine by Mn (VII) and Fe (VI): reaction kinetics and mechanism. *Environmental Science & Technology*, **43**(2), 509–515.

Hübner U., Seiwert B., Reemtsma T. and Jekel M. (2014). Ozonation products of carbamazepine and their removal from secondary effluents by soil aquifer treatment – indications from column experiments. *Water Research*, **49**, 34–43.

Huerta B., Rodriguez-Mozaz S. and Barcelo D. (2012). Pharmaceuticals in biota in the aquatic environment: analytical methods and environmental implications. *Analytical and Bioanalytical Chemistry*, **404**(9), 2611–2624.

Hyland K. C., Dickenson E. R.V., Drewes E., Jr and Higgins C. P. (2012). Sorption of ionized and neutral emerging trace organic compounds onto activated sludge from different wastewater treatment configurations. *Water Research*, **46**(6), 1958–1968.

Ivashechkin P., Corvini P. F. X., Fahrbach M., Hollender J., Konietzko M., Meesters R., Schroder H. F. and Dohmann M. (2004). Comparison of the Elimination of Endocrine Disrupters in Conventional Wastewater Treatment Plants and Membrane Bioreactors. Proceedings of the 2nd IWA Leading-Edge Conference on Water and Wastewater Treatment Technologies – Prague. Part Two: Wastewater Treatment.

Jacob M., Guigui C., Cabassud C., Darras H., Lavison G. and Moulin L. (2010). Performances of RO and NF processes for wastewater reuse: tertiary treatment after a conventional activated sludge or a membrane bioreactor. *Desalination*, **250**(2), 833–839.

Jelic A., Cruz-Morató C., Marco-Urrea E., Sarrà M., Perez S., Vicent T., Petrović M. and Barcelo D. (2012). Degradation of carbamazepine by *Trametes versicolor* in an air pulsed fluidized bed bioreactor and identification of intermediates. *Water Research*, **46**(4), 955–964.

Jin X., Shan J., Wang C., Wei J. and Tang C. Y. (2012). Rejection of pharmaceuticals by forward osmosis membranes. *Journal of Hazardous Materials*, **227–228**, 55–61.

Johnson A. C. and Sumpter J. P. (2001). Removal of endocrine-disrupting chemicals in activated sludge treatment works. *Environmental Science & Technology*, **35**(24), 4697–4703.

Johnson A. C., Belfroid A. and Di Corcia A. (2000). Estimating steroid oestrogen inputs into activated sludge treatment works and observations on their removal from the effluent. *Science of the Total Environment*, **256**(2–3), 163–173.

Joss A., Andersen H., Ternes T., Richle P. R. and Siegrist H. (2004). Removal of estrogens in municipal wastewater treatment under aerobic and anaerobic conditions: consequences for plant optimization. *Environmental Science & Technology*, **38**(11), 3047–3055.

Joss A., Keller E., Alder A. C., Gobel A., McArdell C. S., Ternes T. and Siegrist H. (2005a). Removal of pharmaceuticals and fragrances in biological wastewater treatment. *Water Research*, **39**(14), 3139–3152.

Joss A., Keller E., Alder A. C., Göbel A., McArdell C. S., Ternes T. and Siegrist H. (2005b). Removal of pharmaceuticals and fragrances in biological wastewater treatment. *Water Research*, **39**(14), 3139–3152.

Joss A., Zabczynski S., Gobel A., Hoffmann B., Loffler D., McArdell C. S., Ternes T. A., Thomsen A. and Siegrist H. (2006). Biological degradation of pharmaceuticals in municipal wastewater treatment: proposing a classification scheme. *Water Research*, **40**(8), 1686–1696.

Jury K. L., Khan S. J., Vancov T., Stuetz R. M. and Ashbolt N. J. (2011). Are sewage treatment plants promoting antibiotic resistance. *Critical Reviews in Environmental Science and Technology*, **41**(3), 243–270.

Kang A.-R., Choi H.-T. and Song H.-G. (2008). Optimization of bisphenol A biodegradation by *Trametes versicolor*. *The Korean Journal of Microbiology*, **44**(1), 37–42.

Kim S., Eichhorn P., Jensen J. N., Weber A. S. and Aga D. S. (2005). Removal of antibiotics in wastewater: effect of hydraulic and solid retention times on the fate of tetracycline in the activated sludge process. *Environmental Science & Technology*, **39**(15), 5816–5823.

Kim S. D., Cho J., Kim I. S., Vanderford B. J. and Snyder S. A. (2007). Occurrence and removal of pharmaceuticals and endocrine disruptors in South Korean surface, drinking, and waste waters. *Water Research*, **41**(5), 1013–1021.

Kimura K., Toshima S., Amy G. and Watanabe Y. (2004). Rejection of neutral endocrine disrupting compounds (EDCs) and pharmaceutical active compounds (PhACs) by RO membranes. *Journal of Membrane Science*, **245**(1), 71–78.

Kimura K., Hara H. and Watanabe Y. (2005). Removal of pharmaceutical compounds by submerged membrane bioreactors (MBRs). *Desalination*, **178**(1–3 SPEC. ISS.), 135–140.

Kimura K., Hara H. and Watanabe Y. (2007). Elimination of selected acidic pharmaceuticals from municipal wastewater by an activated sludge system and membrane bioreactors. *Environmental Science & Technology*, **41**(10), 3708–3714.

Kloepfer A., Jekel M. and Reemtsma T. (2005). Occurrence, sources, and fate of benzothiazoles in municipal wastewater treatment plants. *Environmental Science & Technology*, **39**(10), 3792–3798.

Kontchou C. Y. and Gschwind N. (1999). Biodegradation ofs-triazine compounds by a stable mixed bacterial community. *Ecotoxicology and Environmental Safety*, **43**(1), 47–56.

Kosjek T., Andersen H. R., Kompare B., Ledin A. and Heath E. (2009). Fate of carbamazepine during water treatment. *Environmental Science & Technology*, **43**(16), 6256–6261.

Kosjek T., Heath E. and Kompare B. (2007). Removal of pharmaceutical residues in a pilot wastewater treatment plant. *Analytical and Bioanalytical Chemistry*, **387**(4), 1379–1387.

Kraigher B., Kosjek T., Heath E., Kompare B. and Mandic-Mulec I. (2008). Influence of pharmaceutical residues on the structure of activated sludge bacterial communities in wastewater treatment bioreactors. *Water Research*, **42**(17), 4578–4588.

Kreuzinger N., Clara M., Strenn B. and Kroiss H. (2004). Relevance of the sludge retention time (SRT) as design criteria for wastewater treatment plants for the removal of endocrine disruptors and pharmaceuticals from wastewater. *Water Science and Technology*, **50**(5), 149–156.

Kümmerer K. and Al-Ahmad A. (1997). Biodegradability of the anti-tumour agents 5-fluorouracil, cytarabine, and gemcitabine: impact of the chemical structure and synergistic toxicity with hospital effluent. *Acta Hydrochimica et Hydrobiologica*, **25**(4), 166–172.

Kuruto-Niwa R., Nozawa R., Miyakoshi T., Shiozawa T. and Terao Y. (2005). Estrogenic activity of alkylphenols, bisphenol S, and their chlorinated derivatives using a GFP expression system. *Environmental Toxicology and Pharmacology*, **19**(1), 121–130.

Laera G., Chong M. N., Jin B. and Lopez A. (2011). An integrated MBR–TiO_2 photocatalysis process for the removal of carbamazepine from simulated pharmaceutical industrial effluent. *Bioresource Technology*, **102**(13), 7012–7015.

Layton A. C., Gregory B. W., Seward J. R., Schultz T. W. and Sayler G. S. (2000). Mineralization of steroidal hormones by biosolids in wastewater treatment systems in Tennessee, U. S.A *Environmental Science & Technology*, **34**(18), 3925–3931.

Leclercq M., Mathieu O., Gomez E., Casellas C., Fenet H. and Hillaire-Buys D. (2009). Presence and fate of carbamazepine, oxcarbazepine, and seven of their metabolites at wastewater treatment plants. *Archives of Environmental Contamination and Toxicology*, **56**(3), 408–415.

Leclercq M., Mathieu O., Gomez E., Casellas C., Fenet H. and Hillaire-Buys D. (2010). Presence and fate of carbamazepine, oxcarbazepine, and seven of their metabolites at wastewater treatment plants. *Archives of Environmental Contamination and Toxicology*, **56**(3), 408–415.

Lee H. B. and Liu D. (2002). Degradation of 17β-estradiol and its metabolites by sewage bacteria. *Water, Air, and Soil Pollution*, **134**(1–4), 351–366.

Lee C. O., Howe K. J. and Thomson B. M. (2012). Ozone and biofiltration as an alternative to reverse osmosis for removing PPCPs and micropollutants from treated wastewater. *Water Research*, **46**(4), 1005–1014.

Lee W., Kang S. and Shin H. (2003). Sludge characteristics and their contribution to microfiltration in submerged membrane bioreactors. *Journal of Membrane Science*, **216**(1–2), 217–227.

Leet J. K., Gall H. E. and Sepulveda M. S. (2011). A review of studies on androgen and estrogen exposure in fish early life stages: effects on gene and hormonal control of sexual differentiation. *Journal of Applied Toxicology*, **31**(5), 379–398.

Lesjean B., Gnirss R., Buisson H., Keller S., Tazi-Pain A. and Luck F. (2005). Outcomes of a 2-year investigation on enhanced biological nutrients removal and trace organics elimination in membrane bioreactor (MBR). *Water Science and Technology*, **52**(10–11), 453–460.

Lester J. N. (1999). Microbiology and Chemistry for Environmental Scientists and Engineers. 2nd ed., Spon Press [Imprint], New York.

Li F., Yuasa A., Obara A. and Mathews A. P. (2005). Aerobic batch degradation of 17-β estradiol (E2) by activated sludge: effects of spiking E2 concentrations, MLVSS and temperatures. *Water Research*, **39**(10), 2065–2075.

Li X., Hai F. I. and Nghiem L. D. (2011a). Simultaneous activated carbon adsorption within a membrane bioreactor for an enhanced micropollutant removal. *Bioresource Technology*, **102**(9), 5319–5324.

Li X., Hai F. I., Tadkaew N., Gilbertson S. and Nghiem L. D. (2011b). Strategies to enhance the removal of the persistent pharmaceutically active compound carbamazepine by membrane bioreactors. *Desalination and Water Treatment*, **34**(13), 402–407.

Liao B. Q., Allen D. G., Droppo I. G., Leppard G. G. and Liss S. N. (2001). Surface properties of sludge and their role in bioflocculation and settleability. *Water Research*, **35**(2), 339–350.

Libra J. A., Borchert M. and Banit S. (2003). Competition strategies for the decolorization of a textile-reactive dye with the white-rot fungi *Trametes versicolor* under non-sterile conditions. *Biotechnology and Bioengineering*, **82**(6), 736–744.

Linares R. V., Yangali-Quintanilla V., Li Z. and Amy G. (2011). Rejection of micropollutants by clean and fouled forward osmosis membrane. *Water Research*, **45**(20), 6737–6744.

Lipp P., Groay H.-J. and Tiehm A. (2012a). Improved elimination of organic micropollutants by a process combination of membrane bioreactor (MBR) and powdered activated carbon (PAC). *Desalination and Water Treatment*, **42**(1–3), 65–72.

Lipp P., Groß H.-J. and Tiehm A. (2012b). Improved elimination of organic micropollutants by a process combination of membrane bioreactor (MBR) and powdered activated carbon (PAC). *Desalination and Water Treatment*, **42**(1–3), 65–72.

Lishman L., Smyth S. A., Sarafin K., Kleywegt S., Toito J., Peart T., Lee B., Servos M., Beland M. and Seto P. (2006). Occurrence and reductions of pharmaceuticals and personal care products and estrogens by municipal wastewater treatment plants in Ontario, Canada. *Science of the Total Environment*, **367**(2–3), 544–558.

Liu C. and Huang X. (2008). Enhanced atrazine removal using membrane bioreactor bioaugmented with genetically engineered microorganism. *Frontiers of Environmental Science & Engineering in China*, **2**(4), 452–460.

Liu Z. H., Ogejo J. A., Pruden A. and Knowlton K. F. (2011). Occurrence, fate and removal of synthetic oral contraceptives (SOCs) in the natural environment: a review. *Science of the Total Environment*, **409**(24), 5149–5161.

Lloret L., Eibes G., Feijoo G., Moreira M. T. and Lema J. M. (2012). Degradation of estrogens by laccase from *Myceliophthora thermophila* in fed-batch and enzymatic membrane reactors. *Journal of Hazardous Materials*, **213–214**, 175–183.

Lloret L., Eibes G., Feijoo G., Moreira M. T. and Lema J. M. (2013). Application of response surface methodology to study the removal of estrogens in a laccase-mediated continuous membrane reactor. *Biocatalysis and Biotransformation*, **31**(4), 197–207.

Loke M.-L., Tjørnelund J. and Halling-Sørensen B. (2002). Determination of the distribution coefficient (log K_d) of oxytetracycline, tylosin A, olaquindox and metronidazole in manure. *Chemosphere*, **48**(3), 351–361.

Luo W., Hai F. I., Price W. E., Guo W., Ngo H. H., Yamamoto K. and Nghiem L. D. (2014). High retention membrane bioreactors: challenges and opportunities. *Bioresource Technology*, **167**, 539–546.

Luo W., Hai F. I., Kang J., Price W. E., Nghiem L. D. and Elimelech M. (2015). The role of forward osmosis and microfiltration in an integrated osmotic-microfiltration membrane bioreactor system. *Chemosphere*, **136**, 125–132.

Luo W., Phan H. V., Xie M., Hai F. I., Price W. E., Elimelech M. and Nghiem L. D. (2017). Osmotic versus conventional membrane bioreactors integrated with reverse osmosis for water reuse: biological stability, membrane fouling, and contaminant removal. *Water Research*, **109**, 122–134.

Maeng S. K., Sharma S. K., Lekkerkerker-Teunissen K. and Amy G. L. (2011). Occurrence and fate of bulk organic matter and pharmaceutically active compounds in managed aquifer recharge: a review. *Water Research*, **45**(10), 3015–3033.

Mak Y. L., Taniyasu S., Yeung L. W.Y., Lu G. H., Jin L., Yang Y. L., Lam P. K.S., Kannan K. and Yamashita N. (2009). Perfluorinated compounds in tap water from China and several other countries. *Environmental Science & Technology*, **43**(13), 4824–4829.

Marco-Urrea E., Gabarrell X., Sarrà M., Caminal G., Vicent T. and Reddy C. A. (2006). Novel aerobic perchloroethylene degradation by the white-rot fungus *Trametes versicolor*. *Environmental Science & Technology*, **40**(24), 7796–7802.

Marco-Urrea E., Pérez-Trujillo M., Blánquez P., Vicent T. and Caminal G. (2010). Biodegradation of the analgesic naproxen by *Trametes versicolor* and identification of intermediates using HPLC-DAD-MS and NMR. *Bioresource Technology*, **101**(7), 2159–2166.

Matozzo V., Gagne F., Marin M. G., Ricciardi F. and Blaise C. (2008). Vitellogenin as a biomarker of exposure to estrogenic compounds in aquatic invertebrates: a review. *Environment International*, **34**(4), 531–545.

Maurer M., Escher B. I., Richle P., Schaffner C. and Alder A. C. (2007). Elimination of β-blockers in sewage treatment plants. *Water Research*, **41**(7), 1614–1622.

Miao X.-S., Yang J.-J. and Metcalfe C. D. (2005). Carbamazepine and its metabolites in wastewater and in biosolids in a municipal wastewater treatment plant. *Environmental Science & Technology*, **39**(19), 7469–7475.

Morgan-Sagastume F. and Allen D. G. (2003). Effects of temperature transient conditions on aerobic biological treatment of wastewater. *Water Research*, **37**(15), 3590–3601.

Moriyama K., Matsufuji H., Chino M. and Takeda M. (2004). Identification and behavior of reaction products formed by chlorination of ethynylestradiol. *Chemosphere*, **55**(6), 839–847.

Musson S. E., Campo P., Tolaymat T., Suidan M. and Townsend T. G. (2010). Assessment of the anaerobic degradation of six active pharmaceutical ingredients. *Science of the Total Environment*, **408**(9), 2068–2074.

Navaratna D., Elliman J., Cooper A., Shu L., Baskaran K. and Jegatheesan V. (2012a). Impact of herbicide Ametryn on microbial communities in mixed liquor of a membrane bioreactor (MBR). *Bioresource Technology*, **113**, 181–190.

Navaratna D., Shu L., Baskaran K. and Jegatheesan V. (2012b). Treatment of ametryn in wastewater by a hybrid MBR system: a lab-scale study. *Water Science and Technology*, **66**(6), 1317–1324.

Nghiem L. D. and Coleman P. J. (2008). NF/RO filtration of the hydrophobic ionogenic compound triclosan: transport mechanisms and the influence of membrane fouling. *Separation and Purification Technology*, **62**(3), 709–716.

Nghiem L. D., Schäfer A. I. and Elimelech M. (2005). Nanofiltration of hormone mimicking trace organic contaminants. *Separation Science and Technology*, **40**(13), 2633–2649.

Nghiem L. D., Schäfer A. I. and Elimelech M. (2006). Role of electrostatic interactions in the retention of pharmaceutically active contaminants by a loose nanofiltration membrane. *Journal of Membrane Science*, **286**(1), 52–59.

Nguyen L. N., Hai F. I., Kang J., Price W. E. and Nghiem L. D. (2012). Removal of trace organic contaminants by a membrane bioreactor – granular activated carbon (MBR-GAC) system. *Bioresource Technology*, **113**, 169–173.

Nguyen L. N., Hai F. I., Kang J., Nghiem L. D., Price W. E., Guo W., Ngo H. H. and Tung K.-L. (2013a). Comparison between sequential and simultaneous application of activated carbon with membrane bioreactor for trace organic contaminant removal. *Bioresource Technology*, **130**(0), 412–417.

Nguyen L. N., Hai F. I., Kang J., Price W. E. and Nghiem L. D. (2013b). Removal of emerging trace organic contaminants by MBR-based hybrid treatment processes. *International Biodeterioration & Biodegradation*, **85**, 474–482.

Nguyen L. N., Hai F. I., Yang S., Kang J., Leusch F. D., Roddick F., Price W. E. and Nghiem L. D. (2013c). Removal of trace organic contaminants by an MBR comprising a mixed culture of bacteria and white-rot fungi. *Bioresource Technology*, **148**, 234–241.

Nguyen L. N., Hai F. I., Price W. E., Leusch F. D., Roddick F., McAdam E. J., Magram S. F. and Nghiem L. D. (2014a). Continuous biotransformation of bisphenol A and diclofenac by laccase in an enzymatic membrane reactor. *International Biodeterioration & Biodegradation*, **95**, 25–32.

Nguyen L. N., Hai F. I., Price W. E., Leusch F. D.L., Roddick F., McAdam E. J., Magram S. F. and Nghiem L. D. (2014b). Continuous biotransformation of bisphenol a and diclofenac bylaccase in an enzymatic membrane reactor. *International Biodeterioration and Biodegradation*, **95**(PA), 25–32.

Nguyen L. N., Hai F. I., Yang S., Kang J., Leusch F. D.L., Roddick F., Price W. E. and Nghiem L. D. (2014c). Removal of pharmaceuticals, steroid hormones, phytoestrogens, UV-filters, industrial chemicals and pesticides by *Trametes versicolor*: role of biosorption and biodegradation. *International Biodeterioration & Biodegradation*, **88**, 169–175.

Nguyen L. N., Hai F. I., Price W. E., Kang J., Leusch F. D., Roddick F., van de Merwe J. P., Magram S. F. and Nghiem L. D. (2015). Degradation of a broad spectrum of trace organic contaminants by an enzymatic membrane reactor: complementary role of membrane retention and enzymatic degradation. *International Biodeterioration & Biodegradation*, **99**, 115–122.

Nguyen L. N., Hai F. I., Dosseto A., Richardson C., Price W. E. and Nghiem L. D. (2016a). Continuous adsorption and biotransformation of micropollutants by granular activated carbon-bound laccase in a packed-bed enzyme reactor. *Bioresource Technology*, **210**, 108–116.

Nguyen L. N., van de Merwe J. P., Hai F. I., Leusch F. D., Kang J., Price W. E., Roddick F., Magram S. F. and Nghiem L. D. (2016b). Laccase–syringaldehyde-mediated degradation of trace organic contaminants in an enzymatic membrane reactor: removal efficiency and effluent toxicity. *Bioresource Technology*, **200**, 477–484.

Onesios K., Yu J. and Bouwer E. (2009). Biodegradation and removal of pharmaceuticals and personal care products in treatment systems: a review. *Biodegradation*, **20**(4), 441–466.

Oppenheimer J., Stephenson R., Burbano A. and Liu L. (2007). Characterizing the passage of personal care products through wastewater treatment processes. *Water Environment Research*, **79**(13), 2564–2577.

Ottmar K. J., Colosi L. M. and Smith J. A. (2012). Fate and transport of atorvastatin and simvastatin drugs during conventional wastewater treatment. *Chemosphere*, **88**(10), 1184–1189.

Oulton R. L., Kohn T. and Cwiertny D. M. (2010). Pharmaceuticals and personal care products in effluent matrices: a survey of transformation and removal during wastewater treatment and implications for wastewater management. *Journal of Environmental Monitoring*, **12**(11), 1956–1978.

Paetkau M. R. (2011). Comparison of Ethinylestradiol and Nitrogen Removal in a Conventional and Simultaneous Nitrification-Denitrification Membrane Bioreactor. M.Sc., Biosystems Engineering, University of Manitoba, Manitoba, Canada. Accessed from http://hdl.handle.net/1993/4519 on March 2019.

Park N., Vanderford B. J., Snyder S. A., Sarp S., Kim S. D. and Cho J. (2009). Effective controls of micropollutants included in wastewater effluent using constructed wetlands under anoxic condition. *Ecological Engineering*, **35**(3), 418–423.

Perez S., Eichhorn P. and Aga D. S. (2005). Evaluating the biodegradability of sulfamethazine, sulfamethoxazole, sulfathiazole, and trimethoprim at different stages of sewage treatment. *Environmental Toxicology and Chemistry*, **24**(6), 1361–1367.

Phan H. V., McDonald J. A., Hai F. I., Price W. E., Khan S. J., Fujioka T. and Nghiem L. D. (2016). Biological performance and trace organic contaminant removal by a side-stream ceramic nanofiltration membrane bioreactor. *International Biodeterioration & Biodegradation*, **113**, 49–56.

Plumlee M. H., Larabee J. and Reinhard M. (2008). Perfluorochemicals in water reuse. *Chemosphere*, **72**(10), 1541–1547.

Pollice A., Laera G., Cassano D., Diomede S., Pinto A., Lopez A. and Mascolo G. (2012). Removal of nalidixic acid and its degradation products by an integrated MBR-ozonation system. *Journal of Hazardous Materials*, **203–204**, 46–52.

Post G. B., Louis J. B., Cooper K. R., Boros-Russo B. J. and Lippincott R. L. (2009). Occurrence and potential significance of perfluorooctanoic acid (PFOA) detected in New Jersey public drinking water systems. *Environmental Science & Technology*, **43**(12), 4547–4554.

Prado M., Borea L., Cesaro A., Liu H., Naddeo V., Belgiorno V. and Ballesteros F. (2017). Removal of emerging contaminant and fouling control in membrane bioreactors by combined ozonation and sonolysis. *International Biodeterioration & Biodegradation*, **119**(Suppl. C), 577–586.

Prasse C., Wagner M., Schulz R. and Ternes T. A. (2011). Biotransformation of the antiviral drugs acyclovir and penciclovir in activated sludge treatment. *Environmental Science and Technology*, **45**(7), 2761–2769.

Protzman R. S., Lee P.-H., Ong S. K. and Moorman T. B. (1999). Treatment of formulated atrazine rinsate by *Agrobacterium radiobacter* strain J14a in a sequencing batch biofilm reactor. *Water Research*, **33**(6), 1399–1404.

Qin J.-J., Kekre K. A., Tao G., Oo M. H., Wai M. N., Lee T. C., Viswanath B. and Seah H. (2006). New option of MBR-RO process for production of NEWater from domestic sewage. *Journal of Membrane Science*, **272**(1–2), 70.

Quintana J. B., Weiss S. and Reemtsma T. (2005). Pathways and metabolites of microbial degradation of selected acidic pharmaceutical and their occurrence in municipal wastewater treated by a membrane bioreactor. *Water Research*, **39**(12), 2654–2664.

Radjenovic J., Petrovic M. and Barceló D. (2007). Analysis of pharmaceuticals in wastewater and removal using a membrane bioreactor. *Analytical and Bioanalytical Chemistry*, **387**(4), 1365–1377.

Radjenović J., Petrović M., Ventura F. and Barceló D. (2008). Rejection of pharmaceuticals in nanofiltration and reverse osmosis membrane drinking water treatment. *Water Research*, **42**(14), 3601–3610.

Radjenovic J., Petrovic M. and Barceló D. (2009a). Fate and distribution of pharmaceuticals in wastewater and sewage sludge of the conventional activated sludge (CAS) and advanced membrane bioreactor (MBR) treatment. *Water Research*, **43**(3), 831–841.

Radjenovic J., Petroviic M. and Barcelo D. (2009b). Fate and distribution of pharmaceuticals in wastewater and sewage sludge of the conventional activated sludge (CAS) and advanced membrane bioreactor (MBR) treatment. *Water Research*, **43**(3), 831–841.

Reddy S., Iden C. R. and Brownawell B. J. (2005). Analysis of steroid conjugates in sewage influent and effluent by liquid chromatography-tandem mass spectrometry. *Analytical Chemistry*, **77**(21), 7032–7038.

Reemtsma T., Zywicki B., Stueber M., Kloepfer A. and Jekel M. (2002). Removal of sulfur-organic polar micropollutants in a membrane bioreactor treating industrial wastewater. *Environmental Science & Technology*, **36**(5), 1102–1106.

Reif R., Suairez S., Omil F. and Lema J. M. (2008). Fate of pharmaceuticals and cosmetic ingredients during the operation of a MBR treating sewage. *Desalination*, **221**(1 - 3), 511–517.

Reungoat J., Macova M., Escher B. I., Carswell S., Mueller J. F. and Keller J. (2010). Removal of micropollutants and reduction of biological activity in a full scale reclamation plant using ozonation and activated carbon filtration. *Water Research*, **44**(2), 625–637.

Ribeiro A. R., Afonso C. M., Castro P. M. L. and Tiritan M. E. (2013). Enantioselective biodegradation of pharmaceuticals, alprenolol and propranolol, by an activated sludge inoculum. *Ecotoxicology and Environmental Safety*, **87**, 108–114.

Rodarte-Morales A., Feijoo G., Moreira M. and Lema J. (2011). Degradation of selected pharmaceutical and personal care products (PPCPs) by white-rot fungi. *World Journal of Microbiology and Biotechnology*, **27**(8), 1839–1846.

Rosenberger S., Krüger U., Witzig R., Manz W., Szewzyk U. and Kraume M. (2002). Performance of a bioreactor with submerged membranes for aerobic treatment of municipal waste water. *Water Research*, **36**(2), 413–420.

Rotchell J. M. and Ostrander G. K. (2003). Molecular markers of endocrine disruption in aquatic organisms. *Journal of Toxicology and Environmental Health-Part B-Critical Reviews*, **6**(5), 453–495.

Sahar E., Messalem R., Cikurel H., Aharoni A., Brenner A., Godehardt M., Jekel M. and Ernst M. (2011). Fate of antibiotics in activated sludge followed by ultrafiltration (CAS-UF) and in a membrane bioreactor (MBR). *Water Research*, **45**(16), 4827–4836.

Schroder H. F., Tambosi J. L., Sena R. F., Moreira R. F. P. M., Jose H. J. and Pinnekamp J. (2012). The removal and degradation of pharmaceutical compounds during membrane bioreactor treatment. *Water Science and Technology*, **65**(5), 833–839.

Serrano D., Suárez S., Lema J. M. and Omil F. (2011). Removal of persistent pharmaceutical micropollutants from sewage by addition of PAC in a sequential membrane bioreactor. *Water Research*, **45**(16), 5323–5333.

Shariati F. P., Mehrnia M. R., Salmasi B. M., Heran M., Wisniewski C. and Sarrafzadeh M. H. (2010). Membrane bioreactor for treatment of pharmaceutical wastewater containing acetaminophen. *Desalination*, **250**(2), 798–800.

Shon H. K., Vigneswaran S. and Snyder S. A. (2006). Effluent organic matter (EfOM) in wastewater: constituents, effects, and treatment. *Critical Reviews in Environmental Science and Technology*, **36**(4), 327–374.

Slack R. J., Zerva P., Gronow J. R. and Voulvoulis N. (2005). Assessing quantities and disposal routes for household hazardous products in the United Kingdom. *Environmental Science & Technology*, **39**(6), 1912–1919.

Snyder S. A., Adham S., Redding A. M., Cannon F. S., DeCarolis J., Oppenheimer J., Wert E. C. and Yoon Y. (2007). Role of membranes and activated carbon in the removal of endocrine disruptors and pharmaceuticals. *Desalination*, **202**(1–3), 156–181.

Spain J. C., Pritchard P. H. and Bourquin A. W. (1980). Effects of adaptation on biodegradation rates in sediment/water cores from estuarine and freshwater environments. *Applied and Environmental Microbiology*, **40**(4), 726–734.

Spengler P., Karner W. and Metzger J. W. (2001). Substances with estrogenic activity in effluents of sewage treatment plants in southwestern Germany. 1. Chemical analysis. *Environmental Toxicology and Chemistry*, **20**(10), 2133–2141.

Spring A. J., Bagley D. M., Andrews R. C., Lemanik S. and Yang P. (2007). Removal of endocrine disrupting compounds using a membrane bioreactor and disinfection. *Journal of Environmental Engineering and Science*, **6**(2), 131–137.

Stasinakis A. S. (2012). Review on the fate of emerging contaminants during sludge anaerobic digestion. *Bioresource Technology*, **121**, 432–440.

Stasinakis A. S., Kordoutis C. I., Tsiouma V. C., Gatidou G. and Thomaidis N. S. (2010). Removal of selected endocrine disrupters in activated sludge systems: effect of sludge retention time on their sorption and biodegradation. *Bioresource Technology*, **101**(7), 2090–2095.

Stevens-Garmon J., Drewes J. E., Khan S. J., McDonald J. A. and Dickenson E. R.V. (2011). Sorption of emerging trace organic compounds onto wastewater sludge solids. *Water Research*, **45**(11), 3417–3426.

Strenstrom M. K., Cardinal L. and Libra J. (1989). Treatment of hazardous substances in wastewater treatment plants. *Environmental Progress*, **8**(2), 107–112.

Suarez S., Carballa M., Omil F. and Lema J. M. (2008). How are pharmaceutical and personal care products (PPCPs) removed from urban wastewaters. *Reviews in Environmental Science and Biotechnology*, **7**(2), 125–138.

Suarez S., Lema J. M. and Omil F. (2010). Removal of harmaceutical and personal care products (PPCPs) under nitrifying and denitrifying conditions. *Water Research*, **44**(10), 3214–3224.

Suarez S., Reif R., Lema J. M. and Omil F. (2012). Mass balance of pharmaceutical and personal care products in a pilot-scale single-sludge system: influence of T, SRT and recirculation ratio. *Chemosphere*, **89**(2), 164–171.

Sui Q., Huang J., Deng S., Chen W. and Yu G. (2011). Seasonal variation in the occurrence and removal of pharmaceuticals and personal care products in different biological wastewater treatment processes. *Environmental Science and Technology*, **45**(8), 3341–3348.

Tadkaew N., Sivakumar M., Khan S. J., McDonald J. A. and Nghiem L. D. (2010). Effect of mixed liquor pH on the removal of trace organic contaminants in a membrane bioreactor. *Bioresource Technology*, **101**(5), 1494–1500.

Tadkaew N., Hai F. I., McDonald J. A., Khan S. J. and Nghiem L. D. (2011). Removal of trace organics by MBR treatment: the role of molecular properties. *Water Research*, **45**(8), 2439–2451.

Taheran M., Brar S. K., Verma M., Surampalli R. Y., Zhang T. C. and Valéro J. R. (2016). Membrane processes for removal of pharmaceutically active compounds (PhACs) from water and wastewaters. *Science of the Total Environment*, **547**, 60–77.

Tam L. S., Tang T. W., Lau G. N., Sharma K. R. and Chen G. H. (2007). A pilot study for wastewater reclamation and reuse with MBR/RO and MF/RO systems. *Desalination*, **202**(1–3), 106–113.

Tambosi J. L., de Sena R. F., Favier M., Gebhardt W., Jose H. J., Schroder H. F. and Moreira R. d. F. P.M. (2010). Removal of pharmaceutical compounds in membrane bioreactors (MBR) applying submerged membranes. *Desalination*, **261**(1–2), 148–156.

Tanghe T., Devriese G. and Verstraete W. (1998). Nonylphenol degradation in lab scale activated sludge units is temperature dependent. *Water Research*, **32**(10), 2889–2896.

Tauxe-Wuersch A., De Alencastro L. F., Grandjean D. and Tarradellas J. (2005). Occurrence of several acidic drugs in sewage treatment plants in Switzerland and risk assessment. *Water Research*, **39**(9), 1761–1772.

Ternes T. A. (1998). Occurrence of drugs in German sewage treatment plants and rivers. *Water Research*, **32**(11), 3245–3260.

Ternes T. A., Stumpf M., Mueller J., Haberer K., Wilken R. D. and Servos M. (1999). Behavior and occurrence of estrogens in municipal sewage treatment plants – I. Investigations in Germany, Canada and Brazil. *Science of the Total Environment*, **225**(1–2), 81–90.

Ternes T. A., Herrmann N., Bonerz M., Knacker T., Siegrist H. and Joss A. (2004a). A rapid method to measure the solid-water distribution coefficient (K d) for pharmaceuticals and musk fragrances in sewage sludge. *Water Research*, **38**(19), 4075–4084.

Ternes T. A., Joss A. and Siegrist H. (2004b). Peer reviewed: scrutinizing pharmaceuticals and personal care products in wastewater treatment. *Environmental Science & Technology*, **38**(20), 392A–399A.

Terzic S., Matosic M., Ahel M. and Mijatovic I. (2005). Elimination of aromatic surfactants from municipal wastewaters: comparison of conventional activated sludge treatment and membrane biological reactor. *Water Science and Technology*, **51**(6–7), 447–453.

Thompson A., Griffin P., Stuetz R. and Cartmell E. (2005). The fate and removal of triclosan during wastewater treatment. *Water Environment Research*, **77**(1), 63–67.

Topalova Y., Kozuharov D. and Dimkov R. (1999). Adaptive response of trivial activated sludge towards toxic effect of oNP, PCP and combination oNP/PCP. *Waste Management*, **19**(4), 247–254.

Tran N. H., URASE T. and KUSAKABE O. (2010). Biodegradation characteristics of pharmaceutical substances by whole fungal culture *Trametes versicolor* and its laccase. *Journal of Water and Environment Technology*, **8**(2), 125–140.

Urase T., Kagawa C. and Kikuta T. (2005). Factors affecting removal of pharmaceutical substances and estrogens in membrane separation bioreactors. *Desalination*, **178**(1–3), 107–113.

Vader J. S., van Ginkel C. G., Sperling F. M. G. M., de Jong J., de Boer W., de Graaf J. S., van der Most M. and Stokman P. G. W. (2000). Degradation of ethinyl estradiol by nitrifying activated sludge. *Chemosphere*, **41**(8), 1239–1243.

Valladares L. R., Yangali-Quintanilla V., Li Z. and Amy G. (2011). Rejection of micropollutants by clean and fouled forward osmosis membrane. *Water Research*, **45**(20), 6737–6744.

Verliefde A. R., Cornelissen E. R., Heijman S., Petrinic I., Luxbacher T., Amy G., Van der Bruggen B. and Van Dijk J. (2009). Influence of membrane fouling by (pretreated) surface water on rejection of pharmaceutically active compounds (PhACs) by nanofiltration membranes. *Journal of Membrane Science*, **330**(1–2), 90–103.

Verliefde A. R. D., Heijman S. G. J., Cornelissen E. R., Amy G., Van der Bruggen B. and van Dijk J. C. (2007). Influence of electrostatic interactions on the rejection with NF and assessment of the removal efficiency during NF/GAC treatment of pharmaceutically active compounds in surface water. *Water Research*, **41**(15), 3227–3240.

Vieno N. M., Tuhkanen T. and Kronberg L. (2005). Seasonal variation in the occurrence of pharmaceuticals in effluents from a sewage treatment plant and in the recipient water. *Environmental Science & Technology*, **39**(21), 8220–8226.

Vieno N. M., Tuhkanen T. and Kronberg L. (2007). Elimination of pharmaceuticals in sewage treatment plants in Finland. *Water Research*, **41**(5), 1001–1012.

Visvanathan C., Aim R. B. and Parameshwaran K. (2000). Membrane separation bioreactors for wastewater treatment. *Critical Reviews in Environmental Science and Technology*, **30**(1), 1–48.

Yang W., Paetkau M. and Cicek N. (2010). Improving the performance of membrane bioreactors by powdered activated carbon dosing with cost considerations. *Water Science and Technology*, **62**(1), 172–179.

Wang S. and Wang J. (2017). Carbamazepine degradation by gamma irradiation coupled to biological treatment. *Journal of Hazardous Materials*, **321**, 639–646.

Weiss S. and Reemtsma T. (2008). Membrane bioreactors for municipal wastewater treatment – A viable option to reduce the amount of polar pollutants discharged into surface waters? *Water Research*, **42**(14), 3837–3847.

Wick A., Marincas O., Moldovan Z. and Ternes T. A. (2011). Sorption of biocides, triazine and phenylurea herbicides, and UV-filters onto secondary sludge. *Water Research*, **45**(12), 3638–3652.

Wijekoon K. C., Fujioka T., McDonald J. A., Khan S. J., Hai F. I., Price W. E. and Nghiem L. D. (2013). Removal of N-nitrosamines by an aerobic membrane bioreactor. *Bioresource Technology*, **541**, 248–252. DOI: 10.1016/j.biortech.2013.01.057.

Wijekoon K. C., Hai F. I., Kang J., Price W. E., Cath T. Y. and Nghiem L. D. (2014a). Rejection and fate of trace organic compounds (TrOCs) during membrane distillation. *Journal of Membrane Science*, **453**, 636–642.

Wijekoon K. C., Hai F. I., Kang J., Price W. E., Guo W., Ngo H. H., Cath T. Y. and Nghiem L. D. (2014b). A novel membrane distillation–thermophilic bioreactor system: biological stability and trace organic compound removal. *Bioresource Technology*, **159**, 334–341.

Williams M., Ong P. L., Williams D. B. and Kookana R. S. (2009). Estimating the sorption of pharmaceuticals based on their pharmacological distribution. *Environmental Toxicology and Chemistry*, **28**(12), 2572–9.

Wu Q.-Y., Hu H.-Y., Zhao X. and Sun Y.-X. (2009). Effect of chlorination on the estrogenic/antiestrogenic activities of biologically treated wastewater. *Environmental Science & Technology*, **43**(13), 4940–4945.

Xia K., Bhandari A., Das K. and Pillar G. (2005). Occurrence and fate of pharmaceuticals and personal care products (PPCPs) in biosolids. *Journal of Environmental Quality*, **34**(1), 91–104.

Xiao Y., Yaohari H., De Araujo C., Sze C. C. and Stuckey D. C. (2017). Removal of selected pharmaceuticals in an anaerobic membrane bioreactor (AnMBR) with/without powdered activated carbon (PAC). *Chemical Engineering Journal*, **321**, 335–345.

Xie M., Nghiem L. D., Price W. E. and Elimelech M. (2012a). Comparison of the removal of hydrophobic trace organic contaminants by forward osmosis and reverse osmosis. *Water Research*, **46**(8), 2683–2692.

Xie M., Price W. E. and Nghiem L. D. (2012b). Rejection of pharmaceutically active compounds by forward osmosis: role of solution pH and membrane orientation. *Separation and Purification Technology*, **93**, 107–114.

Xu J., Wu L. S. and Chang A. C. (2009). Degradation and adsorption of selected pharmaceuticals and personal care products (PPCPs) in agricultural soils. *Chemosphere*, **77**(10), 1299–1305.

Xu Y., Zhou Y., Wang D., Chen S., Liu J. and Wang Z. (2008). Occurrence and removal of organic micropollutants in the treatment of landfill leachate by combined anaerobic-membrane bioreactor technology. *Journal of Environmental Sciences*, **20**(11), 1281–1287.

Xue W., Wu C., Xiao K., Huang X., Zhou H., Tsuno H. and Tanaka H. (2010). Elimination and fate of selected micro-organic pollutants in a full-scale anaerobic/anoxic/aerobic process combined with membrane bioreactor for municipal wastewater reclamation. *Water Research*, **44**(20), 5999–6010.

Yamamoto T. and Yasuhara A. (2002). Chlorination of bisphenol A in aqueous media: formation of chlorinated bisphenol A congeners and degradation to chlorinated phenolic compounds. *Chemosphere*, **46**(8), 1215–1223.

Yang S., Hai F. I., Nghiem L. D., Nguyen L. N., Roddick F. and Price W. E. (2013a). Removal of bisphenol A and diclofenac by a novel fungal membrane bioreactor operated under non-sterile conditions. *International Biodeterioration & Biodegradation*, **85**, 483–490.

Yang S., Hai F. I., Nghiem L. D., Price W. E., Roddick F., Moreira M. T. and Magram S. F. (2013b). Understanding the factors controlling the removal of trace organic contaminants by white-rot fungi and their lignin modifying enzymes: a critical review. *Bioresource Technology*, **141**, 97–108.

Yeo B. J., Goh S., Zhang J., Livingston A. G. and Fane A. G. (2015). Novel MBRs for the removal of organic priority pollutants from industrial wastewaters: a review. *Journal of Chemical Technology & Biotechnology*, **90**(11), 1949–1967.

Yi T. and Harper W. F., Jr (2007a). The effect of biomass characteristics on the partitioning and sorption hysteresis of 17-α -ethinylestradiol. *Water Research*, **41**(7), 1543–1553.

Yi T. and Harper W. F., Jr (2007b). The link between nitrification and biotransformation of 17α-ethinylestradiol. *Environmental Science and Technology*, **41**(12), 4311–4316.

Yi T., Harper W. F., Jr, David H. R. and Love N. G. (2006). Role of particle size and ammonium oxidation in removal of 17 α-ethinylestradiol in bioreactors. *Journal of Environmental Engineering*, **132**(11), 1527–1529.

Ying G.-G., Williams B. and Kookana R. (2002). Environmental fate of alkylphenols and alkylphenol ethoxylates – a review. *Environment International*, **28**(3), 215–226.

Yoon Y., Westerhoff P., Snyder S. A. and Wert E. C. (2006). Nanofiltration and ultrafiltration of endocrine disrupting compounds, pharmaceuticals and personal care products. *Journal of Membrane Science*, **270**(1), 88–100.

Zaviska F., Drogui P., Grasmick A., Azais A. and Heran M. (2013). Nanofiltration membrane bioreactor for removing pharmaceutical compounds. *Journal of Membrane Science*, **429**, 121–129.

Zeng Q., Li Y. and Gu G. (2009). Nitrate-dependent degradation of 17α-ethinylestradiol by acclimated activated sludge under anaerobic conditions. *Journal of Chemical Technology and Biotechnology*, **84**(12), 1841–1847.

Zhang B., Yamamoto K., Ohgaki S. and Kamiko N. (1997). Floc size distribution and bacterial activities in membrane separation activated sludge processes for small-scale wastewater treatment/reclamation. *Water Science and Technology*, **35**(6), 37–44.

Zhang L. Q., Urase T. and Feng L. (2008). Removal of carbamazepine in an enhanced membrane bioreactor with small dose addition of powdered activated carbon. Proceedings of the International Conference on Advances in Chemical Technologies for Water and Wastewater Treatment, 199–205.

Zhang T., Liu Y. and Fang H. H. P. (2005). Effect of pH change on the performance and microbial community of enhanced biological phosphate removal process. *Biotechnology and Bioengineering*, **92**(2), 173–182.

Zhang Y. and Geißen S.-U. (2012). Elimination of carbamazepine in a non-sterile fungal bioreactor. *Bioresource Technology*, **112**, 221–227.

Zhou P., Su C., Li B. and Qian Y. (2006). Treatment of high-strength pharmaceutical wastewater and removal of antibiotics in anaerobic and aerobic biological treatment processes. *Journal of Environmental Engineering*, **132**(1), 129–136.

Zwiener C. and Frimmel F. H. (2003). Short-term tests with a pilot sewage plant and biofilm reactors for the biological degradation of the pharmaceutical compounds clofibric acid, ibuprofen, and diclofenac. *Science of the Total Environment*, **309**(1–3), 201–211.

Chapter 14

Commercial MBR technologies and selected case studies

Graeme Pearce

Membrane Consultancy Associates Ltd., Little Eastfield, Eastfield Lane, Whitchurch, Reading RG8 7EJ, UK

Correspoding author: graemekpearce@btinternet.com

ABSTRACT

This chapter reviews the current status of membrane bioreactors and examines the diversity of material, format and configuration used in commercial products. The chapter also considers the common features of product design and operational method. Though the first commercial products were introduced about 50 years ago, it is only in the last 15–20 years that there has been widespread acceptance and take-up of the technology across industrial and municipal applications. This commercial success can be put down to the introduction of the immersed configurations introduced in the early 1990s in both flat sheet and hollow fibre formats. The energetically extravagant side-stream configuration was then consigned to difficult applications operating at small scale. However, interestingly, the side-stream configuration has made a comeback in the last few years through the use of air lift concept. Also, side-stream systems have reduced their reliance on cross-flow velocity by using backwash where possible, which has significantly reduced their energy usage.

The market is now dominated by relatively few companies, though there are very many specialists with a focus on niche applications or specific geographic regions. All of the companies have made substantial progress in making their offerings more competitive, particularly in terms of energy requirements. The chapter concludes with a review of case studies illustrating key features of the different formats, and highlighting design and operational improvements from various suppliers. End-user experience is provided to show how different product features have worked out in the field.

 Membrane Biological Reactors: Theory, Modeling, Design, Management and Applications to Wastewater Reuse
Editor(s): Faisal I. Hai, Kazuo Yamamoto and Chung-Hak Lee
doi: 10.2166/9781780409177_0469

14.1 INTRODUCTION TO COMMERCIAL PRODUCTS

14.1.1 Background

Membrane bioreactors (MBRs) were first commercialized about 50 years ago (Wiesmann *et al.*, 2007). The concept was based on the simple premise that membranes could be used synergistically with the activated sludge process to enhance biological oxidation in several ways (Leiknes, 2010). Firstly, the membrane would provide a barrier for the biomass of the biological reaction stage which would enable the solids concentration in the bioreactor to be increased, thereby reducing footprint. Also, the permeate quality would be improved since the membrane would provide an absolute barrier to particles above the membrane pore size, which would improve filtrate quality dramatically. Finally, the air used for the bioreactor could serve the dual purpose of scouring the membrane surface to provide stable performance. Subsequent experience showed that getting a cost-effective system was not straightforward (Pollet *et al.*, 2008) and it took many years before commercial products were adopted widely.

While the advantages of increased solids and better filtrate quality have been relatively easy to realize, issues around membrane fouling and clogging and the efficient use of air have been much more difficult to optimize and have been responsible for the relatively slow uptake of MBR (Brepols, 2012; Gil, 2011). Today's MBR products have significant diversity which reflects the different approaches that have been taken primarily to achieve stable operation and an efficient use of air. In this respect, MBR is somewhat similar to the general field of membrane filtration (MF), which even within the water sector is characterized by diversity in the product offering since different approaches have been taken to achieve optimization of the module and process design (Pearce, 2011). This is in distinct contrast to RO where a common approach has been taken to design and operation, resulting in a commoditized market with interchangeable products (Wilf *et al.*, 2007).

The first wave of commercialization took place in Japan and started during the 1980s as small flat sheet systems were used for treating the wastewater from individual buildings as well as apartment and office blocks. The 1990s saw the uptake of hollow fibre-based systems developed with the idea that MBR could potentially be applied at a larger scale (Lesjean *et al.*, 2008). Alongside these two immersed formats, other companies used tubular systems, initially with crossflow and eventually utilizing air to assist with recirculation, the so-called air lift (AL) design (Pollet *et al.*, 2008).

The upsurge of uptake for MBR gathered pace soon after 2000 as both municipal and industrial users became convinced of the merits of the technology (Judd, 2008). Growth was initially limited due to the perception of the technology as only being applicable for niche opportunities. However, the development of Asian economies with less existing infrastructure for conventional wastewater treatment, transfer, and distribution, as well as increasing interest in treatment for reuse (Katz *et al.*, 2016) has given rise to a sharper uptake and a period of strong sustained growth. Since 2007, a steady trickle of large-scale projects has occurred for indirect potable reuse (IPR) and direct potable reuse (DPR) schemes, particularly in North America and China, and this has massively increased the installed capacity of MBR (Pankratz, 2017). The high flowrate of most of the reuse schemes (e.g., Katz & Grundestam, 2016) has made hollow fibre offerings completely dominant in terms of the share of installed capacity, but flat sheet systems are still preferred at small scale.

The current range of MBR products continues to reflect the initial diversity of product format and system configuration evident in the original developments (Judd *et al.*, 2011). Although products still mainly rely on supported polymeric membranes (e.g., Suez and Mitsubishi), unsupported polymerics (such as those from Asahi and Memstar) have recently increased their share and ceramics are also making in-roads, thus making the market even more diverse (see Section 14.2.2 for detailed product descriptions).

Application requirements in wastewater treatment vary depending on biological and chemical loading of the feed, as well as solids concentration and plant scale. The early products addressed market needs in a variety of ways, but each had a strong niche, which has had the effect of preserving diversity as products have been optimized within the constraints of the original concept. Indeed, different formats have substantial cross-over, and compete closely for main stream municipal wastewater applications at certain plant sizes or within constraints of footprint, energy use, etc.

The first section of this chapter will address the general issues of membrane materials, product format and system configuration, discussing the range of technology and the reasons for such diversity. The chapter will then continue with a review of the products. The MBR market is dominated by relatively few long established leading players who have been present since the inception of the market and have largely been responsible for both developing the technology and gaining end-user acceptance. The leading hollow fibre suppliers are Suez (formerly GE) with the Zenon product range and the relative newcomer from China, Beijing Origin Water (BOW). Together, these two companies account for well over 50% market share in terms of treated water capacity (see Section 14.2.1). The supplier with the largest number of installations is Kubota, with their flat sheet system. Outside of the short list of dominant players, there is a very extensive list of suppliers, many of whom are application specialists or who have a regional focus. Many have their own technology differentiation, but there are an increasing number of examples of interchangeable products such as BOW, Memstar, Koch and Econity, which all target retrofits of each other and of Suez and Mitsubishi (see further discussion in Section 14.2.2).

The product review in this chapter focuses on 13 of the market leading MBR companies, reflecting market share and influence, as well as representing the range of products and formats. Examples from these suppliers are used in the final section of the chapter which presents some selected case studies to describe the activity and innovation of the most influential companies in the market while also illustrating features of all the main format and configuration options.

14.1.2 Membrane materials

14.1.2.1 Polymeric membranes

Membranes for MF applications need to balance several characteristics, some of which may be in conflict (Pearce, 2011). The membrane needs to have a good level of permeability, which requires a high pore density and an open structure, but it also needs to have strength so that it has a good level of integrity, does not break, and has a long life. If used in a process operation with air bubbles or an air scouring action, the membrane needs to have flexibility so that it does not suffer from fatigue damage and can withstand the stress at the potting interface. The membrane should have surface properties that resist fouling, and needs a high level of chemical resistance so that it can be cleaned with aggressive chemicals.

Five polymer families are used for fabricating ultrafiltration (UF) and microfiltration (MF) membranes in water and wastewater applications. However, the choice of polymer families for MBR is more limited than for the general field of MF, due to the onerous nature of the wastewater treatment duty, and the rigorous operational environment. Flexibility is required for the large majority of MBR duties due to the importance of air scour for immersed formats, whereas this characteristic is only desirable for just over half of MF duties, i.e. those using an outside feed flow configuration which utilizes air scour in backwash (Ng *et al.*, 2005).

For MBR, strength and flexibility are more important than permeability, and Cl_2 resistance is seen to be essential since it is used for both cleaning and disinfection. The MBR membrane therefore needs to be mechanically robust, chemically resistant to high Cl_2 concentrations used in cleaning, and non-biodegradable. Clean water permeability is not as important in an MBR as in MF applications, since the membrane transport properties will be strongly influenced by the accumulation of foulant particles at the membrane surface. However, process flux in treating a wastewater feed is important since it will directly affect capital cost, due to its effect on membrane area and footprint, and operating costs due to the effect of membrane area on chemical and air use.

The physical demands of MBR has meant that many products use a supported membrane, which is virtually unheard of in general MF duties since the support adds cost and reduces packing density. However, the additional cost is more acceptable in an MBR, and the issue of packing density is not as critical. The low fluxes used in MBR allow different fibre dimensions to be considered than for general UF/MF.

If the membrane is in a hollow fibre format, it is normally possible to reverse the flow even for a supported membrane, which allows a backwash cycle to be performed. However, some supported hollow fibre membrane products do not use backwash, even though they could so, in order to simplify the process sequence, reduce downtime and increase recovery. Most flat sheet formats and tubular formats do not allow reverse flow, since there is a danger of de-lamination in which the active layer of the membrane would be parted from the support if a backwash was undertaken. This limitation on the possible use of backwash is an important consideration in selecting a suitable format, since backwash can be a simple and efficient means of ensuring stable performance. Formats which do not allow backwash compensate by using different cleaning cycles.

Unsupported membranes are strongly preferred for general water treatment duties due to better packing density, higher permeability, and lower cost (Pearce, 2011), but historically were viewed as insufficiently robust for MBR (Judd *et al.*, 2011). However, in recent years there has been an increasing tendency to also use these membranes for MBR, potentially with an increased wall thickness, or especially if using a particularly robust symmetric structure. Whereas the market share of these products used to be very low, they have now overtaken the flat sheet format in terms of installed capacity, largely due to the fact that they have been used as an option in the booming reuse market in China.

An important distinction between the membranes used for general UF/MF and those used for MBR is the pore-size rating. Water treatment duties mainly use UF at around 0.02 μm or fine MF at around 0.04 μm. When membranes were first considered for water duties, there were some advocates of MF membranes with a 0.2 μm rating. However, the relatively low solid concentrations and low particle sizes in the feed water meant that MF membranes could give unstable performance due to pore plugging with rapidly rising trans-membrane pressure (Mulder, 1996). This would need to be countered by aggressive and frequent backwashing and cleaning. There are still some water treatment products at true MF ratings of up to 0.1 μm, but these products are not as common as UF.

Either UF or MF can be a good choice for wastewater treatment. The high solids concentration and larger particle size often found in wastewater feeds means that the filtration mechanisms are different. The wastewater solids readily form a pre-coat at the membrane surface, protecting the actual pores, and maintain stable permeability, especially at low operating pressures. UF can still have some advantages by providing a virus barrier, but MF achieves stable performance and is preferred by a significant proportion of suppliers. For MBR, many products utilize MF, especially those with flat sheets. Immersed hollow fibre products may offer either UF or MF depending on the supplier, but tubular products are more likely to use UF.

The increase in interest in reuse and the need to meet regulatory standards has shone a spotlight on the effect of membrane ageing on the ability of an MBR to consistently achieve the required level of pathogen removal (Branch & Le-Cech, 2015; Hai *et al.*, 2014; Mosher & Vartanian, 2018). Supported membranes generally work well initially, but nicks and abrasions in operation make it impossible to maintain the log removal value (LRV) credit in a pressure decay test (PDT) using a clean water feed in a laboratory test. Surface damage creates the potential for leak paths through the support layer. However, in reality, removal efficiency does not suffer due to the self-repairing nature of the MBR challenge, thus a supported membrane system does achieve a consistent LRV in situ, as demonstrated in a multi-site survey (Katz *et al.*, 2017). The study even included LRV test results after intentional membrane damage to confirm the ability of the system to self-repair. The findings have been endorsed by an ageing study, which showed that the LRV was effectively unchanged over a 10 year operational time period on a single test site (Branch *et al.*, 2015). Unsupported membranes, especially if using a hollow fibre with a symmetric cross-section, would not pose the same theoretical concern and would be expected to maintain LRV even in laboratory test conditions.

14.1.2.2 Ceramic membranes

While ceramic membranes have not been widely applied yet in either general MF applications in the water industry or specifically for MBR, there are signs that the advantages of this type of material are beginning to overcome the cost barriers (Judd, 2017). In general water treatment, ceramics offer the potential to use much higher flux combined with longevity and aggressive cleaning options. In MBR, the main advantages claimed are longevity and a low fouling propensity. The hydrophilic nature of the material makes a normal backwash highly efficient and reduces the frequency of chemical cleaning.

There are a few examples of commercial ceramic MBR products, using both flat sheet and tubular approaches, such as Meiden and Cembrane, Pall Membralox, and Tami. Ceramic membranes are normally made by coating successively finer layers on an open support. Alumina is the most common support material and the active layer could be alumina, titania, or zirconia. Silicon carbide is also used both as a support and as an active layer.

The multi-layer manufacturing process used for ceramic membranes tends to result in a lower permeability for UF compared to MF in long-term operation with a real feed. In contrast, polymeric UF and MF have similar permeability when operated with real feeds even though pure water permeability would be much higher for MF. As a consequence, ceramic MF tends to be used more often than UF, though both ratings are found in ceramic MBR products. Section 14.2 will provide brief comments on various ceramic products.

14.1.2.3 Membrane materials selected for commercial products

The MBR industry first developed in Japan with the use of chlorinated polyethylene (PE) flat sheet membrane by Kubota, and PE fibres by Mitsubishi (formerly Mitsubishi Rayon Engineering (MRE) then Mitsubishi Rayon Company (MRC)). PE has reasonable strength and flexibility, and the different modifications made by the two companies to the basic polymer improved properties like wettability and resistance to chlorine. PE is normally made as an MF membrane, and even then, it has relatively low permeability, so process fluxes of PE membranes tend to be relatively low. The main polymer options for MBR are summarized in Table 14.1.

In the 1990s, polyvinylidene difluoride (PVDF) became established in MBRs, firstly as a reinforced capillary fibre in Zenon's ZW 500 product range. The disadvantage of PVDF is that it is significantly more expensive as a polymer, but it has impressive performance in terms of strength and flexibility, and overcomes price disadvantage by achieving higher flux. In addition, PVDF membranes may be more durable, which has an important impact on operating cost, and hence on total water cost. Many companies offer PVDF products in both capillary and flat sheet formats, with the result that PVDF is now by far the dominant membrane polymer in the MBR market.

Table 14.1 Polymer selection.

PVDF	Limited ability to modify properties
	Excellent strength, flexibility, and permeability
	Coarse supported UF or MF in hollow fibres, flat sheets, and tubes
PE	Inexpensive, with reasonable strength and good flexibility
	Can be made hydrophilic
	Susceptible to oxidation, but chlorination improves resistance
	Unsupported MF hollow fibre or supported MF flat sheet
PES	Ability to modify properties through polymer blend
	Good strength and permeability
	Coarse supported UF produced for MBR
	Unsupported UF option used in AL

The final significantly used membrane polymer in MBR is polyethersulphone (PES) which may be reinforced, though it is much less widely used than the previous two materials. Companies offering PES-based products include Pentair-Xflow and Koch-Puron (though Koch now focuses on their reinforced PVDF offering). PES is also offered by some small speciality flat sheet MBR suppliers. PES is an important polymer in water treatment, but in wastewater applications, its lack of flexibility limits the possibility of using air scour, particularly due to the effect of stress at the potting interface. Reinforcing the capillary, for example with a polyester braid, does allow air scour, though at the expense of permeability.

There is some use of other polymers in MBR, but these tend to be used by just one supplier. One example is the use of polyethylene tetrafluoride (PTFE) in the Sumitomo poreflon membrane. PTFE has exceptionally high strength and chemical compatibility, but is well known to be strongly hydrophobic, an unwelcome feature in water contact applications. However, Sumitomo uses a hydrophilic surface modification.

14.1.3 Module format

Whereas membrane performance is critical in MF, MBRs are more heavily reliant on module and process design, the method of operation, and the performance of the bioreactor (Judd, 2005). This section examines the format options that have emerged in MBR for design and operation from the various suppliers.

In MBR technology, two broad trends have emerged, namely submerged MBRs or immersed MBRs (iMBR) and side-stream MBRs (sMBR). Immersed technologies tend to be more cost effective for larger scale lower wastewater strength applications, and side-stream technologies are favoured for smaller scale higher strength applications. Although 'submerged' is the preferred term used in general MF, 'immersed' is more widely used in MBR, partly because of the nomenclature clash with the abbreviation for side-stream, i.e. sMBR. The original side-stream MBR (sMBR) was based on crossflow, which gave high performance and high flux, but at a significant energy cost. The sMBR envelope has been extended in recent years by the development of the AL concept, which bridges the gap between iMBR and crossflow sMBR, and may have the potential to challenge immersed systems in larger scale applications (see Section 14.2.4.2).

The immersed and side-stream approaches are each commercially available in two configurations making a total of four MBR formats, as shown in Figure 14.1. For example, the two immersed formats from Suez (with the Zenon product offering) and Kubota, respectively, use hollow fibre membranes (sometimes referred to as capillary) and flat sheet. The side-stream approach of Pentair and other smaller suppliers also divides into two formats. These are the long-established traditional method of crossflow, now used only for the most difficult feeds, and the newer but now well-established concept of AL, which uses air to recirculate the feed and thereby significantly reduces energy demand. Both side-stream formats use tubular membranes, though Pentair has recently introduced an option with a 3 mm diameter tube, which is more properly described as hollow fibre.

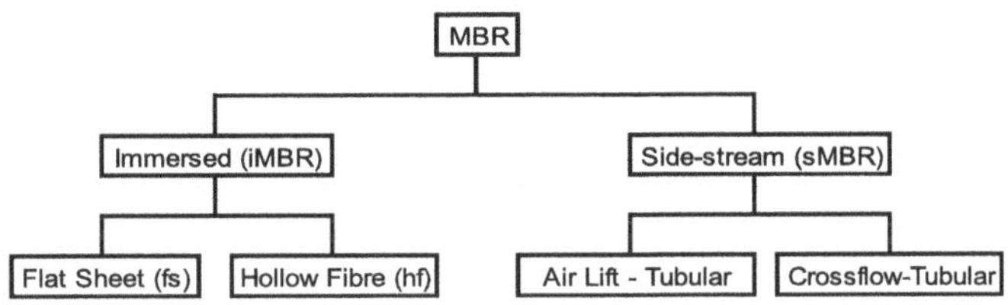

Figure 14.1 MBR formats.

Figure 14.2 shows schematic representation of the two immersed modules, with a flat sheet system on the left and hollow fibre on the right. The flat sheet systems normally use a series of membrane panels, in which a sandwich of membrane sheets, with the active layer facing outwards, captures a permeate channel, which is sealed from the feed stream. The spacing between panels is normally around 6–7 mm, which creates a feed channel.

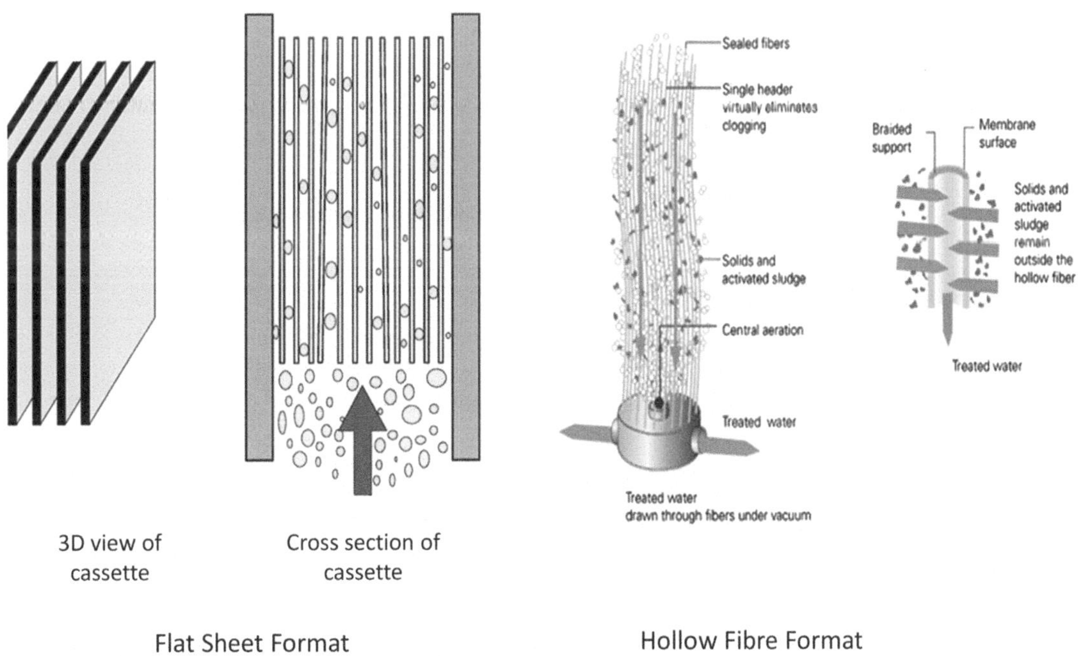

Figure 14.2 Flat sheet and hollow fibre immersed formats.

This relatively wide spacing ensures that clogging problems are minimized. The downside of using a wide spacing is that packing density is rather low, since the channel dimension is akin to that found in a tubular product. Coarse bubbles are used to keep the feed channel clean, and as indicated in the diagram, the module design has been developed to capture the air and utilize it as efficiently as possible by forcing it to enter the channel spacing. Air efficiency can be enhanced by stacking flat sheet modules on top of each other, so double stack designs are common. Flat sheet systems are particularly cost effective at the small to medium scale, and with feeds that could cause clogging. In recent years, the optimum envelope has been extended in some cases to very large-scale duties with straightforward municipal wastewater feeds, due to improved air scour efficiency (see Section 14.2.3.1).

Hollow fibre systems tend to use a much narrower effective channel dimension due to the close packing of fibres with a typical outside diameter of 2–3 mm. Also, fluxes are similar to or even higher than those of flat sheet systems, so this confers an advantage in both footprint and air use. Furthermore, bubble size can be optimized to improve cleaning efficiency, since the bubble both acts as a mechanical scouring agent, but also importantly moves the fibre as it rises through the bundle to remove foulant particles by a shaking action.

The downside of hollow fibres can be that they are prone to clogging, with foulant becoming stuck between the fibres, especially near the potting interfaces (Stefanski *et al.*, 2011). Fibrous matter and hair is a particular problem for this format, and so close attention needs to be paid to pre-treatment with a carefully selected screen designed for the removal of a fibrous challenge. Capital costs can be higher than for flat sheet systems due to pre-treatment requirements and the effects of the economies of scale, so hollow fibre-based MBRs are often chosen for medium-to-large-scale duties with a feed that is easily treatable and less prone to clogging.

Figure 14.3 illustrates how the different technological approaches are most cost effectively applied. The figure represents a spectrum of applications, with high flow, easily treated feeds on the left, moving towards the more difficult feeds and lower flow applications on the right. Municipal feeds tend to be easier to treat than industrial feeds, since they normally have lower biological and chemical oxygen demand (BOD and COD), and are generally more consistent. However, in some countries, storm run off can dilute the wastewater burden and increase flowrates several fold, thereby significantly changing feed characteristics.

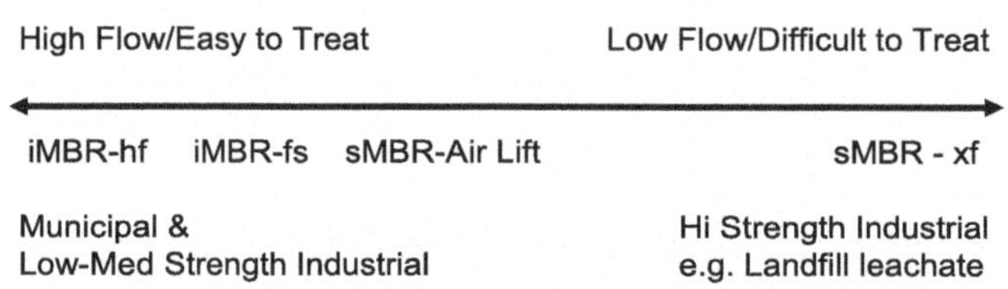

Figure 14.3 Application characteristics of different MBR format options.

Industrial feeds vary widely from one application to another, and may combine several sources. In addition, applications such as landfill leachate are notoriously challenging, due to high loadings, especially COD, and the presence of recalcitrant organics that are difficult to breakdown by biological oxidation (Wiesmann *et al.*, 2007).

In general terms, iMBR formats based on immersed hollow fibres (iMBR-hf) have been found to provide the most cost-effective solution for large scale, easy-to-treat applications. Technology has been developed with optimized packing density and aeration bubble size to achieve stable performance at minimum energy use (Fane *et al.*, 2005). However, the iMBR-hf format can experience operational difficulties due to fibres becoming matted close to the potted ends, so pre-treatment and removal of hairs and fibres is essential.

iMBR formats based on flat sheets (iMBR-fs) have been found to be cost effective for similar types of feed (Adham *et al.*, 2007), but due to higher air use and lower capex, tend to be selected for small-to-medium scale duties. The flat sheet format has operational advantages in terms of plugging and cleaning, and has been used with somewhat more difficult feeds.

The AL format is the most recent development of the MBR formats. It has been developed as a low-energy alternative to the energetically expensive crossflow side-stream format, which has been used historically for the most difficult feeds. The energy cost of crossflow effectively rules it out as a treatment option for any application other than small scale or where there is no other treatment option. However, the AL can have low-energy use, and may even undercut the energy requirements of the immersed options, due to the advantage of containment of the feed and air bubbles inside the tubular membrane. Since AL eliminates operator contact and has good operational characteristics, it could make an impact for medium scale applications in the MBR market in the longer term. AL has also been used with tubular ceramic membranes, which though expensive, show lower fouling propensity than polymeric membranes (Judd *et al.*, 2004). Recent developments of the AL concept show promise in broadening its place on the spectrum in Figure 14.3, as will be discussed further in the commercial product review in Section 14.2.

14.1.4 System configuration

The original MBR designs were based on the simple idea of installing the immersed membrane modules into the biological aeration tank and using the air required for biological oxidation to serve a dual purpose. Air bubbles would both provide the oxidation for the biological reaction as well as scouring the membrane surface and preventing clogging of the feed channels. Since these early beginnings, significant thought has been put into designing systems that operate effectively, and use the minimum amount of air possible, since the energy cost of aeration turns out to be a key component of total treatment costs.

Figure 14.4 shows a general arrangement of the various MBR process flow options for both immersed and side-steam formats (Melin *et al.*, 2006). The initial designs used a single tank with a coarse bubble diffuser located under the membrane modules. Movement of the air bubbles through the module scoured the surface and created a circulation of fluid. Air was provided in excess to feed the biological reaction requirements.

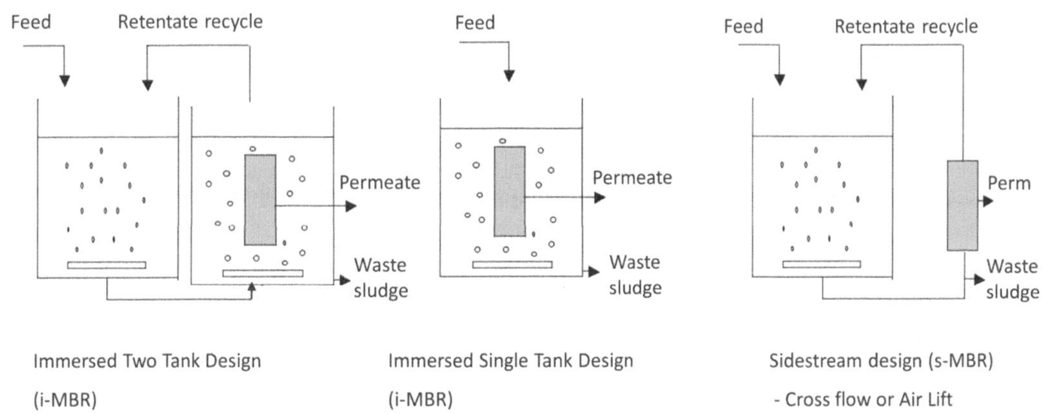

Figure 14.4 System configuration options for different formats.

Figure 14.4 indicates that while a fine bubble size is normally preferred in the oxidation tank to facilitate oxygen uptake, a coarse bubble is preferred in the membrane tank to enhance the cleaning action. The bubble size and air flow can be optimized independently for the two tank arrangement. Furthermore, whereas the air used in the membrane tank needs to be carefully directed and contained to achieve maximum effect, it is beneficial for the air in the reaction tank to be dispersed and distributed as widely as possible. The manufacturers' product review will demonstrate that there are different approaches to air use, distribution and bubble size, but all designs have one thing in common, which is to minimize the use of air as far as possible.

A key advantage of MBR is allowing operation at higher MLSS than conventional treatment, since there is a membrane barrier preventing loss of solids from the system. High MLSS increases the process intensity and reduces footprint. However, a feed with a high MLSS has high viscosity and will require relatively high aeration rates. If the MLSS and fluxes are kept lower, energy efficiency will improve (Verrecht *et al.*, 2008). There are, therefore, several design considerations which need to be balanced to optimize both capital and operating costs, and the two tank design improves the opportunity for optimization considerably. Sizing the biological reaction tank allows the HRT and SRT to be precisely controlled, while the membrane tank size is minimized to ensure minimum hold up which reduces cleaning chemical inventories and improves recovery.

Since 2000, the energy efficiency of MBR has evolved dramatically, and energy use is now 15 times less than the early commercial products (Peeters *et al.*, 2013), largely due to optimization of air usage and bubble size. Several companies offer MBR with an energy use of <0.1 kWhr/m^3, even as low as 0.05 kWhr/m^3. However, it is still higher than conventional activated sludge (CAS) since MBR is effectively penalized when flow is reduced. Recent developments in process automation have reduced aeration proportionately with flow at low plant utilization to address the disadvantage. Even without flow proportional aeration, MBR can have a lower life cycle cost than CAS in scenarios requiring enhanced nutrient removal and/or water reuse (Young *et al.*, 2013).

Even though there is diversity both within and between formats with different dimensions and porting arrangements, as well as different operational procedures, there is some opportunity for retrofitting with a different supplier within the same format. Thus hollow fibre products from several suppliers have dimensions which are close enough for retrofits, but there are others which are basically unique in design. As the market has matured, there has been a gradual consolidation around preferred dimensions. Furthermore, newer entrants from China such as BOW and Memstar, have ensured that their products imitate existing market leaders closely enough that retrofit projects have become a realistic option and there are significant number of retrofit projects. There are fewer flat sheet suppliers, but here too there is some opportunity for retrofits. It is probably true that there are more retrofits in MBR than in MF in general, but it is still the exception rather than the rule.

An important unit operation not shown in Figure 14.4 is provided by the pre-treatment strainers. It is much more likely that MBR problems will be the result of clogging of feed channels or of the underside of the tube-sheet (i.e., the potting interface), than a problem with the membrane such as permeability or integrity loss. For hollow fibre systems, it is normal to use a fairly fine screen of 1 or 2 mm, though coarser ratings have also been considered. The screen is normally preceded by a coarse 6 mm pre-screen at the Works inlet. Screen ratings are determined by the nature of the feed contaminants with fibres and hairs providing a particular challenge to products with a hollow fibre format, since they are prone to becoming trapped below the top header. Flat sheet systems often use a significantly coarse strainer of 3 mm, perhaps eliminating it altogether for some duties. However, there are questions marks over the wisdom of using too coarse a screen for flat sheets regardless of duty; feeds with fibres or hair are particularly challenging for both formats (Ewert, 2014; Stefanski *et al.*, 2011).

The final process flow arrangement shown on the right-hand side of Figure 14.4 is for side-stream formats. This design enshrines the separation of the biological reaction stage from the membrane stage. If not for the recirculation line, the arrangement would resemble an activated sludge process followed by MF, and indeed there are some commercial MBR offerings which are closely related to or even identical to straightforward MF products. The 13 market leading products described in this review will focus on commercial offerings which have been specifically designed for MBR applications.

14.1.5 Product nomenclature

The use of terminology can be confusing, since different terms can be used to describe essentially the same part of the system, and sometimes the same term is used by different suppliers for a different part. At the heart of the system is the membrane. This is packaged into a component, and the component is further assembled into a product to become operable in a system. The component is often called an element in hollow fibre and a cartridge or panel in flat sheet. The product could be called a module, but this term is also used at the component level. Other common terms for the product are cassette, unit, or stack, but terms are also taken from MF such as rack. A summary of terms used for the two main formats are shown in Figure 14.5.

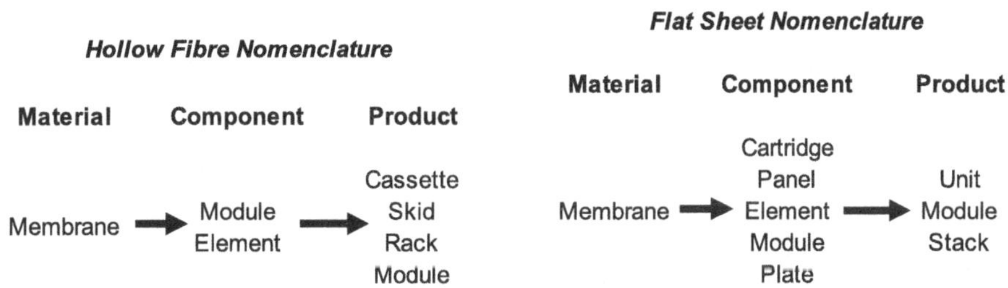

Figure 14.5 Nomenclature for components and products of the two main MBR formats.

Tubular formats tend to use the words tube or sometimes hollow fibre to describe the membrane. Generally, the term tubular is used for a capillary with an outside diameter above 3 mm and hollow fibre for $\leq$3 mm. The module is used to describe the encapsulated tubes or fibres in a housing or pressure vessel. The terms for an assembly of modules are similar to those in MF and include racks, skids, or banks.

14.2 MANUFACTURERS' REVIEW

14.2.1 Overview

The MBR web site www.thembrsite.com, has around 80 companies registered as MBR product or technology suppliers, and although fairly comprehensive, not all suppliers are represented on this list. For example, a report on the China MBR Market Outlook (Frost & Sullivan, 2011) suggests that there are 150 MBR suppliers in China alone, though presumably many of these companies do not manufacture their own membranes.

The MBR web site lists nine of the 80 companies as being providers of the world's largest MBR installations with a capacity >100 MLD. In 2011 (Judd, 2011), the combined treatment capacity of the largest plants was 1900 MLD, but by 2018, this had increased more than seven-fold. Figure 14.6 uses a slightly broader definition of large plants as those being >76 MLD (20 MGD) and illustrates the dramatic growth in the market since the first 'large' plant installed by Suez (i.e., GE-Zenon at that time) at Cleveland Bay, Australia in 2007.

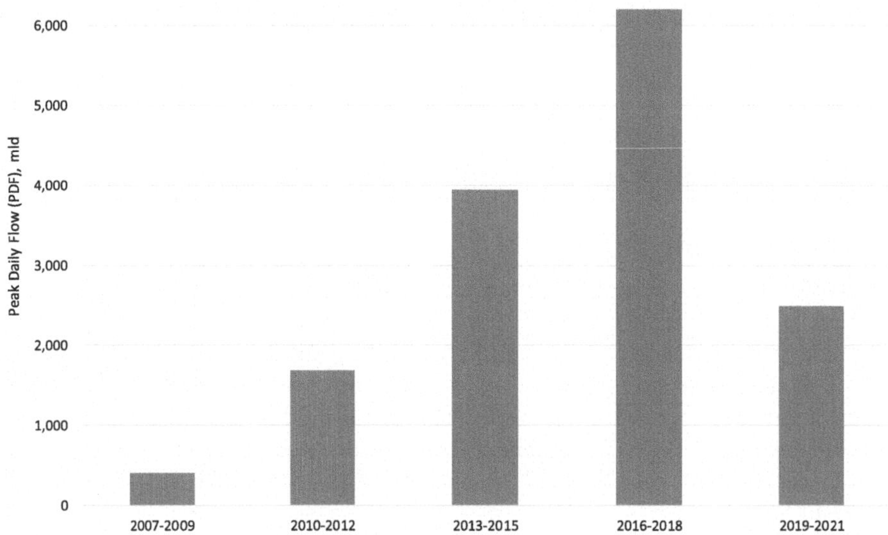

Figure 14.6 Flow capacity of the 20 largest MBR installations, MLD.

The current reference list used for the analysis in this section has been largely drawn from the MBR web site, but has also been expanded using supplier references to include a number of additional plants not listed by the site, thereby providing a comprehensive review of the growth of large systems. The list includes 82 plants >76 MLD installed or under construction from 2007 to date, and has a combined capacity of almost 15,000 MLD.

Figure 14.7 shows where the regional growth has occurred. The width of the columns shows the increasing size of the market in terms of installed flow capacity, while the height on the bar shows the split between the main continental regions. China has been separated from the rest of Asia to illustrate the importance of this market and the dramatic growth that has been experienced. In the last 6 years, China has accounted for 69% of the total installed flow capacity of large plants.

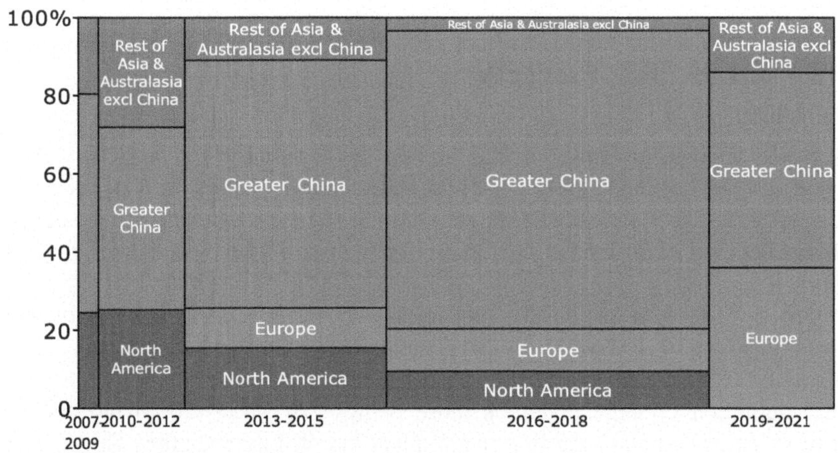

Figure 14.7 Regional trends for MBR installations >76 MLD.

The world's largest plant is currently stated to be the 864 MLD Suez plant at Henriksdal near Stockholm (Katz & Grundestam, 2016), but this will soon be over taken by the 1040 MLD Beijing Origin plant at Beihu, Hubei. An even larger plant of 1200 MLD has recently been announced at Tuas, Singapore, but this is not included in the list since the MBR supplier has not yet been selected.

Interestingly, Figure 14.7 shows that Europe has begun to implement large MBR schemes for reuse whereas in North America, MBR growth has stalled in favour of schemes using CAS followed by MF, due to the easier path to regulatory compliance for in-situ testing. Although this stance is challenged by MBR suppliers (Katz, 2017), several large-scale US reuse schemes have failed to select MBR. However, there is a general industry expectation that MBR will be considered for future potential schemes on an equal footing with CAS-MF/UF (Erdal, 2017) and that regulatory hurdles for proof-testing will soon be overcome.

Figure 14.8 shows the supplier share for these large plants. The picture has changed fundamentally from the large plant review presented in the 1st edition of this book. At that time, Suez had 60% share of both treatment capacity and number of plants from the large plant list. The installed capacity and number of plants of Suez have both now fallen to about 25% of the 82 plants and the Suez share of large capacity plants has been significantly exceeded by BOW. Memstar has also increased its share significantly, and together with BOW, these two companies dominate the huge market for MBR in China. Up to 2013, BOW acted as a system integrator largely supplying systems with either Mitsubishi or Asahi MBR. However, since then, the company has changed to become a vertically integrated player like Suez, and progressively used its own MBR technology, giving it an overall share of the large plant market of 43%. Between them, BOW, Suez, and Memstar account for around 85% of installed capacity of the worldwide MBR market for large plants, with the remaining capacity split fairly evenly between the other six players. Note, however, that the share of Mitsubishi is also substantial when account is taken of projects won through BOW, in which BOW has acted simply as an EPC with Mitsubishi as the specialist MBR supplier.

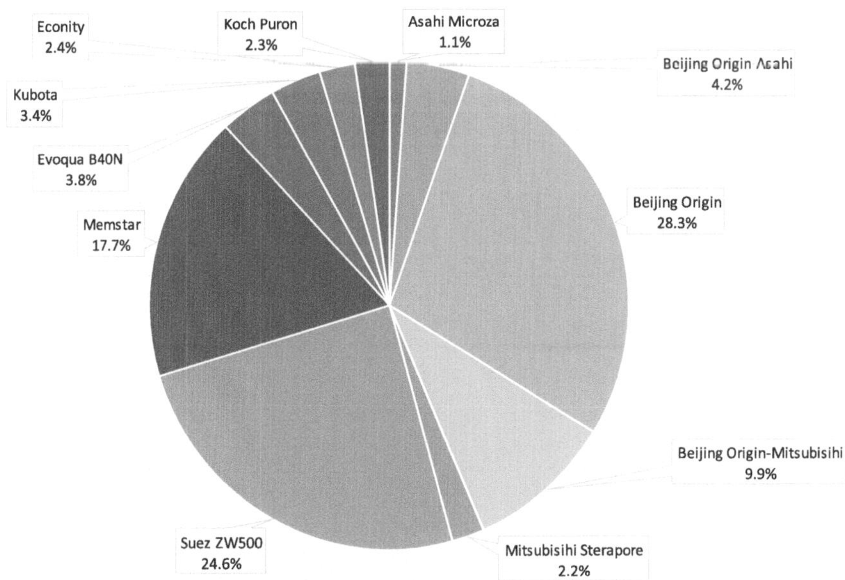

Figure 14.8 Supplier share for MBR installations >76 MLD, 2007–2019.

The supplier review used in this section shows that the four companies with the longest established commercial MBR products, i.e., Kubota, Mitsubishi, Suez (Zenon), and Econity (formerly Korea Membrane Separations) have more than 14,000 references between them. The total installed capacity of MBR of all suppliers for any plant size is thought to approach 25,000 MLD with well over half of that capacity being added since the 1st edition of this book (i.e., within approximately the last five years). Suez is still the dominant player overall accounting for over 40% of the total cumulative installed capacity at all plant sizes.

Figure 14.9 is based on data from the Global MBR Market Report (Frost & Sullivan, 2013). When compared with the latest information presented in Figure 14.8, Figure 14.9 provides a perspective of how market share has changed since the 1st edition of this book. The chart shows that prior to the explosion of reuse projects in China, which occurred around 2011, the market was more diverse with a single dominant player, i.e., the former GE Water Technologies, now Suez. The chart presents share in terms of value of all products sold, rather than being based on the installed capacity share as in Figure 14.8. Two features are apparent. Firstly, Chinese suppliers then had a small share (note that 70% of the BOW share at this time was based on Mitsubishi). Secondly, in value terms, small- and medium-scale MBR made up an important proportion of market value with 'Others' having a significant contribution to the market share.

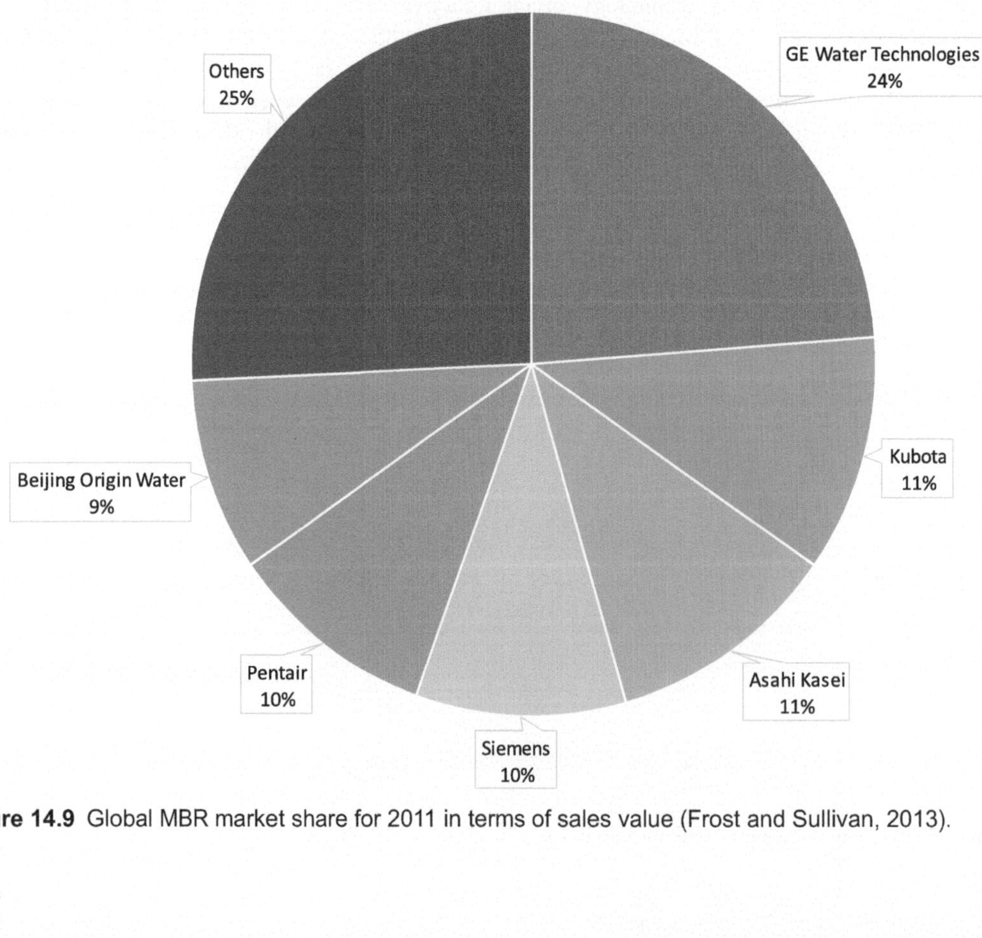

Figure 14.9 Global MBR market share for 2011 in terms of sales value (Frost and Sullivan, 2013).

Significant companies which will be included in the product review which do not feature in the large plant list include Pentair-Xflow, Berghof, Toray, and Meiden. Though not yet having a major market share, Pentair-Xflow is the leading exponent of the AL concept using a tubular format and this has gradually gained some degree of traction as an alternative option to the entrenched position of the immersed formats. A brief mention is also made of some tubular crossflow suppliers, primarily Berghof, even though the market share of this format is relatively low.

Toray has a flat sheet offering with similarities to Kubota and has a significant number of small-scale references. The product review also has a limited discussion of ceramics for MBR with a focus on Meiden. Although the technology has yet to make a significant impact, it does have a reference position and could offer significant potential for the future. Finally, mention is made of innovative or unusual products which are outside of the mainstream offerings and have a unique and/or promising approach which could provide differentiation.

Some of these technologies achieve innovation through module design and others through the mode of operation. Table 14.2 provides a summary of the market leading group in this review, with the overall dominant suppliers in each of the two immersed formats, i.e., Suez and Kubota, at the top. Hollow fibre offerings are then listed alphabetically. The final grouping contains the two main suppliers offering the tubular format.

The information provided throughout this section is based on submissions from the manufacturers solicited specifically for this book, supplemented by recent standard product literature and website downloads (see Appendix 1, A1.2). Where a company has two distinctly different membrane offerings, two entries are provided in the table. The table shows that PVDF is the most common membrane polymer for MBR and the most common format is hollow fibre. Mitsubishi initially offered only PE, but in recent years have put significantly more emphasis on reinforced PVDF. Similarly, Koch's original offering was PES but there is now an almost exclusive focus on the new PVDF product, introduced in 2010. Where the table includes two entries for one supplier, both products are still actively promoted.

Although hollow fibre has an overwhelming share of annual total installed capacity, flat sheet systems are more numerous since they are preferred at small scale. Also, flat sheet has been established longer, so the total cumulative treated capacity is more significant. Since flat sheet systems are often favoured by new start-ups and niche players for small duties, the schemes may fail to be counted in the overall market data.

14.2.2 Immersed hollow fibre

14.2.2.1 Suez

Prior to the acquisition by Suez in 2017, the ZeeWeed membrane, developed by Zenon and later owned by GE, had already become a leading product in wastewater treatment (Zsirai, 2012). Zenon's initial involvement in wastewater dates back to the 1980s when the Permaflow crossflow tubular system was introduced. However, Zenon has been better known for the development of the ZeeWeed immersed hollow fibre membrane. This was first introduced as a small product in 1993, namely the ZW-145, which was followed shortly after by the ZW-150. The company was one of the three early movers in the market, together with Kubota and Mitsubishi Rayon Engineering (now MRC). In 1997, Zenon introduced the ZW-500A module, which was followed by improved and larger modules in quick succession up to the introduction of the ZW-500D in 2002. This module still forms the basis of the current product line, though it now incorporates numerous design and operational improvements, which have reduced energy cost substantially.

Table 14.2 Summary of leading manufacturers[a].

	Membrane	Pore Size mm	Membrane Format	Module Format	Product Name	Flux lmh
Suez	PVDF	0.04	hf	iMBR	ZW 500d, LEAPmbr	25–30 (55)
Asahi Kasei	PVDF	0.1	hf	iMBR	Microza	
BOW	PVDF	0.3	hf	iMBR	RF-III, MBRU	Up to 25
Econity	PE	0.4	hf	iMBR	C-66D (C series)	13–21
Econity	PVDF	0.1	hf	iMBR	E-66D (E series)	17–24
Evoqua-Memcor	PVDF	0.04	hf	iMBR	MemPulse B40	17–24
KMS	PVDF	0.03	hf	iMBR	PURON, PULSION	25–50
Memstar/CEL	PVDF	0.04	hf	iMBR	SMM	15–25
Mitsubishi	PVDF	0.4	hf	iMBR	STERAPORE	Up to 33
Kubota	Cl_2 PE	0.4	fs	iMBR	EK510	17–24
Toray	PVDF	0.08	fs	iMBR	Toray	29
Meiden	Ceramic	0.1	fs	iMBR	CH250-1000E02K-5NA	25–60
Berghof	PES & PVDF	0.006–0.03	tub	sMBR	HyMem, B-SMART	45–150
Pentair-Xflow	PVDF	0.03	tub	sMBR	AirLift F4835	50–60 (80)
Pentair-Xflow	PES	0.02	tub	sMBR	Megablock	Up to 65

Abbreviation: CEL – Citic Envirotech Ltd.
[a]See Appendix 1, A1.1.

The ZW-500 membrane is a reinforced PVDF hollow fibre mounted vertically in a module and potted at both ends. Figure 14.10 shows a photo of the ZW-500 cassette and Table 14.3 shows the features of the membrane, module and cassettes for the Suez product family based around the ZW-500D membrane. The ZW-500 membrane is effectively unchanged since the product was first launched, and has been found to be resilient with a long membrane life of up to 20 years. Indeed, an early ZeeWeed reference at Thetis Lake trailer park in Canada has only just had its first membrane re-fit after slightly more than 20 years and many systems have exceeded a 10 year life (e.g., Angiel *et al.*, 2018; Natvik *et al.*, 2015).

Figure 14.10 ZW-500 MBR cassette (courtesy of Suez).

Suez has continued to develop and improve the ZeeWeed 500 product range. The company's approach has been to improve the cost effectiveness and operability of the MBR product through a steady evolution of engineering design and operating procedures. The key advance has been the introduction of intermittent aeration. Using air scour continuously not only requires a lot of energy, but is not fully effective. Furthermore, the size of air bubbles makes a major difference to efficiency with large slug bubbles much more effective than smaller ones (Sepideh & Bérubé, 2013).

Soon after 2000, intermittent or cyclic aeration was introduced, initially based on a 10 second on/10 second off cycle, which halved the energy requirement for air scour. Over the next few years, it was found that this could be halved again, especially for easily treatable feeds, by extending the off period to 30 seconds. Finally, in 2011, the company introduced LEAPmbr to reduce aeration to 7% of the original continuous aeration system through a combination of improvements. The acronym LEAP is derived from a description of the product benefits, i.e., 'low energy aeration process'. Some of the improvements are process related and have been achieved by optimizing the bubble size, using moderate size coarse bubbles interspersed with occasional large or slug bubbles (Berube, 2007). Others are physical changes to the air distribution system and design of the air release method.

The LEAPmbr improvements have allowed energy to be reduced by 30% in comparison with the previous benchmark. At the same time, the lower fouling rates from the use of slug bubbles has increased productivity by 15% through the use of higher fluxes. Suez claims a very low specific aeration demand (SADm) of about 0.081 Nm^3/h per m^2 membrane. The energy for air scour is now of the order of 0.1 $kWhr/m^3$, which is comparable to the energy required for recirculation. In addition, LEAPmbr has

Table 14.3 Suez product details[a].

Membrane		Module				Cassette			
Brand Name	**ZW-500**	**Product Name**	**ZW-500**			**Product Name**	**ZW-500**		
UF or MF	UF	Model identifier	500M	500S	500D	No. of modules/identifier	4M	16M	52M
Polymer	PVDF, supported	Ht × width, mm	712 × 131	1660 × 264	2198 × 49	Ht × width, mm	2149 × 738	1870 × 729	2561 × 1744
Hydrophilicity	Moderate	Fibre length, m	~0.3	~1.55	~1.9	Fibre length, m	~1.55	~1.55	~1.9
Pore size, pm	0.04	Length or depth, mm	116	210	830	Length or depth, mm	1744	1828	2136
Fibre I.D, mm	0.8	Membrane area, m^2	0.93	27.9	34.4	Membrane area, m^2	446	up to 446	up to 1788
Fibre O.D, mm	1.9	Max TMP, bar	0.55	0.55	0.55	Max TMP, bar	0.55	0.55	0.55
Cl_2 resistance, ppm	1000	Temp. rating, °C	40	40	40	Temp. rating, °C	40	40	40
Cleaning pH range	2–10.5	Shipping weight, kg	3	22	28	Shipping weight, kg	721	295	1863

[a]See Appendix 1, A1.2.2.1.

25% lower footprint, and also simplified design with less maintenance since some cyclic valves are eliminated.

The next development by Suez was an anaerobic MBR which was launched in 2013 (Zsirai, 2012). Anaerobic MBRs are sometimes considered for high strength feeds, for example from the food industry. They have the potential of generating biogas while simultaneously producing a treated wastewater for reuse (see Chapter 11 for details).

Note that in 2015, Suez introduced a membrane-aerated biofilm reactor (MABR) known as ZeeLung (Peeters *et al.*, 2015). Although the product physically resembles the ZeeWeed MBR, the MABR is not a version of an MBR and will not be covered in any detail in this section. In MABR, the membrane does not fulfil a filtration function, but serves to supply oxygen to a biofilm. The membrane performs the function of efficiently passing oxygen to a biofilm that grows on the outside of the fibres, thereby acting as an air diffuser and a support for bio-film growth. The drivers for MABR are energy and footprint saving. The technology should be considered to be a low operating cost alternative to CAS, or as a means of process intensification for expanding a CAS facility which is space constrained. The product was initially tested by the Metropolitan Water Reclamation District in Chicago and by Thames Water in the UK and now several full-scale installations are in operation.

14.2.2.2 Asahi Kasei

Asahi Kasei has a well-established MF membrane which was introduced for water treatment applications in the late 1990s as a pressure-driven outside feed module under the Microza brand name. Outside of Asia, Asahi has marketed the MF Microza product through a marketing arrangement with Pall, and the modules have been widely applied both in drinking water and in recent years for wastewater reuse following conventional activated sludge. In 1999, Asahi Kasei introduced an unencapsulated version of the same unsupported PVDF membrane for MBR applications, also using the Microza brand name, and this is marketed directly across the world, and has also been marketed through BOW in China (see Section 14.2.2.3). Membranes are potted in small bundles with several bundles potted in a circular end cap as shown in Figure 14.11. The original 620A product is no longer supplied but it is compatible with the current 620AllT. Details of the current product (620AllT) are shown in Table 14.4.

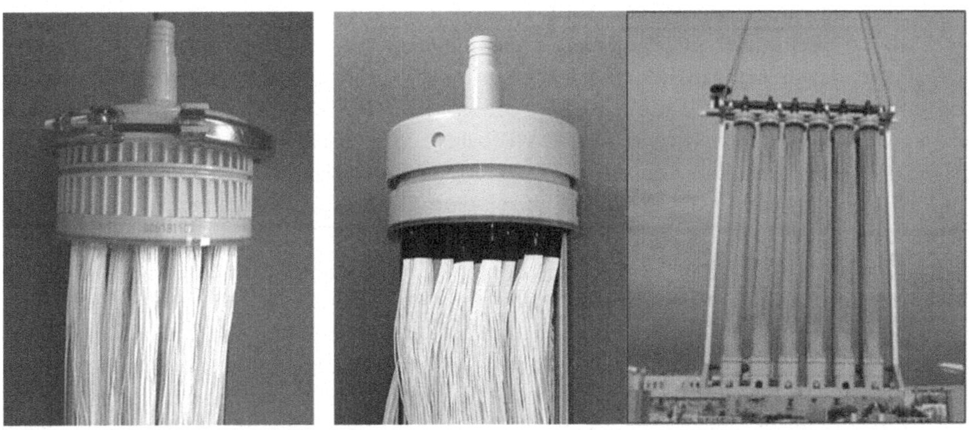

Figure 14.11 Microza™ MUNC-620A (original) left, MUNC-620AllT (current) centre, 12 module skid right (see Appendix 1, A1.1.2.2).

Asahi uses a PVDF fibre produced using a thermal-induced phase separation process (TIPS). This creates different characteristics to the products from Suez and Evoqua, which are made by a non-solvent-induced phase separation process (NIPS). TIPS fibres tend to have a lower degree of asymmetry, with the result that it is challenging (though not impossible) to obtain a UF rating with sufficient permeability for commercial use in water or wastewater applications. However, TIPS fibres are known for displaying excellent mechanical properties both in terms of strength and flexibility and are perfectly satisfactory as an MF membrane. Thus even though unsupported, the fibre can achieve low fibre breakage rates and long life. In fact, the Asahi fibre is close to symmetric, having a uniform distribution of 0.1 μm pores through the fibre wall cross-section. This confers an important benefit in MBR since surface scratches and abrasion have very little impact on membrane integrity.

The composite bundle has a diameter of 167 mm and an active length of around 2 m. The membrane area of each bundle is 25 m^2, which suggests that a relatively low packing density is used compared to the water treatment product. Four bundles are mounted on a frame for small-scale units with 12 or 24 bundles used for the standard commercial skid. The module is potted in polyurethane. Aeration is introduced as for other designs in a skirted end at the base of the module, forcing air bubbles via a series of 10 mm holes and through the bundle as shown in Figure 14.12. Asahi claims reasonable efficiency at low air scour rates with a specific aeration demand (SADm) of about 0.24–0.28 Nm^3/h per m^2 membrane.

The typical process sequence for the Asahi system is 9 minutes filtration followed by 1 minute backwash, alternated with a filtration sequence having 1 minute relaxation. There is a continuous aeration requirement of 6–7 Nm^3/h per module. An advantage of the TIPS fibre is that chemical resistance is improved compared to NIPS PVDF fibres. Asahi typically uses a maintenance clean with 0.1% NaOCl once per week with a 90 minute soak and oxalic or citric acid depending on foulants. A recovery clean with 0.5% NaOCl and 1% caustic might be used twice per year, also with oxalic or citric acid as required. A photo of the membrane tank for an Asahi-Kasei municipal wastewater MBR is shown in Figure 14.13.

Asahi recommends a pre-treatment strainer with a 1 mm rating. Modules can be immersed in the biological reactor, or in a separate membrane tank. A recirculation velocity of at least 3Q, i.e. recirculation of at least 3 times the filtrate production is recommended to ensure stable operation.

14.2.2.3 Beijing Origin Water

BOW, sometimes simply referred to as Origin Water (OW), was established in 2001, and achieved its first large-scale MBR project in 2009. The company claims a cumulative installed capacity in the Chinese water and wastewater market of 20,000 MLD, of which more than 6,000 MLD is in large-scale MBR, according to the survey of large plants >76 MLD discussed in Section 14.2.1. BOW claims a market share of 70% of the Chinese water market. This is consistent with the MBR share of large plants in which, as a system integrator, it also has a 70% share of the market in China. Up to 2013, the company mainly supplied large-scale plants based on MBR membranes from Asahi or Mitsubishi, with Mitsubishi accounting for 70% of the share of BOW MBR sales. Since then, the company has progressively introduced its own MBR.

Standard specification sheets are not readily available for download from the BOW website, so the description below is based on material from conference presentations. The BOW membrane is a 0.3 μm PVDF hollow fibre, which is assumed to be supported, i.e., resembling the Mitsubishi product (see Section 14.2.2.8). The BOW MBR unit (MBRU) has physical similarities to products from both Mitsubishi and Suez (see Section 14.2.2.1). The BOW product has a unit (equivalent to a module for Mitsubishi and cassette for Suez) comprising a multiplicity of thin elements (referred to as 'membrane cases' by BOW). As is also the case for Mitsubishi, there are three element or membrane case sizes available for small-, medium- and large-scale plants, but the surface areas are different. The largest

Table 14.4 Asahi-Kasei product details.

Membrane		Module		Skid		
Brand name	Microza	Product name	MUNC-620AIIT	Product name	M UNC-620AIIT	M UNC-620AIIT
UF or MF	MF	Model identifier		No. of modules/identifier	12	24
Polymer	PVDF	Ht × width, mm	2,163 × 167	Ht × width, mm	2,800 × 460	3,100 × 920
Hydrophilicity	Moderate	Fibre length, m	2.0	Fibre length, m	2.0	2.0
Pore size, mm	0.1	Length or depth, mm		Length or depth, mm	1,400	1,400
Fibre I.D, mm	0.7	Membrane area, m^2	25	Membrane area, m^2	300	600
Fibre O.D, mm	1.3	Max TMP, bar	0.6	Max TMP, bar	0.6	0.6
Cl_2 resistance, ppm	Up to 5,000″	Temp rating, °C	40	Temp rating, °C	40	40
Cleaning pH range	1–12[x]	Shipping Weight, kg	14 (wet)	Shipping weight, kg		

Note: [x]Asahi PVDF has relatively high chlorine and caustic resistance.

Table 14.5 Econity PE product details (C series)[a].

Membrane		Cartridge Module		Cassette			
Brand Name	**Econity**	**Product Name**	**C Series**	**Product Name**	**KSMBR & eMBR**		
UF or MF	MF	Model identifier		No. of modules/identifier	CF-23S	CF35-D	CF-66D
Polymer	HDPE	Ht × width, mm		Ht × width, mm	605 × 784	1,210 × 1146	1300 × 2264
Hydrophilicity	Hydrophilized	Fibre length, m	0.3	Fibre length, m	~0.6	~0.6	~0.6
Pore size, μm	0.4	Length or depth, mm		Length or depth, mm	1,520	2,320	2,810
Fibre I.D, mm	0.41	Membrane area, m^2		Membrane area, m^2	101	504	1,210
Fibre O.D, mm	0.65	Max TMP, bar	0.6	Max TMP, bar	0.6	0.6	0.6
Cl_2 resistance, ppm	10^6 ppm-hrs	Temp rating, °C	35	Temp rating, °C	35	35	35
Cleaning pH range	2–13	Shipping weight, kg		Shipping weight, kg	95	432	1120

[a]See Appendix 1, A1.2.2.4.

membrane case has a membrane area of 35 m^2, which is almost identical to Suez element, but significantly higher in area than the largest Mitsubishi element which has an area of 25 m^2. The 35 m^2 membrane case has a product identifier RF-lll. Up to 60 RF-lll membrane cases can be incorporated in a 'unit', thus is slightly larger than the Suez equivalent, which has 52 modules. The dimensions of the BOW unit are similar to the Mitsubishi modules, but have a significantly higher membrane area. The Suez cassette is lower in height and wider but is more similar in membrane area. BOW claims to operate at a nominal flux of up to 25 lmh, which is at the bottom end of the range of flux for the normal average daily flux for Suez (Figure 14.14).

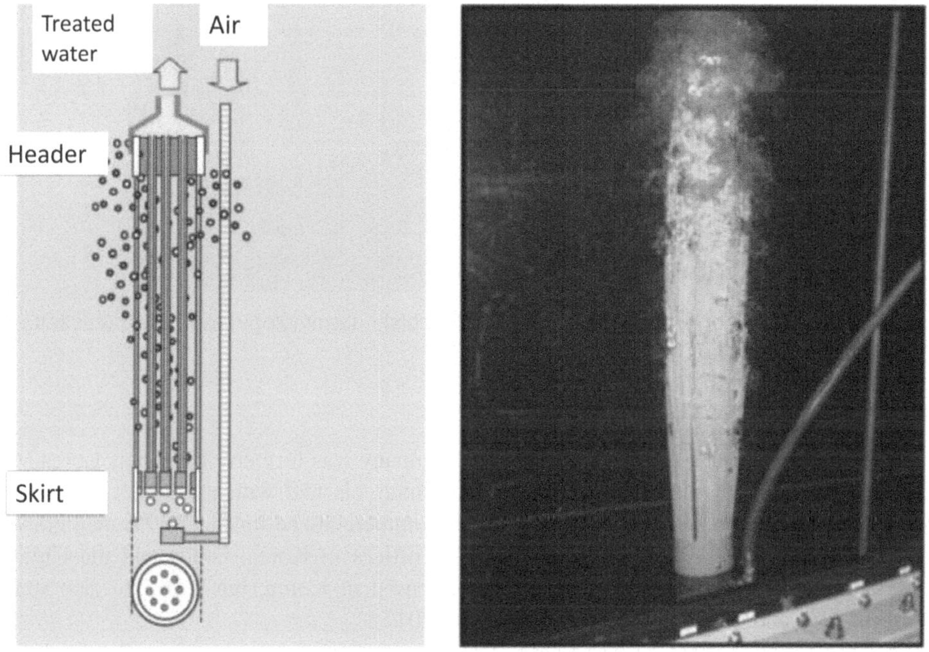

Figure 14.12 Aeration scheme for the Asahi module.

Figure 14.13 Asahi-Kasei MBR at Paju City, Korea, installed by GS E&C/Silver Star Engineering, 2012.

Figure 14.14 BOW membrane case (element in Mitsubishi terminology) and unit (cassette in Suez terminology) (see Appendix 1, A1.2.2.3).

14.2.2.4 Econity

Econity is the leading supplier of MBR in Korea. The company was formerly known as Korea Membrane Separations (KMS), but for international markets the initials clashed with the longer established Koch Membrane Systems. Econity was founded in 1998 and commercialized its current PE membrane in 2003 and PVDF membrane in 2014. The company now has offices in Korea, China and the United States. Most of the long reference list of 2,100 references is based in Korea, but there are also some recent installations in China and the United States (Kim *et al.*, 2017).

Econity has two product series. The original C series uses an unsupported 0.4 μm PE stretch membrane with an asymmetric structure, hydrophilized by post treatment. Both membrane and product modelled closely on the original Mitsubishi product mentioned briefly in Section 14.2.2.8 known as the STERAPORE SUR unit. Modules are mounted horizontally in cassettes to form systems with a high surface area and substantial treatment capacity. Details of the C series PE product are shown in Table 14.5.

Modules are placed in the cassette frame either in a single depth, which has the last letter of the identifier as an 'S', or more commonly as a double layer, in which case the identifier is a 'D'. Table 14.10 provides details of three of the eight standard cassette models, of which two have a single layer and six, a double layer. A short path length is required since the internal diameter of the fibre is low, and operational fluxes are also relatively low. Systems normally use a separate membrane tank. There is a relatively low recirculation rate of between 2Q and 4Q, dependent on flux and other process design parameters. The standard high performances system is designated KSMBR and has a typical specific aeration demand of 0.15–0.21 $Nm^3/hr/m^2$.

Econity has a second product range known as the E series based on an unsupported 0.1 μm asymmetric PVDF membrane, details of which are shown in Table 14.6. The membrane is made by a TIPS process combined with stretching to produce elliptical pores. Details are readily available for download from the company web site.

Table 14.6 Econity PVDF product details (E series).

Membrane		Cartridge Module		Cassette			
Brand Name	**Econity**	**Product Name**	**E Series**	**Product Name**	**KSMBR & eMBR**		
UF or MF	MF	Model ident fier		No. of modules/identifier	CF-23S	CF35-D	CF-66D
Polymer	PVDF	Ht × width, mm		Ht × width, mm	655 × 886	1,260 × 1570	1,300 × 2274
Hydrophilicity	Hydrophilized	Fibre length, m	0.3	Fibre length, m	~0.6	~0.6	~0.6
Pore size, pm	0.1	Length or depth, mm		Length or depth, mm	1,600	2,360	2,800
Fibre I.D, mm	0.7	Membrane area, m^2		Membrane area, m^2	126	630	1,512
Fibre O.D, mm	1.2	Max TMP, bar	0.6	Max TMP, bar	0.6	0.6	0.6
Cl_2 resistance, ppm	1.5×10^6 ppm-hrs	Temp rating, °C	35	Temp rating, °C	35	35	35
Cleaning pH range	2–13	Shipping weight, kg		Shipping weight, kg	95	432	1,120

The more recent E series membrane products use an end-free module design which requires less energy. The energy demand is reduced by changing bubble size during the processing steps. Coarse bubbles are used during scour in the relaxation step if there are high load feeds, and fine bubbles are used during the filtration cycle if solids load is low. The module is stated to have easy handling for the operator.

Econity systems use a 1 or 2 mm screen, which is typical for immersed hollow fibre formats (see Section 14.1.4). Cassettes have side panels to ensure efficient utilization of air. A unique diffuser design below the cassette ensures an even distribution of air. The system is automatically de-sludged by using a seven degree angle on the inclined collector plate below the cassette to prevent sludge accumulating and facilitate recirculation. A typical process sequence uses a filtration cycle of 7–14 min, with 1 min backwash and relaxation. Air scour is used during the filtration and backwash cycles.

14.2.2.5 Evoqua-Memcor

The current MBR product from Evoqua Water Technologies (EWT) is called MemPulse and provides an energy efficient development to the original MBR product, MemJet. A summary of the product is shown in Table 14.7.

MemPulse was introduced in 2008 and used a stronger more durable unsupported PVDF fibre than before with higher permeability, which allowed fluxes to be increased by about 20%. The low-energy usage of the MemPulse system have been successfully demonstrated at many plants including Rimini, Italy (see case study, Section 14.3.2) and Modesto, California, in which the original B40N installation is still in operation and pathogen removal is consistently maintained. Typical power requirements are highly competitive at $<0.08\ kWh/m^3$. A key feature of the MemPulse MBR promoted by Evoqua, which is shared with other hollow fibre systems using unsupported membranes, is that it can readily use a pressure decay test (PDT) to monitor integrity (see Section 14.1.2.1).

Figure 14.15 shows the module and rack for the MemPulse system. Notable features are the square section used for the module and the use of a relatively low packing density with space between individual sub-bundles of fibre. This ensures efficient distribution of air and facilitates the shaking action to counter fouling and clogging. The square section eliminates unused space from the rack by allowing modules to be packed efficiently.

The slug flow innovation, shown for the EWT design in Figure 14.10 and as also used by others (e.g., see Section 14.2.2.1) underlines the importance of the principle scouring mechanism of aeration. The air bubbles do not simply knock foulant particles from the surface as they rise through the fibre bundle, but shake the fibre causing foulant to drop off. The energy of the larger slug bubble is efficient enough to achieve this whereas the shaking effect is less for the more normally sized coarse bubbles (Figure 14.16).

14.2.2.6 Koch Membrane Systems

Koch Membrane Systems (KMS), a subsidiary of Koch Industries, is one of the longest established players in MF starting with its acquisition of Abcor in the 1960s. This was followed by the purchase of the UF specialist, Romicon, in 1991, the NF/RO pioneer, Fluid Systems, in 1998, and the German MBR start up, Puron, in 2004.

The original PURON MBR product utilized a supported PES fibre. The product has developed under the Koch-PURON brand name and now incorporates a PVDF fibre reinforced with a polyester braid. Featuring a high-performance morphology with high porosity and very narrow pore size distribution, as seen in Figure 14.17, the PURON fibre provides high sustainable fluxes and fouling resistance.

Table 14.7 Details of Evoqua Water Technologies (EWT), Memcor products.

Membrane		Module		Rack	
Brand Name	**B40N**	**Product Name**	**MemPulse**	**Product Name**	**Mem Pulse**
UF or MF	fine MF	Model identifier	B40N	No. of modules/identifier	4–24
Polymer	PVDF	Ht × width, mm	1,600 × 203	Ht × width, mm	2,260 × 203
Hydrophilicity	Moderate	Fibre length, m	~1.45	Fibre length, m	~1.45
Pore size, pm	0.04	Length or depth, mm	203	Length or depth, mm	3,960
Fibre I.D, mm		Membrane area, m^2	40	Membrane area, m^2	608
Fibre O.D, mm	1.3	Max TMP, bar	0.75	Max TMP, bar	0.75
Cl_2 resistance, ppm	600-MC; 1500-RC	Temp rating, °C	40	Temp rating, °C	40
Cleaning pH range	2–10.5	Shipping weight, kg	19	Shipping weight, kg	~400

Notes: (1) The module uses a polyurethane (PU) potting compound. (2) Long-term Cl_2 contact is not recommended at pH<6.5. Long-term resistance >1,000,000 ppm hours. (3) Typical number of modules in a rack is 16. (4) Competitor retrofit AM series rack sets also supplied using the B40N+MemPulse building block.

Figure 14.15 MemPulse module and 16 module rack (see Appendix 1, A1.2.2.5).

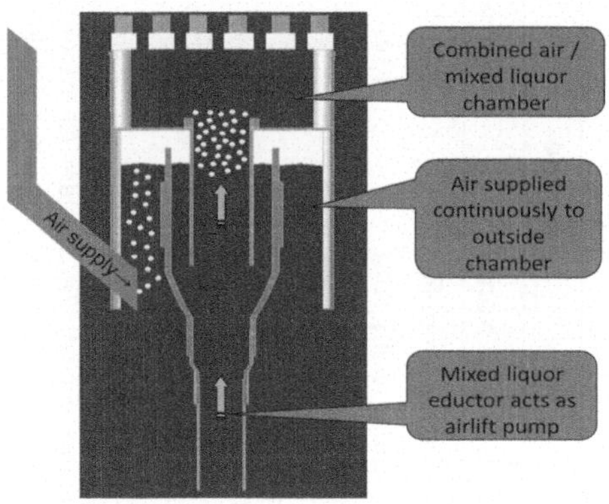

Figure 14.16 Operation of the MemPulse air distribution system.

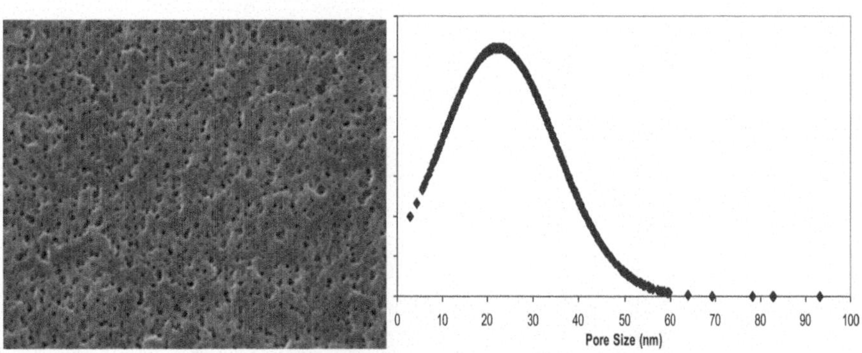

Figure 14.17 SEM of membrane surface (left) and pore-size distribution curve (right) for the Koch hollow fibre.

Good penetration of the membrane into the braid support structure is achieved to enable backwash to be performed without loss of integrity. Details of the membrane fibre can be found in Table 14.8. Figure 14.18 illustrates the epoxy potting of the fibre bundles, and the unique feature of this product, i.e., it is only potted at the bottom. The open bundle top ensures a low clogging tendency since debris, which has a tendency to accumulate on the underside of the top potting is freely released from the bundle. The PURON PSH module has several interesting features which ensure efficient aeration of the fibre bundle without excessive energy use. The left-hand side of Figure 14.18 shows that the fibres are spaced to form five channels. There are relatively few fibres in the cross section of the bundle, typically just five or six to the centre, to ensure intimate contact between individual fibres and the central air distribution port which provides aeration using a patented design. The bubble size is coarse, to give an effective cleaning action, and every 3–4 hours, a power-flush valve operates to avoid clogging of the aeration orifice. The right-hand side of the figure shows the top of the fibre bundle with the end of the fibre sealed and coated. Details of the various modules of the product range are shown in Table 14.9.

Table 14.8 PURON membrane and module details.

Membrane	
Brand name	PURON® SMP3
Filtration class	UF
Polymer	Supported PVDF
Hydrophilicity	Hydrophilic
Pore size (μm)	0.03
Fibre OD (mm)	2.6
Cl2 resistance (ppm-hrs)	1,000,000
Cleaning pH	2–10.5

Figure 14.18 Segmented fibre bundles (left) sealed fibre tips (centre), PSH-1800 module (right) (courtesy of Koch).

In 2015, KMS introduced its latest MBR technology with PULSION MBR. While leveraging the benefits of a unique design combination of a reinforced fibre, single-header, and central aeration, PULSION MBR incorporates large bubble continuous aeration into individual chambered fibre bundles. The result is a highly efficient piston-like pumping effect that drastically improves MLSS recirculation within the membrane bundles and increases the effectiveness of membrane air scouring. As a consequence, PULSION MBR provides lower aeration flow rates, reduced energy, higher sustainable fluxes, and reduced installed membrane area and footprint requirements. The features of PUSION MBR are illustrated in Figure 14.19.

Table 14.9 PURON PSH module details[a].

PURON® PSH Modules							
Product Model	**PSH-95**	**PSH-330**	**PSH-440**	**PSH-520**	**PSH-660**	**PSH-1650**	**PSH-1800**
Dimensions L × W × H (m)	0.45 × 0.89 × 2.06	0.91 × 0.89 × 2.50	0.73 × 1.76 × 2.14	0.73 × 1.76 × 2.47	1.7 × 0.89 × 2.53	2.24 × 1.76 × 2.43	2.2 × 1.8 × 2.53
Fibre Length (m)	1.5	2	1.5	1.8	2	1.8	2
Membrane area (m^2)	93	328	434	525	656	1,650	1,804
Max TMP (bar)	0.6	0.6	0.6	0.6	0.6	0.6	0.6
Temp. rating (°C)	40	40	40	40	40	40	40
Weight (kg)	164	350	461	525	654	1,479	1,570

[a]See Appendix 1, A1.2.2.6.

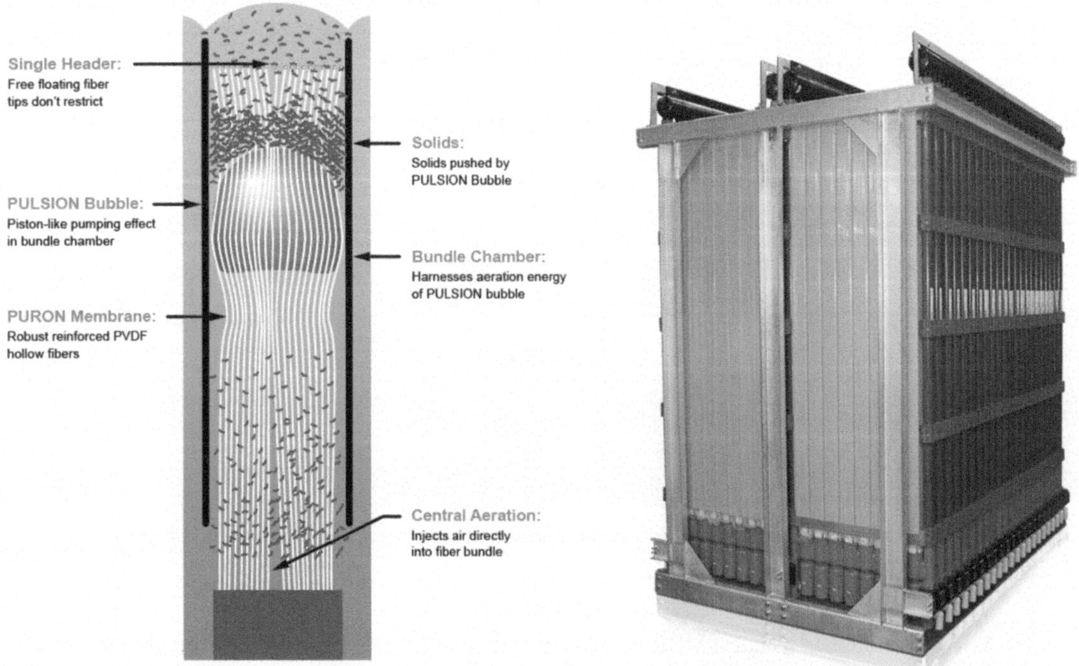

Figure 14.19 PULSION MBR Features (left) PULSION LE-44 module (right) (courtesy of Koch).

Details of module configurations offered can be found in Table 14.10. As with the other hollow fibre suppliers, PURON and PULSION MBR generally favour a separate membrane tank, and would expect to operate with a recirculation rate of 3–5Q. Pre-treatment is typically provided by a 2–3 mm punched hole drum screen.

Table 14.10 PULSION MBR module details.

PULSION® Modules			
Product Model	**LE-8**	**LE-16**	**LE-44**
Dimensions L × W × H (m)	0.93 × 0.94 × 2.68	1.68 × 0.94 × 2.68	2.26 × 1.78 × 2.68
Fibre length (m)	2	2	2
Membrane area (m^2)	348	696	1,914
Max TMP (bar)	0.6	0.6	0.6
Temp. rating (deg. C)	40	40	40
Weight (kg)	590	1135	2754

A typical filtration cycle lasts for 10–12 minutes, with a typical flux range of 25–50 lmh. Backwash is conducted at 30 lmh with air scour for 30–60 s. The relaxation step favoured by some suppliers is not used, presumably to minimize downtime. Koch uses a unique system to minimize aeration, with the effect that air scour energy is just 0.07–0.10 $kWhr/m^3$ for PURON PSH and 0.05–0.08 $kWhr/m^3$ for PULSION. Air scour frequency is adjusted dependent on flux as follows.

- PURON PSH:
  - Low flux: 33 s on/99 s off (at 100% intensity)
  - High flux: 120 s on/20 s off (at 50% intensity)
- PULSION
  - Low flux: continuous at 50% intensity
  - High flux: continuous at 100% intensity

Therefore, as the diurnal flow varies, the amount of air scour used can be adjusted in line with the output of the plant to optimize overall efficiency. Chemical cleaning is carried out in-situ using similar chemicals and cleaning frequencies as for the other hollow fibre suppliers:

- Maintenance clean: (CEB)
  - Sodium hypochlorite: daily to weekly, 30 min, 125–200 ppm
  - Citric acid (50%): 1×/week to 1×/12 weeks, 30 min, 1,000–2,000 ppm
- Recovery clean: (Intensive soak, 1–4× per year)
  - Sodium hypochlorite (12.5%): 4–12 h, 1,000–2,000 ppm
  - Citric acid (50%): 4–10 h, 1,000–2,000 ppm

For difficult filtration applications (other than MBR) in water or wastewater treatment duties with high feed solids, KMS offers a high packing density version of the PURON product (PURON HF). In these cases, modules have a similar configuration to those listed in Table 14.10, but without the bundle gaps and 47% more membrane area within the same footprint.

14.2.2.7 Memstar

Memstar is not as long established as most of the membrane makers in this review, but has had a major impact in a short time, quickly becoming one of the biggest suppliers of installations in China, where it has 30% of the cumulative large plant capacity (see Section 14.2.1). The company has headquarters and a manufacturing unit in Singapore, as well as manufacturing in China. Many of the Memstar references are listed as CITIC Envirotech Limited (formerly, United Envirotech), a Chinese system integrator and major WWTP operator in China. Memstar is a wholly owned subsidiary of CITIC Envirotech Limited. The Memstar membrane is similar to Asahi, i.e., it is an unsupported PVDF hollow fibre, with highly crystalline structure and similar dimensions. The membrane is hydrophilic and has excellent chemical resistance (cleaning pH tolerance between 1 and 14).

The membrane fibres are captured in bundles and potted with multiple bundles in a module. Modules are then mounted in a frame to form a skid, as shown in Figure 14.20. Three-dimensional arrays are commonly used, which can provide skids with a high membrane area. It is also common to use double height stacks to improve the efficiency of air use. A separate membrane tank is strongly preferred so that biology and membrane operation is separately optimized. Recirculation between the membrane tank and the bio-reactor can be anything from 1 to 4 Q. Table 14.11 provides a summary of membrane module and skid specifications. Three different module lengths are used depending on the project height restrictions and flow requirements.

Figure 14.20 Memstar modules in a frame (see Appendix 1, A1.2.2.7).

The typical pre-treatment strainer specification is <2 mm. There are two modes of aeration: variable intensity or energy saving, with a larger bubble size. For variable intensity, the aeration supply to

Table 14.11 Memstar product details.

Membrane		Module				Skid	
Brand Name	**Memstar**	**Product Name**	**SMM**	**SMM**	**SMM**	**Product Name**	**SMM2030-MBR100**
UF or MF	MF	Model identifier	SMM-1015	SMM-1522	SMM-2030	No. of modules/identifier	100
Polymer	PVDF	Ht × width, mm	1,040 × 45	1,540 × 45	2,040 × 45	Ht, mm	4,730
Hydrophilicity	Hydrophilic	Fibre length, m	940	1440	1940	Fibre length, m	1,280
Pore size, μm	0.1					Length or depth, mm	1,280 × 2,085
Fibre I.D, mm	0.7	Membrane area, m^2	15	22	30	Membrane area, m^2	3,000
Fibre O.D, mm	1.3	Max TMP, bar	0.5	0.5	0.5	Max TMP, bar	0.5
Cl_2 resistance	>1,000,000 ppm-hrs	Temp rating, °C	45	45	45	Temp rating, °C	45
Cleaning pH range	1–14	Shipping weight, kg	9	11	13	Weight, kg	3,600

each train is via two sets of air pipes to allow different intensity of air scouring during operation and relaxation. The average air flow is 1.4–3.9 Nm^3 air/m^3 filtrate (i.e. SADp). A typical process sequence is a filtration time of 8–9 minutes followed by relaxation for 1 minute. An unusual feature of Memstar as a hollow fibre product is that the system does not use a regular backwash, instead relying on relaxation and air scour. Typical chemical cleaning is by use of a chemical enhanced backwash with 100 to 300 ppm of NaOCl every 3–7 days. For domestic wastewater, recovery cleans are performed every 6–12 months with 1000–2000 ppm NaOCl and with 0.3–1% of citric acid once every 1–2 years. Cleaning regimes for industrial wastewater depend on the constituents present; high pH caustic cleaning and/or other acids may be used.

14.2.2.8 Mitsubishi (formerly Mitsubishi Rayon Corporation (MRC))

Mitsubishi Chemical, formerly known as Mitsubishi Rayon, is one of the longest established MBR companies with activity dating back to the 1980s. In particular, the company has a strong position in the relatively mature MBR market in Japan and Korea, and in the past decade has been very active in China. The original Mitsubishi product was based on an unsupported PE membrane, and has been used in both water and wastewater MF applications, as well as in MBR. There are a large number of references for this technology in Asia, but relatively few installations elsewhere, though Ionics did promote the technology in North America around 2004 just prior to its acquisition by GE (later sold to Suez). Despite having a regional focus, Mitsubishi is one of the main Worldwide MBR players alongside Suez, Kubota, and Econity in terms of number of installations due to the length of time as a supplier and the importance of the Asian market.

In the early 2000s, Mitsubishi introduced a new reinforced PVDF hollow fibre, specifically developed for MBR applications, designated STERAPORE™. The PE fibre is horizontally mounted whereas the PVDF fibre is vertically mounted as shown in Figure 14.21. The PVDF product was sold for MBR alongside the original PE product for a time, and was initially targeted at larger scale duties, for which the importance of energy cost is greater. Though a more expensive fibre with a lower packing density, it turned out that the PVDF fibre was more competitive in most cases since it could use higher fluxes which reduced membrane area and aeration requirements. Also, it had better membrane life which reduced replacement costs. Eventually, PVDF has essentially displaced the PE product, which is now only used in minor niches, for replacements, and for non-MBR applications. The MBR product uses the product brand name, STERAPORE™, and this now primarily describes a vertical element with reinforced PVDF fibre. Table 14.12 shows details of the reinforced PVDF membrane available in three element sizes, as well as details of the PE element option.

The table features the most commonly used MBR fibre which has a 0.4 µm MF rating. Mitsubishi Chemical also produces a 0.05 µm fibre, which is used in the STERAPORE™ 5600–5700 series products, which has similar dimensions to the 5000 series but a higher packing density and hence larger membrane area.

The Mitsubishi element contains discrete bunches of fibres, potted in polyurethane and captured in a frame, which can be readily mounted into a module as shown in Figure 14.21. Flow is removed from the module from both the top and bottom header. Using a side panel on the module enhances aeration efficiency. Details of the 5600 and 5700 series modules are provided in Table 14.13. Note that Mitsubishi uses a different nomenclature to the previous supplies described in this section. Mitsubishi recommends a finer pre-treatment strainer than most other suppliers with a 0.5–1 mm aperture. The approach to single or two tank design and to specification of recirculation flows is similar to other suppliers.

Figure 14.21 Mitsubishi's PVDF membrane element (left), module (centre), and element flow path (right) (see Appendix 1, A1.2.2.8).

Table 14.12 Mitsubishi membrane and element details.

Membrane Material	**5600 Series**	**5700 Series**	**5000 Series**	**PE Series**
	Reinforced PVDF			**PE**
Membrane type/ Pore size	UF/0.05 μm		MF/0.4 μm	
Application	MBR, tertiary treatment			
Membrane surface area	40 m^2/Element 400–2,400 m^2/Module	5 m^2/Element 50–200 m^2/Module	6, 15, 25 m^2/Element 60–1,500 m^2/Module	3 m^2/Element 210 m^2/Module
Flux*	Up to 33LMH	Up to 40LMH	Up to 33LMH	Up to 10 LMH
Outer diameter	1.65 mm	1.65 mm	2.8 mm	0.5 mm
Storage condition	Dry			

The process flow for Mitsubishi is similar to Memstar relying just on a short filtration cycle of around 7 min, interspersed with a relaxation period of 1 min combined with air scour, and no backwash. A maintenance cleaning procedure is carried out about once per week with an effective chlorine concentration of 300–1,000 ppm. Recovery cleaning using an effective chlorine concentration of 3,000 ppm for 30–90 min is performed every 3 months or when the suction pressure has exceeded the set value. If trans-membrane pressure is still high after recovery cleaning, a chemical soak is performed with acid (oxalic acid: 1%, citric acid: 1%, sulphuric acid or hydrochloric acid: 0.1–0.5 M).

Table 14.13 A Mitsubishi module details (5600–5700 Series).

		5600 Series				
PN (Module)		**56M0400FF**	**56M0800FF**	**56M1200FF**	**56M1600FF**	**56M2400FF**
PN (Element)		56E0040SA(40 m^2/Element)				
Number of element		10	20	30	40	*60*
Membrane surface area	m^2	400	800	1,200	1,600	2,400
capacity*	m3/d	200	400	600	800	1,200
Size	$D \times W \times H$ mm	1,524 × 940 × 2,798	1,524 × 1,390 × 2,798	1,524 × 1,870 × 2,798	1,524 × 2,320 × 2,798	1,524 × 3,416 × 2,798
Min. WL	M	3				

		5700 Series			
PN (Module)		**57M0050FF**	**57M0100FF**	**57M0150FF**	**57M0200FF**
PN (Element)		57E0005SM(5 m^2/Element)			
Number of element		10	20	30	40
Membrane surface area	m^2	50	100	150	200
capacity*	m^3/d	25	50	75	100
Size	D × W × H mm	668 × 675 × 1,600	668 × 967 × 1,602	670 × 1,245 × 1,602	670 × 1,520 × 1,602
Min. WL	M	1.7			

14.2.2.9 Others

This section will briefly mention other hollow fibre products with unique features. The great majority of hollow fibre products use PVDF due to the importance of chlorine cleaning in MBR. There are also several examples of PE-based products, and a few with PP and PES.

However, PTFE is an unusual choice, and the company that has promoted this membrane with some degree of market impact, is Sumitomo, which is sold in North America through its US subsidiary. The first plant was installed in 2003, and since then there has been a steady uptake of the technology, primarily for industrial applications of <10 MLD, particularly in food waste, as well as having a recent push into the municipal sector. The membrane, marketed under the brand name POREFLON™, has a 0.2 µm pore size with high porosity giving high flux. The fibre has an outer diameter of 2.3 mm and a lumen of 1.1 mm (see Appendix 1, A1.2.2.9). It is claimed to have six times the strength of PVDF, allowing it to be used at a TMP of up to 0.9 bar. Furthermore, the membrane is hydrophilic, temperature resistant to 50°C, and has exceptionally high chemical resistance to chlorine, acids and bases.

Sumitomo standard package plants are marketed as MBR-ISAM. They contain between 10 and 120 modules, each with a membrane area of 10 m^2, to make packages that looks similar to the Mitsubishi unit. The system is claimed not to require a pre-treatment strainer.

14.2.3 Immersed flat sheet

14.2.3.1 Kubota

Kubota was one of the early pioneers of the MBR concept, encouraged by a Japanese Government initiative in the 1980s. As with the Zenon product range now owned by Suez, the original Kubota MBR was based on a sidestream crossflow tubular system. Kubota has the highest number of references of any MBR supplier with 5,500 systems as of 2017 for applications ranging from very low to very high flowrates. 1,500 of these MBR's are used for industrial wastewater applications. A high proportion of the remaining 4,000 units are used in the residential/commercial market in Japan. Kubota MBR is now widely established worldwide.

Kubota started in the MBR business in 1986 and first introduced its submerged flat sheet MBR product in 1991. Units for medium or large-scale projects are based on one of two cartridge (panel) sizes, 0.5 m wide and either 1.0 or 1.5 m in height. The cartridges are mounted vertically in a frame, as shown in Figure 14.22. For low flow systems, a single cartridge deck or stack is used, whereas higher flow systems use a double deck to improve the efficiency of air utilization. This arrangement, first introduced in 2001, was developed to overcome the low packing density and relatively high footprint inherent to the flat sheet system, and has enabled Kubota to compete for projects across the flow rate scale. The assembled system is referred to as a Kubota Submerged Membrane Unit (SMU).

Kubota utilizes a partly chlorinated PE membrane with an average pore size of 0.2 µm and a maximum pore size of 0.4 µm, a rating similar to several of the hollow fibre MBR membranes. The membrane is cast on a non-woven backing material. In common with most supported flat sheets, it is not suitable for backwashing or flow reversal, since this can cause delamination of the active layer. Product details are given in Table 14.14.

Flat sheet membranes have an advantage of relatively low manufacturing cost compared to hollow fibres, which can give lower capital costs (Judd *et al.*, 2011). Another important advantage is that the system is better able to resist clogging with particle or fibrous material and can potentially use a pre-treatment strainer with a much higher size rating, typically 3 mm compared with 1 or 1.5 mm for hollow fibre. The disadvantages tend to be the high footprint due to low packing density, and high air use, which leads to high operating costs.

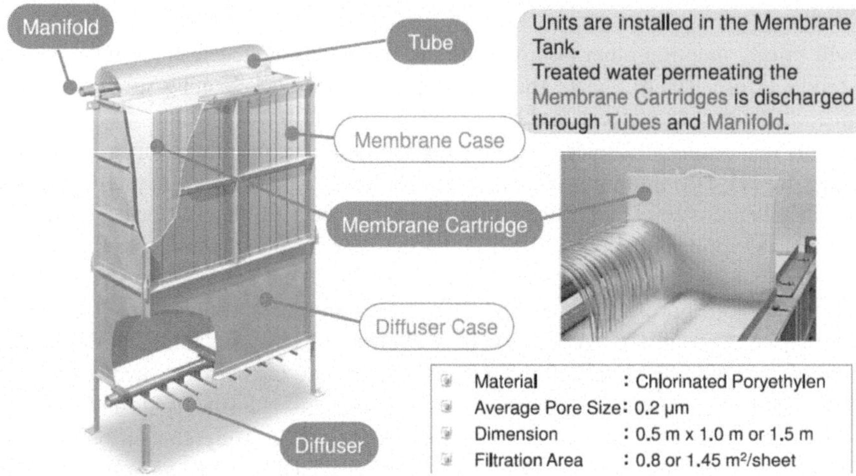

Figure 14.22 Kubota panel or cartridge arrangement in a stack (courtesy of Kubota).

The Kubota membrane is permanently hydrophilic and highly chlorine resistant. A similar operational cycle is used as for the Mitsubishi and Memstar hollow fibre products, i.e., a 9 min filtration cycle with 1 min relaxation using continuous aeration, but no backwash (thereby eliminating the backwash pump). Original designs used an air distribution system with 10 mm holes, but this has now been reduced to 3 or 4 mm to decrease air usage. An important operational challenge for systems with finer air distribution systems is to avoid blocking of the holes. Most Kubota units use coarse bubble aeration, but the SP units use a medium bubble size. An automated cleaning sequence is used once per day to clean the air diffuser by reversing the liquid flow for 5 min.

The PE membrane has good chemical resistance and can tolerate acid cleaning such as 1% oxalic acid, citric or HCl. The membrane also tolerates caustic, but this is not usually used. Chlorine is the most commonly employed cleaning agent. A significant advantage for flat sheet systems over hollow fibre is that cleaning frequency is much less frequent, since lower TMP tends to be used (<0.2 bar). Also the relatively wide 7 mm flow channels are effectively cleaned by the physical air scouring process, with very little tendency for solids to accumulate as can occur near the top potting of hollow fibres.

A typical chemical cleaning frequency for Kubota is once every 2–6 months, conducted automatically, and taking about 3 h. Membrane life is typically 5–10 years. One of the largest early MBRs in Europe was installed by Wessex Water at Porlock wastewater treatment works in 1997/1998. After 10 years, 93.6% of the original membrane panels were still in use (Churchouse *et al.*, 2008). During this period, permeability has been stable for comparable conditions, declining from around 350 to 500 lmh bar when cleaned to 250 lmh bar when the next cleaning was required. The typical cleaning frequency was steady at 6 months. The system did not require an acid clean.

The simplicity and operational robustness of the Kubota system makes it particularly suitable and cost effective for small-scale applications, and there are many adherents for this approach to MBR for industrial applications. However, the importance of energy cost and the efficiency of air utilization in the municipal sector has forced Kubota and other flat sheet suppliers to address various shortcomings. Developments in the last few years have focused on making the product applicable throughout the flow range through improved use of air. In fact, in a dramatic demonstration of progress in this regard, Ovivo used Kubota in their bid for the 333 MLD (88 MGD) peak daily flow plant at the Canton WWTP in

Table 14.14 Kubota product details[a].

Membrane		Cartridge (or Panel)			Unit				
Brand Name	**Kubota Flat Sheet**	**Product Name**	**Kubota SMU**		**Product Name**	**Kubota Submerged Membrane Unit (SMU)**			
UF or MF	**MF**	**Model Identifier**	**510**	**515**	**Model Identifier**	**ES 200**	**EK 400**	**RM 200**	**RW 400**
Polymer	Chlorinated PE, supportec	Ht × width, mm	1,000 × 500	1,500 × 500	Decks	Single	Single	Double	Double
Hydrophilicity	Hydrophilic	Channel length, m	~1.0	~1.5	Ht × width, mm	2,030 × 620	3,500 × 620	2,490 × 575	4,290 × 575
Pore size, pm	0.4	Panel spacing, mm	7	7	Length or depth, mm	2,920	2,920	2,930	2,930
Cl2 resistan ce, ppm	Up to 0.6%	Membrane area, m^2	0.8	1.45	Membrane area, m^2	160	320	290	580
Operational pH	5–10	Rec Max TMP, bar	0.2	0.2	Max TMP, bar	0.2	0.2	0.2	0.2
Cleaning pH range	E.g. 1% oxalic	Temp rating, °C	40	40	Temp rating, °C	40	40	40	40
		Shipping weight, kg	*5*	*11*	Shipping weight, kg	860	1,670	1,390	2,590

[a]See Appendix 1, A1.2.3.1.

2014 featured as a case study in Section 14.3.3.2. There are also two other Kubota MBRs amongst the 82 large plant references discussed in Section 14.2.1, so although not commonplace, flat sheet may be worth considering for high flowrate projects in some circumstances.

14.2.3.2 Toray

Toray introduced a flat sheet MBR in the early 2000s and has established the product as a competitor to Kubota using the brand name MEMBRAY™. Many product details are similar to Kubota, but not identical, as shown in Table 14.15. The Toray membrane is a 0.08 µm PVDF flat sheet on a PET backing, available in two different element or panel sizes. The elements have a slightly greater area than Kubota and are packed slightly more closely. The operational sequence is very similar with a 9 min filtration cycle followed by relaxation, all with continuous aeration, but without a backwash cycle. The system appears to use a finer bubble size than Kubota. A similar pre-treatment specification of a 3 mm screen is suggested and a separate membrane tank is generally preferred.

Chemical cleaning concentrations and frequency are almost the same as Kubota. As with Kubota, Toray quotes double deck systems for larger flows and has considered using triple deck designs. There are extensive references for Toray's MBR in Asia, but the majority of installed plants have relatively low flow and there are no plants in the large flow list in Section 14.2.1.

14.2.3.3 Meiden

Meidensha of Japan offers a flat sheet MBR system using ceramic plates rather than polymeric panels. The 6 mm thick plate has internal permeate channels, as illustrated in Figure 14.23, thereby creating a panel in a single piece. The surface of the Meidensha ceramic membrane is hydrophilic and smooth, to prevent or at least minimize foulant adhesion. Compared with polymeric membranes, Meidensha's product has several advantages such as high mechanical stability and excellent chemical resistance. Scouring air, backwash and chemical cleaning in place can be used to clean the plate if fouled, giving stable permeability.

Since 2012, Meiden has provided more than 70 installations in Singapore, Japan, China, Korea, Taiwan, Middle East and North America. Many projects are for industrial waste water treatment but also some are for municipal waste water treatment and for difficult drinking water applications which can exploit the advantages of a ceramic material. In 2017, Siemens announced that the company planned to use the Meidensha ceramic system in its PACT MBR and had applied the technology for an oil refinery wastewater. PACT MBR is a powdered activated carbon-based wastewater treatment system that provides biological/physio-chemical/and UF in a single, compact system.

In 2013, Meiden conducted operation and testing of a 4 MLD (1 MGD) MBR demonstration facility using the ceramic flat sheet membrane at Jurong Water Reclamation Plant in Singapore (Noguchi *et al.*, 2018). The objective of the demonstration plant was to treat industrial used water with low-energy consumption and produce a high-quality permeate water for reuse as industrial water. The demonstration did not require a recovery clean until after 18 months of operation. After the cleaning, the membrane operated at a good level of flux for a challenging feed of 25 lmh, though this may not be sufficient to overcome the cost penalty of selecting a ceramic membrane. Figure 14.24 shows a double stack flat sheet module assembly from Meiden which resembles mainstream polymetric competitors.

In 2017, a full-scale ceramic MBR system was installed at Changi Water Reclamation Plant (CWRP) as part of an MBR retrofit project to increase treatment capacity without expanding the plant's footprint (Niwa *et al.*, 2018). The ceramic MBR system has a nominal capacity of 15 MLD and uses a domestic sewage feed. During the initial operation the plant has not required cleaning and has operated at a flux in the range 30–60 lmh.

Table 14.15 Toray product details[a].

Membrane		Element			Module				
Brand Name	**TMR**	**Product Name**	**Membray Element**		**Product Name**	**Membray Module**			
UF or MF	MF	Model identifier	TSP-50100	TSP-50150	Model identifier	TMR090-50S	TMR140-050S	TMR140-200W	TMR140-200D
Polymer	PVDF on PET	Ht × width, mm	1059 × 515	1608 × 515	Decks	Single	Single	Single	Double
Hydrophilicity	Moderate	*Channel length, m*	~1.0	~1.5	Ht × width, mm	1,474 × 711	2,100 × 810	2,100 × 840	4,160 × 810
Pore size, pm	0.08	Panel spacing, mm	6	6	Length or depth, mm	1,016	950	3,260	1,620
Cl_2 resistance	Up to 0.5%	Membrane area, m^2	0.9	1.4	Membrane area, m^2	45	70	280	280
Cleaning conc	0.5% oxalic	Rec max TMP, bar	0.2	0.2	Rec max TMP, bar	0.2	0.2	0.2	0.2
		Temp rating, °C	40	40	Temp rating, °C	40	40	40	40
		Shipping weight, kg	3	5	Shipping weight, kg	265	400	1,430	1,365

[a]See Appendix 1, A1.2.3.2.

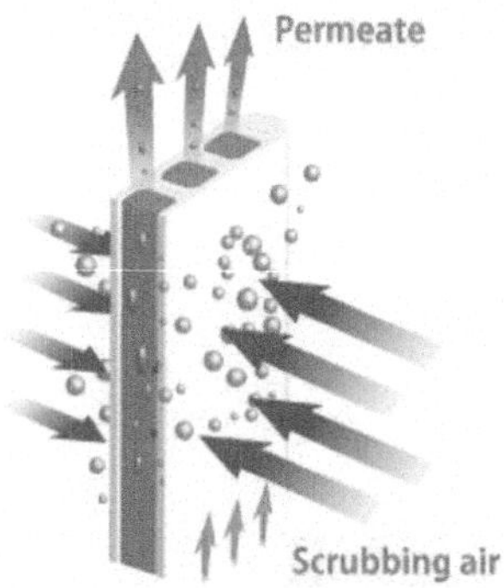

Figure 14.23 Meidensha MBR flat sheet schematic (see Appendix 1, A1.2.3.3).

Figure 14.24 Meidensha MBR double stack module (courtesy of Meidensha).

Although ceramic prices are high, membrane life is long giving a low replacement cost. However, supported polymeric membranes also have a good life span, so arguably the differential between ceramics and polymerics is not as much as in MF applications. However, it does appear that ceramic MBR achieves stable performance with a low chemical cleaning requirement. Any operating cost advantage in air scour and/or chemical cleaning could therefore translate into an important whole life cost advantage.

14.2.3.4 Others

This section will briefly review other flat sheet products with unique features. There are numerous small companies that have developed flat sheet MBR products, since these products tend to be suitable for small industrial niche applications. Interestingly, the majority of these companies are German, which perhaps reflects the first commercialization of MF technology in the early years of the last century. Most of the other flat sheet MBR companies are from China, but so far, the impact of the Chinese companies in international markets has been very limited. The two companies that will be discussed in this section are from Germany, namely Huber, which has a unique product design, and Microdyn-Nadir, with its unique operational method.

Huber has three MBR products. The unique design is the vacuum rotation module (VRM), which is shown in Figure 14.25, which has a trapezoid shape of membrane sheet. The other products are more conventional rectangular modules, designated BioMem and smartMBR.

Figure 14.25 Huber Vacuum Rotation Module (VRM) (see Appendix 1, A1.2.3.4).

All three products use a coarse UF membrane made from PES with a 0.04 µm rating. The BioMem and smartMBR products have a polyester backing, whereas the VRM module has a polypropylene backing. All three of the systems operate with a maximum TMP of 0.4 bar. The VRM unit uses a two-tank design and a coarse bubble size for air scour, whereas the other units can use either single- or two-tank designs and use a medium bubble size. The membranes have a good temperature and chemical resistance, tolerating 50°C and a cleaning pH range of 1–14. However, chlorine resistance is less than for other MBR polymers at 100,000 ppm-hours.

The VRM units use a slow rotation speed of 1 rpm (or sometimes up to 2 rpm) to create a cleaning action which minimizes air scour requirements and reduces cleaning frequency to typically once or twice per year. Huber claims that the energy cost of the rotational system is much lower than for systems relying solely on air scour. The system comes with two types of module plates, the VRM 20, for systems of up to 900 m^2, and the VRM 30 for systems of up to 3840 m^2. The operational sequence for all of the Huber MBRs is typical of flat sheet systems with a 9 min filtration time, 1 min relaxation and no backwash. As with other flat sheet systems, a 3 mm pre-treatment strainer is used. Huber tends to focus its product range on small- to medium-scale duties, such as industrial applications and decentralized municipal systems.

The second system featured in this section is from Microdyn Nadir and is known as BIO-CEL®. The modules look like a conventional flat sheet configuration, but use a laminated PES membrane with a coarse UF rating of 0.04 μm that can be backwashed.

There are two significant differences to the operation. The first of these is that the system uses fine bubble aeration, whereas most other systems favour the more aggressive coarse bubble. However, the main innovation is the second difference, which is that polyurethane beads with a 4–5 mm diameter are dosed into the feed and recirculated to create a continuous scouring action. It is claimed that the particles do not suffer damage or damage the membrane, are not prone to having a bio-film form on the surface, and can be recycled with high efficiency. The use of the beads allows flux to be increased significantly for a given feed with the same air scour rate, for example from 25 to 40 LMH. This has the effect of reducing specific energy use considerably. Furthermore, the membrane is able to maintain high permeability of 400 LMH.bar during operation, which ensures operation at low TMP. The concept can also be used in cross-flow designs.

BIO-CEL® was first introduced in 2007, it was originally available in small units of 10, 50 and 100 m^2 of membrane area, but has gradually increased the module size to make the product more competitive for larger installations, especially in the municipal sector. Two new module sizes added in the last few years have surface areas of 416 and 1920 m^2, potentially allowing the company to address major schemes. The company has a number of references, and field experience has shown very low chemical cleaning intervals of greater than a year. In 2015 Microdyn Nadir announced a partnership with Ovivo, the most influential independent MBR system provider in the US market (see Ovivo case study in Section 14.3.3.2). In 2016, Trisep was acquired by Microdyn-Nadir and in 2017, the company announced the building a second US MBR manufacturing facility.

14.2.4 External

14.2.4.1 Crossflow

When MBR was first commercialized, many systems started with the use of tubular modules mounted externally to the main biological reactor. Indeed, the most influential MBR companies today, Suez and Kubota, cut their teeth with side-stream tubular systems before focusing on immersed systems from the 1990s onwards. Today, there are many providers of side-stream systems, some of which make their own tubular products, e.g., Berghof, but there are many more who build systems using bought-in tubular modules, potentially from different suppliers, e.g., Wehrle. Features of the external cross-flow design provided in this section will be illustrated through a review of one of the longest established and widely used tubular module makers, Berghof Membrane Technology, GmbH, which has more than 700 references worldwide. Berghof has been making modules since 1975 and also provides MBR systems. Some of the tubular modules are shown in Figure 14.26.

Berghof uses hydrophilic PES and PVDF membranes with a polyester support and available in five pore sizes ranging from fine UF at 5 kDa to coarse UF at 0.03 μm. The membrane tubes are available in two different diameters of 5 and 10 mm, selected dependent on solids loading and cross-flow velocity. The modules have a high-pressure rating of 10 bar and high-temperature rating of 60°C, making them suitable for harsh environments if required. There are several different diameters ranging from 75 to 250 mm and two different lengths of 3 and 4 m to give surface area varying from 4 to 67 m^2 per module, dependent on the membrane tube diameter. Chemical cleaning contact is recommended at a maximum of 500 ppm hypochlorite with a contact of up to 250,000 ppm-hours and a pH range of 2–10.

Figure 14.26 Berghof tubular products (see Appendix 1, A1.2.4.1).

Berghof uses two systems designs called B-SMART and B-SMART Pro. The B-SMART system can be customized to the customer needs and can deliver 50–150 LMH at cross-flow velocity of 1–4 m/s. This system design can provide stable performance at high flux, and was designed to lower upfront CAPEX. On the other hand, the B-SMART Pro system uses a much lower cross-flow of 1–2 m/s and a lower flux of 45–60 LMH. The system has the flexibility of utilizing backwash for highly fouling feeds and this flexibility can be used to ensure a cost-effective total treatment cost for an easier feed, consequently lowering overall OPEX. Alternatively, the flexible system can be used for a duty with variable feed quality characteristics or flowrates.

Another supplier of cross-flow systems which also manufactures tubular modules is Pentair-Xflow, though the company is better known for the AL technology described in the next section. The cross-flow membranes are based on 0.03 μm hydrophilic PVDF on a polyester support. Tubular dimensions are similar to some of the Berghof options with 5.2 and 8 mm tube diameter and a 3 m module length. Module diameters are available at 75, 200 and 300 mm, which give a membrane area range from 4 to 75 m^2, dependent on the tube diameter. The normal module has a pressure rating of 8 bar and a temperature rating of 40°C, but a special housing version is also available with ratings of 12.5 bar and 60°C. The membrane can operate at a TMP of up to 5 bar. Details of the Pentair crossflow product offerings are included in the next section with airlift options. The 8 mm tubular products are invariably used in a crossflow mode whereas the 5.2 mm products could be used in either an AL or crossflow design.

Several ceramic suppliers of tubular or hollow fibre membranes also discuss use of their products in MBR, for example Pall and Hyflux. Ceramics can have particular advantage in cross-flow tubular systems for wastewater feeds potentially containing oily contaminants.

14.2.4.2 Air lift

The best-known air lift (AL) MBR system is provided by Pentair-Xflow, and was first introduced into the market in the early 2000s as a low-energy development of a conventional cross-flow system. At this time, development of the AL design was also being undertaken by Wehrle using tubular membrane from Berghof. The concept of AL is that air is introduced at the bottom of a vertically orientated module to provide recirculation and air scour. A schematic of the process flow and a rack are shown in Figure 14.27.

Figure 14.27 AirLift MBR module (left) and the 35 MLD system at Naxcivan, Azerbeidjan (2017) (right) (see Appendix 1, A1.2.4.2).

The buoyancy of the air provides both a motive force to encourage recirculation back to the bio-reactor tank as well as an air scouring action as the air bubbles rise through the tubular membranes. The efficiency of the air scour is relatively high since the bubbles tend to form slugs in the tubes, and it turns out that this type of bubble is the most efficient to enhance mass transfer at the surface and optimize the cleaning action (Cabassud *et al.*, 2001; Sepideh *et al.*, 2013). Unfortunately, if air alone is used to meet the complete requirements for recirculation, the energy usage is excessive. The system has therefore developed to use less air, with a low-energy pump to boost recirculation. Table 14.16 summarizes key features of the original Xflow AirLift product.

The AL product started out by using the same tubular membranes as used for cross-flow, though it is common for the narrower 5.2 mm tube to be used for most applications since the velocity and therefore pressure drop along the length of the tube is low. The product uses a 0.03 μm UF membrane made from PVDF supported by polyester. The membrane is sufficiently hydrophilic to be wettable and has a clean water permeability in excess of 1000 LMH.bar. At 3 m, modules are longer than in normal MF applications due to the wider internal diameter. The most commonly used module is the Compact 33, which has a 200 mm housing diameter and a membrane area of 33 m^2 if used with 5.2 mm tubes. The typical packing density for this module is 308 m^2/m^3.

The membrane is usually operated $\leq$0.5 bar, with a typical flux of 50 LMH. The liquid velocity during AL operation is 0.5 m/s with an average air velocity 0.3–0.5 m/s. Backwash and occasional chlorinated CEB are used to maintain stable permeability. Under cross-flow conditions, the liquid velocity would be increased to between 1 and 6 m/s allowing higher fluxes of up to 80 LMH to be used. A significant advantage of external tubular designs is that cleaning chemicals are encapsulated within the housings during CEB and CIP which limits the hold-up volume.

Figure 14.28 shows a photo of the standard Pentair Xflow MegaBlock package plant and module. The MegaBlock was launched in 2011 in response to the increasing interest in large-scale MBR in the municipal water sector. The MegaBlock is made up of sections each containing a 6 × 6 module block with a footprint of about 3.2 × 3 m. Up to six of these blocks can then by joined together to create a standard package plant with a treatment capacity of up to 10 MLD at a typical operating flux of 50 LMH.

Table 14.16 Pentair AirLift and Crossflow product detail.

Membrane		Module						
Brand Name	**AirLift**	**Product Name**	**AirLift MBR Module**					
UF or MF	UF	Model identifier	Compact 5.1	Compact 4	Compact 33	Compact 27	Compact 75	Compact 60
Polymer	PVDF	Ht × diameter, m	3 × 0.075	3 × 0.075	3 × 0.2	3 × 0.2	3 × 0.3	3 × 0.3
Support	Polyester	Tube length, m	~2.9	~2.9	~2.9	~2.9	~2.9	~2.9
Hydrophilicity	Moderate	Tube diameter, mm	5.2	8	5.2	8	5.2	8
Pore size, μm	0.03	Membrane area, m^2	5.1	4	33	27	75	60
Cl_2, ppm-hrs	250,000	Typical TMP, bar	<0.5	<0.5	<0.5	<0.5	<0.5	<0.5
Cleaning pH	1–11.5	Temp rating, °C	40	40	40	40	40	40

Figure 14.28 AirLift MBR Megablock standard package plant (left) and module (right).

A summary of the product details is provided in Table 14.17. In the early days of MF in the water and wastewater sector, a 3 mm fibre was deemed to be the limit for economic spinning for an unsupported UF membrane from a single polymer dope. Beyond this diameter, the fibre would be too weak, have too low a permeability, or both. However, the development of spinning techniques has now made the goal of a larger diameter fibre achievable. The MegaBlock PES/PVP fibre actually uses an asymmetric 0.03 μm UF fibre with a 3 mm I.D and 4.5 mm O.D. Though unsupported, the fibre has a relatively thick wall to provide robustness for the fairly aggressive AL process. A municipal feed that is relatively easy to treat is suitable for a 3 mm tube if operated in AL at a modest cross-flow velocity. As with the 5.2 and 8 mm tube products, the process operates with a regular backwash that can occur at a frequency of once every 5–10 minutes, and has a duration of 10–20 s. The typical chemical cleaning frequency is once every 25 days.

Table 14.17 Pentair MegaBlock AirLift product details

Membrane		Module		Rack		
Brand Name	**AirLift**	**Product Name**	**Compact**	**MegaBlock**		
UF or MF	UF	Model identifier	Compact 32V	No. of segments	1	6
Polymer	PES	Htx internal diam, m	3.0 × 0.2	No. of modules	36	216
Support	None	Fibre length, m	~2.9	Rack length, m	3	18
Hydrophilicity	Moderate	Fibre I.D /O.D, mm	3.0 / 4.5	Rack width, m	3.2	3.2
Pore size, μm	0.03	Membrane area, m^2	32	Membrane area, m^2	1,152	6,912
Cl_2, ppm-hrs	200,000	Typical TMP, bar	<0.5	Typical TMP, bar	<0.5	<0.5
Cleaning pH	1.5–13	Temp rating, °C	40	Temp rating, °C	40	40

A recent development by Pentair is the *Helix* enhancement of the tubular design, as shown in Figure 14.29. The helical-shaped ridge in the membrane tube, which is made from the membrane material itself, creates additional turbulence to improve performance in all high solids applications. As with spacers in spiral wound elements, the ridge creates flow vortices and promotes turbulence through the tube, and especially close to the membrane surface where foulants tend to accumulate. Standard and helix options are available for the complete tubular product range. Another compatible development which can be used together with the helix membrane enhancement is the use of flow reversal functionality combined with variable cross flow rates, namely lean cross flow (LCF). This procedure is aimed at enhancing mass transfer while minimizing energy utilization.

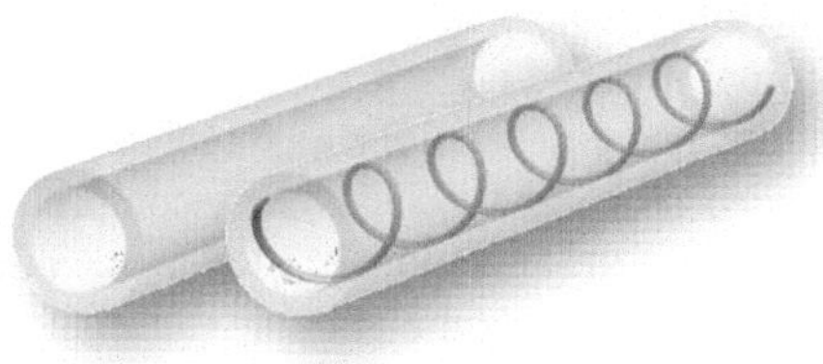

Figure 14.29 Standard tubular membrane on left vs tubular membrane with *Helix* enhancement on the right.

The side stream approach is suitable for anaerobic MBRs, and this development has been a focus of Pentair's X-Flow membranes (see Chapter 11). The helix technology is particularly applicable to this development since feed streams can be highly fouling.

14.3 CASE STUDIES

14.3.1 Immersed hollow fibre case studies

14.3.1.1 Suez: Energy optimization studies at Ulu Pandan, Singapore

The first case study presents results from energy optimization work conducted by the PUB using the Suez MBR (Tao *et al.*, 2010, 2018). The initial work discussed in this section was performed between 2003 and 2010 with a 300 m^3/d pilot and with a 23 MLD demonstration plant. PUB then continued studies with the original demonstration units and a full scale 68 MLD plant.

In 2015, PUB announced the decision to install a 1200 MLD PDF (800 MLD ADF) plant at nearby Tuas (about 20 km away), scheduled to be commissioned in 2025. Accordingly, a decision was made to install a second demonstration plant at Ulu Pandan with a peak capacity of 18.75 MLD (ADF 12.5 MLD). This second demonstration was initially commissioned in early 2017 and some preliminary discussion of the initial results is also presented in this section. Confusingly, the plant is referred to by some as the 'Ulu Pandan demonstration plant' even though it is distinct from the original demonstration and uses a different process flowsheet. The full official name for the second plant is the Tuas Integrated Validation and Demonstration plant (IVP) located at Ulu Pandan.

The original pilot plant was commissioned in March 2003 at the Bedok Water Reclamation Plant (Tao *et al.*, 2005). Primary settled sewage, predominantly domestic in nature, was used as the feed source and the bioreactor parameters were identified using the Modified Ludzack-Ettinger (MLE) model. A baseline study was carried out based on recommendations for the Zenon product. The membrane flux was increased gradually in phase 1 and air supply to the aeration tank and membrane tank was varied in phase 2. This allowed the operator to achieve an optimum oxygen level in the aeration tank and an optimum scouring impact on membranes while maintaining the frequency of the cleaning cycle of three months without compromising the filtrate quality. The performance of the Suez pilot was compared to two other pilots. The air/permeate ratios of three systems were reduced from an initial ratio of 24–50 m^{-3}/m^3 down to 21–34 m^{-3}/m^3 in phase 1 and further reduced to 16–28 m^{-3}/m^3 in phase 2. The specific energy consumption was reduced from 1.3–1.7 to 1.0–1.3 kWh/m^3 in phase 1 and further reduced to 0.8–1.1 kWh/m^3 in phase 2. An energy consumption rate of 0.7 kWh/m^3 was set as the specific energy guarantee figure of the large-scale MBR demonstration plant.

The commissioning of Ulu Pandan MBR Demonstration Plant in December 2006 allowed for more studies to be carried out. The plant also treats primary settled sewage which is predominantly domestic in nature. The net design flux is 25.3 LMH. A baseline study was carried out first to identify the major energy consumption components. Through an energy audit, the key energy consumption components were identified including process aeration, membrane scouring rate, sludge retention time (SRT) and the mixed liquor suspended solids (MLSS) level. A comprehensive system optimization was then carried out in turn, which included optimization of sludge retention time, MLSS levels, MLSS recirculation ratios, process aeration, and membrane scouring rates. Reducing the hydraulic retention time (HRT) down to 3 hours was also tested to assess the impact on product water quality and energy consumption.

Without compromising product quality and quantity, the Ulu Pandan MBR Demonstration Plant was able to reduce energy consumption significantly. Setting a specific energy guarantee figure enabled the contractor to select energy-efficient equipment. This reduced specific energy consumption from 0.7 to 0.59 kWh/m^3 under design conditions. Optimization of the sludge retention time and MLSS level enabled the energy consumption to be reduced to 0.549 kWh/m^3. Optimization of MLSS recirculation resulted in a further reduction to 0.535 kWh/m^3. Optimization of process aeration enabled the energy consumption to be reduced to 0.475 kWh/m^3. Finally, optimization of the membrane scouring rate allowed the energy consumption to be further reduced to 0.37 kWh/m^3. In all the pilot and demonstration studies, there was no observable adverse impact on the MBR filtrate quality.

Through comprehensive and systematic MBR optimization studies, the specific energy consumption was reduced from 1.3 kWh/m^3 to less than 0.8 kWh/m^3 by increasing membrane flux and reducing aeration in the 300 m^3/d pilot-scale plants. This was further reduced to 0.37 kWh/m^3 with the optimization of sludge retention time and MLSS level, MLSS recirculation, process aeration, membrane scouring rate, and energy efficient equipment selection for the 23 MLD MBR demonstration plant.

Since PUB reported the pilot and demonstration studies in 2010, Suez has introduced the LEAPmbr (see Section 14.2.2.1), which offers a further significant reduction of energy as well as other operating and capital cost benefits for the type of feed tested in Singapore. By 2013, data indicated an energy range of 0.148–0.183 $kWhr/m^3$ as the total energy consumption for the MF stage, i.e., air scour and TMP. This value is approximately equivalent to the energy for recirculation which would indicate that it should be possible to further reduce the 0.37 $kWhr/m^3$ final optimized value from the demonstration plant studies reported above to around 0.3 $kWhr/m^3$.

The Tuas IVP demonstration plant has implemented a number of important energy-saving innovations, which have improved the overall process efficiency and significantly reduced energy even further. While the process flowsheet for the original Ulu Pandan demonstration resembled the standard set up for MBR (see Chapter 2, Section 2.6.4), the Tuas IVP Demo Plant has used a radically different approach to minimize energy usage. This has been accomplished by enhancing the primary stages through the use of biosorption and a compact lamella plate primary clarifier, both of which reduce the nutrient loading to the MBR stage (Tao *et al.*, 2018). It also eliminates unnecessary energy usage in the MBR by using multiple anoxic and aerobic zones in a multi-step MBR layout, with five steps, to reduce the reliance on recirculation. This arrangement enables the average MLSS to be reduced to 2.5 g/l, which in itself improves aeration efficiency, and allows nitrification/denitrification treatment goals to be achieved with a lower recirculation range of 1–2 compared to a conventional MBR set up which would require 3–5.

Preliminary results from the first year of operation of the Tuas IVP Demo since mid-2017 have been very encouraging. The process flowsheet changes have reduced process aeration requirements to <0.1 kWhr/m^3 which is about half the lowest value discussed for the original demonstration plant. Overall MBR energy needs are further reduced due to the low recirculation rates. The MBR has met treatment goals with a filtrate of <0.1 NTU, ammonia <1 mg/L, nitrite 'N' <1 mg/L, and nitrate 'N' <8 mg/L. A critical aspect of the optimization has been the use of advanced process automation to provide feedback from the instrumentation and analysers to the process control system. This ensures that aeration provision can be matched with requirements in real time.

14.3.1.2 Evoqua: Rationalization and expansion of the Santa Giustina WWTP, Rimini, Italy

In 2015, HERA, one of the largest municipal operators in Italy, decided to enhance the wastewater treatment plant at Rimini, by implementing an upgrade based on the Evoqua MemPulse MBR (Boveri, 2016). The new MBR system runs alongside an existing CAS. At the same time, some nearby older facilities were shutdown thereby centralizing wastewater treatment for the area in a single enhanced and optimized facility. Following the MBR upgrade, the capacity at the Santa Giustina WWTP has been increased from a population equivalent of 220,000–560,000 (Figure 14.30).

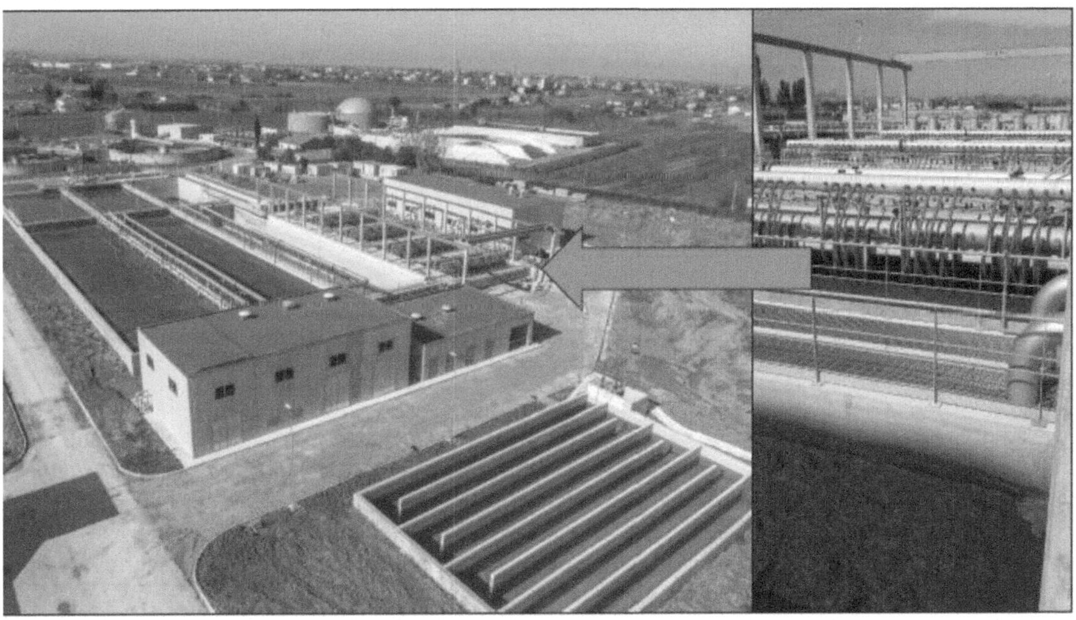

Figure 14.30 MBR plant at the Santa Giustina WWTP, Rimini.

Rimini is an important tourist area in Italy and receives a major influx of visitors in the summer months. The plant is located close to the coast and needs to achieve stringent discharge standards to meet the EU Bathing Water Directive, since the treated effluent is discharged to a local river estuary which is close to the main beach area. The beaches in this region are closely monitored since the coastline is somewhat enclosed due to its location at the northern end of the Adriatic Sea. The new works has a maximum average daily flow in the summer of about 125 MLD, which is 2.5 times the flow in months outside of the tourist season. The plant has been designed to provide a PDF of 188 MLD. The wide range of flow requirements has necessitated a design which maximizes flexibility between the original CAS and the new MBR to meet varying flow and treatment targets.

In this plant, the membranes are protected by 1.5 mm screens. At low flow, a proportion of the flow is recirculated through the screens to improve the removal of fibrous material and control solids accumulation in the biological tanks. Existing primary clarifiers at the site were initially used as settlement tanks, but were later by-passed as they reduced the carbon levels too far which restricted the subsequent denitrification stage. Total nitrogen (TN) needed to be reduced from an influent level of 50–60 ppm to provide an effluent of $\leq$15 ppm.

The MBR membranes are arranged in eight separate tanks fed from a common channel. The process cycle comprises a 12 min filtration cycle followed by a relaxation step with air scour. In common with several other MBR technologies, the system does not use backwash. The bubble size is controlled by a pulsing device which has no moving parts and ensures low-energy use and a highly efficient cleaning action (see Figure 14.16).

The system uses a regular maintenance clean (MC) once per week, and an occasional CIP, once every 3–6 months. Similar chemicals are used (e.g. 200–300 ppm of sodium hypochlorite) for the two procedures, but the latter has a longer soak time, e.g., 6 hours for CIP compared to 1 hour for MC. Occasional acid cleaning can also be used during the CIP. The spent CIP chemical is recirculated to the feed and re-processed.

Evoqua utilizes a 0.04 μm fibre for the MBR which resembles the fibre used for general water treatment applications. The normal design MLSS for the system is 9.1 g/ml, but the fine pore size rating enables the system to operate with lower MLSS than can be used by coarser MF membranes, e.g. 3 g/ml, since it is less dependent on the pre-coat effect (see Section 14.1.2.1).

Since coming into operation in mid-2015, the plant has achieved effluent treatment standards, and met the challenges of variable flow and quality. The average specific energy consumption has been <0.35 kWhr/m^3 of treated water, including usage related to screens, blowers, mixers, valves, pumps, and sludge thickening.

14.3.1.3 Koch: Rationalization and expansion of the Carrée de Réunion WWTP, France

The Carrée de Réunion plant is located adjacent to the Palace of Versailles about 20 km south west of Paris. The treated water is discharged into the Ru de Gally creek located in the palace gardens, and eventually enters the Seine River. The plant is owned and operated by SMAROV (Syndicat mixte d'Assainissement de la Region Ouest de Versailles).

Prior to the upgrade, the pre-existing system comprised a conventional activated sludge (CAS) plant with primary sedimentation, aeration tanks, and secondary sedimentation. The capacity had to be increased as the master plan called for centralization of several smaller WWTPs. As these plants were phased out, the Carrée de Réunion plant needed to take up the load. The main challenges associated with this upgrade were to achieve the capacity increase within the limited available area of the plant using an architectural concept for the plant which did not disturb the view from the famous Versailles palace and palace gardens.

The feed is standard municipal wastewater with a temperature ranging from 6 to 25°C. The average daily flow in rainy conditions is 96.7 MLD with peak flow of 144 MLD, and an annual average capacity of 46 MLD. The treated water quality requirements are very stringent as set out in Table 14.18.

Table 14.18 Treated water quality requirements.

Parameter	Value, mg/L
COD	≤30
BOD	≤5
TKN	≤1
Total N	≤8
Total P	≤0.3
TSS	≤3

The system uses the module specification PURON® PSH1800 with 1,800 m^2 per module (each module contains 44 rows of 40.9 m^2 for each individual unit), as shown in Figure 14.31. The plant comprises 10 trains each containing 9xPSH1800 modules, i.e., 90 modules in total to give a membrane area of 162,000 m^2. Each train has space for an additional module. The plant also has the capability to install two additional trains (for a total of 12) to accommodate future potential capacity expansion. The plant is designed to treat the full peak hourly flow (PHF) of 6,000 m^3/h with one membrane train off-line (N–1). The design flux for the system is as follows:

ADF design flux=11.8 L/m^2 h
PHF design flux=37.0 L/m^2 h
PHF (N–1) design flux=41.1 L/m^2 h

The proposed schedule for maintenance and recovery cleaning (MC and RC) was based on sodium hypochlorite MC using 125 mg/L applied on a daily basis, with one to two RC applied per year per train using 1,000–2,000 mg/L NaOCl and citric acid at a pH of 2.0–2.7.

Figure 14.31 Installation of one of the PURON® MF modules.

Consistent effluent quality is especially important for Carrée de Réunion, because the treated water is discharged into the Ru de Gally creek located in the palace gardens, and eventually enters the Seine River. As a result, the effluent phosphorus is limited to ≤0.3 mg/L, TSS to ≤3 mg/L and TN to ≤8 mg/L

Commissioning started at the end of September 2015 and was phased to accept 100% of the incoming load by end of November 2016. The plant has operated at design flow rates (up to 6,000 m^3/h) according to the nominal design parameters. The process runtime is 10 min followed by a 30 second backflush. The filtration plant has demonstrated stable performance since commissioning. The plant performance was tested relatively early with adverse conditions featuring minimum temperatures and peak hydraulic capacity in winter with one train offline (see Figure 14.32). The permeability of the membranes is still close to the initial values when testing with clear water and varies between values of 400 LMH (beginning of filtration cycle) down to 300 LMH at the end of each filtration cycle under peak flow conditions. The transmembrane pressure was still below 200 mbar even in peak flow conditions with very low temperatures.

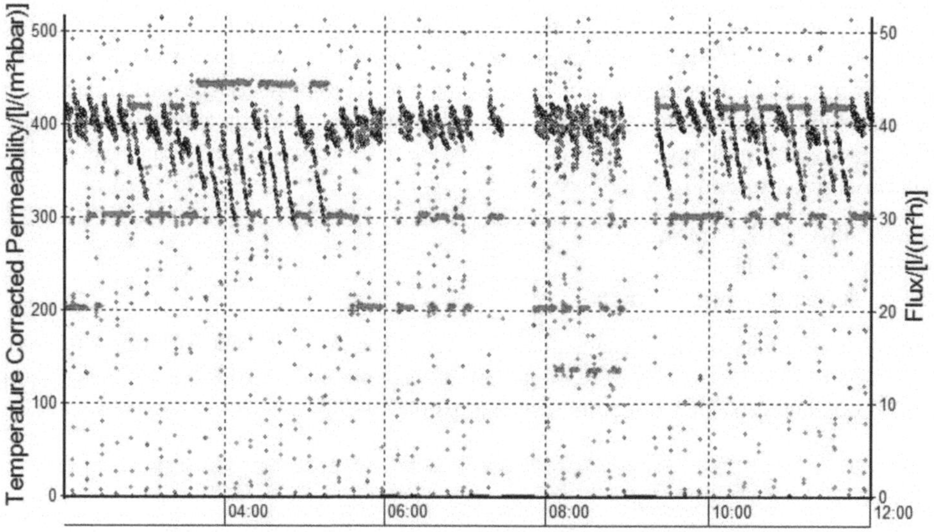

Figure 14.32 Permeability (black) as a function of flux (red) at low temperature (12°C) during a peak event.

14.3.1.4 Memstar: Expansion and upgrade of the Guangzhou STP, China

Guangzhou, China's third largest city with a population of 14 million, required an increase in its sewage treatment capacity in the lead up to the Asian games held there in 2010. It was decided to use MBR due to space constraints and the need to meet tight effluent treatment standards. In fact, space is at such a premium the plant has been installed underground, and claims to be the world's largest underground STP. The EPC for the project was Memstar's parent, United Envirotech (now CITIC Envirotech). Influent and effluent quality for the plant are indicated in Table 14.19.

Table 14.19 Influent wastewater quality and regulated effluent levels.

	BOD5	COD	SS	$NH4^+$-N	TN	TP
Influent (mg/L)	160	270	220	30	35	4.5
Effluent Regulation (mg/L)[a]	10	50	10	5	15	0.5

[a]Based on China's Class IA effluent water quality standards.

The Memstar design uses an innovative scheme to optimize aeration energy and minimize the under-utilization of dissolved oxygen (DO) in case of excessive levels of air scour. In the block flow diagram in Figure 14.33, the three-stage wastewater treatment process comprising anaerobic, anoxic and oxic stages is dubbed A^2O. The standard flow recirculation path which was originally adopted at the Guangzhou STO was for mixed liquor from the membrane tank to be recirculated to the anoxic tank. However, significant energy savings were obtained by recirculating to the oxic (or aerobic tank) instead of to the anoxic tank, thereby leveraging on excess DO to reduce overall aeration demands. In this arrangement, the surplus DO can be consumed along the length of the oxic tanks by the aerobic and facultative bacteria, thereby reducing aeration needs directly. Additionally, when nitrate-rich mixed liquors with lower oxygen levels are recirculated from the end of the oxic zone to the anoxic tanks, denitrification performance can be enhanced thus achieving higher TN removals.

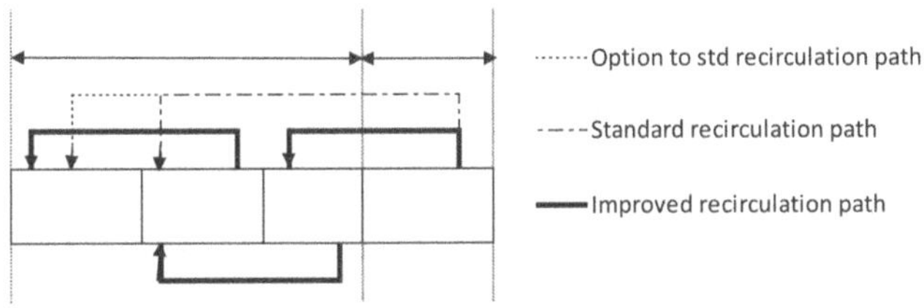

Figure 14.33 Block flow diagram for the Guangzhou STP.

Long-term operation has confirmed that the improved recirculation path protocol has resulted in reduced energy for the MBR blower from 0.22 to 0.18 $kWhr/m^3$, i.e., an 18% saving. Furthermore, the performance of the denitrification stage has been improved with lower levels of DO in the anoxic tank and longer sludge retention times, allowing lower TN levels to be achieved in the effluent.

The original design used a flux of 15 LMH. However, subsequent increased demand meant that the operator required fluxes of up to 25 LMH by utilizing an increased maintenance cleaning (MC) frequency. Three MC procedures were examined, as illustrated in Figure 14.34. Although it was possible to maintain the higher flux with the most aggressive cleaning protocol, namely MC3, the operating costs were more than 1.5 times the costs of protocol MC1. Accordingly, the economic viability of this approach was felt to be questionable.

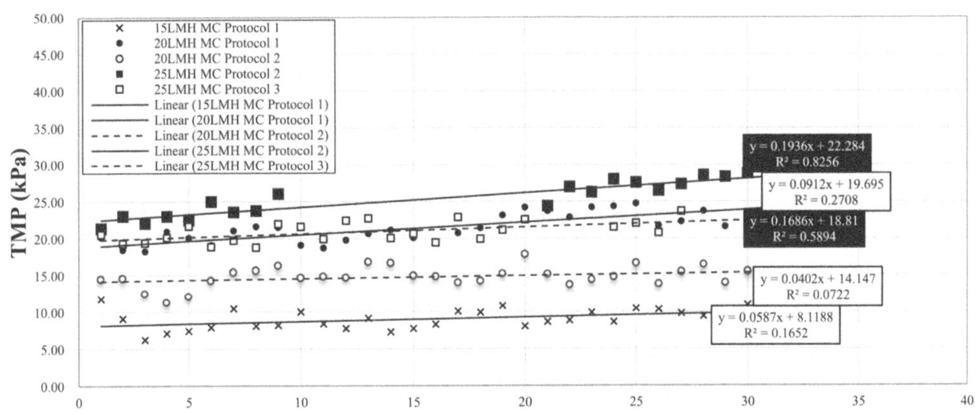

Figure 14.34 TMP response curves for different fluxes and MC protocols.

14.3.2 Flat sheet case studies

14.3.2.1 Kubota: Retrofit MBR at Canton to increase flow and improve effluent quality

In 2015, the world's largest MBR at that time was under construction at the Canton Water Reclamation Facility (WRF) in Ohio, though by 2019, it had been surpassed by four other plants, two in China and two in Europe. The plant was especially notable since it was the only very large plant utilizing the flat sheet format. As discussed in Section 14.2.1, flat sheet is rarely utilized for very large facilities, and Canton is one of only two MBRs in operation from the list of 80 large plants operating in 2019. Normally, the additional operating expense of aeration for flat sheet outweighs other factors in the life cycle assessment. However, other capital cost factors came into play at Canton which resulted in flat sheet technology providing the most attractive option (Gellner, 2016).

The original CAS facility at Canton, which had been in operation since the 1970s, provided an ADF of 125 MLD. In 2010, various options were evaluated to enhance and/or replace this facility to expand the ADF to 148 MLD and improve effluent to meet a standard of TN $\leq$8 ppm, $P \leq 0.7$ ppm. The plant was specified to provide a significant peak flow capability with a PDF of 333 MLD and an instantaneous flow of 416 MLD. MBR was one of six options evaluated and was selected as the most cost effective with a life cycle cost (LCC) at least 20% less than any of the other conventional alternatives. Having decided to proceed with MBR, Ovivo won the MBR system supply contract since their offering, based on the Kubota flat sheet format, was able to utilize existing tanks and other equipment from the original CAS facility in the most cost-effective manner.

Figure 14.35 shows the plant layout at the Canton WRF, indicating the parts of the original system that were deleted or made available for other uses, and those parts that were reused or modified in converting the plant from CAS to MBR. The MBR did not require any new concrete structures or replace existing process equipment. For example, coarse screens and pumps for the influent were reused, and the aeration blowers were re-conditioned rather than being replaced. New preliminary treatment facilities were constructed in the existing pre-aeration tanks and a two-stage set of fine screens were provided with a 6 mm rating followed by a 2 mm rating. A new building was added to handle solids and grease from the screens, and included solids washing.

Figure 14.35 Retrofit and upgrade of the Canton WRF.

The new plant layout shows that the extensive tank farm providing primary and secondary clarifiers were not required by the MBR, and these tanks have been made available for alternative uses. The advantage of not utilizing a primary clarification stage is that it increases the carbon concentration to the biological reaction stage in the MBR system enabling improved denitrification performance and a lower TN specification. P removal is achieved primarily by biological processes and supplemented, if required, by ferric dosing. The key to the selection of the Kubota flat sheet format for Canton has been the flexibility with which it could be accommodated within the constraints and limitations of existing tanks and process equipment, thereby reducing capital cost.

14.3.3 External case studies

14.3.3.1 Pentair-Xflow AL MBR: Energy optimization at Ootmarsum

The WWTP Ootmarsum is owned by the water authority Regge & Dinkel (WRD) and is situated in the municipality of Dinkelland (in the eastern part of the Netherlands). The WWTP needed to be modernized because it discharges the treated wastewater into a water system with considerable ecological vulnerability. In cooperation with Grontmij engineering consultants a design study was carried out resulting in the choice of a hybrid system. This combined a conventional system followed by a sand filter with an MBR to reduce costs compared to an MBR only. After an intensive selection procedure the Xflow AirLift MBR system was chosen for the project. The construction work of the 650 m^3/h plant started in mid-2005 with start-up of the full-scale plant in late 2007.

Table 14.20 summarizes the main process parameters of a pilot study conducted at the site before start-up of the main plant to optimize the energy consumption. The pilot study showed that the energy consumption could be reduced to a level comparable to submerged systems at around 0.25 kWh/m^3. The decrease in energy consumption has been achieved by several improvements such as a better cleaning procedure of the module and better hydraulics of the aerator. Both have resulted in an increase in the net permeate flux and a significant reduction in the air flow rate. The pilot showed that stable permeability could be obtained at fluxes of up to 65 LMH, but the main plant was designed at 50 LMH.

Table 14.20 Process parameters for the Ootmarsum pilot.

Date	Oct-05	Mar-06	May-06	Jul-07
Nominal net flux (lmh)	35	45	50	50
Aeration				
Air flow rate (Nm^3/h/module)	20	20	15	9
Absolute inlet pressure (bar)	1	1	1	1
Absolute outlet pressure (bar)	1.55	1.55	1.55	1.55
Blower efficiency (%)	60	60	60	60
Energy consumption aeration (kWh/m^3)	**0.42**	**0.32**	**0.22**	**0.13**
Recirculation:				
Sludge flow rate (m^3/h/module)	20	20	20	20
Head (m)	2	2	2	2
pump efficiency (%)	85	85	85	85
Energy consumption recirculation (kWh/m^3)	**0.13**	**0.1**	**0.09**	**0.09**

To date, the plant is still using 85% of its original membranes on 5 of the 6 skids, i.e., a membrane life exceeding 11 years. In 2015, one of the skids was upgraded and used as the Helix version of the same module (see Section 14.2.4.2), operating under similar conditions, but at higher flux of 70–80 LMH.

14.3.3.2 Aquabio/Berghof: Wastewater treatment at Kanes food, UK and Holland Malt, Nl

In 2001 Kanes Foods in Evesham, UK, installed an Aquabio 'first generation' AMBR™ plant combined with reverse osmosis (RO) and UV disinfection for wastewater reuse to potable water standards. In doing so, the company became the first food plant in Europe to recycle treated wastewater on a large scale at this treatment standard. Due to factory growth and continuing limitations on water supply and discharge options, Kanes subsequently installed a second AMBR LE™ plant on the same site to increase the treatment capacity and recycle volume. The 2001 'first generation' AMBR™ process treatment scheme comprises the following stages:

- Upstream screening
- Flow balancing
- Dissolved air flotation (DAF) treatment for fine vegetable solids removal
- The AMBR™ itself
- Downstream treatment by RO followed by UV disinfection

This system yields potable quality water suitable for blending with mains water for use within the factory. The process comprises two 250 m^3 bioreactors with four banks of crossflow membrane modules. The maximum biomass solids (MLSS) concentration employed has been as high as 22 g/L, but the bioreactor is generally operated at around 10 g/L. Sludge production is calculated as being 0.14 kg DS/kg COD removed at a sludge age of over 100 days.

Each membrane bank comprises four UF membranes. The membranes operate at an average flux of 153 LMH normalized to 25°C. The permeate water has average TSS, BOD, and COD concentrations of only 4, 7 and 16 mg/L, respectively. The UF permeate then passes to a two-stage RO plant which achieves an overall recovery of 75%. The reject stream is discharged to sewer and the permeate, which typically has a conductivity of 40–100 μS/cm, is passed to the UV disinfection unit and on to the client's water supply tank.

The original plant has now been in continuous operation for about 15 years and has performed consistently in terms of biological treatment, membrane performance and final reuse water quality. For the majority of the time membrane performance has been better than design allowing one bank to be maintained as a standby and so offering more process flexibility and lower energy use. Occasional reductions in membrane flux have been linked to poor biomass health, which has been rectified by closer management of the process.

The latest developments with AMBR LE™ have allowed the end user to expand plant capacity while at the same time achieving significant energy savings. Due to continued production expansion and the consequent pressure on incoming water supply and discharge consent, a second wastewater treatment plant was supplied by Aquabio in 2010. This new plant provides an additional 1.435 MLD of biological/MBR treatment capacity and an additional 0.6 MLD of re-use water. The plant replicates the successful process scheme of the original plant but uses lower energy in operation. New common inlet screening and flow balancing facilities are provided, which allows the original balance tank to be used to balance outgoing flows to sewer. Again, DAF treatment is employed to remove fine vegetable solids.

A single bioreactor of 765 m^3, allowing for MLSS concentrations up to 12 g/L, is aerated by a blower-assisted slot-type aeration system, with high alpha factor characteristics.

The AMBR LE™ UF membrane system shown in Figure 14.36 comprises 4 banks of 4 m long membranes modules containing 8 mm tubes supplied by Berghof. Each bank has 320 m^2 of membrane area. Low-energy performance is achieved by the use of backflushing to control membrane fouling thus allowing much reduced cross flow velocities and hence significantly lower energy use. The combination of backflushing frequency and variable cross flow velocity is optimized to give the lowest energy use for the required throughput. Significant process flexibility is offered by the inclusion of variable speed recirculation pumps and optional permeate pumping to control the TMP and hence the flux performance. Stable performance is currently being achieved at typical energy use of 0.4–0.8 kWh/m^3 of treated permeate for the biomass separation, a significant improvement on the original plant.

Figure 14.36 Aquabio AMBR LE™ system at Kanes Foods.

The Berghof tubular membrane has also been used at Holland Malt in the Netherlands, where the end user needed to double production capacity and did so by recycling process water. As at Kane Foods, the feed has a high degree of variability both in terms of flowrate and COD. Berghof used the B-SMART self-regulating system to cater for the variability. The plant comprises 18 modules which are 250 mm diameter and 4 m long, and contain 8 mm tubular membranes. The system operates at an average permeate flow of 1.584 MLD with a flux normally in the range 65–75 LMH and a crossflow velocity of 2.0 m/s. Table 14.21 shows feed and filtrate quality.

Figure 14.37 shows operating data for a 5 month period soon after start up in 2016. The figure shows that the MLSS can vary between 12 and 18 g/ml. In mid-October, the MLSS increased and the system performed an automatic CIP to restore the original flux. The B-SMART software automatically adjusts crossflow velocity, backwash and CIP frequency in response to TMP. Accordingly, average energy usage was limited to 0.6 kWhr/m^3, which was similar to that at Kanes Foods.

Table 14.21 Feedwater and filtrate quality

Parameter	Feed, mg/L	Fíltrate, mg/L
COD	1,500–5,000	133
BOD	1,000–3,000	<2
Total N	60–70	4.69
NH_4-n	5–20	0.1
Total P	15–30	0.2
PO_4-P	10–20	0.29
Plate count	$>10^6$ CFU/100 mL	<500 CFU/100 mL

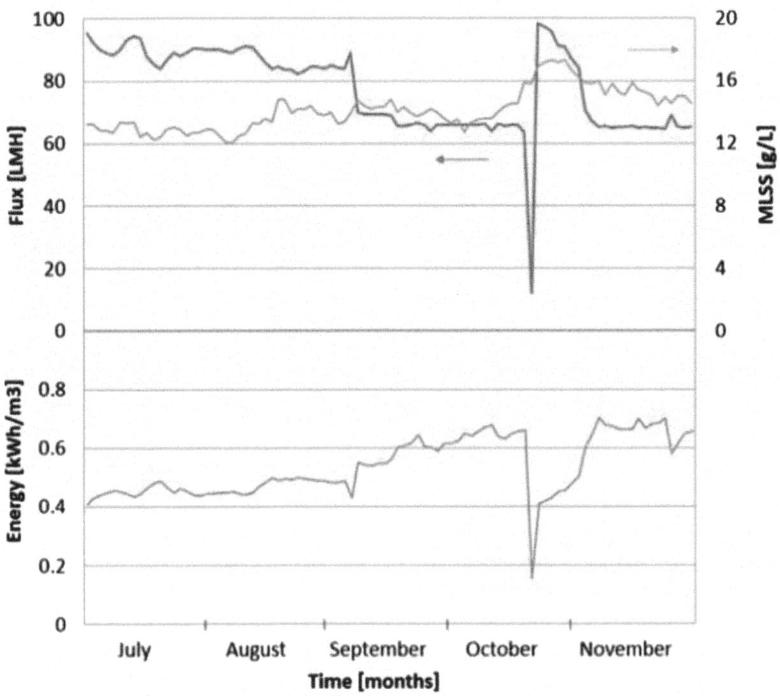

Figure 14.37 Operation results over a period of 5 months.

14.4 SUMMARY AND FUTURE OUTLOOK

The first commercial MBR products were introduced around 50 years ago, initially driven by the fact that they provided excellent treated water quality. In addition, the physical barrier for the biomass meant that a bio-reactor could operate at higher solids loading, which reduced the footprint. Through the next three decades, various systems were offered using a side-stream configuration operating in cross-flow. However, the most significant development for commercial products occurred in the 1990s with the introduction of immersed products based on either flat sheet or hollow fibre formats. These designs reduced capital costs, since membranes could be immersed in the bio-reactor. Also operating cost was reduced since the aeration system required for the bio-reactor could be used to clean the membranes.

The last 15–20 years has seen the concept of MBR become commercially accepted, with an evolution of the product base. Indeed, since 2007 there has been a substantial uptake of large schemes >76 MLD (i.e., >20 MGD) peak daily flow, which has transformed the perception of MBR technology. There are now 82 of these large plants with a cumulative peak capacity of almost 15,000 MLD (4,000 MGD). Most of the schemes have been in China, but the United States and continental Europe has also adopted the technology for major indirect potable reuse projects.

Although current products are still clearly recognizable from the original developments, operational efficiency and cost have now been significantly improved and product sizes have become much larger. Today's commercial products have maintained diversity, with several different materials in use, though PVDF is the clear market leader. As the market has matured, alternative membrane materials are offered less and some have been removed from product portfolios resulting in consolidation for PVDF offerings. Different MBR formats are used, though immersed is much more commonplace than side-stream. For immersed formats, hollow fibre is more widely used than flat sheet, especially for large plants in which hollow fibre has an overwhelming but not exclusive share.

A common theme among all suppliers has been to improve operational simplicity and reliability, by eliminating valves and pumps wherever possible. Equally as important, products now utilize much less air than they did originally, by controlling bubble size, the method of distribution, and the amount of air supplied. These improvements have had the effect of reducing energy use dramatically, making MBR competitive with conventional wastewater treatment. The recent trend has been to reduce air wastage and enhance process automation to ensure that air is used at a rate which is proportional to flow and is potentially modified in response to changes in the feed analysis, resulting in lower energy usage.

There are more than 100 companies worldwide supplying MBR, but only around nine companies dominate the market in terms of the number of plants supplied and the total treatment capacity. Of these, two companies are the most influential players in the market, i.e., Suez, with an immersed hollow fibre product, and Kubota with an immersed flat sheet. Both of these companies were early market entrants and have been pioneering proponents of their chosen formats. Suez has the greatest share of treated water capacity across all plant sizes and regions, having been selected for 25% of the world's largest plants. Conversely, Kubota has by far the greatest number of plants, with a product offering that is particularly cost effective at smaller scale.

However, new players in China have caught up dramatically in just a few years to take the major share of the booming Chinese market which has seen an explosion of projects at large scale. The two main beneficiaries of this growth have been BOW and Memstar. The market share leader, BOW, initially acted as an EPC using imported membranes from Asahi and Mitsubishi, but in the last few years has used its own membranes. The second dominant player is Memstar providing MBR products to its EPC parent, Citic Environment Ltd (CEL). The Chinese market looks set to continue in its use of MBR and these two EPCs are likely to continue to dominate in their own market, which should provide a strong platform for overseas growth. Indeed, in 2018, Memstar opened a US manufacturing facility.

There are several other hollow fibre suppliers with similar products to market leading Suez, each with unique beneficial features, which has created strong niches and the potential for competitive advantage. Though not identical, some of these products can be retrofitted with each other. Indeed, as project size has grown, the number of options for closely sized replacements has grown, and it seems likely that a competitive retrofit market will develop. It should be noted however, that membrane life is often long, with many installations reaching a 10-year life before membrane replacement, and some even reaching 20 years. The retrofit market is therefore potentially not as attractive for MBR as for the general MF duties which do not benefit from self-repairing mechanisms. Where problems occur, it is likely to be due to clogging in the module rather than a fundamental failure or degradation of the membrane, i.e., an issue of module design and/or operation.

There are fewer flat sheet options amongst the leading group of companies, but this format is a popular choice for small-scale applications and/or difficult duties. Overall, the marketplace has retained considerable diversity among product offerings, both in terms of the format choice and membrane material, but also in terms of product dimensions and operational details. It is unlikely therefore that a fully interchangeable product market will become established in the near term, but sufficient similarity is evolving to enable several retrofit options for most products. All of the market leading companies have improved the energy and operational efficiency of their products to a considerable extent in the last few years to improve their competitiveness against conventional technology and each other. Further gains will now become more difficult to achieve.

For the tubular format, recent developments have seen the original side-stream concept make something of a comeback to address the issue of high-energy use. This has included the use of backwash and reverse flushing to reduce the need for continuous high cross-flow velocity. Of wider significance, there has also been a development of the side-stream AL concept, which shows the potential of rivalling the low-energy use now being achieved by immersed format, while retaining operational and maintenance benefits.

A final development that could have an impact in the future is that of ceramic membranes since the low fouling characteristics are potentially beneficial and long membrane life may counter the high initial capital cost. However, the impact of ceramics so far in the MBR field has been limited. Though ceramics achieve long life, polymerics, especially the supported membrane products, have established a good track record, which has reduced the potential advantage of ceramics. Ceramics however, do achieve stable performance in the long term with very low chemical cleaning rates, and both of these factors provide an important operating cost advantage. It is possible therefore that ceramics will gain share, but the initial capital cost will probably have to become relatively close to polymerics (i.e. certainly less than twice as expensive) before these gains are made.

The case studies reviewed in this chapter have shown that the operational experience of MBR has been good, and that recent developments carried out by the suppliers featured has resulted in better operation and performance stability at lower energy cost than would have been achieved with previous product offerings. Different cases have illustrated dramatic energy improvements, the use of MBR in very large schemes both hollow fibre and flat sheet, the advantages of ceramics, and the improvements accessible in the tubular format. It is likely that MBR will continue to develop and improve, and continue to achieve a strong sustained growth.

REFERENCES

Adham S., DeCarolis J., Hirani Z., Wasserman L. and Tran N. (2007). Evaluation of New MBR Systems for Water Reuse. Proceedings of the IWA Conference, Harrogate, UK, paper 7b.

Angiel J. D., Lothman S., Yi P., Kirkpatrick R., Rumke M. and Walker T. (2018). 10 years of MBR Operation; the Broad Run Water Reclamation Facility Experience. Proceedings of AMTA/AWWA Membrane Technology Conference, March 2018, West Palm Beach, Florida.

Berube P. (2007). Shear Forces in an Air Sparged Submerged Hollow Fibre Membrane System. Conference Proceedings – IWA Membrane, UK, 10b.

Boveri L. (2016). Advanced Technologies and Innovative Solutions in Full Scale MBR Installations; Rimini WWTP. Proceedings of AMTA/AWWA Membrane Technology Conference, February 2016, San Antonio, Texas.

Branch A. and Le-Cech P. (2015). National Validation Guidelines for Water Recycling: Membrane Bioreactors, Australian Water Recycling Centre of Excellence. UNSW, ISBN: 978-1-922202-67-3, http://vuir.vu.edu.au/id/eprint/32073.

Branch A., James G., Trinh T., Ta T. M., Leslie G. and Le-Clech P. (2015). Membrane Aging and Replacement – Impact on Pathogen Removal in Full-Scale MBR. Proceedings of Weftec Conference, September 2015, Chicago.

Brepols C. (2012). Performance improvement of full scale membrane bioreactors. In: Water Energy Interaction of Water Reuse, V. Lazarova, K.-H. Choo and P. Cornel (eds.), IWA Publishing, London, UK, ISBN 978-184-33954-16, pp. 127–137.

Cabassud C., Laborie S., Durand-Bourlier L. and Laine J. M. (2001). Air sparing in UF hollow fibers: relationship between flux enhancement, cake characteristics and hydrodynamic parameters. *J Membrane Science*, **181**, 57–69.

Churchouse S., Warren S. and Floyd M. (2008). Feedback from the Porlock MBR Plant in its 10th Year of Operation:- An Analysis of the Flux, Effluent Quality and Membrane Lifetime Data to date. Proceedings of the Conference Membranes in Drinking Water Production and Wastewater Treatment, 20–22 October 2008, Toulouse, France, http://kubota-freeair.com/files/Proceedings%20paper%20by%20SJC%20Oct08%20Aachen.pdf.

Erdal U. (AECOM) (2017). Can MBR Replace MF/UF in a Potable Reuse Train? Concerns and Limitations. Proceedings of the 11th IWA Conference on Water Reclamation & Reuse, July 2017, Long Beach, California.

Ewert J. (2014). Flat Sheet Membrane Bioreactor Operational Experiences – a New Zealand Perspective. http://www.wioa.org.au/conference_papers/2014_vic/documents/Jason_Ewert.pdf

Fane A. G., Yeo A., Law A., Parameshwaran K., Wicaksana F. and Chen V. (2005). Low pressure membrane processes – doing more with less energy. *Desalination*, **185**, 1585–1591.

Frost and Sullivan. (2011). China Membrane Bioreactor (MBR) Market Outlook – Ambitious Water Reuse Targets to Boost Local Membrane Industry and Fuel Exponential Growth. August 2011, UK, M700–15.

Frost and Sullivan. (2013). Global MBR market.

Gellner T. (2016). North America's Largest MBR WWTP; Breaking it Down to Make it Cost Effective and a Small Plant Mentality. Proceedings of AMTA/AWWA Membrane Technology Conference, Th-03-01, February 2016, San Antonio, Texas.

Gil J. A. (2011). MBR performance: operational problems in industry. *Filtration + Separation*, **Nov/Dec**, 36–41.

Hai F., Riley T., Shawkat S., Magram S. and Yamamoto K. (2014) Removal of pathogens by membrane bioreactors: A review of the mechanisms, influencing factors and reduction in chemical disinfectant dosing. *Water*, **6**(12), 3603–3630, https://doi.org/10.3390/w6123603.

Judd S. (2005). Fouling control in submerged membrane bioreactors. *Water Science &Technology*, **51**(6), 27–34.

Judd S. (2008). The status of membrane bioreactor technology. *Trends in Biotechnology*, **26**(2), 109–116.

Judd S. (2011). The Top 20 Largest MBRs in the World? http://www.thembrsite.com/feature_largest_plants.php.

Judd S. (2017). https://www.thembrsite.com/blog/the-cost-of-ceramic-membranes-for-membrane-bioreactors/.

Judd S., Robinson T., Holdner J., Alavarez-Vazquez H. and Jefferson B. (2004). Impact of Membrane Material on Membrane Bioreactor Permeability. Proceedings of Water Environ. Membrane Technology Conference, Seoul.

Judd S. *et al.* (2011). The MBR Book (2nd Edition): Principles and Applications of Membrane Bioreactors for Water and Wastewater Treatment. Publ. Elsevier, UK, ISBN: 978-0-08-096682-3, pp. 289–357.

Katz S. and Grundestam J. (2016). The Largest MBR in the World; Upgrading the Henriksdal WWTP. Proceedings of the SIWW Conference, July 2016, Singapore.

Katz S., Owerdieck C., Arnsten B. and Kendrick S. (2016). MBR's Role in Potable Reuse. Proceedings of AMTA/AWWA Membrane Technology Conference, February 2016, San Antonio, Texas.

Katz S., Salveson A., Fontaine N., Bucher B. and Berrryhill J. (2017). Giving Credit Where Credit is Due – MBR for Potable Water Reuse. Proceedings of Weftec Conference, September 2017, Chicago.

Kim D., Cho Y., McKennon K., Lowe G., Jung S. and Kim J. (2017). A successful transformation of conventional SBR to MBR at Indian casino and facility: A case study of Chumash Casino and Resort Water Reclamation Facility (WRF), water reuse perspectives. *Environ Earth Sci*, **76**, 578.

Leiknes T.-O. (2010). Membrane bioreactors. In: Membrane for Water Treatment, K.-V. Peinemann and S. P. Nunes Wiley-VCH Verlag, Weinheim, Germany, ISBN 978-3-527-31483-6, pp. 193–223.

Lesjean B., Ferre V., Vonghia E. and Moeslang H. (2008). Market and Design Considerations of the 32 Larger MBR Plants in Europe. IWA World Water Congress, 7–12 September, Vienna, Austria.

Melin T., Jefferson B., Bixio D., Thoeye C., De Wilde W., De Koning J., van der Graaf J. and Wintgens T. (2006). Membrane bioreactor technology for wastewater treatment and reuse. *Desalination*, **187**, 271–282.

Mosher J. and Vartanian G. (2018). Guidance Framework for Direct Potable Reuse in Arizona, NWRI Report, NWRI Publication Number: NWRI-2017-09.

Mulder M. (1996). Basic Principles of Membrane Technology. Kluwer Academic, Dordrecht, ISBN 0-7823-4247-X.

Natvik O., Zaghi E., McDermott L., Gauld G., Atkinson R., Totten G. and Calderon A. (2015). Oxford MBR Membrane Longevity; How Many More Years?. Proceedings of AMTA/AWWA Membrane Technology Conference, March 2015, Orlando, Florida.

Ng C. A., Sun D., Zhang J., Chua H. C., Bing W., Tay S., Fane A. G. and Bucher B. (2005). Strategies to Improve the Sustainable Operation of Membrane Bioreactors. Proceedings of the IDA Conference, Singapore, SP05-179.

Niwa T., Yin R., Oo M. H., Noguchi H., Watanabe T., Razali L. Y., Pug H. Y., Lay W. C. L., Ong K. A. and Alom M. (2018). Performance of a full-scale ceramic MBR system to treat domestic sewage. *Water Practice and Technology*, **13**(3), 589–593.

Noguchi H., Niwa T., Nakamura Y., Agrawal J. K., Ang W. S., Kekre K. and Tao G. (2018). Treatment of industrial used water by UASB+CMBR process to produce industrial grade water for reuse. *Water Practice & Technology*, **13**(2), 424–430.

Pankratz T. (2017). Breaking down the barriers for MBR. *GWI Magazine*, **41**, 42–47.

Pearce G. K. (2011). UF/MF Membrane Water Treatment: Principles and Design. Water Treatment Academy, Bangkok, Thailand, ISBN 978-616-90836-3-4, pp. 95–128.

Peeters J., GE WPT, Côté P., (COTE Membrane Separation), Young T., (GHD) and Smoot S., GHD (2013). Energy Consumption of MBR for Municipal Wastewater Treatment: Current Situation and Potential. Proceedings of the WEF Energy & Water Conference, Nashville, paper 12B.

Peeters J., Adams N., Hong Y., Koops G.-H. and Côté P. (2015). A New Membrane Aerated Bio-Film Reactor (MASBR) for Wastewater Treatment. Proceedings of AMTA/AWWA Membrane Technology Conference, March 2015, Orlando, Florida.

Pollet S., Guigui S. C. and Cabassud C. (2008). Fouling and its reversibility in relation to flow properties and module design in aerated hollow fibre modules for membrane bioreactors. *Water Science & Technology*, **57**(4), 629–636.

Sepideh J. and Bérubé P. R. (2013). Efficiency of Coarse and Pulse Bubble Sparging in Terms of Fouling Control in Submerged Hollow Fiber Membrane Systems. Proceedings of IWA Membrane Technology Conference, Toronto, Ontario.

Stefanski M., Kennedy S. and Judd S. (2011). The determination and origin of fibre clogging in membrane bioreactors. *Journal of Membrane Science*, **375**(1–2), 198–203.

Tao G. H., Kekre K., Zhao W., Lee T. C., Viswanath B. and Seah H. (2005). Membrane bioreactors for water reclamation. *Water Science & Technology*, **51**(6–7), 431–440.

Tao G., Kekre K., Htun Oo M., Lew D. and Seah H. (2010). Energy Reduction and Optimisation in Membrane Bioreactor Systems. Proceedings of the Water Convention, 28th June to 2nd July, Singapore.

Tao G. H., Yen J. T., Li J., Htoo C., Ang A. M., Chua S. C., Wah Y. L., Ooi K. E. and Seah H. (2018). Commissioning and Demonstration Study of Low Energy Step-Feed Membrane Bioreactor for Water and Resource Recovery. Proceedings of the Water Convention, 8th to 12th July, Singapore.

Verrecht B., Judd S., Guglielmi G., Brepols C. and Mulder J. M. (2008). An aeration energy model for an immersed membrane bioreactor. *Water Research*, **42**, 4761–4770.

Wiesmann U., Choi I. S. and Dombrowski E.-M. (2007). Fundamentals of Biological Wastewater Treatment. Wiley-VCH Verlag, Weinheim, Germany, ISBN 978-3-527-31219-1, pp. 195–222, 291–327.

Wilf M, Awerbuch L., Bartels C., Mickley M., Pearce G., Voutchkov N. (2007). The Guidebook to Membrane Desalination Technology. Desalination Publications, L'Aquila, Italy, ISBN 0-86689-065-3, pp. 21–40.

Young T., Smoot S., Peeters J. and Côté P. (2013). When does building an MBR make sense? How variations of local construction and operating cost parameters impact overall project economics.

Zsirai I (2012). 30 Years of Commercial MBR, GE Global Membrane Summit, 16/17th October, Tata, Hungary.

APPENDIX 1: STANDARD PRODUCT LITERATURE AND WEBSITE DOWNLOADS

A1.1 INTRODUCTION

A summary of supplier information and links to supplier web sites can be found at the following address: https://www.thembrsite.com/.

A1.1.1 Overview

Supplier web site links used in this review together with product literature details can be found in the supplier specific sections below.

A1.1.2 Immersed (submerged) hollow fibre

A1.1.2.1 Suez (formerly GE-Zenon)

Spec sheet: ZeeWeed500SpecSheetModule
Spec sheet: ZeeWeedLEAP500SpecSheetCassette
Web site: https://www.suezwatertechnologies.com/products/membrane-bioreactor-mbr

A1.1.2.2 Asahi-Kasei

Web site: http://www.asahi-kasei.co.jp/membrane/microza/en/customer_wastewater.html

A1.1.2.3 Beijing Origin Water

Web site: http://www.originwater.com

A1.1.2.4 Econity

Data sheet: ECONITY MBR CF-E Series
Web site: http://econity.com/en/business/products_tab2.html

A1.1.2.5 Evoqua-Memcor

Module Spec sheet: MC-B40N-MODULE-AU-PS-A4-V07_0418
System Spec sheet: MCMEMPULSEBR0416-A4
Web site: http://www.evoqua.com/en/brands/Memcor/Pages/MBR.aspx

A1.1.2.6 Koch Membrane Systems

Web site: https://www.kochmembrane.com/Products/Hollow-Fiber-Membranes/PURON-MBR-and-PULSION-MBR-Modules

A1.1.2.7 Memstar

Web site: http://www.unitedenvirotech.com/article/membrane-products/submerged-membrane-modules/

A1.1.2.8 Mitsubishi chemical (formerly Mitsubishi Rayon Company (MRC))

Data sheet: Specifications of FF Module (STERAPORE™ 5600 & 5700 Series)
Data sheet: Specifications of FF Module (STERAPORE™ 5000 Series)
Web site: https://www.m-chemical.co.jp/sterapore/en/

A1.1.2.9 Others

Data sheet: http://www.sumitomoelectricusa.com/poreflon/wp-content/uploads/July-2017-Newsletter-1.pdf
Web site: https://global-sei.com/poreflon/module/02.html

A1.1.3 Immersed (submerged) flat sheet

A1.1.3.1 Kubota

Data sheet: http://www.kubota-membrane.com/uploads/2018/09/05/Kubota%20Brochure.pdf
Web site: https://www.kubota-mbr.com/en/Mbr

A1.1.3.2 Toray

Web site: http://www.toraywater.com/products/mbr/index.html

A1.1.3.3 Meiden (ceramic)

Brochure: Ceramic flat sheet membrane
Web site: http://www.meidensha.com/products/water/prod_06/prod_06_01/index.html

A1.1.3.4 Others

Web site: http://www.huber.co.uk/products/membrane-filtration-mbr.html
Web site: www.microdyn-nadir.com

A1.1.4 External

A1.1.4.1 Crossflow – Berghof

Data sheet: Berghof Technical Data sheet P.1/1 and P.1/2
Web site: https://www.berghofmembranes.com/products-services/

A1.1.4.2 Air lift – Pentair

Web site: https://advancedfiltration.pentair.com/en/products/airlift-mbr
Web site: https://advancedfiltration.pentair.com/en/products/crossflow-mbr

Chapter 15

MBR design calculations

Muhammad B. Asif and Faisal I. Hai

Strategic Water Infrastructure Laboratory, School of Civil, Mining and Environmental Engineering, The University of Wollongong, Wollongong, Australia

Corresponding author: faisal@uow.edu.au

EXAMPLE 15.1

A membrane bioreactor (MBR) plant at a small resort town comprises two membrane skids each containing 10 membrane modules, i.e., bundles of hollow fibres. The modules were operated at an average flux of 20 LMH (i.e., L/m^2 h). Each module (bundle) consisted of 6100 pieces of fibres. The fibres were each 1.52 m long and had an outer diameter of 1.2 mm. Calculate the wastewater treatment capacity of the MBR plant.

Solution

Considering that hollow fibres are cylindrical in shape,

$$\text{Total surface area per module} = \pi \times D(\text{diameter}) \times L(\text{length}) \times n(\text{number of hollow fibres})$$

$$= \pi \times (1.2 \times 10^{-3}\,\text{m}) \times 1.52\,\text{m} \times 6100 = 35\,\text{m}^2$$

$$\text{Total number of molecules} = 2\,\text{skids} \times 10\,\text{modules/skids} = 20$$

$$\text{Total membrane surface area available in the plant} = 35 \times 20 = 700\,\text{m}^2$$

$$\text{Flow rate, i.e., wastewater treatment capacity} = \text{Total membrane surface area} \times \text{Average flux}$$

$$= 700\,\text{m}^2 \times 20\,\text{LMH} = 700\,\text{m}^2 \times (24 \times 20)\text{L/m}^2\text{d}$$

$$= 336{,}000\,\text{L/d} = 336\,\text{m}^3\text{/d} = 0.336\ \text{Mega litre per day (MLD)}$$

Editor(s): Faisal I. Hai, Kazuo Yamamoto and Chung-Hak Lee
doi: 10.2166/9781780409177_0535

Note: Detailed information on membrane module dimensions and membrane surface area per module from different suppliers is available in Chapter 14.

As noted, this is an example of a small MBR plant. The world's largest MBR plants shown in Figure 1.3 of Chapter 1 have peak daily flow capacities of 40 to over 400 MLD.

EXAMPLE 15.2

You have been assigned to design an MBR plant to treat the wastewater from a locality consisting of 600 individual family residences. The average occupancy in that locality has been estimated to be 2.7 persons per residence. For design wastewater flow, base your calculation on a wastewater production rate of 230 L/person d. The hydraulic retention time of the aerated biological tank will be 12 h. Hollow fibre membranes will be submerged in a separate (but interconnected) membrane tank with a hydraulic retention time of 0.5 h. Consider a design average flux of 20 LMH (i.e., L/m^2 h). The total membrane surface area and volume of each module are 78 m^2 and 0.24 m^3, respectively.

Calculate (a) the volume of the aerated biological tank and the membrane tank, (b) the number of membrane modules required for this plant, (c) the membrane displacement ratio in % (i.e., volume of membrane over volume of the membrane tank).

Solution

(a) Volume of tanks

$$\text{Design capacity of MBR } (Q) = 600 \text{ houses } \times 2.7 \frac{\text{persons}}{\text{house}}$$

$$\times\ 230 \frac{L}{\text{capita. } d} \times \frac{1\,\text{m}^3}{1{,}000\,\text{L}} = \mathbf{372.6} \frac{\boldsymbol{m^3}}{\boldsymbol{d}}$$

$$\text{Volume of aeration tank}(V) = \text{HRT} \times Q = 12\,\text{h} \times 372.6 \frac{\text{m}^3}{d} \times \frac{1\,d}{24\,\text{h}} = \mathbf{186.3}\,\boldsymbol{m^3}$$

$$\text{Volume of membrane tank} = \text{HRT} \times Q = 0.5\,\text{h} \times 372.6 \frac{\text{m}^3}{d} \times \frac{1\,d}{24\,\text{h}} = \mathbf{7.8}\,\boldsymbol{m^3}$$

(b) Number of membrane modules

$$\text{Permeate flux} = \frac{\text{Flow rate}}{\text{Total surface area of membrane}}$$

$$\text{Total surface area of membrane} = \frac{\text{Flow rate}}{\text{Permeate flux}}$$

$$= \frac{372.6(\text{m}^3/d) \times (1\,\text{d}/24\,\text{h}) \times (1{,}000\,\text{L}/1\,\text{m}^3)}{20(\text{L}/\text{m}^2\,h)} = \mathbf{776.25}\,\boldsymbol{m^2}$$

$$\text{No. of membrane modules} = \frac{776.25\,\text{m}^2}{78\,(\text{m}^2/\text{module})} \approx \textbf{10 membrane modules}$$

(c) Membrane displacement ratio

$$\text{Volume of membrane modules} = 10 \times 0.24 = 2.4\,\text{m}^3$$

$$\text{Membrane displacement ratio} = (2.4\,\text{m}^3/7.8\,\text{m}^3) \times 100 = \mathbf{31\%}$$

Note: In this example, the design wastewater flow was estimated only considering a wastewater production rate of 230 L/person d. There are established guidelines for wastewater treatment facility design flow determinations. Consult Chapter 3 of this book and other standard wastewater engineering references such as Metcalf & Eddy *et al.* (2002) for detailed information.

EXAMPLE 15.3

A full-scale MBR plant is planned to be installed for the treatment of municipal wastewater originating from a town. The current population of the town is 20,000, and average water consumption rate is 200 L/person day. The MBR plant will consist of an activated sludge-based bioreactor, and a separate tank will be used to submerge a hollow-fibre membrane with a total filtration area of 75 m^2/module. Estimate:

(a) Design flow for MBR plant if the project life is 25 years (assume an annual geometric population growth rate of 1.2%).
(b) Volume of activated sludge-based bioreactor and membrane tank with an HRT of 10 and 1 h, respectively
(c) Number of membrane modules required for achieving an average flux of 20 LMH (i.e., L/m^2 h).

Solution

(a) Design capacity of MBR
It is given that the current population (P_o) and its average geometric growth rate (k) is 20,000 people and 1.2%/yr, respectively. We need to estimate the population of the town at the end of 25 yr for calculating the capacity of MBR as follows:

$$P_t = P_o.\,e^{kt} = 20{,}000 \times e^{25\times(1.2/100)} = 26{,}997 \approx 27{,}000\,\text{people}$$

Assuming wastewater production rate to be same as water consumption rate (*Note*: This is a conservative assumption. A rule of thumb is to assume wastewater discharged into sewer to be 60–70% of the volume of water consumption per day),

$$\text{The design capacity of MBR (Q)} = 27{,}000\ \text{people} \times 200\frac{L}{\text{person.d}} \times \frac{1\,\text{m}^3}{1000\,\text{L}} = \mathbf{5400}\frac{\mathbf{m^3}}{\boldsymbol{d}}$$

(b) Volume of activated sludge-based bioreactor and membrane tank

Volume of activated sludge-based bioreactor(V)

$$= \text{HRT} \times Q = 10\,\text{h} \times 5{,}400\frac{\text{m}^3}{d} \times \frac{1\,\text{d}}{24\,\text{h}} = \mathbf{2{,}250}\,\boldsymbol{m^3}$$

Volume of membrane tank

$$= \text{HRT} \times Q = 1\,\text{h} \times 5{,}400 \frac{\text{m}^3}{d} \times \frac{1\,\text{d}}{24\,\text{h}} = \mathbf{225}\,\boldsymbol{m^3}$$

(c) Number of membrane modules

$$\text{Permeate flux} = \frac{\text{Flow rate}}{\text{Total surface area of membrane}}$$

$$\text{Total surface area of membrane} = \frac{\text{Flow rate}}{\text{Permeate flux}}$$

$$= \frac{5{,}400(\text{m}^3/d) \times (1\,\text{d}/24\,\text{h}) \times (1{,}000\,\text{L}/1\,\text{m}^3)}{20(\text{L}/\text{m}^2\,\text{h})}$$

$$= 11{,}250\,\text{m}^2$$

$$\text{No. of membrane modules} = \frac{11{,}250\,\text{m}^2}{75(\text{m}^2/\text{module})} = \textbf{150 membrane modules}$$

Note: In this example, the design wastewater flow was estimated based on only wastewater production rate. The wastewater production rate was estimated from the water consumption rate. The wastewater production rate was adjusted to consider the population growth over the life of the treatment plant. However, there are additional considerations that may play a role in determining wastewater treatment capacity of a plant. Consult Chapter 3 of this book and other standard wastewater engineering references such as Metcalf and Eddy *et al.* (2002) for detailed information.

Also note that usually the membrane system is designed with enough membrane area to accommodate the peak daily flow condition with one membrane train out of service. Additionally, some spare space (e.g., 10%) is provided per membrane train to allow for the installation of additional membrane area if needed in the future for unexpected membrane fouling conditions.

EXAMPLE 15.4

Permeate flux of microfiltration (MF), ultrafiltration and nanofiltration (NF) membranes was monitored at different trans-membrane pressures (TMP) using ultrapure Milli-Q water at room temperature (20°C) as shown in Table 15.1. Estimate water permeability of each membrane.

Solution

Water permeability of a membrane can be estimated by drawing a graph of permeate flux vs TMP. The graph should be linear with R^2 values of greater than 0.99. The slope of the line is the water permeability of the membrane.

MF membrane: The graph of permeate flux vs TMP is given below (Figure 15.1) and the values of R^2 and slope are also shown. The permeability of MF membrane is 252 LMH/bar.

Similarly, by drawing the graph of permeate flux vs TMP (graph not shown), permeability of UF and NF membranes is 101 and 2.6 LMH/bar, respectively.

Table 15.1 Permeate flux (L/m^2 h or LMH) of different membranes at varying TMPs.

TMP	MF	UF	NF
bar		LMH	
0	0	0	0
0.5	120	60	1.4
1	275	110	2.53
1.5	350	145	3.88
2	515	210	5.25

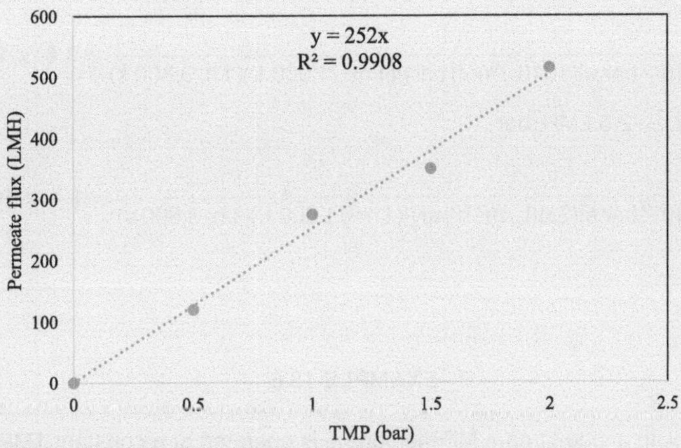

Figure 15.1 Permeate flux of MF membrane at different TMPs.

EXAMPLE 15.5

For the conditions given in Example 15.4, estimate the resistance of each membrane. Assume that the viscosity of water at room temperature (20°C) is 1.002×10^{-7} bar s.

Solution

The following expression is used for measuring the membrane resistance (see Equation 2.28 in Chapter 2):

$$J = \frac{\Delta P}{\mu \times R_T}$$

where J, ΔP, μ and R_T are the permeate flux (LMH), change in TMP (bar), viscosity of water and total resistance of membrane, respectively.

Since ultrapure water was used for measuring the permeate flux in Example 15.4, the total resistance of membrane (R_T) is equal to the intrinsic resistance (R_m) of the membrane. On the other hand, $J/\Delta P\ (=L_p)$ is the permeability of membrane, and its value for each membrane are determined in Example 15.4. Hence, the above equation can be rewritten as

$$R_m = \frac{1}{\mu \times (J/\Delta P)} = \frac{1}{\mu \times L_p}$$

MF membrane: $L_p = 252$ LMH/bar

$$R_m = \frac{1}{(1.002 \times 10^{-7}\ \text{bar s})(252\ (L/(\text{m}^2\ \text{h bar})))(1\ \text{m}^3/1{,}000\ \text{L})(1\ \text{h}/3{,}600\ \text{s})} = \mathbf{1.4 \times 10^{11}}\ \boldsymbol{m^{-1}}$$

UF membrane: $L_p = 101$ LMH/bar

$$R_m = \frac{1}{(1.002 \times 10^{-7}\text{bar s})(101(\text{L}/\text{m}^2\ \text{h bar}))(1\ \text{m}^3/1{,}000\ \text{L})(1\ \text{h}/3{,}600\ \text{s})} = \mathbf{3.6 \times 10^{11}\ m^{-1}}$$

NF membrane: $L_p = 2.6$ LMH/bar

$$R_m = \frac{1}{(1.002 \times 10^{-7}\text{bar s})(2.6(\text{L}/\text{m}^2\ \text{h bar}))(1\ \text{m}^3/1{,}000\ \text{L})(1\ \text{h}/3{,}600\ \text{s})} = \mathbf{1.4 \times 10^{13}\ m^{-1}}$$

EXAMPLE 15.6

An MBR equipped with a side-stream MF membrane is operated at a constant TMP of 0.5 bar for the treatment of municipal wastewater. At an operating pressure of 0.5 bar, the membrane shows an average clean water flux of 125 LMH, but the average permeate flux was reduced to 35 LMH during filtration of municipal wastewater. Estimate the fouling resistance of the membrane if viscosity of water at room temperature (20°C) is 1.002×10^{-7} bar s.

Solution

Fouling resistance (R_f) can be estimated by subtracting the total hydraulic resistance (R_T) from intrinsic membrane resistance (R_m). Hence, we will separately calculate R_m and R_T by using the following equation:

$$J = \frac{\Delta P}{\mu \times R_T}$$

Intrinsic membrane resistance (R_m)

$$R_m = \frac{0.5\ \text{bar}}{(1.002 \times 10^{-7}\text{bar s})(125(\text{L}/\text{m}^2\ \text{h}))(1\ \text{m}^3/1{,}000\ \text{L})(1\ \text{h}/3{,}600\ \text{s})} = \mathbf{1.44 \times 10^{11}\ m^{-1}}$$

Total hydraulic resistance (R_T)

$$R_T = \frac{0.5\ \text{bar}}{(1.002 \times 10^{-7}\text{bar s})(35\ (\text{L}/\text{m}^2\ \text{h}))(1\ \text{m}^3/1{,}000\ \text{L})(1\ \text{h}/3{,}600\ \text{s})} = \mathbf{5.13 \times 10^{11}}\ \boldsymbol{m^{-1}}$$

Fouling resistance (R_T)

$R_T = R_m + R_f$

$R_f = R_T - R_m = 5.13 \times 10^{11} - 1.44 \times 10^{11} = \mathbf{3.7 \times 10^{11}}\ m^{-1}$

EXAMPLE 15.7

An investigation was carried out by operating three MBRs (i.e., MBR_1, MBR_2 and MBR_3) in parallel. Each of the MBRs consisted of a 6 m^3 aeration tank and a 25 m^2 submerged membrane module. All three MBRs were operated at a constant MLSS concentration of 8 g/L. The MBRs were operated at an average permeate flux of 20 LMH. The total organic carbon (TOC) concentration in the influent to MBR_1, MBR_2 and MBR_3 was 500, 1000 and 1500 mg C/L, respectively. During continuous operation, the membranes in MBR_1, MBR_2 and MBR_3 showed an increase in transmembrane pressure (TMP) at the rates of 2.03, 2.4 and 3.5 kPa/d, respectively, over the first 30 days of operation. Calculate the F/M ratio for each MBR and comment on the influence of F/M ratio on membrane fouling rate.

Solution

$$\frac{F}{M}\text{ratio} = \frac{\text{organic loading rate per day}}{\text{MLSS concentration}}$$

$$\text{Flow rate}\,(Q) = 20\text{ LMH} \times 25\,\text{m}^2 = 500\frac{\text{L}}{\text{h}} = 0.5\frac{\text{m}^3}{h}$$

In case of a feed TOC of 500 $\dfrac{\text{mg TOC}}{\text{L}}$,

$$\frac{F}{M}\text{ratio} = \frac{500(\text{g/m}^3) \times 0.5(\text{m}^3/\text{h}) \times (1/6{,}000\,\text{L}) \times (24\,\text{h/d})}{8(\text{g/L})} = 0.125/\text{d}$$

In case of a feed TOC of 1,000 $\dfrac{\text{mg TOC}}{\text{L}}$,

$$\frac{F}{M}\text{ratio} = \frac{1{,}000(\text{g/m}^3) \times 0.5(\text{m}^3/\text{h}) \times (1/6{,}000\text{ L}) \times (24\,\text{h/d})}{8\ (\text{g/L})} = 0.25/\text{d}$$

In case of a feed TOC of 1,500 $\dfrac{\text{mg TOC}}{\text{L}}$,

$$\frac{F}{M}\text{ratio} = \frac{1{,}500(\text{g/m}^3) \times 0.5(\text{m}^3/\text{h}) \times (1/6{,}000\,\text{L}) \times (24\,\text{h/d})}{8(\text{g/}L)} = 0.375/d$$

The membranes were fouled at faster rates (indicated by increasing TMP) when operated at higher ((F/M) ratio).

The ((F/M) ratio) is a controllable parameter found to significantly influence membrane fouling in MBR.

Note: This is a simplified example showing the impact of one of the MBR operating parameters on membrane performance. In fact, SRT, HRT and F/M ratio are interconnected parameters, which affect membrane fouling. Their impact on membrane fouling has been studied intensively (Grelier *et al.*, 2006; Ng *et al.*, 2006; Patsios & Karabelas, 2011; Van den Broeck *et al.*, 2012; Wu *et al.*, 2013).

As an example, Table 15.2, based on data from Grelier *et al.* (2006), demonstrates the intricate relationship of membrane fouling with SRT, HRT and F/M ratio.

Table 15.2 Intricate relationship of membrane fouling with SRT, HRT and F/M ratio (data from Grelier *et al.*, 2006).

	SRT 8 d		SRT 15 d		SRT 40 d	
	Phase 1	**Phase 2**	**Phase 1**	**Phase 2**	**Phase 1**	**Phase 2**
Reactor volume (L)	70	25	70	34	70	70
MF flux (LMH)	20	17.5	20	17.5	20	17.5
Volumetric loading rate (kg COD/m^3 d)	1.0	2.7	1.0	1.9	1.0	0.9
F/M ratio, (kgCOD/kg MLSS. d)	0.3	0.3	0.2	0.2	0.1	0.1
MLSS concentration (g/L)	3.2	8	4.9	8	7.7	8
HRT (h)	12	4.5	12	6.0	12	12
Fouling rate (m^{-1}d^{-1})	2.3×10^{11}	$\mathbf{4.1 \times 10^{11}}$	1.4×10^{11}	1.1×10^{11}	0.4×10^{11}	0.6×10^{11}

EXAMPLE 15.8

During the operation of a full-scale MBR for high-strength industrial wastewater treatment, MF membrane modules with a total filtration area of 750 m^2 were backwashed *in situ* at a flux of 30 LMH using 500 mg/L NaOCl solution for 30 s after each filtration cycle. A filtration cycle comprised 10 min of permeation and 1.5 min of relaxation period (i.e., cease of filtration). As noted in Chapter 4, relaxation periods are applied to reduce membrane fouling. (a) Estimate the mass of NaOCl required for membrane backwashing annually. (b) If the net design water flux through the membranes in this plant is 20 LMH, what is the instantaneous flux taking into account the time the membranes spend in non-productive modes of operation (i.e., relaxation and backwashing).

Solution

One filtration cycle = 10 min filtration + 1.5 min relaxation + 0.5 min backwashing = 12 min/cycle (i.e., 5 cycles/h)

(a) Annual requirement of NaOCl for membrane cleaning:

$$\text{Mass of NaOCl} = 30\ (\text{L/m}^2\ \text{h}) \times 750\ \text{m}^2 \times 500\ (\text{mg/L}) \times \frac{1\ \text{kg}}{1{,}000{,}000\ \text{mg}} = 11.25\frac{\text{kg}}{\text{h}}$$

In this example, backwashing is carried out for a period of 2.5 min/h (60 min/d or 1 h/d) and, thus, 11.25 kg of NaOCl is required daily.

$$\text{Annual requirement of NaOCl} = 11.25 \times 365 = 4{,}106\ \text{kg/yr} = \mathbf{4.1\ tonne/yr}.$$

(b) Instantaneous flux:
Taking into account the relaxation and backwashing period (i.e., 2 min/filtration cycle, or 10 min every hour)

$$\text{Instantaneous flux} = 20\ \text{LMH} \times \frac{1}{50\ \text{min}} \times \frac{60\ \text{min}}{1\ \text{h}} = 24\ \textbf{LMH}$$

Note: This example illustrates that because membrane operation includes non-productive period (such as time spent in relaxation and backwashing), the actual instantaneous flux during membrane operation must be larger than the design value estimated to treat the full influent flow. The instantaneous flux must also be temperature-corrected. For example, if the historical data indicate that the minimum weekly influent temperature at a site is 10°C, then this is the minimum design temperature for the membrane system. In this case, the instantaneous flux at 10°C can be converted into that at 20°C by $J_{\text{instantaneous, 20°C}} = J_{\text{instantaneous, 10°C}} \times 1.025^{20-10}$ (Rautenbach & Albrecht, 1989). During design, the temperature that corresponds to a flux value provided by the vendor or assumed in a design calculation must be verified.

EXAMPLE 15.9

The membranes in an MBR plant undergo flux recovery cleaning for 8 h once in 6 months. This involves emptying the mixed liquor from the membrane tank in 30 min, then filling it with sodium hypochlorite (NaOCl) solution for cleaning and soaking the membranes, then draining the spent chemical cleaning waste from the membrane tanks and, finally, returning the mixed liquor. The membrane system has two trains or subunits that are independently chemically cleaned. Each subunit is housed in a tank with a volume of 100 m^3. The MBR package plant vendor recommended chemical cleaning by 1 g/L NaOCl. Assuming that NaOCl is delivered to the treatment plant with a concentration of 12.5% by weight, and has a specific gravity of 1.175, calculate the volume of bulk chemical required per recovery clean of each membrane subunit.

Solution

Mass of NaOCl required per subunit recovery clean

$$= \frac{1\ \text{g NaOCl}}{\text{L}} \times 100\ \text{m}^3 \times \frac{1{,}000\ \text{L}}{\text{m}^3} = 100{,}000\ \text{g NaOCl}$$

Mass of bulk chemical required per subunit recovery clean

$$= 100{,}000\ \text{g NaOCl} \times \frac{100\ \text{g bulk chemical}}{12.5\ \text{g NaOCl}}$$

$$= 800{,}000\ \text{g bulk chemical}$$

Volume of bulk chemical required per subunit recovery clean

$$= 800{,}000\ \text{g} \times \frac{1\ \text{mL}}{1.175\ \text{g}} \times \frac{1\ \text{L}}{1{,}000\ \text{mL}} = 680\ \textbf{L bulk chemical}$$

EXAMPLE 15.10

An MBR was installed to treat wastewater from a fabric dyeing unit. An MF membrane with a nominal pore size of 0.4 μm was used in the MBR. In separate filtration investigations, pure dye solution in water was observed to pass through the membrane. However, during the operation of the MBR, a certain portion of the dye was observed to be retained within the bioreactor. This was evident from the fact that the sample taken from the bioreactor was more coloured than the membrane-permeate, i.e., treated effluent. The absorbance of different water samples at the peak wavelength of the dye was as follows: raw wastewater = 1.0, bioreactor supernatant = 0.35 and membrane-permeate = 0.03. A closer observation confirmed the presence of a sludge cake layer on the membrane surface. Assuming a linear relationship between the sample absorbance and the dye concentration, estimate the contributions of biodegradation, adsorption and membrane retention in total colour (dye) removal by the MBR. Briefly mention how a soluble dye was retained by the MBR.

Solution

The overall dye removal from the aqueous phase by the MBR

$$= \frac{(1 - 0.03)}{1} \times 100\% = 97\%$$

This removal occurred via (i) adsorption on mixed liquor suspended solids (MLSS), (ii) adsorption on the cake layer over the membrane and (iii) subsequent biodegradation of the adsorbed mass (Hai *et al.*, 2006, 2008, 2009, 2011, 2012, 2013).

Dye removal achieved by (i) dye adsorption on MLSS and (ii) biodegradation of dye in bulk liquid and on MLSS

$$= \frac{(1 - 0.35)}{1} \times 100\% = 65\%$$

Comparing the above dye removal values, it is evident that a significant level of dye was retained by the sludge cake layer over the membrane. A MF membrane with a nominal pore size of 0.4 μm is not expected to retain a dye which is soluble in water. However, in this case the cake layer formed on the membrane surface enhanced dye removal by the MBR.

Note: MBR-permeate is suspended solids-free. Additionally, colloids and some soluble compounds can be retained within MBR after being adsorbed onto MLSS. This additionally retained organics may be eventually biodegraded.

EXAMPLE 15.11

Average wastewater generation rate in a bakery product manufacturing unit is 2,500 m^3/day. Assume that sucrose ($C_{12}H_{22}O_{11}$, molecular weight = 342 g/mol), at a concentration of 7,000 mg COD/L, is the major component of wastewater originating from this unit. The owner of the facility plans to install an aerobic MBR plant at the site so that sucrose in wastewater is completely oxidized to carbon dioxide and water, and the treated effluent can be reused for non-potable purpose within the bakery product manufacturing facility. Estimate the theoretical oxygen demand of the MBR plant by using the microbial stoichiometric equation, if the sludge (biomass) yield is 0.4 g MLVSS/g sucrose. Also, calculate the MLVSS production rate. Use the following biochemical equation (Park *et al.*, 2015):

$$C_{12}H_{22}O_{11} + 1.2NH_3 + 6O_2 \rightarrow 1.2C_5H_7O_2N + 6CO_2 + 8.6H_2O \ldots \quad (15.11.1)$$

where, $C_5H_7O_2N$ (molecular weight = 113 g/mol) is the composition of MLVSS in MBR, and NH_3 is the source of nitrogen for microbial reactions.

Note that chemical reaction of oxidation of sucrose to carbon dioxide and water:

$$C_{12}H_{22}O_{11} + 12O_2 \rightarrow 12CO_2 + 11H_2O \ldots \quad (15.11.2)$$

Solution

Since sucrose concentration in wastewater is given in the form of chemical oxygen demand (COD), it is required to be changed to mass unit for estimating the theoretical oxygen demand from the following equation:

$$C_{12}H_{22}O_{11} + 1.2NH_3 + 6O_2 \rightarrow 1.2C_5H_7O_2N + 6CO_2 + 8.6H_2O \ldots \quad (15.11.1)$$

For this purpose, we have to consider the chemical equation for sucrose degradation:

$$C_{12}H_{22}O_{11} + 12O_2 \rightarrow 12CO_2 + 11H_2O \ldots \quad (15.11.2)$$

which indicates that 12 mole O_2 is required for oxidizing 1 mole of sucrose, i.e. (12 × 32) g of O_2 is required to oxidize 342 g sucrose, or 1.12 g O_2 per g sucrose is required.

Hence, a sucrose solution exhibiting a COD of 7,000 mg O_2/L is equivalent to

$$\left(7000 \frac{\text{mg Oxygen as OD}}{\text{L}}\right) \times \frac{1}{1.12\,\text{mg oxygen per mg sucrose}} = 6250 \frac{\text{mg sucrose}}{\text{L}}$$

From Equation 15.11.1, the theoretical oxygen demand is

$$\left(\frac{6{,}250\,\text{mg sucrose}}{\text{L}}\right)\left(\frac{1\,\text{g sucrose}}{1{,}000\,\text{mg sucrose}}\right)\left(\frac{6 \times 32\,\text{g oxygen}}{1 \times 342\,\text{g sucrose}}\right)$$
$$\left(\frac{1\,\text{kg oxygen}}{1{,}000\,\text{g oxygen}}\right)\left(\frac{2{,}500\,\text{m}^3}{\text{d}}\right)\left(\frac{1{,}000\,\text{L}}{1\,\text{m}^3}\right) = \mathbf{8{,}772}\ \frac{\textbf{kg oxygen}}{\textbf{d}}$$

MLVSS production rate can be estimated by comparing the mass of sucrose available for oxidation and MLVSS generation in MBR by using Equation 15.11.1. Therefore, excess MLVSS production rate is

$$\left(\frac{6{,}250\,\text{mg sucrose}}{\text{L}}\right)\left(\frac{1\,\text{g sucrose}}{1{,}000\,\text{mg sucrose}}\right)\left(\frac{1.2 \times 113\,\text{g MLVSS}}{1 \times 342\,\text{g sucrose}}\right)$$
$$\left(\frac{1\,\text{kg MLVSS}}{1{,}000\,\text{g MLVSS}}\right)\left(\frac{2{,}500\,\text{m}^3}{\text{d}}\right)\left(\frac{1{,}000\,\text{L}}{1\,\text{m}^3}\right) = \mathbf{6195}\ \frac{\textbf{kg MLVSS}}{\textbf{d}}$$

EXAMPLE 15.12

Composition of wastewater is diverse and usually complex, thereby making it difficult to develop a stoichiometric model for estimating oxygen demand. For the conditions described in Example 15.11, estimate the theoretical oxygen demand and MLVSS production rate *without* using the balanced stoichiometric relationship (i.e., Equation 15.11.1).

Solution

Continuous supply of oxygen is necessary for the oxidation of organic matters in MBR because oxygen acts as the terminal electron acceptor. It is important to note that overestimation of oxygen demand can significantly increase the operating cost of MBR, while underestimation can result in inadequate oxidation of organic pollutants. Therefore, oxygen demand should be carefully estimated.

One way of estimating oxygen demand is to use the following empirical relationship (Metcalf & Eddy *et al.*, 2002; Park *et al.*, 2015)

$$\text{Theoretical oxygen demand(TOD)} = Q(S_o - S) - 1.42\,P_{x,bio} \ldots \quad (15.12.1)$$

where,
Q is the flow rate (m^3/d),
S_o is the influent concentration of organic matter (mg/L),
S is the effluent concentration of organic matter (mg/L),
$P_{x,bio}$ is the MLVSS production rate (g/day).

The MLVSS production rate is estimated as follows (Metcalf & Eddy *et al.*, 2002; Park *et al.*, 2015):

$$P_{x,bio} = YQ(S_o - S) \ldots (15.12.2)$$

Here, Y is the sludge (biomass) growth yield in MBR.

In this example, the influent concentration of sucrose (S_o) is 6,250 mg/L (as calculated in Example 15.11), while effluent sucrose concentration is zero (i.e., complete oxidation of sucrose).

Hence, the MLVSS production rate is

$$P_{x,bio} = \left\{\left(\frac{2{,}500\,m^3}{d}\right)\left(\frac{1{,}000\,L}{1\,m^3}\right)\right\}\left(\frac{0.4\,g\,MLVSS}{g\,sucrose}\right)$$
$$\left[\left(6{,}250\frac{mg\,sucrose}{L} - 0\frac{mg\,sucrose}{L}\right)\left(\frac{1\,g\,sucrose}{1{,}000\,mg\,sucrose}\right)\right].$$
$$= 6{,}250{,}000\frac{g\,MLVSS}{day} = 6{,}250\frac{kg\,MLVSS}{day}$$

$$TOD = \left[\left\{\left(\frac{2{,}500\,m^3}{d}\right)\left(\frac{1{,}000\,L}{1\,m^3}\right)\right\}\left\{\left(7{,}000\frac{mg\,oxygen}{L} - 0\,\frac{mg\,oxygen}{L}\right)\left(\frac{1\,g\,oxygen}{1{,}000\,mg\,oxygen}\right)\right\}\right]$$
$$-\left(1.42\,\frac{g\,oxygen}{g\,MLVSS}\right)\left(6{,}250{,}000\frac{g\,MLVSS}{day}\right)$$
$$= 8{,}625{,}000\frac{g\,O_2}{day} = \mathbf{8{,}625}\frac{kg\,O_2}{day}$$

Note: Given that empirical equations were used in Example 15.12, it is not possible to obtain identical values for theoretical oxygen demand and MLVSS production rate as in Example 15.11. However, if we compare the values obtained in Examples 15.11 and 15.12, the values for the same parameter from these examples are close to each other.

EXAMPLE 15.13

An MBR plant has been installed with the design capacity of 2,000 m^3/d for the treatment of municipal wastewater. Hollow-fibre membrane modules with a nominal pore size of 0.1 μm and total filtration area of 3,740 m^2 are submerged in an activated sludge-based bioreactor with a working volume of 1,040 m^3. Assuming an oxygen transfer efficiency (OTE) of 0.5, compare the aeration requirements for (a) microbial metabolism; and (b) membrane scouring based on the data given below:

Influent total organic carbon, TOC (mg/L)	=65
Chemical oxygen demand, COD (mg O_2/L)	=140
Sludge yield, *Y* (g MLVSS/g TOC)	=0.25
Average permeate flux (LMH)	=30
Specific air demand based on membrane area, SAD_m (N m^3/m^2 h)	=0.41
Specific air demand based on permeate volume, SAD_p (N m^3/m^3 permeate)	=15
TOC removal by biodegradation (%)	=74%
COD removal by biodegradation (%)	=85%
Density of air to be supplied for aeration	=1.225 g/L

Solution

(a) *Aeration requirement for microbial metabolism*

To estimate the aeration requirement for microbial metabolism, theoretical oxygen demand can be determined using the following equation (Metcalf & Eddy *et al.*, 2002; Park *et al.*, 2015):

$$\text{Theoretical oxygen demand (TOD)} = Q(S_o - S) - 1.42P_{x,bio} \ldots \quad (15.13.1)$$

where,

Q is the flow rate (m^3/d),
S_o is the influent concentration of organic matter (mg/L),
S is the effluent concentration of organic matter (mg/L),
$P_{x,bio}$ is the MLVSS production rate (g/day)

The MLVSS production rate is estimated as follows (Metcalf & Eddy *et al.*, 2002; Park *et al.*, 2015):

$$P_{x,bio} = YQ(S_o - S) \ldots \quad (15.13.2)$$

Here, Y is the sludge (biomass) growth yield in MBR

$$P_{x,bio} = \left\{\left(\frac{2{,}000\,m^3}{d}\right)\left(\frac{1{,}000\,L}{1\,m^3}\right)\right\}\left(\frac{0.25\,g\,MLVSS}{g\,TOC}\right)\left[\left(65\frac{mg\,TOC}{L} - 16.9\frac{mg\,TOC}{L}\right)\left(\frac{1\,g\,TOC}{1{,}000\,mg\,TOC}\right)\right]$$

$$= 24{,}050\,\frac{g\,MLVSS}{day} \approx \mathbf{24}\,\frac{\mathbf{kg\,MLVSS}}{\mathbf{day}}$$

Therefore, the theoretical oxygen demand from Equation 15.13.1 will be

$$TOD = \left[\left\{\left(\frac{2{,}000\,m^3}{d}\right)\left(\frac{1{,}000\,L}{1\,m^3}\right)\right\}\left\{\left(140\,\frac{mg\,oxygen}{L} - 21\,\frac{mg\,oxygen}{L}\right)\left(\frac{1\,g\,oxygen}{1{,}000\,mg\,oxygen}\right)\right\}\right]$$
$$- \left(1.42\,\frac{g\,oxygen}{g\,MLVSS}\right)\left(24050\,\frac{g\,MLVSS}{day}\right)$$
$$= 203{,}849\,\frac{g\,O_2}{day} \approx \mathbf{204}\,\frac{kgO_2}{day}$$

Because oxygen transfer efficiency (OTE) is 0.5, actual TOD will be equal to 408 kg O_2/day. Required volumetric flowrate of air supplied,

$$Q_{air} = \frac{\text{mass flowrate}}{\rho_{air}} = \frac{408\,(kg/d)}{1.225\,(g/L)} \times \frac{1{,}000\,g}{1\,kg} \times \frac{1\,m^3}{1{,}000\,L} = 333\,m^3/d$$

(b) *Aeration requirement for membrane scouring*
Aeration requirement for membrane scouring can be estimated using specific air flowrate (SAD_m or SAD_p) as presented below (Judd, 2010):

$$SAD_m = \frac{Q_{air}}{A_m}$$

where Q_{air} is the flowrate of air (m^3/d), A_m is the total area of membrane (m^2); and SAD_m is specific air flow rate based on membrane area (N m^3/m^2 h)

$$Q_{air} = SAD_m \times A_m = 0.41\frac{N\,m^3}{m^2\,h} \times 3{,}740\,m^2 = 1{,}533\frac{m^3}{h} = \mathbf{36{,}792}\,\frac{m^3}{d}$$

$$SAD_p = \frac{Q_{air}}{Flux \times A_m}$$

where Q_{air} is the flowrate of air (m^3/d), A_m is the total area of the membrane (m^2); and SAD_p is specific air flowrate based on permeate volume (N m^3/m^3 permeate)

$$Q_{air} = SAD_p \times (Flux \times A_m) = SAD_p \times Q = 15\frac{N\,m^3}{m^3} \times 2{,}000\frac{m^3}{d} = \mathbf{30{,}000}\,\frac{m^3}{d}$$

Note: In the expression of SAD_m or SAD_p, the term 'Nm^3' means the air volume at normal state, i.e., at 0°C and 1 atmospheric pressure. In most submerged MBR plants, SAD_p exceeds 10 and can be as high as 50 at some sites (Judd, 2010). This is because the SADp strongly depends on module configuration and packing density and frequency of chemical cleanings.

EXAMPLE 15.14

An MBR with a design capacity of 1,000 m^3/day is installed for the treatment of municipal wastewater. The MBR plant comprises two tanks, namely, a 2.5 m^3 aerated bioreactor and 0.7 m^3 filtration tank. The filtration tank contains hollow fibre MF membranes with a nominal pore size of 0.4 μm and a total surface of 13.9 m^2. The MBR is operated at an MLSS concentration of 6 g/L. The COD of the influent wastewater is 790 mg/L. If oxygen required for bacterial metabolism is 0.076 kg O_2/(g COD), estimate the volumetric air flowrate (m^3/h) required for membrane scouring and biological aeration demand. Assume that specific air demand based on membrane area (SAD_m) is 0.5 N m^3/m^2 h.

Solution

Air is supplied to an MBR to facilitate membrane scouring (cleaning) and supply adequate O_2 to activated sludge for microbial metabolism. Membranes are submerged in a separate (but interconnected) tank from the main aerated biological tank. The membrane tank is aerated with coarse aerators, and the biological tank is aerated with fine-bubble diffusers as the latter achieves more efficient transfer of O_2.

In this example, the volumetric flowrate of air supplied for membrane scouring,

$$Q_{\text{air,membrane}} = \text{SAD}_m \times A_m = 0.5\frac{\text{N m}^3}{\text{m}^2\text{ h}} \times 13.9\text{ m}^2 = \mathbf{6.95}\frac{\text{m}^3}{\text{h}}$$

The O_2 requirement for bacterial metabolism ($M_{O_2,\text{bio}}$) can be estimated as follows:

$$\begin{aligned} M_{O_{2,\text{bio}}} &= 0.076\,\frac{\text{kg O}_2}{\text{g COD}} \times 790\frac{\text{mg COD}}{L} \times 1{,}000\frac{\text{m}^3}{d} \times \frac{1\text{ d}}{24\text{ h}} \times \frac{1\text{ g}}{1{,}000\text{ mg}} \times \frac{1{,}000\text{ L}}{1\text{ m}^3} \\ &= 2{,}501\text{ kg O}_2/\text{h} \end{aligned}$$

The relationship of mass flowrate of O_2 supplied (M_{O_2}) and the volumetric air flowrate (Q_{air}) can be written as follows (Martin *et al.*, 2011):

$$M_{O_{2,\text{bio}}} = 0.21 \times 1{,}000 \times \alpha \times \beta \times \gamma \times \text{OTE} \times Q_{\text{air,bio}} \cdots \qquad (15.14.1)$$

where β and γ are the oxygen transfer coefficients. In general, values of β and γ are 0.95 and 0.89, respectively. OTE is the oxygen transfer coefficient, and its value for coarse and fine bubble aeration is 0.02 and 0.05, respectively (Martin *et al.*, 2011). Since fine bubble aeration is mostly used for supplying O_2 to the bioreactor, an OTE value of 0.05 is considered in this example. The factor 'α' takes into account the effect of biomass concentration (MLSS or MLVSS) on aeration, and can be estimated as follows (Martin *et al.*, 2011; Metcalf & Eddy *et al.*, 2002):

$$\alpha = e^{-0.082\times\text{MLSS(g/L)}} = e^{-0.082\,\times\,6\text{(g/L)}} = 0.61$$

So from Equation 15.14.1,

$$M_{O_{2,\text{bio}}} = 2{,}501\,\frac{\text{kg O}_2}{\text{hr}} = 0.21 \times 1{,}000 \times 0.61 \times 0.95 \times 0.89 \times 0.05 \times Q_{\text{air,bio}}$$

$$Q_{\text{air,bio}} = \frac{2{,}501}{0.21 \times 1{,}000 \times 0.61 \times 0.95 \times 0.89 \times 0.05} = \mathbf{462}\frac{\text{m}^3}{\text{h}}$$

Note: The coarse bubble diffusers in the membrane tank to scour the membranes also supply some O_2 for microbial metabolism. In this example, that part has been neglected.

EXAMPLE 15.15

A submerged MBR plant treating 2,000 m^3/d municipal wastewater contains hollow fibre MF membranes with a total surface area of 3,000 m^2. The influent wastewater COD concentration is 800 mg/L and the MBR is operated at an HRT of 12 hr. Hence, the volumetric organic loading rate is fixed. At a fixed volumetric organic loading rate, the MLSS concentration in an MBR can be set at different levels depending on the operating SRT. How would the volumetric air flowrate (m^3/h) required for meeting biological aeration demand vary if the MBR was operated at MLSS concentrations of 5, 7, 10 or 15 g/L. Assume that the specific oxygen requirement for bacterial metabolism is 0.076 kg O_2/(mg COD) and the specific air demand based on membrane area (SAD_m) is 0.5 N m^3/m^2 h.

Solution

The O_2 requirement for bacterial metabolism ($M_{O_2,bio}$) can be estimated as follows:

$$M_{O_{2,bio}} = 0.076\ \frac{\text{kg O}_2}{g\ \text{COD}} \times 800\ \frac{\text{mg}}{L} \times 2{,}000\ \frac{\text{m}^3}{d} \times \frac{1\ \text{d}}{24\ \text{h}} \times \frac{1\ \text{g}}{1{,}000\ \text{mg}} \times \frac{1{,}000\ \text{L}}{1\ \text{m}^3} = \mathbf{5{,}067}\ \text{kg/h}$$

The relationship of mass flowrate of O_2 supplied (M_{O_2}) and the volumetric air flowrate (Q_{air}) can be written as follows (Martin *et al.*, 2011):

$$M_{O_2} = 0.21 \times 1{,}000 \times \alpha \times \beta \times \gamma \times \text{OTE} \times Q_{air} \quad (15.15.1)$$

where β and γ are the oxygen transfer coefficients. In general, values of β and γ are 0.95 and 0.89, respectively. OTE is the oxygen transfer coefficient, and its value for coarse and fine bubble aeration is 0.02 and 0.05, respectively (Martin *et al.*, 2011). The factor 'α' takes into account the effect of biomass concentration (MLSS or MLVSS) on aeration, and can be estimated as follows (Martin *et al.*, 2011; Metcalf & Eddy *et al.*, 2002):

$$\alpha = e^{-0.082 \times \text{MLSS (g/L)}} \quad (15.15.2)$$

Air is supplied to an MBR to facilitate membrane scouring (cleaning) and supply adequate O_2 to activated sludge for microbial metabolism. Membranes are submerged in a separate (but interconnected) tank from the main aerated biological tank. The membrane tank is aerated with coarse aerators, and the biological tank is aerated with fine-bubble diffusers as the latter achieves more efficient transfer of O_2. Although the main purpose of the aerators in the membrane tank is to scour the membrane surface, they do supply some O_2 for biological mechanism. Thus, the total O_2 supplied to the bioreactor is the sum of that supplied by the diffusers in the bioreactor and the diffusers in the membrane tank ($M_{O_{2,M}}$). Taking these into account, the required volumetric air flowrate of the diffusers (Q_{air}) in the bioreactor can be estimated as follows:

$$Q_{air} = \frac{M_{O_{2,bio}} - M_{O_{2,M}}}{0.21 \times 1{,}000 \times \alpha \times \beta \times \gamma \times \text{OTE}} \quad (15.15.3)$$

In this example, the volumetric flowrate of air supplied for membrane scouring,

$$Q_{air,membrane} = \text{SAD}_m \times A_m = 0.5\frac{\text{N m}^3}{\text{m}^2\ \text{h}} \times 3{,}000\ \text{m}^2$$

Now, from Equations 15.15.1 and 15.15.2,

$$M_{O_{2,M}} = 0.21 \times 1{,}000 \times (e^{-0.082 \times \text{MLSS(g/L)}}) \times 0.95 \times 0.89 \times 0.02 \times \left(0.5\ \frac{\text{Nm}^3}{\text{m}^2\ \text{h}} \times 3{,}000\ \text{m}^2\right)$$

After rearrangement, O_2 supplied by the membrane-scouring aerators,

$$M_{O_2,M} = 5{,}326.65 \times (e^{-0.082 \times \text{MLSS (g/L)}})$$

MLSS concentration (g/L)	$M_{O_2,M}$
5	$=5{,}326.65 \times (e^{-0.082\times5\ (g/L)}) = 3{,}535$ kg/h
7	3,000 kg/h
10	2,346 kg/h
15	1,556 kg/h

Thus, using Equation 15.3 and the values in the above table, the required volumetric air flowrate of the diffusers (Q_{air}) in the bioreactor can be calculated.

For MLSS concentration = 5 g/L,

$$Q_{air\ 5g/L} = \frac{M_{O_2,bio} - M_{O_2,M}}{0.21 \times 1{,}000 \times \alpha \times \beta \times \gamma \times OTE} = \mathbf{653}\frac{m^3}{h}$$

For MLSS concentration = 7 g/L, $Q_{air\ 7g/L} = \mathbf{765}\frac{m^3}{h}$

For MLSS concentration = 10 g/L, $Q_{air\ 10g/L} = \mathbf{978}\frac{m^3}{h}$

For MLSS concentration = 15 g/L, $Q_{air\ 15g/L} = \mathbf{1{,}475}\frac{m^3}{h}$

At higher MLSS concentrations, the O_2 transfer efficiency drops and, hence, the aerators require higher volumetric air flowrate capacities (Q_{air}). The relationship between MLSS concentration and Q_{air} is depicted in Figure 15.2.

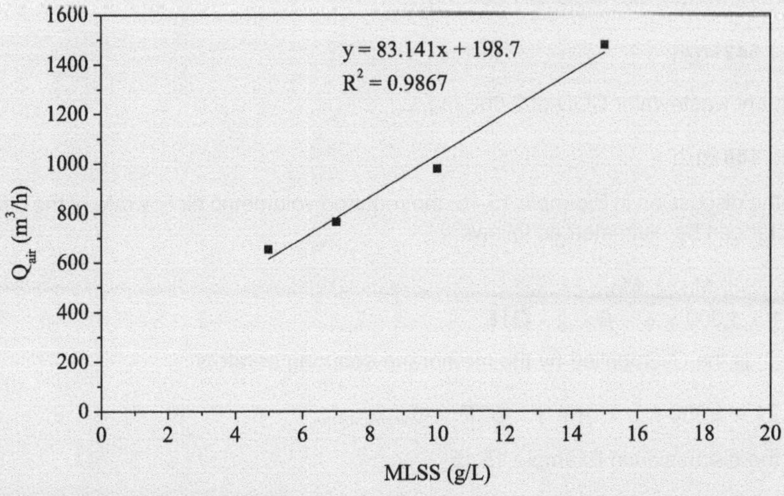

Figure 15.2 Relationship between MLSS concentration and the required volumetric air flowrate (Q_{air}) of the aerators in the bioreactor.

EXAMPLE 15.16

During a pilot-scale investigation, the membrane tank of an MBR housed UF hollow-fibre membrane modules with a total surface area of 150 m^2. The membranes were operated at an average flux of 20 LMH. The COD concentration of the influent wastewater was set at 500, 1,000, 1,500 and 2,000 mg/L at phase 1, phase 2, phase 3 and phase 4 of the investigation. Estimate the energy required for biological aeration as well as membrane scouring at different phases of the investigation. The MLSS concentration was controlled at 8 g/L in all phases. Assume that the specific oxygen requirement for bacterial metabolism is 0.076 kg O_2/(mg COD) and the specific air demand based on membrane area (SAD_m) is 0.4 N m^3/m^2 h. Inlet and outlet pressure of the blower used for aeration are 106,000 and 156,000 Pa, respectively.

Solution

$$\text{Flow rate}(Q) = \text{Flux} \times \text{membrane surface area} = 20\frac{L}{m^2\,h} \times 150\,m^2 \times \frac{1\,m^3}{1{,}000\,L} = \mathbf{3}\frac{m^3}{h}$$

The O_2 requirement for bacterial metabolism ($M_{O_2,\text{bio}}$) can be estimated as follows:

$$M_{O_{2,\text{bio}}} = 0.076\,\frac{\text{kg}\,O_2}{\text{gCOD}} \times \text{Influent COD} \times 3\frac{m^3}{h} \times \frac{1\,g}{1{,}000\,mg} \times \frac{1{,}000\,L}{1\,m^3}$$

For an influent wastewater COD of 500 mg/L,

$$M_{O_{2,\text{bio},500}} = \mathbf{114}\,\text{kg/h}$$

For an influent wastewater COD of 1,000 mg/L,

$$M_{O_{2,\text{bio},1{,}000}}\,\mathbf{228}\,\text{kg/h}$$

For an influent wastewater COD of 1,500 mg/L,

$$M_{O_{2,\text{bio},1{,}500}} = \mathbf{342}\,\text{kg/h}$$

For an influent wastewater COD of 2,000 mg/L,

$$M_{O_{2,\text{bio},2{,}000}} = \mathbf{456}\,\text{kg/h}$$

Following the discussion in Example 15.15, the required volumetric air flowrate of the diffusers (Q_{air}) in the bioreactor can be estimated as follows:

$$Q_{\text{air}} = \frac{M_{O_{2,\text{bio}}} - M_{O_{2,M}}}{0.21 \times 1{,}000 \times \alpha \times \beta \times \gamma \times \text{OTE}} \cdots \qquad (15.16.1)$$

Here, $M_{O_{2,M}}$ is the O_2 supplied by the membrane-scouring aerators.

$$M_{O_{2,M}} = 0.21 \times 1{,}000 \times \alpha \times \beta \times \gamma \times \text{OTE} \times Q_{\text{air,membrane}}$$

Following the discussion in Example 15.15,

$$\alpha = e^{-0.082 \times 8\,(g/L)}, \quad \beta = 0.95, \quad \gamma = 0.89, \quad \text{OTE} = 0.02$$

and, $Q_{air,membrane}$ = the volumetric flowrate of air supplied for membrane scouring

$$= SAD_m \times A_m = 0.4 \frac{N\,m^3}{m^2\,h} \times 150\,m^2$$

Therefore, from Equation 15.16.2, $M_{O_{2,M}}$

$$= 0.21 \times 1{,}000 \times (e^{-0.082 \times 8 (g/L)}) \times 0.95 \times 0.89 \times 0.02 \times \left(0.4 \frac{N\,m^3}{m^2\,h} \times 150\,m^2\right)$$

$$= \mathbf{110.8 \frac{kg}{h}}$$

From Equation 15.16.1, the required volumetric air flowrate of the diffusers (Q_{air}) in the bioreactor,

$$Q_{air} = \frac{M_{O_{2,bio}} - 110.8}{0.21 \times 1{,}000 \times \alpha \times \beta \times \gamma \times OTE}$$

Plugging in the $M_{O_{2,bio}}$ values calculated for different influent COD values,

For an influent wastewater COD of 500 mg/L, $Q_{air,500} \mathbf{2.17} \frac{m^3}{h}$

For an influent wastewater COD of 1,000 mg/L, $Q_{air,1000} \mathbf{63.7} \frac{m^3}{h}$

For an influent wastewater COD of 1,500 mg/L, $Q_{air,1500} \mathbf{125.7} \frac{m^3}{h}$

For an influent wastewater COD of 2,000 mg/L, $Q_{air,2000} \mathbf{187.6} \frac{m^3}{h}$

Energy required for biological aeration:

The power requirements associated with biological aeration ($E_{air,bio}$) can be obtained from the power consumption of a blower delivering the corresponding air flow. Energy required for biological aeration can be determined by using the following expression (Judd, 2014):

$$E_{air,bio} = KP_{a,in}\left(\left(\frac{P_{a,out}}{P_{a,in}}\right)^{0.283} - 1\right)Q_{air} \ldots \tag{15.16.2}$$

where K is the blower energy equation constant, and its value is 6.5 in SI units. $P_{a,in}$ and $P_{a,out}$ are the blower inlet and outlet pressure.

Plugging different values of Q_{air} corresponding to different influent COD concentrations into Equation 15.16.2,

For an influent wastewater COD of 500 mg/L,

$$E_{air,bio,500} = 6.5 \times 106{,}000 \left(\left(\frac{156{,}000}{106{,}000}\right)^{0.283} - 1\right) \times 2.17 = 173{,}600\,Ws = \mathbf{0.05}\,kWh$$

For an influent wastewater COD of 1,000 mg/L, $E_{air,bio,1,000} = \mathbf{1.42}$ kWh

For an influent wastewater COD of 1,500 mg/L, $E_{air,bio,15,000} = \mathbf{2.8}$ kWh

For an influent wastewater COD of 2,000 mg/L, $E_{air,bio,2,000}$

Energy required for membrane scouring:
For membrane scouring, energy requirement can be estimated as follows:

$$E_{air,membrane} = K.P_{a,in}\left(\left(\frac{P_{a,out}}{P_{a,in}}\right)^{0.283} - 1\right).Q_{air,membrane} = \mathbf{1.33}\,\text{kWh}$$

Note: In MBR, energy required for membrane operation comprises energy required for membrane scouring to prevent fouling as well as to withdraw permeate (i.e., treated effluent). Energy required for permeation ($E_{permeation}$) can be calculated by using the following expression (Martin *et al.*, 2011): $E_{permeation} = Q_p \times \text{TMP}$, where, Q_p (m^3/s) is the permeate flow rate and TMP is the transmembrane pressure (Pa).

Recent studies report breakdown of energy demand and cost for different components of MBRs (Bertanza *et al.*, 2017; Fenu *et al.*, 2010; Gil *et al.*, 2010; Lo *et al.*, 2015; Verrecht *et al.*, 2010).

EXAMPLE 15.17

As noted in Chapters 12 and 13, when the MF or UF membrane used in classical MBR is replaced by a high retention membrane such as forward osmosis, NF or membrane distillation process, it is called a high retention MBR. In a laboratory investigation, the membrane distillation (MD) process was combined with an enzymatic bioreactor, thus constituting a 'high retention enzymatic MBR'. During MD process, water in vapour form moves *via* diffusion through a microporous hydrophobic membrane from higher temperature feed solution to lower temperature permeate solution. Since the water moves across the membrane in vapour form, MD can theoretically provide complete removal of non-volatile pollutants. The lab-scale MD-MBR comprised a 6 L bioreactor, and was operated at an HRT of 24 h. A synthetic municipal wastewater containing a mixture of five non-volatile pharmaceutical compounds was continuously fed to the MBR. Duplicate samples were collected from the feed, bioreactor and permeate of the MD-MBR after 5 d of continuous operation, and were analysed by GC/MS. Estimate the removal of the pharmaceutical compounds *via* different mechanisms such as biodegradation, membrane retention and so forth based on the data given in Table 15.3. Assume that the permeate flux was stable throughout the investigation period.

Table 15.3 Concentration of the compounds after 5 d of continuous operation.

Pharmaceutical Compound	Concentration (ng/L)		
	Influent	Bioreactor	Permeate
A	4,085	3,720	125
B	5,420	20,520[a]	30
C	4,160	11,850[a]	25
D	4,105	1,975	0
E	4,500	17,040[a]	5

[a]Higher concentration in bioreactor than influent means that the compound was effectively retained by the MD membrane but biodegradation was slow.

Solution

The overall removal, biodegradation and membrane retention of the compounds can be estimated using the following expressions:

$$R_{\text{overall removal}} = \frac{C_{\text{influent}} - C_{\text{permeate}}}{C_{\text{influent}}} \times 100\ldots \quad (15.17.1)$$

$$R_{\text{biodegradation}} = \frac{\text{massenzymatically biodegraded}}{\text{mass}_{\text{influent}}} \times 100\ldots \quad (15.17.2)$$

$$R_{\text{membrane retention}} = R_{\text{overall removal}} - R_{\text{biodegradation}}\ldots \quad (15.17.3)$$

where C_{influent} and C_{permeate} are the concentrations of a specific compound in the feed and permeate, respectively. m_{influent} is the mass of a specific pharmaceutical in the feed and bioreactor, respectively. The mass of pharmaceuticals biodegraded can be calculated as follows:

Total mass fed = Mass accumulated in bioreactor + Mass lost through permeate + Mass biodegraded

Therefore,

$$C_{\text{influent}} \times V_{\text{influent}} = (C_{\text{bioreactor}} \times V_{\text{bioreactor}}) + (C_{\text{permeate}} \times V_{\text{permeate}}) + \text{biodegradation/biotransformation}\ldots \quad (15.17.4)$$

where V_{influent}, $V_{\text{bioreactor}}$ and V_{permeate} represents the volume of feed, enzymatic bioreactor and permeate, respectively.

For pharmaceutical compound 'A':

$$R_{\text{overall removal}} = \frac{4{,}085 - 125}{4{,}085} \times 100 = 96.9\% \quad (15.17.1)$$

From Equation 15.17.4:

$$\frac{4{,}085\,\text{ng}}{L} \times \left(5\,dx\,6\frac{L}{d}\right) = \left(\frac{3{,}720\,\text{ng}}{L} \times 6\,L\right) + \left(\frac{125\,\text{ng}}{L} \times 30\,L\right) + \text{biodegraded mass}$$

Therefore, biodegraded mass $\asymp$ 96,370 ng

$$R_{\text{biodegradation}} = \frac{96{,}370}{122{,}520} \times 100 = 78.7\% \quad (15.17.2)$$

$$R_{\text{membrane retention}} = 96.9 - 78.7 = 17.9\% \quad (15.17.3)$$

The complete set of calculated data is listed in Table 15.4.

Table 15.4 Removal of pharmaceutical compounds by the MD-MBR.

Pharmaceutical Compound	Removal (%)[a]		
	Overall Removal	Biodegradation	Membrane Retention
A	97	79	18
B	100	24	76
C	99	41	57
D	100	90	10
E	100	24	76

[a]Rounded to nearest whole number.

Note: Pharmaceuticals are one of the groups of emerging hazardous trace organic contaminants that have been discussed in Chapter 13. For other examples of trace organic contaminant removal by high retention MBRs, readers can consult the following references: Asif *et al.* (2018a, b); Luo *et al.* (2017) and Wijekoon *et al.* (2014). For examples of trace organic contaminant removal by classical MBRs, i.e., those equipped with MF or UF membranes, readers can consult the following references: Nguyen *et al.* (2013), Wijekoon *et al.* (2013) and Yang *et al.* (2013).

REFERENCES

Asif M. B., Hai F. I., Dhar B. R., Ngo H. H., Guo W., Jegatheesan V., Price W. E., Nghiem L. D. and Yamamoto K. (2018a). Impact of simultaneous retention of micropollutants and laccase on micropollutant degradation in enzymatic membrane bioreactor. *Bioresource Technology*, **267**, 473–480.

Asif M.B., Hai F.I., Kang J., van de Merwe J.P., Leusch F.D.L., Price W.E. and Nghiem L.D. (2018b). Biocatalytic degradation of pharmaceuticals, personal care products, industrial chemicals, steroid hormones and pesticides in a membrane distillation-enzymatic bioreactor. *Bioresource Technology*, **247**, 528–536.

Bertanza G., Canato M., Laera G., Vaccari M., Svanström M. and Heimersson S. (2017). A comparison between two full-scale MBR and CAS municipal wastewater treatment plants: techno-economic-environmental assessment. *Environmental Science and Pollution Research*, **24**(21), 17383–17393.

Fenu A., Roels J., Wambecq T., De Gussem K., Thoeye C., De Gueldre G. and Van De Steene B. (2010). Energy audit of a full scale MBR system. *Desalination*, **262**(1), 121–128.

Gil J.A., Túa L., Rueda A., Montaño B., Rodríguez M. and Prats D. (2010). Monitoring and analysis of the energy cost of an MBR. *Desalination*, **250**(3), 997–1001.

Grelier P., Rosenberger S. and Tazi-Pain A. (2006). Influence of sludge retention time on membrane bioreactor hydraulic performance. *Desalination*, **192**(1), 10–17.

Hai F.I., Yamamoto K. and Fukushi K. (2006). Development of a submerged membrane fungi reactor for textile wastewater treatment. *Desalination*, **192**(1–3), 315–322.

Hai F.I., Yamamoto K., Nakajima F. and Fukushi K. (2008). Removal of structurally different dyes in submerged membrane fungi reactor – Biosorption/PAC-adsorption, membrane retention and biodegradation. *Journal of Membrane Science*, **325**(1), 395–403.

Hai F.I., Yamamoto K., Nakajima F. and Fukushi K. (2009). Factors governing performance of continuous fungal reactor during non-sterile operation – The case of a membrane bioreactor treating textile wastewater. *Chemosphere*, **74**(6), 810–817.

Hai F.I., Yamamoto K., Nakajima F. and Fukushi K. (2011). Bioaugmented membrane bioreactor (MBR) with a GAC-packed zone for high rate textile wastewater treatment. *Water Research*, **45**(6), 2199–2206.

Hai F.I., Yamamoto K., Nakajima F. and Fukushi K. (2012). Application of a GAC-coated hollow fiber module to couple enzymatic degradation of dye on membrane to whole cell biodegradation within a membrane bioreactor. *Journal of Membrane Science*, **389**, 67–75.

Hai F.I., Yamamoto K., Nakajima F., Fukushi K., Nghiem L.D., Price W.E. and Jin B. (2013). Degradation of azo dye acid orange 7 in a membrane bioreactor by pellets and attached growth of *Coriolus versicolour*. *Bioresource Technology*, **141**, 29–34.

Judd S. (2010). The MBR Book: Principles and Applications of Membrane Bioreactors for Water and Wastewater Treatment, 2nd edn. Butterworth-Heinemann, Oxford.

Judd S. (2014). Industrial MBRs: Membrane Bioreactors for Industrial Wastewater Treatment. IWA Publishing, London.

Lo C.H., McAdam E. and Judd S. (2015). The cost of a small membrane bioreactor. *Water Science and Technology*, **72**(10), 1739–1746.

Luo W., Phan H.V., Xie M., Hai F.I., Price W.E., Elimelech M. and Nghiem L.D. (2017). Osmotic versus conventional membrane bioreactors integrated with reverse osmosis for water reuse: biological stability, membrane fouling, and contaminant removal. *Water Research*, **109**, 122–134.

Martin I., Pidou M., Soares A., Judd S. and Jefferson B. (2011). Modelling the energy demands of aerobic and anaerobic membrane bioreactors for wastewater treatment. *Environmental Technology*, **32**(9), 921–932.

Metcalf & EddyTchobanoglous G., Burton F.L. and Stensel H.D. (2002). Wastewater Engineering: Treatment and Reuse, 4th edn. McGraw-Hill Higher Education, New York, USA.

Ng H.Y., Tan T.W. and Ong S.L. (2006). Membrane fouling of submerged membrane bioreactors: impact of mean cell residence time and the contributing factors. *Environmental Science & Technology*, **40**(8), 2706–2713.

Nguyen L.N., Hai F.I., Yang S., Kang J., Leusch F.D.L., Roddick F., Price W.E. and Nghiem L.D. (2013). Removal of trace organic contaminants by an MBR comprising a mixed culture of bacteria and white-rot fungi. *Bioresource Technology*, **148**, 234–241.

Park H.-D., Chang I.-S. and Lee K.-J. (2015). Principles of Membrane Bioreactors for Wastewater Treatment. CRC Press, Florida, USA.

Patsios S.I. and Karabelas A.J. (2011). An investigation of the long-term filtration performance of a membrane bioreactor (MBR): the role of specific organic fractions. *Journal of Membrane Science*, **372**(1), 102–115.

Rautenbach R. and Albrecht R. (1989). Membrane Processes. Wiley, Chichester, New York.

Van den Broeck R., Van Dierdonck J., Nijskens P., Dotremont C., Krzeminski P., van der Graaf J.H.J.M., van Lier J.B., Van Impe J.F.M. and Smets I.Y. (2012). The influence of solids retention time on activated sludge bioflocculation and membrane fouling in a membrane bioreactor (MBR). *Journal of Membrane Science*, **401–402**, 48–55.

Verrecht B., Maere T., Nopens I., Brepols C. and Judd S. (2010). The cost of a large-scale hollow fibre MBR. *Water Research*, **44**(18), 5274–5283.

Wijekoon K.C., Hai F.I., Kang J., Price W.E., Guo W., Ngo H.H. and Nghiem L.D. (2013). The fate of pharmaceuticals, steroid hormones, phytoestrogens, UV-filters and pesticides during MBR treatment. *Bioresource Technology*, **144**, 247–254.

Wijekoon K.C., Hai F.I., Kang J., Price W.E., Guo W., Ngo H.H., Cath T.Y. and Nghiem L.D. (2014). A novel membrane distillation–thermophilic bioreactor system: biological stability and trace organic compound removal. *Bioresource Technology*, **159**, 334–341.

Wu B., Kitade T., Chong T.H., Uemura T. and Fane A.G. (2013). Impact of membrane bioreactor operating conditions on fouling behavior of reverse osmosis membranes in MBR–RO processes. *Desalination*, **311**, 37–45.

Yang S., Hai F.I., Nghiem L.D., Nguyen L.N., Roddick F. and Price W.E. (2013). Removal of bisphenol A and diclofenac by a novel fungal membrane bioreactor operated under non-sterile conditions. *International Biodeterioration & Biodegradation*, **85**, 483–490.

Index

A

Acclimatization, 365, 427–428
Activated carbon, 61, 68, 72–73, 77–78, 133, 135, 138, 185, 197, 199, 235, 325–326, 362, 371, 393, 418, 438–441, 508
Activated Sludge Models, 84, 269, 270, 273
Advance Oxidation Process, 174, 197, 236, 394, 441
Aeration demand, 6–7, 64, 90, 137, 275, 284, 485, 488, 492, 523, 549–550
Aerobic granular sludge, 388, 407
Air lift, 5, 10, 11, 134, 397, 407, 469, 470, 513
Air scouring, 62–63, 94–96, 106, 135, 137, 217, 295, 297, 403, 471, 498, 502, 506, 514
Ammonia nitrogen, 44, 271, 282
Anaerobic membrane bioreactor (AnMBR), 164, 343, 344
Anammox process, 273, 297, 401
Asahi Kasei, 484, 487–489, 491
Asymmetric cellulose acetate membrane, 2–3
Atomic force microscopy (AFM), 123, 355
Automation, 14, 67, 98–99, 105–106, 478, 519, 529

B

Backwashing, 8, 26, 60–63, 94–95, 103–105, 116, 126, 135, 137, 216, 355, 371, 402, 472, 505, 542–543
Bacteriophage, 139, 144–145, 171, 191, 216–221
Bardenpho process, 43, 163
Bioaugmentation, 315, 401–402, 444, 453
Biocatalytic membrane biological reactors, 2
Biochemical conversion, 2, 305–307, 330
Biochemical oxygen demand (BOD), 40, 188, 189, 210, 238
Biodegradation, 2, 7, 59, 135, 136, 139, 216, 315, 319, 326, 354, 393, 401, 415, 417–418, 421–422, 426–439, 447, 449, 451–453, 544, 547, 554–556
Biofilm, 2, 68, 102–103, 128, 130, 133, 139–141, 143–144, 171–172, 285, 305–308, 310–318, 324, 326, 330–331, 369–370, 373, 388–391, 396, 401, 404–405, 407, 435, 487
Biogas production, 256, 352, 359, 364–365, 372–373
Biological activated carbon (BAC), 393, 438
Biological nutrient removal (BNR), 43–44, 161
Biosolids separation membrane biological reactors, 2

C

Cake layer fouling, 90, 94, 99, 369
Capillary suction time (CST), 25, 117, 213
Capital and operating costs, 11, 14, 27, 196, 358, 478
Carbon nanotubes, 388, 394

Ceramic membrane, 9, 50, 82, 350, 396, 473, 477, 508, 530
Chemical cleaning, 8, 26, 55, 59, 61–62, 91, 94, 103, 116, 118, 129, 133–135, 144, 196, 218, 233, 349, 355, 367, 372, 402, 473, 499, 502, 506, 508, 510, 512, 516, 530, 543, 548
Chemical oxygen demand (COD), 9, 190, 210, 272, 347, 348, 364, 417, 476, 545, 547
Cleaning in place (CIP), 57, 94, 106, 135, 508
Cometabolism, 432–433, 444
Computational Fluid Dynamics (CFD), 270, 295, 316
Cost–benefit analysis, 26, 227, 228
Compound annual growth rate (CAGR), 16, 22
Commercial MBR products, 46, 64, 190, 482, 528
Confocal laser scanning microscopy (CLSM), 123, 286, 355
Conventional activated sludge (CAS), 2, 35, 40, 69, 117, 168, 170, 185, 234, 269, 317, 415, 478, 487, 520
Critical flux, 117–118, 122, 126, 133, 138, 291, 293–294, 367–368, 371
Crossflow velocity, 527

D

Decentralized MBR, 15, 21, 74
Denaturing Gradient Gel Electrophoresis (DGGE), 164, 393
Denitrification, 41–42, 44, 59, 70, 73, 77, 80, 87–88, 95, 156, 161–164, 168, 211, 248, 256, 272–273, 275–278, 281, 283–284, 291, 306–307, 310–311, 314–315, 330, 367, 372–373, 391, 400, 436–437, 518, 520, 523, 525
Desalination, 156, 188, 328, 331, 397
Dewatering, 6, 45, 46, 61, 81, 104, 239, 245, 256, 257, 329, 374, 433
Disinfection Byproducts, 157
Dissolved Oxygen (DO), 96, 163, 147, 188, 223, 271, 275, 283, 284, 288, 290, 311, 370, 401, 434, 523
Dissolved organic matter (DOM), 2, 438
Drivers for MBR market
Dynamic viscosity of mixed liquor, 90

E

Ease of operation, 27, 46, 117, 446
Electrochemical membrane biological reactors, 2, 305–306
Electron withdrawing groups (EWG), 426
Electrospun nanofibers, 388
Endocrine Disrupting Compounds (EDC), 175, 414, 445
Endogenous (autolytic) metabolism, 8
Energy usage, 7, 57, 469, 494, 514, 518, 527, 529
Energydispersive Xray spectroscopy (EDX), 356
Enhanced biological phosphorus removal (EBPR), 6, 166
Enhanced capital allowance (ECA), 23
Environmental biotechnology, 2, 306, 331
Eutrophication, 18, 155–156, 228, 232–236, 245–246, 249, 251, 256, 258, 261–262
Extracellular polymeric substances (EPS), 121, 123, 135, 140, 144, 269, 285, 354–355, 369, 391, 420
Extractive membrane biological reactors, 2

F

Filterability of mixed liquor, 116–117, 119, 121
Filtration resistance, 6, 90, 363, 366, 369
Flat sheet membrane, 56, 118, 126, 308, 353, 356, 473, 505, 508
Flux, 3, 9–12, 15, 24, 26, 47, 51–53, 57, 59–60, 74, 77, 90–92, 99, 103, 105, 116–118, 122, 126, 129, 133–135, 137–138, 144, 155, 156, 168, 216–218, 233, 235, 286, 291, 293–294, 310, 319, 344, 348, 349, 352, 354, 358–363, 365–371, 373, 375, 396, 398–399, 402, 441, 472–474, 484, 491–492, 499, 503, 505, 508, 512–514, 517–518, 521–523, 525–527, 535–543, 547, 552, 554
Food to microorganism ratio (F/M ratio), 41, 275
Fluorescence in situ hybridization (FISH), 129, 164, 357
Foot print, 6
Fourier transform infrared spectroscopy (FTIR), 127
Forward osmosis, 236, 388, 441, 447, 554
Functional groups, 128, 132, 192, 327, 395–396, 421–422, 426, 448

G

Gas sparging, 235–236, 359–361, 367, 370, 373
Gas-diffusion membrane biological reactors, 2
GEZenon, 2
Global warming potential, 235, 247, 259, 373
Greenhouse gases, 234, 248
Grey water recycling, 197
Groundwater replenishment/recharge, 23, 76, 77, 158, 159, 161, 167, 185, 189, 191, 198, 372

H

Heterotrophic bacteria, 40–42, 59, 215, 285, 437
Hollowfiber membrane, 125–126, 136, 141, 251, 307–308, 310, 312, 314, 317, 319, 321, 324–326, 402
Human toxicity potential, 260
Humic substances, 126–128, 132, 166, 369
Hydraulic retention time (HRT), 57, 76, 274, 352, 518
Hydrodynamic conditions, 47, 117–118, 269, 358, 366–367

Hydrophobicity, 51, 59, 120, 293, 358, 421–422, 426–427, 431, 433–434, 447–448

I

Immersed (submerged) MBR (iMBR), 25
Impact of hazardous events, 216
Indirect potable reuse (IPR), 14, 159, 161, 186, 188–189, 196–198, 223, 414, 438, 470, 529
Inflow equalization, 82

K

KochPuron, 474, 494
Kubota, 4–5, 16, 20, 72, 167, 190, 352, 354, 471, 473–474, 482–484, 502, 505–508, 512, 524–525, 529

L

Legislation, 12, 17, 19, 27, 101, 197, 246, 393
Life cycle assessment (LCA), 227–228, 231–232, 524
Life cycle costs (LCC), 15, 101, 374
Life Cycle Inventory (LCI), 231, 242–243, 252–253
Log Removal, 6, 170, 172, 191, 220

M

Market trends, 17
Mega litre per day (MLD), 535
Membrane biofilm reactor (MBfR), 306–307
Membrane bioreactor (MBR), 1–2, 35, 46, 67, 115, 133–134, 137, 155–156, 164, 183, 185, 207, 209, 227–228, 269, 278, 284–286, 291, 293–297, 305–306, 345, 349, 352, 387–388, 390, 404, 413, 469–470, 535
Membrane distillation, 388, 397, 399, 407, 447, 554
Membrane filtration after CAS process (CASMF/UF), 7
Membrane flux enhancer, 371
Membrane fouling, 2, 6, 8, 14, 25–26, 51, 53, 57–62, 64, 93, 103, 116–117, 120–123, 125–126, 128–138, 144–145, 164, 196, 216, 269, 286, 290–291, 295, 322–323, 330, 343, 345, 352, 354–356, 358–360, 362–371, 373–375, 389, 391, 393–395, 402–404, 407, 420, 441, 470, 527, 538, 541–542
Membrane integrity, 14, 62–63, 156, 172, 212, 216, 220, 223, 488
Memstar, 167, 470–471, 478, 481, 484, 500–503, 506, 522–523, 529
Mesophilic, 310, 347, 351–352, 356, 365, 368, 370
Metabolic pathways and fate,
Methanogenic bacteria, 344, 366
Microbial community, 122, 128–130, 132, 156, 164, 175, 211, 310, 357, 360, 428–429, 445, 453
Microbial fuel cell, 327, 388, 405, 407
Microfiltration, 2–4, 15, 48, 78, 163, 185, 199, 233, 238, 321, 325–326, 330, 348, 358, 394, 407, 471, 538
Micropollutant, 25, 78, 234, 237, 251, 397
Mitsubishi Rayon, 3–5, 167, 473, 483, 502
Mixed liquor suspended solids (MLSS), 6, 40, 171, 220, 235, 367, 422, 518, 544
Mixed liquor volatile suspended solids (MLVSS), 427
Modeling, 229, 270, 273–276, 278–281, 283–286, 290–292, 295, 297, 316
Modified LudzackeEttinger (MLE), 517
Modular design, 12
Molecular weight cutoff, 48, 320–321
Moving bed or biofilm bioreactors (MBBR), 102
Multiple barriers, 222

N

Nanofiltration, 48, 166, 196–197, 358, 388, 396–397, 407, 439, 538
Natural organic matter (NOM), 123
Nephelometric Turbidity Unit,
Net Environmental Benefit (NEB), 234, 241, 260
Nitrification, 41–42, 54, 59, 73, 82, 87–88, 95, 156, 161–164, 168, 211, 215, 248, 251, 256, 272–276, 278, 281–284, 291, 297, 306–307, 310–312, 314, 329–330, 367, 372–373, 391, 400, 403, 419–420, 436–437, 518
Nitrifying bacteria, 7, 37, 59, 82, 87, 161–162, 166, 215, 281, 372, 419, 429
Nonwoven filters, 363
Nutrient Recovery, 169

O

Organic loading rate (OLR), 344, 375, 550
Oxidationreduction potential, 174, 446

P

Packing density, 47, 57, 92, 134, 295, 297, 472, 475–476, 488, 494, 499, 502, 505, 514, 548
Pathogens, 6, 21, 26, 68, 78, 81, 156, 166, 170–173, 175, 183, 185–186, 191, 210, 216, 220, 223, 233, 372
Peak daily flow (PDF) capacity, 13
PentairXflow, 474, 483–484, 513–514, 525
Pharmaceutical and Personal Care Products (PPCPs), 228
Phosphorus accumulating organisms (PAOs), 271–272
Polymeric membranes, 9, 49, 51, 316, 363, 394, 470–471, 477, 508, 510
Polyethylsulphone (PES) membrane, 9
Polyethylene (PE) membrane, 3, 9–10, 138, 166, 322–323, 359, 362–363, 473

Polypropylene (PP) membrane, 9–10, 51, 138, 316, 319, 326, 363
Polytetrafluoroethylene (PTFE) membrane, 9–10, 51, 362
Polymerase chain reaction (PCR), 129, 164
Polyvinylidene fluoride (PVDF) membrane, 51, 135, 166, 363, 396
Population equivalent (P.E), 21, 71–72, 82, 105, 519
Pore blocking, 52, 59–61, 293, 355, 363–364, 391
Prescreening, 8, 210
Pressure retarded osmosis, 388
Protein/polysaccharide ratio, 366
Public perception, 14, 160
Pyrosequencing, 129, 132, 164

Q

Quorum sensing/quenching, 128, 133, 139–142, 401–402

R

Relaxation, 8, 61, 94, 99, 105, 137, 217, 221, 359–362, 367, 370–371, 402, 488, 494, 499, 502–503, 506, 508, 511, 520, 542–543
Retrofit, 6, 16, 70, 100, 471, 478, 495, 508, 524, 529–530
Return activated sludge (RAS), 163
Reverse osmosis, 2, 6, 48, 68, 166, 197, 199, 228, 328, 352, 358, 398, 438–439, 526
Risk assessment, 191, 208–209, 223

S

Scanning electron microscopy (SEM), 50, 122, 283, 355
Sequencing batch reactors (SBR), 199, 235, 391, 431
Settling unit, 2
Sewage treatment plants (STPs), 414
Shock loadings, 80, 85, 169, 210
Sidestream MBR (sMBR), 3
SiemensMemcor,
Size exclusion chromatography, 126, 356
Sludgewater partition coefficient (Kp), 433
Sludge production, 5–9, 54, 84–85, 89, 139, 162, 175, 256, 259–261, 269, 276, 278–280, 297, 343, 374, 391, 420, 526
Sludge volume index (SVI), 117
Solids retention time (SRT), 6, 186, 415, 428
Soluble microbial products (SMP), 54, 121, 269, 274, 285, 288, 355, 369, 438
Specific aeration demand (SADm), 90, 485, 488, 492
Staff training, 15
Struvite, 170, 330, 357
Supervisory Control and Data Acquisition (SCADA), 99, 223

T

Technical challenges, 14, 64
Tertiary treatment, 2, 7, 11, 15, 39, 68, 72, 101, 187, 229, 234, 236–238, 241, 245, 415, 503
Thermophilic, 310, 347, 352, 356, 364–365, 368, 399
Toray, 167, 483–484, 508–509
Total Kjeldahl nitrogen (TKN), 44, 215, 352
Total phosphorous (TP), 11, 190, 433
Trace organic contaminant, 25, 175, 183, 185, 372, 393, 413–414, 556
Transmembrane pressure (TMP), 47, 49, 99, 142, 216, 291, 354, 522, 541, 554
Treatment efficiency, 6, 84, 88, 101, 162, 188, 210, 366, 374–375, 418, 439
Tubular membrane, 51, 56–57, 79, 117–118, 137, 295, 350, 477, 513, 514, 517, 527

U

Ultrafiltration, 2–4, 15, 23, 48, 67, 144, 166, 185, 198–199, 212, 228, 236, 239, 286, 319–323, 325, 345, 348, 351, 353, 358, 374, 394, 418, 451, 471, 538
Upflow anaerobic sludge blanket bioreactor (UASB), 348, 351, 362

V

Volatile fatty acids (VFA), 329, 373

W

Wastewater reuse, 3, 14, 22, 26, 67–68, 106, 183, 185–193, 196–198, 372, 487, 526
Wastewater treatment plant (WWTP), 6–7, 18, 21, 39, 44, 64, 156, 164, 172, 173, 196, 227, 231, 262, 273, 329, 331, 372, 415, 427, 438, 519, 526
Water reclamation plant (WRP), 21, 23, 70, 73, 168, 437, 508, 517
Water scarcity, 17–19, 21, 27, 70, 156, 387
Wet weather flow, 18, 64

X

Xenobiotics, 2, 305, 311, 319–326, 419

IWA Publishing's authorised EU representative for General Product Safety Regulations is Diane D'Arras, 15 rue Duret, 75116 Paris, France, e-mail: safety@iwap.co.uk.

Printed and bound by CPI Group (UK) Ltd, Croydon, CR0 4YY
12/05/2026
02108866-0015